SEVENTH EDITION

Understanding Human Sexuality

Janet Shibley Hyde

University of Wisconsin-Madison

John D. DeLamater

University of Wisconsin-Madison

Boston   Burr Ridge, IL   Dubuque, IA   Madison, WI   New York   San Francisco   St. Louis
Bangkok   Bogotá   Caracas   Lisbon   London   Madrid
Mexico City   Milan   New Delhi   Seoul   Singapore   Sydney   Taipei   Toronto

McGraw-Hill Higher Education

A Division of The **McGraw-Hill** *Companies*

UNDERSTANDING HUMAN SEXUALITY, SEVENTH EDITION

This book is printed on acid-free paper.

2 3 4 5 6 7 8 9 0 VNH/VNH 0 9 8 7 6 5 4 3 2 1 0

ISBN 0–07–366204–6

Editorial director: *Jane E. Vaicunas*
Executive editor: *Mickey Cox*
Developmental editor: *Sharon Geary*
Senior marketing manager: *James Rozsa*
Project manager: *Susan J. Brusch*
Production supervisor: *Enboge Chong*
Designer: *K. Wayne Harms*
Senior photo research coordinator: *Carrie K. Burger*
Supplement coordinator: *Sandra M. Schnee*
Compositor: *GTS Graphics, Inc.*
Typeface: *9.5/12 Utopia*
Printer: *Von Hoffmann Press, Inc.*

Interior/cover design: *Kim Rokusek*
Photo research: *Mary Regg Photo Research*

The credits section for this book begins on page 680 and is considered an extension of the copyright page.

Library of Congress Cataloging-in-Publication Data

Hyde, Janet Shibley.
    Understanding human sexuality / Janet Shibley Hyde, John D. DeLamater. — 7th ed.
      p.   cm.
    Includes bibliographical references and index.
    ISBN 0–07–366204–6
    1. Sex. 2. Sex customs. 3. Hygiene, Sexual. 4. Sex (Psychology) 5. Mass media and sex. 6. Computer sex. I. DeLamater, John D. II. Title.
HQ12.H82   2000
306.7—dc21
                                   99–22743
                                      CIP

www.mhhe.com

Author Biographies

Janet Shibley Hyde, Professor and Chair of the Department of Psychology, and Evjue-Bascom Professor of Women's Studies at the University of Wisconsin–Madison, received her education at Oberlin College and the University of California, Berkeley. She has taught a course in human sexuality since 1974, first at Bowling Green State University, then at Denison University, and now at the University of Wisconsin. Her research interests are in gender differences and gender-role development in children. Author of the textbook *Half the Human Experience: The Psychology of Women,* she is currently President of the Society for the Scientific Study of Sexuality and is a Fellow of the American Psychological Association. She has received many other honors, including an award for excellence in teaching at Bowling Green State University, the Chancellor's Award for teaching at the University of Wisconsin, and the Kinsey Award from the Society for the Scientific Study of Sexuality for career contributions to sex research. She is married to John DeLamater.

John D. DeLamater, Professor of Sociology at the University of Wisconsin–Madison, received his education at the University of California, Santa Barbara and the University of Michigan. He created the Human Sexuality course at the University of Wisconsin in 1975 and has taught it regularly since then. His research and writing are focused on social and psychological influences on human sexuality; his recent work is in the areas of HIV/AIDS and STD prevention. He is the coauthor of the textbook *Social Psychology.* He is a Fellow of the Society for the Scientific Study of Sexuality and editor of *The Journal of Sex Research.* He has received awards for excellence in teaching from the Department of Sociology and the Interfraternity Council-Panhellenic Association and is a member of the Teaching Academy at the University of Wisconsin. He regularly teaches a seminar for graduate students on teaching undergraduate courses. He is married to Janet Hyde.

Contents in Brief

Contents

CHAPTER 11 Sexuality and the Life Cycle: Childhood and Adolescence 284

CHAPTER 20 Sexually Transmitted Diseases 521

Preface

The sexual scene seems to be in a constant state of flux. Every year there are new advances in the prevention and treatment of AIDS. New methods of contraception are developed and made available. New discoveries are made about causes and treatment of sexual disorders. Many things have changed in the new edition of this textbook as well; at the same time, we have retained the features of this book that users have praised over more than two decades. First we will describe the successful features that we have kept in place. Then we will provide an overview of new features in this edition.

This text has a unique combination of three features that are of utmost importance in a textbook: a writing style that is accessible and appealing to the student; coverage that is comprehensive and interdisciplinary; and excellent scholarship. Our goal in this text is to provide the best in all three of these features—accessibility, comprehensiveness, and scholarship. This approach has been well-received in previous editions, and we have worked to strengthen these features in the seventh edition.

Plan of the Book

First and foremost, we have tried to keep in mind at all times that students *want* to learn about sexuality and that our job as writers is to help them learn. We have covered topics completely, in as clear a presentation as possible, and have made a special effort to use language that will enlighten rather than intimidate; because students so often know only slang terminology regarding sex, we have included slang terms in parentheses following definitions of scientific sexual terms, to connect the two terminologies. In the selection and preparation of illustrations for the book, the goal has always been to convey as much information as possible, simply and clearly.

The book assumes no prior college courses in biology, psychology, or sociology. It is designed as an introduction following the three major objectives of our own courses in human sexuality:

1. To provide practical information needed for everyday living (information about sexual anatomy, contraception, and sexually transmitted diseases, for example) and to deal with problems in sexual functioning (such as erection problems or inability to have an orgasm).

2. To help students feel more comfortable with thinking and talking about sex, both to minimize their own personal anguish about a tension-causing topic and to help them become responsible decision-makers in an important aspect of their lives.

3. To familiarize students with methods used in research on sexuality, and particularly with problems inherent in some of these methods, so that they can read research reports critically and intelligently.

Our own courses are surveys, designed to provide students with a broad range of information about sexuality. Reflecting that approach, this book is intended to be complete and balanced in its coverage, so that students will want to save it after the course for use as a reference in future years.

The background of the Hyde and DeLamater team is quite compatible with this interdisciplinary approach. Janet Hyde's original graduate training was in psychology, with specialties in behavior genetics and statistics; later her interests expanded to include psychology of women and gender roles. As a result, her expertise is in the biological and psychological viewpoints. John DeLamater's graduate training was in psychology and sociology, so his expertise includes sociological and cultural viewpoints. The team's goal has been to cover all aspects of sexuality with integrity.

Nonetheless, for instructors who feel they lack the time to deal with all the material or who are not prepared to cover certain topics, the chapters have been written to be fairly independent. For example, any of the following chapters could be omitted without loss of continuity: Chapter 13, "Attraction, Intimacy, and Love"; Chapter 17, "Sexual Coercion"; Chapter 21, "Ethics, Religion, and Sexuality"; Chapter 22, "Sex and the Law"; and Chapter 23, "Sexuality Education."

Certainly there are some aspects of sexuality today that are very serious. Nonetheless, it is our belief that, in modern American culture, we are in danger of taking some aspects of sexuality far too seriously. We may not be serious about it in the same way as were our Victorian ancestors, but we are serious nonetheless—serious about whether we are using the best and most up-to-date sexual techniques, serious about whether our partners are having as many orgasms as possible, and so on. To counteract this tendency, we have tried to use a light touch, with occasional bits of humor, in this book. We are not advocating that we treat sex in a flippant or frivolous manner, or that we ignore serious issues such as STDs and sexual coercion, but rather that we keep it all in perspective and remember that there are some very funny things about sex.

One thing we are serious about is the quality of research. The quality of sex research is highly variable, to put it mildly. Journalists think they are sex researchers if they have interviewed 10 people and written a book about it! We see other sexuality textbooks that cite an article from the local newspaper with equal authority to a refereed journal article

from *The New England Journal of Medicine* or *The Journal of Sex Research*—and readers have to do a lot of detective work to find out what the real source is for a statement. We believe that it is our responsibility as textbook authors to sift through available studies and *not* cite all of them, but rather present those that are of the best quality. We are pleased to observe that the quality of sex research improves every decade. In this edition we were able to prune out many studies of lesser quality and rely much more on recent studies that are of excellent quality, in terms of sampling, research design, and measurements.

The Seventh Edition

The seventh edition represents a major revision. What's new about the seventh edition? Two changes are the most substantial and pervasive: a focus on **sexuality and the media** and integration of the **World Wide Web.** Each of these changes is detailed in a section below.

Beyond these new themes, data and theories have been updated throughout. There is a new Focus box in Chapter 5 on the debate over the treatment of intersex individuals. Chapter 8, "Contraception and Abortion," and Chapter 20, "Sexually Transmitted Diseases," have been updated thoroughly with new statistics and new research findings. In the discussion of sexual disorders and therapy, a cognitive-physiological model is emphasized and cognitive-behavioral therapy is presented, in contrast to older methods of behavior therapy pioneered by Masters and Johnson. The discussion of communication (Chapter 10) now includes a section on gender differences in communication.

Sexuality and the Media

A new focus throughout the book is on sexuality in the media. We begin in Chapter 1 by including the media, alongside religion and science, as having a major impact on sexuality today. In Chapter 3, on methodology in sex research, we have added a major section on media content analysis, supplying students with the methods needed to study the media scientifically. Chapter 11 contains a Focus box on "The Impact of Mass Media on Adolescent Sexuality." As another example, a new Focus box has been added to Chapter 15, "Bisexual Characters in Film." Media exercises for students have been added to the Instructor's Manual. Today we need to be teaching media literacy, and no arena provides better material for the analysis of media coverage than sexuality.

The World Wide Web

A particular kind of media, the World Wide Web, has dramatically changed many Americans' ways of thinking and conducting their work. The web is loaded with sex, some of it good and some of it not so good. The web has fantastic potential for doing good things—for example, providing accurate sex information for those with no other access to it. It also has the potential for harm—for example, conveying inaccurate information or increasing the market for kiddie porn. We have therefore integrated web material throughout the book. Informational websites are listed at the end of each chapter and on the text's web site for easy hot-linking: http://www.mhhe.com/hyde7. Chapter 18, "Sex for Sale," has a section on sex on the Internet. As another example, Chapter 19, "Sexual Disorders and Sex Therapy," includes a discussion of the pros and cons of sex therapy on-line. Exercises in the Instructor's Manual help students learn how to distinguish high-quality websites from the rest.

Multicultural/Multiethnic Perspectives

Multicultural perspectives have been a part of this book beginning with the very first edition, published in 1978. The multicultural/multiethnic perspective has come into full flower in this edition, in which we have been able to integrate into nearly every chapter studies on sexuality in various cultures around the world, as well as on various ethnic groups in the United States. Here are a few examples:

- Chapter 1 contains one Focus box on sex in China and another on sexuality in three preliterate cultures.
- Chapter 3, "Sex Research," has a section on methodological issues in conducting research with persons who are members of ethnic minorities.
- Chapter 4, "Sexual Anatomy," includes a discussion of female genital cutting.

- Chapter 6, "Menstruation and Menopause," contains a Focus box on the menstrual experience among indians in South Africa.
- Chapter 8, "Contraception and Abortion," includes a Focus box on abortion in cross-cultural perspective.
- Chapter 14, "Gender Roles," covers research on Native American two-spirits.
- Chapter 15, "Sexual Orientation," contains a section on ethnicity and sexual orientation.
- Chapter 23, "Sexuality Education," includes a new section devoted to multicultural sexuality education.

One of the most important improvements in multicultural coverage was made possible by the publication of the University of Chicago NHSLS (Laumann et al., 1994). This well-sampled survey had large numbers of African Americans and Latinos, and smaller numbers of Asian Americans and Native Americans. As a result, most of our tables on sexual behavior in the United States now include data for these groups separately, allowing students to obtain a much better knowledge of sexuality and ethnicity in the United States.

The fabric art used as the basis of the design elements throughout this book are called Molas. This layered fabric art is made by Cuna Indians in the San Blas Islands off the eastern coast of Panama.

Learning Resources

This book also emphasizes learning resources for the student. There is a running glossary of terms, with pronunciations. Chapter outlines appear at the opening of each chapter. Since research in cognitive psychology indicates that learning and memory are improved considerably if the learner knows the organization of the material in advance, the chapter outlines are designed to facilitate this learning. A comprehensive Summary concludes each chapter. There are Review Questions and Questions for Thought, Discussion, and Debate at the end of each chapter. These questions are designed to help students review for exams as well as stimulate them to think beyond the material presented in the text. Questions that ask students to apply what they have learned are in each set of Questions for Thought, Discussion, and Debate.

Special Features

An Appendix—"A Directory of Resources in Human Sexuality"—lists the names, addresses, and functions of many major organizations in the field of human sexuality, on topics ranging from birth control to toll-free hotlines on sexually transmitted diseases, to scholarly journals. We hope that this listing will serve as a useful reference for both instructors and students. We personally have used the one in the previous edition a great deal.

We hope you will visit our web site at http://www.mhhe.com/hyde7. The site will feature author newsletters, a directory of on-line information resources, sample study guide exercises and test questions, and more.

Ancillary Package

The supplements listed here may accompany *Hyde and DeLamater, Understanding Human Sexuality,* seventh edition. Please contact your local McGraw-Hill representative for details concerning policies, prices, and availability as some restrictions may apply.

For the Instructor

Instructor's Manual by Herb Laube, University of Minnesota

This extensively revised and expanded manual provides the necessary tools for both the seasoned professor and professors new to the human sexuality course. The new edition of the Instructor's Manual contains for each chapter, learning objectives, a chapter outline, lecture suggestions, classroom activities, discussion questions, guest speaker suggestions, and multimedia resources. The learning objectives are integrated throughout each of the print supplements for consistent learning. Lecture suggestions and classroom activities focus on the major concepts of the text as well as topics identified as difficult for students by our panel of reviewers. In addition, with the strong emphasis on sexuality and the media in this edition, the Instructor's Manual includes media-specific exercises. Discussion questions provide interesting and controversial topics to help promote in-class participation and debate about text coverage. Guest speaker suggestions are included to help bring real world applications into the classroom. Multimedia resources include a listing of pertinent films/videos and overhead transparencies for each chapter of the text. Finally, each chapter contains website resources that direct instructors to the McGraw-Hill Human Sexuality Website where hot links to related sites are available. Internet exercises will also be included to help students learn how to distinguish high-quality websites from the rest.

Test Bank
by Gary Hampe, University of Wyoming

The Test Bank was extensively reviewed and revised in this edition to include over 3100 questions specifically related to the main text. Questions include a wide range of multiple-choice, true-false, short answer, and essays from which instructors can create their test material. Page references are also included for easier instructor reference.

Computerized Test Bank
by Gary Hampe, University of Wyoming

This computerized test bank contains all of the questions in the print version and is available in both Macintosh and Windows platforms.

The McGraw-Hill Human Sexuality Psychology Image Database, Overhead Transparencies and CD-ROM

This set of over 160 images was developed using the best selection of our human sexuality art and tables and is available in both a print overhead transparency set as well as in a CD-ROM format with a fully functioning editing feature. Instructors can add their own lecture notes to the CD-ROM as well as organize the images to correspond to their particular classroom needs.

Videocases in Human Sexuality

The package includes a series of four 35-minute videotapes with spontaneous, unrehearsed interviews. A video guide with case vignettes and classroom discussion questions is also included. Please see your McGraw-Hill sales representative for further information on policy, price, and availability.

PowerPoint Slide Presentation

This set of PowerPoint slides follows the chapter organization of **Understanding Human Sexuality,** seventh edition, and includes related images for a more effective lecture presentation.

The AIDS Booklet

The fourth edition by Frank D. Cox of Santa Barbara City College is a brief but comprehensive introduction to acquired immune deficiency syndrome, which is caused by HIV (human immunodeficiency virus) and related viruses.

Annual Editions—Human Sexuality 1999/00

Published by Dushkin/McGraw-Hill, this is a collection of articles on topics related to the latest research and thinking in human sexuality from over 300 public press sources. These editions are updated annually and contain helpful features including a topic guide, an annotated table of contents, unit overviews, and a topical index. An instructor's guide containing testing materials is also available.

Sources: Notable Selections in Human Sexuality

This is a collection of articles, books excerpts, and research studies that have shaped the study of human sexuality and our contemporary understanding of it. The selections are organized topically around major areas of study within human sexuality. Each selection is preceded by a headnote that establishes the relevance of the article or study and provides biographical information on the author.

Taking Sides: Clashing Views on Controversial Issues in Human Sexuality

This is a debate-style reader designed to introduce students to controversial viewpoints on the field's most crucial issues. Each issue is carefully framed for the student, and the pro and con essays represent the arguments of leading scholars and commentators in their fields. Instructor's guide containing testing materials is also available.

Web Site

Please visit our human sexuality web site for additional information on this title as well as text–specific resources and web links for both instructors and students. Our web site address is http://www.mhhe.com/hyde.

For The Student

Student Study Guide
by Pearl Hawe, New Mexico State University

An exciting new addition to the supplement package for the seventh edition is our Student Study Guide. This study guide has been written specifically for students using Hyde/DeLamater, **Understanding Human Sexuality,** seventh edition. The Study Guide provides a complete introduction to students for successful understanding and applications of human sexuality as told by Hyde and DeLamater. Features include for each chapter, learning objectives from the Instructor's Manual, a brief outline, key term exercise, short answer and essay questions, self-assessment exercises, sample quizzes that include multiple choice and true/false questions, and Internet resources. The Learning Objectives are also included in the Instructor's Manual to assure students that they are on the right track to understanding the relevant material in each chapter. The outline and key terms put the chapter into perspective. Short answer and essay questions will encourage critical thinking and offer suggestions for paper topics. Self-assessment exercises give students a personal understanding of the material. The sample quizzes are modeled after the Test Bank to provide the opportunity for maximum performance on exams. Finally, Internet sites are provided so students know where to look for additional information about topics that interest them.

Acknowledgments

Over the course of the first six editions, numerous reviewers contributed to the development of *Understanding Human Sexuality.* We don't have the space to cite them all, but their contributions remain and we are grateful to them.

In addition, we are enormously grateful to the following reviewers who helped shape this revision: **Manuscript Reviews:** Elaine Baker, *Marshall Univeristy;* Larry J. Bloom, *Colorado State University;* Catherine Schuster, *Western Kentucky University;* Michael R. Stevenson, *Ball State University;* Susan L. Woods, *Eastern Illinois University;* David N. Yarbrough, *University of Southwestern Louisiana.* **Reviewed Fifth edition or Sixth edition:** Charles Chase, *West Texas A&M;* Priti Gupta, *Purdue University;* Morton G. Harmatz, *University of Massachusetts at Amherst;* Matt Hobson, *University of Northern Iowa;* Larry M. Lance, *The University of North Carolina at Charlotte;* Susan Lyman, *University of Southwestern Louisiana.* **Expert Reviews:** Milton Diamond, *University of Hawaii at Manoa;* Suzanne G. Frayser, *Colorado College;* Charlene Muehlenhard, *University of Kansas;* Scott J. Spear, *University of Wisconsin–Madison.* **Video Case Series Reviews:** Harold Herzog, *Western Carolina University;* Louis Janda, *Old Dominion University.*

Wendy Theobald, our library researcher, deserves special thanks for tracking down hundreds of articles and books at the numerous libraries spread around the University of Wisconsin. She was creative and thorough and often found studies that even we were not aware of. This book is noticeably better because of her work.

Finally, we owe many thanks to the editors and staff at McGraw-Hill: Mickey Cox, psychology editor, who has offered consistent support and encouragement and many excellent ideas for the book; Sharon Geary, developmental editor, who made many improvements; and Susan Brusch, production supervisor, who did an excellent job carrying the book through to completion.

We love teaching our human sexuality courses and we've loved writing and rewriting this text for it. We hope that you will enjoy reading it, learning from it, and teaching with it.

Janet Shibley Hyde
John D. DeLamater

chapter

1

SEXUALITY IN PERSPECTIVE

CHAPTER HIGHLIGHTS

Sex and Gender

Understanding Sexuality: Religion, Science, and the Media
Religion
Science
The Media

Cross-Cultural Perspectives on Sexuality

Cross-Species Perspectives on Sexuality

"You're incredible," he said as the dress fell away from her. "Oh Page . . ." He devoured her with his lips, his hands, and slowly she undressed him, until at last they stood naked together in the moonlight. He lifted her gently onto the bed, and caressed her with his lips until she moaned in pleasure, arched toward him, and then led him toward her. Their union was a powerful one, throbbing, arching for what they had both longed for, until at last they both exploded in unison, and lay spent in each other's arms, stunned by the force of what they felt for each other.*

Human sexual behavior is a diverse phenomenon. It occurs in different physical locations and social contexts, consists of a wide range of specific activities, and is perceived differently by different people. An individual engages in sexual activity on the basis of a complex set of motivations and organizes that activity on the basis of numerous external factors and influences. Thus, it is unlikely that the tools and concepts from any single scientific discipline will suffice to answer all or even most of the questions one might ask about sexual behavior. . . . Much of the previous scientific research on sexuality has been conducted by biologists and psychologists and has thus focused on sexual behavior purely as an "individual level" phenomenon. . . . Human sexual behavior is only partly determined by factors originating within the individual. In addition, a person's socialization into a particular culture, his or her interaction with sex partners, and the constraints imposed on him or her become extremely important in determining his or her sexual activities.†

Odd though it may seem, both of the quotations above are talking about the same thing—sex. The first quotation is from a romance novel. It stimulates the reader's fantasies and arousal. The second is from a scholarly book about sex. It stimulates the brain but not the genitals. From these two brief excerpts we can quickly see that the topic of sexuality is diverse, complex, and fascinating.

Introductory textbooks on most subjects generally begin with a section designed to motivate students to study whatever topic the text is about. No such section appears in this book; the reason people want to study sex is obvious, and your motivation for studying it is probably already quite high. Sex is an important force in many people's lives, so there are practical reasons for wanting to learn about it. Most people are curious about sex, particularly because exchanging sexual information is somewhat taboo in our culture, so curiosity also motivates us to study sex. Finally, most of us at various times experience problems with our sexual functioning or wish that we could function better, and we hope that learning more about sex will help us. This book is designed to meet all those needs. So let us consider various perspectives on sexuality—that is, the effects of religion, science, and culture on our understanding of sexuality. This will give you a glimpse of the forest before you study the trees: sexual anatomy and physiology (the "plumbing" part), and sexual behavior (the "people" part), which is discussed in later chapters. First we must draw an important distinction, between sex and gender.

*Danielle Steel. (1994). *Accident*. New York: Dell.

†Edward O. Laumann, John H. Gagnon, Robert T. Michael, and Stuart Michaels. (1994). *The social organization of sexuality: Sexual practices in the United States.* Chicago: University of Chicago Press.

Sex and Gender

Sometimes the word "sex" is used ambiguously. Sometimes it refers to being male or female, and sometimes it refers to sexual behavior or reproduction. In most cases, of course, the meaning is clear from the context. If you are filling out a job application form and one item says, "Sex: ," you do not write, "I like it" or "As often as possible." It is clear that your prospective employer wants to know whether you are a male or female. In other cases, though, the meaning is ambiguous. For example, when a book has the title *Sex and Temperament in Three Primitive Societies,* what is it about? Is it about the sexual practices of primitive people and whether having sex frequently gives them pleasant temperaments? Or is it about the kinds of personalities that males and females are expected to have in those societies? Not only does this use of "sex" create ambiguities, but it also clouds our thinking about some important issues.

To remove—or at least reduce—this ambiguity, the term "sex" will be used in this book in contexts referring to sexual anatomy and sexual behavior, and the term **gender** will be used to refer to the state of being male or female.

This is a book about sex, not gender; it is about sexual behavior and the biological, psychological, and social forces that influence it. Of course, although we are arguing that sex and gender are conceptually different, we would not try to argue that they are totally independent of each other. Certainly gender roles—the ways in which males and females are expected to behave—exert a powerful influence on the way people behave sexually, and so one chapter will be devoted to gender roles and their effects on sexuality.

How should we define "sex," aside from saying that it is different from "gender"? A biologist might define sexual behavior as "any behavior that increases the likelihood of gametic union [union of sperm and egg]" (Bermant & Davidson, 1974). This definition emphasizes the reproductive function of sex. However, particularly in the last few decades, technologies such as the birth control pill have been developed that allow us to separate reproduction from sex. Most Americans now use sex not only for procreation but also for recreation.[1]

The noted sex researcher Alfred Kinsey defined "sex" as behavior that leads to orgasm. Although this definition has some merits (it does not imply that sex must be associated with reproduction), it also presents some problems. If a wife has intercourse with her husband but does not have an orgasm, was that not sexual behavior for her?

To try to avoid some of these problems, **sexual behavior** will be defined in this book as *behavior that produces arousal and increases the chance of orgasm.*[2]

Understanding Sexuality: Religion, Science, and the Media

Religion

Throughout most of recorded history, at least until about 100 years ago, religion (and rumor) provided most of the information that people had about sexuality. Thus the ancient Greeks openly acknowledged both heterosexuality and homosexuality in their society and explained the existence of the two in a myth in which the original humans were double creatures with twice the normal number of limbs and organs; some were double males, some were double females, and some were half male and half female (LeVay, 1996). The gods, fearing the power of these creatures, split them in half, and forever after each one continued to search for its missing half. Heterosexuals were thought to have resulted from the splitting of the half male, half female; male homosexuals, from the splitting of the double male; and female homosexuals, from the splitting of the double female. People were thought to long to unite with their other half. It was through this mythology that the ancient Greeks understood sexuality and sexual desire.

---

[1]Actually, even in former times sex was not always associated with reproduction. For example, a man in 1850 might have fathered 10 children; using a very conservative estimate that he engaged in sexual intercourse 1500 times during his adult life (once a week for the 30 years from age 20 to age 50), one concludes that only 10 in 1500 of those acts, or less than 1 percent, resulted in reproduction.

[2]This definition, though an improvement over some, still has its problems. For example, consider a woman who feels no arousal at all during intercourse. According to the definition, intercourse would not be sexual behavior for her. However, intercourse would generally be something we would want to classify as sexual behavior. It should be clear that defining "sexual behavior" is very difficult. The definition given is good, though not perfect.

Gender: The state of being male or female.
Sexual behavior: Behavior that produces arousal and increases the chance of orgasm.

The fifteenth-century Christian believed that "wet dreams" (nocturnal emissions) resulted from intercourse with tiny spiritual creatures called *incubi* and *succubi,* a notion put forth in a papal bull of 1484 and a companion book, the *Malleus Maleficarum* ("witch's hammer"); the person who had wet dreams was guilty of sodomy (see Chapter 21) as well as witchcraft.

The Muslim believed that sexual intercourse was one of the finest pleasures of life, reflecting the teachings of the great prophet Muhammad. Moreover, extraordinary sexual prowess is attributed to Muhammad, who had several wives.

People of different religions hold different understandings of human sexuality, and these religious views often have a profound impact on people growing up with them. A detailed discussion of religion and sexuality is provided in Chapter 21.

Science

It was against this background of religious understandings of sexuality that the scientific study of sex began in the nineteenth century, although, of course, religious notions continue to influence our ideas about sexuality to the present day. In addition, the groundwork for an understanding of the biological aspects of sexuality had already been laid by the research of physicians and biologists. The Dutch microscopist Anton van Leeuwenhoek (1632–1723) and his student John Ham had discovered sperm swimming in human semen. In 1875 Oscar Hertwig first observed the actual fertilization of the egg by the sperm in sea urchins, although the ovum in humans was not directly observed until the twentieth century.

A major advance in the scientific understanding of the psychological aspects of human sexuality came with the work of the Viennese physician Sigmund Freud (1856–1939), founder of psychiatry and psychoanalysis. His ideas will be discussed in detail in Chapter 2.

It is important to recognize the cultural context in which Freud and the other early sex researchers began their research and writing. They began their

(a)

(b)

Figure 1.1   Two important early sex researchers. *(a)* Sigmund Freud. *(b)* Henry Havelock Ellis.

Figure 1.2 The Victorian era, from which Freud and Ellis emerged, was characterized by extreme sexual repression. Here are some apparatuses that were sold to prevent onanism (masturbation).

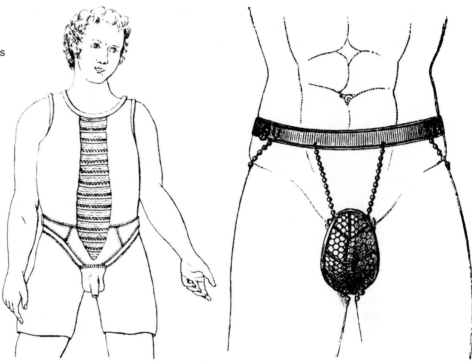

Fig. 350. Korsett von Lajade-Lafond zur Verhinderung der Onanie. Nach Fleck: Die Verirrungen des Geschlechtstriebes. Stuttgart 1830.

Fig. 351.
Onaniebandage   Onaniebandage
für weibliche Patienten.   für männliche Patienten.

work in the Victorian era, the late 1800s, both in the United States and Europe. Norms about sexuality were extraordinarily rigid and oppressive. Historian Peter Gay characterized this repressive aspect of Victorian cultural norms as

> . . . a devious and insincere world in which middle-class husbands slaked their lust by keeping mistresses, frequenting prostitutes, or molesting children, while their wives, timid, dutiful, obedient, were sexually anesthetic and poured all their capacity for love into their housekeeping and their child-rearing. (Gay, 1984, p. 6)

Certainly traces of these Victorian attitudes remain with us today. Yet at the same time the actual sexual behavior of Victorians was sometimes in violation of societal norms (see Focus 1.1). In his history of sexuality in the Victorian era, Gay documents the story of Mabel Loomis Todd, who, though married, carried on a sensual and lengthy affair with Austin Dickinson, a community leader in Amherst, Massachusetts. Many people actually knew about the "secret" affair, yet Mrs. Loomis did not become an outcast (Gay, 1984). Doubtless, this wide discrepancy between Victorian sexual norms

Focus 1.1
A Victorian Sex Survey

In the late 1800s Queen Victoria reigned in England, and both there and in the United States the ideal was to repress sexuality as much as possible. Women, particularly, were to have no sexual desires. And standards of modesty were so great that pianos had "limbs" rather than vulgar "legs."

Out of the Victorian environment emerged a remarkable woman, Dr. Clelia Mosher. Born in Albany, New York, in 1863, she began college at Wellesley and finished at Stanford. For her master's degree from Stanford, she collected data to debunk a popular myth of the time: that women could breathe only high in the chest, whereas men breathed deeply. Mosher concluded, quite reasonably, that any differences resulted purely from women being laced into tight-fitting corsets. Mosher began medical school at Johns Hopkins when she was 32 and earned her M.D. degree four years later. Interestingly, Gertrude Stein, the famous author, entered the same medical school one year after Mosher, but never finished her degree.

Over a period of 30 years, beginning when she was an undergraduate, Mosher conducted a sex survey of Victorian women, most of whom were born around the time of the Civil War. In all, she administered her nine-page questionnaire to 47 women. (This sample is, admittedly, small and nonrandom; the criteria for valid survey research are discussed in Chapter 3.) Many of the women were faculty wives at universities, or women from Mosher's medical practice, and surely they were a select sample to agree to answer the questions; 81 percent had attended college, a high level of education for women in those days. Nonetheless, the survey is remarkable because—despite well-known ideas about Victorian women—this is the only actual survey of those women known to exist. Here are some interesting findings from the study:

- Despite the Victorian stereotype that women felt no sexual desire, 80 percent of the women who answered the question said that they felt a desire for sexual intercourse.

- Thirty-four of the women (72 percent) indicated that they experienced orgasm. Interestingly, Mosher worded the question "Do you always have a venereal orgasm?" thus assuming that orgasm was to be expected.

- Mosher suspected that women's relative slowness at reaching orgasm might be a cause of marital conflicts. Many of her respondents supported this idea. One said that sex had been unpleasant to her for years because of her "slow reaction," but "orgasm [occurs] if time is taken." Another complained that "Men have not been properly trained." And for some, not reaching orgasm was psychologically devastating (one can't help thinking that things haven't changed so much from the 1890s to the present).

- At least 30 of the women (64 percent) used some form of birth control. Douching was the most popular method, followed by withdrawal and "timing." Several women's husbands used a "male sheath," and two women used a "rubber cap over the uterus." One woman used cocoa butter. She did not explain how or why.

Clelia Mosher's survey is fascinating because it demonstrates that despite the Victorian era's repressive teachings, some women still managed to enjoy sex. True, some were affected by Victorian mores; three of the women said their ideal would be to abstain from intercourse entirely. But the majority of the women still expressed sexual desires, experienced orgasms, and seemed to enjoy sex with their husbands.

Source: Kathryn A. Jacob. (1981). The Mosher report. *American Heritage*, pp. 57–64.

and actual behavior created a great deal of personal tension. That tension probably propelled a good many people into Dr. Freud's office, providing data for his theory, which emphasizes sexual tensions and conflict.

An equally great—though not so well known—early contributor to the scientific study of sex was Henry Havelock Ellis (1859–1939). A physician in Victorian England, he compiled a vast collection of information on sexuality—including medical and anthropological findings, as well as case histories—which was published in a series of volumes entitled *Studies in the Psychology of Sex* beginning in 1896. Havelock Ellis was a remarkably objective and tolerant scholar, particularly for his era. He believed that women, like men, are sexual creatures. A sexual reformer, he believed that sexual deviations from the norm are often harmless, and he urged society to accept them. In his desire to collect information about human sexuality rather than to make judgments about it, he can be considered the forerunner of modern sex research (for an autobiography, see Ellis, 1939; numerous biographies also exist).

Another important figure in nineteenth-century sex research was Richard von Krafft-Ebing (1840–1902). His special interest was "pathological" sexuality, and he managed to collect over 200 case histories of pathological individuals, which appeared in his book entitled *Psychopathia Sexualis*. His work tended to be neither objective nor tolerant. One of his case histories is presented in Chapter 16.

One other early contributor to the scientific understanding of sexuality deserves mention, the German Magnus Hirschfeld (1868–1935). He founded the first sex research institute and administered the first large-scale sex survey, obtaining data from 10,000 people on a 130-item questionnaire. Unfortunately, most of the information he amassed was destroyed by Nazi hoodlums. Hirschfeld also established the first journal devoted to the study of sex, established a marriage counseling service, worked for legal reforms, and gave advice on contraception and sex problems. His special interest, however, was homosexuality. Doubtless some of his avant-garde approaches resulted from the fact that he was himself both homosexual and a transvestite. His contributions as a pioneer sex researcher cannot be denied (Bullough, 1994).

In the twentieth century, major breakthroughs in the scientific understanding of sex came with the massive surveys of human sexual behavior in the United States conducted by Alfred Kinsey and his colleagues in the 1940s and with Masters and Johnson's investigations of sexual disorders and the physiology of sexual response. At about the same time that the Kinsey research was being conducted, some anthropologists—most notably Margaret Mead and Bronislaw Malinowski—were beginning to collect data on sexual behavior in other cultures. Other, smaller investigations also provided important information. By the 1990s we had a rich array of sex research, including major national surveys (e.g., Laumann et al., 1994), detailed investigations of sexual disorders and sexual orientation, and studies of the biological processes underlying sexual response.

The scientific study of sex has not emerged as a separate, unified academic discipline like biology or psychology or sociology. Rather, it tends to be interdisciplinary—a joint effort by biologists, psychologists, sociologists, anthropologists, and physicians (see Figure 1.3). In a sense, this is a major virtue in our current approach to understanding sexuality, since it gives us a better view of humans in all their sexual complexity.

The Media

In terms of potency of influence, the mass media in America today may play the same role that religion did in previous centuries. The television set is on about 7 hours per day in the average American home (Brown & Steele, 1996). Primetime television programs feature an average of 10 instances of sexual behavior per hour. In a study of sex on the soaps, the most frequent sexual activity was heterosexual intercourse between unmarried persons (Greenberg & Busselle, 1996)—although this pattern is far from what occurs in the real world, where marital sex is most frequent. References to safe sex—both for STD prevention and pregnancy prevention—are rare. A typical viewer sees about 25 instances of sexual behavior for every one instance of preventive behavior or discussion of prevention (Brown & Steele, 1996).

Figure 1.3 The history of scientific research on sex.

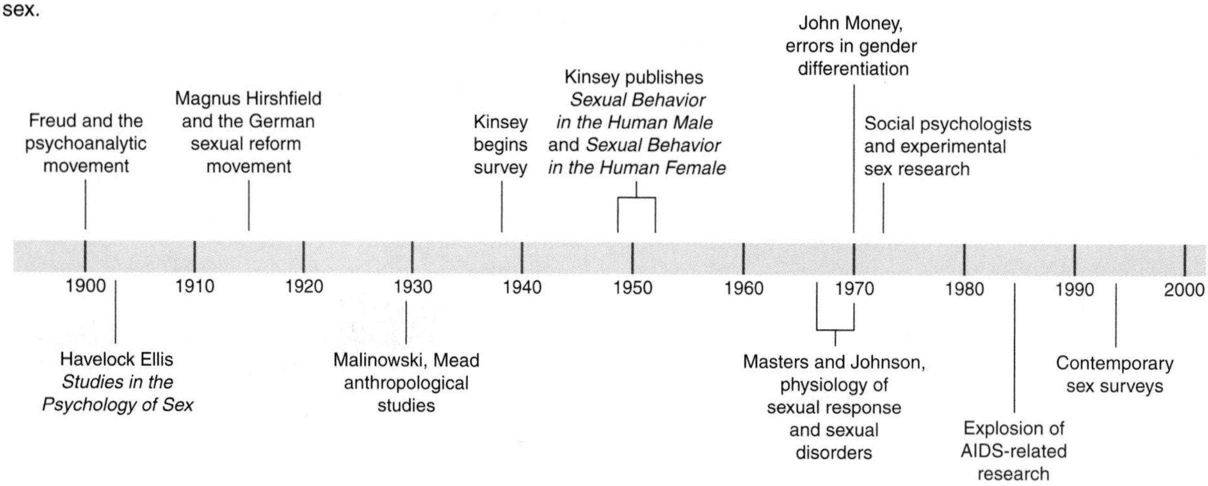

							John Money, errors in gender differentiation			
Freud and the psychoanalytic movement	Magnus Hirshfield and the German sexual reform movement		Kinsey begins survey	Kinsey publishes *Sexual Behavior in the Human Male* and *Sexual Behavior in the Human Female*			Social psychologists and experimental sex research			
1900	1910	1920	1930	1940	1950	1960	1970	1980	1990	2000
Havelock Ellis *Studies in the Psychology of Sex*			Malinowski, Mead anthropological studies			Masters and Johnson, physiology of sexual response and sexual disorders		Explosion of AIDS-related research	Contemporary sex surveys	

Figure 1.4 Sexual portrayals in the media have become much more explicit. In this 1998 scene from the prime-time show *Melrose Place*, Lexi seduces Ryan, an employee of a competing advertising firm. The scene ends with sex on his boss's desk.

In short, the average American's views about sexuality are likely to be much more influenced by the mass media than by scientific findings. Communications theorists believe that the media can have three types of influence (Brown & Steele, 1996). The first, called **mainstreaming**, refers to the notion that people begin to think that what they see on television and other media really represents the mainstream of what happens in our culture. For example, college students who watch the soaps are more likely than nonviewing students to overestimate the incidence of divorce. The second influence is agenda-setting. News reporters select what to report and what to ignore and, within the stories they report, what to emphasize. For example, in 1998 the media chose to highlight the sexual dalliances of President Bill Clinton, letting the public know that these matters were important. In contrast, the illicit sexual activities of President John F. Kennedy were not revealed during his presidency. The media in many ways tell us what the agenda is to which we should pay attention. The third influence is social learning, a theory we will take up in

Mainstreaming: In communications theory, the view that exposure to the mass media makes people think that what they see there represents the mainstream of what really occurs.

detail in Chapter 2. The contention here is that characters on television, in the movies, or in romance novels may serve as role models whom we imitate, perhaps without even realizing it. Research has found, for example, that teens who watch more sexy television engage in first intercourse earlier than do other teens (Brown & Steele, 1996).

In the chapters that follow, we will examine the content of the media on numerous sexual issues and we will consider what the effects of exposure to this media content might have on viewers.

Let us now consider the perspectives on sexuality that are provided by scientific observations of humans in a wide variety of societies.

Cross-Cultural Perspectives on Sexuality

Ethnocentrism tends to influence our understanding of human sexual behavior. Most of us have had experience with sexuality in only one culture—the United States, for example—and we tend to view our sexual behavior as the only pattern in existence, and certainly as the only "natural" pattern. But anthropologists have discovered that there are wide variations in sexual behavior and attitudes from one culture to the next. Considering these variations should help us to put our own sexual behavior in perspective.

One source of data on the enormous cross-cultural variations in sexual behavior is the classic study of anthropologist Clellan Ford and psychologist Frank Beach (1951), who surveyed sexual behavior in 190 societies around the world and also sexual behavior across various animal species. Other major sources are massive cross-cultural surveys of sexual practices done by anthropologists Edgar Gregersen (1983) and Suzanne Frayser (1985).

Before we proceed, it is worth noting that Ford and Beach concluded, in 1951, that the United States culture was a relatively sexually restrictive one; the majority of the others studied were far more permissive. This may help us put our own standards in better perspective. On the other hand, most people would agree—and scientific research supports the notion (see Chapters 11 and 12)—that sexual behavior and attitudes among Americans have undergone substantial changes in the

Figure 1.5   Margaret Mead, an anthropologist who contributed much to the early cross-cultural study of sexuality.

half century since Ford and Beach wrote their book. Modern America has been described as an "eroticized" culture (Gregor, 1985).

The major generalization that emerges from cross-cultural studies is that all societies regulate sexual behavior in some way, though the exact regulations vary greatly from one culture to the next (DeLamater, 1987). Apparently no society has seen fit to leave sexuality totally unregulated, perhaps because social disruption would result. As an example, **incest taboos** are nearly universal: Sex is regulated in that intercourse between blood relatives is prohibited (Leavitt, 1989). Most societies also condemn forced sexual relations, such as rape.

Ethnocentrism: The tendency to regard one's own ethnic group and culture as superior to others and to believe that its customs and way of life are the standards by which other cultures should be judged.

Incest taboo: A regulation prohibiting sexual interaction between blood relatives, such as brother and sister or father and daughter.

Focus 1.2
Sexuality in Three Societies

Inis Beag

Inis Beag is a small island off the coast of Ireland. It is probably one of the most naive and sexually repressive societies in the world.

The people of Inis Beag seem to have no knowledge of a number of sexual activities such as French kissing, mouth stimulation of the breast, or hand stimulation of the partner's penis, much less oral sex, or homosexuality. Sex education is virtually nonexistent; parents do not seem to be able to bring themselves to discuss such embarrassing matters with their children, and they simply trust that, after marriage, nature will take its course.

Menstruation and menopause are sources of fear for the island women because they have no idea of their physiological significance. It is commonly believed that menopause can produce insanity; in order to ward off this condition, some women have retired from life in their mid-forties, and a few have confined themselves to bed until death years later.

The men believe that intercourse is hard on one's health. They will desist from sex the night before they are to do a job that takes great energy. They do not approach women sexually during menstruation or for months after childbirth; a woman is considered dangerous to the male at these times.

The islanders abhor nudity. Only babies are allowed to bathe while nude. Adults wash only the parts of their bodies that extend beyond their clothing—face, neck, lower arms, hands, lower legs, and feet. The fear of nudity has even cost lives. Sailors who never learned to swim because it involved wearing scanty clothing have drowned when their ships have sunk.

Premarital sex is essentially unknown. In marital sex, foreplay is generally limited to kissing and rough fondling of the buttocks. The husband invariably initiates the activity. The male-on-top is the only position used, and both partners keep their underwear on during the activity. The man has an orgasm quickly and falls asleep immediately. Female orgasm either is believed not to exist or is considered deviant.

Mangaia

In distinct contrast to Inis Beag is Mangaia, an island in the South Pacific. For the Mangaians, sex—for pleasure and for procreation—is a principal interest.

The Mangaian boy hears of masturbation when he is about 7, and he may begin to masturbate at age 8 or 9. At around age 13 he undergoes the superincision ritual (in which a slit is made on the top of the penis, along its entire length). This ritual initiates him into manhood; more important, however, the expert who performs the superincision gives him sexual instruction. He shows the boy how to perform oral sex, how to kiss and suck breasts, and how to bring his partner to orgasm several times before he has his own orgasm. About

Beyond this generalization, though, regulations vary greatly from one society to the next, and sexual behavior and attitudes vary correspondingly (see Focus 1.2). Let's look at the ways in which various societies treat some key areas of human sexual behavior.

Variations in Sexual Techniques

Kissing is one of the most common sexual techniques in our culture. It is also very common in most other societies (Gregersen, 1983). There are a few societies, though, in which kissing is unknown.

two weeks after the operation, the boy has intercourse with an experienced woman, which removes the scab. She provides him with practice in various acts and positions and trains him to hold back until he can have simultaneous orgasms with his partner.

After this, the Mangaian boy actively seeks out girls, or they seek him out; soon he has coitus every night. The girl, who has received sexual instruction from an older woman, expects demonstration of the boy's virility as proof of his desire for her. What is valued is the ability of the male to continue vigorously the in-and-out action of coitus over long periods of time while the female moves her hips "like a washing machine." Nothing is despised more than a "dead" partner who does not move. A good man is expected to continue his actions for 15 to 30 minutes or more.

The average "nice" girl will have three or four successive boyfriends between the ages of 13 and 20; the average boy may have 10 or more girlfriends. Mangaian parents encourage their daughters to have sexual experiences with several men. They want her to find a marriage partner who is congenial.

At around age 18, the Mangaians typically have sex most nights of the week, with about three orgasms per night. By about age 48, they have sex two or three times per week, with one orgasm each time.

All women in Mangaia apparently learn to have orgasms. Bringing his partner to orgasm is one of the man's chief sources of sexual pleasure.

Mehinaku

Between Inis Beag, where there is little sex and plenty of anxiety, and Mangaia, where there is plenty of sex and little anxiety, is Mehinaku, where there is plenty of sex and plenty of anxiety.

In the central Brazilian village of Mehinaku, sex is believed to be very fascinating and the culture is highly eroticized. There is an openness with children about sex, and children can easily list the names of their parents' extramarital lovers, who are typically many. The men have a very high libido, leading them to compete with each other for women's sexual favors by bringing small gifts such as fish.

On the other hand, the culture is very gender-segregated. There is a men's house and if a woman enters it and sees what she is forbidden to see, she is taken to the woods and gang raped, in a culture that is otherwise very nonviolent. Women are believed to have a much weaker sex drive than men, and there seems to be no recognition of female orgasm. Women's menstruation is believed to be dangerous.

The dreams and mythic stories told by the people testify to their sexual anxieties—for example, those in myths who engage in extramarital sex typically die in fantastic ways. In reality the people continue with a great deal of sexual activity while feeling intense ambivalence and anxiety about it.

Sources: John C. Messenger. Sex and repression in an Irish folk community. In D. N. Suggs and A. W. Miracle (Eds.), (1993). *Culture and human sexuality.* Pacific Grove, CA: Brooks/Cole. Donald S. Marshall. Sexual behavior on Mangaia. In D. S. Marshall and R. C. Suggs (Eds.). (1971). *Human sexual behavior.* New York: Basic Books. Thomas Gregor. (1985). *Anxious pleasures: The sexual lives of an Amazonian people.* Chicago: University of Chicago Press.

For example, when the Thonga of Africa first saw Europeans kissing, they laughed and said, "Look at them; they eat each other's saliva and dirt." There is also some variation in techniques of kissing. For example, among the Kwakiutl of Canada and the Trobriand Islanders, kissing consists of sucking the lips and tongue of the partner, permitting saliva to flow from one mouth to the other. Many Americans might find such a practice somewhat repulsive, but other peoples find it sexually arousing.

Cunnilingus (mouth stimulation of the female genitals) is fairly common in our society, and it occurs in a few other societies as well, especially in the South Pacific. A particularly interesting variation is reported on the island of Ponape; the man places a fish in the woman's vulva and then gradually licks it out prior to coitus.

Inflicting pain on the partner is also a part of the sexual technique in some societies. The Apinaye woman of South America may bite off bits of her partner's eyebrows, noisily spitting them aside. Ponapean men usually tug at the woman's eyebrows, occasionally yanking out tufts of hair. The Trukese women of the South Pacific poke a finger into the man's ear when they are highly aroused. People of various societies bite their partners to the point of drawing blood and leaving scars; most commonly both men and women mutually inflict pain on each other (Frayser, 1985).

The frequency of intercourse for married couples varies considerably from one culture to the next. The lowest frequency seems to be among the Irish natives of Inis Beag, discussed in Focus 1.2, who engage in intercourse perhaps only once or twice a month; however, the anthropologists who studied them were unable to determine exactly how often couples did have sex because so much secrecy surrounds the act. At the opposite extreme, the Mangaians, also described in Focus 1.2, have intercourse several times a night, at least among the young. The Santals of southern Asia copulate as often as five times per day every day early in marriage (Gregersen, 1983). Surveys of United States sexuality in the 1990s indicate that our frequency of intercourse is about in the middle compared with other societies (e.g., Laumann et al., 1994).

Very few societies actually encourage people to engage in sexual intercourse at particular times (Frayser, 1985). Instead, most groups have restrictions that forbid intercourse at certain times or in certain situations. For example, almost every society has a postpartum sex taboo, that is, a prohibition on sexual intercourse for a period of time after a woman has given birth to a baby, although the taboo typically lasts less than six months (Frayser, 1985).

Masturbation

Attitudes toward **masturbation**, or sexual self-stimulation of the genitals, vary widely across cultures. Some societies tolerate or even encourage masturbation during childhood and adolescence, whereas others condemn the practice at any age. Almost all human societies express some disapproval of adult masturbation, ranging from mild ridicule to severe punishment (Gregersen, 1983). On the other hand, at least some adults in all societies appear to practice it.

Female masturbation certainly occurs in other societies. The African Azande woman does it with a phallus made of a wooden root; however, if her husband catches her masturbating, he may beat her severely. The following is a description of the Lesu of the South Pacific, who are one of the few societies that express no disapproval of adult female masturbation:

> A woman will masturbate if she is sexually excited and there is no man to satisfy her. A couple may be having intercourse in the same house, or near enough for her to see them, and she may thus become aroused. She then sits down and bends her right leg so that her heel presses against her genitalia. Even young girls of about six years may do this quite casually as they sit on the ground. The women and men talk about it freely, and there is no shame attached to it. It is a customary position for women to take, and they learn it in childhood. They never use their hands for manipulation. (Powdermaker, 1933, pp. 276–277)

Premarital and Extramarital Sex

Societies differ considerably in their rules regarding premarital sex (Frayser, 1985). At one extreme are the Marquesans of eastern Polynesia. Both boys and girls in that culture have participated in a wide range of sexual experiences before puberty. Their first experience with intercourse occurs with a heterosexual partner who is 30 to 40 years old. Mothers are proud if their daughters have many lovers. Only later does marriage occur. In contrast are the Egyptians of Siwa. In this culture a girl's clitoris is removed at age 7 or 8 in order to decrease her potential for sexual excitement and intercourse. Premarital intercourse is believed to bring shame on the family. Marriage usually occurs around the age of 12 or 13, short-

---

Masturbation: Self-stimulation of the genitals to produce sexual arousal.

ening the premarital period and any temptations it might contain.

These two cultures are fairly typical of their regions. According to one study, 90 percent of Pacific Island societies permit premarital sex, as do 88 percent of African and 82 percent of Eurasian societies; however, 73 percent of Mediterranean societies prohibit premarital sex (Frayser, 1985).

Extramarital sex is more complex and conflicted for most cultures. Extramarital sex ranks second only to incest as the most strictly prohibited type of sexual contact. One study found that it was forbidden for one or both partners in 74 percent of the cultures surveyed (Frayser, 1985). Even when extramarital sex is permitted, it is subjected to regulations; the most common pattern of restriction is to allow extramarital sex for husbands but not wives.

Sex with Same-Gender Partners

There is a wide range of attitudes toward same-gender sexual expression—what we in the United States call "homosexuality"—in various cultures. At one extreme are societies that strongly disapprove of same-gender sexual behavior for people of any age. In contrast, some societies tolerate the behavior for children but disapprove of it in adults. Still other societies actively encourage all their male members to engage in some same-gender sexual behavior, usually in conjunction with puberty rites. In Africa, prominent Siwan men lend their sons to one another, and they discuss their same-gender love affairs as openly as they discuss the love of women. A few societies have a formalized role for the adult gay man that gives him status and dignity.

While there is wide variation in attitudes toward homosexuality and in same-gender sexual behavior, four general rules do seem to emerge (Ford & Beach, 1951; Whitam, 1983): (1) No matter how a particular society treats homosexuality, the behavior always occurs in at least some individuals—that is, same-gender sexuality is found universally in all societies; (2) males are more likely to engage in same-gender sexual behavior than females; and (3) same-gender sexual behavior is never the predominant form of sexual behavior for adults in any of the societies studied.

In the United States and other Western nations, we hold an unquestioned assumption that people have a sexual identity, whether gay, lesbian, bisexual, or heterosexual. Yet sexual identity, as an unvarying, life-long characteristic of the self, is unknown or rare in some cultures, such as Indonesia (Stevenson, 1995). In those cultures, the self and individualism, so prominent in American culture, are downplayed. Instead, a person is defined in relation to others and behavior is seen as much more the product of the situation than of lifelong personality traits. In such a culture, having a "gay identity" just doesn't compute.

Standards of Attractiveness

In all human societies physical characteristics are important in determining whom one chooses as a sex partner. What is considered attractive varies considerably, though. For example, the region of the body that is judged for attractiveness varies from one culture to the next. For some peoples, the shape and color of the eyes are especially significant. For others, the shape of the ears is most important. Some societies go directly to the heart of the matter and judge attractiveness by the appearance of the external genitals. In a few societies, elongated labia majora (the pads of fat on either side of the vaginal opening in women) are considered sexually attractive, and it is common practice for a woman to pull on hers in order to make them longer. Elongated labia majora among the Nawa women of Africa are considered a mark of beauty and are quite prominent.

Our society's standards are in the minority in one way: in most cultures, a plump woman is considered more attractive than a thin one.

One standard does seem to be a general rule: A poor complexion is considered unattractive in the majority of human societies.

Social-Class and Ethnic-Group Variations in the United States

The discussion so far may have seemed to imply that there is one uniform standard of sexual behavior in the United States and that all Americans behave alike sexually. In fact, though, there are large variations in sexual behavior within our culture. Some of these subcultural variations can be classified as social-class differences and some as ethnic differences.

Social Class and Sex

Table 1.1 shows data on some social-class variations in sexuality. Education is used as an indicator of social class. The more educated the respon-

Table 1.1 Social-Class Variations in Sexual Behavior in the United States (Education is used as an indicator of social class)

	Less Than High School	High School Graduates	Some College	College Graduates	Graduate Degree
Percent who masturbated in last year					
Men	45%	55%	67%	76%	81%
Women	25	32	49	52	59
Percent who have performed oral sex					
Men	59	75	80	84	81
Women	41	60	78	79	79
Percent of conceptions terminated by abortion	5	10	11	11	14
Percent who have had two or more sexual partners in the last 12 months	17	15	18	19	13

Source: Edward O. Laumann, John H. Gagnon, Robert T. Michael, & Stuart Michaels. (1994). *The social organization of sexuality: Sexual practices in the United States.* Chicago: University of Chicago Press, Table 3.1 (p. 82), Table 3.6 (p. 98), Table 12.6A (p. 458), Table 5.1A, (p. 177).

dents, the more likely they are to have masturbated within the last year. This trend is true for both men and women, and the differences are large. Those who have advanced degrees are approximately twice as likely to have masturbated as those who did not finish high school.

The pattern for oral sex is more complex. Among men, for example, there are no important social-class variations, except that those who did not graduate from high school are less likely to

have performed oral sex. Oral sex has become a standard sexual technique in the United States, and about three-quarters of all people engage in it. Most of those who don't are in the lowest educational groups.

Among women, the percentage of conceptions that are terminated by abortion rises steadily from 5 percent for those who did not complete high school to 14 percent for those who hold advanced degrees. These findings raise the interesting possi-

(a)

(b)

Figure 1.6 Cross-cultural differences, cross-cultural similarities. *(a)* Woman of Labe Guinea, West Africa. *(b)* Three beauty queens in a Cinco de Mayo parade, San Jose, California. The custom of female adornment is found in most cultures, although the exact definition of beauty varies from culture to culture.

bility that, especially for women, social class and sexuality exert mutual influence on each other. That is, thus far we have assumed that one's social class affects one's sexual behavior. But it may also be true that sexuality influences one's social class. In this case, being able to afford an abortion may allow women to continue their education.

Finally, there are some social-class similarities. The percentage of people who have had two or more sex partners in the past 12 months (which puts them in a high-risk category for acquiring a sexually transmitted disease) is about the same at all educational levels (Table 1.1).

In summary, there are some social-class variations in sexuality. For example, the percentage of people who masturbate rises steadily with the level of education. At the same time, there are some social-class similarities.

Ethnicity and Sexuality in the United States

The United States population is composed of many ethnic groups, and there are some variations among these groups in sexual behavior. These vari-ations are a result of different cultural heritages, as well as of current economic and social conditions. Here we will discuss the cultural heritages and their influence on sexuality of four groups: African Americans, Latinos, Asian Americans, and whites. We would like to have included Native Americans as well, but the major sex surveys have not re-ported data on them because there were too few of them in the samples. A summary of some ethnic-group variations in sexuality is shown in Table 1.2.

In examining these data on ethnic-group varia-tions in sexuality, it is important to keep in mind two points: (1) There are ethnic-group variations, but there are also ethnic-group similarities. The sexuality of these four groups is not totally differ-ent. (2) Cultural context is the key. The sexuality of any particular group can be understood only by understanding the cultural heritage of that group as well as its current social and economic condi-tions. In the following sections, we briefly discuss the cultural contexts for African Americans, Lati-nos, and Asian Americans and examine how these cultural contexts are reflected in their sexuality.

Table 1.2 Comparison of the Sexuality of Whites, African Americans, Latinos, and Asian Americans

	Whites	African Americans	Latinos	Asian Americans
Percent who masturbated in the last year				
Men	67%	40%	67%	61%
Women	44%	32%	35%	NA*
Percent who have performed oral sex				
Men	81%	51%	71%	64%
Women	75%	34%	60%	NA*
Sex ratio (number of males per 100 females), 30- to 34-year-olds[†]	101	85	98	NA*
Percent of 30- to 34-year-olds who have never been married[†]				
Men	25%	45%	25%	NA*
Women	14%	35%	17%	NA*
Percent of conceptions terminated by abortion	10%	9%	11%	21%
Had sex with a same-gender partner				
Men	10%	8%	8%	3%
Women	5%	3%	4%	0
Percent who have had two or more sexual partners in the last 12 months	15%	27%	20%	8%

*NA means not available because the number of respondents in this category was too small for statistics to be computed.

[†]*Source:* U.S. Bureau of the Census (1990).

Source: Edward O. Laumann, John H. Gagnon, Robert T. Michael, & Stuart Michaels. (1994). *The social organization of sexuality: Sex-ual practices in the United States.* Chicago: University of Chicago Press, Table 3.6 (p. 98), Table 12.6A (p. 458), Table 8.2 (p. 305), Table 5.1A (p. 177).

African Americans.   The sexuality of African Americans is influenced by many of the same factors influencing the sexuality of white Americans, such as the legacies of the Victorian era and the influence of the Judeo-Christian tradition. In addition, at least three other factors act to make the sexuality of blacks somewhat different from that of whites: (1) the African heritage (Savage & Tchombe, 1994), (2) the forces that acted upon blacks during slavery, and (3) current economic and social conditions (Sudarkasa, 1997).

Table 1.2 shows some data comparing the sexuality of African Americans with whites, Latinos, and Asian Americans. In some cases, differences between blacks and whites are striking. For example, black men are considerably less likely than white men to have masturbated in the last year, and both black men and black women are less likely than whites to have performed oral sex. These differences, though, must be balanced against the similarities. For example, African American women and white women are about equally likely to terminate a conception with an abortion.

Notice that the marriage rate is lower for African Americans. This is due to a number of factors. First, there is not an equal gender ratio among blacks. As shown in Table 1.2, the gender ratio is nearly equal among whites and among Latinos; that is, there are about 100 men for every 100 women. Among African Americans, however, there are only about 85 men for every 100 women. This creates lower marriage rates among African American women because there are simply not enough black men to go around [among both blacks and whites, 98 percent of marriages are between two people of the same race (Thornton & Wason, 1995)]. Second, lower marriage rates among African American men are also due to the obstacles that they have encountered in seeking and maintaining jobs necessary to support a family. Since World War II, the number of manufacturing jobs, which once were a major source of employment for working-class black men, has declined dramatically. The result has been a decline in the black working class and an expansion of the black underclass. Joblessness among black men further contributes to low marriage rates.

In later chapters, we will discuss other issues having to do with race/ethnicity and sexuality, always bearing in mind the cultural context that shapes and gives meaning to different sexual patterns.

Latinos.   **Latinos** are people of Latin American heritage; therefore, the category includes many different cultural groups, such as Mexican Americans, Puerto Ricans, and Cuban Americans. "Latinos" can be used to refer to the entire group or specifically to men; the term "Latinas" refers exclusively to women of Latin American origin.

Latinos have a cultural heritage distinct from that of both African Americans and Anglos, although forces such as the Judeo-Christian religious tradition affect all three groups. In traditional Latin American cultures, gender roles are rigidly defined (Comas-Diaz, 1987). Such roles are emphasized early in the socialization process for children. Boys are given greater freedom and are encouraged in sexual exploits. Girls are expected to be passive, obedient, and weak. Latinos in the United States today have a cultural heritage that blends these traditional cultural values with the contemporary values of the dominant Anglo culture.

The rigid gender roles of the traditional Latino culture are epitomized in the concepts of *machismo* and *marianismo* (Comas-Diaz, 1987). The term *machismo,* or *macho,* has come to be used loosely in American culture today. Literally, *machismo* means "maleness" or "virility." More generally, it refers to the "mystique of manliness" (Ruth, 1990). The cultural code of machismo among Latin Americans mandates that the man must be responsible for the well-being and honor of his family, but in extreme forms it also means that men's sexual infidelities should be tolerated. *Marianismo,* the female counterpart of machismo, derives from Roman Catholic worship of Mary, the virgin mother of Jesus. Thus motherhood is highly valued while virginity until marriage is closely guarded. In Mexico, for example, 62 percent of males have had premarital intercourse by age 19, compared with only 14 percent of females (Liskin, 1985).

The data in Table 1.2 reveal a striking similarity in the data for Latinos and whites. The incidence of masturbation is approximately the same for both groups, although Latinas have a slightly lower incidence than white women. Marriage rates and abortion rates are also approximately equal between the two groups.

Latinos: People of Latin American heritage.

(a)

(b)

Figure 1.7   The sexuality of members of different ethnic groups is profoundly shaped by their cultures. *(a)* Roman Catholicism has a powerful impact on Latinos. *(b)* There is strong emphasis ●n the family among Asian Americans.

Focus 1.3

Sex in China

For the first 4000 years of recorded Chinese history, there was a yin-yang philosophy and open, positive attitudes about human sexuality, including a rich erotic literature. Indeed, the oldest sex manuals in the world came from China, dating from approximately 200 B.C. The most recent 1000 years, however, have been just the opposite, characterized by repression of sexuality and censorship.

A major philosophical concept in Chinese culture is yin and yang, originating around 300 B.C. and found in important writings on Confucianism and Taoism. According to yin-yang philosophy, all objects and events are the products of two elements: yin, which is negative, passive, weak, and destructive; and yang, which is positive, active, strong, and constructive. Yin is associated with the female, and yang with the male. For several thousand years, the Chinese have used yin and yang in words dealing with sexuality. For example, *yin fu* (the door of yin) means "vulva," and *yang ju* (the organ of yang) means "penis." *Huo yin yang* (the union of yin and yang) is the term used for sexual intercourse. This philosophy holds that the harmonious interaction between the male and female principles is vital, creating positive cultural attitudes toward sexuality.

Of the three major religions of China—Confucianism, Taoism, and Buddhism—Taoism is the only truly indigenous one, dating from the writings of Chang Ling around A.D. 143. Taoism is one of the few religions to advocate the cultivation of sexual techniques for the benefit of the individual. To quote from a classic Taoist work, *The Canon of the Immaculate Girl,*

Said P'eng, "One achieves longevity by loving the essence, cultivating the spiritual, and partaking of many kinds of medicines. If you don't know the ways of intercourse, taking herbs is of no benefit. The producing of man and woman is like the begetting of Heaven and Earth. Heaven and Earth have attained the method of intercourse and, therefore, they lack the limitation of finality. Man loses the method of intercourse and therefore suffers the mortification of early death. If you can avoid mortification and injury and attain the arts of sex, you will have found the way of nondeath. (Ruan, 1991, p. 56)

The tradition of erotic literature and openness about sexuality began to change about 1000 years ago, led by several famous neo-Confucianists, so that negative and repressive attitudes became dominant. In 1422 there was a ban on erotic literature, and a second major ban occurred in 1664. A commoner involved in printing a banned book could be beaten and exiled.

When the communist government founded the People's Republic of China in 1949, it imposed a

---

Asian Americans.   The broad category of Asian Americans includes many different cultural groups, such as Japanese Americans, Chinese Americans, and Indian Americans, as well as the relative newcomer groups such as Vietnamese Americans and the Hmong. As discussed in Focus 1.3, traditional Asian cultures, particularly Chinese culture, have been repressive about sexuality. Asian Americans today tend to be the sexual conservatives of the ethnic groups. For example, they have the lowest incidence of multiple sexual partners in the last 12 months (see Table 1.2). A similar pattern has been documented for Asians in Canada (Meston et al., 1996).

Table 1.2 shows that Asian Americans have the lowest incidence of same-gender sexual experience. This is consistent with the widespread denial by Chinese that homosexuality even exists in China. Moreover, it is consistent with the Asian emphasis on family, defined as a mother, father, and children. Gays and lesbians, then, are seen not only as violating a sexual norm but also as violating the paramount cultural norm of the family.

strict ban on all sexually explicit materials. The policy was quite effective in the 1950s and 1960s. By the late 1960s, however, erotica were produced much more in Western nations and in China there was increased openness to the West. By the late 1970s, X-rated videotapes were being smuggled into China from Hong Kong and other countries, and they quickly became a fad. Small parties were organized around the viewing of these tapes. The government reacted harshly, promulgating a new antipornography law in 1985. According to the law, "Pornography is very harmful, poisoning people's minds, inducing crimes . . . and must be banned. . . . The person who smuggled, produced, sold, or organized the showing of pornography, whether for sale or not, shall be punished according to the conditions, by imprisonment or administrative punishment" (Ruan, 1991, p. 100). Publishing houses were given stiff fines, and by 1986, 217 illegal publishers had been arrested and 42 publishing houses had been forced to close. In one incident, a Shanghai railway station employee was sentenced to death for having organized sex parties on nine different occasions, during which pornographic videotapes were viewed and he engaged in sexual activity with women.

Male homosexuality is recognized in historical writings in China as early as 2000 years ago. Homosexuality was so widespread among the upper classes in ancient China that the period is known as the Golden Age of Homosexuality in China. One historical book on the Han dynasty contained a special section describing the emperors' male sexual partners. There were tolerant attitudes toward lesbianism. But with the founding of the People's Republic in 1949, homosexuality, like all other sexuality, was severely repressed. Most Chinese today claim that they have never known a homosexual and that there must be very few in Chinese society.

In the early 1980s China was characterized by a puritanism that probably far exceeded that observed by the original Puritans. It was considered scandalous for a married couple to hold hands in public. Prostitution, premarital sex, homosexuality, and variant sexual behaviors were all illegal, and the laws were enforced. Even sexuality in marriage was given little encouragement.

A moderate sexual liberation began in the 1980s. Open displays of affection, such as holding hands in public, are no longer treated as signs of promiscuity. High schools now include sex education in the curriculum. The rationale is that a scientific understanding of sexual development is essential to the healthy development of young people and to the maintenance of high moral standards and well-controlled social order. Sexual images are found much more often in the media, although censors continue to monitor the content. There is open discussion of the importance of women's sexual pleasure. It will be interesting to see whether liberalization continues, or an eventual swing back to repression occurs.

Sources: Ruan, Fang-fu. (1991). *Sex in China.* New York: Plenum Press. Evans, Harriet (1995). Defining differences: The "scientific" construction of sexuality and gender in the People's Repubiic of China, *Signs, 20,* 357–394.

Among the ethnic groups shown in Table 1.2, Asian Americans have the highest rate of terminating conceptions by abortion. Again, this is consistent with cultural heritage. In China, for example, abortion is coznsidered a reasonable backup method of birth control, the IUD being the primary method. China's severe overpopulation problem has led to a policy of permitting only one child per family, and the survival of all Chinese is thought to rest on everyone complying with this policy. Under these circumstances, abortion is considered ethical and, in fact, desirable in order to avoid mass starvation and to ensure the survival of the Chinese people. In addition, most Asians are not Christian, and therefore are not part of that religious tradition, which questions the morality of abortion. In this context, it is not surprising that Asian Americans, the largest subgroup of whom are Chinese Americans, are more accepting of abortion than are other ethnic groups.

The Significance of the Cross-Cultural Studies

What relevance do the cross-cultural data have to an understanding of human sexuality? They are important for two basic reasons. First, they give us a notion of the enormous variation that exists in human sexual behavior, and they help us put our own standards and our own behavior in perspective. Second, these studies provide us with impressive evidence concerning the importance of culture and learning in the shaping of our sexual behavior; they show us that human sexual behavior is not completely determined by biology or drives or instincts. For example, a woman of Inis Beag and a woman of Mangaia presumably have vaginas that are similarly constructed and clitorises that are approximately the same size and have the same nerve supply. But the woman of Inis Beag never has an orgasm, and all Mangaian women orgasm.[3] Why? Their cultures are different, and they and their partners learned different things about sex as they were growing up. Culture is a major determinant of human sexual behavior.

The point of studying sexuality in different cultures is *not* to teach you that there are a lot of exotic people out there doing exotic things. Rather, the point is to remind you that each group has its own culture, and this culture has a profound influence on the sexual expression of the women and men who grow up in it. We will return with more examples in many of the chapters that follow.

Cross-Species Perspectives on Sexuality

Humans are just one of many animal species, and all of them display sexual behavior. To put our own sexual behavior in evolutionary perspective, it is helpful to explore the similarities and differences between our own sexuality and that of other species. There is one other reason for this particular discussion. Some people classify sexual behaviors as "natural" or "unnatural" depending on whether other species do or do not exhibit those behaviors. Sometimes, though, the data are twisted to suit the purposes of the person making the argument, and so there is a need for a less biased view. Let us see exactly what some other species do!

Masturbation

Humans are definitely not the only species that masturbates. Masturbation is found among many species of mammals, and it is particularly common among the primates (monkeys and apes). Male monkeys and apes in zoos can be observed masturbating, often to the horror of the proper folk who have come to see them. At one time it was thought that this behavior might be the result of the unnatural living conditions of zoos. However, observations of free-living primates indicate that they, too, masturbate. Techniques include hand stimulation of the genitals or rubbing the genitals against an object. In terms of technique, monkeys and nonhuman apes have one advantage over humans: their bodies are so flexible that they can perform mouth-genital sex on themselves. A unique form of male masturbation is found among red deer; during the rutting season they move the tips of their antlers through low-growing vegetation, producing erection and ejaculation (Beach, 1976).

Female masturbation is also found among many species besides our own. The prize for the most inventive technique probably should go to the female porcupine. She holds one end of a stick in her paws and walks around while straddling the stick; as the stick bumps against the ground, it vibrates against her genitals (Ford & Beach, 1951). Human females are apparently not the first to enjoy vibrators.

Same-Gender Sexual Behavior

Same-gender behavior is found in many species besides our own (Vasey, 1995; Wallen & Parsons, 1997). Indeed, observations of other species indicate that our basic mammalian heritage is bisexual; it is composed of both heterosexual and homosexual elements (Ford & Beach, 1951).

Males of many species will mount other males, and anal intercourse has been observed in some

---

[3]We like to use the word "orgasm" not only as a noun but also as a verb. The reason is that alternative expressions, such as "to *achieve* orgasm" and "to *reach* orgasm," reflect the tendency of Americans to make sex an achievement situation (an idea to be discussed further in Chapter 10). To avoid this, we use "to have an orgasm" or "to orgasm."

(a)   *(b)*

Figure 1.8   The sexual behavior of primates. (*a*) Females have various ways of expressing choice. Here a female Barbary macaque presents her sexual swelling to a male. He seems to be interested. (*b*) Bonobos mating in the male-on-top position, which is rare among nonhumans. Bonobos are among the most sexual of primates. Females use sex for nonsexual purposes, such as gaining food, transferring into new groups, and forming alliances. They engage in sexual activity with males and with females. (Photographs by Frans de Waal.)

male primates (Wallen & Parsons, 1997). Male porpoises have also been observed repeatedly attempting to insert their penis into the anus of an intended partner, even though females were available (McBride & Hebb, 1948). Among domestic sheep, 9 percent of adult males strongly prefer other males as sex partners (Price et al., 1988; Ellis, 1996). In a number of primate species, including bonobos and Japanese macaques, females mount other females (Small, 1993).

In species that form long-term bonds or relationships, long homosexual bonds have been observed. For example, Konrad Lorenz (1966) reported a long-term relationship between two male ducks.

Sexual Signaling
Female primates engage in sexual signaling to males, saying, in effect, "Come on over, big guy" (Dixson, 1990). For example, females in one species of macaque engage in "parading" in front of males to signal interest. Among baboons, spider monkeys, and orangutans, the female makes eye contact with

the male. The female patas monkey puffs out her cheeks and drools. The parading and eye contact sound very familiar—they could easily be observed among women at a singles bar. The puffing and drooling probably wouldn't play so well, though.

Human Uniqueness
The general trend, as we move from lower species such as fish or rodents to higher species such as primates, is for sexual behavior to be more hormonally (instinctively) controlled among lower species and to be controlled more by the brain (and therefore by learning) in the higher species (Beach, 1947). Thus environmental influences are much more important in shaping primate—especially human—sexual behavior than they are in shaping the sexual behavior of other species.

An illustration of this fact is provided by studies of the adult sexual behavior of animals that have been raised in deprived environments. If mice are reared in isolation, their adult sexual behavior will nonetheless be normal (Scott, 1964). But if rhesus monkeys are reared in isolation, their adult sexual behavior is severely disturbed, to the point where they may be incapable of reproducing (Harlow et al., 1963). Thus environmental experiences are crucial in shaping the sexual behavior of the higher species, particularly humans; for us, sexual behavior is a lot more than just "doin' what comes naturally."

Female sexuality provides a particularly good illustration of the shift in hormonal control from lower to higher species. Throughout most of the animal kingdom, female sexual behavior is strongly controlled by hormones. In virtually all species, females do not engage in sexual behavior at all except when they are in "heat" (estrus), which is a particular hormonal and physiological state. In contrast, human females are capable of engaging in sexual behavior—and actually do engage in it— during any phase of their hormonal (menstrual) cycle. Thus the sexual behavior of the human female is not nearly so much under hormonal control as that of females of other species.

Traditionally it was thought that female orgasm is unique to humans and does not exist in other species. Then some studies found evidence of orgasm in rhesus macaques (monkeys), although under very artificial laboratory conditions involving stimulation of the female by a mechanical penis (Burton, 1970; Zumpe & Michael, 1968). Another study showed the same physiological responses indicative of orgasm in human females—specifically, increased heart rate and uterine contractions—in stump-tailed macaques as a result of female same-gender sexual activity, and perhaps for heterosexual activity as well (Goldfoot et al., 1980). Thus it seems that humans can no longer claim that they have a corner on the female orgasm market. This fact has interesting implications for understanding the evolution of sexuality. Perhaps the higher species, in which the females are not driven to sexual activity by their hormones, have the pleasure of orgasm as an incentive.

The Nonsexual Uses of Sexual Behavior

Two male baboons are locked in combat. One begins to emerge as the victor. The other "presents" (the "female" sexual posture, in which the rump is directed toward the other and is elevated somewhat).

Two male monkeys are members of the same troop. Long ago they established which one is dominant and which one is subordinate. The dominant one mounts (the "male" sexual behavior) the subordinate one.

These are examples of the fact that animals sometimes use sexual behavior for nonsexual purposes (Small, 1993; Jensen, 1976b). Commonly this behavior is done to signal the end of a fight, as in the first example. The loser indicates surrender by presenting, and the winner signals victory by mounting. Sexual behaviors can also symbolize an animals' rank in a dominance hierarchy. Dominant animals mount subordinate ones. As another example, male squirrel monkeys sometimes use an exhibitionist display of their erect penis as part of an aggressive display against another male, something that is called *phallic aggression* (Wickler, 1973).

All this is perfectly obvious when we observe it in monkeys. But do humans ever use sexual behavior for nonsexual purposes? Consider the rapist, who uses sex as an expression of aggression against and power over a woman (Holmstrom & Burgess, 1980), or over another man in the case of homosexual rape. Another example is the exhibitionist, who uses the display of his erect penis to shock and frighten women, much as the male squirrel monkey uses this display to shock and frighten his opponent. Humans also use sex for economic purposes; the best examples are prostitutes and **gigolos.**

There are also less extreme examples. Consider the couple who have a fight and then make love to signal an end to the hostilities.[4] Or consider the woman who goes to bed with an influential— though unattractive—politician because this gives her a vicarious sense of power.

You can probably think of other examples of the nonsexual use of sexual behavior. Humans, just like members of other species, can use sex for a variety of nonsexual purposes.

---

[4]It has been our observation that this practice does not always mean the same thing to the man and to the woman. To the man it can mean that everything is fine again, but the woman can be left feeling dissatisfied and not at all convinced that the issues are resolved. Thus this situation can be a source of miscommunication between the two.

Gigolo (JIH-guh-lo): A man who sells his sexual services to women.

SUMMARY

"Sexual behavior" is behavior that produces arousal and increases the chance of orgasm. "Sex" (sexual behavior and anatomy) is distinct from "gender" (being male or female).

Throughout most of human history, religion was the main source of information concerning sexuality. In the late 1800s and early 1900s, important contributions to the scientific understanding of sex were made by Sigmund Freud, Havelock Ellis, Richard von Krafft-Ebing, and Magnus Hirschfeld. These early researchers emerged from the Victorian era, in which sexual norms were highly rigid—although many people's actual behavior violated these norms. By the 1990s, major, well-conducted sex surveys were available. Today, the mass media are a powerful influence on most people's understanding of sexuality.

Studies of various human cultures around the world provide evidence of the enormous variations in human sexual behavior. For example, frequency of intercourse may vary from once a week in some cultures to three or four times a night in others. One generalization that does emerge is that all societies regulate sexual behavior in some way. Attitudes regarding premarital and extramarital sex, masturbation, same-gender sexual behavior, and gender roles vary considerably from one culture to the next. The great variations provide evidence of the importance of learning in shaping sexual behavior.

Within the United States, sexual behavior varies with social class and ethnic group. For example, African Americans are less likely to perform oral sex than whites are. In other areas, though, blacks and whites are quite similar. Traditional Latino cultures are characterized by rigid gender roles and restrictions on female sexuality, but not on male sexuality. Asian cultures tend to be conservative about sexuality for both males and females.

Studies of sexual behavior in various animal species show that masturbation, mouth-genital stimulation, and same-gender sexual behavior are by no means limited to humans. They also illustrate how sexual behavior may be used for a variety of nonsexual purposes, such as expressing dominance.

REVIEW QUESTIONS

1. A sex survey of Victorian women found that none of them reported having orgasms during marital intercourse. True or false?
2. On the soaps, the most frequent sexual activity is _____ .
3. On Inis Beag, an island off the coast of Ireland, premarital intercourse is common and many babies are born out of wedlock. True or false?
4. African American women and white women are about equally likely to terminate a pregnancy with an abortion. True or false?
5. The wide variations in sexual behavior found cross-culturally provide evidence of the importance of learning and culture in shaping human sexual behavior. True or false?
6. Same-gender sexual behavior is unknown among animals. True or false?
7. In traditional Latino cultures, males are far more likely to engage in premarital sex than females are. True or false?
8. This textbook uses the term "sex" to refer to sexual anatomy and sexual behavior, and the term _____ to refer to the state of being male or female.
9. The early sex researcher Richard von Krafft-Ebing focused on "pathological" sexual behavior. True or false?
10. All societies regulate sexuality in some way, although the exact regulations vary from culture to culture. True or false?

(The answers to all review questions are at the end of the book, after the Glossary.)

QUESTIONS FOR THOUGHT, DISCUSSION, AND DEBATE

1. In the wide spectrum of sexual practices in different cultures, from the conservatism of Inis Beag to the permissiveness of Mangaia, where would you place the United States today? Are we permissive, restrictive, or somewhere in between? Why?

2. Research indicates that masturbation, mouth-genital stimulation, and same-gender sexual behaviors are present in other species besides humans. What is the significance of that finding?

3. Sally is a newspaper reporter in Captown, the state capital. She learns, through a trusted friend, that the highly popular governor, George Smith, has been having a long-term affair with a local woman attorney, despite the fact that each of them is apparently happily married. Should Sally break the news in an article? In terms of communications theories, what effects might she expect her article to have on the public's attitudes about sexuality?

4. How does premarital sexual behavior in the United States compare with premarital sex in Mangaia and Inis Beag (see Focus 1.2)? Which of these three cultures do you think it is best for adolescents to grow up in? Why? In which of those three cultures do you think adults are likely to have the most positive feelings about their sexuality? Why?

SUGGESTIONS FOR FURTHER READING

Bullough, Vern L. (1994). *Science in the bedroom: A history of sex research.* New York: Basic Books. Bullough is one of the most knowledgeable historians of sexuality, and this book provides fascinating details about sex research from the ancient Greeks to Hirschfeld, Ellis, and Freud and brings us all the way to the 1990s.

Greenberg, Bradley S., Brown, Jane D., & Buerkel-Rothfuss (Eds.) (1993). *Media, sex, and the adolescent.* This is a marvelous collection of articles on the sexual content of the media to which adolescents are exposed, and how adolescents respond.

Gregersen, Edgar. (1983). *Sexual practices. The story of human sexuality.* New York: Franklin Watts. Gregersen, an anthropologist, has compiled a vast amount of information about sexuality in cultures around the world. The book also includes a treasure trove of fascinating illustrations.

Kon, Igor S. (1995). *The sexual revolution in Russia: From the age of the Czars to today.* New York: Free Press. Kon, a prominent Russian sex researcher, traces Russian sexuality, including repression in the Soviet Union and then sudden liberation with its fall.

Small, Meredith. (1993). *Female choices: Sexual behavior of female primates.* Ithaca, NY: Cornell University Press. Written for a general audience, this book provides a delightful journey through the sexual world of female primates.

WEB RESOURCES

http://www.blackgirl.org
    African American Women resources

http://www.lib.uci.edu/rrsc/asiamer.html
    Asian American studies resources

http://clnet.ucr.edu/research/educ.html
    Chicano/a/Latino/a resources

http://www.umiacs.umd.edu/users/sawweb/sawnet/
    South Asian Women's Network

http://songweaver.com/info/bonobos.html
    An article describing bonobos, social organization

chapter 2

THEORETICAL PERSPECTIVES ON SEXUALITY

CHAPTER HIGHLIGHTS

Evolutionary Perspectives
Sociobiology
Evolutionary Psychology

Psychological Theories
Psychoanalytic Theory
Learning Theory
Cognitive Theories

Sociological Perspectives
The Sociological Approach: Levels of Analysis
Social Institutions
Sexual Scripts
Reiss's Sociological Theory of Sexuality

"One of the discoveries of psychoanalysis consists in the assertion that impulses, which can only be described as sexual in both the narrower and the wider sense, play a peculiarly large part, never before sufficiently appreciated, in the causation of nervous and mental disorders. Nay, more, that these sexual impulses have contributed invaluably to the highest cultural, artistic, and social achievements of the human mind.*

On the face of it, sex is at best a peculiar way to reproduce; at worst, it seems profoundly self-defeating.[†]

Imagine, for a moment, a heterosexual couple making love. Imagine, too, that sitting with you in the room, thinking your same thoughts, are Freud, E. O. Wilson (a leading sociobiologist), Albert Bandura (a leading social learning theorist), and John Gagnon (a proponent of script theory in sociology). The scene you are imagining may evoke arousal and nothing more in you, but your imaginary companions would have a rich set of additional thoughts as they viewed the scene through the specially colored lenses of their own theoretical perspectives. Freud might be marveling at how the biological sex drive, the *libido,* expresses itself so strongly and directly in this couple. Wilson, the sociobiologist, would be thinking how mating behavior in humans is similar to mating behavior in other species of animals, and how it is clearly the product of evolutionary selection for behaviors that lead to successful reproduction. Bandura might be thinking how sexual arousal and orgasm act as powerful positive reinforcers that will lead the couple to repeat the act frequently, and how they are imitating a technique of neck nibbling that they saw in an X-rated film last week. Finally, Gagnon's thoughts might be about the social scripting of sexuality; this couple begins with kissing, moves on to petting, and finishes up with intercourse, following a script written by society.

Some of the major theories in the social sciences have had many—and different—things to say about sexuality, and it is these theories that we consider in this chapter.

*Sigmund Freud. (1924). *A general introduction to psychoanalysis.* New York: Permabooks, 1953 (Boni & Liveright edition, 1924), pp. 26–27.
[†]D. P. Barash. (1982). *Sociobiology and behavior.* New York: Elsevier, p. 216.

Evolutionary Perspectives

Sociobiology

Sociobiology is a highly controversial theory. It was heralded by Harvard biologist E. O. Wilson's book *Sociobiology: The New Synthesis* (1975). **Sociobiology** is defined as the application of evolutionary biology to understanding the social behavior of animals, including humans (Barash, 1982). Sexual behavior, of course, is a form of social behavior, and so the sociobiologists, often through observations of other species, try to understand why certain patterns of sexual behavior have evolved in humans. Donald Symons has applied sociobiological thinking to human sexuality in his book *The Evolution of Human Sexuality* (1979; Symons, 1987).

Before we proceed, we should note that in terms of **evolution,** what counts is producing lots of healthy, viable offspring who will carry on one's genes. Evolution occurs via **natural selection,** the process by which those animals that are best adapted to their environment are more likely to survive, reproduce, and pass their genes on to the next generation.

How do humans select mates? One major criterion is the physical attractiveness of the person

Sociobiology: The application of evolutionary biology to understanding the social behavior of animals, including humans.
Evolution: A theory that all living things have acquired their present forms through gradual changes in their genetic endowment over successive generations.
Natural selection: A process in nature resulting in greater rates of survival of those plants and animals that are adapted to their environment.

(see Chapter 13). The sociobiologist argues that many of the characteristics we evaluate in judging attractiveness—for example, physique and complexion—are indicative of the health and vigor of the individual. These in turn are probably related to the person's reproductive potential; the unhealthy are less likely to produce many vigorous offspring. Natural selection would favor those individuals preferring mates who would have maximum reproductive success. Thus, perhaps our concern with physical attractiveness is a product of evolution and natural selection. (See Barash, 1982, for an extended discussion of this point and the ones that follow.) We choose an attractive, healthy mate who will help us produce many offspring. Can you guess why the sociobiologist thinks most men are attracted to women with large breasts?

From this viewpoint, dating, going steady, getting engaged, and similar customs are much like the courtship rituals of other species (See Fig. 2.1). For example, many falcons and eagles have a flying courtship in which objects are exchanged between the pair in midair. The sociobiologist views this courtship as an opportunity for each member of the prospective couple to assess the other's fitness. For example, any lack of speed or coordination would be apparent during the airborne acrobatics. Evolution would favor courtship patterns that permitted individuals to decide on mates who would increase their reproductive success. Perhaps that is exactly what we are doing in our human courtship rituals. The expenditure of money by men on dates indicates their ability to support a family. Dancing permits the assessment of physical prowess, and so on.

Sociobiologists can also explain why the nuclear family structure of a man, a woman, and their offspring is found in every society. Once a man and woman mate, there are several obstacles to reproductive success; two of these are infant vulnerability and maternal death. Infant vulnerability is greatly reduced if the mother provides continuing physical care, including breast feeding. It is further reduced if the father provides resources and security from attack for mother and infant. Two

(a)

(b)

Figure 2.1   *(a)* The courtship rituals of great egrets. *(b)* Dancing is a human dating custom. According to sociobiologists, human customs of dating and becoming engaged are biologically produced and serve the same functions as courtship rituals in other species: They allow potential mates to assess each other's fitness.

mechanisms that facilitate these conditions are a *pair-bond* between mother and father, and *attachment* between infant and parent (Miller & Fishkin, 1997). Thus, an offspring's chances of survival are greatly increased if the parents bond emotionally, that is, love each other, and if the parents have a propensity for attachment. Further, an emotional bond might lead to more frequent sexual interaction; the pleasurable consequences of sex in turn will strengthen the bond. Research with small mammals, including mice and voles, demonstrates the advantages of biparental care of offspring, and the critical role of bonding (Morell, 1998).

In addition to natural selection, Darwin also proposed a mechanism that is not so much a household word, **sexual selection** (Gangestad & Thornhill, 1997). Sexual selection is selection that results from differences in traits affecting access to mates. It consists of two processes: (1) competition among members of one gender (usually males) for mating access to members of the other gender *(intrasexual selection)*; and (2) preferential choice by members of one gender (usually females) for certain members of the other gender *(intersexual selection)*. In other words, in many, though not all, species, males compete among themselves for the right to mate with females; and females, for their part, prefer certain males and mate with them while refusing to mate with other males.[1] Researchers are currently testing with humans some of the predictions that come from the theory of sexual selection. For example, the theory predicts that men should compete with each other in ways that involve displaying material resources that should be attractive to women, and men should engage in these displays more than women (Buss, 1988). Examples might be giving impressive gifts to potential mates, flashy showing of possessions (e.g., cars, stereos), or displaying personality characteristics that are likely to lead to the acquisition of resources (e.g., ambition). Research shows that men engage in these behaviors significantly more than women do, and that both men and women believe these tactics are effective (Buss, 1988).

One of the most controversial ideas is the argument by some sociobiologists that rape is the product of evolutionary selection because it is adaptive for the male (Barash, 1977b; Shields & Shields, 1983; see Palmer, 1991, for a critical review). They have even suggested that human males are larger than human females as a result of evolutionary selection that made large males more successful at rape. They have found evidence that males accomplish copulation by force in a number of other species. The argument is that human males could reproduce either by courtship, bonding with a female, and then reproducing, or else by rape. Rape would be predicted for those males that lacked the resources to carry on a successful courtship and win a consenting female, or in response to a female's rejection of his advances. One problem with this argument is that if rape is punished, it may lead to less reproduction, not more. For example, if there is capital punishment for rape, it becomes a poor reproductive strategy; the single act of rape produces a certain probability (but certainly not 100 percent) of impregnating the female, but if it leads to the man's death he cannot reproduce any more—scarcely a good strategy for passing on one's genes to the next generation. There are also problems internally with the sociobiological arguments about rape which are too complicated to discuss here (Palmer, 1991). This example, though, will give you a flavor of some theorizing by sociobiologists.

Evolutionary Psychology

A somewhat different approach is taken by **evolutionary psychology,** which focuses on psychological mechanisms that have been shaped by natural selection (Buss, 1991). If behaviors evolved in response to selective pressures, it is plausible to argue that cognitive or emotional structures evolved in the same way. Thus, a man who accurately judged whether a woman was healthy and fertile would be more successful in reproducing. If his offspring exhibited the same ability to judge accurately, they in turn would have a competitive advantage.

Sexual selection: Selection that results from differences in traits affecting access to mates.
Evolutionary psychology: The study of psychological mechanisms that have been shaped by natural selection.

[1] Sociobiologists use this mechanism to explain gender differences.

Recent research has concentrated on *sexual strategies* (Buss & Schmitt, 1993). According to this theory, females and males face different adaptive problems in short-term, or casual, mating and in long-term mating and reproduction. These differences lead to different strategies, or behaviors designed to solve these problems. In short-term mating, a female may choose a partner who offers her immediate resources, such as food or money. In long-term mating, a female may choose a partner who appears able and willing to provide resources for the indefinite future. A male may choose a sexually available female for a short-term liaison, and avoid such females when looking for a long-term mate. Many specific predictions about human sexual strategies based on this theory have been supported by research data (Buss, 1994).

Many criticisms of evolutionary perspectives have been made. Some protest that sociobiology ignores the importance of culture and learning in human behavior. However, most contemporary evolutionary theorists acknowledge that behavior is the result of an interaction between evolved mechanisms and environmental input (Buss & Schmitt, 1993). Other critics resent the biological determinism that it introduces. Sociobiology has been criticized for resting on an outmoded version of evolutionary theory that modern biologists consider naive (Gould, 1987). For example, sociobiology has focused mainly on individuals' struggle for survival, whereas modern biologists focus on more complex issues such as the survival of the species and the evolution of a successful adaptation between a species and its environment. Further, sociobiologists assume that the central function of sex is reproduction; this may have been true historically, but it probably is not at present. Finally, evolutionary perspectives cannot be evaluated empirically in the same way that we evaluate other theories. We cannot manipulate evolution and measure the outcome.

Psychological Theories

Three of the major theories in psychology are relevant to sexuality: psychoanalytic theory, learning theory, and cognitive theory.

Psychoanalytic Theory

Freud's **psychoanalytic theory** has been one of the most influential of all psychological theories. Because Freud saw sex as one of the key forces in human life, his theory gives full treatment to human sexuality.

Freud termed the sex drive or sex energy **libido,** and he saw it as one of the two major forces motivating human behavior (the other being *thanatos,* or the death instinct).

Id, Ego, and Superego

Freud described the human personality as being divided into three major parts: the id, the ego, and the superego. The **id** is the basic part of personality and is present at birth. It is the reservoir of psychic energy (including libido) and contains the instincts. Basically it operates on the *pleasure principle;* it cannot tolerate any increase in psychic tensions, and so it seeks to discharge these tensions.

While the id operates only on the pleasure principle and can thus be pretty irrational, the **ego** operates on the *reality principle* and tries to keep the id in line. The ego functions to make the person have realistic, rational interactions with others.

Finally, the **superego** is the conscience. It contains the values and ideals of society that we learn, and it operates on *idealism.* Thus its aim is to inhibit the impulses of the id and to persuade the ego to strive for moral goals rather than realistic ones.

To illustrate the operation of these three components of the personality in a sexual situation, consider the case of the president of the corporation who is at a meeting of the board of directors; the meeting is also attended by her gorgeous, muscular colleague, Mr. Hunk. She looks at Mr. Hunk,

Psychoanalytic theory: A psychological theory originated by Freud; it contains a basic assumption that part of human personality is unconscious.
Libido (lih-BEE-doh): In psychoanalytic theory, the term for the sex energy or sex drive.
Id: According to Freud, the part of the personality containing the libido.
Ego: According to Freud, the part of the personality that helps the person have realistic, rational interactions.
Superego: According to Freud, the part of the personality containing the conscience.

and her id says, "I want to throw him on the table and make love to him immediately. Let's do it!" The ego intervenes and says "We can't do it now because the other members of the board are also here. Let's wait until 5 P.M., when they're all gone, and then do it." The superego says, "I shouldn't make love to Mr. Hunk at all because I'm a married woman." What actually happens? It depends on the relative strengths of this woman's id, ego, and superego.

The id, ego, and superego develop sequentially. The id contains the set of instincts present at birth. The ego develops later, as the child learns how to interact realistically with his or her environment and the people in it. The superego develops last, as the child learns moral values.

Erogenous Zones

Freud saw the libido as being focused in various regions of the body known as **erogenous zones.** An erogenous zone is a part of the skin or mucous membrane which is extremely sensitive to stimulation; touching it in certain ways produces feelings of pleasure. The lips and mouth are one such erogenous zone, the genitals are a second, and the rectum and anus are a third. Thus sucking produces oral pleasure, defecating produces anal pleasure, and rubbing the genital area produces genital pleasure.

Stages of Psychosexual Development

Freud believed that the child passes through a series of stages of development. In each of these stages, a different erogenous zone is the focus of the child's striving.

The first stage, lasting from birth to about 1 year of age, is the *oral stage.* The child's chief pleasure is derived from sucking and otherwise stimulating the lips and mouth. Anyone who has observed children of this age knows how they delight in putting anything they can into their mouths. The second stage, which occurs during approximately the second year of life, is the *anal stage.* During this stage, the child's interest is focused on elimination.

The third stage of development, lasting from age 3 to perhaps age 5 or 6, is the *phallic stage.* The boy's interest is focused on his phallus (penis), and he derives great pleasure from masturbating.[2] Per-

haps the most important occurrence in this stage is the development of the **Oedipus complex,** which derives its name from the Greek story of Oedipus, who killed his father and married his mother. In the Oedipus complex, the boy loves his mother and desires her sexually. He hates his father, whom he sees as a rival for the mother's affection. The boy's hostility toward his father grows, but eventually he comes to fear that his father will retaliate by castrating him—cutting off his prized penis. Thus the boy feels *castration anxiety.* Eventually the castration anxiety becomes so great that he stops desiring his mother and shifts to identifying with his father, taking on the father's gender role and acquiring the characteristics expected of males by society. Freud considered the Oedipus complex and its resolution to be one of the key factors in human personality development.

As might be expected from the name of this stage, the girl will have a considerably different, and much more difficult, time passing through it, since she has none of what the stage is all about. For a girl, the phallic stage begins with her traumatic realization that she has no penis, perhaps after observing that of her father or her brother. She feels envious and cheated, and she suffers from *penis envy,* wishing that she too had a wonderful wand. (Presumably she thinks her own clitoris is totally inadequate, or she is not even aware that she has it.) She believes that at one time she had a penis but that it was cut off, and she holds her mother responsible. Thus she begins to hate her mother and shifts to loving her father, forming her version of the Oedipus complex, sometimes called the **Electra complex.** In part, her incestuous desires for her father result from a desire to be impregnated by him, to substitute for the unobtainable penis. Unlike the boy, the girl does not have a strong motive of castration anxiety for resolving the Oedipus complex; she has already lost her penis. Thus the girl's resolution of the Electra complex is not so complete as the boy's resolution of

---

[2]Masturbation to orgasm is physiologically possible at this age, although males are not capable of ejaculation until they reach puberty (see Chapter 5).

Erogenous zones (eh-RAH-jen-us): Areas of the body that are particularly sensitive to sexual stimulation.
Oedipus complex (EH-di-pus): According to Freud, the sexual attraction of a little boy for his mother.
Electra complex (eh-LEK-tra): According to Freud, the sexual attraction of a little girl for her father.

the Oedipus complex, and for the rest of her life she remains somewhat immature compared with men.

Freud said that following the resolution of the Oedipus or Electra complex, children pass into a prolonged stage known as *latency,* which lasts until adolescence. During this stage, the sexual impulses are repressed or are in a quiescent state, and so nothing much happens sexually. The postulation of this stage is one of the weaker parts of Freudian theory, because it is clear from the data of modern sex researchers that children do continue to engage in behavior with sexual components during this period.

With puberty, sexual urges reawaken, and the child moves into the *genital stage.* During this stage, sexual urges become more specifically genital, and the oral, anal, and genital urges all fuse together to promote the biological function of reproduction. Sexuality becomes less narcissistic (self-directed) than it was in childhood and is directed toward other people as appropriate sexual objects.

According to Freud, people do not always mature from one stage to the next as they should. A person might remain permanently fixated, for example, at the oral stage; symptoms of such a situation would include incessant cigarette smoking and fingernail biting, which gratify oral urges. Most adults have at least traces of earlier stages remaining in their personalities.

Freud on Women

In recent years a storm of criticism of Freudian theory has arisen from feminists. Let us first review what Freud had to say about women and then discuss what feminists object to in his theory (Lerman, 1986; Millett, 1969).

Essentially, Freud assumed that the female is biologically inferior to the male because she lacks a penis. He saw this absence as a key factor in her personality development. The penis envy she feels as a result of her biological deficiency causes her to develop the Electra complex. Yet she never adequately resolves this complex, and she is left, throughout her life, with feelings of jealousy and inferiority, all because of her lack of a penis. As Freud said, "Anatomy is destiny."

Freud believed that female sexuality is inherently passive (as opposed to the active aggressiveness of the male). He also felt that female sexuality is masochistic; in seeking intercourse, the female is trying to bring pain on herself, since childbirth, which may result, is painful and since intercourse itself may sometimes be painful. The following quotation from Marie Bonaparte, a follower of Freud, illustrates the psychoanalytic position:

> Throughout the whole range of living creatures, animal or vegetable, passivity is characteristic of the female cell, the ovum whose mission is to *await* the male cell, the active mobile spermatozoan to come and *penetrate* it. Such penetration, however, implies infraction of its tissue, but infraction of a living creature's tissue may entail destruction: death as much as life. Thus the fecundation of the female cell is initiated by a kind of wound; in its way, the female cell is primordially "masochistic." (1953, p. 79)

Freud also originated the distinction between *vaginal orgasm* and *clitoral orgasm* in women. During childhood little girls rub their clitorises to produce orgasm (clitoral orgasm). Freud believed, though, that as they grow to adulthood women need to shift their focus to having orgasm during heterosexual intercourse, with the penis stimulating the vagina (vaginal orgasm). Thus, not only did he postulate two kinds of orgasm for women, but he also maintained that one kind was better (more mature) than the other. The evidence that Masters and Johnson have collected on this issue will be reviewed in Chapter 9; suffice it to say for now that there seems to be little or no physiological difference between the two kinds of orgasm. The assertion that the vaginal orgasm is more mature is not supported by Masters and Johnson's and others' findings that most adult women orgasm as a result of clitoral stimulation.

Feminists understandably object to several aspects of Freud's theory. A chief objection is to the whole notion that women are anatomically inferior to men because they lack a penis. What is so intrinsically valuable about a penis that makes it better than a clitoris, a vagina, or a pair of ovaries? In a creative approach, psychoanalyst Karen Horney (1926/1973) coined the concept of "womb envy," arguing that males have a powerful envy of females' reproductive capacity, more than females envy the penis. Similarly, feminists find the assertion that women are inherently passive, masochistic, and narcissistic to be offensive. Is it not likelier that men simply have higher status in our culture than women do and that women's feelings of jealousy or inferiority result from these cultural status differences? Feminists argue that psychoanalytic theory is essentially a male-centered theory which may cause harm to women, particularly those who seek psychotherapy from therapists who use a psychoanalytic approach.

Evaluation of Psychoanalytic Theory

From a scientific point of view, one of the major problems with psychoanalytic theory is that most of its concepts cannot be evaluated scientifically to see whether they are accurate. Freud postulated that many of the most important forces in personality are unconscious, and thus they cannot be studied by any of the usual scientific techniques.

Another criticism is that Freud derived his data almost exclusively from his work with patients who sought therapy from him. Thus, his theory may provide a view not so much of the human personality as of *disturbances* in the human personality.

Finally, many modern psychologists feel that Freud overemphasized biological determinants of behavior and instincts and that he gave insufficient recognition to the importance of the environment and learning.

Nonetheless, Freud did make some important contributions to our understanding of human behavior. He managed to rise above the sexually repressive Victorian era of which he was a part and teach that the libido is an important part of personality (although he may have overestimated its importance). His recognition that humans pass through stages in their psychological development was a great contribution. Perhaps most important from the perspective of this text, Freud took sex out of the closet; he brought it to the attention of the general public and suggested that we could talk about it and that it was an appropriate topic for scientific research.

Learning Theory

While psychoanalytic and sociobiological theories are based on the notion that much of human sexual behavior is biologically controlled, it is also quite apparent that much of it is learned. Some of the best evidence for this point comes from studies of sexual behavior across different human societies, which were considered in Chapter 1. Here the various principles of modern learning theory will

be reviewed, because they can help us understand our own sexuality (for a more detailed discussion, see McConaghy, 1987).

Classical Conditioning

Classical conditioning is a concept usually associated with the work of the Russian scientist Ivan Pavlov (1849–1936). Think of the following situations: You salivate in response to the sight or smell of food, you blink in response to someone poking a finger in your eye, or you experience sexual arousal in response to stroking the inner part of your thigh. In all these cases, an unconditioned stimulus (US, for example, appealing food) automatically, reflexively elicits an unconditioned response (UR, for example, salivation). The process of learning that occurs in classical conditioning takes place when a new stimulus, the conditioned stimulus (CS, for example, the sound of a bell) repeatedly occurs paired with the original unconditioned stimulus (food). After this happens many times, the conditioned stimulus (ringing bell) can eventually be presented without the unconditioned stimulus (food) and will evoke the original response, now called the conditioned response (CR, salivation).

As an example, suppose that Nadia's first serious boyfriend in high school always wears Eau de Male aftershave when they go out. As they advance in their sexual intimacy, they have many pleasant times in the back seat of the car, where he strokes her thighs and other sexually responsive parts of her body and she feels highly aroused, always with the aroma of Eau de Male in her nostrils. One day she enters an elevator full of strangers in a department store, and someone is wearing Eau de Male. Nadia instantly feels sexually aroused, although she is not engaged in any sexual activity. From the point of view of classical conditioning, this makes perfect sense, although Nadia may wonder why she is feeling so aroused in the elevator. The thigh-stroking and sexy touching were the US. Her arousal was the UR. The aroma of Eau de Male, the

Classical conditioning: The learning process in which a previously neutral stimulus (conditioned stimulus) is repeatedly paired with an unconditioned stimulus that reflexively elicits an unconditioned response. Eventually the conditioned stimulus itself will evoke the response.

CS, was repeatedly paired with the US. Eventually, the Eau de Male aroma occurred by itself, evoking arousal, the CR.

Classical conditioning of sexual arousal has been demonstrated in an experiment using male students as participants (Lalumiere & Quinsey, 1998). All of the participants were first shown 20 slides of partially clothed women; a slide rated as 5 on a scale of sexual attractiveness which ranged from 1 to 10 was selected as the target slide. Ten participants were shown the target slide, followed by a 40-second segment of a sexually explicit videotape, for 11 trials; men in the control group saw only the target slide for 11 trials. Arousal was measured by a penile strain gauge, which measures the extent of engorgement or erection of the penis (see Chapter 14). Each man then rated the 20 original slides again. In the experimental group, the target slide was associated with an increase in arousal as measured by the strain gauge; in the control group, men were less aroused by the target slide following the repeated exposure.

Classical conditioning is useful in explaining a number of phenomena in sexuality. One example is fetishes, as explained in Chapter 16.

Operant Conditioning

Operant conditioning, a concept that is often associated with the psychologist B. F. Skinner, refers to the following process. A person performs a particular behavior (the operant). That behavior may be followed by either a reward (positive reinforcement) or a punishment. If a reward follows, the person will be likely to repeat the behavior again in the future; if a punishment follows, the person will be less likely to repeat the behavior. Thus if a behavior is repeatedly rewarded, it may become very frequent, and if it is repeatedly punished, it may become very infrequent or even be eliminated.

Some rewards are considered to be primary reinforcers; that is, there is something intrinsically rewarding about them. Food is one such primary reinforcer, and sex is another. Rats, for example, can be trained to learn a maze if they find a willing sex partner at the end of it. Thus sexual behavior plays dual roles in learning theory: It can itself be a positive reinforcer, but it can also be the behavior that is rewarded or punished.

Simple principles of operant conditioning can help explain some aspects of sex (McGuire et al.,

1965). For example, if a woman repeatedly experiences pain when she has intercourse (perhaps because she has a vaginal infection), she will probably want to have sex infrequently or not at all. In operant conditioning terms, sexual intercourse has repeatedly been associated with a punishment (pain), and so the behavior becomes less frequent.

Another principle of operant conditioning that is useful in understanding sexual behavior holds that consequences, whether reinforcement or punishment, are most effective in shaping behavior when they occur immediately after the behavior. The longer they are delayed after the behavior has occurred, the less effective they become. As an example, consider a young man who has had gonorrhea three times yet continues to have unprotected sexual intercourse. The pain associated with gonorrhea is certainly punishing, so why does he persist in having sex without a condom? The delay principle suggests the following explanation. Each time he engages in intercourse, he finds it highly rewarding; this immediate reward maintains the behavior; the punishment, gonorrhea, does not occur until several days later and so is not effective in eliminating that behavior.

A third principle that has emerged in operant conditioning studies is that, compared with rewards, punishments are not very effective in shaping behavior. Often, as in the case of the child who is punished for taking an illicit cookie, punishments do not eliminate a behavior but rather teach the person to be sneaky and engage in it without being caught. As an example, some parents, as many commonly did in earlier times in our culture, punish children for masturbating; yet most of those children continue to masturbate, perhaps learning instead to do it under circumstances (such as in a bathroom with the door locked) in which they are not likely to be caught.

One important difference between psychoanalytic theory and learning theory should be noticed. Psychoanalytic theorists believe that the determinants of human sexual behavior occur in early

Operant (OP-ur-unt) conditioning: The process of changing the frequency of a behavior (the operant) by following it with reinforcement (which will make the behavior more frequent in the future) or punishment (which should make the behavior less frequent in the future).

childhood, particularly during the Oedipal complex period. Learning theorists, in contrast, believe that sexual behavior can be learned and changed at any time in the lifespan—in childhood, in adolescence, in young adulthood, or later. When we try to understand what causes certain sexual behaviors and how to treat people with sex problems, this distinction between the theories will have important implications.

Behavior Modification

Behavior modification involves a set of techniques, based on principles of classical or operant conditioning, that are used to change (or modify) human behavior. These techniques have been used to modify everything from problem behaviors of children in the classroom to the behavior of schizophrenics. In particular, these methods can be used to modify problematic sexual behaviors—that is, sexual disorders such as orgasm problems (see Chapter 19) or deviant sexual behavior such as child molesting. Behavior modification methods differ from more traditional methods of psychotherapy such as psychoanalysis in that the behavioral therapist considers only the problem behavior and how to modify it using learning-theory principles; the therapist does not worry about a depth analysis of the person's personality to see, for example, what unconscious forces might be motivating the behavior.

One example of a technique used in modifying sexual behavior is *olfactory aversion therapy* (Abel et al., 1992). In aversion therapy, the problematic behavior is punished using an aversive stimulus. Repeated pairing of the behavior and the aversive stimulus should produce a decline in the frequency of the behavior, or its extinction. In olfactory aversion therapy, the problematic sexual behavior is punished using an unpleasant odor, such as the odor of spirits of ammonia, as the aversive stimulus. With the help of a therapist, the patient first identifies the behavior chain or sequence that leads up to the problem behavior. Then the patient imagines one event in the chain and is simultaneously exposed to the odor. The odor can be administered by the patient, using a breakable inhaler. This form of therapy not only punishes the behavior but also creates the perception in the patient that the behavior is under his or her control.

Figure 2.2 According to social learning theory, children learn about sex and gender in part by imitation. These children may be imitating their parents or a scene they have watched on TV.

Social Learning

Social learning theory (Bandura, 1977; Bandura & Walters, 1963) is a somewhat more complex form of learning theory. It is based on principles of operant conditioning, but it also recognizes two other processes at work: *imitation* and *identification.* These two processes are useful in explaining the development of gender identity, or one's sense of maleness or femaleness. For example, it seems that a little girl acquires many characteristics of the female role by identifying with her mother and imitating her, as when she plays at dressing up after observing her mother getting ready to go to a party. Because most sexual behaviors in our society are kept rather private and hidden, imitation and identification have less of a chance to play a part. However, some of the more open forms of sexuality may be learned through imitation. In high school, for example, the sexiest girl in the senior class may find that other girls are imitating her behaviors and the way she dresses. Or a boy might see a movie in which the hero's technique seems to "turn women on"; then he tries to use this technique with his own dates.

Behavior modification: A set of operant conditioning techniques used to modify human behavior.

Focus 2.1

Learning Theory and Sexual Orientations in a Non-Western Society

The Sambia are a tribe living in Papua New Guinea in the South Pacific, which has been extensively studied by anthropologists (see Focus 15.3 in Chapter 15). The Sambia are interesting for a number of reasons, the chief one being that young males are expected to spend 10 or more years of their lives in exclusively homosexual relations. During this time they are taught to fear women and believe that women have polluting effects on them. After that stage of their lives, they are expected to marry women. They do, and their sexual behavior becomes exclusively heterosexual. These observations defy our Western notions that sexual orientation is a permanent characteristic throughout one's life. Indeed, the very concepts of having a "heterosexual identity" or "homosexual identity" are not present in Sambia culture.

Can social learning theory explain these patterns of sexual behavior? It can, according to the analysis of John and Janice Baldwin. The thing that is puzzling is how the Sambia male, who has had years of erotic conditioning to homosexual behavior just at puberty when he is most easily aroused and sensitive to conditioning, would then switch to heterosexual behavior and do so happily.

According to the Baldwins' analysis, several factors in social learning theory explain this switch. First, positive conditioning in the direction of heterosexuality occurs early in life. The boy spends the first 7 to 10 years of his life with his family. He has a close, warm relationship with his mother. In essence, he has been conditioned to positive feelings about women.

Second, observational learning occurs. In those same first 7 to 10 years, the boy observes closely the heterosexual relationship between two adults, his mother and father. This observational learning can be used a decade later when it is time for him to marry and form a heterosexual relationship.

Third, the boy is provided with much cognitive structuring, a notion present in social learning theory as well as cognitive psychology. He is instructed that a boy must pass through a series of stages to become a strong, masculine man. This includes first becoming the receptive partner to fellatio, then being the inserting partner to fellatio, marrying, defending himself from his wife's first menstruation (girls are usually married before puberty and undergo no homosexual stage of development), and then fathering a child by her. Essentially he is given all the cognitive structures necessary to convince him that it is perfectly natural, indeed desirable, to engage in sex with men for 10 years and then switch to women. Finally, there is some aversive conditioning to the homosexual behavior that leads it to be not particularly erotic. The boy performs fellatio for the first time after several days of initiation, when he is exhausted. The activities are staged so that the boy feels fearful about it. He must do it in darkness with an older boy who may be an enemy, and he is required to do it with many males in succession. In essence, unpleasantness or punishment is associated with homosexual behavior.

In summary, then, social learning theory provides a sensible explanation of the seemingly puzzling shift that Sambia males make from exclusively homosexual expression to exclusively heterosexual expression.

Sources: Baldwin, John D., & Baldwin, Janice I. (1989). The socialization of homosexuality and heterosexuality in a non-Western society. *Archives of Sexual Behavior, 18,* 13–29. Herdt, Gilbert H. (Ed.). (1984). *Ritualized homosexuality in Melanesia.* Berkeley: University of California Press.

Once a behavior is learned, the likelihood of its being performed depends on its consequences. The young man who imitates actor Brad Pitt's romantic technique may not succeed in arousing his companion. If the behavior is not reinforced, he will stop performing it. If it is reinforced, he will repeat it. Successful experiences with an activity over time create a sense of competence at performing the activity, or **self-efficacy** (Bandura, 1982). If a woman feels efficacious at using the female condom, she will expend more effort (going to the drugstore to buy one) and will show greater persistence in the face of difficulty (continuing to adjust it until it fits properly) than she did before. The concept of self-efficacy has been widely used in designing health intervention programs such as those that encourage individuals to use condoms to prevent transmission of sexually transmitted diseases and HIV infection (e.g., DeLamater, Wagstaff, & Havens, 1994). These programs provide opportunities for participants to practice the behaviors and be successful.

Cognitive Theories

Over the last two decades, a "cognitive revolution" has swept through psychology. In contrast to the older behaviorist tradition (which insisted that psychologists should study only behaviors that could be directly observed), cognitive psychologists believe that it is very important to study people's thoughts—that is, the way people perceive and think.

Cognition and Sexuality

Cognitive psychology is increasingly being applied in understanding human sexuality (Walen & Roth, 1987). A basic assumption is that what we think influences what we feel. If we think happy, positive thoughts, we will tend to feel better than if we think negative ones. Therapists using a cognitive approach believe that psychological distress is often a result of unpleasant thoughts that are usually not tuned to reality and include misconceptions, distortion, exaggerations of problems, and unreasonably negative evaluations of events.

Self-efficacy: A sense of competence at performing an activity.
Schema (SKEE-muh): A general knowledge framework that a person has about a particular topic.

To the cognitive psychologist, how we perceive and evaluate a sexual event makes all the difference in the world (Walen & Roth, 1987). For example, suppose that a man engaged in lovemaking with his wife does not get an erection. Starting from that basic event, his thoughts might take one of two directions. In the first, he thinks that it is quite common for men in his age group (fifties) not to get an erection every time they have sex; this has happened to him a few times before, once every two or three months, and it's nothing to worry about. At any rate, the fellatio was fun, and so was the cunnilingus, and his wife had an orgasm from that, so all in all it was a nice enough encounter. In the second possibility, he began the lovemaking thinking that he had to have an erection, had to have intercourse, and had to have an orgasm. When he didn't get an erection, he mentally labeled it *impotence* and imagined that he would never again have an erection. He thought of the whole episode as a frustrating disaster because he never had an orgasm.

As cognitive psychologists point out, perception, labeling, and evaluating events are crucial. In one case, the man perceived a slight problem, labeled it a temporary erection problem, and evaluated his sexual experience as pretty good. In the other case, the man perceived a serious problem, labeled it impotence, and evaluated the experience as horrible.

We shall see cognitive psychology several times again in this book, as theorists use it to understand the cycle of sexual arousal (see Chapter 9), the causes of some sexual variations such as fetishes (see Chapter 16), and the causes and treatment of sexual disorders (see Chapter 19). Before we leave cognitive psychology, however, we will look at one cognitive theory, schema theory, that has been used especially to understand issues of sex and gender.

Gender Schema Theory

Psychologist Sandra Bem (1981) has proposed a schema theory to explain gender-role development and the impact of gender on people's daily lives and thinking. "Schema" is a term taken from cognitive psychology. A **schema** is a general knowledge framework that a person has about a particular topic. A schema organizes and guides perception; it helps us remember, but it sometimes also distorts our memory, especially if the information is inconsistent with our schema. Thus, for example, you might have a "football game schema," the set of ideas you have about what elements should be present at the game (two teams, spectators, bleachers, etc.) and what kinds of activities should occur (opening kickoff, occasional touchdown, band playing at half-time, and so on).

It is Bem's contention that all of us possess a *gender schema*—the set of attributes (behaviors, personality, appearance) that we associate with males and females. Our gender schema, according to Bem, predisposes us to process information on the basis of gender. That is, we tend to want to think of things as gender-related and to want to dichotomize them on the basis of gender. A good example is the case of the infant whose gender isn't clear when we meet him or her. We eagerly seek out the information or feel awkward if we don't, because we seem to need to know the baby's gender in order to continue to process information about it.

Bem (1981) has done a number of experiments that provide evidence for her theory, and there are confirming experiments by other researchers as well, although the evidence is not always completely consistent (Ruble & Stangor, 1986). In one of the most interesting of these experiments, 5- and 6-year-old children were shown pictures like those in Figure 2.3, showing males or females performing either stereotype-consistent activities (such as a girl baking cookies) or stereotype-inconsistent activities (such as girls boxing) (Martin & Halverson, 1983). One week later the children were tested for their recall of the pictures. The results indicated that the children distorted information by changing the gender of people in the stereotype-inconsistent pictures, while not making such changes for the stereotype-consistent pictures. That is, children tended to remember a picture of girls boxing as having been a picture of boys boxing. These results are just what would be predicted by gender schema theory. The schema helps us remember schema-consistent (stereotype-consistent) information well, but it distorts our memory of information that is inconsistent with the schema (stereotype-inconsistent).

Figure 2.3 Pictures like these were used in the Martin and Halverson research on gender schemas and children's memory. *(a)* A girl engaged in a stereotype-consistent activity. *(b)* Girls engaged in a stereotype-inconsistent activity. In a test of recall a week later, children tended to distort the stereotype-inconsistent pictures to make them stereotype-consistent; for example, they remembered that they had seen boys boxing.

(a)   (b)

One of the interesting implications of gender schema theory is that stereotypes—whether they are about males and females, or heterosexuals and homosexuals, or other groups—may be very slow to change. The reason is that our schema tends to filter out stereotype-inconsistent (that is, schema-inconsistent) information so that we don't even remember it.

Sociological Perspectives

Sociologists are most interested in the ways in which society or culture shapes human sexuality. (For a detailed articulation of the sociological perspective, see DeLamater, 1987; the arguments that follow are taken from that source.)

Sociologists approach the study of sexuality with three basic assumptions: (1) Every society regulates the sexuality of its members. (For a discussion of reasons why, see Horrocks, 1997.) (2) Basic institutions of society (such as religion and the family) affect the rules governing sexuality in that society. (3) The appropriateness or inappropriateness of a particular sexual behavior depends on the culture in which it occurs.

The Sociological Approach: Levels of Analysis

The sociological approach to understanding influences on sexuality is shown in Figure 2.4. As you look down the left side of the diagram, you will notice that sociologists view societal influences on

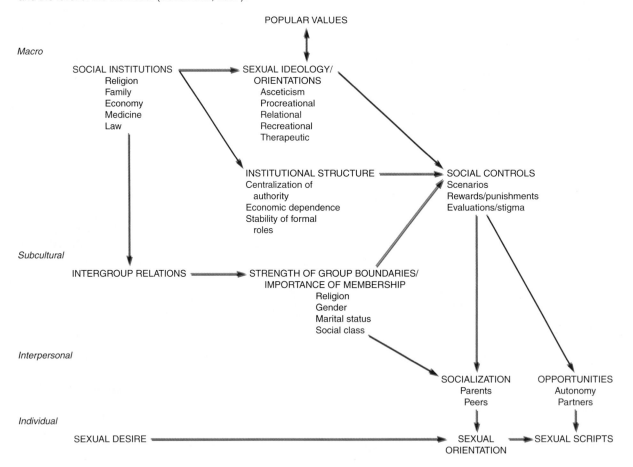

Figure 2.4 A sociological approach to understanding human sexuality. Sociologists focus on four levels of analysis: the macro level, the subcultural level, the interpersonal level, and the level of the individual (DeLamater, 1987).

Figure 2.5 According to sociologists, a culture's economy may have a profound effect on patterns of sexuality, marriage, and childbearing. High rates of male unemployment may lead to an increase in the number of female-headed households.

human sexuality as occurring on four levels: the macro-level or society as a whole; the subcultural level, at which one's social class or ethnic group may have an impact on one's sexuality; the interpersonal level, in which interactions with parents, peers, or lovers influence us; and finally the individual level, at which each of us has her or his own level of sexual desire, a sexual orientation, and a set of sexual scripts stored in memory. Our discussion here emphasizes the macro-level of influence.

Social Institutions

According to the sociological perspective, at the macro-level our sexuality is influenced by powerful social institutions, including religion, the family, the economy, medicine, and the law. [3]

[3]Each major institution supports a sexual ideology, or discourse, about sexual activity.

Religion

In our culture, the Judeo-Christian religious tradition has been a powerful shaper of sexual norms. A detailed discussion of that religious tradition and its teachings on sexuality is provided in Chapter 21. Suffice it to say here that the Christian religion has contained within it a tradition of asceticism, in which abstinence from sexual pleasures—especially by certain persons such as monks and priests—is seen as virtuous. The tradition, at least until recently, has also been oriented toward procreation—that is, a belief that sexuality is legitimate only within traditional heterosexual marriage and only with the goal of having children, a procreational ideology. This view has created within our culture a set of norms, or standards for behavior, that say, for example, that premarital sex, extramarital sex, and homosexual sex are wrong.

The Economy

The nature and structure of the economy is another macro-level influence on sexuality. Before the industrial revolution, most work was done in the family unit in the home or farm. This kind of togetherness permitted rather strict surveillance of family members' sexuality and thus strict norms could be enforced. However, with the industrial revolution, people—most frequently men—spent many hours per day at work away from the home. Thus they were under less surveillance, and scenarios such as extramarital affairs and homosexual sex could be acted out more often.

Today we see much evidence of the extent to which economic conditions, and especially the unemployment rate, can affect the structure of the family and thus sexuality. For example, when a group of men—such as lower-class black men—have less access to jobs and have a high unemployment rate, they are understandably reluctant to marry, knowing that they cannot support a family. The result is many female-headed households, with sexuality occurring outside marriage and children born without a legal father, although the father may be present in the household, providing care for the children. The point is that a culture's economy may have a profound effect on patterns of sexuality, marriage, and childbearing.

The Family

The family is a third macro-level factor influencing sexuality. As we noted earlier, before the industrial revolution the family was an important economic unit, producing the goods necessary for survival. As that function waned after the industrial revolution, there was increased emphasis on the quality of interpersonal relationships in the family. At the same time, love was increasingly seen as an important reason for marriage. By 1850, popular American magazines sang the praises of marriage based on romantic love (Lantz et al., 1975). Thus a triple linkage between love, marriage, and sex was formed. Ironically, the linkage eventually became a direct one between love and sex (removing marriage as the middleman) so that, by the 1970s, some people were arguing that premarital sex, if in the context of a loving relationship, was permissible, as was homosexual sex, again if the relationship was a loving one. This is the relational ideology.

The family exerts a particularly important force on sexuality through *socialization* of children. That is, parents socialize their children—teach them appropriate norms for behavior—in many areas,

including sexuality. Others, of course, such as the peer group, also have important socializing influences. Sociologists see these socializing influences occurring not at the macro-level, but at the interpersonal level (see Figure 2.4).

Medicine

The institution of medicine is another macro-level factor that, particularly in the last 100 years, has influenced our sexuality. Physicians tell us what is healthy and what is unhealthy. In the late 1800s, physicians warned that masturbation could cause various pathologies. Today sex therapists tell us that sexual expression is natural and healthy and sometimes even "prescribe" masturbation as a treatment. Because we tend to have great confidence in medical advice, these pronouncements, based on a therapeutic ideology, have an enormous impact on sexuality.

The Law

The legal system is another institution influencing sexuality at the macro-level. A detailed discussion of laws relating to sexuality is provided in Chapter 22. The point to be made here is that from a sociological perspective, the law influences people's sexual behaviors in a number of ways. First, laws determine norms. Generally we think that what is legal is right and what is illegal is wrong. Thus, a society in which prostitution is illegal will have much different views of it than a society in which it is legal.

Second, laws are the basis for mechanisms of social control. They may specify punishments for certain acts and thus discourage people from engaging in them. An example is public sexual acts such as exhibitionism, or even nudity on beaches. One wonders how many people would prefer to be nude at the beach if they did not fear arrest because the behavior is generally illegal and if they did not fear possible embarrassing publicity such as having their names in the paper as a result of an arrest.

Third, the law reflects the interests of the powerful, dominant groups within a society. In part, the law functions to confirm the superiority of the ideologies of these dominant groups. Consider Mormons in the United States. In the past, their religion approved of polygyny (a man having several wives). Mormons did not become the dominant group in American society—although they might have, with the kind of reproduction one could achieve with polygyny. Rather, the Judeo-Christian tradition was the ideology of the dominant group, and the Judeo-Christian tradition takes a very dim view of polygyny. Accordingly, polygyny is illegal in the United States and Mormons have been arrested for their practice. Also, one wonders what kinds of laws we would have on prostitution or sexual harassment if the composition of the state legislatures were 90 percent women rather than 90 percent men. Would prostitution, for example, be legal? Or would it still be illegal, but would the male customer be held as guilty as or guiltier than the female prostitute?

In summary, then, the sociological perspective focuses on how society or culture shapes and controls our sexual expression, at levels from institutions such as religion and the law to the interpersonal level of socialization by family and peers.

Sexual Scripts

The outcome of all these social influences is that each of us learns a set of *sexual scripts* (Gagnon & Simon, 1973; Gagnon, 1977, 1990). The idea is that sexual behavior (and virtually all human behavior, for that matter) is scripted much as a play in a theater is. That is, sexual behavior is a result of elaborate prior learning that teaches us an etiquette of sexual behavior. According to this concept, little in human sexual behavior is spontaneous. Instead, we have learned an elaborate script that tells us who, what, when, where, and why we do what we do sexually. For example, the "who" part of the script tells us that sex should occur with someone of the other gender, of approximately our own age, of our own race, and so on. Even the sequence of sexual activity is scripted. *Scripts,* then, are plans that people carry around in their heads for what they are doing and what they are going to do; they are also devices for helping people remember what they have done in the past (Gagnon, 1977, p. 6).

How could we study these scripts? How could we find out if there are widely shared beliefs about how one should behave in a specific situation? One way is to ask people to describe what one should do in such a situation. Researchers asked male and female college students to make a list of the things that a man or woman would typically do on a first date (Rose & Frieze, 1993). The hypothetical script written by many of the participants included a core

Figure 2.6   According to some people's sexual scripts, a man taking a woman to dinner is one scene of the first act in a sexual script that features intercourse as the finale.

sequence of acts: dress, pick up, get to know, evaluate, eat, make out, and go home. This is the first-date script. This script is also influenced by contemporary views of gender roles. Males were portrayed as proactive, taking the initiative: pick up the female, pay for the date, attempt to make out, and ask for another date. Females were portrayed as reactive: be picked up, be treated, accept or reject male's attempt to make out, and accept or reject invitation to another date. The widely shared nature of this script enables relative strangers to interact smoothly the first time they are together.

One study attempted to identify the sequence of sexual behaviors that is scripted for males and females in a heterosexual relationship in our culture (Jemail & Geer, 1977). People were given 25 sentences, each describing an event in a heterosexual interaction. They were asked to rearrange the sentences in a sequence that was "most sexually arousing" and then to do it again to indicate what was "the most likely to occur." There was a high degree of agreement among people about what the sequence should be. There was also high

Figure 2.7 According to Reiss's sociological theory, sex is important to us because it is associated with great physical pleasure and self-disclosure.

agreement between males and females. The standard sequence was kissing, hand stimulation of the breasts, hand stimulation of the genitals, mouth-genital stimulation, intercourse, and orgasm. Does it sound familiar? Interestingly, not only is this the sequence in a sexual encounter, it is also the sequence that occurs as a couple progresses in a relationship. These results suggest that there are culturally defined sequences of behaviors that we all have learned, much as the notion of "script" suggests.

Can you imagine a young man, on the first date, attempting oral stimulation of the young lady's genitals before he has kissed her? The idea seems amazing, perhaps humorous or shocking. Why? Because the young man has performed Act IV before Act I.

Scripts also tell us the meaning we should attach to a particular sexual event (Gagnon, 1990). Television programs and films frequently suggest but do not show sexual activity between people.

How do we make sense out of these implicit portrayals? A study of how women interpret such scenes in films found that they utilize scripts. If the film showed a couple engaging in two actions that are part of the script for sexual intercourse (e.g., kissing and undressing each other) and then faded out, viewers inferred that intercourse had occurred (Meischke, 1995).

Reiss's Sociological Theory of Sexuality

Ira Reiss (1986) has proposed a sociological theory of human sexuality. Borrowing from script theory discussed earlier in this section, he defines sexuality as "erotic and genital responses produced by the cultural scripts of a society" (p. 37). As he points out, a sociological theory, because it focuses on societal influences on sexuality, must be able to account for both cross-cultural variations in sexuality, as well as cross-cultural universals in sexuality.

One cross-cultural universal is that all societies believe that sexuality is important. Even in those cultures that are sexually repressive, sexuality is still accorded great importance as something that is dangerous and must be controlled. Why is sexuality so important? Many theorists have claimed that it is the link of sexuality to making babies, which is undeniably important for any society. Reiss argues against this notion, however, citing instances of societies that do not understand the link between sex and reproduction, yet still find sex important. Indeed, in North America today effective methods of contraception have allowed us to separate much of sexuality from reproduction, yet we still think sex is important, even nonreproductive sex.

Reiss's explanation for the universal importance of sexuality has two components: (1) sexuality is associated with great physical pleasure, and (2) sexual interactions are associated with great personal self-disclosure, involving not only disclosure of one's body, but in an intimate interaction, of one's thoughts and feelings as well. Humans seem to find intrinsic value in the physical pleasures of sex and in the psychic satisfaction of the self-disclosures associated with sex—therefore its importance.

According to Reiss, sexuality is linked to the structures of any society in three areas: the kinship system, the power structure, and the ideology of the society.

First, because sexuality is the source of reproduction, it is always linked to *kinship,* and all societies seek to maintain social order through stable kinship systems. This linkage then becomes the explanation for sexual jealousy, which is universal cross-culturally although it exists in varied forms. Jealousy is a way of setting boundaries on a relationship that is considered very important, important enough so that it should not be breached. Marriage is typically such a relationship, and jealousy in marriage about extramarital affairs exists in all human societies. Kin define what relationships are and are not acceptable and enforce the resulting rules. Furthermore, all societies have structured ways of dealing with such jealousy. Even in societies that practice polygyny, rituals develop to minimize jealousy among the wives—for example, the husband must sleep one night with one wife, the next with another, and so on, and has violated norms if he spends two nights with the same wife! Reiss argues, therefore, that no society will be able to eliminate sexual jealousy because jealousy is a statement of the value or importance kinship groups and individuals attach to a particular relationship such as marriage.

Second, sexuality is always linked to the *power structure* of a society. Reiss defines power as the ability to influence others and achieve one's objectives even if there is opposition from the other person. Powerful groups in any society generally seek to control the sexuality of the less powerful. Males are more powerful than females in most societies, so sexuality becomes linked to gender roles and males exercise control over female sexuality. Cross-cultural research shows, interestingly, that the closer females are in power to males in a society, the more sexual freedom women have; in societies in which women have little power, their sexuality is greatly restricted.

Third, sexuality is closely linked to the *ideologies* of a culture. Reiss defines "ideology" as fundamental assumptions about human nature. Societies define carefully what sexual practices are normal and abnormal, and which are right and wrong. Some cultures define homosexuality as abnormal, whereas others define it as normal—but the point is that all cultures define it one way or the other. Similarly, some cultures take a permissive attitude toward premarital sex both for males and females, some are permissive for males but not for females, and some are permissive for neither. A culture's ideologies define what is right and wrong sexually.

In summary, Reiss's sociological theory of sexuality argues that societies regard sexuality as important because it is associated with physical pleasure and with self-disclosure. Sexuality is tightly woven into the fabric of any society because of its links to the kinship system, to power structures, and to the ideologies of the society.

SUMMARY

Sociobiologists view human sexual behaviors as the product of natural selection in evolution, and thus view these behavioral patterns as being genetically controlled. Contemporary evolutionary theorists view behavior as the result of an interaction between evolved mechanisms and environmental influence.

Among the psychological theories, Freud's psychoanalytic theory views the sex energy, or libido, as a major influence on personality and behavior. Freud introduced the concepts of erogenous zones and psychosexual stages of development. Learning theory emphasizes how sexual behavior is learned and modified through reinforcements and punishments according to principles of operant conditioning. Behavior modification techniques—therapies based on learning theory—are used in treating sexual variations and sexual disorders. Social learning theory adds the concepts of imitation, identification, and self-efficacy to learning theory. Cognitive psychologists focus on people's thoughts and perceptions—whether positive or negative—and how these influence sexuality.

Sociologists study the ways in which society influences our sexual expression. At the macro-level of analysis, sociologists investigate the ways in which institutions such as religion, the economy, the family, medicine, and the law influence sexuality. At the subcultural level, our social class or ethnic group may shape our sexuality. At an interpersonal level we are socialized by others—such as parents and peers—as to the appropriate norms for sexual expression. The resulting sexual scripts provide us with guidelines for behavior in many situations. Reiss's sociological theory argues that all societies regard sexuality as important because it is associated with great physical pleasure and self-disclosure.

REVIEW QUESTIONS

1. _____ is the term for the theory that applies evolutionary theory to social behaviors such as sexual behavior.

2. Freud termed the sex energy or sex drive _____ .

3. According to Freud, the three major parts of the personality are the id, the ego, and the libido. True or false?

4. Freud believed that women could have either of two kinds of orgasms, vaginal orgasm or clitoral orgasm. True or false?

5. According to learning theory, teenagers are in the genital stage of development. True or false?

6. According to social learning theory, imitation is a powerful force shaping our sexual behaviors. True or false?

7. According to cognitive psychologists, how we perceive and evaluate a sexual event is at least as important as what occurred. True or false?

8. Sociologists study social influence at four levels of analysis: _____, _____, _____, and _____ .

9. According to the notion of scripts, we have been socialized to believe that the appropriate sequence of social behaviors is kissing before engaging in intercourse. True or false?

10. According to Reiss's sociological theory, sex is important in any society because it involves physical pleasure and _____ .

(The answers to all review questions are at the end of the book, after the Glossary.)

QUESTIONS FOR THOUGHT, DISCUSSION, AND DEBATE

1. Compare and contrast how a sociobiologist, a psychoanalyst, and a sociologist would explain why people engage in premarital intercourse.

2. Of the theories described in this chapter, which do you think provides the most insight into human sexuality? Why?

3. You are visiting your married sister and her family for the holidays. Her six-year-old asks you a question: "Why do mommies stay home with kids?" Drawing on the knowledge you gained from Chapter 2, how would you answer this question? Which ideas about parental investment, social learning, or social norms/scripts would you use to explain why, in most cultures, women perform the child-rearing activities?

4. From a sociologist's point of view, how have social institutions controlled female sexuality?

5. Could script theory explain the pattern of homosexual and heterosexual behavior of Sambia males described in Focus 2.1? Explain your answer.

6. In Chapter 1, we identified a number of differences in the sexual behavior of whites, African Americans, Latinos, and Asian Americans (see Table 1.2, page 15). Which of the theories discussed in Chapter 2 can most easily explain these differences? Pick one specific difference and construct an explanation for it using that theory. How well does the theory help you to understand cultural differences in sexuality?

SUGGESTIONS FOR FURTHER READING

Baker, Robin (1996). *Sperm Wars: The evolutionary logic of love and lust.* New York: Basic Books. Baker uses a series of vignettes describing sexual interactions to illustrate the many ways evolutionary mechanisms may influence our selection of mates, frequency of intercourse, success and failure to conceive, and infidelity.

Buss, David M. (1994). *The evolution of desire.* New York: Basic Books. A detailed presentation of sexual strategy theory and data that are consistent with the theory.

Freud, Sigmund. (1943). *A general introduction to psychoanalysis*. Garden City, NY: Garden City Publishing. (Original in German, 1917.) Good for the reader who wants a basic introduction to Freud. For a good one-chapter summary, see

Hall, C. S., and Lindzey, G. (1970). *Theories of personality* (2nd ed.). New York: Wiley.

Gagnon, John H., & Simon, William. (1987). The sexual scripting of oral-genital contacts. *Archives of Sexual Behavior, 16*, 1–26. An interesting discussion and application of script theory.

WEB RESOURCES

http://www.behavior.net/
Behavior Online Page, including threads about various theoretical perspectives.

http://www.psych.ucsb.edu/research/cep/primer.html
The Evolutionary Primer, by Lea Cosmides and John Tooby; a must read.

http://www.runet.edu/~lridener/courses/SOCBIO.HTML
A description of the sociobiological perspective by E. O. Wilson.

http://www.humanitas.ucsb.edu/users/steen/CogWeb/Sociobiology.html
A critique of sociobiology.

chapter

3

SEX RESEARCH

CHAPTER HIGHLIGHTS

Issues in Sex Research

The Major Sex Surveys

Studies of Special Populations

Media Content Analysis

**Laboratory Studies Using Direct
  Observations of Sexual Behavior**

Participant-Observer Studies

Experimental Sex Research

Some Statistical Concepts

A ccording to the Kinsey Report
  Every average man you know,
Much prefers to play his favorite sport,
When the temperature is low,
But when the thermometer goes way up,

And the weather is sizzling hot,
  Mr. Adam
For his madam
Is not.
    'Cause it's too darn hot.*

In the last few decades, sex research has made increasing advances, and the names of Kinsey and Masters and Johnson are household words. How do sex researchers do it? How valid are their conclusions?

There are many different types of sex research, but basically the techniques vary in terms of the following: (1) whether they rely on people's self-reports of their sexual behavior or whether the scientist observes the sexual behavior directly; (2) whether large numbers of people are studied (surveys) or whether a small number or just a single individual is studied (in laboratory studies or case studies); (3) whether the studies are conducted in the laboratory or in the field; and (4) whether sexual behavior is studied simply as it occurs naturally or whether some attempt is made to manipulate it experimentally.

Examples of studies using all these techniques will be considered and evaluated later in the chapter. First some issues in sex research—objections frequently made to studies that have been done—will be discussed.

It is important to understand the techniques of sex research and their strengths as well as their limitations. This knowledge will help you evaluate the studies that are cited as evidence for various conclusions in later chapters and will also help you decide how willing you are to accept these conclusions. Perhaps more important, this knowledge will help you evaluate future sex research. Much sex research has been conducted already, but much more will be done in the future. The information in this chapter should help you understand and evaluate sex research that appears 10 or 20 years from now.

*Kiss Me Kate, a musical comedy. (1953). Music and lyrics by Cole Porter; book by Sam and Bella Spewack. New York: Knopf.

Issues in Sex Research

Sampling

One of the first steps in conducting sex research is to identify the appropriate **population** of people to be studied. Does the population in question consist of all adult human beings, all adults in the United States, all adolescents in the United States, all people guilty of sex crimes, or all married couples who engage in swinging? Generally, of course, the scientist is unable to get data for all the people in the population, and so a **sample** is taken.

At this point, things begin to get sticky. If the sample is a **random sample** or representative sample of the population in question and if it is a reasonably large sample, then results obtained from it can safely be generalized to the population that was originally identified. That is, if a researcher has really randomly selected 1 out of every 50 adolescents in the United States, then the results obtained from that sample are probably true of all adolescents in the United States. One technique that is sometimes used to get such a sample is **probability sampling.**[1] But if the sample consists

[1]A detailed discussion of probability sampling is beyond the scope of this book. For a good description of this method as applied to sex research, see Cochran et al. (1953). A random sample is one example of a probability sample.

Population: A group of people a researcher wants to study and make inferences about.
Sample: A part of a population.
Random sampling: An excellent method of sampling in research, in which each member of the population has an equal chance of being included in the sample.
Probability sampling: An excellent method of sampling in research, in which each member of the population has a known probability of being included in the sample.

only of adolescents with certain characteristics—for example, only those whose parents would agree to let them participate in sex research—then the results obtained from that sample may not be true of all adolescents. Sampling has been a serious problem in sex research.

Typically, sampling proceeds in three phases: the population is identified, a method for obtaining a sample is adopted, and the people in the sample are contacted and asked to participate. The scientific techniques of the second phase—obtaining a sample—are by now fairly well developed and should not be a problem in future research, provided investigators use them. What is perhaps the thorniest problem, though, occurs in the last phase: getting the people identified for the sample to participate. If any of the people refuse to participate, then the great probability sample is ruined. This is called the **problem of refusal (or nonresponse).** As a result, the researcher is essentially studying volunteers, that is, people who do agree to be in the research. The outcomes of the research may therefore contain **volunteer bias.** In casually conducted research such as the Hite report (Hite, 1976, 1981), the response rate was only 3 percent, making it impossible to reach any conclusions about the population based on the sample. The problem of refusal in sex research is difficult, since there is no ethical way of forcing people to participate when they do not want to.

The problem of volunteer bias would not be so great if those who refused to participate were identical in their sexual behavior to those who partici-

pated. But it seems likely that those who refuse to participate differ in some ways from those who agree to, and that means the sample is biased. Evidence suggests that volunteers who participate in sex research hold more permissive attitudes about sexuality and are more sexually experienced than those who don't; for example, they masturbate more frequently and have had more sexual partners (Morokoff, 1986; Strassberg & Lowe, 1995; Wiederman, 1993; Wiederman et al., 1994). In addition, women are less likely to volunteer for sex research than men are (e.g., Wiederman et al., 1994), so that female samples are even more highly selected than male samples. In sum, volunteer bias is potentially a serious problem when we try to reach conclusions based on sex research.

Table 3.1 shows how different the results of sex surveys can be, depending on how carefully the sampling is done (Greeley, 1994). The table shows results from the Janus report (Janus & Janus, 1993), which used sampling methods so haphazard that the researchers ended up with what some call a "convenience sample." It included volunteers who came to sex therapists' offices and friends re-

---

Problem of refusal or nonresponse: The problem that some people will refuse to participate in a sex survey, thus making it difficult to have a random sample.
Volunteer bias: A bias in the results of sex surveys that arises when some people refuse to participate, so that those who are in the sample are volunteers who may in some ways differ from those who refuse to participate.

Table 3.1 The Percentage of People Reporting Having Sex at Least Once a Week: Comparing a Convenience Sample with a Probability Sample

| Age | Men | | Women | |
	Convenience Sample (Janus Report)	Probability Sample (General Social Survey)	Convenience Sample (Janus Report)	Probability Sample (General Social Survey)
18–26	72%	57%	68%	58%
27–38	83	69	78	61
39–50	83	56	68	49
51–64	81	43	65	25
Over 65	69	17	74	6

Source: Andrew M. Greeley. (1994). The Janus Report. *Contemporary Sociology, 23,* 221–223.

cruited by the original volunteers. This report contrasts with the probability sample obtained in the General Social Survey conducted by the University of Chicago. Notice that a considerably higher level of sexual activity is reported by the convenience sample in the Janus report, compared with the probability sample. This difference is especially pronounced among the elderly. Convenience samples simply do not give us a very good picture of what is going on in the general population.

Reliability of Self-Reports of Sexual Behavior

Most sex researchers have not directly observed the sexual behavior of their research participants. Instead, most have relied on respondents' self-reports of their sexual practices. The question is: How accurately do people report their own sexual behavior? There are several ways in which inaccuracies may occur, and these inaccuracies are problems for sex surveys. These problems are discussed next.

Purposeful Distortion

If you were an interviewer in a sex research project and a 90-year-old man said that he and his wife made love twice a day, would you believe him, or would you suspect that he might be exaggerating slightly? If a 35-year-old woman told you that she had never masturbated, would you believe her, or would you suspect that she had masturbated but was unwilling to admit it?

Respondents in sex research may, for one reason or another, engage in **purposeful distortion,** intentionally giving self-reports that are distortions of reality. These distortions may be in either of two directions. People may exaggerate their sexual activity (a tendency toward "enlargement"), or they may minimize their sexual activity or hide the fact

Purposeful distortion: Purposely giving false information in a survey.

Figure 3.1   The reliability of self-reports of sexual behavior. if you were interviewing this man in a sex survey and he said that he had never masturbated, would you believe him, or would you think that he was concealing a taboo behavior?

that they have done certain things ("concealment"). Unfortunately, we do not know whether most people tend toward enlargement or concealment.

Distortion is a basic problem when using self-reports. To minimize distortion, participants must be impressed with the fact that because the study will be used for scientific purposes, their reports must be as accurate as possible. They must also be assured that their responses will be completely anonymous; this is necessary, for example, so that a politician would not be tempted to hide an extra-marital affair or a history of sex with animals for fear that the information could be used to blackmail him.

But even if all respondents were very truthful and tried to give as accurate information as possible, two factors might still cause their self-reports to be inaccurate: memory and difficulties with estimates.

Memory

Some of the questions asked in sex surveys require respondents to recall what their sexual behavior was like many years before. For example, some of the data we have on sexual behavior in childhood come from the Kinsey study, in which adults were asked about their childhood sex behavior. This might involve asking a 50-year-old man to remember at what age he began masturbating and how frequently he masturbated when he was 16 years old. It may be difficult to remember such facts accurately. The alternative is to ask people about their current sexual behavior, although getting data like this from children raises serious ethical and practical problems.

Difficulties with Estimates

One of the questions sex researchers have asked is: How long, on the average, do you spend in pre-coital foreplay? If you were asked this question, how accurate a response do you think you could give? It is rather difficult to estimate time to begin with, and it is even more difficult to do so when engaged in an absorbing activity. The point is that in some sex surveys people are asked to give estimates of things that they probably cannot estimate very accurately. This may be another source of inaccuracy in self-report data.

Evidence on the Reliability of Self-Reports

Scientists have developed several methods for assessing how reliable or accurate people's reports are

(Catania et al., 1995). One is the method of **test-retest reliability,** in which the respondent is asked a series of questions and then is asked the same set of questions after a period of time has passed, for example, a week or a month. The correlation[2] between answers at the two times (test and retest) measures the reliability of responses. If people answer identically both times, the correlation would be 1.0, meaning perfect reliability. If there were absolutely no relationship between what they said the first time and what they said the second time (a situation that never actually occurs), the correlation would be 0, meaning that the responses are not at all reliable.

In one study, heterosexual college students were asked to estimate their frequency of vaginal intercourse for a one-month period; the test-retest reliability was .89, which is excellent (Catania et al., 1990a). However, when they were asked their frequency of intercourse in a six-month period of time, the test-retest reliability fell to .65, and when they were asked about frequency during a one-year period, reliability was only .36. Respondents can give their best estimates about short, recent time intervals.

Kinsey, too, used the test-retest method, re-interviewing 162 men with a minimum of 18 months between the first two interviews. The results indicated a high degree of agreement between the first and second interviews. Correlations greater than .95 between the first and second interviews were obtained for reports of incidence of masturbation, extramarital coitus, and homosexual activity.

Another method for assessing reliability involves obtaining independent reports from two different people who share sexual activity, such as husbands and wives. Kinsey was the first to use this method, interviewing 706 couples. Reports of objective facts—such as the number of years they had been married, how long they were engaged, and how much time elapsed between their marriage and the birth of the first child—showed per-

[2]The statistical concept of correlation is discussed in the last section of this chapter.

Test-retest reliability: A method for testing whether self-reports are reliable or accurate; participants are interviewed (or given a questionnaire) and then interviewed a second time sometime later to determine whether their answers are the same both times.

fect or near-perfect agreement. However, as noted earlier, some other questions required much more subjective responses or estimates of things that might be difficult to estimate. On such items there was less agreement between spouses. For example, the correlation between the average husband's and wife's estimate of the average frequency of intercourse early in marriage was only about .50, which was the lowest correlation obtained. Even with these subjective reports, though, husbands and wives showed a fairly high degree of agreement.

More recently, we used this same method of checking for agreement between spouses (Hyde et al., 1996). On a simple item such as whether they had engaged in intercourse in the last month, there was 93 percent agreement. When reporting on something that requires somewhat more difficult estimation, the number of times they had intercourse in the past month, the correlation was .80, which still indicates good reliability.

Interviews Versus Questionnaires

In the large-scale sex surveys, three methods of collecting data have been used: the face-to-face interview, the phone interview, and the written questionnaire. Each of these methods has some advantages when compared with the others (Catania et al., 1995).

The advantage of the interview, particularly the face-to-face interview, is that the interviewer can establish rapport with the respondent and, it is hoped, convince that person of the research's worth and of the necessity for being honest. An interviewer can also vary the sequence of questions depending on the person's response. For example, if a person mentioned having had a homosexual experience, this response would be followed by a series of questions about the experience; however, those questions would be omitted if the person reported having had no homosexual experiences. It is hard to get this kind of flexibility in a printed questionnaire. Finally, interviews can be administered to persons who cannot read or write.

Questionnaires are much less costly, since they do not require hiring interviewers to spend the many hours necessary to interview respondents individually. It is also possible that respondents would be more honest in answering a questionnaire because they are more anonymous.

A recent innovation is the computer-assisted self-interview method (CASI), which can be combined with an audio component so that the respondent not only reads but hears the questions. This method offers the privacy of the written questionnaire while accommodating poor readers. The computer can be programmed to follow varying sequences of questions depending on respondents' answers, just as a human interviewer does.

What do the data say about which method works best for sex research? Several researchers have compared the results obtained through use of two different methods. For example, in one study, the rate of reporting rape was nearly double (11 percent) in a face-to-face interview compared with a telephone interview (6 percent) (Koss et al., 1994, p. 174). This finding seems to indicate that, when respondents are reporting very sensitive information, interviewers can establish rapport and trust better in person than over the telephone. In a study assessing risky sexual behavior among gay men, both face-to-face interviews and written questionnaires were used (Siegel et al., 1994). Riskier behaviors were more likely to be reported on the questionnaire than in the interview. People evidently feel a bit freer to report particularly sensitive information on the more private written questionnaire than to an interviewer. Many experts in sex research recommend that a face-to-face interview to build rapport be combined with a written questionnaire administered during the interview to tap particularly sensitive information (Laumann et al., 1994; Siegel et al., 1994).

Self-Reports Versus Direct Observations

As we noted earlier, one of the major ways of classifying techniques of sex research is according to whether the scientist relied on people's self-reports of their behavior or observed the sexual behavior directly.

The problems of self-reports have been discussed above. In a word, self-reports may be inaccurate. Direct observations—such as those done by Masters and Johnson in their work on the physiology of sexual response—have a major advantage over self-reports in that they are accurate. No purposeful distortion or inaccurate memory can intervene. On the other hand, direct observations have their own set of problems. They are expensive and time-consuming, with the result that generally a rather small sample is studied. Furthermore, obtaining a random or probability sample of the

Figure 3.2   Sex researchers have tried to revise some ingenious methods for overcoming the problems of self-reports.
©Punch/Rothco

"Another one of those damned sex surveys, I suppose."

population is even more difficult than in survey research. While some people are reticent about completing a questionnaire concerning their sexual behavior, even more would be unwilling to come to a laboratory where their sexual behavior would be observed by a scientist or where they would be hooked up to recording instruments while they engaged in sex. Thus results obtained from the unusual group of volunteers who would be willing to do this might not be generalizable to the rest of the population. One study showed that volunteers for a laboratory study of male sexual arousal were less guilty, less sexually fearful, and more sexually experienced than nonvolunteers (Farkas et al., 1978; for similar results with females, see Wolchik et al., 1983).

Direct observations of sexual behavior in the laboratory, such as those made by Masters and Johnson, involve one other problem: Is sexual behavior in the laboratory the same as sexual behavior in the privacy of one's own bedroom? For example, might sexual response in the laboratory be somewhat inhibited?

Extraneous Factors

Various extraneous factors may also influence the outcomes of sex research. In interviews about sexual behavior, for example, there is some indication that both male and female respondents prefer a female interviewer (DeLamater & MacCorquodale, 1979). Thus such extraneous factors as the gender, race, or age of the interviewer may influence the outcome of sex research. Questionnaires do not get around these problems, since such simple factors as the wording of a question may influence the results. In one study, respondents were given either standard or supportive wording of some items (Catania et al., 1995). For the question about extramarital sex, the standard wording was as follows:

At any time while you were married during the past 10 years, did you have sex with someone other than your (husband/wife)?

The supportive wording was as follows:

Many people feel that being sexually faithful to a spouse is important, and some do not. However, even those who think being faithful is important have found themselves in situations where they ended up having sex with someone other than their (husband/wife). At any time while you were married during the past 10 years, did you have sex with someone other than your (husband/wife)?

The supportive wording significantly increased reports of extramarital sex from 12 percent with the standard wording to 16 percent with the supportive wording, if the interviewer was of the same gender as the respondent; the wording made no difference when the interviewer and respondent were of different genders. Sex researchers must be careful to control these extraneous factors so that they influence the results as little as possible.

Ethical Issues

There is a possibility of ethical problems involved in doing research. Ethical problems are particularly difficult in sex research, because people are more likely to feel that their privacy has been invaded when you ask them about sex than when you ask them to name their favorite presidential candidate or memorize a list of words. The ethical standards of most scientific organizations—such as the American Psychological Association and the American Sociological Association—involve two basic principles: informed consent and protection from harm (see, for example, American Psychological Association, 1973).

Informed Consent

According to the principle of **informed consent,** participants have a right to be told, before they participate, what the purpose of the research is and what they will be asked to do. They may not be forced to participate or be forced to continue. An investigator may not coerce people to be in a study, and it is the scientist's responsibility to see to it that all participants understand exactly what they are agreeing to do. In the case of children who may be too young to give truly informed consent, it is usually given by the parents.

The principle of informed consent was adopted by scientific organizations in the 1970s. It was violated in some of the older sex studies, as will be discussed later in this chapter.

Protection from Harm

Investigators should minimize the amount of physical and psychological stress to people in their research. Thus, for example, if an investigator must shock participants during a study, there should be a good reason for doing this. Questioning people about their sexual behavior may be psychologically stressful to them and might conceivably harm them in some way, so sex researchers must be careful to minimize the stress involved in their procedure. The principle of *anonymity* of response is important to ensure that participants will not suffer afterward for their participation in research.

A Cost-Benefit Approach

Considering the possible dangers involved in sex research, is it ever ethical to do such research? Officials in universities and government agencies sponsoring sex research must answer this question for every proposed sex research study. Typically they use a **cost-benefit approach.** That is, stress to research participants should be minimized as much as possible, but some stresses will remain; they are the cost. The question then becomes: Will the benefits that result from the research be greater than the cost? That is, will the participants benefit in some way from participating, and will science and society in general benefit from the knowledge resulting from the study? Do these benefits outweigh the costs? If they do, the research is justifiable; otherwise, it is not.

As an example, Masters and Johnson considered these issues carefully and they feel that their research participants benefited from being in their research; they have collected data from former participants that confirm this belief. Thus a cost-benefit analysis would suggest that their research is ethical, even though their participants may be temporarily stressed by it. Even in a study as ethically questionable as Laud Humphreys's study of the tearoom trade (discussed later in this chapter), the potential cost to the participants must be weighed against the benefits that accrue to society

> **Informed consent:** An ethical principle in research, in which people have a right to be informed, before participating, of what they will be asked to do in the research.
> **Cost-benefit approach:** An approach to analyzing the ethics of a research study, based on weighing the costs of the research (the subjects' time, stress to subjects, and so on) against the benefits of the research (gaining knowledge about human sexuality).

Focus 3.1
Alfred C. Kinsey

Alfred C. Kinsey was born in 1894 in New Jersey, the first child of uneducated parents. In high school he did not date, and a classmate recalled that he was "the shyest guy around girls you could think of."

His father was determined that Kinsey become a mechanical engineer. From 1912 to 1914 he tried studying mechanical engineering at Stevens Institute, but he showed little talent for it. At one point he was close to failing physics, but a compromise was reached with the professor, who agreed to pass him if he would not attempt any advanced work in the field! In 1914 Kinsey made the break and enrolled at Bowdoin College to pursue his real love: biology. Because this went against his father's wishes, Kinsey was put on his own financially.

In 1916 he began graduate work at Harvard. There he developed an interest in insects, specializing in gall wasps. While still a graduate student he wrote a definitive book on the edible plants of eastern North America.

In 1920 he went to Bloomington, Indiana, to take a job as assistant professor of zoology at Indiana University. That fall he met Clara McMillen, whom he married six months later. They soon had four children.

With his intense curiosity and driving ambition, Kinsey quickly gained academic success. He published a high school biology text in 1926, and it received enthusiastic reviews. By 1936 he had published two major books on gall wasps; they established his reputation as a leading authority in the field and contributed not only to knowledge of gall wasps but also to genetic theory.

Kinsey came to the study of human sexual behavior as a biologist. His shift to the study of sex began in 1938, when Indiana University began a "marriage" course; Kinsey chaired the faculty committee teaching it. Part of the course included individual conferences between students and faculty, and these were Kinsey's first sex interviews. When confronted with teaching the course, he also became aware of the appalling lack of information on human sexual behavior. Thus his research resulted in part from his realization of the need of people, especially young people, for sex information. In 1939 he made his first field trip to collect sex histories in Chicago. His lifetime goal was to collect 100,000 sex histories.

His work culminated with the publication of the Kinsey reports in 1948 (*Sexual Behavior in the Human Male*) and 1953 (*Sexual Behavior in the Human*

from being informed about this aspect of sexual behavior.

The Major Sex Surveys

In the major sex surveys, the data were collected from a large sample of people by means of questionnaires or interviews. The best known of these studies is the one done by Kinsey, so we will consider it first. The data were collected in the late 1930s and 1940s, and thus the results are now largely of historical interest. However, Kinsey documented his methods with extraordinary care, so his research is a good example to study for both the good and the bad points of surveys.

The Kinsey Report

The Sample

Kinsey (see Focus 3.1) and his colleagues interviewed a total of 5300 males, and their responses were reported in *Sexual Behavior in the Human Male* (1948); 5940 females contributed to *Sexual*

Female). While the scientific community generally received them as a landmark contribution, they also provoked hate mail.

In 1947 he founded the Institute for Sex Research (known popularly as the Kinsey Institute) at Indiana University. It was financed by a grant from the Rockefeller Foundation and, later, by book royalties. But in the 1950s Senator Joseph McCarthy, the communist baiter, was in power. He made a particularly vicious attack on the Institute and its research, claiming that its effect was to weaken American morality and thus make the nation more susceptible to a communist takeover. Under his pressure, support from the Rockefeller Foundation was terminated.

Kinsey's health began to fail, partly as a result of the heavy work load he set for himself, partly because he was so involved with the research that he took attacks personally, and partly because he saw financial support for the research collapsing. He died in 1956 at the age of 62 of heart failure, while honoring a lecture engagement when his doctor had ordered him to convalesce.

By 1957 McCarthy had been discredited, and the grant funds returned. The Institute was then headed by Paul Gebhard, an anthropologist who had been a member of the staff for many years. The Institute continues to do research today; it also houses a large library on sex and an archival collection including countless works of sexual art.

In a recent, highly publicized, tell-all biography of Kinsey, James Jones (1997) argued that, although Kinsey's public self was a stable, married man, Kinsey was in fact homosexual (more accurately, bisexual) and practiced masochism. According to Jones, this discredits Kinsey's research. Jones's logic is poor, though, because one can evaluate the quality of the research methods independent of Kinsey's personal sex life.

Sources: P. Gebhard. The Institute. In M. S. Weinberg (Ed.). (1976). *Sex research: Studies from the Kinsey Institute.* New York: Oxford University Press, pp. 10–22. C. V. Christensen. (1971). *Kinsey: A biography.* Bloomington: Indiana University Press. J. H. Jones (1997), *Alfred C. Kinsey: A public-private life.* New York: Norton.

Figure 3.3 Alfred C. Kinsey (second from right, holding the folder), with colleagues Martin, Gebhard, and Pomeroy.

Behavior in the Human Female (1953). Though some blacks were interviewed, only interviews with whites were included in the publications. The interviews were conducted between 1938 and 1949.

Initially, Kinsey was not much concerned with sampling issues. His goal was simply to collect sex histories from as wide a variety of people as possible. He began conducting interviews on his university campus and then moved on to large cities, such as Chicago.

Kinsey later became more concerned with sampling issues and developed a technique called *100 percent sampling*. In this method he contacted a group, obtained its cooperation, and then got every one of its members to give a history. Once the cooperation of a group had been secured, peer pressure assured that all members would participate. Unfortunately, although he was successful in getting a complete sample from such groups, the groups themselves were by no means chosen randomly. Thus among the groups from which 100 percent samples were obtained were 2 sororities, 9 fraternities, and 13 professional groups.

In the 1953 volume, Kinsey said that he and his colleagues had deliberately chosen not to use probability sampling methods because of the problems of nonresponse. This is a legitimate point. But as a result, we have almost no information on how adequate the sample was. As one scholar observed, the sampling was haphazard but not random (Kirby, 1977). For example, there were more respondents from Indiana than from any other state. Generally, the following kinds of people were overrepresented in the sample: college students, young people, well-educated people, Protestants, people living in cities, and people living in Indiana and the northeast. Underrepresented groups included manual laborers, less well-educated people, older people, Roman Catholics, Jews, members of racial minorities, and people living in rural areas.

The Interviews

Although scientists generally regard Kinsey's sampling methods with some dismay, his face-to-face interviewing techniques are highly regarded. Over 50 percent of the interviews were done by Kinsey himself, and the rest by his associates, whom he trained carefully. The interviewers made every attempt to establish rapport with the people they spoke to, and they treated all reports matter-of-factly. They were also skillful at phrasing questions in language that was easily understood. Questions were worded so as to encourage people to report anything they had done. For example, rather than asking "Have you ever masturbated?" the interviewers asked "At what age did you begin masturbating?" They also developed a number of methods for cross-checking a person's report so that false information would be detected. Wardell Pomeroy recounted an example:

> Kinsey illustrated this point with the case of an older Negro male who at first was wary and evasive in his answers. From the fact that he listed a number of minor jobs when asked about his occupation and seemed reluctant to go into any of them [Kinsey] deduced that he might have been active in the underworld, so he began to follow up by asking the man whether he had ever been married. He denied it, at which Kinsey resorted to the vernacu-

lar and inquired if he had ever "lived common law." The man admitted he had, and that it had first happened when he was 14.

> "How old was the woman?" [Kinsey] asked.

> "Thirty-five," he admitted, smiling.

> Kinsey showed no surprise. "She was a hustler, wasn't she?" he said flatly.

> At this the subject's eyes opened wide. Then he smiled in a friendly way for the first time, and said, "Well, sir, since you appear to know something about these things, I'll tell you straight."

> After that, [Kinsey] got an extraordinary record of this man's history as a pimp. . . . (Pomeroy, 1972, pp. 115–116)

Strict precautions were taken to ensure that responses were anonymous and that they would remain anonymous. The data were stored on IBM cards, but using a code that had been memorized by only a few people directly involved in the project and that was never written down. The research team had even made contingency plans for destroying the data in the event that the police tried to demand access to the records for the purposes of prosecuting people.

Put simply, the interviewing techniques were probably very successful in minimizing purposeful distortion. However, other problems of self-report remained: the problems of memory and of inability to estimate some of the numbers requested.

How Accurate Are the Kinsey Statistics?

When all is said and done, how accurate are the statistics presented by Kinsey? The American Statistical Association appointed a blue-ribbon panel to evaluate the Kinsey reports (Cochran et al., 1953; for other evaluations, see Terman, 1948; Wallin, 1949). While the panel members generally felt that the interview techniques had been excellent, they were dismayed by Kinsey's failure to use probability sampling, and they concluded, somewhat pessimistically:

> In the absence of a probability-sample benchmark, the present results must be regarded as subject to systematic errors of unknown magnitude due to selective sampling (via volunteering and the like). (Cochran et al., 1953, p. 711)

However, they also felt that this was a nearly insoluble problem for sex research; even if a probability sample were used, refusals would still create serious problems.

The statisticians who evaluated Kinsey's methods felt that one aspect of his findings might have been particularly subject to error: the generally high levels of sexual activity, and particularly the high incidence of homosexual behavior. These conclusions might have been seriously influenced by discrepancies between reported and actual behavior and by sampling problems, particularly Kinsey's tendency to seek out persons with unusual sexual practices.

Kinsey's associates felt that the most questionable statistic was the incidence of male homosexuality. Wardell Pomeroy commented, "The magic 37 percent of males who had one or more homosexual experiences was, no doubt, overestimated." (1972, p. 466).

In sum, it is impossible to say how accurate the Kinsey statistics are; they may be very accurate, or they may contain serious errors. Probably the single most doubtful figure is the high incidence of homosexuality. Also, at this point the Kinsey survey is approximately 50 years old; for accurate information about sexuality today, we need to look to more recent research.

The NHSLS

Since the time of the Kinsey report, many sex surveys have been conducted. Some have used slipshod sampling methods; others have been more carefully done studies of special populations such as teenagers between the ages of 15 and 19. What was needed was a large-scale, national survey of sexuality using probability sampling methods to tell us what Americans' patterns of sexual behavior are today. Such a study appeared in 1994. The research team was headed by Edward Laumann, a distinguished sociologist at the University of Chicago, and was conducted by the National Opinion Research Center (NORC), one of the best-respected survey organizations in the country. The survey was called the National Health and Social Life Survey; to keep things simple, we will call this study the NHSLS (Laumann et al., 1994; Michael et al., 1994).

The sampling method involved a probability sampling of households in the United States. This excluded less than 3 percent of Americans, but did exclude people living in institutions (e.g., prisons, college dormitories) and the homeless. People were eligible if they were adults between the ages of 18 and 59.

The researchers obtained an impressive 79 percent cooperation rate. Apparently, the great majority of people are willing to respond to a carefully conducted sex survey. The response rate is particularly impressive in view of the fact that today, even surveys of more neutral topics such as political opinions generally have a response rate of only about 75 percent.

The researchers had originally planned to obtain a sample of 20,000 people. However, federal funding for the project was blocked by the same political processes described in Focus 3.2. They were able to obtain funding from private foundations, but only enough to interview a sample of 3432.

Figure 3.4 Sociologist Edward Laumann, head of the research team that conducted the NORC study.

Focus 3.2
Politics Versus Sex Research

Although major well-sampled surveys of adult sexual behavior have recently been published, we do not have comparable studies of teenage sexuality, and we need them. Many of the surveys we have tend to be out-of-date (the Kantner and Zelnik studies) or poorly sampled (studies of college students). The surveys that use good sampling techniques typically ask only a limited set of questions about sex, or are surveys of adults only. As a result, we have a knowledge gap. Scientists do not know with any degree of accuracy, for example, what percentage of teenage males in the United States have engaged in homosexual interactions. We cannot hope to educate people about sex, much less fight the AIDS epidemic, with such a glaring lack of information.

Recognizing this serious problem, the National Institutes of Health (NIH) decided to fund a major survey of teenage sexuality. The grant was won by a team of researchers from the University of North Carolina, headed by Dr. Ronald Rindfuss and co-directed by the eminent sex researcher Dr. J. Richard Udry. The research was a five-year project studying teenagers' sexual behavior, the focus of which was a survey of 20,000 children in grades 7 through 11 whose parents had given consent to participate. The project had been evaluated by scientific review panels at the NIH and had received glowing reviews. Funding was awarded in May 1991.

Then the Secretary of Health and Human Services (HHS), Louis Sullivan, who had been appointed by President George Bush, heard about the study, as did members of Congress. NIH is under the jurisdiction of Health and Human Services. Senator Jesse Helms (R-NC) and Representative William Dannemeyer (R-CA) quickly initiated legislation to force NIH to stop funding the study. Senator Helms said that the real purpose of the study "is to 'cook the books,' so to speak, in terms of presenting 'scientific facts'—in order to do what? To legitimize homosexual lifestyles of course."

Meanwhile Sullivan, seeing the direction the political winds were blowing, terminated the study. This was the first time in the history of federal funding for scientific research that a political appointee, the Secretary of HHS, interfered with the scientific process and cancelled a study that had met the high standards of the scientific peer review process and been approved.

These actions raise a number of serious questions. How can we fight the AIDS epidemic when we lack current, accurate information on teenagers' sexual behavior? Can a federal agency arbitrarily terminate a contract it has made with a university? Should politicians be permitted to interfere with the integrity of the scientific process?

Sources: Charrow, Robert P. (1991). (November). Sex, politics, and science. *The Journal of NIH Research, 3,* 80–83. Udry, J. Richard. (1993). The politics of sex research. *Journal of Sex Research, 30,* 103–110.

The data were obtained in face-to-face interviews supplemented by brief written questionnaires, which were handed to the respondents for particularly sensitive topics (such as masturbation) and sealed in a "privacy envelope" when they had been completed. The researchers chose the face-to-face interview because they felt that it would yield a higher response rate than a written questionnaire alone, and it allowed the researchers to ask more complex, detailed sequences of questions than would have been possible with just a written questionnaire or a phone interview.

Laumann's team was careful to obtain the respondents' informed consent. About a week before an interviewer went to a household, a letter was sent explaining that the purpose of the survey was to help "doctors, teachers, and counselors better understand and prevent the spread of diseases like AIDS and better understand the nature and extent of harmful and of healthy sexual behavior in our country" (Laumann et al., 1994, p. 55). The purpose was therefore clearly and honestly described to participants before their consent was requested. In order to protect confidentiality, all identifying information about the respondent was destroyed after the interview. Each respondent was paid $35 for the interview, which lasted, on the average, 90 minutes.

The NHSLS is the best sex survey of the general population of the United States that we have today, and its findings will be referred to in many chapters in this book. The researchers made outstanding efforts to use the best sampling methods and interview methods.

Nonetheless, the study has some limitations. The sample did not include enough people from some statistically small minority groups—in particular, Native Americans—to compute reliable statistics for them, so their data were omitted from most tables of the results. This problem would probably not have occurred if there had been funding for the full sample of 20,000. On the other hand, for other ethnic minority groups—African Americans, Latinos, and Asian Americans—there are lots of interesting findings. No doubt some respondents engaged in concealment and perhaps also in enlargement, because self-reports were used. The skill of interviewers and their ability to build rapport is crucial in overcoming such problems. The researchers reported training the interviewers extensively, but nonetheless the extent of concealment, or *underreporting,* remains unknown.

One criticism of the NHSLS is that some of the interviews took place with a third person present, obviously not an ideal situation for honest reporting about sexuality. Fully 79 percent of the interviews were conducted privately; 6 percent of the interviews took place with the spouse or partner present, and the remaining 15 percent took place with some other person present, usually the respondent's child (Laumann et al., 1994, p. 568). The researchers analyzed the data to see whether the presence of another person was associated with different patterns of reporting that might suggest concealment. In fact, 17 percent of respondents interviewed privately reported 2 or more sex partners in the past year, whereas only 5 percent of respondents interviewed with another person present admitted to 2 or more sex partners. However, before we leap to the conclusion that the results are seriously distorted in this subsample with a third person present for the interview, it is important to realize that the presence of the third person was likely associated with other factors. Specifically, in many cases a woman was being interviewed with her preschool child present when she could not obtain child care. It may therefore be that mothers with preschool children are simply less likely to have had multiple partners in the last year. Laumann's analyses support this view: when simple social variables were controlled, only a few random reporting differences remained between those with a third person present and those whose interview was private. It seems unlikely that this methodological problem caused major difficulties with the results.

Ironically, the most controversial statistic in the study was the same as in Kinsey's research—the incidence of homosexuality. In Kinsey's case, people thought the numbers were too high. In the case of the NHSLS, some people thought that they were too low. We will return to this issue in Chapter 15.

Sexual Behavior in France and Britain

Again stimulated by a need for far better information about sexual behavior in order to deal with the AIDS crisis, a team of French researchers, called the ACSF Investigators, conducted a major French sex survey (ACSF Investigators, 1992). The data were collected in 1991 and 1992. These researchers chose the method of telephone interviews, preceded by a letter notifying potential respondents that they had been identified for the representative sample. The response rate was 76.5 percent. The result was a sample of 20,055 adults ranging in age from 18 to 69.

At the time of this writing only preliminary results of the survey have been released. They indicate, for example, that 13 percent of French men, compared with 6 percent of French women, had multiple sex partners (two or more) in the past 12 months and therefore were at higher risk of HIV

Figure 3.5   Research conducted among racial and ethnic minority groups in the United States must be culturally sensitive. Ideally, for example, interviewers should be of the same cultural background as research participants.

infection. Other results from this study will be discussed in later chapters of this book.

A similar British survey was conducted by Anne Johnson and her colleagues (1992), yielding data for 18,876 men and women aged 16 to 59, living in England, Wales, and Scotland. The data indicated that, for the entire sample, 14 percent of the men and 7 percent of the women had multiple sex partners in the past 12 months, figures quite similar to the French survey. However, when one looks just at 16- to 24-year-olds, 27 percent of the men and 16 percent of the women had two or more partners in the past 12 months—a considerably greater proportion.

Many of the statistics in the French and British surveys match quite closely those from the United States NHSLS.

A Sex Survey of African American and Hispanic Youth

Kathleen Ford and Anne Norris (1997) conducted a major sex survey of African American and Latino youth between the ages of 15 and 24 years, residing in low-income areas of Detroit. They used excellent sampling methods, contacting a probability sample of households in low-income neighborhoods. More than 95 percent of the 60 interviewers were themselves ethnic minority residents of Detroit. Interviewers were given extensive training on skills for interviewing people about sensitive topics such as sexuality.

The study focused on the respondents' networks of sexual relationships. Although both Latinos and African Americans formed sexual partnerships with another member of their ethnic group the majority of the time, Latinos were far more likely than African Americans to have a partner outside their ethnic group. Among both African Americans and Latinos, women were likely to have partners older than themselves. We see in these results a pattern of both ethnic-group differences and ethnic-group similarities.

In other writings, Ford and Norris (1991) have provided extensive discussions of methodological points that must be kept in mind as researchers move beyond studying all or mostly white samples and focus increasingly on ethnic minorities in the United States (see also Matsumoto, 1994). Here are some of the issues they raise.

Respondents should be interviewed by an interviewer of the same gender and ethnic background as themselves. This practice is important for building rapport and establishing trust during the interview, both of which are critical in obtaining honest answers.

Language is another important issue in constructing interviews. Many people, Anglos included, do not know scientific terms for sexual concepts. Interviewers therefore have to be ready with a supply of slang terms so that they can switch to these if a respondent does not understand a question. The problem becomes more complex when one interviews people whose first language is not English. Ford and Norris had the interview questions translated into Spanish by experts. It was then checked by both Mexican and Puerto Rican persons.

Cultural differences may mean that people from different ethnic groups attach different meanings to sexual concepts. An example is the apparently simple question, "Would you classify yourself as homosexual, heterosexual, or bisexual?" In Latino cultures, men who engage in anal intercourse with other men are typically not categorized as homosexual as long as they are the inserting partner (the receiving partner is definitely categorized as homosexual) (see Chapter 15). Therefore, a Latino male who frequently has sex with other men but is always the inserter might quite truthfully answer that he does not classify himself as homosexual.

The Hispanic women presented some special challenges for these researchers. Hispanic culture is characterized by sharply defined gender roles, a high valuation on virginity for unmarried women, and protection of women from sexual discussions and sexual knowledge. In keeping with these cultural traditions, the Latinas in this sample seemed to have the least sexual knowledge of all the groups and were especially sensitive about sexual topics, making it difficult to interview them.

In conclusion, doing sex research with ethnic minorities in the United States requires more than just administering the same old surveys to samples of minorities. It also requires revisions to methodology that are culturally sensitive on issues such as the ethnicity of the interviewer, the language used in the interview, and the special sensitivity of some groups regarding some topics.

Magazine Surveys

Many large-scale sex surveys have been conducted through magazines. Often the survey is printed in one issue of the magazine, and readers are asked to respond. The result can be a huge sample—perhaps 20,000 people—which sounds impressive. But are these magazine surveys really all they claim to be?

Sampling is just plain out of control with magazine surveys. The survey is distributed just to readers of the magazine, and different magazines have different clienteles. No one magazine reaches a random sample of Americans. If the survey appeared in *Redbook*, it would go to certain kinds of women; if it were in *Ladies' Home Journal*, it would go to others. It would be risky to assume that women who read *Redbook* have the same sexual patterns as those who read *Ladies' Home Journal*. To make matters worse, the response rate is unknown. We can't know how many people saw the survey and did not fill it out, compared with the number who did. The response rate may be something like 3 percent. One does not, therefore, even have a random sample of readers of that magazine.

As an example, let's consider a survey that was reported in the August 1998 issue of *Cosmopolitan* (Moritz, 1998). In an earlier issue they had printed their "Lust Survey." A boxed insert claims "1000 *Cosmo* readers confess," but the text of the article reports that "thousands of you sent in answers." Which was it? And how can we (or the editors at *Cosmopolitan*) know the response rate? Among the respondents, how many were married? Single? What about their ethnic backgrounds? How old were they? Of course these details are not the sort of thing that *Cosmo* probably thinks will entertain its readers. Nonetheless, they could have printed the information in a small box at the end of the article. More importantly, these details mean everything in terms of whether one can take their claims seriously.

For example, in response to a question about what qualities of a man turn you on, the most frequent response, from 41 percent of the women, was "looks" (other common responses were "general attitude," endorsed by 37 percent, and "personality," endorsed by 32 percent). From this, can we conclude that 41 percent of American women are most turned on by a man's appearance? That conclusion would require a leap of logic that is too big for safety. *Cosmo* isn't even close to a random sample in this study. Perhaps the sample was all white. The results, then, might be totally different for African Americans or Asian Americans or Native Americans. Perhaps all the respondents were in their 20s; then the results might have been totally different for women in their 50s.

For all these reasons, it would not be legitimate to infer that these statistics characterize American women in general. We could continue with more examples of magazine surveys, but the general conclusion should be clear by now. Although they may appear impressive because of the large number of respondents, magazine sex surveys actually are poor in quality because the sample is generally seriously biased.

Studies of Special Populations

In addition to the large-scale studies of the U.S. population discussed earlier, many studies of special populations have been done. Two examples are given here: the George and Weiler study of sexuality in middle and late life, and the Bell, Weinberg, and Hammersmith study of gays, lesbians, and heterosexuals.

George and Weiler: Sexuality in Middle and Late Life

Linda George and Stephen Weiler of Duke University conducted an important study of the sexuality of middle-aged and elderly persons (George & Weiler, 1981). With the "graying of America," there is increased interest in the behavior, including the sexual behavior, of older persons. Previous research, such as Kinsey's, had shown substantial declines in sexual expression among the elderly.

George and Weiler studied a broadly based sample of volunteers from the community who were between the ages of 46 and 71 at the beginning of the study. One important feature of their study is that they used a longitudinal design to study developmental changes with aging. Developmental researchers typically use one of two kinds of designs: cross-sectional or longitudinal. With the *cross-sectional design,* one studies several samples of people of different ages, collecting all the data within a short period of time. There is a problem with the cross-sectional design, however. Suppose we find that the average frequency of intercourse is twice a month in a sample of 50-year-olds, and once a month in a sample of 80-year-olds. Can we conclude that sexual activity declines in that 30-year period? No, we can't. The reason is that we had two different samples of people at the two different ages. Those two samples may differ in many ways other than their age. For example, the 80-year-olds grew up 30 years before the 50-year-olds, and socialization about sexuality might have been much more rigid during childhood for the 80-year-olds. The way to overcome this problem is by doing *longitudinal research,* in which the same people are studied repeatedly over the years. Obviously this kind of research takes considerably more time than cross-sectional research, but it allows the investigator to reach conclusions about actual changes in people's lives as they age.

George and Weiler were also careful to restrict their sample to 278 people who were married throughout the four testing times of the study (1969, 1971, 1973, and 1975). This is an important control because widows often cease engaging in sexual intercourse simply because no partner is available.

George and Weiler's results indicated very little decline in sexual activity across the 6 years of the study, in contrast to the findings of previous researchers. Probably George and Weiler got different results because of their longitudinal design. On the other hand, they looked at only a 6-year period. If they had extended the study to 10 or 20 years (an ambitious project), they might have obtained different results.

Bell, Weinberg, and Hammersmith: Homosexuals and Heterosexuals

Under the sponsorship of the Kinsey Institute at Indiana University, Alan Bell, Martin Weinberg, and Sue Hammersmith conducted a major survey of homosexuals and heterosexuals, reporting the results in their book *Sexual Preference* (1981). As the title of the book implies, their goal was to find out what factors determine people's sexual orientation, whether heterosexual or homosexual.

The data came from face-to-face interviews with 979 homosexual women and men and 477 heterosexual women and men, all living in the San Francisco Bay Area. Although the sampling has an obvious geographical limitation, Bell and his colleagues justified their choice of working in the Bay Area as permitting them to obtain a large sample of homosexuals who could be open enough to cooperate and participate in the research. In order to find prospective participants, recruiters—half of them gay themselves—visited locations such as gay bars; recruiting was also done by posters, ads in local newspapers, television spots, and referral by persons already interviewed for the study. Heterosexual participants were obtained by a random sampling technique.

The interview contained 200 questions and took three to five hours. The questions focused on a wide variety of events in the person's childhood and adolescence, the goal being to use the responses to test the various theories that have been proposed to explain why people become homosexual or heterosexual. In several cases, two questions on the same topic appeared at different places in the interview, so that an individual's answers could be checked for reliability. Some of the respondents were reinterviewed six months after the original interview, again to check for the reliability of the self-reports. Bell and his colleagues reported no specific results of these reliability checks, but they appeared to be satisfied with the results.

They used a statistical technique called path analysis to analyze the data. We would have to go too far afield from our discussion of sexuality to explain path analysis here, but briefly, it is a statistical technique that allows one to make conclusions about causal factors from correlational data. Bell and his colleagues wanted to test various hypotheses about experiences and environmental factors that might cause homosexuality, yet their survey data were clearly correlational, so path analysis was a good solution. The results of this study will be discussed in detail in Chapter 15; to summarize them briefly here, they found that the usual environmental factors that have been used to explain homosexuality—parent-child relationship, parental identification, early heterosexual trauma—were not confirmed by their data.

In evaluating this study, one can see that it was done more carefully and according to better scientific standards than many other surveys. The interviewing seems to have met the same high standards as the earlier Kinsey report. There were internal checks for reliability, although the method of self-report was still used, and there might have been problems of memory for events that occurred years ago, in childhood. Although the sampling was done carefully, it is still problematic. The sample could not be considered random or representative of all gays in the United States. It omits all those who are outside the "San Francisco scene." And it omits covert gays, those who do not frequent bars or parties and are not willing to participate in a research interview. This research raises the point that studies of special populations defined by their sexual behavior—such as homosexuals or bisexuals or fetishists—are essentially impossible to do in any kind of representative fashion. It is impossible to identify all the people in the population in the first place, and therefore it is impossible to sample them

properly. In contrast, samples of the general population, such as the NHSLS, are more feasible to obtain, though not easy.

Media Content Analysis

To this point we have focused on methods used to analyze people's responses. Yet we have also recognized the profound impact of the mass media on Americans' sexuality. To be able to understand this impact, we need to be able to analyze the media, and the standard technique is called **content analysis** (Reinharz, 1992; Weber, 1990).

Content analysis refers to a set of procedures used to make valid inferences about text. The "text" might be romance novels, advice columns in *Cosmopolitan* magazine, lyrics from rap music, or prime-time television programs. As it turns out, many of the same methodological issues discussed earlier also come into play with content analysis.

Sampling is an issue. Suppose that you want to do a content analysis of advice columns in *Playboy* magazine. First, you need to define the population. Do you just want to sample from *Playboy*, or do you want to sample from all sex-oriented magazines? If you want to focus only on *Playboy*, then you will surely want to collect your sample of columns from more than one issue. You will have to define the span of years of magazine issues in order to define the population. Finally, you will have to decide whether you will analyze all advice columns from those years, sample from only certain years, or sample some columns in some issues.

The next step is to create a coding scheme. First, you must define the recording unit—is it the word, the sentence, the entire text, or perhaps themes that run across several sentences? Then, perhaps most importantly, you must define the coding categories. Defining the coding scheme involves defining the basic content categories, the presence or absence of which will be recorded, for example, in the advice columns. The coding categories will depend on the question you want to ask. For example,

Content analysis: A set of procedures used to make valid inferences about text.

Figure 3.6 Precise methods have been developed for analyzing the content of the media.

suppose your question about prime-time television shows is, What is the frequency of nonmarital compared with marital sex? In creating a coding scheme, you would have to define carefully what observable behaviors on television count as "sex." Suppose you include kissing, fondling of the breasts or genitals, sexual intercourse actually shown, and implied sexual intercourse. You could then code each of these behaviors as they occurred on a sample of prime-time shows, and, for each act, indicate whether it was between married persons or unmarried persons.

The reliability of the data must be demonstrated in a content analysis just as they must in research with human participants. Without a demonstration of reliability, a critic might accuse you of bias, for example seeing far more acts of sex on the programs than actually occurred. Usually a measure called **intercoder reliability** is used. The researcher trains another person in the exact use of the coding scheme. Then the researcher and the trained coder each independently code a sample of the texts in the study—for example, 20 of the advice columns or 20 of the prime-time shows. The researcher then computes a correlation or percent of agreement between the two coders' results, and that gives the measure of intercoder reliability. If the two coders agree exactly, the correlation will be 1.0.

Content analysis is a powerful scientific technique that allows us to know how the media portray sexuality.

Laboratory Studies Using Direct Observations of Sexual Behavior

The numerous problems associated with using self-reports of sexual behavior in scientific research have been discussed. The major alternative to using self-reports is to make direct observations of sexual behavior in the laboratory. These direct observations overcome the major problems of self-reports: purposeful distortion, inaccurate memory, and inability of people to estimate correctly or describe certain aspects of their behavior. The pioneering example of this approach is Masters and Johnson's work on the physiology of sexual response.

Masters and Johnson: The Physiology of Sexual Response

William Masters began his research on the physiology of sexual response in 1954. No one had ever studied human sexual behavior in the laboratory before, so he had to develop all the necessary research techniques from scratch. He began by interviewing 188 female prostitutes, as well as 27 male prostitutes working for a homosexual clientele. They gave him important preliminary data in which they "described many methods for elevating and controlling sexual tensions and demonstrated innumerable variations in stimulative techniques," some of which were useful in the later program of therapy for sexual disorders.

Meanwhile, Masters began setting up his laboratory and equipping it with the necessary instruments: an electrocardiograph to measure changes in heart rate over the sexual cycle, an electromyograph to measure muscular contractions in the body during sexual response, and a pH meter to measure the acidity of the vagina during the various stages of sexual response.

Sampling

Masters made a major breakthrough when he decided that it was possible to recruit normal participants from the general population and have them

Intercoder reliability: In content analysis, the correlation or percent of agreement between two coders independently rating the same texts.

engage in sexual behavior in the laboratory, where their behavior and physiological responses could be carefully observed and measured. This approach had never been used before, as even the daring Kinsey had settled for people's verbal reports of their behavior.

Most of the research participants were obtained from the local community simply by word of mouth. Masters let it be known in the medical school and university community that he needed volunteers for laboratory studies of human sexual response. Some people volunteered because of their belief in the importance of the research. Some, of course, came out of curiosity or because they were exhibitionists; they were weeded out in the initial interviews. Participants were paid for their hours in the laboratory, as is typical in medical research, and so many medical students and graduate students participated because it was a way to earn money.

Initially, all prospective participants were given detailed interviews by the Masters and Johnson team. People who had histories of emotional problems or who seemed uncomfortable with the topic of sex either failed to come back after this interview or were eliminated even if they were willing to proceed. Participants were also assured that the anonymity and confidentiality of their participation would be protected carefully. In all, 694 people participated in the laboratory studies reported in *Human Sexual Response*. The men ranged in age from 21 to 89, while the women ranged from 18 to 78. A total of 276 married couples participated, as well as 106 women and 36 men who were unmarried when they entered the research program. The unmarried persons were helpful mainly in the studies that did not require sexual intercourse, for example, studies of the ejaculatory mechanism in males and of the effects of sexual arousal on the positioning of the diaphragm in the vagina.

Certainly the group of people Masters and Johnson studied were not a random sample of the population of the United States. In fact, one might imagine that people who would agree to participate in such research would be rather unusual. The data indicate that they were more educated than the general population, and the sample was mostly white, with only a few blacks participating. Paying the participants probably helped since it attracted some people who simply needed the money. The sample omitted two notable types of people: those who are not sexually experienced or do not respond to sexual stimulation and those who are unwilling to have their sexual behavior studied in the laboratory. Therefore, the results Masters and Johnson obtained might not generalize to such people.

Just exactly how critical is this sampling problem to the validity of the research? Masters and Johnson were not particularly concerned about it because they assumed that the processes they were studying are normative; that is, they work in essentially the same way in all people. This assumption is commonly made in medical research. For example, a researcher who is studying the digestive process does not worry that the sample is composed of all medical students, since the assumption is that digestion works the same way in all human beings. If this assumption is also true for the physiology of sexual response, then all people respond similarly, and it does not matter that the sample is not random. Whether this assumption is correct remains to be seen (see Chapter 9 for further critiques). The sampling problem, however, does mean that Masters and Johnson cannot make statistical conclusions on the basis of their research; for example, they cannot say that X percent of all women have multiple orgasms. Any percentages they calculate would be specific to their sample and could not be generalized to the rest of the population.

In defense of their sampling techniques, even if they had identified an initial probability sample, they would still almost surely have had a very high refusal rate, probably higher than in survey research, and the probability sample would have been ruined. At present, this seems to be an unsolvable problem in this type of research.

Data Collection Techniques

After they were accepted for the project, participants then proceeded to the laboratory phase of the study. First, they had a "practice session," in which they engaged in sexual activity in the laboratory in complete privacy, with no data being recorded and with no researchers present. The purpose of this was to allow the participants to become comfortable with engaging in sexual behavior in a laboratory setting.

The physical responses of the participants were

then recorded during sexual intercourse, masturbation, and "artificial coition." Masters and Johnson made an important technical advance with the development of the artificial coition technique. In it, a female participant stimulates herself with an artificial penis constructed of clear plastic; it is powered by an electric motor, and the woman can adjust the depth and frequency of the thrust. There is a light and a recording apparatus inside the artificial penis, so the changes occurring inside the vagina can be photographed.

Measures such as these avoid the problems of distortion that are possible with self-reports. They also answer much different questions. That is, it would be impossible from such measures to tell whether the person had had any homosexual experiences or how frequently he or she masturbated. Instead, they ascertain how the body responds to sexual stimulation, with a kind of accuracy and detail that would be impossible to obtain through self-reports.

One final potential problem also deserves mention. It has to do with the problems of laboratory studies: Do people respond the same sexually in the laboratory as they do in the privacy of their own homes?

Ethical Considerations

Masters and Johnson were attentive to ethical principles. They were careful to use informed consent. Potential participants were given detailed explanations of the kinds of things they would be required to do in the research, and they were given ample opportunity at all stages to withdraw from the research if they so desired. Furthermore, Masters and Johnson eliminated people who appeared too anxious or distressed during the preliminary interviews.

It is also possible that participating in the research itself might have been harmful in some way to people. Masters and Johnson were particularly concerned with the long-term effects of participating in the research. Accordingly, they made follow-up contacts with the participants at five-year intervals. In no case did a participant report developing a sexual disorder (for example, impotence). In fact, many of the couples reported specific ways in which participating in the research enriched their marriages. Thus the available data seem to indicate that such research does not harm the participants and may in some ways benefit them, not to mention the benefit to society that results from gaining

information in such an important area.

In sum, direct observations of sexual behavior of the type done by Masters and Johnson have some distinct advantages but also some disadvantages, compared with survey-type research. The research avoids the problems of self-reports and is capable of answering much more detailed physiological questions than self-reports could. But the research is costly and time-consuming, making large samples impractical; furthermore, a high refusal rate is probably inevitable, so probability samples are impossible to obtain.

Participant-Observer Studies

A research method used by anthropologists and sociologists is the **participant-observer technique.** In this type of research, the scientist actually becomes a part of the community to be studied, and she or he makes observations from inside the community. In the study of sexual behavior, the researcher may be able to get direct observations of sexual behavior combined with interview data.

Examples of this type of research are studies of sexual behavior in other cultures, such as those done in Mangaia, Mehinaku, and Inis Beag, which were discussed in Chapter 1. Two other examples are Laud Humphreys's study of the tearoom trade and Bartell's study of swinging.

Humphreys: The Tearoom Trade

Sociologist Laud Humphreys (1970) conducted a participant-observer study of impersonal sex between men in public places. The study is discussed in detail in Focus 15.2 in Chapter 15. Briefly, Humphreys acted as a lookout while men engaged in sex acts in public rest rooms ("tearooms"); his job was to sound a warning if police or other intruders approached. This permitted Humphreys to make direct observations of the sexual behavior. He also got the license-plate numbers of the men involved, traced them, and later interviewed them in their

Participant-observer technique: A research method in which the scientist becomes part of the community to be studied and makes observations from inside the community.

homes under the pretext of conducting a routine survey.

Humphreys obtained a wealth of information from the study, but in so doing he violated several ethical principles of behavioral research. He had no informed consent from his subjects; they were never even aware of the fact that they were participants in research, much less of the nature of the research. Thus this study was quite controversial.

Bartell: Swinging

Anthropologist Gilbert Bartell (1970) did a participant-observer study of swinging (a married couple having sex with another person or another couple). He and his wife recruited research participants by responding to ads that swingers had placed in newspapers. They took the role of "baby swingers"—a couple who are swinging for the first time—but they also stated that they did not misrepresent themselves and informed all subjects that they were anthropologists interested in knowing more about swinging. In addition, they attended a large number of swingers' parties and large-scale group sexual activities and made observations in that context.

Sampling is often a problem in studies of special groups such as this one, since it is difficult to get any kind of a random sample of people who engage in a particular kind of sexual behavior. In this particular case the problem was made somewhat less difficult because many such contacts are made through newspaper ads, to which the researchers could respond. This method permitted studying only current swingers; it did not allow sampling of people who had engaged in swinging but had stopped for one reason or another. There is also some question as to how honest the respondents were, since swingers are doing something that they take great pains to hide; in fact, Bartell cited specific examples of distortion in reports of information on such things as age and interests.

Ethically, it was easy to preserve anonymity, since swingers go to great lengths to preserve their own anonymity; even with one another, they use first names only. Bartell obtained some kind of informed consent, since he told his subjects that he was doing research; in other, similar studies, however, researchers have posed as swingers, have not divulged the fact that data were being collected, and therefore have seriously misled their subjects.

Experimental Sex Research

All the studies discussed so far had one thing in common: they all were studies of people's sexual behavior as it occurs naturally, conducted by means of either self-reports or direct observations. Such studies are **correlational** in nature; that is, at best the data they obtain can tell us that certain factors are related. They cannot tell us what *causes* various aspects of sexual behavior.

As an example, suppose we conduct a survey and find that women who masturbated to orgasm before marriage are more likely to have a high consistency of orgasm in marriage than women who did not. From this it would be tempting to conclude that practice in masturbating to orgasm causes women to have more orgasms in heterosexual sex. Unfortunately, this is not a legitimate conclusion to draw from the data, since many other factors might also explain the results. For example, it could be that some women have a higher sex drive than others; this high sex drive causes them to masturbate and also to have orgasms in heterosexual sex. Therefore, the most we can conclude is that masturbation experience is related to (or correlated with) orgasm consistency in marital sex.

An alternative method that does allow researchers to determine the causes of various aspects of behavior is the **experiment.** According to the technical definition of "experiment," one factor must be manipulated while all other factors are held constant. Thus any differences among the groups of people who received different treatments on that one factor can be said to be caused by that factor. For obvious reasons, most experimental research is conducted in the laboratory.

Correlational study: A study in which the researcher does not manipulate variables but rather studies naturally occurring relationships (correlations) among variables.
Experiment: A type of research study in which one variable (the independent variable) is manipulated by the experimenter while all other factors are held constant; the research can then study the effects of the independent variable on some measured variable (the dependent variable); the researcher is permitted to make causal inferences about the effects of the independent variable on the dependent variable.

Figure 3.7   An innovation in surveys of children is the use of "Talking Computers" to ask questions, with the child entering her answers using the mouse or the keyboard.

As an example of an experiment, let us consider a study that investigated whether being interviewed face-to-face causes children to underreport their sexual experiences (Romer et al., 1997). The participants were approximately 400 low-income children between the ages of 9 and 15. Some were assigned to a face-to-face interview with an experienced adult interviewer of their own gender. Others were assigned to be interviewed by a "talking computer," which had the same questions programmed into it. The questions appeared on the screen and, simultaneously, came through headphones, for those who were not good readers. Presumably in the talking computer condition, the child feels more of a sense of privacy and anonymity and therefore responds more truthfully.

Among 13-year-old boys interviewed by the talking computer, 76 percent said they had "had sex," compared with only 50 percent of the boys in the face-to-face interview. Forty-eight percent of

13-year-old girls interviewed by computer said they had had sex, compared with 25 percent of those interviewed by a human. The children clearly reported more sexual activity to the computer than to a human interviewer.

In the language of experimental design, the *independent variable* (manipulated variable) was the type of interview (computer or human interviewer). The *dependent variable* (the measured variable) was whether they had had sex (there were a number of other dependent variables as well, but a discussion of them would take us too far afield).

The results indicated that those interviewed by humans reported significantly less sexual activity than those interviewed by computer. Because the research design was experimental, we can make causal inferences. We can say confidently that the type of interview influenced the amount of reporting. That is, we can say that the type of interview had an effect on children's answers. We might also say that a face-to-face interview causes children to underreport their activity. That statement is a bit shakier than the previous one, because it assumes that the answers given to the talking computer were "true." It is possible that children overreported or exaggerated in responding to the computer and that their answers to the human interviewer were accurate, although this interpretation seems rather far-fetched.

Experimental sex research permits us to make much more powerful statements about the causes of various kinds of sexual phenomena. As for disadvantages, much of the experimental sex research, including the study described here, still relies on self-reports. Experimental sex research is time-consuming and costly, and it can generally be done only on small samples of participants. Sometimes in their efforts to control all variables except the independent variable, researchers control too much. Finally, experiments cannot address some of the most interesting, but most complex, questions in the field of sexual behavior, such as what factors cause people to develop heterosexual or homosexual orientations.

Some Statistical Concepts

Before you can understand reports of sex research, you must understand some basic statistical concepts.

Average

Suppose we get data from a sample of married couples on how many times per week they have sexual intercourse. How can we summarize the data? One way to do this is to compute some average value; this will tell us how often, on the average, these people have intercourse. In sex research, the number that is usually calculated is either the mean or the median; both of these give us an indication of approximately where the average value for that group of people is. The **mean** is simply the average of the scores of all the people. The *median* is the score that splits the sample in half, with half the respondents scoring below that number and half scoring above it.

Variability

In addition to having an indication of the average for the sample of respondents, it is also interesting to know how much variability there was from one respondent to the next in the numbers reported. That is, it is one thing to say that the average married couple in a sample had intercourse 3 times per week, with a range in the sample from 2 to 4 times per week, and it is quite another thing to say that the average was 3 times per week, with a range from 0 to 15 times per week. In both cases the mean is the same, but in the first there is little variability, and in the second there is a great deal of variability. These two alternatives are shown graphically in Figure 3.8. There is great variability in virtually all kinds of sexual behavior.

Average Versus Normal

It is interesting and informative to report the average frequency of a particular sexual behavior, but this also introduces the danger that people will confuse "average" with "normal." That is, there is a tendency, when reading a statistic like "The average person has intercourse twice per week," to think of one's own sexual behavior, compare it with that average, and then conclude that one is abnormal if one differs much from the average. If you read that statistic and your frequency of intercourse is only once a week, you may begin to worry that you are undersexed or that you are not getting as much as you should. If you are having intercourse seven times per week, you might begin worrying that you are oversexed or that you are wearing out your sex organs. Such conclusions are a mistake, first because they can make you miserable and second because there is so much variability in sexual behavior that any behavior (or frequency or length of time) within a wide range is perfectly normal. Don't confuse average with normal.

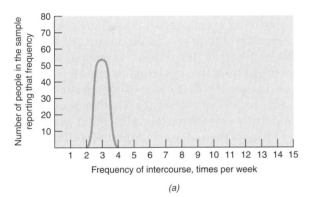

(a)

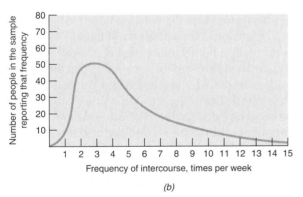

(b)

Figure 3.8 Two hypothetical graphs of the frequency of intercourse for married couples in a sample. In both, the average frequency is about three times per week, but in (b) there is little variability (almost everyone has a frequency between two and four times per week), whereas in (b) there is great variability (the frequency ranges from zero to fifteen or twenty times per week). The graph for most sexual behavior looks like (b); there is great variability.

Incidence Versus Frequency

In sex statistics, the terms "incidence" and "frequency" are often used. **Incidence** refers to the

Mean: The average of respondents' scores.
Incidence: The percentage of people giving a particular response.
Frequency: How often a person does something.

percentage of people who have engaged in a certain behavior. **Frequency** refers to how often people do something. Thus we might say that the incidence of masturbation among males is 92 percent (meaning that 92 percent of all males masturbate at least once in their lives), whereas the average frequency of masturbation among males between the ages of 16 and 20 is about once per week.

A closely related concept is that of cumulative incidence. If we consider a sexual behavior according to the age at which each person in the sample first engaged in it, the *cumulative incidence* refers to the percentage of people who have engaged in that behavior before a certain age. Thus the cumulative incidence of masturbation in males might be 10 percent by age 11, 25 percent by age 12, 80 percent by age 15, and 95 percent by age 20. Graphs of cumulative incidence always begin in the lower left-hand corner and move toward the upper right-hand corner. An example of a cumulative-incidence curve is shown in Figure 3.9.

Correlation

In this chapter the concept of correlation has already been mentioned several times—for example, test-retest reliability is measured by the correlation between people's answer to a question with their answer to the same question a week or two later—and the concept of correlation will reappear in later chapters of the book.

The term "correlation" is used by laypeople in contexts such as the following: "There seems to be a correlation here between how warm the days are and how fast the corn is growing." But what do statisticians mean by the term "correlation"? A **correlation** is a number that measures the relationship between two variables. A correlation can be positive or negative. A positive correlation occurs when there is a positive relationship between the two variables, that is, people who have high scores on one variable tend to have high scores on the other variable; low scores go with low scores. A negative correlation occurs when there is an opposite relationship between the two variables; that is, people with high scores on one variable tend to have low scores on the other variable. We might want to know, for example, whether there is a correlation between the number of years a couple has been married and the frequency with which they have sexual intercourse. In this case we might expect that there would be a negative correlation, and that

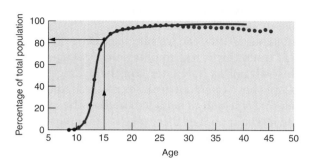

Figure 3.9   A cumulative-incidence curve for masturbation in males. From the graph, you can read off the percentage of males who report having masturbated by a given age. For example, about 82 percent have masturbated to orgasm by age 15.

is just what researchers have found. That is, the *greater* the number of years of marriage, the *lower* the frequency of intercourse. As another example, we might want to know whether there is a correlation between people's sexual attitudes and their sexual behavior, specifically whether people who hold more permissive attitudes about premarital sex have more premarital partners. In this case we expect a positive correlation in the sense that the people who score high on the measure of permissive attitudes are expected to have more partners and that people who score low on the measure of permissiveness are expected to have few partners.

Correlations range between +1.0 and −1.0. A correlation of +1.0 indicates a perfect positive relationship between two variables, meaning that the person in the sample who scores highest on one variable also has the highest score on the other variable, the person with the second highest score on the first variable also has the second highest score on the other variable, and so on. A correlation of 0 indicates no relationship between the two variables. Knowing a person's score on one variable tells us nothing about whether the person will have a high or low score on the other variable. Positive correlations between 0 and +1.0—for example, +.62—say that the relationship is positive but not a perfect relationship. A correlation of −1.0 means that there is a perfect negative correlation between the two variables. That is, the person in the sample with the highest score on

Correlation: A number that measures the relationship between two variables.

variable X has the lowest score on variable Y, the person with the second highest score on variable X has the second lowest score on variable Y, and so on.

Returning to the example of test-retest reliability discussed earlier in the chapter, suppose we administer a questionnaire to a sample of adults. One of the questions asks, "How many times did you masturbate to orgasm during the month of September?" We ask this question of the sample on October 1 and again on October 8. If each person in the sample gives us exactly the same answer on October 1 and on October 8, the correlation between the two variables (the number given on October 1 and the number given on October 8) would be +1.0 and the test-retest reliability would be a perfect +1.0. In fact, test-retest reliabilities for questions about sex typically range between +.60 and +.90, indicating that people's answers on the two occasions are not identical but are very similar.

SUMMARY

Knowledge of the major methods that have been used in sex research and of the problems and merits associated with each is necessary for understanding and evaluating sex research.

Ideally, sex research should employ probability sampling techniques.

Large-scale surveys of sexual behavior generally rely on people's self-reports, which may be inaccurate because of purposeful distortion, problems of memory, or inability to estimate some of the information requested. Direct observations of sexual behavior avoid these problems, but they lead to an even more restricted sample. They also answer questions that are somewhat different from those answered by surveys.

In all behavioral research, the ethical principles of informed consent and protection from harm must be observed, although historically some sex researchers did not do this.

One major sex survey was Kinsey's large-scale interview study of the sexual behavior of Americans, done during the 1940s. The NHSLS is a recent, large-scale survey of the sexual behavior of Americans; it was based on probability sampling. Other large surveys have been done, including some through magazines; the samples in magazine surveys, however, are so restricted that we cannot draw any general conclusions from them.

Studies of special populations include George and Weiler's study of sexuality in later life and the Bell, Weinberg, and Hammersmith study of homosexuals and heterosexuals.

In media content analysis, researchers use systematic coding categories to analyze representations in the media, such as television, romance novels, or magazine ads.

In participant-observer studies, the scientist becomes a part of the community to be studied, and he or she uses a combination of direct observations and interviewing. Examples are studies of sexual behavior in other cultures, Humphreys's study of the tearoom trade, and Bartell's study of swinging.

In experimental sex research, the goal is to discover what factors cause various aspects of sexual behavior. The researcher manipulates an independent variable and measures a dependent variable.

The following statistical terms were introduced: "average," "variability," "incidence," "frequency," and "correlation."

REVIEW QUESTIONS

1. Sex researchers can study only those people who agree to participate. This introduces the problem of _____.

2. Purposeful distortion and memory problems may create problems with the reliability of self-reports in sex research. True or false?

3. One alternative to self-reports that helps overcome some of their problems is the method of _____, which was used by Masters and Johnson.

4. Kinsey's sampling techniques were excellent; his interviewing techniques were also excellent. True or false?

5. The NHSLS used a convenience sample. True or false?

6. Magazine surveys provide a good, rich source of information about human sexuality. True or false?

7. If we wanted to find whether articles in men's magazines or women's magazines are more likely to be about sex, we would use the research technique called _____ .

8. _____ is the term for the type of study in which the researcher becomes a part of the community being studied and thus observes it from the inside, as in the studies of swingers.

9. The ACSF Investigators' representative sample of French adults' sexuality would properly be termed an experiment. True or false?

10. If someone says "Approximately 67 percent of people engage in premarital intercourse," this is a statement about frequency. True or false?

(The answers to all review questions are at the end of the book, after the Glossary.)

QUESTIONS FOR THOUGHT, DISCUSSION, AND DEBATE

1. Find a recent sex survey in a magazine. Evaluate the quality of the study, using concepts you have learned in this chapter.

2. Of the research techniques in this chapter—surveys, laboratory studies using direct observations, media content analysis, participant-observer studies, experiments—which do you think is best for learning about human sexuality? Why?

3. You want to conduct a survey, using face-to-face interviews, to determine whether there are differences between Asian American and white American teenagers (ages 15 to 19) in their sexual behavior and attitudes. In what ways would you tailor the research methods in order to make them culturally sensitive?

4. Imagine that you have been hired by your college or university to produce a report on the patterns of sexual behavior of the students there, with the goal of helping the administration to do better planning in areas such as health services and counseling services. You are given a generous budget for data collection. How would you go about collecting the data you would need in order to produce a truly excellent report?

SUGGESTIONS FOR FURTHER READING

Catania, Joseph A., et al. (1990). Methodological problems in AIDS behavioral research: Influences on measurement error and participation bias in studies of sexual behavior. *Psychological Bulletin, 108,* 339–362. This article takes up where the present chapter leaves off and offers an excellent analysis of methodological issues in sex research.

Harry, Joseph. (1986). Sampling gay men. *Journal of Sex Research, 22,* 21–34. This article provides an interesting discussion of some of the methodological problems that occur when sampling special populations of people, such as gay men.

Matsumoto, David. (1994). *Cultural influences on research methods and statistics.* Pacific Grove, CA: Brooks/Cole. This concise book, written for undergraduates, explains principles of cross-cultural research and how one should modify research methods depending on the culture being studied.

Michael, Robert T., Gagnon, John H., Laumann, Edward O., & Kolata, Gina. (1994). *Sex in America: A definitive survey.* Boston: Little, Brown. This book reports the results of the NHSLS, and is written for the general public.

WEB RESOURCE

www.indiana.edu/~kinsey/index.html.
    The Kinsey Institute Home Page.

chapter 4

SEXUAL ANATOMY

CHAPTER HIGHLIGHTS

Female Sexual Organs
External Organs
Internal Organs
The Breasts

Male Sexual Organs
External Organs
Internal Organs

Cancer of the Sex Organs
Breast Cancer
Cancer of the Cervix
Cancer of the Prostate
Cancer of the Testes

Men and women, all in all, behave just like our basic sexual elements. If you watch single men on a weekend night they really act very much like sperm—all disorganized, bumping into their friends, swimming in the wrong direction.
    "I was first."

"Let me through."
"You're on my tail."
"That's my spot."
We're like the Three Billion Stooges.
    But the egg is very cool: "Well, who's it going to be? I can divide. I can wait a month. I'm not swimming anywhere."*

The women's health movement has emphasized that women need to know more about their bodies. Actually, that is a good principle for everyone to follow. The current trend is away from the elitist view that only a select group of people—physicians—should understand the functioning of the body and toward the view that everyone needs more information about his or her own body. The purpose of this chapter is to provide basic information about the structure and functions of the parts of the body that are involved in sexuality and reproduction. Some readers may anticipate that this will be a boring exercise. Everyone, after all, knows what a penis is and what a vagina is. But even today, we find some bright college students who think a woman's urine passes out through her vagina. And how many of you know what the epididymis and the seminiferous tubules are? If you don't know, keep reading. You may even find out a few interesting things about the penis and the vagina that you were not aware of.

Female Sexual Organs

The female sexual organs can be classified into two categories: the *external organs* and the *internal organs*.

External Organs

The external genitals of the female consist of the clitoris, the mons pubis, the inner lips, the outer lips, and the vaginal opening (see Figure 4.1). Col-

lectively, they are known as the **vulva** ("crotch"; other terms such as "cunt" and "pussy" may refer either to the vulva or to the vagina, and some ethnic groups use "cock" for the vulva—slang, alas, is not so precise as scientific language). "Vulva" is a wonderful term but, unfortunately, it tends to be underused—the term, that is. The appearance of the vulva varies greatly from one woman to another (see Figure 4.2).

The Clitoris

The **clitoris** is a sensitive organ that is exceptionally important in female sexual response (Figure 4.3). It consists of the tip, a knob of tissue situated externally in front of the vaginal opening and the urethral opening; a shaft consisting of two corpora cavernosa (spongy bodies similar to those in the male's penis) that extends perhaps an inch into the body; and two crura, longer spongy bodies that lie deep in the body and run from the tip of the clitoris to either side of the vagina, under the major lips (Clemente, 1987). Some refer to the entire structure as having a "wishbone" shape. Close to the crura are the vestibular bulbs, which will be discussed in the section on internal organs.

As will be discussed in Chapter 5, female sexual organs and male sexual organs develop from similar tissue before birth; thus we can speak of the organs of one gender as being *homologous* (in the sense of developing from the same source) to the organs of the other gender. The female's clitoris is homologous to the male's penis; that is, both

*Jerry Seinfeld. (1993). *SeinLanguage.* New York: Bantam Books, p. 17.

Vulva (VULL-vuh): The collective term for the external genitals of the female.
Clitoris (KLIT-or-is): A small, highly sensitive sexual organ in the female, found in front of the vaginal entrance.

Figure 4.1   The vulva: The external genitals of the female.

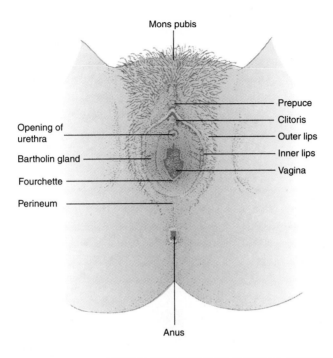

Mons pubis

Prepuce

Clitoris

Opening of urethra

Outer lips

Inner lips

Bartholin gland

Vagina

Fourchette

Perineum

Anus

Figure 4.2   The shape of the vulva varies widely from one woman to the next.

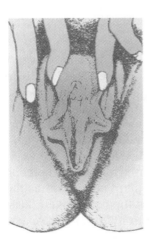  

Figure 4.3   Structure of the clitoris

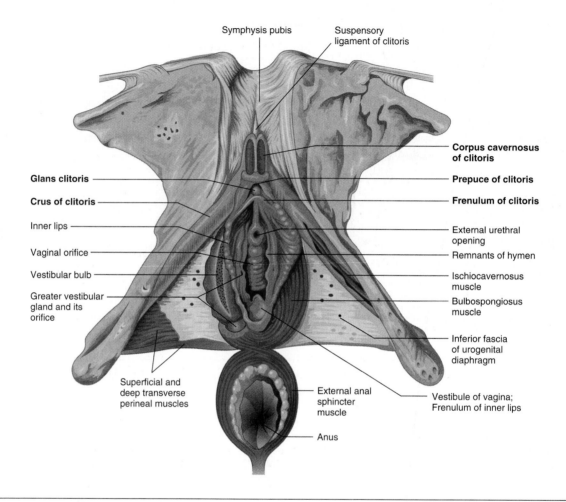

develop from the same embryonic tissue. The clitoris has a structure similar to that of the penis in that both have corpora cavernosa. The clitoris varies in size from one woman to the next, much as the penis varies in size from man to man. Also, the clitoris, like the penis, is erectile. Its erection is possible because its internal structure contains corpora cavernosa that fill with blood, as the similar structures in the penis do. The corpora cavernosa and the mechanism of erection will be considered in more detail in the discussion of the male sexual organs. Like the penis, the clitoris has a rich supply of nerve endings, making it very sensitive to stroking. Most women find it to be more sensitive to erotic stimulation than any other part of the body.

The clitoris is unique in that it is the only part of the sexual anatomy with no known reproductive function. All the other sexual organs serve sexual and reproductive functions. For example, not only is the vagina used for sexual intercourse, but it also receives the sperm and serves as the passageway through which the baby travels during childbirth. The penis not only produces sexual arousal and pleasure but also is responsible for ejaculation and

Figure 4.4 Twenty-eight African nations practice some form of ritualized genital cutting of young girls as an initiation into womanhood. Clitoridectomy is also practiced in Muslim countries outside Africa, and was practiced in the United States during the Victorian era.

impregnation. The clitoris clearly has an important function in producing sexual arousal. Unlike the other sexual organs, however, it appears to have no direct function in reproduction.

The Mons

In outward appearance, the more obvious parts of the vulva are the mons pubis, the inner lips, and the outer lips. The **mons pubis** (also called the *mons* or the *mons veneris,* for "mountain of Venus") is the rounded, fatty pad of tissue, covered with pubic hair, at the front of the body. It lies on top of the pubic bones (which come together in the center at a point called the *pubic symphysis*).

The Labia

The **outer lips** (or *labia majora,* for "major lips") are rounded pads of fatty tissue lying along both sides of the vaginal opening; they are covered with pubic hair. The **inner lips** (or *labia minora,* for "minor lips") are two hairless folds of skin lying between the outer lips and running right along the edge of the vaginal opening. Sometimes they are folded over, concealing the vaginal opening until they are spread apart. The inner lips extend forward and come together in front, forming the clitoral hood. The inner and outer lips are well supplied with nerve endings and thus are also important in sexual stimulation and arousal.

Mons pubis (PYOO-bis): The fatty pad of tissue under the pubic hair.
Outer lips: Rounded pads of fatty tissue lying on either side of the vaginal entrance.
Inner lips: Thin folds of skin lying on either side of the vaginal entrance.

Focus 4.1
Female Genital Cutting

Today a worldwide sexual health controversy rages over female genital cutting (FGC, also known as female genital mutilation or FGM). FGC is practiced in 28 African nations as well as in several countries in the Middle East and among Muslim populations in Indonesia and Malaysia. It is estimated to affect between 80 million and 110 million women worldwide. Typically girls are subjected to the procedure between the ages 4 and 10, although sometimes it is done during infancy. Often the procedure is done by a native woman, with no anesthetic and under unsanitary conditions. If these practices seem remote to many North Americans, it is important to recognize that immigrant women from these countries now reside in North America. Moreover, some of these practices were performed in Britain and North America in the 1800s, during the Victorian era, to cure various "female weaknesses" such as masturbation.

FGC is practiced in several forms, depending on the customs of the particular culture. The mildest form is called *sunna* and refers to the removal of just the clitoral hood (prepuce), not the clitoris; it is the only procedure that could legitimately be called "female circumcision," analogous to male circumcision. In its mildest form, it may involve only making a slit in the prepuce, not removing it. A second form of FGC is *excision* (sometimes known as *clitoridectomy*), which refers to complete removal of the clitoris and perhaps some of the inner lips. The most extreme form is *infibulation* or pharaonic circumcision (named for the pharaohs of Egypt, under whose reign the practice is said to have originated thousands of years ago). It involves the removal of the clitoris, all of the inner lips, and part of the outer lips; the raw edges of the outer lips are then stitched together to cover the urethral opening and the vaginal entrance, with only a small opening left for the passage of urine and menstrual fluid.

Infibulation poses severe health problems for women. Hemorrhaging may occur, leading to shock and even to bleeding to death. Because of the unsanitary conditions present during the procedure, tetanus and other infections are risks. Because the same instrument may be used on multiple girls, HIV and hepatitis B can be transmitted. A common problem is that the pain of the wound is so severe and the stitching so tight that the girl avoids urinating or cannot urinate properly, leading to urinary infections and other complications. A tightly infibulated woman can only urinate drop by drop, and her menstrual period may take 10 days and be extremely painful.

The sexual and reproductive health consequences are no less severe. Infibulation is an effective method for ensuring virginity until marriage, but on the wedding night the man must force an opening through stitching and scar tissue. This is painful and may take days; a midwife may be called to cut open the tissue. Orgasm would be only a remote possibility for a woman whose clitoris has been removed. Infibulated women have a substantial risk of complications during childbirth. The scar must always be cut open to permit delivery and, if done too late, the baby may die.

In a recent study of the Sudan, 90 percent of the women said that they had had or would have these procedures performed on their daughters, and about half the women favored the most severe procedures (Williams & Sobieszczyk, 1997). If these procedures are so harmful, why do they persist? Why do girls submit to them, and even ask for them, and why do their parents permit it, even encourage it? The answer lies in the complex and powerful interplay of culture and gender. Being infibulated indicates a woman's loyalty to and identification with her culture and its traditions, a particularly sensitive issue for people who had long been dominated by European colonizers. A woman who is not infibulated is not marriageable in these cultures in which marriage is the only acceptable way of life for an adult woman. When those are the rules of the game, it is less surprising that girls submit or even want to be circumcised and that their parents require them to do it. Particular communities may also hold certain beliefs that make these procedures seem necessary. For example, in some areas there is a belief that the clitoris contains poison and can harm men during sexual intercourse and kill children during birth. Some Muslims mistakenly believe that it is required by their faith, although it is not mentioned in the Koran.

Infibulation raises a number of dilemmas for North Americans. In universities, we generally encourage the approach of "cultural relativism," an openness to and appreciation of the customs of other cultures. FGC is definitely a custom of other cultures. If we apply standards of cultural relativism, we should say, "Great, if that's what those people want." But should there be limits to cultural relativism? Can one oppose certain practices that pose well-documented, serious health risks? Medical personnel in North America face difficult dilemmas. A physician may cut open an infibulated woman to permit childbirth. What if the woman requests that she be restitched? Should the physician comply, knowing the risks? Immigrant women whose daughters are born in North America may request that a physician perform an excision or infibulation, knowing that the procedure will be far safer if done by a physician in a hospital. Should the physician comply, knowing that the realistic alternative is that the procedure will be performed by someone, possibly untrained and inexperienced, from that culture under unsanitary conditions?

On a more hopeful note, a grass-roots movement of women has sprung up in a number of African nations, including Kenya, Gambia, Sudan, Somalia, and Nigeria, that is dedicated to eliminating these practices. Furthermore, to put matters into perspective, only about 15 percent of cultures that practice FGC do the severe form, infibulation. The remaining cultures practice the milder forms ranging from a slit in the prepuce to clitoridectomy. To add additional perspective, far more cultures practice male genital modifications than female genital modifications. An example is the United States, where male circumcision is widespread but FGC is extremely rare and occurs almost exclusively among a subgroup of immigrants, those from Islamic cultures.

Sources: Council on Scientific Affairs (1995); Gregersen (1983); Hicks, (1996); Horowitz & Jackson (1997); Kiragu (1995); Lightfoot-Klein, (1989); Schroeder, (1994); Toubia (1994, 1995); Williams & Sobieszczyk (1997).

A pair of small glands, the **Bartholin glands,** lie just inside the inner lips (Figure 4.1). Their function is unknown, and they are of interest only because they sometimes become infected.[1] *Skene's glands* are located nearby and, similarly, usually attract attention only when infected.

A few more landmarks should be noted (Figure 4.1). The place where the inner lips come together behind the vaginal opening is called the *fourchette.* The area of skin between the vaginal opening and the anus is called the **perineum.** The vaginal opening itself is sometimes called the **introitus.** Notice also that the urinary opening lies about midway between the clitoris and the vaginal opening. Thus urine does not pass out through the clitoris (as might be expected from analogy with the male) or through the vagina, but instead through a separate pathway, the *urethra,* with a separate opening.

Self-Knowledge

One important difference between the male sex organs and the female sex organs—and a difference that has some important psychological consequences—is that the female's external genitals are much less visible than the male's. A male can view his external genitals directly either by looking down at them or by looking into a mirror while naked. Either of these two strategies for the female, however, will result at best in a view of the mons. The clitoris, the inner and outer lips, and the vaginal opening remain hidden. Indeed, many adult women have never taken a direct look at their own vulva. This obstacle can be overcome by the simple method of using a mirror. The genitals can be viewed either by putting a mirror on the floor and sitting in front of it or by standing up and putting one foot on the edge of a chair, bed, or something similar and holding the mirror up near the genitals (see Figure 4.5). We recommend that all women use a mirror to identify on their own bodies all the parts shown in Figure 4.1.

The Hymen

Before the internal structures are discussed, one other external structure deserves mention: the hymen. The **hymen** ("cherry," "maidenhead") is a thin membrane which, if present, partially covers the vaginal opening. The hymen may be one of a number of different types (see Figure 4.6), although it generally has some openings in it; otherwise the menstrual flow would not be able to pass

Figure 4.5 Body education: The mirror exercise lets women see their own genitals.

out.[2] At the time of first intercourse, the hymen may be broken or stretched as the penis moves into the vagina. This may cause bleeding and possibly some pain. Typically, though, it is an untraumatic occurrence and goes unnoticed in the excitement of the moment. For a woman who is very concerned about her hymen and what will happen to it at first coitus, there are two possible approaches. A physician can cut the hymen neatly so that it will not tear at the time of first intercourse, or the woman herself can stretch it by repeatedly inserting a finger into the vagina and pressing on it.

[1]And there is a limerick about them:

There was a young man from Calcutta
Who was heard in his beard to mutter,
"If her Bartholin glands
Don't respond to my hands,
I'm afraid I shall have to use butter."

Actually, there is a biological fallacy in the limerick. Can you spot it? If not, see Chapter 9.

[2]The rare condition in which the hymen is a tough tissue with no opening is called *imperforate hymen* and can be corrected with fairly simple surgery.

Bartholin glands: Two tiny glands located on either side of the vaginal entrance.
Perineum (pair-ih-NEE-um): The skin between the vaginal entrance and the anus.
Introitus: Another word for the vaginal entrance.
Hymen (HYE-men): A thin membrane that may partially cover the vaginal entrance.

Figure 4.6   There are several types of hymens.

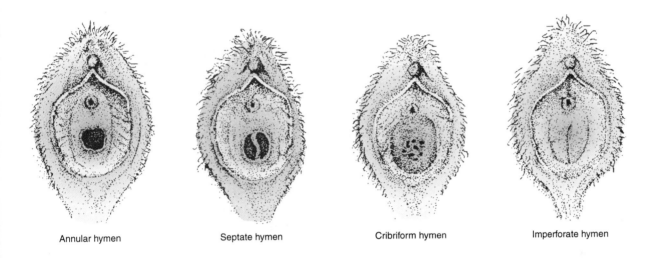

| Annular hymen | Septate hymen | Cribriform hymen | Imperforate hymen |

The hymen, and its destruction at first intercourse, has captured the interest of people in many cultures. In Europe during the Middle Ages, the lord might claim the right to deflower a peasant bride on her wedding night before passing her on to her husband (the practice is called *droit du seigneur* in French and *jus primae noctis* in Latin). The hymen has been taken as evidence of virginity. Thus bleeding on the wedding night was proof that the bride had been delivered intact to the groom; the parading of the bloody bed sheets on the wedding night, a custom of the Kurds of the Middle East, is one ritual based on this belief.

Such practices rest on the assumption that a woman without a hymen is not a virgin. However, we now know that this is not true. Some females are simply born without a hymen, and others may tear it in active sports such as horseback riding. Unfortunately, this means that some women have been humiliated unjustly for their lack of a hymen.

Internal Organs

The internal sex organs of the female consist of the vagina, the uterus, a pair of ovaries, and a pair of fallopian tubes (see Figure 4.7).

The Vagina

The **vagina** is the tube-shaped organ into which the penis is inserted during coitus, and it receives the ejaculate. It is also the passageway through which a baby travels during birth, so it is sometimes called the *birth canal*. In the resting or unaroused state, the vagina is about 8 to 10 centimeters (3 to 4 inches) long and tilts slightly backward from the bottom to the top. At the bottom it ends in the vaginal opening, or *introitus*. At the top it connects with the cervix (the lower part of the uterus). It is a very flexible organ that works somewhat like a balloon. In the resting state its walls lie against each other like the sides of an uninflated balloon; during arousal it expands like an inflated balloon, allowing space to accommodate the penis.

The walls of the vagina have three layers. The inner layer, the *vaginal mucosa,* is a mucous membrane similar to the inner lining of the mouth. The middle layer is muscular, and the outer layer forms a covering. The walls of the vagina are extremely elastic and are capable of expanding to the extent necessary during intercourse and childbirth, although with age they become thinner and less flexible.

The nerve supply of the vagina is mostly to the lower one-third, near the introitus. That part is

Vagina (vuh-JINE-uh): The tube-shaped organ in the female into which the penis is inserted during coitus and through which a baby passes during birth.

Figure 4.7   Internal sexual and reproductive organs of the female from a side view and from a front view.

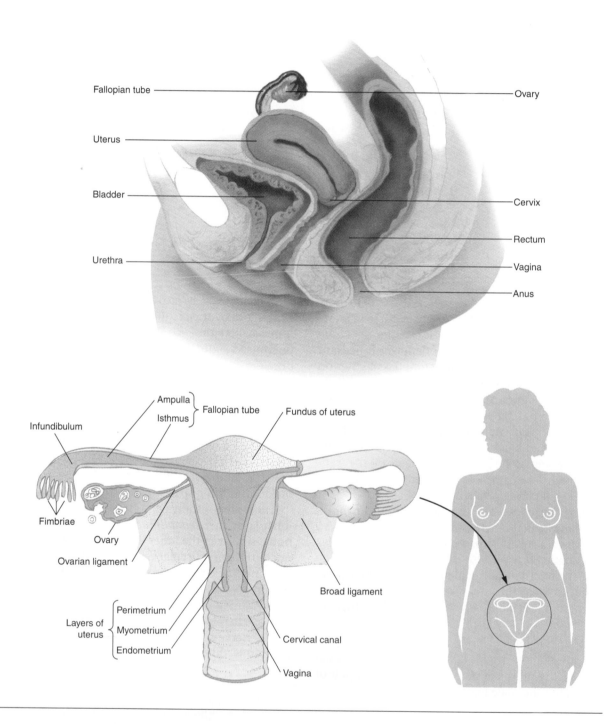

Figure 4.8 Appearance of the vulva of a woman who is a virgin; a woman who has had intercourse but has not had a baby (nulliparous); and a woman who has had a baby (parous).

Virginal   In the woman who   In the woman
                            has had intercourse   who has had a baby

sensitive to erotic stimulation. The inner two-thirds of the vagina contains almost no nerve endings and is therefore relatively insensitive except to feelings of deep pressure. Some women have a spot on the front wall of the vagina that is more sensitive than the rest of the vagina, but even it is not nearly so sensitive as the inner lips, outer lips, or clitoris (Schultz et al., 1989). This spot is referred to by some as the G-spot (see Chapter 9).

The number of slang terms for the vagina (for example, "beaver," "cunt") and the frequency of their usage testify to its power of fascination across the ages. One concern has been with size: whether some vaginas are too small or too large. As noted earlier, though, the vagina is highly elastic and expandable. Thus, at least in principle, any penis can fit into any vagina. The penis is, after all, not nearly so large as a baby's head, which manages to fit through the vagina. The part of the vagina which is most responsible for the male's sensation that it is "tight," "too tight," or "too loose" is the introitus. One of the things that can stretch the introitus is childbirth; indeed, there is a considerable difference in the appearance of the vulva of a woman who has never had a baby *(nulliparous)* and the vulva of a woman who has *(parous)* (see Figure 4.8).

Surrounding the vagina, the urethra, and the anus is a set of muscles called the *pelvic floor muscles.* One of these muscles, the **pubococcygeus muscle,** is

particularly important. It may be stretched during childbirth, or it may simply be weak. However, it can be strengthened through exercise, and this is recommended by sex therapists (see Chapter 19) as well as by many popular sex manuals and magazines.

The Vestibular Bulbs

The **vestibular bulbs** (or bulbs of the clitoris) are two organs about the size and shape of a pea pod (Figure 4.3). They lie on either side of the vaginal wall, near the entrance, under the inner lips (O'Connell et al., 1998). They are erectile tissue and lie close to the crura of the clitoris.

The Uterus

The **uterus** (womb) is about the size of a fist and is shaped somewhat like an upside-down pear. It is usually tilted forward and is held in place by ligaments. The narrow lower third is called the *cervix* and opens into the vagina. The top is the *fundus,* and the main part is the *body.* The entrance to the

> **Pubococcygeus muscle (pyoo-bo-cox-ih-GEE-us):** A muscle around the vaginal entrance.
> **Vestibular bulbs:** Erectile tissue running under the inner lips.
> **Uterus (YOO-tur-us):** The organ in the female in which the fetus develops.

uterus through the cervix is very narrow, about the diameter of a drinking straw, and is called the *os* (or cervical canal). The major function of the uterus is to hold and nourish a developing fetus.

The uterus, like the vagina, consists of three layers. The inner layer, or *endometrium,* is richly supplied with glands and blood vessels. Its state varies according to the age of the woman and the phase of the menstrual cycle. It is the endometrium which is sloughed off at menstruation and creates most of the menstrual discharge. The middle layer, the *myometrium,* is muscular. The muscles are very strong, creating the powerful contractions of labor and orgasm, and also highly elastic; they are capable of stretching to accommodate a nine-month-old fetus. The outer layer—the *perimetrium* or *serosa*—forms the external cover of the uterus.

The Fallopian Tubes

Extending out from the sides of the upper end of the uterus are the **fallopian tubes,** also called the *oviducts* ("egg ducts") or *uterine tubes* (see Figure 4.7). The fallopian tubes are extremely narrow and are lined with hairlike projections called *cilia.* The fallopian tubes are the pathway by which the egg leaves the ovaries and the sperm reach the egg. Fertilization of the egg typically occurs in the infundibulum, the section of the tube closest to the ovary; the fertilized egg then travels the rest of the way through the tube to the uterus. The infundibulum curves around toward the ovary; at its end are numerous fingerlike projections called *fimbriae* which extend toward the ovary.

The Ovaries

The **ovaries** are two organs about the size and shape of unshelled almonds; they lie on either side of the uterus. The ovaries have two important functions; they produce eggs (ova), and they manufacture the female sex hormones, *estrogen* and *progesterone.*

Each ovary contains numerous follicles. A *follicle* is a capsule that surrounds an egg (not to be confused with hair follicles, which are quite different).

Fallopian tube (fuh-LOW-pee-un): The tube extending from the uterus to the ovary; also called the oviduct.
Ovaries: Two organs in the female that produce eggs and sex hormones.

It is estimated that a female is born with about 400,000 immature eggs. Beginning at puberty, one or several of the follicles mature during each menstrual cycle. When the egg has matured, the follicle bursts open and releases the egg. The ovaries do not actually connect directly to the fallopian tubes. Rather, the egg is released into the body cavity and reaches the tube by moving toward the fimbriae. If the egg does not reach the tube, it may be fertilized outside the tube, resulting in an abdominal pregnancy (see the section on ectopic pregnancy in Chapter 7). There have also been cases recorded of women who, although they are missing one ovary and the opposite fallopian tube, have nonetheless become pregnant. Apparently, in such cases the egg migrates to the tube on the opposite side.

The Breasts

Although they are not actually sex organs, the *breasts* deserve discussion here because of their erotic and reproductive significance. The breast consists of about 15 or 20 clusters of *mammary glands,* each with a separate opening to the nipple, and of fatty and fibrous tissue which surrounds the clusters of glands (see Figure 4.9). The nipple, into which the milk ducts open, is at the tip of the breast. It is richly supplied with nerve endings and therefore very important in erotic stimulation for many women. The nipple consists of smooth muscle fibers; when they contract, the nipple becomes erect. The area surrounding the nipple is called the *areola.*

There is wide variation among women in the size and shape of the breasts. One thing is fairly consistent, though: few women are satisfied with the size of their breasts. Most women think they are either too small or too large, and almost no woman thinks hers are just right. It is well to remember that there are the same number of nerve endings in small breasts as in large breasts. It follows that small breasts are actually more erotically sensitive per square inch than are large ones.

Breasts may take on enormous psychological meaning; they can be a symbol of femininity or a means of attracting men. Ours is a very breast-oriented culture. Many American men develop a powerful interest in, and attraction to, women's breasts. The social definition of beauty is a compelling force; many women strive to meet the ideal and a few overadapt, going too far in their striving

Figure 4.9   The internal structure of the breast.

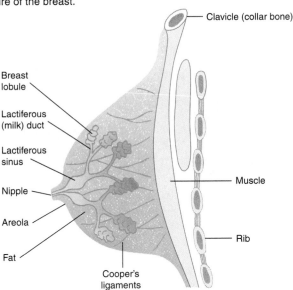

Clavicle (collar bone)

Breast
lobule

Lactiferous
(milk) duct

Lactiferous
sinus

Muscle

Nipple

Areola

Rib

Fat

Cooper's
ligaments

(Mazur, 1986). Breast augmentation surgery has increased steadily since the 1960s, while other women are undergoing breast reduction surgery, in both cases to meet a socially defined standard of beauty.

Male Sexual Organs

Externally, the most noticeable parts of the male sexual anatomy are the penis and the scrotum, or scrotal sac, which contains the testes (see Figure 4.10).

External Organs
The Penis

The **penis** (phallus, "prick," "cock," "johnson," and many other slang terms too numerous to list) serves important functions in sexual pleasure, reproduction, and elimination of body wastes by urination. It is a tubular organ with an end or tip called the *glans*. The opening at the end of the glans is the *meatus,* or *urethral opening,* through which urine and semen pass. The part of the penis that attaches to the body is called the *root,* and the main part of the penis is called the *body* or *shaft.* The raised ridge separating the glans from the

body of the penis is called the *corona,* or *coronal ridge.* While the entire penis is sensitive to sexual stimulation, the glans and corona are the most sexually excitable region of the male anatomy.

Internally, the penis contains three long cylinders of spongy tissue running parallel to the *urethra,* which is the pathway through which semen and urine pass (see Figure 4.11). The two spongy bodies lying on top are called the **corpora cavernosa,** and the single one lying on the bottom of the penis is called the **corpus spongiosum** (the urethra runs through the middle of it). During erection, the corpus spongiosum can be seen as a raised column on the lower side of the penis. As the names suggest, these bodies are tissues filled with many spaces and cavities, much like a sponge. They are richly supplied with blood vessels and nerves. In the flaccid (unaroused, not erect) state, they contain little blood. *Erection,* or *tumescence,* occurs

Penis: The male external sexual organ, which functions both in sexual activity and in urination.
Corpora cavernosa: Spongy bodies running the length of the top of the penis.
Corpus spongiosum: A spongy body running the length of the underside of the penis.

Figure 4.10   The male sexual and reproductive organs from a side view.

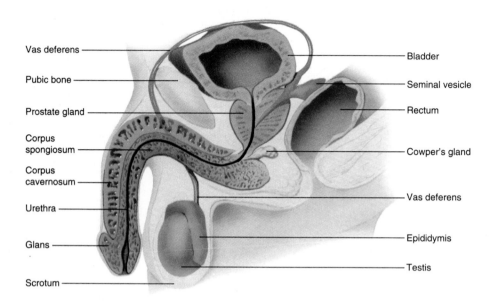

Vas deferens

Pubic bone

Prostate gland

Corpus spongiosum

Corpus cavernosum

Urethra

Glans

Scrotum

Bladder

Seminal vesicle

Rectum

Cowper's gland

Vas deferens

Epididymis

Testis

when they become filled with blood (engorged) and expand, making the penis stiff.

Contrary to popular belief, the penis does not contain a muscle, and no muscle is involved in erection. Erection is purely a vascular phenomenon; that is, it results entirely from blood flow. It is also commonly believed that the penis of the human male contains a bone. This is not true either, although in some other species—for example, dogs—the penis does contain a bone, which aids in intromission (insertion of the penis into the vagina). In human males, however, there is none.

The skin of the penis usually is hairless and is arranged in loose folds, permitting expansion during erection. The **foreskin,** or *prepuce,* is an additional layer of skin which forms a sheathlike covering over the glans; it may be present or absent in the adult male, depending on whether he has been circumcised (see Figure 4.12). Under the foreskin are small glands (Tyson's glands) that produce a substance called *smegma,* which is cheesy in texture. The foreskin is easily retractable,[3] and its retraction is extremely important for proper hygiene. If it is not pulled back and the glans washed thoroughly, the smegma may accumulate, producing a very unpleasant smell.

Circumcision refers to the surgical cutting away or removal of the foreskin. Circumcision is practiced in many parts of the world and, when parents so choose, is done to infants in the United States within a few days after birth.

Circumcision may be done for cultural and religious reasons. Circumcision has been a part of Jewish religious practice for thousands of years. It symbolizes the covenant between God and the Jewish people and is done on the eighth day after birth, according to scriptural teaching (Genesis

Foreskin: A layer of skin covering the glans or tip of the penis in an uncircumcised male; also called the prepuce.
Circumcision: Surgical removal of the foreskin of the penis.

[3]In a rare condition, the foreskin is so tight that it cannot be pulled back; this is called *phimosis* and requires correction by circumcision.

Figure 4.11 The internal structure of the penis.

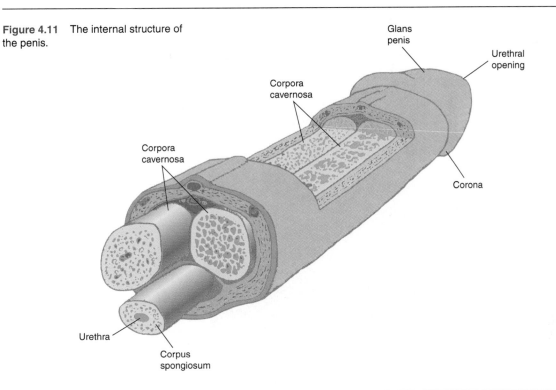

17:9–27). Circumcision is also common in Muslim cultures. In most cultures circumcision may be done at puberty as an initiation ritual, or *rite de passage*. The ability of the young boy to stand the pain may be seen as a proof of manhood.

In the 1980s, an anticircumcision movement gained momentum in the United States. Its proponents argue that circumcision does not have any health benefits and that it does entail some health risk as well as psychological trauma. According to this view, circumcision is nothing more than cruel mutilation. (For a statement of the anticircumcision position, see Wallerstein, 1980.) In fact, as early as 1971 the American Academy of Pediatrics had gone on record saying that there is no medical need for routine circumcision of newborn boys. Reflecting this advice and the growth of controversy about circumcision, only 59 percent of infant boys were circumcised in the United States in 1986, compared with 90 percent in 1970 (Lindsey, 1988).

New evidence accumulated, though, and in 1989 the American Academy of Pediatrics, changing its 1971 statement, declared that there are potential medical benefits and advantages to circumcision as well as some potential risks. The new evidence indicates, for example, that uncircumcised male babies are 11 times more likely to get urinary tract infections than are circumcised babies (Wiswell et al., 1987). There is also some evidence that uncircumcised men have a higher risk of infection with HIV, the AIDS virus (Touchette, 1991). It is thought that the foreskin can harbor HIV and other viruses. In a study of different geographic and ethnic groups in Africa, some of which practice circumcision and some of which do not, it was found that HIV infection rates were very low among groups that practice circumcision and high among those that do not (Moses et al., 1990). However, research in a developed nation, Australia, indicated that there was no association between being uncircumcised and contracting a number of sexually transmitted diseases, including herpes and genital warts (Donovan et al., 1994). Unfortunately, this study did not look at HIV transmission, so we are left wondering about that important question. In our view, the scientific data on the health benefits and risks of circumcision are simply not conclusive at the present time. Consistent with this view, in

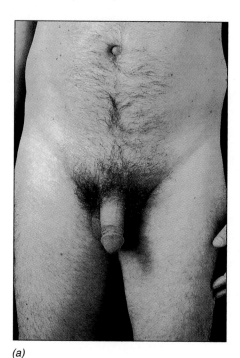

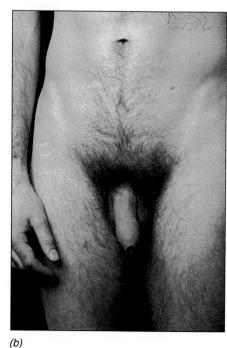

Figure 4.12   (*a*) A circumcised penis and (*b*) an uncircumcised penis, showing the foreskin.

(a)   (b)

1999 the American Academy of Pediatrics returned to its position that there is no medical need for circumcision.

Other arguments have focused on whether the circumcised or the uncircumcised male receives more pleasure from sexual intercourse. In fact, Masters and Johnson (1966) found that there is no difference in excitability between the circumcised and the uncircumcised penis.

Other forms of male genital cutting are done throughout the world. In fact, male genital cutting is done in more cultures than is female genital cutting (Gregersen, 1983). A common form, across most of Polynesia, is **supercision** (also known as *superincision*), which involves making a slit the length of the foreskin on the top, with the foreskin otherwise remaining intact (Gregersen, 1983). With **subincision,** which is common in some tribes in central Australia, a slit is made on the lower side of the penis along its entire length and to the depth of the urethra. Urine is then excreted at the base rather than at the tip of the penis.

To say the least, the penis has been the focus of quite a lot of attention throughout history. In some cultures, the attention has become so pronounced that the male genitals have actually become the object of religious worship (phallic worship). Not surprisingly, the male genitals were often seen as symbols of fertility and thus were worshipped for their powers of procreativity. In ancient Greece, phallic worship centered on Priapus, the son of Aphrodite (the goddess of love) and Dionysus (the god of fertility and wine). Priapus is usually represented as a grinning man with a huge penis.

In contemporary American society, phallic concern often focuses on the size of the penis. It is commonly believed that a man with a large penis is a better lover and can satisfy a woman more than the man with a small penis can. Masters and Johnson (1966), however, have found that this is not true. While there is considerable variation in the length of the penis from one man to the next—the average penis is generally somewhere between 6.4 centimeters (2.5 inches) and 10 centimeters (4 inches) in length when

Supercision: A form of male genital cutting in which a slit is made the length of the foreskin.
Subincision: A form of male genital cutting in which a slit is made on the lower side of the penis along its entire length.

flaccid (not erect)—there seems to be a tendency for the small penis to grow more in erection than the one that starts out large. As a result, there is little correlation between the length of the penis when flaccid and the length when erect. As the saying has it, "Erection is the great equalizer." The average erect penis is about 15 centimeters (6 inches) long, although erect penises longer than 33 centimeters (13 inches) have been measured (Dickinson, 1949). Further, as noted earlier, the vagina has relatively few nerve endings and is relatively insensitive. Hence penetration to the far reaches of the vagina by a very long penis is not essential and may not even be noticeable. Many other factors are more important than penis size in giving a woman pleasure (see Chapters 10 and 19).

Phallic concern has also included an interest in the variations in the shape of the penis when flaccid and when erect, as reflected in this limerick:

There was a young man of Kent
Whose kirp in the middle was bent.
To save himself trouble
He put it in double,
And instead of coming, he went.

Phallic concern has also been expressed in psychological theory, especially in psychoanalytic theory. As we saw in Chapter 2, Freud viewed concern for the penis and a related castration anxiety as the key factors in male psychological development, leading to the resolution of the Oedipus complex, increased independence from parents, and increased psychological maturity. All this from such a small part of the body! (Indeed, the theory even says that the key factor in female psychological development is the lack of a penis.)[4]

The Scrotum

The other major external genital structure in the male is the **scrotum;** this is a loose pouch of skin, lightly covered with hair, which contains the testes ("balls" or "nuts" in slang[5]). The testes themselves are considered part of the internal genitals.

Internal Organs

The **testes** are the *gonads,* or reproductive glands, of the male, and thus they are analogous to the female's ovaries. Like the ovaries, they serve two major functions: they manufacture germ cells

(sperm), and they manufacture sex hormones, in particular *testosterone.* Both testes are about the same size, although the left one usually hangs lower than the right one.

In the internal structure of the testes, two parts are important: the seminiferous tubules and the interstitial cells (see Figure 4.13). The **seminiferous tubules** carry out the important function of manufacturing and storing sperm, a process called *spermatogenesis.* They are a long series of threadlike tubes curled and packed densely into the testes. There are about 1000 of these tubules, and it is estimated that if they were stretched out end to end, they would be several hundred feet in length.

The **interstitial cells** (or *Leydig's cells*) carry out the second important function of the testes, the production of testosterone. These cells are found in the connective tissue lying between the seminiferous tubules. The cells lie close to the blood vessels in the testes and pour the hormones they manufacture directly into the blood vessels. Thus the testes are endocrine glands (hormone-secreting glands).

---

[4]Anxieties about the penis became particularly obvious during the much-publicized case of John Wayne Bobbitt and Lorena Bobbitt and the jokes that poured out. Lorena Bobbitt, apparently after a history of being beaten by her husband, cut off his penis one night. A surgeon later reattached it. The jokes told the real story about the anxieties that were aroused among people around the country.

[5]That brings to mind another limerick:

There once was a pirate named Gates
Who thought he could rhumba on skates.
He slipped on his cutlass
And now he is nutless
And practically useless on dates.

Scrotum (SKROH-tum): The pouch of skin that contains the testes in the male.
Testes: The pair of glands in the scrotum that manufacture sperm and sex hormones.
Seminiferous tubules (sem-ih-NIFF-ur-us): Tubes in the testes that manufacture sperm.
Interstitial cells (int-er-STIH-shul): Cells in the testes that manufacture testosterone.

Figure 4.13 Schematic cross section of the internal structure of the testis.

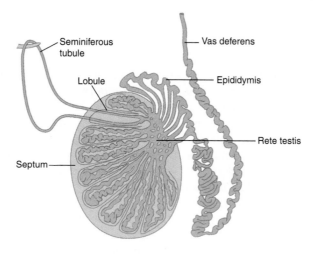

Seminiferous tubule

Vas deferens

Lobule

Epididymis

Septum

Rete testis

Each testis is surrounded by a rigid, whitish sheath (the *tunica albuginea*). This sheath is responsible for the problem of sterility caused by mumps in the adult male. When the virus invades the testes, it causes them to swell. The tunica albuginea is tight, however, and does not permit expansion. The delicate tubules are thus crushed, which impairs their sperm-producing function. Mumps is not a problem in the female because the ovaries are not enclosed in a comparable tight sheath.

One of the clever tricks that the scrotum and testes can perform, as any male will testify, is to move up and close to the body or down and away from the body. These changes are brought about mainly by changes in temperature (although emotional factors may also produce them). If a man plunges into a cold lake, the scrotum will shrivel and move close to the body. If the man is working in an extremely hot place, the scrotum will hang down and away from the body. This mechanism is important because the testes should remain at a fairly constant temperature, slightly lower than normal body temperature. This constancy of temperature is necessary to protect sperm, which may be injured by extremes of temperature. Thus if the air is cold, the testes move closer to the body to maintain warmth, but if the air is too hot, they move away from the body to keep cool. The mechanics of this movement are made possible by the *dartos muscle*, which forms the middle layer of the scrotum. It contracts or relaxes reflexively, thereby moving the testes up or down.

Many people believe that taking hot baths, wearing tight athletic supporters, or having a high fever can cause infertility. Indeed, in some countries the men take long, hot baths as a method of contraception. Such a practice has some basis in biological fact, because sperm can be destroyed by heat. However, as a method of contraception, this practice has not been particularly effective. In one study, it was found that the use of a special jockstrap raised the temperature of the scrotum by nearly 1°C (1.7°F) and that wearing the device daily for seven weeks caused about a 25 percent reduction in the number of sperm produced (Robinson and Rock, 1967). Thus such practices might decrease a man's fertility somewhat, but they are far from 100 percent effective as contraceptives. On the other hand, men with problems of infertility can sometimes cure them by getting out of their tight jockstraps and jockey shorts.

Following initial cell division in the seminiferous tubules, the male germ cells go through several stages of maturation. At the earliest stage, the cell is called a *spermatogonium.* Then it becomes a *spermatocyte* (first primary and then secondary) and then a *spermatid.* Finally when fully mature, it is a *spermatozoan,* or **sperm.** *Spermatogenesis,* the manufacture of sperm, occurs continuously in the adult human male. An average ejaculate contains about 300 million sperm.

A mature sperm is very tiny—about 60 micrometers, or 60/10,000 millimeter (0.0024 inch), long—and consists of a head, a neck, a midpiece, and a tail. A normal human sperm carries 23 chromosomes in the head. This 23 is half the normal number in the other cells of the human body. When the sperm unites with the egg, which also carries 23 chromosomes, the full complement of 46 for the offspring is produced. (See Chapter 7 for a discussion of the sperm's role in conception.)

After the sperm are manufactured in the seminiferous tubules, they proceed into the *rete testes,* a converging network of tubes on the surface of the testis toward the top. The sperm then pass out of the testis and into a single tube, the epididymis. The **epididymis** is a long tube (about 6 meters, or 20 feet, in length) coiled into a small crescent-shaped region on the top and side of the testis. The sperm are stored in the epididymis, in which they ripen and mature, possibly for as long as six weeks.

Upon ejaculation, the sperm pass from the epididymis into the **vas deferens** (it is the vas that is cut in a vasectomy—see Chapter 8). The vas passes up and out of the scrotum and then follows a peculiar circular path as it loops over the pubic bone, crosses beside the urinary bladder, and then turns downward toward the prostate. As the tube passes through the prostate, it narrows, and at this point is called the *ejaculatory duct.* The ejaculatory duct opens into the *urethra,* which has the dual function of conveying sperm and conveying urine; sperm are ejaculated out through the penis via the urethra.

Sperm have little motility (capability of movement) of their own while in the epididymis and vas. Not until they mix with the secretions of the prostate are they capable of movement on their own (Breton et al., 1996). Up to this point, they are conveyed by the cilia and by contractions of the epididymis and vas.

The **seminal vesicles** are two saclike structures that lie above the prostate, behind the bladder, and in front of the rectum. They produce about 70 percent of the seminal fluid, or ejaculate. The remaining 30 percent is produced by the prostate (Spring-Mills & Hafez, 1980). They empty their fluid into the ejaculatory duct to combine with the sperm.

The **prostate** lies below the bladder and is about the size and shape of a chestnut. It is composed of both muscle and glandular tissue. The prostate secretes a milky alkaline fluid that is part of the ejaculate. It is thought that the alkalinity of the secretion provides a favorable environment for the sperm and helps prevent their destruction by the acidity of the vagina. The prostate is fairly small at birth, enlarges at puberty, and typically shrinks in old age. It may become enlarged enough so that it interferes with urination, in which case surgery or drug therapy is required. Its size can be determined by rectal examination.

Cowper's glands, or the *bulbourethral glands,* are located just below the prostate and empty into the urethra. During sexual arousal, these glands secrete a small amount of a clear alkaline fluid, which appears as droplets at the tip of the penis before ejaculation occurs. It is thought that the function of this secretion is to neutralize the acidic urethra, allowing safe passage of the sperm. Generally it is not produced in sufficient quantity to serve as a lubricant in intercourse. The fluid often contains some stray sperm. Thus it is possible (though not likely) for a woman to become pregnant from the sperm in this fluid even though the man has not ejaculated.

Sperm: The mature male reproductive cell, capable of fertilizing an egg.

Epididymis (ep-ih-DIH-dih-mus): Highly coiled tube located on the edge of the testis, where sperm mature.

Vas deferens: The tube through which sperm pass on their way from the testes and epididymis, out of the scrotum, and to the urethra.

Seminal vesicles: Saclike structures that lie above the prostate which produce about 70 percent of the seminal fluid.

Prostate: The gland in the male, located below the bladder, that secretes some of the fluid in semen.

Cowper's glands: Glands that secrete substances into the male's urethra.

Cancer of the Sex Organs

Breast Cancer

Cancer of the breast is the second most common form of cancer in women (exceeded only by skin cancer). About 1 out of every 9 American women has breast cancer at some time in her life. Every year, 44,000 women in this country die of breast cancer (American Cancer Society, 1998). The risk is higher for the woman whose mother, sister, or grandmother has had breast cancer.

Causes

Approximately 10 percent of the cases of breast cancer in women are due to genetic factors (Ezzell, 1994). The remaining cases may be related to a particular virus or to diet. In countries such as Japan and Romania where the diet is low in fat, breast cancer rates are less than half that of the United States, where the diet is high in fat. The evidence also suggests that cigarette smoking increases the risk of breast cancer in women with certain genetic backgrounds (Ambrosone et al., 1996).

There have been great breakthroughs in the last few years in research into the genetics of breast cancer. Scientists have identified two breast cancer genes: BRCA1 (for BReast CAncer 1) on chromosome 17 and BRCA2 on chromosome 13 (Ezzell, 1994; Miki et al., 1994; Shattuck-Eidens et al., 1995). Mutations of these genes seem to cause breast cancer. BRCA1 and BRCA2 mutations also seem to increase susceptibility to ovarian cancer (Miki et al., 1994). In one study, of women having a mutation in BRCA1 or BRCA2, 56 percent eventually developed breast cancer and 16 percent developed ovarian cancer (Struewing et al., 1997). In fact, among men carrying mutations of these genes, 16% developed prostate cancer. These findings may lead to genetic screening tests that will detect BRCA mutations in women who have a family history of breast cancer (Ezzell, 1995; Shattuck-Eidens et al., 1995). If a BRCA mutation was found, the woman could be monitored closely, and if it was not found, she could feel relieved!

Researchers are currently exploring gene therapy (Ezzell, 1996, May). Genetically engineered cells containing normal BRCA1 genes are injected into the woman. Studies with mice indicate that this treatment is successful in shrinking tumors.

Diagnosis

All women should do a breast self-exam regularly (see Focus 4.2). The exam should be done once a month, but not during the menstrual period, when there tend to be natural lumps in the breast. It is important to do the self-exam once a month because the earlier breast cancer is detected, the better the chances of a complete recovery with proper treatment.

Unfortunately, psychological factors can interfere with this process. Many women do not do the self-exam because they are afraid they will find a lump; thus one can go undetected for a long time, making recovery less likely. Some other women do the self-exam, but when they discover a lump, they become so frightened that they do nothing about it. Perhaps knowing a bit more about the realities of breast lumps and the surgery that is performed when they are discovered will help dispel some of the fears surrounding the subject of breast cancer. There are three kinds of breast lumps: *cysts* (fluid-filled sacs, also called *fibrocystic* or *cystic mastitis*), *fibroadenomas,* and *malignant tumors.* The important thing to realize is that 80 percent of breast lumps are cysts or fibroadenomas and are benign—that is, not dangerous. Therefore, if you find a lump in your breast, the chances are fairly good that it is not malignant; of course, you cannot be sure of this until a doctor has performed a biopsy.

Once you have discovered a lump, you should see a doctor immediately. One of several diagnostic procedures may then be carried out. One is *needle aspiration,* in which a fine needle is inserted into the breast; if the lump is a cyst, the fluid in the cyst will be drained out. If the lump disappears after this procedure, then it was a cyst; the cyst is gone, and there is no need for further concern. If the lump remains, it must be either a fibroadenoma or a malignant tumor.

Other techniques are also available for early detection of breast cancer, specifically mammography. Basically, *mammography* involves taking an x-ray of the breast. This technique is highly accurate, although some errors are still made. The major advantage, though, is that it is capable of detecting tumors that are so small that they cannot yet be felt; thus it can detect cancer in very early stages, making recovery more likely. Nonetheless, mammography involves some exposure to radiation,

Focus 4.2

Breast Self-Examination

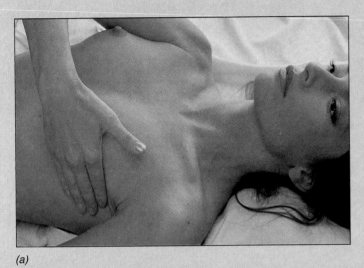

(a)

(b)

Figure 4.14
Techniques used in the
breast self-exam.

By regularly examining her own breasts, a woman is likely to notice any changes that occur. The best time for breast self-examination (BSE) is about a week after your period ends, when your breasts are not tender or swollen. If you are not having regular periods, do BSE on the same day every month.

Lie down with a pillow under your right shoulder and place your right arm behind your head. Use the finger pads of the three middle fingers on your left hand to feel for lumps in the right breast. Press firmly enough to know how your breast feels. A firm ridge in the lower curve of each breast is normal. If you're not sure how hard to press, talk with your doctor or nurse. Move around the breast in a circular, up and down line, or wedge pattern. Be sure to do it the same way every time, check the entire breast area, and remember how your breast feels from month to month. Repeat the exam on your left breast, using the finger pads of the right hand. (Move the pillow to under your left shoulder.) If you find any changes, see your doctor right away. Repeat the examination of both breasts while standing, with your one arm behind your head. The upright position makes it easier to check the upper and outer part of the breasts (toward your armpit). This is where about half of breast cancers are found. You may want to do the standing part of the BSE while you are in the shower. Some breast changes can be felt more easily when your skin is wet and soapy.

For added safety, you can check your breasts for any dimpling of the skin, changes in the nipple, redness, or swelling while standing in front of a mirror right after your BSE each month.

which itself may increase the risk of cancer. The question is: Which is more dangerous—having mammography or not detecting breast cancer until a later stage? Experts have concluded that the benefits outweigh the risks. The American Cancer Society now says that a woman should have her first baseline mammogram at around age 35 and then should have one done regularly at one- or two-year intervals from age 40 to 49, and every year from age 50 on.

Most physicians feel that the only definitive way to differentiate between a fibroadenoma and a malignant tumor is to do a *biopsy.* A small slit is made in the breast, and the lump is removed. A pathologist then examines it to determine whether it is cancerous. If it is simply a fibroadenoma, it has been removed and there is no further need for concern.

Mastectomy

Several forms of surgery may be performed when a breast lump is found to be malignant. Radiation therapy, chemotherapy, and hormone therapy may also be used. The most serious surgery is **radical mastectomy,** in which the entire breast and the underlying muscle (pectoral muscle) and the lymph nodes are removed. Advocates of this procedure argue that if the cancer has spread to the adjoining lymph nodes, the procedure ensures that all the affected tissue is removed. The disadvantages of the procedure are that it is disfiguring and there may be difficulty in arm movement following removal of the pectoral muscles. In *modified radical mastectomy* the breast and lymph nodes, but not the muscles, are removed. In *simple mastectomy* only the breast (and possibly a few lymph nodes) is removed. In **lumpectomy,** only the lump itself and a small bit of surrounding tissues are removed. The breast is thus preserved. Research indicates that in cases of early breast cancer (the cancer has not spread beyond the breast, e.g., to the lymph nodes), lumpectomy followed by radiation therapy is as effective as radical mastectomy (Veronesi et al., 1981;

Radical mastectomy (mast-ECT-uh-mee): A surgical treatment for breast cancer in which the entire breast, as well as underlying muscles and lymph nodes, is removed.
Lumpectomy: A surgical treatment for breast cancer in which only the lump and a small bit of surrounding tissue are removed.

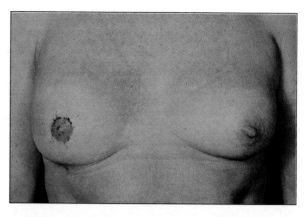

Figure 4.15   Appearance of a breast reconstructed after a mastectomy.

Henahan, 1984) and obviously much preferable. Hormone therapy (tamoxifen) may also be used.

Treatments generally are highly effective. If the cancer has not spread beyond the breast, the survival rate is over 94 percent five years after treatment (American Cancer Society, 1995).

Psychological Aspects

A lot more is at stake with breast cancer and mastectomy than technical details about diagnosis and surgery. The psychological impact of breast cancer and mastectomy can be enormous (for an excellent review see Meyerowitz, 1980). There seem to be two sources of the trauma: Finding out that one has cancer of any kind is traumatic, and the amputation of the breast is additionally stressful.

The typical emotional response of the mastectomy patient is depression, often associated with anxiety and anger. These responses are so common that they can be considered normal. The woman must make a number of physical and psychological adaptations, including different positions for sleeping and lovemaking and, for many women, a change to less revealing clothing. It is common for women to have difficulty showing their incisions to their sexual partners. Marital tensions and sexual problems may increase. Many women experience a fear of recurring cancer and its treatment and of death, as well as concerns about the mutilation of mastectomy and loss of femininity. Our culture is very breast-oriented, and a woman who has defined her identity in terms of her beauty and voluptuous figure may have a more difficult time adjusting.

Often these emotional responses last for a year. However, long-term studies indicate that women gradually adapt to the stresses they have experienced and return to their precancer level of psychological functioning.

It is extremely important for a mastectomy patient and her husband or partner to have some form of counseling available to them. In many towns the American Cancer Society has organized support groups for mastectomy patients.

Cancer of the Cervix

Cancer of the cervix and other portions of the uterus is the third most common form of cancer in women; about 2 to 3 percent of all women develop it at some time in their lives. In 1995, 11,000 American women died of uterine cancer, and another 48,000 new cases were detected (American Cancer Society, 1995). It is encouraging to realize that the death rate from cervical cancer has decreased 74 percent in the last 40 years, mainly as a result of the Pap test and more regular checkups (American Cancer Society, 1998). The death rate from cervical cancer is about twice as high in black women as it is in white women, probably because many black women do not have access to good health care and regular pelvic exams. The typical pelvic exam is described in Focus 4.3.

The exact causes of cervical cancer are unknown, but a virus is suspected. A number of pieces of data lead to this conclusion. There seems to be an association between heterosexual intercourse and cervical cancer; the greater the number of partners, the greater the chances of developing cervical cancer. Teenagers who start having intercourse very early and have multiple partners seem especially susceptible, suggesting that the virus or viruses may be spread by intercourse (National Cancer Institute, 1994). HPV (warts) infection (see Chapter 20) sharply increases the risk of cervical cancer. Research shows that tumor suppressor genes are active in normal cells, preventing them from becoming cancerous. HPV interferes with the activity of those genes (American Cancer Society, 1998).

If detected in the early stages, cervical cancer is quite curable. Fortunately, there is a good, routine test to detect it, the *Pap test,* which is highly accurate and can detect the cancer long before the woman feels any pain from it. Every woman over the age of 20 should have a Pap test annually.

If cervical cancer is confirmed, the treatment is usually a **hysterectomy,** which involves the surgical removal of the entire uterus and cervix (although in some cases, less radical procedures may be possible). The side effects of hysterectomy are minimal, except, of course, for the risks associated with any major surgery. A hysterectomy does not leave the woman "masculinized," with a beard growing and a deep voice developing, because hormone production is not affected; recall that it is the ovaries that manufacture hormones, and they are not removed (except in rare cases when the cancer has spread to them). If cervical cancer is detected very early, it may be treated by less severe procedures such as cryotherapy, which involves destruction just of the abnormal cells using extreme cold, or laser surgery.

Other cancers of the female sexual-reproductive organs include uterine (endometrial) cancer, ovarian cancer, and cancer of the vulva, vagina, and fallopian tubes, but these are all rare in comparison with cervical cancer.

Cancer of the Prostate

Cancer of the prostate is the second most common form of cancer in men (the most common being skin cancer). It is not, however, a major cause of death because it generally affects older men (25 percent of men over 90 have prostate cancer) and because most of the tumors are small and spread (metastasize) only very slowly. On the other hand, a certain percentage of prostate tumors do spread and are lethal. This form of prostate cancer causes 40,000 deaths a year, nearly as many as from breast cancer (American Cancer Society, 1998). A prostate cancer gene (HPC1, for Hereditary Prostate Cancer) has been discovered, but it accounts for only about 3 percent of all cases (Pennisi, 1996; Smith et al., 1996).

Early symptoms of prostate cancer are frequent urination (especially at night), difficulty in urination, and difficulty emptying the bladder. These are also symptoms of benign prostate enlargement,

Hysterectomy (his-tuh-REK-tuh-mee): Surgical removal of the uterus.

Focus 4.3
The Pelvic Exam

All adult women should have a checkup every year that includes a thorough pelvic exam. Among other things, such an exam is extremely important in the detection of cervical cancer, and early detection is the key to cure (see the last section of this chapter). Some women neglect to have the exam because they feel anxious or embarrassed about it or because they think they are "too young" or "too old"; however, having regular pelvic exams can be a matter of life and death. Actually, the exam is quite simple and need not cause any discomfort. The following is a description of the procedures in a pelvic exam (Boston Women's Health Book Collective, 1992).

First, the health care provider inspects the vulva, checking for irritations, discolorations, bumps, lice, skin lesions, and unusual vaginal discharge. Then there is an internal check for *cystoceles* (bulges of the bladder into the vagina) and *rectoceles* (bulges of the rectum into the vagina), for pus in the Skene glands, for cysts in the Bartholin glands, and for the strength of the pelvic floor muscles and abdominal muscles. There is also a test for stress incontinence; the physician asks the patient to cough and checks to see whether urine flows involuntarily.

Next comes the speculum exam. The *speculum* is a plastic or metal instrument that is inserted into the vagina to hold the vaginal walls apart to permit examination. Once the speculum is in place (it should be prewarmed to body temperature if it is metal), the health care provider looks for any unusual signs, such as lesions, inflammation, or unusual discharge from the vaginal walls, and for any signs of infections or damage to the cervix. The health care provider then uses a small metal spatula to scrape a tiny bit of tissue from the cervix for the Pap test for cervical cancer. If done properly, this should be painless. A battery of tests for sexually transmitted diseases can also be run.

If the woman is interested in seeing her own cervix, she can ask the doctor to hold up a mirror so that she can view it through the speculum. Indeed, some women's groups advocate that women learn to use a speculum and give themselves regular exams with it; early detection of diseases would thus be much more likely. (For a more detailed description, see the Boston Women's Health Book Collective, 1992.)

Next, the health care provider does a bimanual vaginal exam. She or he slides the index and middle fingers of one hand into the vagina and then, with the other hand, presses down from the outside on the abdominal wall. The health care provider then feels for the position of the uterus, tubes, and ovaries and for any signs of growths, pain, or inflammation.

which itself may require treatment by surgery or drugs (Oesterling, 1995). These symptoms result from the pressure of the prostate tumor on the urethra. In the early stages, there may be frequent erections and an increase in sex drive; however, as the disease progresses, there are often problems with sexual functioning.

Preliminary diagnosis of prostate cancer is by a rectal examination, which is simple and causes no more than minimal discomfort. The physician (wearing a lubricated glove) simply inserts one finger into the rectum and palpates (feels) the prostate. All men over 40 should have a rectal exam at least once a year. If the rectal exam provides evidence of a tumor, further laboratory tests can be conducted as confirmation. The rectal exam has its disadvantages, though. Some men dislike the discomfort it causes, and it is not 100 percent accurate. A blood test for PSA (prostate-specific antigen) is also available, but it is not as accurate as one might like, and it is therefore controversial.

Treatment often involves surgical removal of some or all of the prostate, plus some type of hormone therapy, radiation therapy, or anticancer

Finally, the health care provider may do a recto-vaginal exam by inserting one finger into the vagina and one into the rectum; this provides further information on the positioning of the pelvic organs and can include a test for colon cancer.

Once again, it is important to emphasize that these are not painful procedures and that having them performed regularly is extremely important to a woman's health.

Figure 4.16 (*a*) The speculum in place for a pelvic exam. The Ayre spatula is used to get a sample of cells for the Pap test. (*b*) The bimanual pelvic exam.

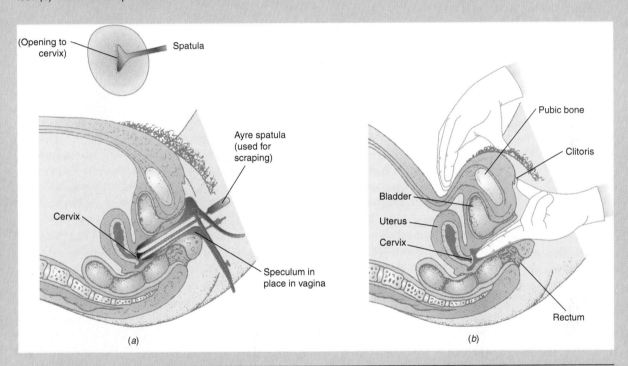

(*a*)   (*b*)

drugs. Because prostate cancer is often a slow-growing cancer, it may be left untreated, particularly if the man is elderly. Surgery may result in erection problems or retrograde ejaculation.

Cancer of the penis is another cancer of the male sexual-reproductive system, but it is rare compared with prostate cancer. It seems to be much more common among uncircumcised men than among circumcised men, suggesting that the accumulation of smegma under the foreskin may be related to its cause. Treatment may consist of surgery or radiation therapy.

Cancer of the Testes

Cancer of the testes is not a particularly common form of cancer. About 7600 new cases are diagnosed each year (American Cancer Society, 1998). However, it tends to be a disease of young men, and it is the most common form of cancer in men between the ages of 29 and 35. The rate for white men is about double that for black men, with rates for Hispanics, Native Americans, and Asian Americans falling in between (American Cancer Society, 1988).

The first sign is usually a painless lump in the testes, or a slight enlargement or change in

Focus 4.4

Testicular Examination

Doctors agree that examination of a man's testicle is an important part of a general physical examination. The American Cancer Society includes the examination in its recommendations for routine cancer-related checkups. The issue of regular testicular self-examination is more controversial. The ACS does not feel that there is any medical evidence to suggest that, for men with average testicular cancer risk, monthly examination is any more effective than simple awareness and prompt medical evaluation. However, the choice of whether or not to perform this examination should be made by each man, so instructions for testicular examination are included in this section.

If you plan to perform the self-exam the best time to do so is during or after a bath or shower, when the skin of the scrotum is relaxed. Stand in front of a mirror and hold the penis out of the way. Examine each testicle separately. Hold the testicle between the thumbs and fingers with both hands and roll it gently between the fingers. Look and feel for any hard lumps or *nodules* (smooth rounded masses) or any change in the size, shape, or consistency of the testes. Contact your doctor if you detect any troublesome signs. Be aware that the testicles contain blood vessels, supporting tissues, and tubes that conduct sperm and that some men may confuse these with a cancer. If you have any doubts, ask your doctor.

These are warning signs of testicular cancer:

1. A painless or an uncomfortable lump on a testicle

2. A testicular enlargement or swelling
3. A sensation of heaviness or aching in the lower abdomen or scrotum.
4. In rare cases, men with germ cell cancer notice breast tenderness or breast growth.

See a physician promptly if you have any of these symptoms.

Figure 4.17   The technique used in the testicular self-exam.

Source: American Cancer Society web site.

consistency of the testes. There may be pain in the lower abdomen or groin. Unfortunately, many men do not discover the tumor, or if they do, they do not see a physician soon, so that in most cases the cancer has spread to other organs by the time a physician is consulted. One study showed that when the lump was reported early to the physician, the chances for survival were better than 90 percent.

However, when the man waited over three months to see a physician, the survival rate dropped to approximately 25 percent ("A Regular Check," 1980).

Diagnosis is made either by a physician's or by the man's examination of the testes (see Focus 4.4) and by ultrasound. Final diagnosis involves surgical removal of the entire testis. This is also the first step in treatment. Fortunately, the other testicle remains, so that hormone production and sexual functioning can continue unimpaired. An artificial, gel-filled testicle can be implanted to restore a normal appearance.

The cause of testicular cancer is not known for certain. An undescended testis has a much greater risk of developing cancer.

SUMMARY

The external sexual organs of the female are the clitoris, the mons, the inner lips, the outer lips, and the vaginal opening. Collectively these are referred to as the vulva. The clitoris is an extremely sensitive organ and is very important in female sexual response. Clitoridectomy and infibulation are rituals that involve cutting of the clitoris and other parts of the vulva and are practiced widely in some African nations and elsewhere. Another external structure is the hymen, which has taken on great symbolic significance as a sign of virginity, although its absence is not a reliable indicator that a woman is not a virgin. The important internal structures are the vagina, which receives the penis during coitus; the uterus, which houses the developing fetus; the ovaries, which produce eggs and manufacture sex hormones; and the fallopian tubes, which convey the egg to the uterus. The breasts of the female also function in sexual arousal and may have great symbolic significance.

The external sexual organs of the male are the penis and the scrotum. The penis contains three spongy bodies which, when filled with blood, produce an erection. Circumcision, or surgical removal of the foreskin of the penis, is a debated practice in the United States but may have some health advantages. The scrotum contains the testes, which are responsible for the manufacture of sperm (in the seminiferous tubules) and sex hormones (in the interstitial cells). Sperm pass out of the testes during ejaculation via the vas deferens, the ejaculatory duct, and the urethra. The seminal vesicles manufacture most of the fluid that mixes with the sperm to form semen. The prostate also contributes secretions.

Breast cancer is the second most common form of cancer in women. All women should do a monthly self-exam because the earlier a lump is detected, the greater the chances of complete recovery. The Pap test is used to detect cervical cancer. Prostate cancer is the second most common form of cancer in men, but it generally affects only older men. Cancer of the testes, although rare, is the most common cancer in men between the ages of 29 and 35. Men should do a monthly testicular self-exam, just as women should do a monthly breast self-exam.

REVIEW QUESTIONS

1. The most sexually sensitive organ in the female is the _____.

2. _____ is a ritual practiced in some African nations; it involves cutting off the clitoris and inner lips and sewing together the outer lips.

3. The pubococcygeal muscle is a muscle that supports the uterus, keeping it in place. True or false?

4. The inner layer of the uterus is termed the _____.

5. The ovaries manufacture the sex hormones _____ and _____.

6. _____ is the term for the surgical removal of the foreskin or prepuce of the penis.

7. Testosterone is manufactured in the _____ in the testes.

8. After passing out of the testes and epididymis, sperm move to the vas deferens. True or false?

9. Approximately _____ percent of breast lumps are benign.

10. Testicular cancer most commonly occurs in men between the ages of 29 and 35. True or false?

(*The answers to all review questions are at the end of the book, after the Glossary.*)

QUESTIONS FOR THOUGHT, DISCUSSION, AND DEBATE

1. Form two groups of students to debate the following: Resolved: Circumcision should not be performed routinely. You can draw on many resources to provide evidence for your debate, including interviews with doctors and nurses, library materials (books and journal articles on the effects of circumcision), and interviews with parents of infants.

2. You are a gynecologist practicing in New York. An immigrant woman from Somalia makes an appointment with you. She wants you to perform infibulation on her 8-year-old daughter. She pleads with you, saying that she wants the girl to have the procedure performed under safe, sanitary conditions in a hospital. She firmly believes that she would betray her culture, to which they will return in 2 years, if she does not carry out this ancient custom with her daughter. What should you do?

3. Clitoridectomy and infibulation are practiced widely today, particularly in East Africa. Scientific evidence indicates that these practices can cause serious negative health consequences for women and girls. Some argue that people throughout the world should work to eradicate this practice, perhaps with the help of an organization such as the World Health Organization. Others argue that these practices are deeply rooted in the cultures of these countries, and that outsiders have no right to judge it, much less try to stop it. What do you think? Why?

SUGGESTIONS FOR FURTHER READING

Boston Women's Health Book Collective. (1996). *The new our bodies, ourselves: A book by and for women*. New York: Simon & Schuster. A good, easy-to-read source on female biology and sexuality.

Lightfoot-Klein, Hanny. (1989). *Prisoners of ritual: An odyssey into female genital circumcision in Africa*. New York: Haworth. An amazing story of the author's investigations into female circumcision practices.

Morgentaler, Abraham. (1993). *The male body: A physician's guide to what every man should know about his sexual health*. New York: Simon & Schuster. An authoritative book on men's health.

WEB RESOURCES

http://www.mayohealth.org/mayo/common/htm/womenpg.htm
Mayo Clinic Women's Health Page.

http://www.mayohealth.org/mayo/common/htm/menspg.htm
Mayo Clinic Men's Health Page.

http://www.mayohealth.org/mayo/common/htm/canhpage.htm
Mayo Clinic Cancer Page.

http://www.medscape.com/
News about a variety of health topics, including cancer, women's health.

http://www.nocirc.org/
National Organization of Circumcision Information Resource Centers Information and resources about male and female circumcision.

chapter

5

Sex Hormones and Sexual Differentiation

CHAPTER HIGHLIGHTS

This Way

I have AIS, I guess,
because there is a god,
and he or she or both,
peered deep into my heart
to see
that all that I can be
is best expressed
in female form.

The alternative for me
would be XY, and I
would be virilized;
so all that's soft and tender
would instead surrender

to a strand of DNA.
In the lie of X and Y
I came to challenge the
immutability
of "he" and the certainty
of "she." Blended and infused,
a ruse of gender
that upends
a different fate.

Non-functioning receptors
have rescued me

Not a failed mess
But a smashing success of nature!*

One of the marvels of human biology is that the complex and different male and female anatomies—males with penis and scrotum; females with vagina, uterus, and breasts—arise from a single cell, the fertilized egg, which varies only in whether it carries two X chromosomes (XX) or one X and one Y (XY). Many of the structural differences between males and females arise before birth, during the **prenatal period,** in a process called *prenatal sexual differentiation.* Further differences also develop during puberty. This process of sexual differentiation—both prenatally and during puberty—will be examined in this chapter. First, however, another biological system, the endocrine (hormonal) system, needs to be considered, paying particular attention to the sex hormones, which play a major role in the differentiation process.

Sex Hormones

Hormones are powerful chemical substances manufactured by the *endocrine glands* and secreted directly into the bloodstream. Because they go into the blood, their effects are felt fairly rapidly and at places in the body quite distant from the place in

which they were manufactured. The most important sex hormones are **testosterone** (one of a group of hormones called **androgens**) and **estrogens** and **progesterone.** The thyroid, the adrenals, and the pituitary are examples of endocrine glands. We are interested here in the gonads, or sex glands: the testes in the male and the ovaries in the female. The **pituitary gland** and a closely related region of the brain, the **hypothalamus,** are also important because the hypothalamus regulates the pituitary, which regulates the other glands, in particular the testes and

---

Prenatal period (pree-NAY-tul): The time from conception to birth.
Hormones: Chemical substances secreted by the endocrine glands into the bloodstream.
Testosterone: A hormone secreted by the testes in the male (and also present at lower levels in the female).
Androgens: The group of "male" sex hormones, one of which is testosterone.
Estrogens (ESS-troh-jens): The group of "female" sex hormones.
Progesterone (pro-JES-tur-ohn): A "female" sex hormone secreted by the ovaries.
Pituitary gland (pih-TOO-ih-tair-ee): A small endocrine gland located on the lower side of the brain below the hypothalamus; the pituitary is important in regulating levels of sex hormones.
Hypothalamus (hy-poh-THAL-ah-mus): A small region of the brain that is important in regulating many body functions, including the functioning of the sex hormones.

---

*Sherri Groveman, an intersex individual with Androgen Insensitivity Syndrome (AIS). In *Hermaphrodites with Attitude,* 1995, p. 2. (Contact Info@isna.org)

ovaries. Because of its importance, the pituitary has been called the "master gland" of the endocrine system. The pituitary is a small gland, about the size of a pea, which projects down from the lower side of the brain. It is divided into three lobes: the anterior lobe, the intermediary lobe, and the posterior lobe. The anterior lobe is the one that interacts with the gonads. The hypothalamus is a region at the base of the brain just above the pituitary (see Figure 5.1); it plays a part in regulating many vital behaviors such as eating, drinking, and sexual behavior,[1] and it is important in regulating the pituitary.

These three structures, then—hypothalamus, pituitary, and gonads (testes or ovaries)—function together. They influence such important sexual functions as the menstrual cycle, pregnancy, the changes of puberty, and sexual behavior. Because these systems are, not surprisingly, somewhat different in males and in females, the sex hormone systems in the male and in the female will be discussed separately.

[1]One psychologist summarized the functions of the hypothalamus as being the four F's: fighting, feeding, fleeing, and, ahem, sexual behavior.

Figure 5.1   The hypothalamus-pituitary-gonad feedback loop in women, which regulates production of the sex hormones.

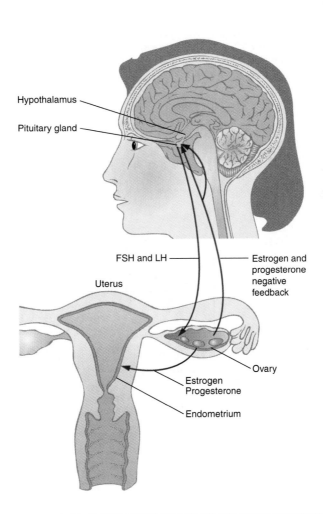

Sex Hormone Systems in the Male

The pituitary and the testes both produce hormones. The important hormone produced by testes is *testosterone.* Testosterone, a "male" or masculinizing sex hormone, has important functions in stimulating and maintaining the secondary sex characteristics (such as beard growth), maintaining the genitals and their sperm-producing capability, and stimulating the growth of bone and muscle.

The pituitary produces several hormones, two of which are important in this discussion: **follicle-stimulating hormone (FSH)** and **luteinizing hormone (LH).** These hormones affect the functioning of the testes. LH controls testosterone production, and FSH controls sperm production.

Testosterone levels in males are relatively constant. These constant levels are maintained because the hypothalamus, pituitary, and testes operate in a negative feedback loop (Figure 5.2). The levels of testosterone are regulated by a substance called **GnRH (gonadotropin-releasing hormone),** which is secreted by the hypothalamus. (FSH levels are similarly regulated by GnRH.) The system comes full circle because the hypothalamus is sensitive to the levels of testosterone present, and thus testosterone influences the output of GnRH.

Follicle-stimulating hormone (FSH): A hormone secreted by the pituitary; it stimulates follicle development in females and sperm production in males.

Luteinizing hormone (LH): A hormone secreted by the pituitary; it regulates estrogen secretion and ovum development in the female and testosterone production in the male.

GnRH (gonadotropin-releasing hormone): A hormone secreted by the hypothalamus that regulates the pituitary's secretion of gonad-stimulating hormones.

Figure 5.2   Schematic diagram of hormonal control of testosterone secretion and sperm production by the testes. The negative signs indicate that testosterone inhibits LH production, both in the pituitary and in the hypothalamus.

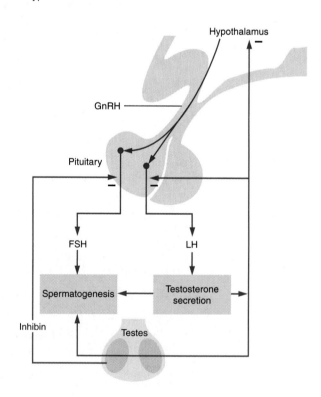

This negative feedback loop operates much like a thermostat-furnace system. If a room is cold, certain changes occur in the thermostat, and it signals the furnace to turn on. The action of the furnace warms the air in the room. Eventually the air becomes so warm that another change is produced in the thermostat, and it sends a signal to the furnace to turn off. The temperature in the room then gradually falls until it produces another change in the thermostat, which then turns on the furnace, and the cycle is repeated. This is a *negative* feedback loop because *rises* in temperature turn *off* the furnace, whereas *decreases* in temperature turn *on* the furnace.

The hypothalamus, pituitary, and testes form a similar negative feedback loop, ensuring that testosterone is maintained at a fairly constant level, just as the temperature of a room is kept fairly constant. The pituitary's production of LH stimulates the testes to produce testosterone. But when testosterone levels get high, the hypothalamus reduces its production of GnRH; the pituitary's production of LH is then reduced, and the production of testosterone by the testes consequently decreases. When it has fallen, the hypothalamus again increases production of GnRH, and the process starts again.

While the level of testosterone in men is fairly constant, there is probably some cycling, with variations according to the time of the day and possibly according to the time of the month (see Chapter 6).

Although it has been clear for some time that there is a negative feedback loop between testosterone levels and LH levels, it has not been clear what regulates FSH levels. **Inhibin** is a substance produced in the testes (by cells called the Sertoli cells) which serves exactly that function—it acts to regulate FSH levels in a negative feedback loop (Plant et al., 1993; Moodbidri et al., 1980).

Interest in inhibin has been intense because it shows great promise, at least theoretically, as a male contraceptive. That is, because inhibin suppresses FSH production, sperm production in turn is inhibited. Future developments in this field should be interesting.

Sex Hormone Systems in the Female

The ovaries produce two important hormones, *estrogen* and *progesterone*. The functions of estrogen include bringing about many of the changes of puberty (stimulating the growth of the uterus and vagina, enlarging the pelvis, and stimulating breast growth). Estrogen is also responsible for maintaining the mucous membranes of the vagina and stopping the growth of bone and muscle, which accounts for the generally smaller size of females as compared with males.

In adult women the levels of estrogen and progesterone fluctuate according to the various phases of the menstrual cycle (see Chapter 6) and during various other stages such as pregnancy and menopause. The levels of estrogen and progesterone are regulated by the two pituitary hormones, FSH and LH. Thus the levels of estrogen and progesterone are controlled by a negative feedback loop of the hypothalamus, pituitary, and ovaries, similar to the negative feedback loop in the male (see Figure 5.3). For example, as shown on the right side of Figure 5.3, increases in the level of GnRH increase the level of LH, and the increases in LH eventually produce increases in the output of estrogen; finally, the increases in the level of estrogen inhibit (decrease) the production of GnRH and LH.

The pituitary produces two other hormones, *prolactin* and *oxytocin,* which play roles in stimulating secretion of milk by the mammary glands after a woman has given birth to a child. Oxytocin also stimulates contractions of the uterus during childbirth.

The female sex hormone system functions much like the male sex hormone system. The ovaries and testes produce many of the same hormones, but in different amounts. The ovaries, like the testes, produce inhibin, and it forms a negative feedback loop with FSH production (Burger, 1993). The functioning of the female sex hormone system and the menstrual cycle will be considered in more detail in Chapter 6.

Prenatal Sexual Differentiation

Sex Chromosomes

As noted above, at the time of conception the future human being consists of only a single cell, the fertilized egg. The only difference between the fertilized egg that will become a female and the fertilized egg

Inhibin: A substance secreted by the testes and ovaries which regulates FSH levels.

that will become a male is the sex chromosomes carried in that fertilized egg. If there are two X chromosomes, the result will typically be a female; if there is one X and one Y, the result will typically be a male. Thus, while incredibly tiny, the sex chromosomes carry a wealth of information which they transmit to various organs throughout the body, giving them instructions on how to differentiate in the course of development. The Y chromosome, because it is smaller, carries less information than the X.

Occasionally, individuals receive at conception a sex chromosome combination other than XX or XY. Such abnormal sex chromosome complements may lead to a variety of clinical syndromes, such as Klinefelter's syndrome. In this syndrome, a genetic male has an extra X chromosome (XXY). As a result, the testes are abnormal, no sperm are produced, and testosterone levels are low (Winter & Couch, 1995).

The single cell divides repeatedly, becoming a two-celled organism, then a four-celled organism, then an eight-celled organism, and so on. By 28 days after conception, the embryo is about 1 centimeter (less than ½ inch) long, but the male and female embryo are still identical, save for the sex chromosomes; that is, the embryo is still in the undifferentiated state. However, by the seventh week after conception, some basic structures have been formed that will eventually become either a male or a female reproductive system. At this point, the embryo has a pair of gonads (each gonad has two parts, an outer cortex and an inner medulla), two sets of ducts (the *Müllerian ducts* and the *Wolffian* ducts), and rudimentary external genitals (the *genital tubercle,* the *genital folds,* and the *genital swelling*) (see Figure 5.4).

Gonads

In the seventh week after conception, the sex chromosomes direct the gonads to begin differentiation. In the male, the undifferentiated gonad develops into a testis at about 7 weeks. In the female, the process occurs somewhat later, with the ovaries developing at around 10 or 11 weeks.

Figure 5.3   Schematic diagram of hormonal control of estrogen secretion and ovum production by the ovaries (during the follicular phase of the menstrual cycle). Note how similar the mechanism is to the one in the male.

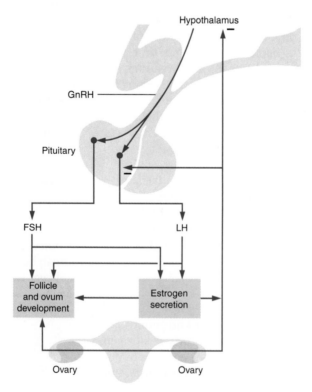

Figure 5.4   Development of the male and female external genitals from the undifferentiated stage. This occurs during prenatal development. Note homologous organs in the female and male.

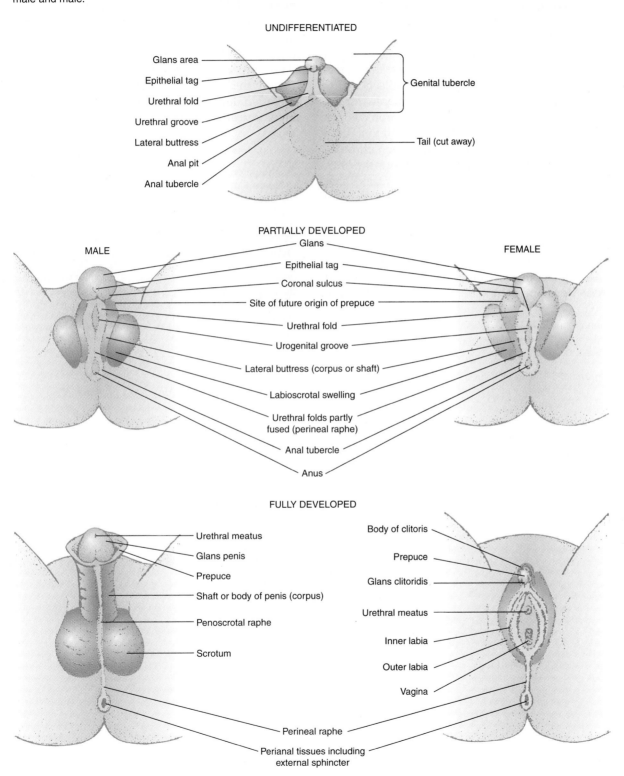

An important gene that directs differentiation of the gonads, located on the Y chromosome, is called **testis-determining factor,** or TDF (Page et al., 1987). It is also called *SRY,* for sex-determining region, Y chromosome. If TDF is present, testes differentiate and male development occurs (see Figure 5.5 for a summary of all the genes that regulate sexual differentiation). Researchers discovered this gene by studying cases of abnormal development—for example, adult women who had XY sex chromosomes and were infertile. The researchers found that these individuals were missing a section of the Y chromosome, precisely the section containing the SRY. The X chromosome carries a number of genes that control normal functioning of the ovaries (Winter & Couch, 1995).

Prenatal Hormones and the Genitals

Once the ovaries and testes have differentiated, they begin to produce different sex hormones, and these hormones then direct the differentiation of the rest of the internal and external genital system (see Figure 5.4).

In the female the Wolffian ducts degenerate, and the **Müllerian ducts** turn into the fallopian tubes, the uterus, and the upper part of the vagina. The tubercle become the clitoris, the folds become the inner lips, and the swelling develops into the outer lips.

In the male SRY encodes a protein called anti-Müllerian hormone (AMH), which is manufactured in the testes (Marx, 1995b). AMH causes the Müllerian ducts to degenerate, while the **Wolffian ducts** turn into the epididymis, the vas deferens,

> **Testis-determining factor (TDF):** A gene on the Y chromosome that causes testes to differentiate prenatally. Also called SRY, for sex-determining region, Y chromosome.
> **Müllerian ducts:** Ducts found in both male and female fetuses; in males they degenerate and in females they develop into the fallopian tubes, the uterus, and the upper part of the vagina.
> **Wolffian ducts:** Ducts found in both male and female fetuses; in females they degenerate and in males they develop into the epididymis, the vas deferens, and the ejaculatory duct.

Figure 5.5   The possible functions of the genes linked so far to sexual differentiation in mammals: SRY, Sox9, AMH, WT1, SF-1, and DAX-1 (Marx, 1995b).

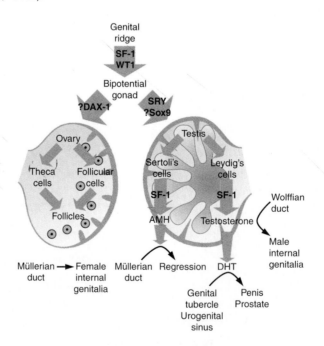

and the ejaculatory duct. The tubercle becomes the glans penis, the folds form the shaft of the penis, and the swelling develops into the scrotum.

The mechanism by which the internal and external genitals differentiate is the subject of much exciting new research. The principle seems to be that the presence or absence of the testes and their products is critical. The testes of the male fetus secrete androgens, which stimulate the development of male external structures, and a second substance, AMH, which suppresses the development of the Müllerian ducts. Thus if the output of the testes is present, male structures develop. If it is absent, female structures develop. At the same time, at least six different genes are involved in prenatal sexual differentiation (Figure 5.5), and a mutation in any one of them can cause an error in development.

By four months after conception, the gender of the fetus is clear from the appearance of the external genitals (Figure 5.4).

Descent of the Testes and Ovaries

As the developmental changes are taking place, the ovaries and testes are changing in shape and position. At first, the ovaries and testes lie near the top of the abdominal cavity. By the tenth week they have grown and have moved down to the level of the upper edge of the pelvis. The ovaries remain there until after birth; later they shift to their adult position in the pelvis.

The male testes must make a much longer journey, down into the scrotum via a passageway called the *inguinal canal.* Normally this movement occurs around the seventh month after conception. After the descent of the testes, the inguinal canal is closed off.

Two problems in this process may occur. First, one or both testes may have failed to descend into the scrotum by the time of birth, a condition known as *undescended testes,* or **cryptorchidism** (Santen, 1995). This occurs in about 2 percent of all males; most frequently, only one testis is undescended, the other being in the normal position. In most of these cases, the testes do descend by puberty, and so only about 1 in 500 adult men has undescended testes. If the testes do not descend spontaneously, however, the condition must be corrected by surgery or hormonal therapy. The optimum time for doing this is before age 5. Other-

wise, if both testes have failed to descend, the man will be sterile because, as discussed in Chapter 4, the high temperature of the testes inside the body inhibits the production of sperm. Undescended testes are also more likely to develop cancer.

The second possible problem occurs when the inguinal canal does not close off completely. It may then reopen later in life, creating a passageway through which loops of the intestine may enter the scrotum. This condition is called *inguinal hernia* and can be remedied by simple surgery.

Brain Differentiation

During the prenatal period, when sex hormones are having a big impact on genital anatomy, are they doing anything to the brain? Might this be a time when gender differences in the brain are created? The results of many experiments with animals indicate that there are differences between male and female brains; these differences tend to be found in the hypothalamus, and in particular in a region of the hypothalamus called the preoptic area (see Hines & Collaer, 1993, for a review). The evidence indicates that differences exist between human male and female brains in the preoptic area of the hypothalamus, although these differences may develop after birth rather than prenatally (Swaab et al., 1995).

Homologous Organs

The preceding discussion of sexual differentiation highlights the fact that although adult men and women appear to have very different reproductive anatomies, their reproductive organs have similar origins. When an organ in the male and an organ in the female both develop from the same embryonic tissue, the organs are said to be **homologous.** When the two organs have similar functions, they are said to be **analogous.** Table 5.1 summarizes the

---

Cryptorchidism: Undescended testes; the condition in which the testes do not descend to the scrotum as they should during prenatal development.
Homologous organs (huh-MOLL-uh-gus): Organs in the male and female that develop from the same embryonic tissue.
Analogous organs (an-AL-uh-gus): Organs in the male and female that have similar functions.

Table 5.1 Homologous and Analogous Organs of the Male and Female Reproductive Systems

Embryonic Source	Homologous Organs		Analogous Organs	
	In the Adult Male	In the Adult Female	In the Adult Male	In the Adult Female
Gonad (medulla plus cortex)	Testes (from medulla)	Ovaries (from cortex)	Testes (from medulla)	Ovaries (from cortex)
Genital tubercle	Glans penis	Clitoris	Glans penis	Clitoris
Genital swelling	Scrotum	Outer lips		
Müllerian duct	Degenerates, leaving only remnants	Fallopian tubes, uterus, part of vagina		
Wolffian duct	Epididymis, vas deferens, seminal vesicles	Degenerates, leaving only remnants		
Urethral primordia	Prostate, Cowper's glands	Skene's glands, Bartholin glands	Prostate, Cowper's glands	Skene's glands, Bartholin glands

major homologies and analogies of the male and female reproductive systems. For example, ovaries and testes are both homologous (they develop from an indifferent gonad) and analogous (they produce gametes and sex hormones).

Atypical Prenatal Gender Differentiation

Gender is not a simple matter, a fact that is apparent from the preceding discussion. Most people, however, assume that it is. That is, people typically assume that if a person is female, she will be feminine; will think of herself as a woman; will be sexually attracted to men; will have a clitoris, vagina, uterus, and ovaries; and will have sex chromosomes XX. The parallel assumption is that all males are masculine; think of themselves as male; are sexually attracted to women; have a penis, testes, and scrotum; and have sex chromosomes XY.

A great deal of research over the last several decades challenges these assumptions and provides a great deal of information about sexuality and gender and their development. Before the results of this research are discussed, however, some background information is necessary.

We can distinguish among the following eight variables of gender (Money, 1987):[2]

1. *Chromosomal gender* XX in the female; XY in the male

2. *Gonadal gender* Ovaries in the female; testes in the male

3. *Prenatal hormonal gender* Testosterone and anti-Müllerian hormone in the male but not the female before birth

3a. *Prenatal and neonatal brain hormonalization* Testosterone present for masculinization, absent for feminization

4. *Internal accessory organs* Fallopian tubes, uterus, and upper vagina in the female; prostate and seminal vesicles in the male

5. *External genital appearance* Clitoris, inner and outer lips, and vaginal opening in the female; penis and scrotum in the male

6. *Pubertal hormonal gender* At puberty, estrogen and progesterone in the female; testosterone in the male

7. *Assigned gender* The announcement at birth, "It's a girl" or "It's a boy," based on the appearance of the external genitals; the gender the parents and the rest of society believe the child to be; the gender in which the child is reared

8. *Gender identity* The person's private, internal sense of maleness or femaleness—which is expressed in ideation, imagery, and behavior—and the integration of this sense with the rest of the personality and with the gender roles prescribed by society

[2]The distinction between the terms "gender" and "sex," discussed in Chapter 1, is being maintained here.

These variables might be subdivided into biological variables (the first six) and psychological variables (the last two).

In most cases, of course, all the variables are in agreement in an individual. That is, in most cases the person is a "consistent" female or male. If the person is a female, she has XX chromosomes, ovaries, a uterus and vagina, and a clitoris; she is reared as a female; and she thinks of herself as a female. If the person is a male, he has the parallel set of appropriate characteristics.

However, as a result of any one of a number of factors during the course of prenatal sexual development and differentiation, the gender indicated by one or more of these variables may disagree with the gender indicated by others. When the contradictions are among several of the biological variables (1 through 6), the person is called an **intersex** or pseudohermaphrodite.[3] Biologically, the gender of such a person is ambiguous; the reproductive structures may be partly male and partly female, or they may be incompletely male or incompletely female.

A number of syndromes can cause intersex, some of the most common being congenital adrenal hyperplasia, progestin-induced pseudohermaphroditism, and the androgen insensitivity syndrome. In **congenital adrenal hyperplasia** (**CAH**, also called "adrenogenital syndrome"), a genetic female develops ovaries normally as a fetus; later in the course of prenatal development, however, the adrenal gland begins to function abnormally (as a result of a recessive genetic condition unconnected with the sex chromosomes), and an excess amount of androgens is produced. Prenatal sexual differentiation then does not follow the normal female course. As a result, the external genitals are partly or completely male in appearance; the labia are partly or totally fused (and thus there is no vaginal opening), and the clitoris is enlarged to the size of a small penis or even a full-sized one. Hence at birth these genetic females are often identified as males.

Progestin-induced pseudohermaphroditism is a similar syndrome which resulted from a drug, progestin, that was at one time given to pregnant women to help them maintain the pregnancy if they were prone to miscarriage. (The drug is no longer prescribed because of the following effects.) As the drug circulated in the mother's bloodstream, the developing fetus was essentially exposed to a high dose of androgens. (Progestin and androgens are quite similar biochemicals, and in the body the progestin acted like androgen.) In genetic females this produced an abnormal, masculinized genital development similar to that found in CAH.

The reverse case occurs in *androgen insensitivity syndrome* (AIS). In this syndrome a genetic male produces normal levels of testosterone; however, as a result of a genetic condition, the body tissues are insensitive to the testosterone, and prenatal development is feminized. Thus the individual is born with the external appearance of a female: a small vagina (but no uterus) and undescended testes. The individual whose poem appeared at the beginning of the chapter has AIS.

Intersex persons provide good evidence of the great complexity of sex and gender and their development. Many variables are involved in gender and sex, and there are many steps in gender differentiation, even before birth. Because the process is complex, it is very vulnerable to disturbances, creating conditions such as intersex. Indeed, the research serves to question our basic notions of what it means to be male or female. In CAH, is the genetic female who is born with male external genitals a male or a female? What makes a person male or female? Chromosomal gender? External genital appearance? Gender identity?

A related phenomenon was first studied in a small community in the Dominican Republic (Imperato-McGinley et al., 1974). Due to a genetic-endocrine problem, a large number of genetic males were born who, at birth, appeared to be females.

---

[3]The term "hermaphrodite" is taken from Hermaphroditos, the name of the mythological son of Hermes and Aphrodite. The latter was the Greek goddess of love.

Intersex: An individual who has a mixture of male and female reproductive structures, so that it is not clear at birth whether the individual is a male or a female. Also called a pseudohermaphrodite.
CAH: Congenital adrenal hyperphasia, a condition in which a genetic female produces abnormal levels of testosterone prenatally and therefore has male-appearing genitals at birth.

Focus 5.1

The Debate Over the Treatment of Intersex Individuals

When Chris was born, her clitoris was 1.7 cm long. That's about half-way between the length of the average newborn clitoris and the average newborn penis. She had a scrotum but no testes in it. As a result, the physician was unsure whether she was a girl or a boy. A blood test revealed that her sex chromosomes were XY. After 24 hours of consultations, during which her parents were in agony over wondering what the problem was, the physician decided that Chris should be a girl because it would be impossible for her to function as a normal boy with such strange genitals. While a baby, she had several surgeries, one of them to remove her testes, which were still in her abdomen. Her clitoris was surgically reduced in size when she was 5, old enough to remember it. Today she is 27 and angry about what she considers the mutilations of her body. She now knows that she has androgen-insensitivity syndrome. So much of her clitoris was removed that she is not able to have an orgasm.

Chris (a composite of several case histories in the scientific literature) is an intersex individual; that is, her genitals have combined male, female, or ambiguous elements. She was treated according to a protocol that became standard beginning in the 1960s and persists to the present day. The protocol is based on the pioneering research of Dr.

John Money and others. According to Money, individuals such as Chris, whom he called "pseudohermaphrodites," could successfully be assigned to either gender, providing that it was done before 18 months of age and that the necessary surgeries and follow-up medical treatments (such as hormone treatment) occurred. His research indicated that individuals treated with the standard protocol grew up to be healthy and well adjusted.

In the last decade, however, intersex individuals have come out of the closet and formed an activist organization, the Intersex Society of North America (ISNA).* Intersex activists argue that they simply have cases of genital *variability,* not genital abnormality. The medical standard is that an infant's organ that is 0.9 cm or less is a clitoris and 2.5 cm or more is a penis. Activists argue that these cutoffs are arbitrary. What is wrong with a clitoris that is 1.7 cm long? Perhaps the only thing wrong with it is that it makes doctors, and perhaps parents, uncomfortable and embarrassed. Issues of medical ethics are raised: Should essentially cosmetic surgery be performed on a baby who clearly cannot give informed consent? Should parents be encouraged to lie to their child and friends?

Sex researcher Milton Diamond conducted long-term follow-ups on several individuals treated

These people are called Guevodoces. The syndrome is called *5-Alpha Reductase Syndrome.* They had a vaginal pouch instead of a scrotum and a clitoris-sized penis. The uneducated parents, according to the researchers, were unaware that there were any problems, and these genetic males were treated as typical females. At puberty, a spontaneous biological change causes a penis to develop. Significantly, their psychological orientation also changed. Despite

rearing as females, their gender identity switched to male, and they developed heterosexual interests.

Anthropologist Gilbert Herdt (1990) is critical of the research and interpretations about the Guevodoces. The major criticism is that the Western researchers assumed that this culture is a two-gender society, like the United States, and that people have to fall into one of only two categories, either male or female. Anthropologists, however,

using Money's standard protocol. He found that, contrary to the glowing picture of perfect adjustment painted by Money and others, these intersex individuals had serious adjustment problems that they traced directly to the medical "management" of their condition. Diamond's research has sparked a debate in the medical community over the proper treatment of intersex individuals. Diamond himself has proposed a protocol in which he urges physicians, in cases of intersex infants, (1) to make their most informed judgment about the child's eventual gender identity (CAH girls, for example, almost invariably have a female gender identity) and counsel the parents to rear the child in that gender; (2) not to perform surgeries that might later need to be reversed; and (3) to provide honest counseling and education to the parents and child as he or she grows up so that the child can eventually make an educated, informed decision regarding treatment.

At this point, we have no experience with the outcomes that would occur if Diamond's recommendations were followed. How would a child with ambiguous genitals function in elementary school? How old would a child have to be in order to make a true informed decision? Until we have answers to these questions, it is difficult to know which treatment plan to follow.

*For information about ISNA and other sexuality organizations, including its Website, see the Directory of Resources at the end of this book, page 633.

Figure 5.6   Cheryl Chase, an intersex activist

Sources: Money & Ehrhardt (1972); Kessler (1990, 1997); Diamond (1996); Diamond & Sigmundson (1997); Meyer-Bahlburg (1998)

have documented the existence of three-gender societies—that is, societies in which there are three, not two, gender categories—and the society in which the Guevodoces grow up is a three-gender society. The third gender is the Guevodoces. Their gender identity is not male or female, but Guevodoce. The 5-Alpha Reductase Syndrome has also been found among the Sambia of New Guinea, and they, too, have a three-gender culture. Again we see

the profound effect of culture on our most basic ideas about sex and gender.

Sexual Differentiation During Puberty

Puberty is not a point in time, but rather a process during which there is further sexual differentiation.

It is the stage in life during which the body changes from that of a child into that of an adult, with secondary sexual characteristics and the ability to reproduce sexually. **Puberty** can be scientifically defined as the time during which there is sudden enlargement and maturation of the gonads, other genitalia, and secondary sex characteristics, leading to reproductive capacity (Tanner, 1967). It is the second important period—the other being the prenatal period—during which sexual differentiation takes place. Perhaps the most important single event in the process is the first ejaculation for the male and the first menstruation for the female, although the latter is not necessarily a sign of reproductive capability since girls typically do not produce mature eggs until a year or two after the first menstruation.

The physiological process that underlies puberty in both genders is a marked increase in level of sex hormones. Thus the hypothalamus, pituitary, and gonads control the changes.

Adolescence is a socially defined period of development which bears some relationship to puberty. Adolescence represents a psychological transition from the behavior and attitudes of a child to the behavior, attitudes, and responsibilities of an adult. In the United States it corresponds roughly to the teenage years. Modern American culture has an unusually long period of adolescence. A century ago, adolescence was much shorter; the lengthening of the educational process has served to prolong adolescence. In some cultures, in fact, adolescence does not exist; the child shifts to being an adult directly, with only a *rite de passage* in between.

Before describing the changes that take place during puberty, we should note two points. First, the timing of the pubertal process differs considerably for males and females. Girls begin the change around 8 to 12 years of age, while boys do so about two years later. Girls reach their full height by about age 16, while boys continue growing until about age 18 or later. The phenomenon of males and females being out of step with each other at this stage creates no small number of crises for the adolescent. Girls are interested in boys long before boys are aware that girls exist. A girl may be stuck with a date who barely reaches her armpits, while a boy may have to cope with someone who is better qualified to be on the basketball team than he is.

Second, there are large individual differences (differences from one person to the next) in the age at which the processes of puberty take place. Thus there is no "normal" time to begin menstruating or growing a beard. Accordingly, we give age ranges in describing the timing of the process.

Changes in the Female

A summary of the physical changes of puberty in males and females is provided in Table 5.2. The first sign of puberty in the female is the beginning of breast development, generally at around 8 to 13 years of age. The ducts in the nipple area swell, and there is growth of fatty and connective tissue, causing the nipples to project forward and the small, conical buds to increase in size. These changes are produced by increases in the levels of the sex hormones by a mechanism that will be described below.

As the growth of fatty and supporting tissue increases in the breasts, a similar increase takes place at the hips and buttocks, leading to the rounded contours that distinguish adult female bodies from adult male bodies. Individual females have unique patterns of fat deposit, so there are also considerable individual differences in the resulting female shapes.

Another visible sign of puberty is the growth of pubic hair, which occurs shortly after breast development begins. About two years later, axillary (underarm) hair appears.

Body growth increases sharply during puberty, during the approximate age range of 9.5 to 14.5 years. The growth spurt for girls occurs about two years before the growth spurt for boys (Figure 5.7). This is consistent with girls' general pattern of maturing earlier than boys. Even prenatally, girls show an earlier hardening of the structures that become bones. One exception to this pattern is in fertility; boys produce mature sperm earlier than girls produce mature ova.

Estrogen eventually applies the brakes to the growth spurt in girls; the presence of estrogen also causes the growth period to end sooner in girls,

Puberty: The time during which there is sudden enlargement and maturation of the gonads, other genitalia, and secondary sex characteristics, so that the individual becomes capable of reproduction.

Table 5.2 Summary of the Changes of Puberty and Their Sequence

Girls			Boys		
Characteristic	Age Range for First Appearance (Years)	Major Hormonal Influence	Characteristic	Age Range for First Appearance (Years)	Major Hormonal Influence
1. Growth of breasts	8–13	Pituitary growth hormone, estrogens, progesterone, thyroxine	1. Growth of testes, scrotal sac	10–13.5	Pituitary growth hormone, testosterone
2. Growth of pubic hair	8–14	Adrenal androgens	2. Growth of pubic hair	10–15	Testosterone
3. Body growth	9.5–14.5	Pituitary growth hormone, adrenal androgens, estrogens	3. Body growth	10.5–16	Pituitary growth hormone, testosterone
4. Menarche	10–16.5	GnRH, FSH, LH, estrogens, progesterone	4. Growth of penis	11–14.5	Testosterone
			5. Change in voice (growth of larynx)	About the same time as penis growth	Testosterone
5. Underarm hair	About two years after pubic hair	Adrenal androgens	6. Facial and underarm hair	About two years after pubic hair	Testosterone
6. Oil- and sweat-producing glands (acne occurs when glands are clogged)	About the same time as under-arm hair	Adrenal androgens	7. Oil- and sweat-producing glands, acne	About the same time as under-arm hair	Testosterone

Source: After B. Goldstein. (1976). *Human sexuality.* New York: McGraw-Hill, pp. 80–81.

thus accounting for the lesser average height of adult women as compared with adult men.

At about 12 to 13 years of age, the **menarche** (first menstruation) occurs. The girl, however, is not capable of becoming pregnant until ovulation begins, typically about two years after the menarche. The first menstruation is not only an important biological event but also an important psychological one. Various cultures have ceremonies recognizing its importance. In some families, it is a piece of news that spreads quickly to the relatives. Girls themselves display a wide range of reactions to the event, ranging from negative ones, such as fear, shame, or disgust, to positive ones, such as pride and a sense of maturity and womanliness.

Some of the most negative reactions occur when the girl has not been prepared for the menarche, which is still the case surprisingly often. Parents who are concerned about preparing their daughters for the first menstruation should remember that there is a wide range in the age at which it occurs. It is not unusual for a girl to start menstruating in the fifth grade, and instances of the menarche during the fourth grade, while rare, do occur.

What determines the age at which a girl first menstruates? One explanation is the *percent body*

Menarche (MEN-ar-key): First menstruation.

Figure 5.7   The adolescent spurt of growth for boys and girls. Note that girls experience their growth spurt earlier than boys do.

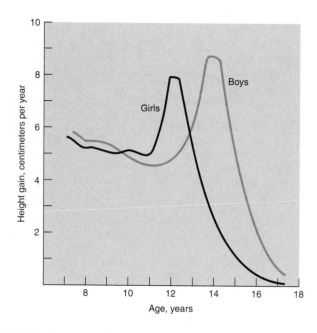

fat hypothesis (Frisch & McArthur, 1974; Hopwood et al., 1990). During puberty, deposits of body fat increase in females. According to the percent body fat hypothesis, the percentage of body weight that is fat must rise to a certain level for menstruation to occur for the first time and for it to be maintained. Thus very skinny adolescent girls would tend to be late in the timing of first menstruation. The percent body fat hypothesis also helps to make sense of two related phenomena: the cessation of menstruation in anorexics and the cessation of menstruation in women distance runners. *Anorexia nervosa* is a condition in which the person—most commonly an adolescent girl—engages in compulsive, extreme dieting, perhaps to the point of starving herself to death. As anorexia progresses, the percentage of body fat declines and menstruation ceases. It is also fairly common for women who are runners, and all women who exercise seriously to the point where their body fat is substantially reduced, to cease menstruating. For both anorexics and female runners, it seems that when the percentage of body fat falls below a critical value, the biological mechanisms that control the menstrual cycle shut down menstruation.[4]

Before leaving the topic of running, we should note that there is some evidence that serious exercise also affects the male reproductive system. One study of male distance runners found that their testosterone levels were only about 68 percent as high, on the average, as a control group's testosterone levels (Wheeler et al., 1984). There are some reports of male long-distance runners complaining of a loss of sex drive, but it is unclear whether it results from reduced testosterone levels or from

[4]On the other hand, programs of moderate, regular aerobic exercise have been shown to reduce menstrual problems such as cramps (Golub, 1992).

the perpetual feelings of fatigue they have from their intensive training (Wheeler et al., 1984).

Other body changes in girls during puberty include a development of the blood supply to the clitoris, a thickening of the walls of the vagina, and a rapid growth of the uterus, which doubles in size between the tenth and the eighteenth years. The pelvic bone structure grows and widens, contributing to the rounded shape of the female and creating a passageway large enough for an infant to move through during birth. There are large individual differences in the shape of the pelvis, and it is important for the physician to identify shapes that may lead to problems in childbirth.

The dramatic changes that occur during puberty are produced, basically, by the endocrine system and its upsurge in sex hormone production during puberty. The process begins with an increase in secretion of FSH by the pituitary gland. FSH in turn stimulates the ovaries to produce estrogen. Estrogen is responsible for many of the changes that occur; it stimulates breast growth and the growth of the uterus and vagina.

Also involved in puberty are the paired **adrenal glands,** which are located just above the kidneys. In the female, the adrenal glands are the major producer of androgens, which exist at low levels in females. Adrenal androgens stimulate the growth of pubic and axillary hair and are related to the female sex drive.

A problem for some girls during puberty is a temporary period of overweight or obesity. If the weight gain is not excessive and lasts for only a short period, there is no cause for concern.

Changes in the Male

As noted above, puberty begins at about 10 or 11 years of age in boys, about two years later than it does in girls. The physical causes of puberty in boys parallel those in girls. They are initiated by increased production of FSH and LH by the pituitary. At the beginning of puberty, the increase in LH stimulates the testes to produce testosterone, which is responsible for most of the changes of puberty in the male.

The first noticeable pubertal change in males is the growth of the testes and scrotal sac, which begins at around 10 to 13 years of age as a result of testosterone stimulation. The growth of pubic hair begins at about the same time. About a year later the penis begins to enlarge, first thickening, and then lengthening. This change also results from testosterone stimulation. As the testes enlarge, their production of testosterone increases even more; thus there is rapid growth of the penis, testes, and pubic hair at ages 13 and 14.

The growth of facial and axillary hair begins about two years after the beginning of pubic-hair growth. The growth of facial hair begins with the appearance of fuzz on the upper lip; adult beards do not appear until two or three years later. Indeed, by age 17, 50 percent of American males have not yet shaved. These changes also result from testosterone stimulation, which continues to produce growth of facial and chest hair beyond 20 years of age.

Erections increase in frequency. The organs that produce the fluid of semen, particularly the prostate, enlarge considerably at about the same time the other organs are growing. By age 13 or 14 the boy is capable of ejaculation.[5] By about age 15, the ejaculate contains mature sperm, and the male is now fertile. The pituitary hormone FSH is responsible for initiating and maintaining the production of mature sperm.

Beginning about a year after the first ejaculation, many boys begin having nocturnal emissions, or "wet dreams." For the boy who has never masturbated, a wet dream may be his first ejaculation.

At about the same time penis growth occurs, the larynx ("voice box") also begins to grow in response to testosterone. As the larynx enlarges, the boy's voice drops, or "changes." Typically the transition occurs at around age 13 or 14. Because testosterone is necessary to produce the change in voice, castration before puberty results in a male with a permanently high voice. This principle was used to produce the castrati, who sang in the great choirs of Europe during the eighteenth century.

---

[5]Note that orgasm and ejaculation are two separate processes, even though they generally occur together, at least in males after puberty. But orgasm may occur without ejaculation, and ejaculation may occur without orgasm.

Adrenal glands (uh-DREE-nul): Endocrine glands located just above the kidneys; in the female they are the major producers of androgens.

(a)

They began as lovely boy sopranos, and their parents or the choirmaster, hating to see their beautiful voices destroyed at puberty, had them castrated so that they remained permanent sopranos. Female bodies are not the only ones that have been mutilated for strange reasons! Contrary to popular belief, castration in adulthood will not produce a high voice, because the larynx has already grown.

A great spurt of body growth begins in males at around 11 to 16 years of age (Figure 5.8). Height increases rapidly. Body contours also change. While the changes in girls involve mainly the increase in fatty tissue in the breasts and hips, the changes in boys involve mainly an increase in muscle mass. Eventually testosterone brings the growth process to an end, although it permits the growth period to continue longer than it does in females.

Puberty brings changes and also problems. One problem is *acne,* which affects boys more frequently than girls. *Acne* is a distressing skin condition that is caused by a clogging of the sebaceous (oil-producing) glands, resulting in pustules, blackheads, and redness on the face and possibly the chest and

(b)

Figure 5.8   There is great variability in the onset of puberty and its growth spurt. Both girls are the same age. All boys are the same age.

(a)

(b)

Figure 5.9 Most cultures celebrate puberty, but cultures vary widely in the nature of the celebration. American Jewish youth celebrate a bar mitzvah (for boys) or bat mitzvah (for girls). The Samburu youth of Kenya celebrate a male circumcision ritual.

back. Generally acne is not severe enough to be a medical problem, although its psychological impact may be great. In order to avoid scarring, severe cases should be treated by a physician, the treatment typically being ultraviolet light, the drug Retin-A, and/or antibiotics. The drug Accutane is highly effective for severe cases. However, it must be used cautiously because it may have serious side effects, including birth defects if taken by a pregnant woman.

Gynecomastia (breast enlargement) may occur temporarily in boys, creating considerable embarrassment. About 80 percent of boys in puberty experience this growth, which is probably caused by small amounts of female sex hormones produced by the testes. Obesity may also be a temporary problem, although it is more frequent in girls than boys.

In various cultures around the world, puberty rites are performed to signify the adolescent's passage to adulthood. In the United States the only remaining vestiges of such ceremonies are the Jewish bar mitzvah for boys and bat mitzvah for girls and, in Christian churches, confirmation. In a sense, it is unfortunate that we do not give more formal recognition to puberty. Puberty rites probably serve an important psychological function in that they are a formal, public announcement of the fact that the boy or girl is passing through an important and difficult period of change. In the absence of such rituals, the young person may think that his or her body is doing strange things and may feel very much alone. This may be particularly problematic for boys, who lack an obvious sign of puberty like the first menstruation (the first ejaculation is probably the closest analogy) to help them identify the stage they are in.

SUMMARY

The major sex hormones are testosterone, which is produced in the male by the testes, and estrogen and progesterone, which are produced in the female by the ovaries. Levels of the sex hormones are regulated by two hormones secreted by the pituitary: FSH (follicle-stimulating hormone) and LH (luteinizing hormone). The gonads, pituitary, and hypothalamus regulate one another's output through a negative feedback loop. Inhibin regulates FSH levels.

At conception males and females differ only in the sex chromosomes (XX in females and XY in males). As the fetus grows, the TDF gene on the Y chromosome directs the gonads to differentiate into the testes. In the absence of the TDF gene, ovaries develop. Different hormones are then produced by the gonads, and these stimulate further differentiation of the internal and external reproductive structures of males and females. A male organ and a female organ that derive from the same embryonic tissue are said to be homologous to each other.

Intersex conditions are generally the result of various syndromes (such as CAH) and accidents that occur during the course of prenatal sexual differentiation. Currently there is a debate over the best medical treatment of these individuals. The Guevodoces provide an interesting case of gender change at puberty.

Puberty is initiated and characterized by a great increase in the production of sex hormones. Pubertal changes in both males and females include body growth, the development of pubic and axillary hair, and increased output from the oil-producing glands. Changes in the female include breast development and the beginning of menstruation. Changes in the male include growth of the penis and testes, the beginning of ejaculation, and a deepening of the voice.

REVIEW QUESTIONS

1. The _____ is the region of the brain that works together with the pituitary and gonads to regulate sex hormone levels.

2. FSH and LH are manufactured by the _____.

3. The hypothalamus produces GnRH. True or false?

4. In the female, during prenatal sexual differentiation, the Wolffian ducts degenerate, leaving the _____ ducts.

5. There are well-documented differences in humans between the male and the female hypothalamus. True or false?

6. The testes in the male are homologous to the _____ in the female.

7. CAH results in an intersex condition. True or false?

8. The Guevodoces of the Dominican Republic grow up in a four-gender society. True or false?

9. "Menarche" refers to a teenage girl who has stopped menstruating. True or false?

10. The percent body fat hypothesis has been used to explain the age at which a girl first menstruates. True or false?

(The answers to all review questions are at the end of the book, after the Glossary.)

QUESTIONS FOR THOUGHT, DISCUSSION, AND DEBATE

1. Of the physical changes of puberty, which are the most difficult to cope with?

2. The society in the Dominican Republic in which the Guevodoces are born (see p. 112) is a three-gender society, unlike the two-gender society of the dominant American culture. What would the United States be like if it was a three-gender society? Who would be classified in the third gender? Would their lives be better or worse as a result? Could we have a four-gender society? Who would be classified in the fourth gender? (For further information, see Herdt, 1990.)

3. Teresa has just given birth to her first baby. The doctor approaches her with a worried expression and says that her baby's genitals are unusual and some decisions will have to be made. The phallus is too big for a clitoris, but too small for a penis. What should Teresa do? What other information should she obtain from the doctor before making a decision?

SUGGESTIONS FOR FURTHER READING

Bancroft, John, and Reinisch, June M. (Eds.) (1990). *Adolescence and puberty.* New York: Oxford University Press. A collection of chapters presenting the most recent research on biological, psychological, and social aspects of puberty.

Fausto-Sterling, Anne (1993, March/April). The five sexes: why male and female are not enough. *The Sciences,* 20–24. Fausto-Sterling, a developmental geneticist, has written a provocative article in which she argues that our two-gender society is at odds with nature's creation of intersex individuals.

Kessler, Suzanne J. (1998). *Lessons from the intersexed.* New Brunswick, NJ: Rutgers University Press. Kessler, a psychologist, reports on her years of research with intersex individuals and the medical and psychological professionals who treat them, and she proposes new approaches in dealing with the condition.

Wilson, Jean D., and Foster, Daniel W. (1992). *Williams textbook of endocrinology.* 8th ed. Philadelphia: Saunders. An outstanding endocrinology text, with a particularly good chapter on sexual differentiation.

WEB RESOURCES

http://www.isna.org/
    The Intersex Society of North America Homepage; resources for intersexuals.

chapter
6

MENSTRUATION AND MENOPAUSE

CHAPTER HIGHLIGHTS

> I t [menstruation] makes me very much aware of the fact that I am a woman, and that's something very important to me. . . . It's also a link to other women. I actually enjoy having my period. I feel like I've been cleaned out inside.
>
> —M. P. W.

> Menstruation is a pain in the vagina.
>
> —S. B. P.

> Menopause, can I get through it without collapse? Men don't have that damned inconvenience and discomfort. God must have been a man—a woman would have done a better job on women's bodies.
>
> —B. B. W.

> Menopause, it's the best form of birth control. Face it graciously and brag about it. It's great.
>
> —S. F.*

Women's sexual and reproductive lives have a rhythm of changes much like that of the seasons. And, as is true of the seasons, there are some tangible signs that mark the shifts, the two most notable being menstruation and menopause. As the women quoted above testify, these events are not only biological but psychological as well, and the psychological responses to them may range from very positive to very negative. This chapter deals with the biology and psychology of menstruation and menopause, as well as with biological cycles in men, including a discussion of "male menopause."

Biology and the Menstrual Cycle

The menstrual cycle is regulated by fluctuating levels of sex hormones which produce certain changes in the ovaries and uterus. The hormone cycles are regulated by means of the negative feedback loops discussed in Chapter 5.

It is important to note that humans are nearly unique among species in having a menstrual cycle. Only a few other species of apes and monkeys also have menstrual cycles. All other species of mammals (for example, horses and dogs) have *estrous* cycles. There are several differences between estrous cycles and menstrual cycles. First, in animals that have estrous cycles, there is no menstruation; there is either no bleeding or only a slight spotting of blood (as in dogs), which is not a real menstruation. Second, the timing of ovulation in relation to bleeding

(if there is any) is different in the two cycles. For estrous animals, ovulation occurs while the animal is in "heat," or estrus, which is also the time of slight spotting. Dogs ovulate at about the time of bleeding. In the menstrual cycle, however, ovulation occurs about midway between the periods of menstruation. A third difference is that female animals with estrous cycles engage in sexual behavior only when they are in heat, that is, during the estrus phase of the cycle. Females with menstrual cycles are capable of engaging in sexual behavior throughout the cycle. It is important to note these differences because some people mistakenly believe that women's cycles are like those of a dog or cat, when in fact the cycles are quite different.

The Phases of the Menstrual Cycle

The menstrual cycle has four phases, each characterized by a set of hormonal, ovarian, and uterine changes (see Figure 6.1). Because menstruation is the easiest phase to identify, it is tempting to call it the first phase; biologically, though, it is actually the last phase (although in numbering the days of the menstrual cycle, day 1 is counted as the first day of menstruation because it is the most identifiable day of the cycle).

Hormones and What Happens in the Ovaries

The first phase of the menstrual cycle is called the **follicular phase** (it is sometimes called the

*Paula Weideger. (1976). *Menstruation and menopause.* New York: Knopf, pp. 4–5.

> **Follicular phase (fuh-LIK-you-lur):** The first phase of the menstrual cycle, beginning just after menstruation, during which an egg matures in preparation for ovulation.

Figure 6.1   The biological events of the menstrual cycle.

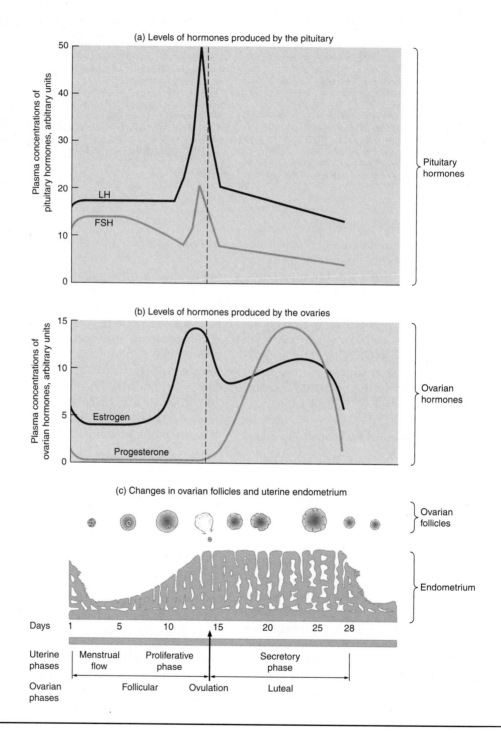

(a) Levels of hormones produced by the pituitary

Pituitary hormones

(b) Levels of hormones produced by the ovaries

Ovarian hormones

(c) Changes in ovarian follicles and uterine endometrium

Ovarian follicles

Endometrium

proliferative phase or the *preovulatory phase*). At the beginning of this phase, the pituitary secretes relatively high levels of FSH (follicle-stimulating hormone). As the name of this hormone implies, its function is to stimulate follicles in the ovaries. At the beginning of the follicular phase, it signals one follicle (occasionally more than one) in the ovaries to begin to ripen and bring an egg to maturity. At the same time, the follicle secretes estrogen.

The second phase of the cycle is **ovulation,** which is the phase during which the follicle ruptures open, releasing the ripened egg (see Figure 6.2). By this time, estrogen has risen to a high level, which inhibits FSH production, and so FSH has fallen back to a low level. The high levels of estrogen also stimulate the hypothalamus to produce GnRH, which causes the pituitary to begin production of LH (luteinizing hormone).[1] A surge of LH and FSH causes ovulation.

The third phase of the cycle is called the **luteal phase** (sometimes also called the *secretory phase* or the *postovulatory phase*). After releasing an egg, the follicle, under stimulation of LH, turns into a glandular mass of cells called the **corpus luteum**[2] (hence the names "luteal phase" and "luteinizing hormone"). The corpus luteum manufactures progesterone; thus progesterone levels rise during the luteal phase. But high levels of progesterone also inhibit the pituitary's secretion of LH, and as LH levels decline, the corpus luteum degenerates. Thus the corpus luteum's output leads to its own eventual destruction. With this degeneration comes a sharp decline in estrogen and progesterone levels at the end of the luteal phase. The falling levels of estrogen stimulate the pituitary to begin production of FSH, and the whole cycle begins again.

The fourth and final phase of the cycle is **menstruation.** Physiologically, menstruation is a shedding of the inner lining of the uterus (the en-

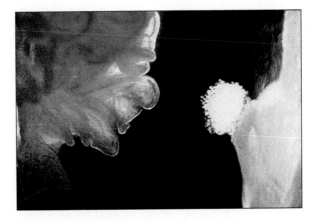

Figure 6.2 Ovulation, showing the egg bursting forth from the wall of the ovary.

dometrium), which then passes out through the cervix and the vagina. During this phase, estrogen and progesterone levels are low, and FSH levels are rising. Menstruation is triggered by the sharp decline in estrogen and progesterone levels at the end of the luteal phase.

What Happens in the Uterus

This brings us to the changes that have been occurring in the uterus while the ovaries and endocrine system were going through the four phases described above. During the first, or follicular, phase, the high levels of estrogen stimulate the endometrium (the inner layer) of the uterus to grow, thicken, and form glands that will eventually secrete substances to nourish the embryo; that is, the endometrium proliferates (hence the alternative name for this first phase, the "proliferative phase"). During the luteal phase, the progesterone secreted by the corpus luteum stimulates the glands of the endometrium to start secreting the nourishing substances (hence the name "secretory phase"). If the egg is fertilized and the

---

[1]This statement may seem to contradict one made in Chapter 5, that high estrogen levels cause a decline in LH. Both of these effects occur, but at different times in the menstrual cycle (Molitch, 1995). There are two centers in the hypothalamus, one of which produces a negative feedback loop between estrogen and LH, the other of which produces a positive feedback loop between the two.

[2]*Corpus luteum* is Latin for "yellow body." The corpus luteum is so named because the mass of cells is yellowish in appearance.

Ovulation: Release of an egg from the ovaries; the second phase of the menstrual cycle.
Luteal phase (LOO-tee-uhl): The third phase of the menstrual cycle, following ovulation.
Corpus luteum: The mass of cells of the follicle remaining after ovulation; it secretes progesterone.
Menstruation: The fourth phase of the menstrual cycle, during which the endometrium of the uterus is sloughed off in the menstrual discharge.

timing goes properly, about six days after ovulation the fertilized egg arrives in a uterus that is well prepared to cradle and nourish it.

The corpus luteum will continue to produce estrogen and progesterone for about 10 to 12 days. If pregnancy has not occurred, its hormone output declines sharply at the end of this period. The uterine lining thus cannot be maintained, and it is shed, resulting in menstruation. Immediately afterward, a new lining starts forming in the next proliferative phase.

The menstrual fluid itself is a combination of blood (from the endometrium), degenerated cells, and mucus from the cervix and vagina. Normally the discharge for an entire period is only about 2 ounces (4 tablespoons). Most commonly the fluid is absorbed with sanitary napkins, which are worn externally, or tampons, which are worn inside the vagina.

Toxic shock syndrome, sometimes abbreviated TSS, is caused by the bacterium *Staphylococcus aureus.* In 1980 a disturbing discovery was made that TSS was associated with tampon use (Price, 1981). It was particularly associated with Rely brand tampons, and they were withdrawn from the market. Tampon use seems to encourage an abnormal growth of the bacteria. Symptoms of toxic shock syndrome include high fever (102°F or greater) accompanied by vomiting or diarrhea; any woman who experiences these symptoms during her period should discontinue tampon use immediately and see a doctor. Toxic shock syndrome leads to death in approximately 10 percent of cases. It is now recommended that women change tampons frequently, at least every six to eight hours during their periods (although the effectiveness of this is debated), that they not use tampons continuously throughout a menstrual period, and that they minimize use of super-absorbent tampons.

The number of cases of TSS fell from 319 in 1983 to 75 in 1987, in a major public health victory, so that the disease is now rare (Golub, 1992). The success has been a result of public education, as well as a change in the chemicals used in tampon manufacture.

Length and Timing of the Cycle

How long is a normal menstrual cycle? Generally anywhere from 20 to 36 or 40 days is considered within the normal range. The average is about 28 days, but somehow this number has taken on more significance than it deserves. There is enormous variation from one woman to the next in the average length of the cycle, and for a given woman there can be considerable variation in length from one cycle to the next.

What is the timing of the various phases of the cycle? In a perfectly regular 28-day cycle, menstruation begins on day 1 and continues until about day 4 or 5. The follicular phase extends from about day 5 to about day 13. Ovulation occurs on day 14, and the luteal phase extends from day 15 to the end of the cycle, day 28 (see Figure 6.1). But what if the cycle is not one of those perfect 28-day ones? In cycles that are shorter or longer than 28 days, the principle is that the length of the *luteal* phase is relatively constant. That is, the time from ovulation to menstruation is always 14 days, give or take only a day or two. It is the follicular phase which is of variable length. Thus, for example, if a woman has a 44-day cycle, she ovulates on about day 30. If she has a 22-day cycle, she ovulates on about day 8.

Some women report that they can actually feel themselves ovulate, a phenomenon called *Mittelschmerz* ("middle pain"). The sensation described is a cramping on one or both sides of the lower abdomen, lasting for about a day, and it is sometimes confused with appendicitis.

It is also true that ovulation does not occur in every menstrual cycle. That is, menstruation may take place without ovulation. When this happens the woman is said to have an *anovulatory cycle.* Such cycles occur once or twice a year in women in their twenties and thirties and are fairly common among girls during puberty and among women during menopause.

Other Cyclic Changes

Two other physiological processes that fluctuate with the menstrual cycle deserve mention: the cervical mucus cycle and the basal body temperature cycle. The cervix contains glands that secrete mucus throughout the menstrual cycle. One function of the mucus is to protect the entrance to the cervix, helping to keep bacteria out. These glands respond to

---

Toxic shock syndrome: A sometimes fatal bacterial infection associated with tampon use during menstruation.

the changing levels of estrogen during the cycle. As estrogen increases at the start of a new cycle, the mucus is alkaline, thick, and viscous. When LH production begins, just before ovulation, the cervical mucus changes markedly. It becomes even more alkaline, thin, and watery. Thus the environment for sperm passage is most hospitable just at ovulation. After ovulation, the mucus returns to its former viscous, less alkaline state. If a sample of mucus is taken just before ovulation and is allowed to dry, the dried mucus takes on a fern-shaped pattern. After ovulation, during the luteal phase, the fernlike patterning will not occur. Thus the "fern test" is one method for detecting ovulation.

A woman's *basal body temperature,* taken with a thermometer, also fluctuates with the phases of the menstrual cycle. Basically, the pattern is the following: The temperature is low during the follicular phase and may take a dip on the day of ovulation; on the day after ovulation it rises noticeably, generally by 0.4°F or more, and then continues at the higher level for the rest of the cycle (Figure 6.3). Progesterone raises body temperature, so the higher temperature during the luteal phase is due to the increased production of progesterone during that time. As the saying goes, "Where there's progesterone, there's heat." This change in basal body temperature is important when a couple are using the rhythm method of birth control (Chapter 8) and when a woman is trying to determine the time of ovulation so that she may become pregnant (Chapter 7).

Menstrual Problems

The most common menstrual problem is **dysmenorrhea** (painful menstruation). Almost every woman experiences at least some menstrual discomfort at various times in her life, but the frequency and severity of the discomfort vary considerably from one woman to the next. Cramping pains in the pelvic region are the most common symptom, and other symptoms may include headaches, backaches, nausea, and a feeling of pressure and bloating in the pelvis.

Although the exact causes of dysmenorrhea are unknown, the current leading hypothesis involves **prostaglandins,** hormonelike substances produced by many tissues of the body, including the lining of the uterus (Golub, 1992). Prostaglandins can cause

> **Dysmenorrhea (dis-men-oh-REE-uh):** Painful menstruation.
> **Prostaglandins:** Chemicals secreted by the uterus that cause the uterine muscles to contract; they are a likely cause of painful menstruation.

Figure 6.3   A basal body temperature graph. Note the dip in temperature, indicating ovulation on day 14.

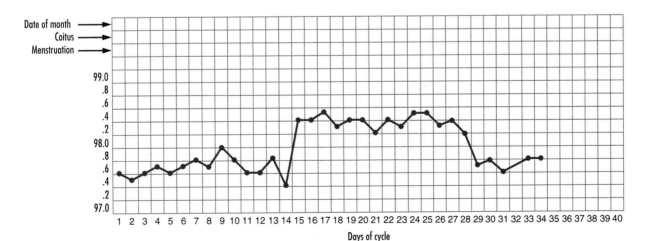

smooth muscle to contract and can affect the size of blood vessels. Women with severe menstrual pain have unusually high levels of prostaglandins. The high levels cause intense uterine contractions, which in turn choke off some of the uterus's supply of oxygen-carrying blood. Prostaglandins may also cause greater sensitivity in nerve endings. The combination of the uterine contractions, lack of oxygen, and heightened nerve sensitivity produces menstrual cramps.

Household remedies for painful menstruation are available and may be helpful to some women. Aspirin appears to be the best, and cheapest, painkiller available, and it can help to relieve menstrual pain. A somewhat more provocative remedy suggested by, among others, Masters and Johnson is masturbation. This makes good physiological sense because part of the discomfort of menstruation—the pressure and bloating—results from pelvic edema (a congestion of fluids in the pelvic region). During sexual arousal and orgasm, pelvic congestion increases. After orgasm, the congestion dissipates (see Chapter 9). Thus orgasm, whether produced by masturbation or some other means, should help to relieve the pelvic edema which causes menstrual discomfort. And it's a lot more fun than taking medicine!

Mefenamic acid (an antiprostaglandin drug) is a powerful and effective drug for use in the treatment of menstrual pain. The drug is sold with brand names such as Naprosyn and Anaprox. About 80 to 85 percent of women who take this drug report significant relief from menstrual pain and symptoms such as nausea, vomiting, dizziness, and weakness (Golub, 1992). Interestingly, aspirin is also an antiprostaglandin.

Dietary changes and aerobic exercise may also be helpful (Golub, 1992; Hatcher et al., 1998). Caffeine should be avoided; a diet high in carbohydrates and low in protein, during the luteal phase, is helpful for some.

A menstrual problem which may be mistaken for dysmenorrhea is **endometriosis.** As noted previously the endometrium is the lining of the uterus; it grows during each menstrual cycle and is sloughed off in menstruation. Endometriosis occurs when the endometrium grows in a place other than the uterus. The place may be the ovaries, fallopian tubes, rectum, bladder, vagina, vulva, cervix, or lymph glands. The symptoms vary, depending on the location of the growth, but very painful periods that last an unusually long time are the most common. En-

dometriosis is fairly serious and should be treated by a physician; if left untreated, it may lead to sterility. Hormones are generally used in treatment, but if the problem is severe, surgery may be required. Laser surgery is a new, experimental treatment.

Another menstrual problem is **amenorrhea,** or the absence of menstruation. It is called *primary amenorrhea* if the girl has not yet menstruated by about age 18. It is called *secondary amenorrhea* if she has had at least one period. Amenorrhea has received considerable attention from physicians because, while rare, it is a symptom of infertility. Some of the causes of amenorrhea include pregnancy, congenital defects of the reproductive system, hormonal imbalance, cysts or tumors, disease, stress, and emotional factors related to puberty. Amenorrhea resulting from programs of strenuous exercise and from anorexia was discussed in Chapter 5.

Psychological Aspects of the Menstrual Cycle

"Why do I get so emotional?" screams the ad for Midol PMS in *Teen* magazine. It is part of the folk wisdom of our culture that women experience fluctuations in mood over the phases of the menstrual cycle. In particular, women are supposed to be especially cranky and depressed just before and during their periods. In France, if a woman commits a crime during her premenstrual phase, she may use the fact in her defense, claiming "temporary impairment of sanity."

What is the scientific evidence concerning the occurrence of such fluctuations in mood, and, if they do occur, what causes them?

Fluctuations in Mood: Do Women Become Extra Emotional?

In 1931, the name "premenstrual tension" was coined for the mood changes that may occur during the three or four days immediately preceding

---

Endometriosis: A condition in which the endometrium grows abnormally outside the uterus; the symptom is unusually painful periods with excessive bleeding.
Amenorrhea: The absence of menstruation.

menstruation (about days 24 to 28 of the cycle). Symptoms of premenstrual tension include depression, anxiety, irritability, fatigue, headaches, and low self-esteem. The term **premenstrual syndrome (PMS)** is used to refer to those cases in which the woman has a particularly severe combination of physical and psychological symptoms premenstrually; these symptoms may include tension, depression, irritability, backache, and water retention (Dalton, 1979). In the last several decades a great many data have been collected on moods during the premenstrual period and on whether moods fluctuate during the cycle.

Research based on women's daily self-reports throughout the cycle generally finds positive moods around the time of ovulation (that is, mid-cycle), and various symptoms, such as anxiety, irritability, depression, fatigue, and headaches, premenstrually (Golub, 1992; Parlee, 1973). However, the fluctuations are not large, on average. In one study, the average depression score of women was 6.84 around ovulation and 9.30 premenstrually, compared with a mean of 16.03 for depressed psychiatric patients (Golub, 1992). Premenstrual women are not ready for the psychiatric ward!

The evidence supporting mood fluctuation and the premenstrual syndrome has not been accepted without challenge. Of the numerous criticisms that have been made (Parlee, 1973), three deserve special mention here. First, much of the evidence depends on subjective reports of mood and symptoms, which are probably not very reliable. Second, the research has not given sufficient consideration to coping mechanisms; most women do not dissolve into tears and confine themselves to bed for the eight premenstrual and menstrual days of each month. Women develop mechanisms for coping with the symptoms. Third, the interpretation of the direction of the differences might be questioned. The typical interpretation is that women show a psychological "deficit" premenstrually, as compared with the "normal" state at ovulation and during the rest of the cycle. However, the opposite interpretation might also be made: that women are "normal" premenstrually and experience unusual well-being psychologically at mid-cycle. What defines "normal" or "average" mood, then? Men's moods? Support for a reinterpretation might come from the statistics on violent crimes committed by women. While it may be true that women are

somewhat more likely to commit crimes during the eight premenstrual and menstrual days, even during this period they are far less likely to commit crimes than men are. Thus women might be considered to experience "normal" or typical moods (comparable to men's moods) during the premenstrual and menstrual days and to have feelings of unusual well-being around ovulation.

Taking into account the criticisms and available evidence, it seems reasonable to conclude the following:

1. Women do, on the average, experience some fluctuations in mood over the phases of the menstrual cycle.

2. Present evidence does not clearly indicate how the direction of the shifts should be interpreted—whether women are unusually "low" premenstrually or unusually "high" around the time of ovulation.

3. There is a great deal of variation from one woman to the next in the size of these shifts and the way they are expressed. Some women experience no shifts or shifts so slight that they are not noticeable, while others may experience large shifts. It would be interesting to know how many do show mood fluctuations and how many do not. Unfortunately, the studies that have tried to provide this information themselves show substantial variation; the percentage reporting symptoms ranges between 20 percent and 95 percent in various studies (Golub, 1992). It is important to make a distinction between women who have full-blown PMS and women who experience no fluctuation or only moderate fluctuations in mood over the cycle.

Fluctuations in Performance: Can a Woman Be President?

So far our discussion has concentrated on fluctuations in psychological characteristics such as depression, anxiety, and low self-confidence. However,

Premenstrual syndrome (PMS): A combination of severe physical and psychological symptoms, such as depression and irritability, occurring just before menstruation.

in some situations performance is of more practical importance than mood. For example, is a woman secretary's clerical work less accurate premenstrually and menstrually? Is a female athlete's coordination or speed impaired during the premenstrual-menstrual period?

Research on performance—such as intellectual performance or athletic performance—generally shows no fluctuations over the cycle. Research has found no fluctuations in academic performance, problem solving, memory, or creative thinking (Golub, 1992).

In one study, 31 percent of female athletes said they believed that they experienced a decline in performance during the premenstrual or menstrual phases; yet when their actual performance was measured, they showed no deficits in strength (weight lifters) or swimming speed (swimming team members) (Quadagno et al., 1991). Thus there is no reliable evidence indicating that the kinds of performance required in a work situation or an athletic competition fluctuate over the menstrual cycle.

Fluctuations in Sex Drive

Another psychological characteristic that has been investigated for fluctuations over the cycle is women's sex drive or arousability. Observations of female animals that have estrous cycles indicate that sexual behavior depends a great deal on cycle phase and the corresponding hormonal state. Females of these species engage in sexual behavior enthusiastically when they are in the estrus, or "heat," phase of the cycle and do not engage in sexual behavior at all during any other phase. This makes good biological sense, since the females engage in sex precisely when they are fertile.

Human females, of course, engage in sexual behavior throughout the menstrual cycle. But might some subtle cycling in drive remain, expressed, perhaps, in fluctuations in frequency of intercourse? Studies have yielded contradictory results. Some have found a peak frequency of intercourse around ovulation, which would be biologically functional. But others have found peaks just before and just after menstruation (reviewed by Zillmann et al., 1994).

Moreover, one should be cautious about using frequency of intercourse as a measure of a woman's sex drive. Intercourse requires some agreement between partners, and thus reflects not only her desires but her partner's as well. One study investigated autosexual activity (masturbation, fantasy, and so on) in addition to intercourse and found that the frequency of autosexual activity actually increased during menstruation, while the frequency of intercourse decreased (Gold & Adams, 1981). Two other studies examined women's sexual arousability to erotic films at the different phases of the cycle, measuring both physiological arousal and self-ratings of arousal (Morrell et al., 1984; Hoon et al., 1982). Both studies concluded that there were no differences in arousability at the different phases.

In the most sophisticated study to date, women kept daily diaries of their moods and sexual interest and provided blood samples for hormone assays (Van Goozen et al., 1997). This study assessed testosterone levels, as well as levels of estrogen and progesterone, an important addition because the evidence is strongest for an association between testosterone and sex drive in women. The results indicated that testosterone levels peaked at ovulation. The women seemed to fall into two subgroups, with different patterns of sexual interest. About half the women reported that they suffered from premenstrual symptoms; their sexual interest peaked at ovulation, exactly when testosterone levels are high. The other half of the women said they did not suffer from premenstrual symptoms and their peak in sexual interest occurred premenstrually, perhaps in anticipation of a deprivation during menstruation.

If there is a link between phase of the menstrual cycle and sexual interest, it most likely will reflect an association between testosterone levels and sexuality. But with humans, psychological and social factors—such as some couples' dislike of intercourse when the woman is menstruating—will play a strong role as well.

What Causes the Fluctuations in Mood: Why Do Some Women Get Emotional?

The answer to the question of what causes mood fluctuations during the menstrual cycle touches off a nature-nurture, or biology-environment, controversy. That is, some investigators argue that the mood fluctuations are caused primarily by biological factors—in particular, fluctuations in levels of hormones—whereas others argue that environmental factors such as menstrual taboos and cultural expectations are the primary cause.

The Biological Explanation

On the biology side, changes in mood appear to be related to changes in hormone levels during the cycle. The fact that depression is more frequent in women premenstrually, at menopause, and postpartum (after having a baby) suggests that there is at least some relationship between sex hormones and depression. The exact hormone-mood relationship is not known, though. Theories of hormonal causes of premenstrual tension involving the following factors have been proposed: (1) absolute amount of estrogen; (2) absolute amount of progesterone; (3) the estrogen-progesterone ratio, or balance; (4) hypersensitivity of some individuals to estrogen levels; and (5) withdrawal reactions to either estrogen or progesterone (during all the premenstrual, postpartum, and menopausal periods, hormone levels drop rapidly). Research has not determined which, if any, of these factors is the real cause. Neither is it known exactly what the mechanism is by which hormones influence mood (Golub, 1992). Research does indicate that the estrogen-progesterone system interacts with the production of the neurotransmitters norepinephrine, serotonin, and dopamine, and neurotransmitter levels are linked to mood disorders such as depression (Halbreich, 1996; Mortola, 1998).

Critics of the hormone point of view note that causality is being inferred from correlational data. That is, the data show a correlation between cycle phase (hormone levels) and mood, but it is unwarranted to infer that the hormone levels cause the mood shifts.

A classic study that partially answers this objection was done by psychologist Karen Paige (1971). She used a manipulation of hormone levels—birth control pills—and studied mood fluctuations resulting from the manipulation. The spoken stories of 102 married women were obtained on days 4, 10, and 16 and 2 days before menstruation during one cycle, and they were scored for the amount of anxiety shown in them. The subjects fell into three groups: (1) those who were not taking oral contraceptives and never had; (2) those who were taking a combination pill (combination pills provide a steady high dose of both estrogen and progestin, a synthetic progesterone, for 20 or 21 days—see Chapter 8 for a more complete discussion); and (3) those who were taking sequential pills (which provide 15 days of estrogen followed by 5 days of estrogen plus progestin, which is a fluctuation similar to the natural cycle but at higher levels). The hormone levels across the cycles of the women in these three groups are shown in Figure 6.4. Paige found that the nonpill women experienced statistically significant variation in their anxiety and hostility levels over the menstrual cycle, which was in agreement with findings from previous studies. Women taking the sequential pill showed the same mood change that nonpill women did. This agrees with the predicted outcome, since their artificial hormone cycle is similar to the natural one. Most important, combination-pill women, whose hormone levels are constant over the cycle, showed no mood shifts over the cycle; their hostility and anxiety levels were constant. This study therefore provides evidence that fluctuations in hormone levels over the menstrual cycle cause mood fluctuations and that when hormone levels are constant, mood is constant.

The Cultural Explanation

Those arguing the other side—that the fluctuations are due to cultural forces—note the widespread cultural expectations and taboos surrounding menstruation (for reviews, see Golub, 1992; Weideger, 1976; also see Focus 6.1). In some nonindustrialized cultures, women who are menstruating are isolated from the community and may have to stay in a menstrual hut at the edge of town during their period. Often the menstrual blood itself is thought to have supernatural, dangerous powers, and the woman's isolation is considered necessary for the safety of the community. Among the Lele of the Congo, for example:

> A menstruating woman was a danger to the whole community if she entered the forest. Not only was her menstruation certain to wreck any enterprise in the forest that she might undertake, but it was thought to produce unfavorable conditions for men. Hunting would be difficult for a long time after, and rituals based on forest plants would have no efficacy. Women found these rules extremely irksome, especially as they were regularly shorthanded and later in their planting, weeding, harvesting, and fishing. (Douglas, 1970, p. 179)

Lest one think that such practices occur only among non-Western people, it should be noted

Figure 6.4   Hormone levels over the menstrual cycle for the three groups of women in Paige's study (see text for further explanation).

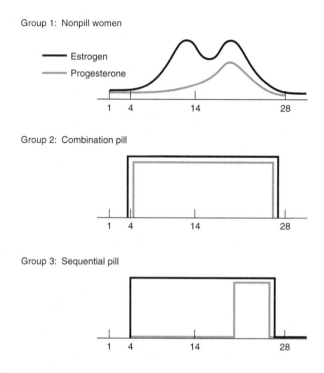

that there is a history of similar practices in our own culture as well. For example, the following passage is from the book of Leviticus in the Bible:

> When a woman has a discharge of blood which is her regular discharge from her body, she shall be in her impurity for seven days, and whoever touches her shall be unclean until the evening. ... And whoever touches her bed shall wash his clothes, and bathe himself in water, and be unclean until the evening; whether it is the bed or anything upon which she sits, when he touches it he shall be unclean until the evening. (Leviticus 15:19–23)

Among the most common menstrual taboos are those prohibiting sexual intercourse with a menstruating woman. For example, the continuation of the passage from Leviticus quoted above is:

> And if any man lies with her, and her impurity is on him, he shall be unclean seven days; and every bed on which he lies shall be unclean. (Leviticus 15:24)

Couples who violated the taboo could be stoned. To this day, Orthodox Jews abstain from sex during the woman's period and for seven days afterward. At the end of this time the woman goes to the mikvah (ritual bath) to be cleansed, and only after this cleansing may she resume sexual relations.

Advocates of the cultural explanation argue, then, that women become anxious and depressed around the time of menstruation because of the many cultural forces, such as menstrual taboos, that create negative attitudes toward menstruation. Furthermore, women's expectations may play a role. It is a well-documented phenomenon that people's expectations can influence their behavior. Our culture is filled with teachings that women are supposed to behave strangely just before and during their periods—for example, the drug company ads that ask "Why am I so emotional?" Thus, according to this line of reasoning, women are taught that they should be depressed around the time of

Focus 6.1

The Menstrual Experience Among Indians in South Africa

Anthropologist Brian du Toit studied women who resided in South Africa but whose ancestors came from India. Most Indian South Africans are Hindu or Muslim. A total of 30 premenopausal and 30 postmenopausal women were interviewed.

Approximately 75 percent of the women reported experiencing the premenstrual syndrome (PMS), although they did not use that term. They complained of a heaviness, a bloated feeling, irritability, general crabbiness, or tension associated with headaches, symptoms that are quite similar to those reported by women in the United States.

Despite these symptoms, menstruation was generally viewed by the women in the study as a cleansing process that leaves the body healthy. "Old blood" and impurities are flushed out. The participants saw advantages to menstruation: it means that a woman can have children, it is an indicator of youthfulness, and it has a cleansing function.

When asked why women menstruate, most said that it is somehow related to cleansing the system. Others thought it is the body's preparation for pregnancy. A few believed that it is God's doing—that it is simply the way He made women.

There are various menstrual taboos in this culture. Menstruating women are not supposed to engage in certain religious rituals, although they may engage in kitchen work such as food preparation (which is prohibited in some other cultures). There was unanimous agreement among the women that sexual intercourse should not occur during the flow, but the older women and the younger women had different reasons for this belief. The older women believed that it was dangerous to the man

or that a sexually transmitted disease could be contracted through such a practice. The younger women simply thought that it was messy.

Among the postmenopausal women, the cessation of menstruation was greeted uniformly by the same emotion: relief. Interestingly, this is exactly the same attitude that researchers have found among American women. The Indian women in South Africa gave several reasons for their feelings of relief. For some, menopause brought an end to unpredictable periods of heavy bleeding; for others, it removed the danger of pregnancy; and for others, it was proof to their husbands that they were now old and should not be expected to engage in sexual intercourse.

The women gave various explanations of why menopause occurs. Some thought that it was because Allah does not want Muslims to become pregnant at that age. Some believed that women produce less and less blood as they age, and finally have no more to shed.

This study is interesting because it shows us both cross-cultural similarities (for example, PMS symptoms and a feeling of relief at menopause) and cross-cultural differences (for example, different explanations for the existence of menstruation) between the Indian culture of South Africa and the dominant American culture.

Source: du Toit, Brian M. (1994). Menstruation: Attitudes and experience of Indian South Africans. In B. M. du Toit (Ed.), *Human sexuality: Cross-cultural readings.* 3d ed. (pp. 11–25). New York: McGraw-Hill.

Figure 6.5   PMS. The media, from drug ads to teen magazine articles, have created negative stereotypes about women and PMS.

menstruation, and because they expect to become depressed, they do become depressed. Psychologist Elaine Blechman (1988) has proposed an additional role for expectations. Women who experience painful cramps may well spend the several days before they know their period is due anticipating the pain, which makes them feel anxious and depressed.

Surely such forces do exist in our culture. But is there any evidence that they really do have an ef-

fect on women's moods and behavior? Psychologist Diane Ruble (1977; see also Klebanov & Jemmott, 1992) did a clever experiment to determine whether women's culturally induced expectations influence their reporting of premenstrual symptoms. College students were tested on the sixth or seventh day before the onset of their next menstrual period. They were told that they would participate in a study on a new technique for predicting the expected date of menstruation using an electroencephalogram (EEG), a method that had already been successfully tested with older women. After the EEG had been run (it actually wasn't), the woman was informed of when her next period was to occur, depending on which of three experimental groups she had randomly been assigned to: (1) the woman was told she was "premenstrual" and her period was due in 1 or 2 days; (2) the woman was told she was "intermenstrual" or "mid-cycle" and her period was not expected for at least a week to 10 days; or (3) she was given no information at all about the predicted date of menstruation (control group). The women then completed a self-report menstrual distress questionnaire. The results indicated that women who had been led to believe they were in the premenstrual phase reported significantly more water retention, pain, and changes in eating habits than did women who had been led to believe they were around mid-cycle. (In fact, women in these groups did not differ significantly in when their periods actually arrived.) There were no significant differences between the groups in ratings of negative moods, however. This study indicates that probably because of learned beliefs, women overstate the changes in body states that occur over the menstrual cycle. When they think they are in the premenstrual phase, they report more problems than when they think they are at mid-cycle.

This nature-nurture argument will not be easily resolved, particularly because there is evidence for both points of view. Perhaps the best solution is to say that some women probably do experience mood shifts caused by hormonal and possibly other physical factors and that for many others, slight biological influences are magnified by psychological and cultural influences. A woman's premenstrual hormonal state may act as a sort of "trigger." It may, for example, provide a state conducive to depression; if the environment provides further stimuli to depression, the woman becomes depressed.

Cycles in Men

The traditional assumption, of both laypeople and scientists, has been that monthly biological and psychological cycles are the exclusive property of women. The corollary assumption has been that men experience no monthly cycles. These assumptions are made, at least in part, because men have no obvious signs like menstruation to call attention to the fact that some kind of periodic change is occurring.

Do men experience subtle monthly hormone cycles and corresponding psychological cycles? An interesting applied study found some evidence of behavioral cycles in men (cited in Ramey, 1972). The Omi Railway Company of Japan operated more than 700 buses and taxis. In 1969 the directors became concerned about the high losses resulting from accidents. The company's efficiency experts studied each man working for the company to determine his monthly cycle pattern of mood and efficiency. Then schedules were adjusted to coincide with the best times for the drivers. The accident rate dropped by one-third.

In another study, the testosterone levels of 20 men were studied for a period of 60 days (Doering et al., 1975). The majority of the men had identifiable testosterone cycles ranging in length from 3 to 30 days, with many clustered around 21 to 23 days. A psychological test to measure the men's moods was also administered (Doering et al., 1978). Although testosterone levels were not correlated with several kinds of moods such as anxiety, high testosterone levels were correlated with depression. Other researchers have also found cycles in men's emotional states (Parlee, 1978). One study, in fact, found no differences between men and women in day-to-day mood changes—men were no more nor less changeable than women (McFarlane et al., 1988; see also McFarlane & Williams, 1994). Undoubtedly there will be increased research on this topic in the next decade.

Menopause

Biological Changes

The *climacteric* is a period lasting about 15 or 20 years (from about ages 45 to 60) during which a woman's body makes the transition from being able to reproduce to not being able to reproduce; the climacteric is marked particularly by a decline in the functioning of the ovaries. But climacteric changes occur in many other body tissues and systems as well. **Menopause** (the "change of life," the "change") refers to one specific event in this process, the cessation of menstruation; this occurs, on the average, over a two-year period beginning at around age 50 (with a normal menopause occurring anywhere between the ages of 40 and 60).

Biologically, as a woman grows older, the pituitary continues a normal output of FSH and LH; however, as the ovaries age, they become less able to respond to the pituitary hormones. In addition, the brain—including the hypothalamus-pituitary unit—ages (Lamberts et al., 1997; Wise et al., 1996). With the aging of the ovaries, there is an accompanying decline in the output of their two major products: eggs and the sex hormones estrogen and progesterone. More specifically, the ovaries become less capable of responding to FSH by maturing and releasing an egg. The hormonal changes of menopause involve a decline in estrogen and progesterone levels and hormonal imbalance.

Physical symptoms of menopause may include "hot flashes" or "hot flushes," headaches, dizziness, heart palpitations, and pains in the joints. For some women, a long-range effect of the decline in estrogen levels is *osteoporosis* (porous and brittle bones). The hot flash is probably the best known of the symptoms. Typically it is described as a sudden wave of heat from the waist up. The woman may get red and perspire a lot; when the flush goes away, she may feel chilled and sometimes may shiver. The flashes may last from a few seconds to half an hour and may occur several or many times a day. They may also occur at night, causing insomnia, and the resulting perspiration can actually soak the sheets.

Do all women experience these menopausal symptoms? The Massachusetts Women's Health Study (Avis & McKinlay, 1995; McKinlay et al., 1992) followed a large sample of middle-aged women for several years, beginning when they were premenopausal and continuing through menopause to the postmenopausal period. The peak in reporting of hot flashes and night sweats occurred just before the onset of menopause, with 50 percent of women reporting them. About one-quarter of the sample (23 percent) did not report a hot flash at any of the interview times. Furthermore, 69 percent of the women reported not being bothered by the hot flashes and night sweats. So, about one-quarter of women do not experience hot flashes, and 50 to 75 percent of women do, but the majority report that they are not bothered by them.

Hormone-replacement therapy (HRT; also called *estrogen-replacement therapy* or ERT) is available and may be helpful to many menopausal women, particularly for relief of physical discomfort such as hot flashes and some of the sexual problems such as lack of vaginal lubrication. The treatment today often includes both estrogen and progesterone. The therapy is somewhat controversial, and its benefits need to be weighed against its potential risks. HRT conveys several substantial benefits. It protects women from cardiovascular disease such as heart attack and stroke (Ross & Stevenson, 1993) and from osteoporosis (which may cause a broken hip leading to death in an elderly woman). However, HRT increases the risk of breast cancer and endometrial cancer (Rosenberg, 1993; Steinberg et al., 1991). On balance, the benefits appear to outweigh the costs; in one study, women using ERT or HRT had a lower rate of mortality than women who did not (Grodstein et al., 1997). A woman's risk of death between the ages of 50 and 94 is 31 percent from coronary heart disease and only 2.8 percent for breast cancer and 2.8 percent from hip fractures (Brinton & Schairer, 1997). Therefore, HRT protects women against the more serious risk.

The latest development in HRT is the so-called "designer estrogens" or "tissue-specific estrogens" (Fuleihan, 1997). They are intended to provide the beneficial effects of traditional HRT without increasing the risk of breast and endometrial cancer. It is too soon to know their long-term effects on coronary heart disease, osteoporosis, and cancer. Nonetheless, a short-term, two-year study of raloxifene (one of these drugs) indicated that it increased bone density and lowered total cholesterol levels, yet it did not stimulate the endometrium—precisely the intended

---

Menopause: The cessation of menstruation in middle age.

effects (Delmas et al., 1997). These drugs hold much promise for the future.

Sexuality and Menopause

During the climacteric, physical changes occur in the vagina. The lack of estrogen causes the vagina to become less acidic, which leaves it more vulnerable to infections. Estrogen is also responsible for maintaining the mucous membranes of the vaginal walls. With a decline in estrogen, there is a decline in vaginal lubrication during arousal, and the vaginal walls become less elastic. Either or both of these may make intercourse painful for the woman. Several remedies are available, including estrogen-replacement therapy and the use of artificial lubricants. Unfortunately, some women do not communicate their discomforts to their partners or physician and instead suffer quietly and develop an aversion to sex. On the other hand, some women report that intercourse is even better after menopause, when the fear of pregnancy no longer inhibits them.

Experts reviewing the research on women's sexuality during and after menopause have reached the following conclusions (McCoy, 1996, 1997; Morokoff, 1988): (1) The majority of women continue to engage in sexual activity and enjoy it both during and after menopause. (2) There is some decline in sexual functioning, on the average, during menopause and particularly after the last period. (3) Estrogen is related to the decline in sexual functioning, in part because low estrogen levels cause vaginal dryness. There is some evidence that higher estrogen levels are associated with better sexual functioning. (4) Testosterone is also important, and a woman's sexual desire may decline if her levels of testosterone have been reduced because the ovaries were removed in a hysterectomy. Testosterone replacement therapy may be helpful in such cases.

The topic of sexuality and the elderly will be discussed in more detail in Chapter 12.

Psychological Changes

Psychological problems that have been associated with menopause include depression, irritability, anxiety, nervousness, crying spells, inability to concentrate, and feelings of suffocation. Moreover, there is a cultural stereotype that women suffer from an "empty-nest syndrome" in the middle years—that is, that they become depressed when all their children have left home and the mother role is at an end for them. However, sociologist Lillian B. Rubin (1979) challenged the whole notion of the empty-nest syndrome and questioned whether there is substantial depression in the midlife period for women. Her results are based on a study of 160 women, a cross section of white mothers aged 35 to 54, from the working, middle, and professional classes. To be included in the sample, they had to have given up work or careers after a minimum of 3 years and to have assumed the traditional role of housewife and mother for at least 10 years after the birth of their first child. Therefore this group should be most prone to the empty-nest syndrome. Typically these women said, "My career was my child." Contrary to the notion of the empty nest, Rubin found that although some women were momentarily sad, lonely, or frightened, they were not depressed in response to the departure of their children. The predominant feeling of every woman except one was a feeling of relief. As one woman put it,

> Lonesome? God, no! From the day the kids are born, if it's not one thing, it's another. After all these years of being responsible for them, you finally get to the point where you want to scream: "Fall out of the nest already, you guys, will you? It's time." (Rubin, 1979, p. 13)

Rather than experiencing an immobilizing depression, most of the women found new jobs and reorganized their daily lives.

In agreement with Rubin's thesis, one authoritative review concluded that menopause does *not* increase the rate of depression (Weissman & Klerman, 1977). And the Massachusetts Women's Health Study found that 85 percent of women were never depressed during the menopausal years, 10 percent were depressed occasionally, and just 5 percent were persistently depressed (Golub, 1992).

In short, the evidence indicates that the incidence of depression is not higher during menopause than at other times in a woman's life.

Figure 6.6   According to Rubin's research, women's predominant feeling when all their children have left home is not empty-nest depression but rather relief!

What Causes the Symptoms?

The difficulties associated with menopause are attributed to biology (in particular, hormones) by some and to culture and its expectations by others.

From the biological perspective, the symptoms of menopause appear to be due to the woman's hormonal state. In particular, the symptoms appear to be related either to low estrogen levels or to hormonal imbalance. The former hypothesis, called the *estrogen-deficiency theory,* has been the subject of the most research. Proponents of this theory argue that the physical symptoms, such as hot flashes, and the psychological symptoms, such as depression, are caused by declining amounts of estrogen in the body—although, as we saw in the last section, there is no evidence of increased depression.

The best evidence for the estrogen-deficiency theory comes from the success of estrogen-replacement therapy. It is effective in relieving low-estrogen menopausal symptoms like hot flashes, sweating, cold hands and feet, and osteoporosis. It also keeps the vagina "young," maintaining the elasticity of its walls and the capacity for lubrication, as well as reducing the risk of vaginitis (Walling et al., 1990). The success of this therapy suggests that low estrogen levels cause menopausal symptoms and that raising estrogen levels relieves the symptoms.

On the other hand, there is a strong cultural bias toward expecting menopausal symptoms. Thus any quirk in a middle-aged woman's behavior is attributed to the "change." It simultaneously becomes the cause of, and explanation for, all the problems and complaints of the middle-aged woman. Given such expectations, it is not surprising that the average person perceives widespread evidence of the pervasiveness of menopausal symptoms. Ironically, idiosyncrasies in women of childbearing age are blamed on menstruation, while problems experienced by women who are past that age are blamed on the *lack* of it.

In summary, the evidence indicates that many, though not all, women experience physical problems such as hot flashes during menopause. These physical symptoms seem to be most closely related to declining estrogen levels and are relieved for many women by estrogen-replacement therapy. On the other hand, there is little evidence that the menopausal years are a bad time psychologically, so we needn't waste our time trying to figure out the causes of psychological problems during that period.

Male Menopause

Biological Changes

In the technical sense, men do not experience a "menopause"; never having menstruated, they can scarcely cease menstruating. However, men do experience a very gradual decline in the manufacture of both testosterone and sperm by the testes, a mild version of the climacteric process in women, although viable sperm may still be produced at age 90. Some refer to this time of life as **andropause,** referring to declining levels of androgens (Lamberts et al., 1997). Some experts argue that a low testosterone syndrome occurs in some older males and that testosterone treatments are beneficial, although this approach is controversial (Morley et al., 1997; Schow et al., 1997; Tserotas & Merino, 1998).

One of the most common physical problems for middle-aged and older men is noncancerous enlargement of the prostate gland, which is believed to be related to changes in hormone levels. Prostate enlargement occurs in 10 percent of men by age 40, and 50 percent of men who reach age 80. Enlargement of the prostate causes urination problems; there is difficulty in voluntarily initiating urination, as well as frequent nocturnal urination. These symptoms are usually remedied by surgery to remove the part of the gland pressing against the urethra. Unfortunately, complications of the surgery include retrograde ejaculation in

about 75 percent of men and erection problems in 5 to 10 percent (Oesterling, 1995). New treatments have recently been developed to avoid these problems, including anti-androgen drugs, and microwave and laser treatments (Oesterling, 1995). Contrary to a number of myths, prostate enlargement is not caused by masturbation or excessive sexual activity. The infection and inflammation resulting from sexually transmitted diseases, though, can cause prostate enlargement.

Psychological Changes

Research has focused on the notion of a "male midlife crisis," whether men experience one in their forties, and what its nature is (Levinson, 1978; Lowenthal et al., 1975). Major themes that emerge from that research are discussed below.

One theme is the *aspiration-achievement gap.* Most human beings have a desire to feel good about themselves based on their achievements. For men, a major source of this positive sense of self comes from their job or career. Around age 40 many men recognize that their achievements have not matched the high aspirations and goals they set for themselves when they were in their twenties. How does a man bridge this aspiration-achievement gap for himself? For many, perhaps the majority, there is a gradual reconciliation, with aspirations being reduced until they are at a realistic, attainable level, and the man emerges feeling good about himself. For some, the reconciliation is not easy and there is a crisis and depression.

According to Erik Erikson (1950), one of the major tasks of adult development is a resolution of the *stagnation versus generativity* conflict. Most people seem to have a deep-seated desire to feel a sense of personal growth, or generativity, in their lives. At age 40, when the only thing that seems to be growing is the waistline, it may be difficult to feel a sense of generativity, and stagnation may set in. This crisis may be resolved positively by having a continual sense of generativity in adulthood, often by finding

Andropause: The time of declining androgen levels in middle-aged men. The male version of menopause.

Figure 6.7 According to Erik Erikson, one of the major tasks of adult development is the resolution of the stagnation-versus-generativity conflict. Middle-aged men may gain a positive sense of generativity, for example, by cultivating enjoyable relationships with their grandchildren.

growth in other sources, such as the growth of one's children or grandchildren. Single men or gay men with no children can gain a sense of generativity in many other ways, such as taking an interest in fostering the careers of their younger coworkers.

Relationships within the man's family shift during the mid-life period (Brim, 1976; Kilmartin, 1994). The children grow and leave home, leaving the husband and wife alone together. Although it is a popular stereotype that this is a difficult time in marriage, causing many divorces, the data actually indicate that married couples on the average rate the postparental period as one of the happiest in their lives (Brim, 1976). The man's own parents may become increasingly dependent on him, requiring a transformation of that relationship. And the man's wife, free from child-care responsibilities, may seek education, a career, or a more active involvement in a job she already has, requiring renegotiation of the marital relationship.

Systematic, well-sampled research indicates that men in their forties—compared with groups of men aged 25 to 39 and 50 to 69—do show significantly higher depression scores and more alcohol and drug use; on the other hand, their levels of anxiety are no higher, and they report no less life satisfaction or happiness (Tamir, 1982). Thus it seems that men in their forties have some problems, but the problems are probably not much worse than the problems men face at other ages.

It is important that the complaints of middle-aged men and women be recognized and that they be considered legitimate complaints. It is also important to note how many of these crises of middle age lead to satisfactory resolution, in which the person makes a positive alteration in her or his life.

SUMMARY

Biologically, the menstrual cycle is divided into four phases: the follicular phase, ovulation, the luteal phase, and menstruation. Corresponding to these phases, there are changes in the levels of pituitary hormones (FSH and LH) and in the levels of ovarian hormones (estrogen and progesterone), as well as changes in the ovaries and the uterus. A fairly common menstrual problem is dysmenorrhea, or painful menstruation.

Research indicates that some, though probably not all, women experience changes in mood over the phases of the menstrual cycle. For those who experience such changes, mood is generally positive around the middle of the cycle (i.e., around ovulation), while negative moods characterized by depression and irritability are more likely just before and during menstruation. These negative moods and physical discomforts are termed the premenstrual syndrome. On the other hand, research indicates that there are no fluctuations in performance over the cycle. There is evidence suggesting that fluctuations in mood are related to changes in hormone levels, but data are also available suggesting that mood fluctuations are related to cultural factors. Research attempting to document whether men experience monthly biological and/or psychological cycles is now in progress.

The climacteric is the period in middle age during which the functioning of the ovaries (both hormone and egg production) declines gradually. One symptom of this process is menopause, the cessation of menstruation. Physical symptoms, such as hot flashes, during this period result from declining levels of estrogen, and may be relieved by estrogen-replacement therapy; contrary to popular belief, research does not show an increased incidence of depression at the time of menopause. Changes in sexual functioning across the menstrual cycle and at menopause are most likely related to changes in levels of testosterone.

Men experience a much more gradual decline in the functioning of their gonads. They may experience a psychological transition in middle age parallel to that of women.

REVIEW QUESTIONS

1. The phase of the menstrual cycle following ovulation is called the _____.

2. Toxic shock syndrome is caused by an abnormal growth of a *Staphylococcus* bacterium in the vagina. True or false?

3. A woman's basal body temperature generally takes a dip on the day of ovulation and then shows a noticeable rise on the day after ovulation. True or false?

4. _____ are the hormonelike substances that are responsible for menstrual cramps.

5. The premenstrual syndrome consists of physical and psychological symptoms, including tension, depression, irritability, backache, and headache. True or false?

6. Research shows that there are fluctuations in women's athletic and intellectual performance over the phases of the menstrual cycle. True or false?

7. Research indicates that some women experience mood shifts over the phases of the menstrual cycle, but other women do not experience such mood shifts. True or false?

8. _____ has been shown to be effective in treating the physical symptoms of menopause, as well as protecting women from osteoporosis.

9. Challenging the notion of the empty-nest syndrome, one sociologist found in interviews with middle-aged women that their predominant feeling about the departure of their children was relief. True or false?

10. One major theme of the male mid-life crisis is the aspiration-achievement gap. True or false?

(The answers to all review questions are at the end of the book, after the Glossary.)

QUESTIONS FOR THOUGHT, DISCUSSION, AND DEBATE

1. Are women's fluctuations in mood over the menstrual cycle caused by biological factors or by environmental/cultural factors?

2. Is there a male mid-life crisis? If so, what are its "symptoms," and what causes it?

3. Among the Yoruba of West Africa, younger women may not travel or engage in trade, but older (postmenopausal) women may. Among the Winnebago of North America, old women sit with men; they are considered to be the same as men. In some cultures, then, women gain in freedom and status after menopause. Does this happen for most women in the United States? Are there variations among ethnic groups—for example, African Americans, Latinos, Asian Americans, whites, and Native Americans—in the way a postmenopausal woman is viewed and treated? (For more information, see Golub, 1992.)

4. Your 10-year-old daughter tells you that she has heard about PMS and wonders whether she will get it when she begins her periods. What would you tell her?

SUGGESTIONS FOR FURTHER READING

Barbach, Lonnie. (1993). *The pause: Positive approaches to menopause.* New York: Dutton. Barbach, a well-known therapist and author, provides many details about menopausal symptoms and positive ways to cope with them.

Golub, Sharon. (1992). *Periods: From menarche to menopause.* Newbury Park, CA: Sage. This book, by a well-known researcher in the field, tells you everything you always wanted to know about menstruation and menopause.

WEB RESOURCES

http://world.std.com/~susan207/
Menopause Matters Homepage; information about menopause and links to related resources.

chapter

7

CONCEPTION, PREGNANCY, AND CHILDBIRTH

CHAPTER HIGHLIGHTS

Conception

Development of the Conceptus

The Stages of Pregnancy

Sex During Pregnancy

Nutrition During Pregnancy

Effects of Drugs Taken During Pregnancy

Birth

Childbirth Options

After the Baby Is Born:
    The Postpartum Period

Breast-Feeding

Problem Pregnancies

Infertility

New Reproductive Technologies

t gave me a sense that I was actually a woman. I had never felt sexy before. I went through a lot of changes. It was a very sexual thing. I felt very voluptuous.

thought it would never end. I was enormous. I couldn't bend over and wash my feet. And it was incredibly hot.*

Chapter 5 described the remarkable biological process by which a single fertilized egg develops into a male or a female human being. This chapter is about some equally remarkable processes involved in creating human beings: conception, pregnancy, and childbirth.

Conception

Sperm Meets Egg: The Incredible Journey

On about day 14 of an average menstrual cycle the woman ovulates. The egg is released from the ovary into the body cavity. Typically it is then picked up by the fimbriae (long fingerlike structures at the end of the fallopian tube—see Figure 7.1) and enters the fallopian tube. It then begins a leisurely trip down the tube toward the uterus, reaching it in about five days, if it has been fertilized. Otherwise, it disintegrates in about 48 hours. The egg, unlike the sperm, has no means of moving itself and is propelled by the cilia (hairlike structures) lining the fallopian tube. The egg has begun its part of the journey toward conception.

Meanwhile, the couple have been having intercourse. The woman's cervix secretes mucus which flushes the passageways to prepare for the arrival of the sperm. The man has an orgasm and ejaculates inside the woman's vagina. The sperm are deposited in the vagina, there to begin their journey toward the egg. Actually they have made an incredible trip even before reaching the vagina. Initially they were manufactured in the seminiferous tubules of the testes (see Chapter 4). They then collected and were stored in the epididymis. During ejaculation they moved up and over the top of the bladder

in the vas deferens; then they traveled down through the ejaculatory duct, mixed with seminal fluid, and went out through the urethra.

The sperm is one of the tiniest cells in the human body. It is composed of a *head,* a *midpiece,* and a *tail* (see Figure 7.2). The head is about 5 micrometers long, and the total length, from the tip of the head to the tip of the tail, is about 60 micrometers (about 2/1000 inch, or 0.06 millimeter). The chromosomal material, which is the sperm's most important contribution when it unites with the egg, is contained in the nucleus, which is in the head of the sperm. The *acrosome,* a chemical reservoir, is also in the head of the sperm. The midpiece contains mitochondria, which are tiny structures in which chemical reactions occur that provide energy. This energy is used when the sperm lashes its tail back and forth. The lashing action (called *flagellation*) propels the sperm forward.

A typical ejaculate has a volume of about 3 milliliters, or about a teaspoonful, and contains about 300 million sperm. Although this might seem to be a wasteful amount of sperm if only one is needed for fertilization, the great majority of the sperm never even get close to the egg. Some of the ejaculate, including one-half of the sperm, will flow out of the vagina as a result of gravity. Other sperm may be killed by the acidity of the vagina, to which they are very sensitive. Of those that make it safely into the uterus, half swim up the wrong fallopian tube (the one containing no egg).

But here we are, several hours later, with a hearty band of sperm swimming up the fallopian tube toward the egg, against the currents that are bringing the egg down. Sperm are capable of swimming 1 to 3 centimeters (about 1 inch) per hour, although it has been documented that sperm may arrive at the egg within 1½ hours after ejaculation, which is much sooner than would be expected, given their swimming rate. It is thought that muscular contractions in the uterus may help speed them along. By the time a

*Boston Women's Health Book Collective. (1992). *The new our bodies, ourselves.* New York: Simon and Schuster, pp. 420, 424.

7.1   Sexual intercourse in the man-on-top position, showing the pathway of sperm and egg from manufacture in the testes and ovary to conception, which typically occurs in the fallopian tube.

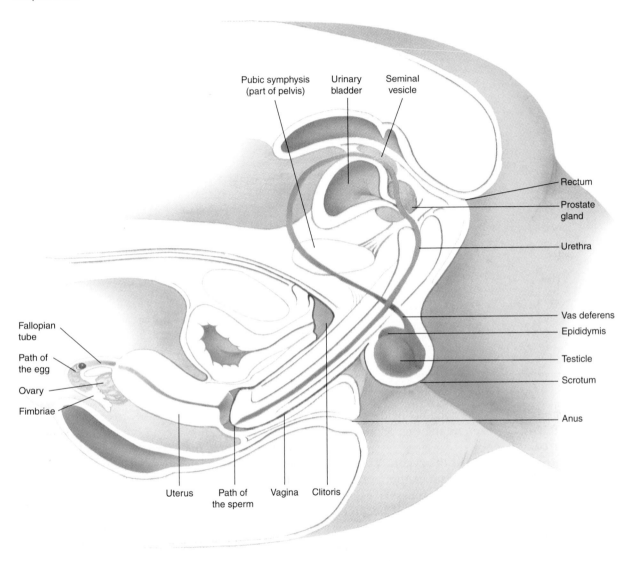

sperm reaches the egg, it has swum approximately 3000 times its own length. This would be comparable to a swim of over 3 miles for a human being.

Contrary to the popular belief that conception occurs in the uterus, typically it occurs in the outer third (the part near the ovary) of the fallopian tube. Of the original 300 million sperm, only about 2000 reach the tube containing the egg. As they approach, a chemical secreted by the egg attracts the sperm to the egg. The egg is surrounded by a thin, gelatinous layer called the *zona pellucida*. Sperm swarm around the egg and secrete an enzyme called **hyaluronidase** (produced by the acrosome located in the head of the sperm—see Figure 7.2);

Hyaluronidase: An enzyme secreted by the sperm that allows one sperm to penetrate the egg.

Figure 7.2   The structure of a mature human sperm.

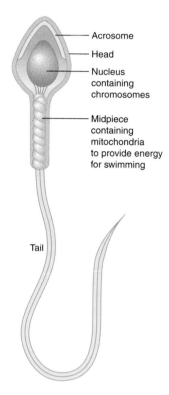

Acrosome

Head

Nucleus
containing
chromosomes

Midpiece
containing
mitochondria
to provide energy
for swimming

Tail

this enzyme dissolves the zona pellucida, permitting one sperm to penetrate the egg.[1] Conception has occurred.

The fertilized egg, called the **zygote,** continues to travel down the fallopian tube. About 36 hours after conception, it begins the process of cell division, by which the original one cell becomes a mass of two cells, then four cells, then eight cells, and so on. About five to seven days after conception, the mass of cells implants itself in the lining of the uterus, there to be nourished and grow. For the first eight weeks of gestation the *conceptus* (product of conception) is called an *embryo;* from then until birth it is called a *fetus.*

[1]Thus while only one sperm is necessary to accomplish fertilization, it appears that it is important for him to have a lot of his buddies along to help him get into the egg. Therefore, maintaining a high sperm count seems to be important for conception.

Improving the Chances of Conception: Making a Baby

While this topic may seem rather remote to a 20-year-old college student, whose principal concern is probably *avoiding* conception, some couples do want to have a baby. The following are points for them to keep in mind.

The whole trick, of course, is to time intercourse so that it occurs around the time of ovulation. To do this, it is necessary to determine when the woman ovulates. If she is that idealized woman with the perfectly regular 28-day cycle, then she ovulates on day 14. But for the vast majority of women, the time of ovulation can best be determined by keeping a *basal body temperature chart.* To do this, the woman takes her temperature every morning immediately upon waking (that means before getting up and moving

Zygote: A fertilized egg.

around or drinking a cup of coffee). She then keeps a graph of her temperature (like the one shown in Figure 6.3). During the preovulatory phase, the temperature will be relatively constant (the temperature is below 98.6°F because temperature is low in the early morning). On the day of ovulation the temperature drops, and on the day following ovulation it rises sharply, by 0.4° to 1.0°F above the preovulatory level. The temperature should then stay at that high level until just before menstruation. The most reliable indicator of ovulation is the rise in temperature the day after it occurs. From this, the woman can determine the day of ovulation, and that determination should be consistent with menstruation occurring about 14 days later. After doing this for a couple of cycles, the woman should have a fairly good idea of the day in her cycle on which she ovulates. Two other methods for determining when a woman is ovulating are the cervical mucus and sympto-thermal methods, described in Chapter 8.

Sperm live inside the woman's body for up to 5 days (Wilcox et al., 1995). The egg is capable of being fertilized for about the first 12 to 24 hours after ovulation. Allowing the sperm some swimming time, this means that intercourse should be timed right at ovulation or one or two days before.

Assuming you have some idea of the time of ovulation, how frequent should intercourse be? While more may be merrier, more is not necessarily more effective. The reason for this is that it is important for the man's sperm count to be maintained. It takes a while to manufacture 300 million sperm—at least 24 hours. And, as was discussed earlier, maintaining a high sperm count appears to be important in accomplishing the task of fertilizing the egg. For purposes of conceiving, then, it is probably best to have intercourse about every 24 to 48 hours, or about four times during the week in which the woman is to ovulate.

It is also important to take some steps to ensure that once deposited in the vagina, the sperm get a decent chance to survive and to find their way into the fallopian tubes. Position during and after intercourse is important. For purposes of conceiving, the best position for intercourse is with the woman on her back (man-on-top, or "missionary," position—see Chapter 10). If the woman is on top, much of the ejaculate may run out of the vagina because of the pull of gravity. After intercourse, she should remain on her back, possibly with her legs pulled up and a pillow under her hips, preferably

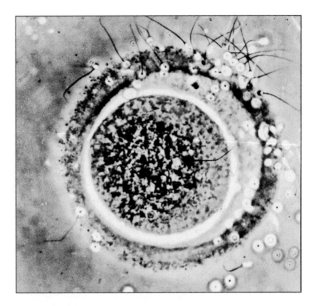

Figure 7.3 The egg is fertilized by one sperm, as many sperm cluster about.

for about a half hour to an hour. This allows the semen to remain in a pool in the vagina, which gives the sperm a good chance to swim up into the uterus. Because sperm are very sensitive to the pH (acidity-alkalinity) of the vagina, this factor also requires some consideration. Acidity kills sperm. Douching with commercial preparations or with acidic solutions (such as vinegar) should be avoided. If anything, the woman may want to douche before intercourse with a slightly basic solution made by adding two or three tablespoons of baking soda to a quart of water. Finally, lubricants and/or suppositories should not be used; they may kill sperm or block their entrance into the uterus.

Development of the Conceptus

For the nine months of pregnancy, two organisms—the conceptus and the pregnant woman—undergo parallel, dramatic changes. The changes that occur in the developing conceptus will be discussed in this section; a later section will be about the changes that take place in the pregnant woman.

Typically the nine months of pregnancy are divided into three equal periods of three months, called *trimesters*. Thus the first trimester is months 1 to 3, the second trimester is months 4 to 6, and the third (or last) trimester is months 7 to 9.

The Embryo and Its Support Systems

We left the conceptus, which began as a single fertilized egg cell, dividing into many cells as it passed down the fallopian tube, finally arriving in the uterus and implanting itself in the uterine wall.

During the embryonic period of development (the first eight weeks), most of the fetus's major organ systems are formed in processes that occur with amazing speed. The inner part of the ball of cells implanted in the uterus now differentiates into two layers, the endoderm and the ectoderm. Later a third layer, the mesoderm, forms between them. The various organs of the body differentiate from these layers. The *ectoderm* will form the entire nervous system and the skin. The *endoderm* differentiates into the digestive system—from the pharynx, to the stomach and intestines, to the rectum—and the respiratory system. The muscles, skeleton, connective tissues, and reproductive and circulatory systems derive from the *mesoderm*. Fetal development generally proceeds in a cephalocaudal order. That is, the head develops first, and the lower body last. For this reason, the head of the fetus is enormous compared with the rest of the body.

Meanwhile, another group of cells has differentiated into the *trophoblast,* which has important functions in maintaining the embryo and which will eventually become the placenta. The **placenta** is the mass of tissues that surrounds the conceptus early in development and nurtures its growth. Later it moves to the side of the fetus. The placenta has a number of important functions, perhaps the most important of which is that it serves as a site for the exchange of substances between the woman's blood and the fetus's blood. It is important to note that the woman's circulatory system and the fetus's circulatory system are completely separate. That is, with only rare exceptions, the woman's blood never circulates inside the fetus; nor does the fetus's blood circulate in the woman's blood vessels. Instead, the fetus's blood passes out of its body through the umbilical cord to the placenta. There it circulates in the numerous *villi* (tiny fingerlike projections in the placenta). The woman's blood circulates around the outside of these villi. Thus there is a membrane barrier between the two blood systems. Some substances are capable of passing through this barrier, whereas others are not. Oxygen and nutrients can pass through the barrier,

and thus the woman's blood supplies oxygen and nutrients to the fetus, providing substitutes for breathing and eating. Carbon dioxide and waste products similarly pass back from the fetal blood to the woman's blood. Some viruses and other disease-causing organisms can pass through the barrier, including those for German measles (rubella) and syphilis. But other organisms cannot pass through the barrier; thus the woman may have a terrible cold, but the fetus will remain completely healthy. Various drugs can also cross the placental barrier, and the woman should therefore be careful about drugs taken during pregnancy (see the section on drugs during pregnancy, later in this chapter).

Another major function of the placenta is that it secretes hormones. The placenta produces large quantities of estrogen and progesterone. Many of the physical symptoms of pregnancy may be caused by these elevated levels of hormones. Another hormone manufactured by the placenta is **human chorionic gonadotropin** (**hCG**). hCG is the hormone that is detected in pregnancy tests.

The **umbilical cord** is formed during the fifth week of embryonic development. The fully developed cord is about 55 centimeters (20 inches) long. Normally, it contains three blood vessels: two arteries and one vein. Some people believe that the fetus's umbilical cord attaches to the woman's navel; actually, the umbilical cord attaches to the placenta, thereby providing for the interchanges of substances described above.

Two membranes surround the fetus, the *chorion* and the *amnion,* the amnion being the innermost. The amnion is filled with a watery liquid called **amniotic fluid,** in which the fetus floats and can readily move. It is the amniotic fluid that is sampled when an amniocentesis is performed (see below). The amniotic fluid maintains the fetus at a

---

Placenta (plah-SEN-tuh): An organ formed on the wall of the uterus through which the fetus receives oxygen and nutrients and gets rid of waste products.
Human chorionic gonadotropin (hCG): A hormone secreted by the placenta; it is the substance detected in pregnancy tests.
Umbilical cord: The tube that connects the fetus to the placenta.
Amniotic fluid: The watery fluid surrounding a developing fetus in the uterus.

constant temperature and, most important, cushions the fetus against possible injury. Thus the woman can fall down a flight of stairs, and the fetus will remain undisturbed. Indeed, the amniotic fluid might be considered the original waterbed.

Fetal Development During the First Trimester

In a sense, the development of the fetus during the first trimester is more remarkable than its development during the second and third trimesters. That's because during the first trimester the small mass of cells implanted in the uterus develops into a fetus with most of the major organ systems present and with recognizable human features.

By the third week of gestation, the embryo appears as a small bit of flesh and is about 0.2 centimeters ($\frac{1}{12}$ inch) long. During the third and fourth weeks, the head undergoes a great deal of development. The central nervous system begins to form, and the beginnings of eyes and ears are visible. The backbone is constructed by the end of the fourth week. A "tail" is noticeable early in embryonic development but has disappeared by the eighth week.

From the fourth to the eighth weeks, the external body parts—eyes, ears, arms, hands, fingers, legs, feet, and toes—develop. By the end of the tenth week they are completely formed. Indeed, by the tenth week, the embryo has not only a complete set of fingers but also fingernails.

By the end of the seventh week, the liver, lungs, pancreas, kidneys, and intestines have formed and have begun limited functioning. The gonads have also formed, but the gender of the fetus is not clearly distinguishable until the twelfth week.

At the end of the twelfth week (end of the first trimester) the fetus is unmistakably human and looks like a small infant. It is about 10 centimeters (4 inches) long and weighs about 19 grams (⅔ ounce). From this point on, development consists mainly of enlargement and differentiation of structures that are already present.

Fetal Development During the Second Trimester

Around the end of the fourteenth week, the movements of the fetus can be detected ("quickening").

By the eighteenth week, the woman has been able to feel movement for two to four weeks, and the physician can detect the fetal heartbeat. The latter is an important point, because it helps the physician determine the length of gestation. The baby should be born about 20 weeks later.

The fetus first opens its eyes around the twentieth week. By about the twenty-fourth week, it is sensitive to light and can hear sounds in utero. Arm and leg movements are vigorous at this time, and the fetus alternates between periods of wakefulness and sleep.

Fetal Development During the Third Trimester

At the end of the second trimester the fetus's skin is wrinkled and covered with downlike hair. At the beginning of the third trimester, fat deposits form under the skin; these will give the infant the characteristic chubby appearance of babyhood. The downlike hair is lost.

During the seventh month the fetus turns in the uterus to assume a head-down position. If this turning does not occur by the time of delivery, there will be a *breech presentation.* Women can do certain exercises to aid the turning (Boston Women's Health Book Collective, 1992). Physicians and midwives can also perform certain procedures to turn the fetus.

The fetus's growth during the last two months is rapid. At the end of the eighth month it weighs an average of 2500 grams (5 pounds, 4 ounces). The average full-term baby weighs 3300 grams (7.5 pounds) and is 50 centimeters (20 inches) long.

The Stages of Pregnancy

The First Trimester (The First 12 Weeks)

Symptoms of Pregnancy

For most women, the first symptom of pregnancy is a missed menstrual period. Of course, there may be a wide variety of reactions to this event. For the teenager who is not married or for the married woman who feels that she already has enough children, the reaction may be negative—depression, anger, and fear. For the woman who has been

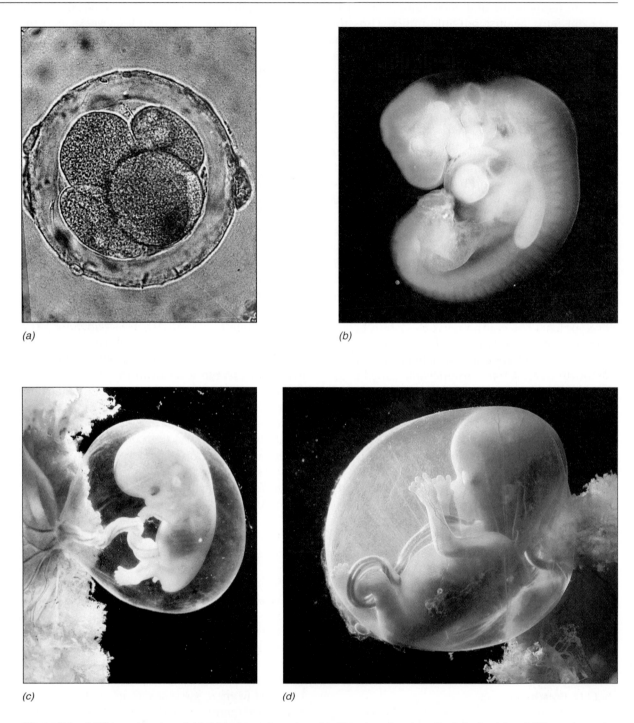

Figure 7.4   *(a)* This embryo has divided into four cells and would still be traveling down the fallopian tube. *(b)* The embryo after 4 weeks of development. The major organs are forming; the bright red, blood-filled heart is just below the lower jaw. *(c)* At 9 weeks the human fetus is recognizable as a primate. Limbs have formed and ears are clearly visible. (d) By about 3 months the fetus is approximately 8 centimeters long and weighs about 28 grams. Muscles have formed and they move the limbs and body.

trying to conceive for several months, the reaction may be joy and eager anticipation.

In fact, there are many other reasons besides pregnancy for a woman to have a late period or miss a period. Illness or emotional stress may delay a period, and women occasionally skip a period for no apparent reason.

It is also true that a woman may continue to experience some cyclic bleeding or spotting during pregnancy. It is not particularly a danger sign, except that in a few cases it is a symptom of a miscarriage.

If the woman has been keeping a basal body temperature chart, this can provide a very early sign that she is pregnant. If her temperature rises abruptly at about the time ovulation would normally occur and then stays up for more than two weeks—say, about three weeks—the chances are fairly good that she is pregnant. The increased temperature results from the high level of progesterone manufactured by the corpus luteum and, later, the placenta.

Other early symptoms of pregnancy are tenderness of the breasts—a tingling sensation and special sensitivity of the nipples—and nausea and vomiting (called "morning sickness," although these symptoms may actually happen anytime during the day). More frequent urination, feelings of fatigue, and a need for more sleep are other early signs of pregnancy.

Pregnancy Tests

It is important that early, accurate pregnancy tests be available and that women make use of them. This is true for several reasons. A woman needs to know that she is pregnant as early as possible so that she can see a physician or midwife and begin getting good prenatal care. She also needs to know so that she can get the nutrition she requires during pregnancy (see the next section on nutrition). And if she does not want to carry the baby to term, she needs to know as soon as possible, because abortions are much safer and simpler when performed in the first trimester than when done in the second trimester.

A pregnancy test may be done by a physician, at a Planned Parenthood or family planning clinic, or at a medical laboratory. The most common pregnancy test is an immunologic test based on detecting the presence of hCG (human chorionic gonadotropin, secreted by the placenta) in the woman's urine. It can be done in a matter of minutes and is very accurate. It involves mixing a drop of urine with certain chemicals, either on a slide or in a tube.

The laboratory tests for pregnancy are 98 to 99 percent accurate. A laboratory test may produce a false negative (tell the woman she is not pregnant when she really is) if it is done too early or if errors are made in processing. Also, some women simply do not show positive signs in the tests or do not do so until the second or third test. The modern urine tests are 98 percent accurate 7 days after implantation (just when a period is missed).

A different type of test, called the *beta-hCG radioimmunoassay,* assesses the presence of beta-hCG in a blood sample. It can detect hCG at very low levels, so it can reliably detect pregnancy 7 days after fertilization. It is much more expensive than the urine tests and is available only in laboratories associated with hospitals or large clinics.

Home pregnancy tests are also available, sold under such brand names as e.p.t. (for "early pregnancy test"), Advance, Answer, and ClearBlue. These are all urine tests designed to measure the presence of hCG; they cost $9 to $13. They can be used about nine days after a missed menstrual period. Their charm lies in their convenience and the privacy of getting the results. The major problem with them is that they have a very high rate—between 25 and 38 percent—of false negatives (Hart, 1990; Doshi, 1986). This compares with an error rate of 1 to 2 percent for laboratory tests. The home pregnancy tests also have a 16 percent rate of false positives (Hatcher et al., 1994). To guard against false negatives, the manufacturers recommend repeating the test one week later if the results are negative the first time, although this increases the cost by $3 to $5. The reason that a high rate of false negatives is so serious is that it leads a pregnant woman to think she is not pregnant, and thus she might take drugs that would harm the fetus, and she will not begin getting prenatal care; such dangerous conditions as ectopic pregnancy might therefore go undetected. Performing the tests also requires a certain amount of coordination and care. All in all, they are probably not as good an idea as they seem, although they may be improved in the future.

The signs of pregnancy may be classified as *presumptive signs, probable signs,* and *positive signs.* Amenorrhea, breast tenderness, nausea, and so on, are presumptive signs. The pregnancy tests discussed previously all provide probable signs. Three signs are interpreted as positive signs, that is, as definite indications of pregnancy: (1) beating of the fetal heart, (2) active fetal movement, and (3) detection of a fetal skeleton by ultrasound. These signs cannot be detected until the fourth month, with the exception of ultrasound, which can be used in the first trimester.

Once the pregnancy has been confirmed, the woman generally is very interested in determining her expected delivery date (called EDC for a rather antiquated expression, "expected date of confinement"). The EDC is calculated using *Nägele's rule*. The rule says to take the date of the first day of the last menstrual period, subtract three months, add seven days, and finally add one year. Thus if the first day of the last menstrual period was September 10, 1999, the expected delivery date would be June 17, 2000: subtracting three months from September 10 gives June 10, adding seven days yields June 17, and adding one year gives June 17, 2000. In cases where the date the last menstrual period began is not known, an ultrasound procedure may be used to determine gestational age (Afriat, 1995).

Physical Changes

The basic physical change that takes place in the woman's body during the first trimester is the large increase in the levels of hormones, especially estrogen and progesterone, that are produced by the placenta. Many of the other physical symptoms of the first trimester arise from these endocrine changes.

The breasts swell and tingle. This results from the development of the mammary glands, which is stimulated by hormones. The nipples and the area around them (areola) may darken and broaden.

There is often a need to urinate more frequently. This is related to changes in the pituitary hormones that affect the adrenals, which in turn change the water balance in the body so that more water is retained. The growing uterus also contributes by pressing against the bladder.

Some women experience morning sickness—feelings of nausea, perhaps to the point of vomiting, and of revulsion toward food or its odor. The nausea and vomiting may occur on waking or at other times during the day. The exact cause of this

is not known. One theory is that the high levels of estrogen irritate the stomach. The rapid expansion of the uterus may also be involved. While these symptoms are quite common, about 25 percent of pregnant women experience no vomiting at all.

Vaginal discharges may also increase at this time, partly because the increased hormone levels change the pH of the vagina and partly because the vaginal secretions are changing in their chemical composition and quantity.

The feelings of fatigue and sleepiness are probably related to the high levels of progesterone, which is known to have a sedative effect.

Psychological Changes

Our culture is full of stereotypes about the psychological characteristics of pregnant women. According to one view, pregnancy is supposed to be a time of happiness and calm. Radiant contentment, the "pregnant glow," is said to emanate from the woman's face, making this a good time for her to be photographed. According to another view, pregnancy is a time of emotional ups and downs. The pregnant woman swings from very happy to depressed and crying, and back again. She is irrational, sending her partner out in a blizzard for kosher dill spears. One study compared 70 pregnant women (planned pregnancies) with 92 nonpregnant women. The researchers assessed both physical and psychological states at three points during the pregnancy (or nine months for the nonpregnant women). On the whole, they found that pregnancy is neither a time of heightened well-being nor heightened emotional turmoil (Striegel-Moore et al., 1996).

Research indicates that the situation is more complex than these stereotypes suggest. A woman's emotional state during pregnancy, often assessed with measures of depression, varies according to several factors. First, her attitude toward the pregnancy makes a difference; women who desire the pregnancy are less anxious than women who do not (Kalil et al., 1993). A second factor is social class. Several studies have found that low income is associated with depression during pregnancy. For example, a study involving interviews with 192 poor, inner-city pregnant women found that they were twice as likely as their middle-income counterparts to be depressed (Hobfoll et al., 1995). This may be due to the economic situation these women face; also, there may be more unwanted pregnancies among low-income women. A third influence is the availability of social support. Women with a supportive partner are less likely to be depressed, perhaps because the partner serves as a buffer against stressful events (Chapman et al., 1997).

During the first trimester, the variables that distinguish pregnant from nonpregnant women are nausea, associated with morning sickness, and fatigue (Streigel-Moore et al., 1996). Depression is not uncommon during this time. Women who led very active lives prior to becoming pregnant may find fatigue and lack of energy especially distressing. Depression during the first trimester is more likely among women experiencing other stressful life events, such as moving, changes in their jobs, changes in relationships, or illnesses (Kalil et al., 1993). In this trimester, women's anxieties often center on concerns about miscarriage. An intensive study of 19 women found that only the four women for whom the pregnancy was unplanned expressed overall negative emotions. The other women were either ambivalent or had overall positive feelings (Leifer, 1980).

The Second Trimester (Weeks 13 to 26)

Physical Changes

During the fourth month, the woman becomes aware of the fetus's movements ("quickening"). Many women find this to be a very exciting experience.

The woman is made even more aware of the pregnancy by her rapidly expanding belly. There are a variety of reactions to this. Some women feel that it is a magnificent symbol of womanhood, and they rush out to buy maternity clothes and wear them before they are even necessary. Other women feel awkward and resentful of their bulky shape and may begin to wonder whether they can fit through doorways and turnstiles.

Most of the physical symptoms of the first trimester, such as morning sickness, disappear, and discomforts are at a minimum. Physical problems at this time include constipation and nosebleeds (caused by increased blood volume).

Edema—water retention and swelling—may be a problem in the face, hands, wrists, ankles, and feet; it results from increased water retention throughout the body.

By about mid-pregnancy, the breasts, under hormonal stimulation, have essentially completed their development in preparation for nursing. Beginning about the nineteenth week, a thin amber or yellow fluid called **colostrum** may come out of the nipple, although there is no milk yet.

Psychological Changes

While the first trimester can be relatively tempestuous, particularly with morning sickness, the second trimester is usually a period of relative calm and well-being. The discomforts of the first trimester are past, and the tensions associated with labor and delivery are not yet present. Fear of miscarriage diminishes as the woman feels fetal movement (Leifer, 1980).

Depression is less likely during the second trimester if the pregnant woman has a cohabiting partner or spouse (Hobfoll et al., 1995). Interestingly, women who have had a previous pregnancy are more distressed during this time than women who have not (Wilkinson, 1995). This may reflect the impact of the demands associated with the care of other children when one is pregnant. Research also indicates that feelings of nurturance, or maternal responsiveness to the infant, increase steadily from the pre-pregnant to the postpartum period (Fleming et al., 1997). This increase does not appear to be related to changes in hormone levels during pregnancy.

Research also indicates that women's beliefs about pregnancy and about how they should behave during pregnancy are influenced by culture. A comparison of Indian women giving birth in London with non-Indian women giving birth there found that culture influenced beliefs about diet during pregnancy, and about the amount and type of medical care they should seek (Woollett et al., 1995).

Edema (eh-DEE-muh): Excessive fluid retention and swelling
Colostrum: A watery substance that is secreted from the breast at the end of pregnancy and during the first few days after delivery.

The Third Trimester (Weeks 27 to 38)

Physical Changes

The uterus is very large and hard now. The woman is increasingly aware of her size and of the fetus, which is becoming more and more active. In fact, some women are kept awake at night by its somersaults and hiccups.

The extreme size of the uterus puts pressure on a number of other organs, causing some discomfort. There is pressure on the lungs, which may cause shortness of breath. The stomach is also being squeezed, and indigestion is common. The navel is pushed out. The heart is being strained because of the large increase in blood volume. Most women feel low in energy (Leifer, 1980).

The weight gain of the second trimester continues. Most physicians recommend about 20 to 26 pounds of weight gain during pregnancy. The average infant at birth weighs 7.5 pounds; the rest of the weight gain is accounted for by the placenta (about 1 pound), the amniotic fluid (about 2 pounds), enlargement of the uterus (about 2 pounds), enlargement of the breasts (1.5 pounds), and the additional fat and water retained by the woman (8 or more pounds). Physicians restrict the amount of weight gain because the incidence of complications such as high blood pressure and strain on the heart is much higher in women who gain an excessive amount of weight. Also, excessive weight gained during pregnancy can be very hard to lose afterward.

The woman's balance is somewhat disturbed because of the large amount of weight that has been added to the front part of her body. She may compensate for this by adopting the characteristic "waddling" walk of the pregnant woman, which can result in back pains.

The uterus tightens occasionally in painless contractions called **Braxton-Hicks contractions.** These contractions are not part of labor. It is thought that they help to strengthen the uterine muscles, preparing them for labor.

In a first pregnancy, around two to four weeks before delivery the baby turns, and the head drops into the pelvis. This is called *lightening, dropping,* or *engagement.* Engagement usually occurs during labor in women who have had babies before.

Some women are concerned about the appropriate amount of activity during pregnancy—whether some things constitute "overdoing it." Traditionally physicians and textbooks warned of the dangers of physical activities and tried to discourage them. It appears now, however, that such restrictions were based more on superstition than on scientific fact. Current thinking holds that for a healthy pregnant woman, moderate activity is not dangerous and is actually psychologically and physically beneficial. Modern methods of childbirth encourage sensible exercise for the pregnant woman so that she will be in shape for labor (see the section on childbirth options, below). The matter, of course, is highly individual.

Psychological Changes

The patterns noted earlier continue into the third trimester. Psychological well-being is greater among women who have social support (often in the form of a cohabiting partner or husband), have higher incomes/are middle class, and experience fewer concurrent stressful life events. One study assessed the relationship between social support and birth outcomes. Women who reported less support during pregnancy were more likely to have low-birth-weight babies (McWilliams, 1994).

What happens to the relationship of the pregnant woman with her husband? A comparison of women pregnant for the first time with women who had experienced previous pregnancies found that first-time mothers reported a significant increase in dissatisfaction with their husbands from the second to the third trimester (Wilkinson, 1995). Another study included 54 women who were pregnant for the first time and their husbands (Zimmerman-Tansella et al., 1994). Wives who reported that higher levels of affection were exchanged between husband and wife reported lower levels of anxiety and of insomnia in the third trimester.

A study of 200 women pregnant for the first time assessed depression in the third trimester, three weeks after the birth, and at 18 months postpartum (Greene et al., 1991). Depression scores were significantly higher in the third trimester, and single women's scores were significantly higher than married women's scores at that time.

Braxton-Hicks contractions: Contractions of the uterus during pregnancy that are not part of actual labor.

The Father's Role in Pregnancy

Couvade

Occasionally some men experience *couvade syndrome*, or male pregnancy symptoms; the symptoms are somatic and include indigestion, gastritis, nausea, change in appetite, and headaches (Kiselica & Scheckel, 1995). A variety of interpretations for this experience have been suggested; one view is that it enables the male to vicariously experience the pregnancy and thus prepare himself for fatherhood (Mason & Elwood, 1995). While a majority of expectant fathers may experience some physiological and behavioral symptoms, only a small percentage experience couvade.

In some cultures this phenomenon takes a more dramatic form, known as *couvade ritual*. In the couvade ritual, the husband retires to bed while his wife is in labor. He suffers all the pains of delivery, moaning and groaning as she does. Couvade is still practiced in parts of Asia, North and South America, and Oceania (Mead & Newton, 1967).

The Father-to-Be

In modern American culture many men expect to devote substantial amounts of time and energy to fathering, unlike the distant fathers of yesteryear. In fact, it has even been claimed that there is a "father instinct" (Biller & Meredith, 1975). Recall our discussion in Chapter 2 of the reproductive advantage of a father-infant bond. Although some men choose to remain in the background, many choose to be actively involved in the pregnancy and the parenting. Early in the pregnancy, first-time fathers may recall their own childhood and relationship to their father, and may need to resolve any mixed emotions they still have about their relationship. In one study, 70 percent of expectant fathers were initially ambivalent about fathering, but their feelings gradually became more positive, in anticipation of satisfactions to be derived from being fathers (Obzrut, 1976). In the same study, the fathers reported engaging in many activities in preparation for becoming fathers. Many reported attending parenting classes, planning father-child

Figure 7.5   Dad changes his daughter's diaper at a "Bootee Camp" in Irvine, California, which helps new or prospective fathers adjust to their new role.

activities, observing and talking to other fathers, and daydreaming about the baby. Most of these activities, of course, parallel those done by expectant mothers. It has been theorized that men who display this active involvement will do best in the father role after the baby is born (Antle, 1978).

Diversity in the Contexts of Pregnancy

There are lots of family contexts in which women have babies these days besides the traditional one of being married to the baby's father. These include living in a stable relationship with the baby's father but not being married; not being married to or living with the baby's father but seeing him regularly; being a single mother-to-be who has no contact with the baby's father; being a single mother-to-be who is pregnant as a result of artificial insemination or other reproductive technologies; and being a woman in a stable relationship with another woman, who is pregnant as a result of artificial insemination or other technologies. Because it is too complicated to mention these alternatives constantly, in the sections that follow our language will be based on a situation in which the woman is married to the baby's father, which is still statistically the most common context in which babies are born in the United States. Readers should keep in mind all these other possible family scenarios, though.

Sex During Pregnancy

Many women and men are concerned about whether it is safe or advisable for a pregnant woman to have sexual intercourse, particularly during the latter stages of pregnancy. Traditionally, physicians believed that intercourse might (1) cause an infection, or (2) precipitate labor prematurely or cause a miscarriage (for a review on sexuality and pregnancy, see Reamy & White, 1987).

Current medical opinion, however, is that—given a normal, healthy pregnancy—intercourse can continue safely until four weeks before the baby is due (Cunningham et al., 1993). There is no evidence that intercourse or orgasm is associated with preterm labor (Lumley & Astbury, 1989). The only exception is a case where a miscarriage or preterm labor is threatened. Whether and how frequently to have intercourse is a matter for a couple to decide, perhaps in consultation with a physician or midwife.

Most pregnant women continue to have intercourse throughout the pregnancy (Reamy & White, 1987). The most common pattern is a significant decline in the frequency of intercourse during the first trimester, no change in the second trimester, and an even greater decline in the third trimester (see Table 7.1). One study included male partners of pregnant women; the men reported the same pattern (Bogren, 1991).

During the latter stages of pregnancy, the woman's shape makes intercourse increasingly awkward. The man-on-top position is probably best abandoned at this time. The side-to-side position (see Chapter 10) is probably the most suitable one for intercourse during the late stages of pregnancy. Couples should also remember that there are many ways of experiencing sexual pleasure and orgasm besides having intercourse—hand-genital

Table 7.1 Percentage of Women Having Various Frequencies of Coitus at Different Stages of Pregnancy

Number of Acts of Coitus per Week	Baseline Prepregnancy, %	12 Weeks Pregnant, %	24 Weeks Pregnant, %	36 Weeks Pregnant, %
None	0	11	8	36
< 1	7	24	25	26
1–3	54	52	55	33
4 or more	40	14	13	5

Source: R. Kumar, H. A. Brant, and K. M. Robson. (1981). Childbearing and maternal sexuality: A prospective study of 119 primiparas. *Journal of Psychosomatic Research, 25,* 373–383.

stimulation or oral-genital sex may be good alternatives.[2] The best guide in this matter is the woman's feelings. If intercourse becomes uncomfortable for her, alternatives should be explored.

Nutrition During Pregnancy

During pregnancy, another living being is growing inside the woman, and she needs lots of energy, protein, vitamins, and minerals at this time. Therefore, diet during pregnancy is extremely important. If the woman's diet is good, she has a much better chance of remaining healthy during pregnancy and of bearing a healthy baby; if her diet is inadequate, she stands more of a chance of developing one of a number of diseases during pregnancy herself and of bearing a child whose weight is low at birth. Babies with low birth weights do not have as good a chance of survival as babies with normal birth weights. According to a study done in Toronto, mothers in a poor-diet group had four times as many serious health problems during pregnancy as a group of mothers whose diets were supplemented with highly nutritious foods. Those with the poor diets had seven times as many threatened miscarriages and three times as many stillbirths; their labor lasted five hours longer on the average (Newton, 1972).

It is particularly important that a pregnant woman get enough protein, folic acid, calcium, magnesium, and vitamin A (Luke, 1994). Protein is important for building new tissues. Folic acid is also important for growth; symptoms of folic acid deficiency are anemia and fatigue. A pregnant woman needs much more iron than usual, because the fetus draws off iron for itself from the blood that circulates to the placenta. Muscle cramps, nerve pains, uterine ligament pains, sleeplessness, and irritability may all be symptoms of a calcium deficiency. Severe calcium deficiency during pregnancy is associated with increased blood pressure, which may lead to a serious condition called eclampsia, discussed later in this chapter (Repke, 1994). Deficiencies of calcium and magnesium are associated with premature birth. Sometimes even

an excellent diet does not provide enough iron, calcium, or folic acid, in which case the pregnant woman should take supplements.

Effects of Drugs Taken During Pregnancy

We are such a pill-popping culture that we seldom stop to think about whether we should take a certain drug. The pregnant woman, however, needs to know that when she takes a drug, not only does it circulate through her body, but it may also circulate through the fetus. Because the fetus develops so rapidly during pregnancy, drugs may produce severe consequences, including serious malformations. Drugs that produce such defects are called **teratogens.**[3] Of course, not all drugs can cross the placental barrier, but many can. The drugs that pregnant women should be cautious in using are discussed below.

Antibiotics

Long-term use of antibiotics by the woman may cause damage to the fetus. Tetracycline may cause stained teeth and bone deformities. Gentamycin, kanamycin, neomycin, streptomycin, and vancomycin all may cause deafness. Nitrofurantoin may cause jaundice. Accutane (isotretinoin), used to treat acne, can cause severe birth defects if taken by a pregnant woman. Some drugs taken by diabetics may cause various fetal anomalies.

Alcohol

WARNING: ACCORDING TO THE SURGEON GENERAL, WOMEN SHOULD NOT DRINK ALCOHOLIC BEVERAGES DURING PREGNANCY BECAUSE OF THE RISK OF BIRTH DEFECTS.

A substantial amount of research has documented the risks to a child of maternal drinking

---

[2]There is, however, some risk associated with cunnilingus for the pregnant woman, as discussed in Chapter 10.

[3]"Teratogen" is from the Greek words *teras*, meaning "monster," and *gen*, meaning "cause."

Teratogen: A substance that produces defects in a fetus.

during pregnancy. Alcohol consumed by the woman circulates through the fetus, so it can have pervasive effects on fetal growth and development.

The effects of prenatal alcohol consumption are dose-dependent; that is, the more alcohol the mother drinks, the larger the number and severity of effects on the child. Consuming the equivalent of one drink or two glasses of wine per day was associated with slower information-processing times in a study of 6-month-old infants (Jacobson et al., 1993). A study of 403 black children at 1 year found that slower reaction times, increased visual fixation, and reduced complexity of play were associated with maternal alcohol consumption during the second and third trimesters of pregnancy (Jacobson et al., 1994). The children of mothers who had taken at least one drink per day in mid-pregnancy had poorer gross and fine motor skills (e.g., catching and throwing a ball) at age 4 (Barr et al., 1990). Deficits of 10 points or more in intelligence test scores at 1 year and lower scores at 4 years of age are associated with prenatal drinking (O'Connor et al., 1993; Barr et al., 1990). Data from the Seattle longitudinal study indicate that maternal binge drinking (five or more drinks per occasion) is associated with numerous academic and behavioral problems among children at age 11 (Olson et al., 1992).

The abuse of alcohol during pregnancy may result in offspring who display a pattern of malformations termed the **fetal alcohol syndrome (FAS)** (for reviews, see Abel, 1984; Clarren & Smith, 1978). Among the characteristics of the syndrome are both prenatal and postnatal growth deficiencies, a small brain, small eye openings, and joint, limb, and heart malformations. Perhaps the most serious effect is mental retardation. About 85 percent of children with the FAS score two or more standard deviations below the mean on intelligence tests—that would be an IQ of about 70 or below. Indeed, FAS is the leading preventable cause of mental retardation (Braun, 1996).

Current research indicates that the fetus is at risk for FAS if the mother's consumption of alcohol is six or more drinks per day. It is also clear that women who drink moderately may have children who are affected to some degree (Abel, 1980). The effects have even been documented in the children when they reach adulthood (Streissguth et al., 1989). Unfortunately, "safe" limits for alcohol consumption during pregnancy have not been established, so probably the best advice about drinking for women who are—or may be—pregnant is "Don't."

Cocaine

Cocaine abuse during pregnancy is associated with an increased risk of premature birth (Handler et al., 1991) and low birth weight (Phibbs et al., 1991). The only regularly noted physical abnormality is smaller head circumference (Cherukuri et al., 1988); cocaine-exposed children are more likely to be microcephalic, which is in turn associated with poorer growth and lower intelligence-test scores in school-age children. Infants exposed to cocaine in utero display neurological deficits and central nervous system anomalies (O'Shea, 1995). At 3 to 6 months of age they are more irritable. At 18 months, they have difficulty focussing their attention. In school, such children have significantly more behavior problems (Vogel, 1997).

Steroids

Synthetic hormones such as progestin can cause masculinization of a female fetus, as discussed in Chapter 5. Corticosteroids are linked with low birth weight, cleft palate, and stillbirth in some studies but not in others (Østensen, 1994). Excessive amounts of vitamin A are associated with cleft palate. Excesses of vitamins D, B_6, and K have also been associated with fetal defects. A potent estrogen, diethylstilbestrol (DES), has been shown to cause cancer of the vagina in girls whose mothers took the drug while pregnant (Herbst, 1972). Long-term exposure to DES is associated with an increased risk of low birth weight (Zhang & Bracken, 1995). At one time, DES was used as a "morning after" treatment, following unprotected intercourse, but such use is no longer approved (Hatcher et al., 1994).

Other Drugs

According to the U.S. Public Health Service, maternal smoking during pregnancy exerts a retarding

Fetal alcohol syndrome (FAS): Serious growth deficiency and malformations in the child of a mother who abuses alcohol during pregnancy.

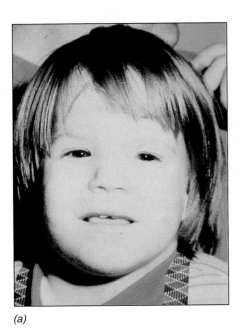

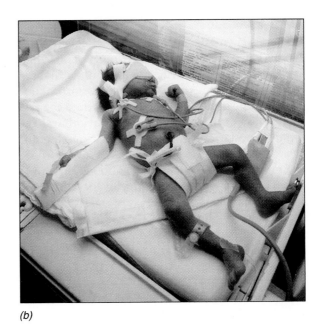

(a)

(b)

Figure 7.6   The effects of prenatal exposure to alcohol and other drugs. *(a)* A child born with fetal alcohol syndrome. *(b)* A baby whose mother used crack cocaine during pregnancy.

influence on fetal growth indicated by decreased infant birth weight and increased incidence of prematurity (for a review, see Coleman et al., 1979). A study of 28,000 children found that children whose mothers smoked heavily during pregnancy were almost twice as likely to be hyperactive and impulsive when they were 6 years old than were the children of nonsmoking mothers; the children of smoking mothers also had lower IQs and less developed motor skills (Dunn et al., 1977).

Some antihistamines may produce malformations. Even plain aspirin may cause blood problems (Cunningham et al., 1989).

Although not classified as a drug, x-rays deserve mention here since they can damage the fetus, particularly during the first 42 days after conception.

The psychoactive chemical in marijuana crosses the placental barrier (Harbison & Mantilla-Plata, 1972; Idänpään-Heikkilä et al., 1969). There is some evidence that marijuana inhibits ovulation (Abel, 1984); thus its use might make it more difficult to become pregnant. In one Canadian study, women who used marijuana during pregnancy were compared with a control group of nonusers. There were no differences between the two groups in rate of miscarriage, complications during birth,

or incidence of birth defects. However, the newborns of the marijuana users had more tremors and had a higher rate of visual problems in the preschool years (Fried, 1986). A longitudinal study of 668 children found that children of users attained lower scores on the Stanford-Binet Intelligence Scale at ages 3 and 6. Six-year-olds who had been exposed to marijuana in utero performed poorly on tasks requiring attention (Day & Richardson, 1994).

Tricyclic antidepressant medications such as amitriptyline and imipramine have been associated with birth defects in some studies but not others. One study of children whose mothers took tricyclics during pregnancy found no significant differences in language development, measured intelligence, mood or temperament at 16 to 86 months of age (Nulman et al., 1997). Use during pregnancy of lithium, especially lithium carbonate, is associated with fetal cardiovascular abnormalities (Cunningham et al., 1993). The effects of Prozac during pregnancy have been studied in animals; no adverse effects have been found.

As with alcohol, the best rule for the pregnant woman considering using a drug is, "When in doubt, don't."

Dads and Drugs

Most research has focused on the effects of drugs taken by the pregnant woman. However, new theorizing suggests that drugs taken by men before a conception may also cause birth defects, probably because the drugs damage the sperm and their genetic contents (Narod et al., 1988). In addition, one study found evidence that mother's smoking during the first trimester of pregnancy increased her offspring's risk of cancer in childhood; but father's smoking during the pregnancy in the absence of mother's smoking also increased the risk of childhood cancer (John et al., 1991). In the future we will surely see more concern and research on the effects of fathers' use of drugs on their children.

Birth

The Beginning of Labor

The signs that labor is about to begin vary from one woman to the next. There may be a discharge of a small amount of bloody mucus (the "bloody show"). This is the mucus plug that was in the cervical opening during pregnancy, its purpose being to prevent germs from passing from the vagina up into the uterus. In about 10 percent of all women the membranes containing the amniotic fluid rupture (the bag of waters bursts), and there is a gush of warm fluid down the woman's legs. Labor usually begins within 24 hours after this occurs. More commonly, the amniotic sac does not rupture until the end of the first stage of labor. The Braxton-Hicks contractions may increase before labor and actually may be mistaken for labor. Typically they are distinct from the contractions of labor in that they are very irregular.

The biological mechanism that initiates and maintains labor is not completely understood. The progesterone-withdrawal theory is the leading hypothesis (Schwartz, 1997). Progesterone is known to inhibit uterine contractions. It has been proposed that some mechanism such as increased production of antiprogesterone reduces the inhibiting effect of progesterone and labor begins.

The Stages of Labor

Labor is typically divided into three stages, although the length of the stages may vary considerably from one woman to the next. The whole process of childbirth is sometimes referred to as *parturition.*

First-Stage Labor

First-stage labor begins with the regular contractions of the muscles of the uterus. These contractions are responsible for producing two changes in the cervix, both of which must occur before the baby can be delivered. These changes are called **effacement** (thinning out) and **dilation** (opening up). The cervix must dilate until it has an opening 10 centimeters (4 inches) in diameter before the baby can be born.

First-stage labor itself is divided into three stages: early, late, and transition. In *early first-stage labor,* contractions are spaced far apart, with perhaps 15 to 20 minutes between them. A contraction typically lasts 45 seconds to a minute. This stage of labor is fairly easy, and the woman is quite comfortable between contractions. Meanwhile, the cervix is effacing and dilating.

Late first-stage labor is marked by the dilation of the cervix from 5 to 8 centimeters (2 to 3 inches). It is generally shorter than the early stage, and the contractions are more frequent and more intense.

The final dilation of the cervix from 8 to 10 centimeters (3 to 4 inches) occurs during the **transition** phase, which is both short and difficult. The contractions are very strong, and it is during this stage that women report pain and exhaustion.

The first stage of labor can last anywhere from 2 to 24 hours. It averages about 12 to 15 hours for a first pregnancy and about 8 hours for later pregnancies. (In most respects, first labors are the hardest, and later ones are easier.) The woman is usually told to go to the hospital when the contractions are 4 to 5 minutes apart. Once there, she is put in the labor room or birthing room for the rest of first-stage labor.

Effacement: A thinning out of the cervix during labor.
Dilation: An opening up of the cervix during labor; also called dilatation.
First-stage labor: The beginning of labor during which there are regular contractions of the uterus; the stage lasts until the cervix is dilated 8 centimeters (3 inches).
Transition: The difficult part of labor at the end of the first stage, during which the cervix dilates from 8 to 10 centimeters (3 to 4 inches).

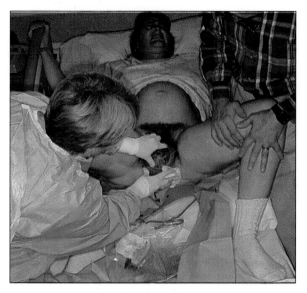

(a)

Figure 7.7 Second-stage labor. *(a)* Baby's head crowning and then *(b)* moving out.

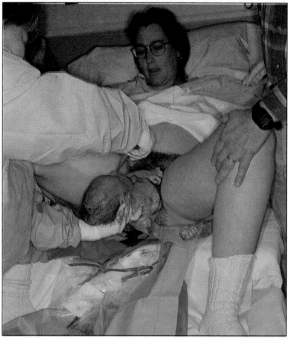

(b)

Second-Stage Labor: Delivery

Second-stage labor (Figure 7.7) begins when the cervix is fully dilated and the baby's head (or whichever part comes first, if the baby is in some other position; see Figure 7.8) begins to move into the vagina, or birth canal. It lasts from a few minutes to a few hours and is generally much shorter than the first stage.

During this stage, many women feel an urge to push or bear down, and if done properly, this may be of great assistance in pushing the baby out. With each contraction the baby is pushed farther along.

When the baby's head has traversed the entire length of the vagina, the top of it becomes visible at the vaginal entrance; this is called *crowning*. It is at this point that many physicians perform an **episiotomy** (see Figure 7.9), in which an incision or slit is made in the perineum, the skin just behind the vagina. Most women do not feel the episiotomy being performed because the pressure of the baby against the pelvic floor provides a natural anesthetic. The incision is stitched closed after the baby is born. The reason physicians give for performing an episiotomy is that if it is not done, the baby's head may rip the perineum; a neat incision is easier to repair than a ragged tear, and the tear may go

much deeper and damage more tissue. But the use of the episiotomy has been questioned; critics claim that it is unecessary and is done merely for the doctor's convenience, while causing the woman discomfort later as it is healing. They note that episiotomies are usually not performed in western European countries, where delivery still takes place quite nicely.

The baby is finally eased completely out of the mother's body. At this point, the baby is still connected to the mother by the umbilical cord, which runs from the baby's navel to the placenta, and the placenta is still inside the mother's uterus. As the baby takes its first breath of air, the functioning of its body changes dramatically. Blood begins to flow to the lungs, there to take on oxygen, and a flap closes between the two atria (chambers) in the heart. This process generally takes a few minutes, during which time the baby changes from a somewhat bluish color

---

Second-stage labor: The stage during which the baby moves out through the vagina and is delivered.
Episiotomy (ih-pee-see-AH-tuh-mee): An incision made in the skin just behind the vagina, allowing the baby to be delivered more easily.

Figure 7.8   Possible positions of the fetus during birth. *(a)* A breech presentation (4 percent of births). *(b)* A transverse presentation (less than 1 percent). *(c)* A normal, head-first or cephalic presentation (96 percent of births).

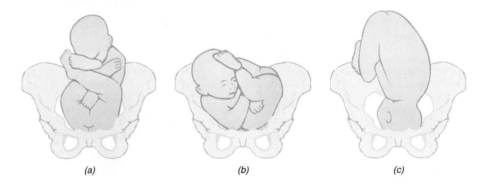

(a)   *(b)*   *(c)*

Figure 7.9   Episiotomy. A Mediolateral or median cut may be performed.

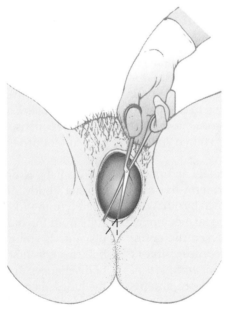

to a healthy, pink hue. At this point, the baby no longer needs the umbilical cord, which is clamped and cut off about 7 centimeters (3 inches) from the body. The stub gradually dries up and falls off.

To avoid the possibility of transmitting gonorrhea or other eye infections from the mother to the baby, drops of silver nitrate or a similar drug are placed in the baby's eyes (see Chapter 20).

Third-Stage Labor
During **third-stage labor,** the placenta detaches from the walls of the uterus, and the afterbirth (placenta and fetal membranes) is expelled. This stage

Third-stage labor: The stage during which the afterbirth is expelled.

may take from a few minutes to an hour. Several contractions may accompany the expulsion of the placenta. The episiotomy and/or any tears are sewn up.

Cesarean Section (C Section)

Cesarean section is a surgical procedure for delivery; it is used when normal vaginal birth is impossible or undesirable. Cesarean section may be required for a number of different reasons: if the baby is too large or if the mother's pelvis is too small to allow the baby to move into the vagina; if the labor has been very long and hard and the cervix is not dilating or if the mother is nearing the point of total exhaustion; if the umbilical cord *prolapses* (moves into a position such that it is coming out through the cervix ahead of the baby); if there is an Rh incompatibility (see below); or if there is excessive bleeding or the mother's or the infant's condition takes a sudden turn for the worse. Another condition requiring a C section is *placenta previa*, in which the placenta is attached to the wall of the uterus over or close to the cervix.

In the cesarean section, an incision is made through the abdomen and through the wall of the uterus. The physician lifts out the baby and then sews up the uterine wall and the abdominal wall.

Cesarean delivery rates increased steadily in the United States in the 1970s and 1980s, peaking at 24 percent in 1989 (Taffel et al., 1991), which is a rate considerably higher than in most western European countries. For example, the cesarean rate is 10 percent of deliveries in the Netherlands, 12 percent of deliveries in Norway, and 10 percent of deliveries in England and Wales (Notzon, 1990). Contrary to popular opinion, it is not true that once a woman has had one delivery by cesarean she must have all subsequent deliveries by the same method. Vaginal births are possible after cesareans (Taffel et al., 1991), although often the same conditions are present in later deliveries that necessitated the first cesarean, making it necessary again in the later deliveries. Rates of cesarean section in the United States have fallen slightly since 1989, to 20.6 percent in 1996 (Guyer et al., 1997). The decline is due almost entirely to an increase in the number of women giving birth vaginally after an earlier cesarean birth.

There is concern about the high U.S. cesarean rates. Reasons that have been proposed to explain them include the following: (a) Physicians make more money performing cesareans. (b) There are more older women giving birth; they may have more difficult labors necessitating cesareans. (c) There are more births to teenagers, who also are at risk for difficult deliveries. (d) Fetal monitors are used increasingly; they can give the physician early warning if the fetus is in distress, necessitating a cesarean to save the fetus.

Programs in two hospitals indicate that cesarean section rates can be decreased dramatically. In one hospital, prenatal education was expanded, guidelines for cesareans were tightened, women were encouraged to remain active during labor (for example, to walk at regular intervals), and physicians were given statistics each month about their own deliveries. In May 1988, when the study began, the rate of cesarean births was 31 percent. For the period January to June 1994, it was 15 percent (Hollander, 1996).

Childbirth Options

Pregnant women and their partners can choose from a variety of childbirth options. Foremost among these is taking childbirth classes to prepare mentally and physically for labor and delivery. In addition, there are several options regarding the use of anesthesia during childbirth. Third, women can often choose to give birth at home, in a birthing or maternity center, or in a hospital delivery suite.

The Lamaze Method

English obstetrician Grantly Dick-Read coined the term "natural childbirth" in 1932. He postulated that fear causes tension and tension causes pain. Thus to attempt to eliminate the pain of childbirth, he recommended a program consisting of education (to eliminate the woman's fears of the unknown) and the learning of relaxation techniques (to eliminate tension).

One of the most widely used methods of *prepared childbirth* was developed by French obstetrician Fernand Lamaze. Classes teaching the Lamaze method or variations of it are now offered in most areas of the country. The **Lamaze method**

Cesarean section (C section): A method of delivering a baby surgically, by an incision in the abdomen.
Lamaze method: A method of "prepared" childbirth involving relaxation and controlled breathing.

involves two basic techniques, *relaxation* and *controlled breathing* (Figure 7.10). The woman learns to relax all the muscles in her body. Knowing how to do this has a number of advantages, including conservation of energy during an event that requires considerable endurance and, more important, avoidance of the tension that increases the perception of pain. The woman also learns a series of controlled breathing exercises, which she will use to help her during each contraction.

Some other techniques are taught as well. One, called *effleurage*, consists of a light, circular stroking of the abdomen with the fingertips. There are also exercises to strengthen specific muscles, such as the leg muscles, which undergo considerable strain during labor and delivery. Finally, because the Lamaze method is based on the idea that fear and the pain it causes are best eliminated through education, the Lamaze student learns a great deal about the processes involved in pregnancy and childbirth.

One other important component of the Lamaze method is the requirement that the woman be accompanied during the classes and during childbirth itself by her partner or some other person, who serves as "coach." The coach plays an integral role in the woman's learning of the techniques and her use of them during labor. He (we shall assume that it is the baby's father) is present during labor and delivery. He times contractions, checks on the woman's state of relaxation and gives her feedback if she is not relaxed, suggests breathing patterns, helps elevate her back as she pushes the baby out, and generally provides encouragement and moral support. Aside from the obvious benefits to the woman, this principle of the Lamaze method represents real progress in that it allows the man to play an active role in the birth of his child and to experience more fully one of the most basic and moving of all human experiences.

One common misunderstanding about the Lamaze method is that the use of anesthetics is prohibited. In fact, the Lamaze method is more flexible than that. Its goal is to teach each woman the techniques she needs to control her reactions to labor so that she will not need an anesthetic; however, her right to have an anesthetic if she wants one is affirmed. The topic of anesthetics in childbirth, which has become quite controversial in recent years, is discussed next.

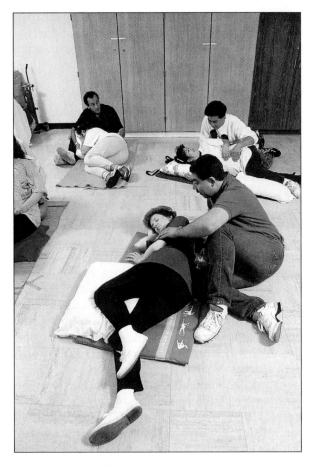

Figure 7.10 Practice in relaxation and breathing techniques is essential in preparing for a Lamaze childbirth.

A number of studies indicate that childbirth training, such as the Lamaze method, has several desirable results. These include reduction in the length of labor, decreased incidence of birth complications, a decrease in the use of anesthetics, a more positive attitude after birth, increased self-esteem, and an increased sense of being in control (e.g., Felton & Segelman, 1978; Zax et al., 1975). Lamaze training is associated with increased tolerance for pain and reduced anxiety both before birth and four weeks after birth (Markman & Kadushin, 1986; McClure & Brewer, 1980; Worthington et al., 1983).

There is no doubt that the Lamaze method has improved the childbirth experiences of thousands of women and men. On the other hand, some Lamaze advocates are so idealistic that they may create unrealistic expectations about delivery, especially for women having their first baby (**primiparas**). The use of the Lamaze method reduces pain in childbirth but does not eliminate it completely. For primiparas, there is often a discrepancy between their positive expectations for delivery and the actual outcomes (Booth & Meltzoff, 1984). Thus, while the Lamaze method produces excellent outcomes and helps women control pain, childbirth still involves some pain, as well as unexpected complications in some cases.

The Use of Anesthetics in Childbirth

Throughout most of human history, childbirth has been "natural"; that is, it has taken place without anesthetics and in the woman's home or other familiar surroundings. The pattern in the United States began to change about 200 years ago, at the time of the Revolutionary War, when male physicians rather than midwives started to assist during birth (Wertz & Wertz, 1977). The next major change came around the middle of the nineteenth century, with the development of anesthetics for use in surgery. When their use in childbirth was suggested, there was some opposition from physicians, who felt they interfered with "natural" processes, and some opposition from the clergy, who argued that women's pain in childbirth was prescribed in the Bible, quoting Genesis 3:16: "In sorrow thou shalt bring forth children." Opposition to the use of anesthetics virtually ceased, however, when Queen Victoria gave birth under chloroform anesthesia in 1853. Since then, the use of anesthetics has become routine (too routine, according to some) and effective. Before discussing the arguments for and against the use of anesthetics, let us briefly review some of the common techniques of anesthesia used in childbirth.

Tranquilizers (such as Valium) or narcotics may be administered when labor becomes fairly intense. They relax the woman and take the edge off the pain. Barbiturates (Nembutal or Seconal) are administered to put the woman to sleep. Scopolamine may sometimes be used for its amnesic effects; it makes the woman forget what has happened, and thus she has no memory afterward of any pain during childbirth. Regional and local anesthetics, which numb only the specific region of the body that is painful, are used most commonly. An example is the pudendal block (named for the pudendum, or vulva), in which an injection numbs only the external genitals. Other examples are spinal anesthesia (a "spinal"), in which an injection near the spinal cord numbs the entire birth area, from the waist down, and the caudal block and epidural anesthesia, which are both administered by injections in the back and produce regional numbing from the belly to the thighs (For more information, see Coustan, 1995a).

The routine use of anesthetics has been questioned by some. Proponents of the use of anesthetics argue that with modern technology, women no longer need to experience pain during childbirth and that it is therefore silly for them to suffer unnecessarily. Opponents argue that anesthetics have a number of well-documented dangerous effects on both mother and infant. Anesthetics in the mother's bloodstream pass through the placenta to the infant. Thus while they have the desired effect of depressing the mother's central nervous system, they also depress the infant's nervous system. Research indicates that babies born under anesthesia have poor sucking ability and retarded muscular and neural development in the first four weeks of life compared with infants born with no anesthesia (Boston Women's Health Book Collective, 1992; MacFarlane et al., 1978). Anesthetics prevent the mother from using her body as effectively as she might to help push the baby out. If administered early in labor, anesthetics may inhibit uterine contractions, slow cervical dilation, and prolong labor. They also numb a woman to one of the most fundamental experiences of her life.

Perhaps the best resolution of this controversy is to say that a pregnant woman should participate in prepared childbirth classes and should use the techniques during labor. If, when she is in labor, she discovers that she cannot control the pain and wants an anesthetic, she should feel free to request it and to do so without guilt; the anesthetic should then be administered with great caution.

Home Birth Versus Hospital Birth

Home birth has become increasingly popular. Either a physician or a nurse-midwife may assist in a

Primipara: A woman having her first baby.

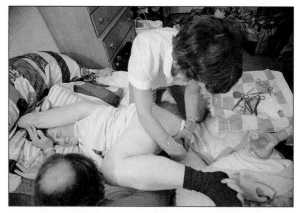

(a)

(b)

Figure 7.11 A home birth

home birth. Advocates of home birth argue that the atmosphere in a hospital—with its forbidding machines, rules and regulations, and general lack of comfort and "homeyness"—is stressful to the woman and detracts from what should be a joyous, natural human experience. Furthermore, hospitals are meant to deal with illness, and the delivery of a baby should not be viewed as illness; hospital births encourage the use of procedures such as forceps deliveries and episiotomies that are themselves dangerous. Birth at home is likely to be more relaxed and less stressful; friends and other children are allowed to be present. There are some studies that indicate that—for uncomplicated pregnancies—home delivery is as safe as hospital delivery (Hahn & Paige, 1980; for a detailed discussion, see Hoff & Schneiderman, 1985).

On the other side of the argument, if unforeseen emergency medical procedures are necessary, home birth may be downright dangerous for the mother, the baby, or both. Furthermore, hospital practices in labor and delivery have changed radically, particularly with the increased popularity of the Lamaze method; thus hospitals are not the forbidding, alien environments they once were. Most hospitals, for example, allow fathers to be present for the entire labor and delivery, and many allow the father to be present in the operating room during cesarean deliveries. Many hospitals have created birthing centers which contain a set of home-like rooms, with comfortable beds and armchairs, that permit labor and delivery to occur in a relaxed atmosphere, while being only a minute away from emergency equipment if it is required.

For any home birth, careful medical screening is essential. Only women with normal pregnancies and anticipated normal deliveries should attempt a home birth. A qualified physician or nurse-midwife must be part of the planning. Finally, there must be access to a hospital in case an emergency arises.

After the Baby Is Born: The Postpartum Period

Physical Changes

With the birth of the baby, the woman's body undergoes a drastic physiological change. During pregnancy the placenta produces high levels of both estrogen and progesterone. When the placenta is expelled, the levels of these hormones drop sharply, and thus the postpartum period is characterized by low levels of both estrogen and progesterone. The levels of these hormones gradually return to normal over a period of a few weeks to a few months. Other endocrine changes include an increase in hormones associated with breast-feeding (discussed next).

In addition, the body undergoes considerable stress during labor and delivery, and the woman may feel exhausted. Discomfort from the episiotomy is common in the first postpartum weeks.

Psychological Changes

For a day or two after parturition, the woman typically remains in the hospital, although there is an increasing national trend toward leaving the hospital less than 24 hours after delivery. For the first two days, women often feel elated; the long pregnancy is over, they have been successful competitors in a demanding athletic event and are pleased with their efforts, and the baby is finally there, to be cuddled and loved.

Within a couple of days after delivery, many women experience depression and periods of crying. These mood swings range from mild to severe. In the mildest type, *maternity blues* or "baby blues," the woman experiences sadness and periods of crying, but this mood lasts only 24 to 48 hours (Hopkins et al., 1984). Between 50 percent and 80 percent of women experience mild baby blues postpartum. Mild to moderate **postpartum depression** is experienced by approximately 13 percent of women and typically lasts 6 to 8 weeks (O'Hara & Swain, 1996; Hopkins et al., 1984). Postpartum depression is characterized by a depressed mood, insomnia, tearfulness, feelings of inadequacy and inability to cope, irritability, and fatigue. Finally, the most severe disturbance is *postpartum depressive psychosis;* fortunately, it is rare, affecting only 0.01 percent of women following birth (Hopkins et al., 1984).

It appears that many factors contribute to this depression. Being in a hospital in and of itself is stressful, as noted previously. Once the woman returns home, another set of stresses faces her. She has probably not yet returned to her normal level of energy, yet she must perform the exhausting task of caring for a newborn infant. For the first several weeks or months she may not get enough sleep, rising several times during the night to tend to a baby that is crying because it is hungry or sick, and she may become exhausted. Clearly she needs help and support from her partner and friends at this time. Some stresses vary depending on whether this is a first child or a later child. The first child is stressful because of the woman's inexperience; while she is in the hospital she may become anxious, wondering whether she will be capable of caring for the infant when she returns home. In the case of later-born children, and some first-borns, the mother may become depressed because she did not really want the baby.

Physical stresses are also present during the postpartum period; hormone levels have declined sharply, and the body has been under stress. Thus it appears that postpartum depression is caused by a combination of physical and social factors.

Another view of the psychological aspects of pregnancy and the postpartum period has been proposed by psychoanalyst Grete Bibring and her colleagues (1961), who studied pregnant women over a 10-year period. Bibring and her colleagues see the pregnancy-postpartum period as a developmental stage, a time of maturational crisis, the resolution of which leads to emotional growth. Perhaps, then, we should begin to think of pregnancy and the postpartum period as normal parts of the process of adult development.

Fathers, too, sometimes experience depression after the birth of a baby. In one study, 89 percent of the mothers and 62 percent of the fathers had experienced the blues during the three months after the birth (Zaslow et al., 1985). The research indicates that less depressed fathers are more likely to be involved with the infant and the parenting role.

Attachment to the Baby

While much of the traditional psychological research focused on the baby's developing attachment to the mother, more recent interest has been about the development of the mother's attachment (bond) to the infant. Leifer's study (1980) showed clearly that this process begins even before the baby is born; during pregnancy, most women in her sample developed an increasing sense of the fetus as a separate individual and developed an increasing emotional attachment to it. In this sense, pregnancy is something of a psychological preparation for motherhood.

In the 1970s, pediatricians Marshall Klaus and John Kennell popularized the idea that there is a kind of "critical period" or "sensitive period" in the minutes and hours immediately after birth, during which the mother and infant should "bond" to each other (Klaus & Kennell, 1976a). These ideas have enthusiastically been incorporated into hospital practices, so that in many progressive hospitals the infant is given to the mother to hold immediately after birth. Ironically, scientists later concluded that there

---

Postpartum depression: Mild to moderate depression in women following the birth of a baby.

is little or no evidence for the sensitive-period-for-bonding hypothesis (e.g., Goldberg, 1983; Lamb, 1982; Lamb & Hwang, 1982; Myers, 1984). That outcome is fortunate. Otherwise, mothers who give birth by cesarean section (and may therefore be asleep under a general anesthetic for an hour or more after the birth) and adoptive parents would have to be presumed to have inadequate bonds with their children. We know that, in both cases, strong bonds of love form between parents and children despite the lack of immediate contact following birth. The hospital policy of immediate mother-infant and father-infant contact is still a good one, if only because it is intensely pleasurable and satisfying.

Sex During Postpartum

The birth of a child has a substantial effect on a couple's sexual relationship. Following the birth, the mother is at some risk of infection or hemorrhage (Cunningham et al., 1993) so the couple should wait at least two weeks before resuming intercourse. When coitus is resumed, it may be uncomfortable or even painful for the woman. If she had an episiotomy, she may experience vaginal discomfort; if she had a cesarean birth, she may experience abdominal discomfort. Fatigue of both the woman and her partner also may influence when they resume sexual activity.

A longitudinal study of the adjustment of couples to the birth of a child collected data from 570 women (and 550 partners) four times: during the second trimester of pregnancy, and at one, four,

and twelve months postpartum (Hyde et al., 1996). Data on the sexual relationship are displayed in Table 7.2. In the month following birth, only 17 percent resumed intercourse; by the fourth month, nine out of ten couples had, the same percentage as reported intercourse during the second trimester. Reports of cunnilingus showed a similar pattern, while reports of fellatio did not indicate a marked decline. Note that although sexual behavior was much less frequent in the month following birth, satisfaction with the sexual relationship remained high. A major influence on when the couple resumed intercourse was whether the mother was breast-feeding. At both 1 month and four months after birth, breast-feeding women reported significantly less sexual activity and lower sexual satisfaction. One reason is that lactation suppresses estrogen production, which in turn results in decreased vaginal lubrication; this makes intercourse uncomfortable. This problem can be resolved by the use of vaginal lubricants.

Breast-Feeding

Biological Mechanisms

Two hormones, both secreted by the pituitary, are involved in lactation (milk production). One, *prolactin,* stimulates the breasts to produce milk. Prolactin is produced fairly constantly for whatever length of time the woman breast-feeds. The other hormone, *oxytocin,* stimulates the breasts to eject milk. Oxytocin is produced reflexively by the pitu-

Table 7.2 Sexual Behaviors in the Last Month, Reported by Mothers During Pregnancy and the Year Postpartum

Behavior	Pregnancy 2nd Trimester	Postpartum 1 month	Postpartum 4 months	Postpartum 12 months
Intercourse	89%	17%	89%	92%
Mean frequency of intercourse/month	4.97	.42	5.27	5.1
Fellatio	43%	34%	48%	47%
Cunnilingus	30%	8%	44%	49%
Satisfaction with sexual relationship	3.76	3.31	3.36	3.53

Satisfaction with the relationship was rated on a scale from 1 (very dissatisfied) to 5 (very satisfied).

Source: J. S. Hyde, J. D. DeLamater, E. A. Plant, & J. M. Byrd. (1996). Sexuality During Pregnancy and the Year Postpartum. *The Journal of Sex Research, 33,* 143–151.

itary in response to the infant's sucking of the breast. Thus sucking stimulates nerve cells in the nipple; this nerve signal is transmitted to the brain, which then relays the message to the pituitary, which sends out the messenger oxytocin, which stimulates the breasts to eject milk. Interestingly, research with animals indicates that oxytocin stimulates maternal behavior (Jenkins & Nussey, 1991).

Actually, milk is not produced for several days after delivery. For the first few days, the breast secretes colostrum, discussed earlier, which is high in protein and is believed to give the baby a temporary immunity to infectious diseases. Two or three days after delivery, true lactation begins; this may be accompanied by discomfort for a day or so because the breasts are swollen and congested.

It is also important to note that, much as in pregnancy, substances ingested by the mother may be transmitted through the milk to the infant. The nursing mother thus needs to be cautious about using alcohol and other drugs.

Physical and Mental Health

The National Institutes of Health strongly encourages mothers to breast-feed, because breast milk is the ideal food for a baby and has even been termed the "ultimate health food." It provides the baby with the right mixture of nutrients, it contains antibodies that protect the infant from some diseases, it is free from bacteria, and it is always the right temperature. Thus there is little question that it is superior to cow's milk and commercial formulas.

The evidence indicates that the percentage of infants who were breast-fed rose steadily during the 1970s. It peaked in 1983 and then declined somewhat in the late 1980s, so that in 1989, 52 percent of infants were breast-fed initially and 18 percent were still breast-fed at five to six months of age (Ryan et al., 1991). By 1994, 58 percent of babies were breast-fed initially. The incidence varies as a function of ethnicity, with 61 percent of Anglo mothers, 67 percent of Hispanic mothers, and 27 percent of African-American mothers breast-feeding during the first three months (National Study for Health Statistics, 1998).

From the mother's point of view, breast-feeding has several advantages. These include a quicker shrinking of the uterus to its normal size and reduced likelihood of becoming pregnant again im-

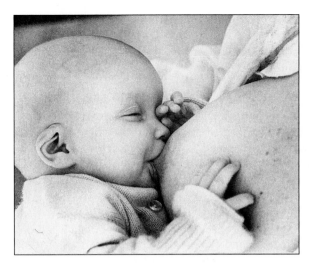

Figure 7.12 Breast-feeding

mediately. The return of normal menstrual cycles is delayed, on the average, in women who breast-feed, compared with those who do not. This provides some period of rest between pregnancies. However, it is important to note that a woman can become pregnant again after parturition but before she has had her first menstrual period. As was discussed in Chapter 6, ovulation precedes menstruation; thus a woman may ovulate and conceive without having a period. Therefore, a woman should not count on breast-feeding as a means of contraception.

Some women report sexual arousal during breast-feeding, and a few even report having orgasms. Unfortunately, this sometimes produces anxiety in the mother, leading her to discontinue breast-feeding. However, there is nothing "wrong" with this arousal, and it appears to stem from activation of hormonal mechanisms. Clearly, from an adaptive point of view, if breast-feeding is important to the infant's survival, it would be wise for Nature to design the process so that it is rewarding to the mother.

The La Leche League is devoted to encouraging women to breast-feed their babies and has helped to spread information on breast-feeding. The organization tends to be a bit militant in its advocacy of breast-feeding, however. A few women are physically unable to breast-feed, while some others feel psychologically uncomfortable with the idea. And breast-feeding can be very inconvenient for the woman who works outside the home. While breast-feeding has some important advantages, as noted previously, long-term studies comparing breast-fed children with bottle-fed children have found no significant differences between them (Schmidt, 1970). What appears to be more important than the method of feeding are the quality of the relationship between the mother and infant and feelings that the mother communicates to the baby:

> A baby raised in a loving home can grow up to be a healthy, psychologically secure individual no matter how he receives his nourishment. While successful nursing is a beautiful, happy experience for both mother and child, the woman who nurses grudgingly because she feels she should will probably do more harm to her baby by communicating her feelings of resentment and unhappiness, than she would if she were a relaxed, loving, bottle-feeding mother (Olds & Eiger, 1973, p. 18).

Problem Pregnancies

Ectopic Pregnancy

An **ectopic pregnancy** (misplaced pregnancy) occurs when the fertilized egg implants somewhere other than the uterus. Most commonly, ectopic pregnancies occur when the egg implants in the fallopian tube (tubal pregnancy; Schenker & Evron, 1983). In rare cases, implantation may also occur in the abdominal cavity, the ovary, or the cervix.

A tubal pregnancy may occur if, for one reason or another, the egg is prevented from moving down the tube to the uterus, as when the tubes are obstructed as a result of a gonorrheal infection. Early in a tubal pregnancy, the fertilized egg implants in the tube and begins development, forming a placenta and producing the normal hormones of pregnancy. The woman may experience the early symptoms of pregnancy, such as nausea and amenorrhea, and think she is pregnant; or she may experience some bleeding which she mistakes for a period, and think that she is not pregnant. It is therefore quite difficult to diagnose a tubal pregnancy early.

A tubal pregnancy may end in one of two ways. The embryo may spontaneously abort and be released into the abdominal cavity, or the embryo and placenta may continue to expand, stretching the tube until it ruptures. Symptoms of a rupture include sharp abdominal pain or cramping, dull abdominal pain and possibly pain in the shoulder, and vaginal bleeding. Meanwhile, hemorrhaging is occurring, and the woman may go into shock and, possibly, die; thus it is extremely important for a woman displaying these symptoms to see a doctor quickly.

The rate of ectopic pregnancy increased 300 percent in the United States from 1970 to 1987, going from 4.5 cases per 1000 reported pregnancies to 16.8 per 1000 pregnancies (Cunningham et al., 1993). Similar increases have been observed in a number of western European nations. It is thought that these changes are due to (1) increased rates of sexually transmitted diseases, some of which lead to blocking of the fallopian tubes; and (2) increased use of contraceptives such as the IUD and progestin-only methods that prevent implantation in the uterus but do not necessarily prevent conception.

Pseudocyesis (False Pregnancy)

In **pseudocyesis,** or *false pregnancy,* the woman believes that she is pregnant and shows the signs and symptoms of pregnancy without really being pregnant. She may stop menstruating and may have morning sickness. She may begin gaining weight, and her abdomen may bulge. The condition may persist for several months before it goes away, either spontaneously or as a result of psychotherapy. In rare cases it persists until the woman goes into labor and delivers nothing but air and fluid.

Pregnancy-Induced Hypertension

Pregnancy may cause a woman's blood pressure to rise to an abnormal level. *Pregnancy-induced hypertension* includes three increasingly serious conditions: (1) hypertension, (2) preeclampsia, and (3) eclampsia. Hypertension refers to elevated

---

Ectopic pregnancy: A pregnancy in which the fertilized egg implants somewhere other than the uterus.
Pseudocyesis: False pregnancy, in which the woman displays the signs of pregnancy but is not actually pregnant.
Preeclampsia: A serious disease of pregnancy, marked by high blood pressure, severe edema, and proteinuria.

blood pressure alone. **Preeclampsia** refers to elevated blood pressure accompanied by generalized edema (fluid retention and swelling) and proteinuria (protein in the urine). The combination of hypertension and proteinuria is associated with an increased risk of fetal death. In severe preeclampsia, the earlier symptoms persist, and the woman may also experience vision problems, abdominal pain, and severe headaches. In *eclampsia,* the woman has convulsions, may go into a coma, and may die (Cunningham et al., 1993).

Preeclampsia usually does not appear until after the twentieth week of pregnancy. It is more likely to occur in women who have not completed a pregnancy before. It is especially common among teenagers. Latina and black women are much more likely than non-Hispanic white women to experience preeclampsia. The possibility of preeclampsia emphasizes the need for proper medical care during pregnancy, especially for teenage and minority women. Hypertension and preeclampsia can be managed well during their early stages. Most maternal deaths occur among women who do not receive prenatal medical care.

Viral Illness During Pregnancy

Certain viruses may cross the placental barrier from the woman to the fetus and cause considerable harm, particularly if the illness occurs during the first trimester of pregnancy. The best-known example is rubella, or German measles. If a woman gets German measles during the first month of pregnancy, there is a 50 percent chance that the infant will be born deaf or mentally deficient or with cataracts or congenital heart defects. The risk then declines, so that by the third month of pregnancy the chance of abnormalities is only about 10 percent. While most women have an immunity to rubella because they had it when they were children, a woman who suspects that she is not immune can receive a vaccination that will give her immunity; she should do this well before she becomes pregnant.

Herpes simplex is also *teratogenic,* that is, capable of producing defects in the fetus. Symptoms of herpes simplex are usually mild: cold sores or fever blisters. Genital herpes (see Chapter 20) is a form of herpes simplex in which sores may appear in the genital region. Usually the infant contracts the disease by direct contact with the sore; delivery by cesarean section can prevent this. Women with herpes genitalis also have a high risk of spontaneously aborting.

Birth Defects

As has been noted, a number of factors, such as drugs taken during pregnancy and illness during pregnancy, may cause defects in the fetus. Other causes include genetic defects (for example, phenylketonuria, PKU, which causes retardation) and chromosomal defects (for example, Down syndrome, which causes retardation).

Two to three percent of all babies born in the United States have a significant birth defect. About one-fourth of miscarried fetuses are malformed. The cause of more than half of these defects is unknown (O'Shea, 1995).

In most cases, families have simply had to learn, as best they could, to live with a child who had a birth defect. Now, however, amniocentesis, chorionic villus sampling, and genetic counseling are available to help prevent some of the sorrow, provided that abortion is ethically acceptable to the parents.

Amniocentesis involves inserting a fine tube through the pregnant woman's abdomen and removing some amniotic fluid, including cells sloughed off by the fetus, for analysis. The technique is capable of providing an early diagnosis of most chromosomal abnormalities, some genetically produced biochemical disorders, and sex-linked diseases carried by females but affecting males (hemophilia and muscular dystrophy), although it cannot detect all defects. If a defect is discovered, the woman may then decide to terminate the pregnancy with an abortion.

Amniocentesis should be performed between the thirteenth and sixteenth weeks of pregnancy. This timing is important for two reasons. First, if a defect is discovered and an abortion is to be performed, it should be done as early as possible (see Chapter 8). Second, there is a 1 percent chance that

---

Amniocentesis (am-nee-oh-sen-TEE-sus): A test done to determine whether a fetus has birth defects; done by inserting a fine tube into the woman's abdomen in order to obtain a sample of amniotic fluid.

the amniocentesis itself will cause the woman to lose her baby, and the risk becomes greater as the pregnancy progresses.

Because amniocentesis itself involves some risk, it is generally thought (although the matter is controversial) that it should be performed only on women who have a high risk of bearing a child with a birth defect. A woman is in that category (1) if she has already had one child with a genetic defect; (2) if she believes that she is a carrier of a genetic defect, which can usually be established through genetic counseling; and (3) if she is over 35, in which case she has a greatly increased chance of bearing a child with a chromosomal abnormality.

Chorionic villus sampling (CVS) (see Figure 7.13) may eventually replace amniocentesis for prenatal diagnosis of genetic defects (Doran, 1990; Kolker, 1989). A major problem with amniocentesis is that it cannot be done until the second trimester of pregnancy; if genetic defects are discovered, there may have to be a late abortion. Chorionic villus sampling, in contrast, can be done in the first trimester of pregnancy, usually around 9 to 11 weeks postconception. Chorionic villus sampling can be performed in one of two ways: transcervically, in which a catheter is inserted into the uterus through the cervix; and transabdominally, in which a needle (guided by ultrasound) is inserted through the abdomen. In either case a sample of cells is taken from the chorionic villi (the chorion is the outermost membrane surrounding the fetus, the amnion, and the amniotic fluid), and these cells are analyzed for evidence of genetic defects. Studies indicate the CVS is as accurate as amniocentesis. Like amniocentesis, it carries with it a slight risk of fetal loss (due, for example, to miscarriage). For amniocentesis, the fetal loss rate is around 1 percent; for CVS it is around 2 percent (Wass et al., 1991).

Amniocentesis and CVS (when followed by abortion) raise a number of serious ethical ques-

> **Chorionic villus sampling (CVS):** A technique for prenatal diagnosis of birth defects, involving taking a sample of cells from the chorionic villus and analyzing them.

Figure 7.13   CVS and amniocentesis are both available for prenatal diagnosis of genetic defects. CVS (shown here) is able to detect chromosomal abnormalities and sex-linked diseases.

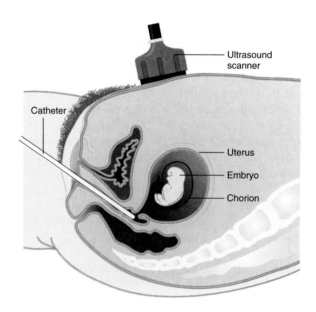

tions, some of which will be discussed in later chapters. However, it is important to note the extreme psychological stress to which families of children with birth defects are often subjected.

Rh Incompatibility

The Rh factor is a substance in the blood; if it is present, the person is said to be Rh positive (Rh+); if it is absent, the person is said to be Rh negative (Rh−). The Rh factor is genetically transmitted, with Rh+ being dominant over Rh−.

The presence or absence of the Rh factor does not constitute a health problem except when an Rh− person receives a blood transfusion or when an Rh− woman is pregnant with an Rh+ fetus (which can happen only if the father is Rh+). A blood test is done routinely early in pregnancy to determine whether a woman is Rh−. Fortunately, about 85 percent of whites and 93 percent of blacks are Rh+; thus the problems associated with being Rh− are not very common.

If some Rh+ blood gets into Rh− blood, the Rh− blood forms antibodies as a reaction against the Rh factor in the invading blood. Typically, as has been noted, there is little interchange between the woman's blood and the fetus's; the placenta keeps them separate. However, during parturition there can be considerable mixing of the two. Thus during birth, the blood of an Rh+ baby causes the formation of antibodies in an Rh− woman's blood. During the next pregnancy, some of the woman's blood enters the fetus, and the antibodies attack the fetus's red cells. The baby may be stillborn, severely anemic, or retarded. Thus there is little risk for an Rh− woman with the first pregnancy because antibodies have not yet formed; however, later pregnancies can be extremely dangerous.

Fortunately, techniques for dealing with this situation have been developed. An injection of a substance called *Rhogam* prevents the woman's blood from producing antibodies. If necessary the fetus or newborn infant may get a transfusion.

Miscarriage (Spontaneous Abortion)

Miscarriage, or *spontaneous abortion,* occurs when a pregnancy terminates through natural causes, before the conceptus is viable (capable of surviving on its own). It is not to be confused with

induced abortion, in which a pregnancy is terminated by mechanical or medicinal means (what is commonly called *abortion*—see Chapter 8), or with *prematurity,* in which the pregnancy terminates early, but after the infant is viable.

It is estimated that 20 percent of all pregnancies end in spontaneous abortion (Frishman, 1995). This is probably an underestimate, since very early spontaneous abortions may not be detected. The woman may not know that she is pregnant and may mistake the products of the miscarriage for a menstrual period. Thus the true incidence may be closer to 40 percent (Cunningham et al., 1993). Most spontaneous abortions (80 percent) occur during the first trimester of pregnancy.

Most spontaneous abortions occur because the conceptus is defective. Studies of spontaneously aborted fetuses indicate that about 60 percent showed abnormalities that were incompatible with life; for example, many had gross chromosomal abnormalities (Singer, 1995). Thus, contrary to popular belief, psychological and physical traumas are not common causes of miscarriage. In fact, spontaneous abortions seem to be functional in that they naturally eliminate many defective fetuses.

Preterm Birth

A major complication during the third trimester of pregnancy is premature labor and delivery of the fetus. When delivery occurs prior to 37 weeks gestation, it is considered *preterm.* Because the date of conception cannot always be accurately determined, preterm birth (prematurity) may be defined in terms of the birth weight of the infant; an infant weighing less than 2500 grams (5½ pounds) is considered to be in the low-birth-weight category. However, this is inappropriate. The principal concern should be the functional development of the infant rather than the weight. It is estimated that about 10 percent of the births in the United States are preterm (Coustan, 1995b).

Preterm birth is a cause for concern because the premature infant is much less likely to survive than the full-term infant. It is estimated that more than

> **Miscarriage:** The termination of a pregnancy before the fetus is viable, as a result of natural causes (not medical intervention).

half of the deaths of newborn babies in the United States are due to preterm birth. Preterm infants are particularly susceptible to respiratory infections, and they must receive expert care. Advances in medical techniques have considerably improved survival rates for preterm infants. Currently 99 percent of infants weighing 2500 grams at birth survive, as do 64 percent of those weighing 1000 grams (Cunningham et al., 1993). However, prematurity may cause damage to an infant who survives.

Maternal factors such as poor health, poor nutrition, heavy smoking, cocaine use, and syphilis are associated with prematurity. Pregnancy-induced hypertension can also lead to preterm birth. Young teenage mothers, whose bodies are not yet ready to bear children, are also very susceptible to premature labor and delivery. However, in over 50 percent of the cases, the cause of prematurity is unknown (Pritchard et al., 1985).

Infertility

Infertility refers to a woman's inability to conceive and give birth to a living child, or a man's inability to impregnate a woman. It is estimated that 14 percent of all couples in the United States have an infertility problem at some time (Sciarra, 1991). When fertile couples are purposely attempting to conceive a child, about 20 percent succeed within the first menstrual cycle, and about 50 percent succeed within the first six cycles (Hatcher et al., 1994). Most doctors consider a couple infertile if they have not conceived after one year of "trying." The term "sterile" refers only to an individual who has an absolute factor preventing conception.

Causes of Infertility

Contrary to popular opinion, a couple's infertility is not always caused by the woman. In about 40 percent of infertile couples, male factors are responsible, and female factors are responsible in an addi-

> **Infertility:** A woman's inability to conceive and give birth to a living child, or a man's inability to impregnate a woman.

tional 40 percent. In the remaining 20 percent, either both have problems or the cause is unknown (for a detailed discussion, see Liebmann-Smith, 1987).

Causes in the Female

The most common cause of infertility in women is pelvic inflammatory disease (PID) caused by a sexually transmitted disease, especially gonorrhea or chlamydia. Other causes include failure to ovulate, blockage of the fallopian tubes, and "hostile mucus," meaning cervical mucus that does not permit the passage of sperm. Less common causes include poor nutrition, eating disorders, exposure to toxic chemicals such as lead, and use of alcohol, narcotics, or barbiturates. Age may also be a factor; fertility declines in women after 35 years of age, the decline being especially sharp after age 40.

Causes in the Male

The most common cause of infertility in men is infections in the reproductive system caused by sexually transmitted diseases. Other causes include low sperm count (often due to varicoceles, which are varicose veins in the testes) and low motility of the sperm, which means that the sperm are not good swimmers. Less common causes include exposure to toxic agents such as lead and pesticides, smoking, alcohol and marijuana use, and use of some prescription drugs (Hatcher et al., 1994).

Combined Factors

In some situations, a combination of factors in both the man and woman causes the infertility. For example, there may be an immunologic response. The woman may have an allergic reaction to the man's sperm, causing her to produce antibodies that destroy the sperm, or the man may produce the antibodies himself. A couple may also simply lack knowledge; for instance, they may not know how to time intercourse correctly so that conception may take place.

Psychological Aspects of Infertility

It is important to recognize the psychological stress to which an infertile couple may be subjected (Liebmann-Smith, 1987). Because the male role is defined partly in our society by the ability to father children, the man may feel that his masculinity or virility is in question. Similarly, the female role is defined largely by the ability to bear children and be a mother, so the woman may feel inadequate.[4] Historically, in most cultures fertility has been encouraged, and indeed demanded; hence, pressures on infertile couples may be high, leading to more psychological stress. As emphasis on population control increases in our society, and as childlessness[5] becomes an acceptable and more recognized option in marriage, the stress on infertile couples may lessen.

Research indicates that, among couples entering infertility treatment programs, the women perceive themselves as experiencing greater emotional and social stress than do the men (unless the man is diagnosed as responsible for the infertility) (Leiblum, 1993). At the same time, women expect to receive more social support in coping with these stresses. Infertility does not significantly reduce marital satisfaction, but it does cause conflict (Andrews & Halman, 1992). It does affect the couple's sexual relationship; it reduces spontaneity (especially for couples in treatment programs which include scheduled intercourse) and is associated with lower sexual satisfaction (Zoldbrod, 1993).

Treatment of Infertility

There are physicians and clinics that specialize in the evaluation and treatment of infertility. An infertility evaluation should include an assessment of the couple's knowledge of sexual behavior and conception, and lifestyle factors such as regular drug use. Infertility caused by such factors can be easily treated.

If the infertility problem stems from the woman's failure to ovulate, the treatment may involve the so-called fertility drugs. The drug of first choice is clomiphene (Clomid). It stimulates the pituitary to produce LH and FSH, thus inducing ovulation. The treatment produces a pregnancy in about half the women who are given it. Contrary to media reports, multiple births occur only about 8 percent of the time with Clomid, compared with 1.2 percent with natural pregnancies. If treatment with Clomid is not successful, a second possibility is injections with HMG (human menopausal gonadotropin).

If the infertility is caused by blocked fallopian tubes, delicate microsurgery can sometimes be effective in removing the blockage.

If the infertility is caused by varicoceles in the testes, the condition can usually be treated successfully by a surgical procedure known as varicocelectomy.

Finally, a number of new reproductive technologies, such as in vitro fertilization, are now available for those with fertility problems, as discussed in the next section.

A Canadian study is helpful in putting issues of the treatment of infertility into perspective. Among infertile couples seeking treatment, 65 percent subsequently achieved a pregnancy with *no treatment* (Rousseau et al., 1983). For some couples conception just takes a bit longer. Thus the risks associated with treatments need to be weighed against the possibility that a pregnancy can be achieved without treatment.

New Reproductive Technologies

Reproductive technologies developed in the last two decades mean that there are many ways to conceive and birth babies besides old-fashioned sexual intercourse and pregnancy.

Artificial Insemination

Artificial insemination involves artificially placing semen in the vagina to produce a pregnancy; thus it is a means of accomplishing reproduction without having sexual intercourse. Artificial insemination in animals was first done in 1776. In 1949, when British scientists successfully froze sperm without any apparent damage to them, a new era of reproductive

---

[4]A survey of infertile couples found that the stress of infertility was associated with reduced self-esteem in both husband and wife (Andrews & Halmm, 1992).

[5]Semantics can make a big difference here. Many couples who choose not to have children prefer to call themselves "child-free" rather than "childless."

Artificial insemination: Procedure in which sperm are placed into the vagina by means other than sexual intercourse.

technology for animals began. Today cattle are routinely bred by artificial insemination.

In humans, there are two kinds of artificial insemination: artificial insemination by the husband (AIH) and artificial insemination by a donor (AID, not to be confused with the disease AIDS). AIH can be used when the husband has a low sperm count. Several samples of his semen are collected and pooled to make one sample with a higher count. This sample is then placed in the woman's vagina at the time of ovulation. AID is used when the husband is sterile. A donor provides semen to impregnate the wife. Estimates are that between 10,000 and 20,000 babies are born every year in the United States as a result of AID.

Sperm Banks

Because it is now possible to freeze sperm, it is possible to store it, and that is just what some people are doing: using frozen human *sperm banks*. The sperm banks open up many new possibilities for various life choices. For example, suppose that a couple decide, after having had two children, that they want a permanent method of contraception. The husband then has a vasectomy. Two years after he has the vasectomy, however, one of their children dies, and they very much want to have another baby. If the man has stored semen in a sperm bank, they can.

Young men can use sperm banks to store sperm before they undergo radiation therapy for cancer. They can later father children without fearing that they will transmit damaged chromosomes (as a result of the radiation) to their offspring.

One of the flashiest projects in this area is that of Robert K. Graham, a wealthy California businessman (*Time,* March 10, 1980). He started a sperm bank to collect "superior" sperm from scientists who have won a Nobel prize. He offers the sperm to young women who have a high IQ. At least five Nobel prizewinners have given donations, including physicist William Shockley. Graham provides a description of each scientist, and bright women applicants may choose whose sperm they want. Some Nobel winners are not amused, however. Burton Richter of Stanford, who received a Nobel prize in physics, reported that his students have asked him if he supplements his salary with stud fees.

Embryo Transfer

With **embryo transfer,** a fertilized, developing egg (embryo) is transferred from the uterus of one woman to the uterus of another woman. Dr. John Buster of UCLA perfected the technique for use with humans, and the first two births resulting from the procedure were announced in 1984 (Brotman, 1984; Associated Press, 1984).

This technique may enable a woman who can conceive but who always miscarries early in the pregnancy to transfer her embryo to another woman who serves as the *surrogate mother,* that is, the person who provides the uterus in which the fetus grows (and whom the media, somewhat callously, have called a "rent-a-womb"). The embryo transfer procedure also essentially can serve as the opposite of artificial insemination. That is, if a woman produces no viable eggs, her husband's sperm can be used to artificially inseminate another woman (who donates her egg), and the fertilized egg is then transferred from the donor to the mother.

Test-Tube Babies

It is possible for scientists to make sperm and egg unite outside the human body (in a "test tube"). The scientific term for this procedure is **in vitro fertilization** or IVF (*in vitro* is Latin for "in glass"). The fertilized egg or embryo can then be implanted in the uterus of a woman and carried to term. This technique can be of great benefit to couples who are infertile because the woman's fallopian tubes are blocked.

A milestone was reached with the birth of Louise Brown, the first test-tube baby, in England in July 25, 1978. Obstetrician Patrick Steptoe and physiologist Robert Edwards had fertilized the mother's egg with her husband's sperm in a laboratory dish and implanted the embryo in the mother's uterus. The pregnancy went smoothly, and Louise was born healthy and normal. The procedure is now per-

Embryo transfer: Procedure in which an embryo is transferred from the uterus of one woman into the uterus of another.
In vitro fertilization: A procedure in which an egg is fertilized by sperm in a laboratory dish.

Figure 7.14

"I ALREADY KNOW ABOUT THE BIRDS AND THE BEES, MOM;
I WANT TO KNOW ABOUT ARTIFICIAL INSEMINATION, IN-
VITRO FERTILIZATION AND SURROGATE MOTHERING!"

Renault/Sacramento Bee, CA/Rothco

formed in a number of countries, with 70 clinics in the United States alone.

According to data collected by the Society for Assisted Reproductive Technology, 70 percent of the procedures performed in the United States in 1995 were IVF (CDC, 1997). A survey of 281 infertility clinics indicated that about 20 percent of all procedures were successful, that is, they resulted in a live birth. The procedure is expensive, around $8000 per attempt, not counting preliminary procedures. It is estimated that the average IVF baby costs between $10,000 and $18,000 to produce.

It is also possible to freeze eggs that have been fertilized in vitro, resulting in frozen embryos. This procedure creates the possibility of donated embryos; the birth of a baby resulting from this procedure was first announced in 1984 in Australia. The legal and moral status of the frozen embryo is a difficult question, and some worry about "embryo wastage."

GIFT

GIFT (for gamate intra-fallopian transfer) is an improvement, in some cases, over IVF. Sperm and eggs (gametes) are collected and then inserted together into the fallopian tube, where natural fertilization can take place, followed by natural implantation. About 6 percent of the procedures performed in 1995 were of this type. The success rate for this procedure is also about 20 percent.

Yet another improvement is ZIFT (zygote intra-fallopian transfer), which involves fertilizing the egg with sperm in a laboratory dish and then placing the developing fertilized egg (zygote) into the

GIFT: Gamete intra-fallopian transfer, a procedure in which sperm and eggs are collected and then inserted together into the fallopian tube.

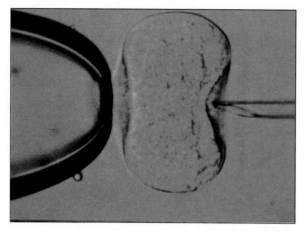

(a)

(b)

Figure 7.15   New reproductive technologies. *(a)* With in vitro fertilization, conception is more likely if the egg is scratched, allowing the sperm to enter more easily. *(b)* Donor and patient sperm to be used in IVF and other infertility procedures are stored using liquid nitrogen.

fallopian tube, again allowing natural implantation. About 2 percent of the procedures performed in 1995 involved this technique.

Assisted reproduction is more likely to be successful if the woman is younger (under 34) and if fresh embryos rather than frozen ones are used. Success rates do not vary by cause of the infertility.

Cloning

Cloning is the reproduction of an individual from a single cell taken from a "donor" or "parent." The technique involves replacing the nucleus of an ovum with the nucleus from a donor, thus producing an embryo that is genetically identical to the donor. Normally, of course, a child has only half of its genes in common with the mother; the other half come from the father. Therefore, children are never genetically identical to either parent. But in cloning, no sperm is necessary and the result is an individual who is genetically identical to the donor.

The concept of cloning raises ugly possibilities in the mind of the general public, such as an army of 100,000 genetically identical individuals. A breakthrough came with the first successful cloning of mice in 1981 (*Time*, January 19, 1981). In February 1997, researchers announced that they had cloned a sheep from a single cell of an adult ewe, a triumph of embryology. This moves us much closer to the possibility of cloning a human being. In June 1997, the United States' National Bioethics Advisory Commission recommended that cloning a human be made a criminal offense.

Gender Selection

There is much interest in techniques that will allow couples to choose whether to have a boy or a girl. Such a technology would be useful to parents who have six girls and really want a boy, or for people who would like to have two children, one of each gender. Problems might arise, though. Some scientists fear that the result of being able to choose gender would be a great imbalance in our population, with many more males than females, because many couples prefer their first child to be a boy.

The commercial aspects of this concept are being capitalized on with the marketing, through drugstores, of a kit called Gender Choice, although there is no scientific evidence that the procedure it uses is effective. Various home methods, such as douching with vinegar before intercourse (supposedly to increase the chances of conceiving a boy) have been discredited. At this point there is no proven method of gender selection.

The technologies discussed here, especially GIFT and ZIFT, require expert practitioners and appropriate facilities. Consequently, they are very expensive (Hatcher et al., 1994). As noted, their success rates are low. For these and other reasons, these procedures raise complex legal questions (discussed in Chapter 22) and ethical questions (discussed in Chapter 21).

SUMMARY

Sperm are manufactured in the testes and ejaculated out through the vas deferens and urethra into the vagina. Then they begin their swim through the cervix and uterus and up a fallopian tube to meet the egg, which has already been released from the ovary. When the sperm and egg unite in the fallopian tube, conception occurs. The single fertilized egg cell then begins dividing as it travels down the tube, and finally it implants in the uterus. Various techniques for improving the chances of conception are available.

The placenta, which is important in transmitting substances between the woman and the fetus, develops early in pregnancy. The most remarkable development of the fetus occurs during the first trimester (first three months), when most of the major organ systems are formed and when human features develop.

For the woman, early signs of pregnancy include amenorrhea, tenderness of the breasts, and nausea. The most common pregnancy tests are designed to detect hCG in the urine or blood. Physical changes during the first trimester are mainly the result of the increasing levels of estrogen and progesterone produced by the placenta. Despite cultural myths about the radiant contentment of the pregnant woman, some women do have negative feelings during the first trimester. During the second trimester the woman generally feels better, both physically and psychologically.

Despite many people's concerns, sexual intercourse is generally quite safe during pregnancy. Nutrition is exceptionally important during pregnancy because the woman's body has to supply the materials to create another human being. Pregnant women also must be very careful about ingesting drugs because some can penetrate the placental barrier and enter the fetus, possibly causing damage.

Labor is typically divided into three stages. During the first stage, the cervix undergoes effacement and dilation. During the second stage, the baby moves out through the vagina. The placenta is delivered during the third stage. Cesarean section is a surgical method of delivering a baby.

The Lamaze method of "prepared" childbirth has become very popular; it emphasizes the use of relaxation and controlled breathing to control contractions and minimize the woman's discomfort. Anesthetics may not be necessary, which seems desirable, since they are potentially dangerous to the infant.

During the postpartum period, hormone levels are very low. Postpartum depression may arise from a combination of this hormonal state and the many environmental stresses on the woman at this time.

Two hormones are involved in lactation: prolactin and oxytocin. Breast-feeding has a number of psychological as well as health advantages, although the nature of the relationship between mother and infant appears to be more important than whether the baby is bottle-fed or breast-fed.

Problems of pregnancy include: ectopic (misplaced) pregnancy, pseudocyesis (false pregnancy), preeclampsia and eclampsia, illness (such as German measles), a defective conceptus, Rh incompatibility, spontaneous abortion, and preterm birth.

The most common cause of infertility in men and women is infection related to sexually transmitted diseases.

New reproductive technologies include artificial insemination, frozen sperm banks, embryo transplants, in vitro fertilization (test-tube babies), and GIFT, all of which are now a reality. These procedures are expensive and have low success rates.

REVIEW QUESTIONS

1. If an egg is not fertilized, it disintegrates about 10 days after ovulation. True or false?

2. It is possible for several sperm to penetrate the egg and fertilize it simultaneously, which is how twins and triplets occur. True or false?

3. The _____ is the mass of tissue lying beside the fetus that serves important functions in allowing nutrients and oxygen to pass from the mother's blood to the baby's blood.

4. During pregnancy, the placenta manufactures a hormone called _____, which is the substance detected in pregnancy tests.

5. Most of the major organ systems of the body develop during the first three months of fetal development. True or false?

6. Drugs taken during pregnancy—including aspirin, antibiotics, and alcohol—may cause damage to the developing fetus. True or false?

7. The _____ method is a technique of prepared childbirth in which the woman learns relaxation techniques and controlled breathing to help control the pain of childbirth.

8. Approximately 90 percent of cases of infertility are caused by problems with the woman's reproductive system. True or false?

9. The technique in which egg and sperm are mixed outside the body in a laboratory dish, in the hope of creating a conceptus, is called _____.

10. _____ refers to a doctor inserting eggs and sperm into the fallopian tube, where natural fertilization may occur, followed by implantation.

(The answers to all review questions are at the end of the book, after the Glossary.)

QUESTIONS FOR THOUGHT, DISCUSSION, AND DEBATE

1. Taking the point of view of a pregnant woman, which would you prefer to have, a home birth or a hospital birth? Why?

2. For those readers who are men, what role would you envision for yourself in parenting if you had a child? Do you feel that you are adequately prepared for that role? For those readers who are women, what role would you ideally like an imaginary husband to take in the parenting of your imaginary children?

3. A close friend confides in you that she is afraid of pregnancy. She has heard that pregnant women experience unpleasant physical (aches and pains, fatigue, illness) and psychological (crying spells, depression) symptoms. She is especially concerned that pregnancy might have a bad effect on her relationship with her partner. What information would you give her about these fears?

SUGGESTIONS FOR FURTHER READING

Dorris, Michael. (1989). *The broken cord.* New York: Harper & Row. The true story of a man and the child he adopted, who turned out to have fetal alcohol syndrome and all the behavior disturbances that go with it.

Kane, Elizabeth. (1988). *Birth mother: The story of America's first legal surrogate mother.* San Diego: Harcourt Brace Jovanovich. An insightful, first-person account by the woman who was the first to have a contract to bear a child for another couple but later developed serious misgivings.

Nilsson, A. L., et al. (1986). *A child is born.* New York: Dell. Contains exceptional photographs of prenatal development.

Zoldbrod, Aline P. (1993). *Men, women and infertility: Intervention and treatment strategies.* New York: Lexington. Explores the impact of infertility on personality, the couple's relationship, and sexuality.

WEB RESOURCES

http://www.childbirth.org/
Information about childbirth and childbirth options from a consumer oriented viewpoint.

www.plannedparenthood.org/
Planned Parenthood: Go to Parenting and Pregnancy Page.

http://www.noah.cuny.edu/pregnancy/fertility.html
Ask NOAH about pregnancy, fertility and infertility. Health information for consumers, links to other resources.

chapter
8

CONTRACEPTION AND ABORTION

CHAPTER HIGHLIGHTS

> For a short time I worked in an abortion clinic. One day I was counseling a woman who had come in for an abortion. I began to discuss the possible methods of contraception she could use in the future (she had been using rhythm), and I asked her what method she planned to use after the abortion. "Rhythm," she answered. "I used it for eleven months and it worked!"*

The average student of today grew up in the pill era and simply assumes that highly effective methods of contraception are available. It is sometimes difficult to remember that this has not always been true and that previously contraception was a hit-or-miss affair at best. Contraception is less controversial than it once was (except for the issue of side effects), and yet the use of contraceptives was illegal in Connecticut until 1965 (the Supreme Court decision in the case of *Griswold v. Connecticut*, 1965, is discussed in Chapter 22).

Today there are a variety of reasons for an individual's use of contraceptives. Many women desire to space pregnancies at least two years apart, knowing that that pattern is better for their health and for the health of their babies. Most couples want to limit the size of their family—usually to one or two children. Unmarried persons typically wish to avoid pregnancy. In some cases a couple know, through genetic counseling, that they have a high risk of having a child with a birth defect and they therefore wish to prevent pregnancy. And in this era of successful career women, many women feel that it is essential to be able to control when and whether they have children.

At the level of society as a whole, there are also important reasons for encouraging the use of contraceptives. There are approximately 1 million adolescent pregnancies annually in the United States (Henshaw, 1997), and they constitute a major social problem. On the global level, the problem of overpopulation is serious (see Figure 8.1). Most experts believe that we must limit the size of the American population as well as assist other countries in limiting theirs (Green, 1992). For a summary of contraceptive practices around the world, see Table 8.1.

In this chapter we discuss various methods of birth control, how each works, how effective they are, what side effects they have, and their relative advantages and disadvantages. We also discuss abortion and advances in contraceptive technology.

The Pill

With **combination birth control pills** (sometimes called *oral contraceptives*) such as Loestrin and Ovcon, the woman takes a pill that contains estrogen and progestin (a synthetic progesterone), both at doses higher than natural levels, for 21 days. Then she takes no pill or a placebo for seven days, after which she repeats the cycle.

How It Works

The pill works mainly by preventing ovulation. Recall from Chapter 6 that in a natural menstrual cycle, the low levels of estrogen during and just after the menstrual period trigger the pituitary to produce FSH, which stimulates the process of ovulation. The woman starts taking the birth control pills on about day 5 of her cycle. Thus just when estrogen levels would normally be low, they are artificially made high. This high level of estrogen inhibits FSH production, and the message to ovulate is never sent out. The high level of progesterone inhibits LH production, further preventing ovulation.

The progestin provides additional backup effects. It keeps the cervical mucus very thick, making it difficult for sperm to get through, and it changes the lining of the uterus in such a way that even if a fertilized egg arrived, implantation would be unlikely.

*Paula Weideger. (1976). *Menstruation and menopause*. New York: Knopf, p. 42.

Combination pills: Birth control pills that contain a combination of estrogen and progestin (progesterone).

Figure 8.1 United Nations projections show that differing patterns of birth control and population growth beginning in 1990 could lead to a world population as high as 19 billion in the year 2100 or as low as 6 billion. The low growth projection assumes that fertility eventually stabilizes at 1.7 children per woman. The medium projection assumes fertility stabilizes at 2.1, and the high projection assumes fertility stabilizes at 2.5. Reducing the population problem is one major reason for using birth control (Green, 1992).

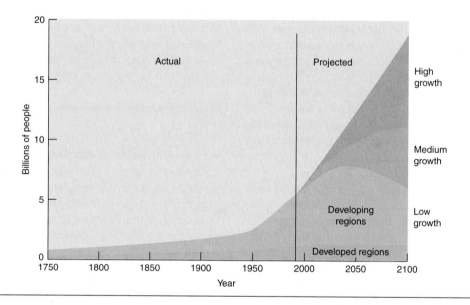

When the estrogen and progestin are withdrawn after day 21, the lining of the uterus disintegrates, and withdrawal bleeding or menstruation occurs, although the flow is typically reduced because the progestin has inhibited development of the endometrium.

Hormonally, the action of the pill produces a condition much like pregnancy, when hormone levels are also high, preventing further ovulation and menstrual periods. Thus it is not too surprising that some of the side effects of the pill are similar to the symptoms of pregnancy.

Effectiveness

Before the effectiveness of the pill is discussed, several technical terms that are used in communicating data on contraceptive effectiveness need to be defined. If 100 women use a contraceptive method for one year, the number of them who become pregnant during that first year of use is called the **failure rate** or *pregnancy rate*. That is, if 5 women out of 100 become pregnant during a year of using contraceptive A, then A's failure rate is 5 percent. *Effectiveness* is 100 minus the failure rate; thus contraceptive A would be said to be 95 percent effective. We can also talk about two kinds of failure rate: the *failure rate for perfect users* and the *failure rate for typical users*. The perfect-user failure rate refers to studies of the best possible use of the method—for example, when the user has been well-taught about the method, uses it with perfect consistency, and so on. The failure rate for typical users is just that—the failure rate when people ac-

Failure rate: The pregnancy rate occurring using a particular contraceptive method; the percentage of women who will be pregnant after a year of use of the method.

Table 8.1 Contraception Around the World, Reported by Currently Married Women
(The great variations reflect differences among cultures in such factors as availability of medical service, people's education about contraception, and gender roles.)

Percentage Using Contraceptive Method

Region, Country	Voluntary Sterilization		Pill	IUD	Condom	Implants & Injectables	Vaginal Methods	Rhythm
	Male	Female						
North America								
United States	11.	28	27	1	20	4	2.3	2
Europe								
Netherlands	11	8	38	10	7	NA	NA	NA
Italy	NA	1	14	2	13	NA	2	9
Norway	2	4	13	28	16	NA	2	3
Africa								
Botswana	0	4	16	6	1	6	0	0
Nigeria	0	0	1	1	0	1	0	1
Asia								
Korea	11	37	3	7	10	NA	2	NA
Thailand	6	22	20	7	1	9	0	1
Latin America								
El Salvador	1	30	8	2	2	1	0	2
Mexico	1	18	11	11	2	3	1	5
Middle East and North Africa								
Egypt	0	1	16	17	3	0	0	1
Jordan	0	6	5	15	1	0	1	4

NA: Statistics not available.

*Includes injections such as Depo-Provera and implants such as Norplant.

†Includes diaphragm, cervical cap, and spermicides.

Sources: Bryant Robey et al. (1992). The reproductive revolution: New survey findings. *Population Reports,* Series M, Number 11. Baltimore: Johns Hopkins University, Population Information Program. Kathy A. London et al. (1985). Fertility and family planning surveys: An update. *Population Reports,* Series M, Number 8, M-291–M-348. Baltimore: Johns Hopkins University, Population Information Program. Linda Piccinino & William D. Mosher (1998). Trends in contraceptive use in the United States. *Family Planning Perspectives, 30*(1), 4–10.

tually use the method, perhaps imperfectly when they forget to take a pill or do not use a condom every time. The good news is that if you are very responsible about contraception, you can anticipate close to the perfect-user failure rate for yourself.

The use of combination pills is one of the most effective methods of birth control. The perfect-user failure rate is 0.1 percent (that is, the method is essentially 100 percent effective), and the typical-user failure rate is 5 percent (Hatcher et al., 1998). Failures occur primarily as a result of forgetting to take a pill for two or more days. If a woman forgets to take a pill, she should take it as soon as she remembers and take the next one at the regular time; this does not appear to increase the preg-

nancy risk appreciably. If she forgets for two days, she should take two pills as soon as she remembers and then two the next day; however, the chances of pregnancy are now increased and a back-up method of contraception should be used. If she forgets for three or more days, she should switch to some other method of birth control for the remainder of that cycle.

Side Effects

You may have seen various reports in the media on the dangerous side effects of birth control pills. Some of these reports are no more than scare stories with little or no evidence behind them. However, there are some well-documented risks associated with the use of the pill, and women who are using it or who are contemplating using it should be aware of them.

Among the serious side effects associated with use of the pill are slight but significant increases in certain diseases of the circulatory system. One of these is problems of blood clotting (thromboembolic disorders). Women who use the pill have a higher chance than nonusers of developing blood clots (thrombi). Often these form in the legs, and they may then move to the lungs. There may also be clotting or hemorrhaging in the brain (stroke). The clots may lead to pain, hospitalization, and (in rare cases) death. Symptoms of blood clots are severe headaches, severe leg or chest pains, and shortness of breath. For some women, the pill can cause high blood pressure; thus it is important to have regular checkups so that this side effect can be detected if it occurs.

There have been many reports in the media of the pill causing cancer. However, the scientific data do not provide evidence that the pill causes cancer of the cervix, uterus, or breast. The good news is that the pill actually protects women from endometrial cancer and ovarian cancer (Hatcher et al., 1998). However, the pill may aggravate already existing cancer. And while some of the studies have looked at effects after fairly long periods of time, there is a need for studies of even longer-term effects, since it is known that cancer-causing agents (carcinogens) may not show their effects for as many as 20 years.

For women who have taken the pill for more than five years, the risk of benign liver tumors increases (Hatcher et al., 1994). These tumors can cause death due to bleeding if they rupture. Although these problems are relatively rare, they underline the importance of the doctor's giving a thorough examination before prescribing birth control pills and of the woman's having regular checkups while using them.

The pill increases the amount of vaginal discharge and the susceptibility to vaginitis (vaginal inflammations such as monilia and trichomonas—see Chapter 20) because it alters the chemical balance of the lining of the vagina. Women on the pill have an increased susceptibility to chlamydia, probably for similar reasons, as well as the fact that they are unlikely to be using condoms or other methods that protect against sexually transmitted diseases. In one study women using the pill had a 73 percent higher rate of acquiring chlamydia and a 70 percent higher rate of acquiring gonorrhea than did comparison groups of women using sterilization or an IUD (Louv et al., 1989).

The pill may cause some nausea, although this almost always goes away after the first month or two of use. Some brands of pills can also cause weight gain, by increasing appetite or water retention, but this side effect can often be reversed by switching to another brand.

Finally, there may be some psychological effects. About 20 percent of women on the pill report increased irritability and depression, which become worse with length of time used. These side effects are probably related to the progesterone in the pill; switching to a different brand may be helpful. There may also be changes in sexual desire. Some women report an increase in sexual interest (McCoy & Matyas, 1996). But other women report a decrease in sexual desire as well as a decrease in vaginal lubrication (Graham et al., 1995). Once again, switching brands may be helpful.

Because of the side effects discussed previously, women in the following groups should *not* use the pill (Hatcher et al., 1998): those with poor blood circulation or blood clotting problems; those who have had a heart attack or who have coronary artery disease; those with liver tumors; those with cancer of the breast; nursing mothers (the pill tends to dry up the milk supply, and the hormones may be transmitted through the milk to the baby); and pregnant women (prenatal doses of hormones, as has been noted, can damage the fetus). Women over 35 who are cigarette smokers should use the pill only with caution, because the risk of heart attack is considerably higher in this group.

After all this discussion, just how dangerous is the pill? It seems that the answer to this question depends on who you are and on how you look at it. If you have blood-clotting problems, the pill is dangerous to you; if you have none of the contraindications listed above, it is very safe (Hatcher et al., 1998). One's point of view and standard of comparison also matter. While a death rate of 1.6 per 100,000 sounds high, it is important to consider that one alternative to the pill is intercourse with no contraceptive, and that can mean pregnancy, which has a set of side effects and a death rate all its own. Thus while the death rate for the pill is 1.6 per 100,000, the death rate for pregnancy and delivery is 9 per 100,000 (Hatcher et al., 1994). And while the pill may precipitate diabetes, pregnancy may do so, too. Thus in many ways the pill is no more dangerous than the alternative, pregnancy, and may even be safer. Another possible standard of comparison is drugs that are commonly taken for less serious reasons. Aspirin, for example, is routinely used for headaches. Recent reports indicate that aspirin has side effects, and thus the birth control pill may be no more dangerous than drugs we take without worrying much.

In short, the pill does have some serious potential side effects, particularly for high-risk individuals, but for many others it is an extremely effective means of contraception that poses little or no danger.

Advantages and Disadvantages

The pill has a number of advantages. It is close to 100 percent effective if used properly. It does not interfere with intercourse, as some other methods—the diaphragm, the condom, and foam—do. It is not messy. Some of its side effects are also advantages; it reduces the amount of menstrual flow and thus reduces cramps. Indeed, it is sometimes prescribed for the noncontraceptive purpose of regulating menstruation and eliminating cramps. Iron-deficiency anemia is also less likely to occur among pill users. The pill can clear up acne, and has a protective effect against some rather serious things, including pelvic inflammatory disease (PID) and ovarian and endometrial cancer (Hatcher et al., 1998).

The side effects of birth control pills, discussed above, are of course major disadvantages. Another disadvantage is the cost, which is about $16 a month (or as little as $2 to $4 per month through a Planned Parenthood clinic) for as long as they are used. They also place the entire burden of contraception on the woman. In addition, taking them correctly is a little complicated; the woman must understand when they are to be taken, and she must remember when to take them and when not to take them. This effort would not be too taxing for the average college student, but for an illiterate peasant woman in a developing nation who thinks the pills are to be worn like an amulet on a chain around the neck or for the mentally retarded (they need contraceptives too), currently available birth control pills are too complicated.

One other criticism of the pill is that for a woman who has intercourse only infrequently (say, once or twice a month or less), it represents contraceptive "overkill"; that is, the pill makes her infertile every day of the month (with the side effects of taking it every day), and yet she needs it only a few days each month. Women in this situation might consider a method, such as the diaphragm, that is used only when needed.

Finally, it is important to recognize that, although it is an excellent contraceptive, the pill provides absolutely no protection against sexually transmitted diseases.

Reversibility

When a woman wants to become pregnant, she simply stops taking pills after the end of one cycle. Some women experience a brief delay (2 or 3 months) in becoming pregnant, but pregnancy rates are about the same as for women who never took the pill.

Drug Interactions

If you are taking birth control pills, you are taking a prescription drug and there may be interactions with other prescription drugs you take (Hatcher et al., 1998). Taking antibiotics such as ampicillin and tetracycline reduces the effectiveness of the pill. Women in this situation should use a backup method in addition, such as rhythm, condoms, or foam.

The pill may also increase the metabolism of some drugs, making them more potent (Hatcher et al., 1988). Examples include some anti-anxiety drugs, corticosteroids used for inflammations, and theophylline (a drug used for asthma and an ingredient in, for example, Primatene). Therefore, women using the pill may require lower doses of these drugs.

Other Kinds of Pills

To this point, the discussion has centered chiefly on the *combination pill,* so named because it contains both estrogen and progestin. This variety of pill is the most widely used, but there are many kinds of combination pills and several kinds of pills other than combination ones.

Focus 8.1

Margaret Sanger—Birth Control Pioneer

Margaret Higgins Sanger (1879–1966) was a crusader for birth control in the United States; to reach her goals, she had to take on a variety of opponents, including the United States government, and she served one jail term.

Sanger was born in Corning, New York, the daughter of a tubercular mother who died young after bearing 11 children. Her father was a free spirit who fought for women's suffrage. After caring for her dying mother, she embarked on a career in nursing and married William Sanger in 1902.

She became interested in women's health and began writing articles on the subject. Later these were published as books entitled *What Every Girl Should Know* (1916) and *What Every Mother Should Know* (1917).

Perhaps her strongest motivation came from her work as a nurse. Her patients were poor maternity cases on New York's Lower East Side. Among these women, pregnancy was a "chronic condition." Margaret Sanger saw them, weary and old at 35, resorting to self-induced abortions, which killed many of them. Frustrated at her inability to help them, she renounced nursing:

> I came to a sudden realization that my work as a nurse and my activities in social service were entirely palliative and consequently futile and useless to relieve the misery I saw all about me.

She determined, instead, to "seek out the root of the evil." Though she was often accused of wanting to lower the birthrate, she instead envisioned families, rich and poor alike, in which children were wanted and given every advantage.

Impeding her work was the Comstock Act of 1873 (see Chapter 22), which classified contraceptive information as obscene and made it illegal to send it through the mail. In 1914 she founded the National Birth Control League, launching the birth control movement in the United States. Though her magazine, *Woman Rebel*, obeyed the letter of the law and did not give contraceptive information, she was nonetheless indicted on nine counts and made liable to a prison term of 45 years.

Margaret Sanger left the United States on the eve of the trial. She toured Europe, and in Holland she visited the first birth control clinics to be established anywhere. There she got the idea of opening birth control clinics in the United States. Meanwhile, the charges against her had been dropped.

She returned to the United States and, in 1916, opened the first U.S. birth control clinic in Brooklyn. The office was closed by the police after nine

---

Combination pills vary from one brand to the next in the dosages of estrogen and progestin. The dose of estrogen is important because higher doses are more likely to induce blood-clotting problems. Most women do well on pills containing no more than 30 to 50 micrograms of estrogen; brands fulfilling that requirement include Ortho-Novum 1/35, Norinyl 1/35, Demulen, Lo-Ovral, and Levlen (Hatcher et al., 1998). Because of concerns about side effects due to the estrogen in the pill, current pills have considerably lower levels of estrogen

than early pills; for example, Ortho-Novum 1/35 has one-third the amount of estrogen of the early pill Enovid 10. High-progestin brands are related to symptoms such as vaginitis and depression. Thus, depending on what side effects the woman wants to avoid, she can choose a brand for its high or low estrogen or progesterone level. (See Hatcher et al., 1994, pp. 241–242, for a list of symptoms related to dosages of estrogen and progestin.)

Another variation of the combination pill is the 28-day pill. There are 28 pills in every package. The

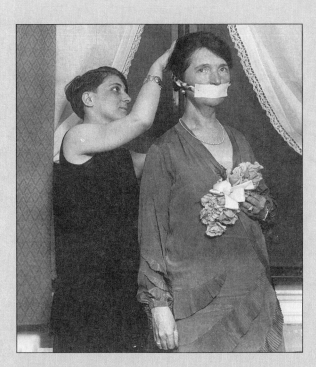

Figure 8.2 Margaret Sanger, a pioneer of the birth control movement, shown here in 1929. She was forbidden by Boston authorities to speak on birth control, so she taped her mouth in protest and wrote on a chalkboard.

days of operation, and Margaret Sanger was put in jail for 30 days. However, on appeal, her side was upheld by the courts, and in 1918, a decision was handed down allowing doctors to give contraceptive information to women for the "cure and prevention of disease."

The birth control movement was gaining followers, and the first National Birth Control Conference was held in 1921 in New York; it was attended by doctors, scientists, and lay supporters. In 1931, the Pope approved the rhythm method for use by Roman Catholics.

Women were also at the forefront of the birth control movement in Canada. In Hamilton, Ontario, Mary Elizabeth Hawkins organized the Hamilton Birth Control Society in 1932. Dr. Elizabeth Bagshaw, one of Hamilton's few female physicians at the time, served the clinic for the next thirty-odd years. Providing information about birth control was technically illegal in Canada, too, at the time, unless it served "the public good," and an Ottawa social worker was actually charged (and acquitted in 1937) for her family planning activity.

Margaret Sanger's role in getting birth control information to American women and in making it legal for them to use the information is unquestioned. Heywood Broun once remarked that Margaret Sanger had no sense of humor. She replied, "I am the protagonist of women who have nothing to laugh at."

Sources: Ellen Chesler. (1992). *Woman of valor: Margaret Sanger and the birth control movement in America.* New York: Anchor Books. *Current Biography.* (1944). P. Van Preagh. (1982). The Hamilton birth control clinic: In response to need. *News/Nouvelles, Journal of Planned Parenthood Federation of Canada, 3*(2).

first 21 pills that the woman takes are regular combination pills, and the last 7 are placebos (they contain no drugs). The purpose of this is simply to help the woman use the pills properly, the idea being that it is easier to remember to take a pill every day than to remember to take one each day for 21 days and then none for 7 days; also, the seven placebos eliminate confusion about when to start taking pills again.

The **triphasic pill** (e.g., Tri-Norinyl) contains a steady level of estrogen like the combination pill does, but there are three phases in the levels of progesterone. The idea is to reduce total hormone exposure and provide a cycle more similar to the natural one.

Triphasic pill: A birth control pill containing a steady level of estrogen and three phases of progesterone, intended to mimic more closely women's natural hormonal cycles.

Figure 8.3   Different types of birth control pills.

Progestin-only pills (such as Micronor, Nor-Q-D, and Ovrette) have also been developed. They are sometimes called *mini-pills.* The pills contain only a low dose of progestin and no estrogen, and they were designed to avoid the estrogen-related side effects of the standard pills. The woman takes one beginning on the first day of her period and every day thereafter, at the same time each day. The effects of progestin-only pills that prevent pregnancy include the following: changes in the cervical mucus such that sperm cannot get through, inhibition of implantation, and inhibition of ovulation (although while taking mini-pills, about 40 percent of women ovulate consistently). Progestin-only pills have a typical-user failure rate of 5 percent, which is higher than that of combination pills, although much of the failure occurs during the first six months of use. Their major side effect seems to be that they produce very irregular menstrual cycles. The mini-pill is probably most useful for women who cannot take combination pills—for example, women over 35 who smoke, or women with a history of high blood pressure or blood-clotting problems.

Progestin-only pills are also useful for women who are breast-feeding and cannot use combination pills because they reduce milk production. Neither kind of pill should be used in the first 6 weeks after birth when breast-feeding, because trace amounts of the hormones can reach the infant through the breast milk. After that time, though, progestin-only pills are a good choice.

Emergency Contraception

Emergency contraception or *postcoital methods* are available in pill form for cases of emergency, such as rape or a condom breaking (Hatcher et al., 1998; von Hertzen & Van Look, 1996). The treatment is most effective if begun within 12 to 24 hours after unprotected intercourse and cannot be delayed longer than 72 hours. Regular birth control pills are taken at higher doses—for example, two doses of two tablets each of Ovral, or four tablets each of Lo-Ovral or Levlen, taken 12 hours apart. Nausea is a common side effect. This treatment reduces by 74 percent the risk of pregnancy following unprotected intercourse (Hatcher et al., 1998). Emergency contraception should be used only under the supervision of a health care provider.

Norplant and Depo-Provera

The first introduction of a major new method of contraception in the United States in years came in 1990 with the introduction of Norplant, a progestin-only implant that lasts for five years.

Norplant comes in six rod-shaped silicone capsules, each about the size of a matchbook match. They are implanted by a medical professional using a minor surgical procedure under local anesthetic. They are usually placed in a fan-shaped configuration on the inside of the woman's upper arm. Once in place, the progestin diffuses slowly but steadily into the body, and the implant lasts for five years.

How Norplant Works

Norplant, like other progestin-only contraceptives (progestin-only pills and injections) works in several ways. It inhibits ovulation; it thickens the cervical mucus, making it difficult for sperm to get through; and it creates a puny endometrium in the uterus so that if a fertilized egg did arrive, it would not have much to implant in.

Effectiveness

Norplant is the most effective method of contraception now available (with the exception of sterilization). The failure rate is only 0.05 percent (Hatcher et al., 1998). That is, it is essentially 100 percent effective. Failure rates go up substantially after five years, but it is only active for five years; a new implant must be obtained after that time has elapsed.

Side Effects

The side effects found so far with Norplant can be regarded as minor rather than major. Menstrual cycle irregularities are the most common side effect. These vary from woman to woman and may include prolonged menstruation during the first few months of use, spotting between periods, and amenorrhea (Darney et al., 1990). There are reports

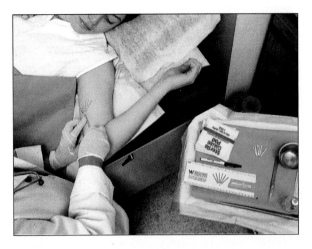

Figure 8.4 Norplant, a progestin contraceptive implanted in the woman's upper arm, is one of the most exciting new developments in contraception of the last 15 years. It was the first major new method of contraception to be introduced in the United States in a decade. It is implanted using a minor surgical procedure, as shown here.

of difficulty with removal, as well as some cases of inflammation or infection at the site of the implant (Hatcher et al., 1998).

At this point there is no evidence of circulatory system side effects (for example, blood-clotting disorders) with Norplant use such as those that have been found with birth control pills. However, Norplant has been introduced so recently that the number of women who have used it is still relatively small, and none have used it for a long period of time. Therefore, future research may document side effects. However, it seems unlikely that these side effects will be very frequent because most of the side effects of combination pills are due to the estrogen, and Norplant uses only progestin.

At the current time, it is recommended that Norplant not be used by pregnant women and those with breast cancer or unexplained vaginal bleeding. The effectiveness of Norplant is reduced considerably if the woman is also taking an antiseizure drug. In the unlikely event that a woman becomes pregnant while using Norplant, the capsules should be removed because it is unwise to expose the fetus to hormones, even at the low dose found in Norplant.

Advantages and Disadvantages

The big advantage of Norplant is that it does not require memory to use it properly. There are no failures due to forgetting to take a pill. In addition, it has all the advantages of the pill, particularly the fact that it does not intrude in lovemaking. Its extremely high effectiveness is another major advantage. Because it contains no estrogen, it has no estrogen-related side effects.

One disadvantage is the high initial cost ($570 for a full-paying client at Planned Parenthood, but less for someone on a smaller income). However, when the cost is considered over five years of use, it reduces to slightly over $100 per year, which is cheaper than birth control pills, and is certainly cheaper than having a baby. The implants may be slightly visible, and some women find them uncomfortable. Menstrual irregularities are a disadvantage for some women, but others find relief from menstrual pain. Finally, Norplant must be inserted and removed by a trained medical professional, so it cannot be used on the spur of the moment.

It is a method best suited for a woman who wants long-term contraception. It is not sensible for someone needing contraception for only a short period of time.

Reversibility

If the woman desires to become pregnant any time during the five years of use, she simply has a trained medical professional remove the capsules. Progestin levels quickly go down, and the woman can become pregnant.

Depo-Provera Injections

Depo-Provera (DMPA) is a progestin administered by injection. Injections must be repeated every three months for maximum effectiveness. Depo-Provera is currently in use in more than 90 nations worldwide (Hatcher et al., 1998). It became available in the United States in 1992. The staff at Planned Parenthood report that it is quickly becoming more popular than Norplant, and they anticipate that it will be even more popular than the pill within a few years.

Depo-Provera works like the other progestin-only methods, by inhibiting ovulation, thickening cervical mucus, and inhibiting the growth of the endometrium. Depo-Provera is highly effective, with a typical-user failure rate of less than 1 percent, making it slightly less effective than Norplant, but more effective than the pill.

Depo-Provera has many advantages. It does not interfere with lovemaking. It requires far less reliance on memory than birth control pills do, although the woman must remember to have a new injection every three months. It is available for women who cannot use the combination pill, such as those over 35 who smoke and those with blood pressure problems. Compared with Norplant, it has a considerably lower initial cost, although the two are probably about equal in cost over a long period of time.

A disadvantage of Depo-Provera is that most users experience amenorrhea (no menstrual periods). However, this may be an advantage. It can relieve anemia due to heavy menstrual periods, and Depo-Provera can be used in the treatment of endometriosis.

No lethal side effects of Depo-Provera have been found, although long-term studies have not yet been done (Hatcher et al., 1998).

The method is reversible simply by not getting another injection. Many women are infertile for 6 to 12 months after stopping its use, but then are able to become pregnant at normal rates (Lande, 1995).

The IUD

The **intrauterine device,** or **IUD,** is a small piece of plastic; it comes in various shapes. Metal or a hormone may also be part of the device. An IUD is inserted into the uterus by a doctor or nurse practitioner and then remains in place until the woman wants to have it removed. One or two plastic strings hang down from the IUD through the cervix, enabling the woman to check to see whether it is in place.

The basic idea for the IUD has been around for some time. In 1909 Richter reported on the use of an IUD made of silkworm gut. In the 1920s the German physician Ernst Grafenberg reported data on 2000 insertions of silk or silver wire rings. In spite of the high effectiveness he reported (98.4 percent), his work was poorly received. Not until the 1950s,

Intrauterine device (IUD): A plastic device sometimes containing metal or a hormone that is inserted into the uterus for contraceptive purposes.

with the development of plastic and stainless-steel devices, did the method gain much popularity. In the 1980s the use of the IUD in the United States was sharply reduced by numerous lawsuits against manufacturers by persons claiming to have been damaged by the device. Some companies stopped producing IUDs, and others declared bankruptcy. As a result, only two IUDs are available in the United States today. Both are T-shaped; one contains copper (the copper T or CuT-380A), and the other contains progesterone (Progestasert). Currently 106 million women worldwide are using IUDs, 40 million of them in the People's Republic of China, and experts predict a resurgence of enthusiasm for them in the United States (Kaunitz, 1997).

How It Works

The IUD works by preventing fertilization. It produces changes in the uterus and fallopian tubes, and in this environment, sperm that reach the uterus are immobilized and cannot move into the fallopian tube (Treiman et al., 1995). The egg may also move more swiftly through the fallopian tube, reducing the chances of fertilization.

The Progestasert releases progesterone directly into the uterus. One effect is to reduce the endometrium. This results in reduced menstrual flow and reduced risk of anemia, thus overcoming two undesirable side effects of other IUDs.

The small amount of copper that is added to the copper T is thought to have an additional contraceptive effect. It seems to alter the functioning of the enzymes involved in implantation. The progestin thickens cervical mucus, disrupts ovulation, and changes the endometrium.

Effectiveness

The IUD is extremely effective; it is fourth behind Norplant, Depo-Provera, and the pill (and sterilization) in effectiveness. The pregnancy rate for the copper T is 0.8 percent for the first year of use; after that, the failure rate is even lower. The copper T is effective for 10 years.

Most failures occur during the first three months of use, either because the IUD is expelled or for other, unknown reasons. Expulsion is most likely in women who have had no children, in younger women, and in women during menstrua-

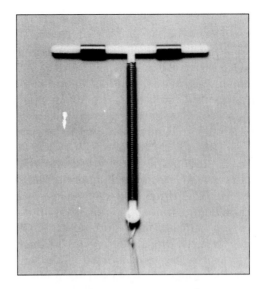

Figure 8.5 Copper-T IUD (shown enlarged).

tion. The expulsion rate is about 1 to 7 percent during the first year (Treiman et al., 1995).

Side Effects

The major serious side effect is **pelvic inflammatory disease,** or **PID** (e.g., uterine or tubal infection). The IUD aggravates already existing pelvic infections, and so it should not be used by a woman who has such an infection or a history of such infections.

The most common side effects of IUDs are increased menstrual cramps, irregular bleeding, and increased menstrual flow. These symptoms occur in 10 to 20 percent of women using IUDs and are most likely immediately after insertion. These side effects are a major reason for requests for IUD removal.

There is no evidence that the IUD causes cancer.

Because of the possible side effects, women with the following conditions should not use an IUD: pregnancy, endometriosis, vaginal or uterine infection (including gonorrhea or chlamydia), or pelvic inflammatory disease.

Pelvic inflammatory disease (PID): Infection of the pelvic organs such as the fallopian tubes.

Advantages and Disadvantages

All the side effects discussed above are disadvantages of the IUD. Another disadvantage is the initial cost, which is about $150 for a full-paying client at Planned Parenthood, for the IUD plus insertion. Even at that rate, though, the IUD is a cheap means of contraception over a long period of use, since the cost is incurred only once and the copper T, for example, lasts 10 years.

The effectiveness of the copper-T IUD is a major advantage. The typical-user failure rate is only 0.8 percent, making it more effective than combination birth control pills and approximately as effective as Norplant and Depo-Provera.

Once inserted, the IUD is perfectly simple to use. The woman has only to check periodically to see that the strings are in place. Thus it has an advantage over methods like the diaphragm or condom in that it does not interrupt intercourse in any way. It has an advantage over the pill in that the woman does not have to remember to use it. The IUD is a method that can be used safely by women after having a baby and while breast-feeding.

Contrary to what some people think, the IUD does not interfere with the use of a tampon during menstruation; nor does it have any effect on intercourse.

Reversibility

When a woman who is using an IUD wants to become pregnant, she simply has a physician remove the device. She can become pregnant immediately.

The Diaphragm and the Cervical Cap

The **diaphragm** is a circular, dome-shaped piece of thin rubber with a rubber-covered rim of flexible metal (see Figure 8.6). It is inserted into the vagina and, when properly in place, fits snugly over the cervix. In order for it to be used effectively, a contraceptive cream or jelly (such as Delfen) must be applied to the diaphragm. The cream is spread on the rim and the inside surface (the surface that fits against the cervix). The diaphragm may be inserted up to 6 hours before intercourse; it must be left in place for at least 6 hours afterward and may be left in for as long as 24 hours. Wearing it longer than that is thought to increase the risk of toxic shock syndrome (see Chapter 6).

Use of the diaphragm was the earliest of the highly effective methods of contraception for women. It was popularized in a paper in 1882 by the German researcher Mensinga. In 1925, Margaret Sanger's husband funded the first U.S. company to manufacture them, and they were the mainstay of contraception until about 1960.

How It Works

The primary action of the diaphragm itself is mechanical; it blocks the entrance to the uterus so that sperm cannot swim up into it. The cream kills any sperm that manage to get past the barrier. Any sperm remaining in the vagina die after about eight hours (this is why the diaphragm should not be removed until at least six hours after intercourse).

Effectiveness

The typical-user failure rate of the diaphragm has been estimated to be about 20 percent. Most failures are due to improper use: the woman may not use it every time, she may not leave it in long enough, or she may not use contraceptive cream or jelly. Even with perfectly proper use, there is still a failure rate. For example, Masters and Johnson found that expansion of the vagina during sexual arousal (see Chapter 9) may cause the diaphragm to slip. To get closer to 100 percent effectiveness, the diaphragm can be combined with a condom around the time of ovulation or throughout the cycle.

Failure rates for the diaphragm and cervical cap are often stated as ranges, for example, 17 to 25 percent. This is done because failure rates for these methods depend so much on the fertility characteristics of the user. For example, a woman under 30 who has intercourse four or more times weekly has twice the average failure rate of a woman over 30 who has intercourse less than four times a week.

Because the fit of the diaphragm is so important

Diaphragm: A cap-shaped rubber contraceptive device that fits inside a woman's vagina over the cervix.

Figure 8.6 The proper use of a diaphragm. *(a)* Spermicide is applied (about 1 tablespoon in center and around the rim). *(b)* The edges are held together to permit easier insertion. *(c)* The folded diaphragm is inserted up through the vagina. *(d)* The diaphragm is placed properly, covering the cervix. To check for proper placement, feel the cervix to be sure that it is completely covered by the diaphragm.

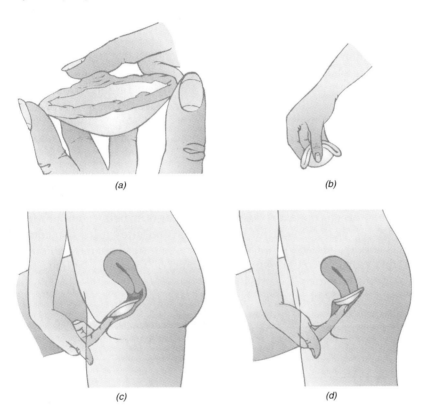

(a)

(b)

(c)

(d)

to its effectiveness, it is important that the woman be individually fitted for one by her physician. She must be refitted after the birth of a child, an abortion, extreme weight gain or loss, or any similar occurrence that would alter the shape and size of the vagina.

Side Effects

The diaphragm has few side effects. One is the possible irritation of the vagina or the penis; this is caused by the cream or jelly and can be relieved by switching to another brand. Another side effect is the rare occurrence of toxic shock syndrome that has been reported in women who left the di-

aphragm in place for more than 24 hours. Therefore users should be careful not to leave the diaphragm in place for much more than the necessary 6 to 8 hours, especially during menstruation.

Advantages and Disadvantages

Some people feel that the diaphragm is undesirable because it must be inserted before intercourse and therefore ruins the "spontaneity" of sex. People with this attitude, of course, should not use the diaphragm as a means of birth control, since they probably will not use it all the time, in which case it will not work. However, a student told us that she and her partner made the preparation and insertion of the diaphragm a ritual part of their foreplay; he inserts it, and they both have a good time! Couples who maintain this kind of attitude are much more likely to use the diaphragm effectively. Moreover, the diaphragm can be inserted an hour or more before sex.

Some women dislike touching their genitals and sticking their fingers into their vagina. Use of the diaphragm is not a good method for them.

The diaphragm requires some thought and presence of mind on the woman's part. She must remember to have it with her when she needs it and to have a supply of cream or jelly. She also needs to avoid becoming so carried away with passion that she forgets about it or decides not to use it.

A disadvantage is that the cream or jelly may leak out after intercourse.

The cost of a diaphragm is about $10 plus the cost of the office visit and the cost of the contraceptive cream. With proper care, a diaphragm should last about two years, and thus it is not expensive.

The major advantages of the diaphragm are that it has few side effects and, when used properly, is very effective. For this reason, women who are worried about the side effects of the pill or the IUD should seriously consider the diaphragm as an alternative. There is also evidence of a reduction in the rate of cervical cancer among long-time users of the diaphragm.

Reversibility

If a woman wishes to become pregnant, she simply stops using the diaphragm. Its use has no effect on her later chances of conceiving.

The Cervical Cap

The **cervical cap** was approved by the FDA in 1988. It is similar to the diaphragm but is somewhat different in shape, fitting more snugly over the cervix (see Figure 8.7). Like the diaphragm, it should be used with a spermicidal cream. One advantage is that it can be left in place longer than the diaphragm. It can remain in place for 48 hours, but experts advise against leaving it in longer because of potential problems with odors and increased risk of toxic shock syndrome. Some women are allergic to the latex or to spermicides.

The cervical cap has approximately the same effectiveness as the diaphragm; that is, it has a typical-user failure rate around 18 percent. However, the rate varies considerably depending on whether the woman is nulliparous (has never had a baby) or is parous (has had a baby). The failure rate

Cervical cap: A method of birth control involving a rubber cap that fits snugly over the cervix.

Figure 8.7   The cervical cap.

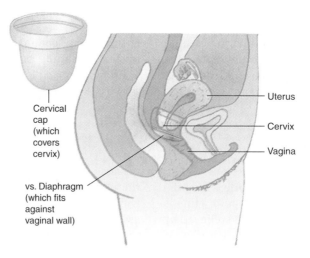

Cervical cap (which covers cervix)

vs. Diaphragm (which fits against vaginal wall)

Uterus

Cervix

Vagina

Figure 8.8   A variety of male condoms.

is 18 percent for nulliparous women, but a whopping 36 percent for parous women.

Protection from STDs

One advantage of both the diaphragm and the cervical cap is that they probably lower the risk of transmission of sexually transmitted diseases, in part because spermicides are fairly effective at killing the organisms that cause these diseases.

The Male Condom

The **male condom** ("rubber," "prophylactic," "safe") is a thin sheath that fits over the penis. It comes rolled up in a little packet and must be unrolled onto the penis before use. It may be made of latex ("rubber"), of polyurethane, or of the intestinal tissue of lambs ("skin"). The widespread use of the modern condom, both for contraception and for protection against diseases, dates from about 1843, when vulcanized rubber was developed; however, the use of a sheath to cover the penis has been known throughout most of recorded history.[1] Casanova (1725–1798) was one of the first to popularize it for its contraceptive value as well as its protective value. Condoms have become increasingly

[1]Condoms have also been the stimulus for humor throughout history, an example being this limerick:

There was a young man of Cape Horn
Who wished he had never been born
   And he wouldn't have been
   If his father had seen
That the end of the rubber was torn.

Male condom: A contraceptive sheath that is placed over the penis.

popular because they help protect against sexually transmitted diseases (STDs). Approximately 450 million are sold each year in the United States (Hatcher et al., 1998).

To be effective, the condom must be used properly. It must be unrolled onto the erect penis before the penis ever enters the vagina—*not* just before ejaculation, since long before then some drops containing a few thousand sperm may have been produced. Condoms come in two shapes: those with plain ends and those with a protruding tip that catches the semen. If a plain-ended one is used, about ½ inch of air-free space should be left at the tip to catch the ejaculate. Care should be taken that the condom does not slip during intercourse. After the man has ejaculated, he must hold the rim of the condom against the base of the penis as he withdraws. It is best to withdraw soon after ejaculation, while the man still has an erection, in order to minimize the chances of leakage. A new condom must be used with each act of intercourse.

Condoms may be either lubricated or unlubricated. Some further lubrication for intercourse may be necessary. A contraceptive foam or jelly works well and provides additional protection. A sterile lubricant such as K-Y jelly may also be used. Many kinds of condoms are available today that are coated with a spermicide containing nonoxynol-9.

How It Works

The condom catches the semen and thus prevents it from entering the vagina. For condoms coated with a spermicide, the spermicide kills sperm and thus provides double protection. Spermicide-coated condoms are not necessarily to be preferred, though. They may create allergies to the spermicide for the man or his partner, and the amount of spermicide is probably not sufficient to be very effective. For couples who want to improve the effectiveness of the condom, it is probably wiser for the woman to use a contraceptive foam.

Effectiveness

Condoms are actually much more effective as a contraceptive than most people think. The perfect-user failure rate is about 3 percent. The typical-user failure rate is about 14 percent, but many failures result from improper or inconsistent use. The FDA controls the quality of condoms carefully, and

thus the chances of a failure due to a defect in the condom itself are small.

Combined with a contraceptive foam or cream or a diaphragm, the condom is close to 100 percent effective.

Side Effects

The condom has no side effects, except that some users are allergic to latex. For them, nonlatex condoms made of polyurethane or other plastics are available.

Advantages and Disadvantages

One disadvantage of the condom is that it must be put on just before intercourse, raising the old "spontaneity" problem again. If the couple can make an enjoyable, erotic ritual of putting it on together, this problem can be minimized.

Some men complain that the condom reduces their sensation and thus lessens their pleasure in intercourse ("It's like taking a shower with a raincoat on"). This can be a major disadvantage. The reduction in sensation, however, may be an advantage for some; for example, it may help the premature ejaculator. Polyurethane condoms are thinner and should provide more sensation.

There are several advantages to condoms. They are the only contraceptive presently available for men except sterilization. They are cheap (around $1.00 to $1.50 for three), they are readily available without prescription at any drugstore and some convenience stores, and they are fairly easy to use, although the man (or woman) must plan ahead so that one will be available when it is needed.

Finally, it is one of the few contraceptive devices that also provide protection against sexually transmitted diseases, an important consideration in our epidemic era. Research indicates that, used consistently and correctly, latex and polyurethane condoms are highly effective in preventing the transmission of most STDs (Hatcher et al., 1998). However, animal-skin condoms allow the HIV, herpes, and hepatitis B viruses to pass through and so are not effective against these STDs.

Reversibility

The method is easily and completely reversible. The man simply stops using them if conception is desired.

Innovations

With the increased popularity of condoms in the 1990s, several innovations have occurred. The new Mentor brand has an adhesive inner surface, the idea being to reduce the risk of slippage and leakage. The marketing of condoms has also changed dramatically. Promotion efforts are now directed increasingly toward women, and there is a new openness in marketing and purchasing. Brands such as Lady Trojan can be found packaged in attractive pastel wrappers in the women's health sections of drugstores. *Consumer Reports* (May, 1995) reviewed the wide variety of brands now available.

The Female Condom

The female condom became available in 1994. It is made of polyurethane and resembles a clear balloon (Figure 8.9). There are two rings in it, one at either end. One ring is inserted into the vagina much like a diaphragm, while the other is spread over the vaginal entrance. The inside is prelubricated, and additional lubrication may be applied if desired. The condom is removed immediately after intercourse, before the woman stands up. The outer ring is squeezed together and twisted to keep the semen inside. A new female condom must be used with each act of intercourse.

How It Works

The female condom works by preventing sperm from entering the vagina and by blocking the entrance to the uterus.

Effectiveness

The female condom is new, so we have less data on its effectiveness compared with other methods. The data we have do not look impressive. The typical-user failure rate is 21 percent (Hatcher et al., 1998), which is unacceptably high for many women. The perfect-user failure rate is 5 percent.

Side Effects

There are few if any side effects with the female condom. A few women experience vaginal irritation and a few men experience irritation of the penis as a result of using it.

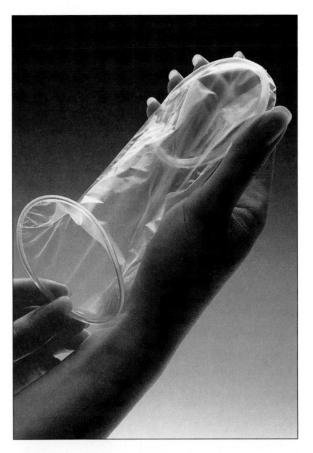

Figure 8.9   The female condom. One ring fits over the cervix, and the other goes outside the body, over the vulva, so that the condom lines the vagina and partly covers the vulva.

Advantages and Disadvantages

The female condom is made of polyurethane, not the latex used in most male condoms. Polyurethane is less susceptible to tearing and does not deteriorate with exposure to oil-based substances in the way that latex does. It does not create the allergic reactions that some people have to latex.

One major advantage is that the female condom is a method that a woman can use herself to reduce her risk of contracting an STD. The polyurethane is impermeable to the HIV virus and to the viruses and bacteria that cause other STDs.

In regard to disadvantages, the spontaneity problem presents itself again. The female condom, at least in its present form (one hopes it will be improved in the near future), is awkward and makes rustling noises while in use. It makes the male condom seem sophisticated and unobtrusive by comparison. Also, it is the least effective of the methods discussed so far in this chapter.

Another disadvantage is the cost, about $2 per condom, which is considerably higher than the cost for male condoms.

Reversibility

The method is easily and completely reversible. The woman simply stops using the condom.

Spermicides

Contraceptive foams (Delfen, Emko), creams, and jellies are all classified as **spermicides,** that is, sperm killers. They come in a tube or a can, along with a plastic applicator. The applicator is filled and inserted into the vagina; the applicator's plunger is then used to push the spermicide out into the vagina near the cervix. Thus the spermicide is inserted much as a tampon is. It must be left in for 6 to 8 hours after intercourse. One application provides protection for only one act of intercourse.

Spermicides are not to be confused with the various feminine hygiene products (vaginal deodorants) on the market. The latter are not effective as contraceptives.

How They Work

Spermicides consist of a spermicidal chemical in an inert base, and they work in two ways: chemical and mechanical. The chemicals in them kill sperm, while the inert base itself mechanically blocks the entrance to the cervix so that sperm cannot swim into it.

Effectiveness

Failure rates for spermicides can be as high as 25 percent. Put simply, they are not very effective. Foams tend to be more effective, and creams and

Spermicide (SPERM-ih-side): A substance that kills sperm.

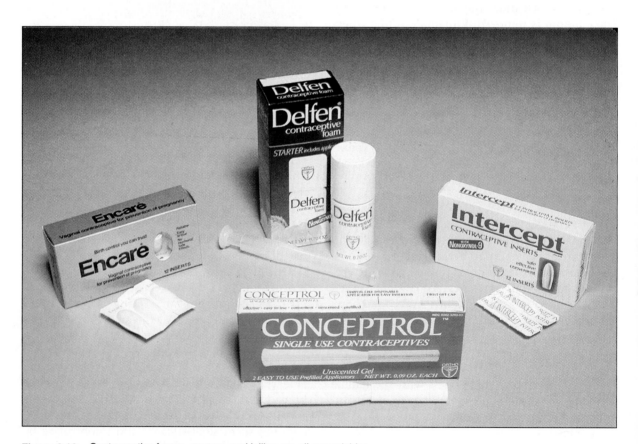

Figure 8.10   Contraceptive foams, creams, and jellies are all spermicides.

jellies less so. Spermicidal tablets and suppositories are also available, but they are the least effective. Spermicides are highly effective only when used with a diaphragm or a condom.

Side Effects
Some people experience an allergic reaction—irritation of the vagina or penis—to spermicides. Because we couldn't find any scientific studies on the incidence of these allergies, we surveyed our sexuality classes. We found that, of those students who had used spermicides, about 2 percent of the men reported an allergic reaction, as did 26 percent of the women.

Advantages and Disadvantages
The major advantage of spermicides is that they are readily available, without a prescription, in any drugstore. Thus they can be used as a stopgap method until the woman can see a physician and get a more effective contraceptive. Their failure rate is so high, though, that we cannot recommend using them by themselves; always combine them with a second method such as a condom.

Spermicides provide some—not a lot—of protection against bacterial STDs such as chlamydia and gonorrhea. There is no evidence, though, that they protect against HIV.

Their major disadvantage is that by themselves, they are not very effective. They also interrupt the spontaneity of sex, although only very briefly. Some women dislike the sensation of the spermicide leaking out after intercourse, and some are irritated by the chemicals. Finally, some people find that they taste terrible, and so their use interferes with oral sex.

Douching

Some people mistakenly believe that **douching** (flushing the vagina with a liquid) with any one of a variety of solutions is an effective contraceptive technique. A popular rumor among teenagers is that douching with Coca-Cola after intercourse will prevent pregnancy. Unfortunately, while it is true that acidic solutions will kill sperm, it takes only seconds for some of the sperm to reach the cervical mucus; once there, they are free to continue moving up into the uterus, and no douching solution will reach them. The woman would have to be a championship sprinter to get herself up and douched

Figure 8.11 Male responsibility is a key issue in birth control.

soon enough. And the douche itself may even push some sperm up into the uterus. Douching, therefore, is just not effective as a contraceptive method.

Withdrawal

Withdrawal (*coitus interruptus*, "pulling out") is probably the most ancient form of birth control. (A reference to it is even found in Genesis 38:8–9, in the story of Onan; hence it is sometimes called *onanism*, although this term is also sometimes used for masturbation.) Withdrawal is still widely used throughout the world. The man withdraws

Douching (DOOSH-ing): Flushing out the inside of the vagina with a liquid.
Withdrawal: A method of birth control in which the man withdraws his penis from his partner's vagina before he has an orgasm.

his penis from his partner's vagina before he has an orgasm and thus ejaculates outside the vagina. To be effective as contraception, the ejaculation must occur completely away from the woman's vulva.

Effectiveness

Withdrawal is not very effective as a method of birth control. The failure rate is around 19 percent. Failures occur for several reasons: The few drops of fluid that come out of the penis during arousal may carry enough sperm for conception to occur; if ejaculation occurs outside the vagina but near or on the vulva, sperm may still get into the vagina and continue up into the uterus; and sometimes the man simply does not withdraw in time.

Side Effects

Withdrawal produces no direct physical side effects. However, over long periods of time, it may contribute to sexual dysfunctions in the man, such as premature ejaculation, and also sexual dysfunction in the woman.

Advantages and Disadvantages

The major advantage of withdrawal is that it is the only last-minute method; it can be used when nothing else is available, although if the situation is this desperate, one might consider abstinence or some other form of sexual expression such as mouth-genital sex as alternatives. Obviously, withdrawal requires no prescription, and it is completely free.

One major disadvantage is that withdrawal is not very effective. In addition, it requires exceptional motivation on the part of the man, and it may be psychologically stressful to him. He must constantly maintain a kind of self-conscious control. The woman may worry about whether he really will withdraw in time, and the situation is certainly less than ideal for her to have an orgasm.

Fertility Awareness (Rhythm) Methods

Rhythm (fertility awareness) **methods** are the only form of "natural" birth control and are therefore the only methods officially approved by the Roman Catholic Church. They require abstaining from intercourse during the woman's fertile period (around

ovulation). There are several rhythm methods, in each of which the woman's fertile period is determined in a different way.

The Calendar Method

The **calendar method** is the simplest rhythm method. It is based on the assumption that ovulation occurs about 14 days before the onset of menstruation. It works best for the woman with the perfectly regular 28-day cycle. She should ovulate on day 14, and almost surely on one of days 13 to 15. Three days are added in front of that period (previously deposited sperm may result in conception), and 2 days are added after it (to allow for long-lasting eggs); thus the couple must abstain from sexual intercourse from day 10 to day 17. Therefore, even for the woman with perfectly regular cycles, 8 days of abstinence are required in the middle of each cycle. Recent research shows that sperm can live up to 5 days inside the female reproductive tract and eggs live less than a day (Wilcox et al., 1995).

The woman who is not perfectly regular must keep a record of her cycles for at least 6 months, and preferably a year. From this she determines the length of her shortest cycle and the length of her longest cycle. The preovulatory safe period is then calculated by subtracting 18 from the number of days in the shortest cycle, and the postovulatory safe period is calculated by subtracting 11 from the number of days in the longest cycle (see Table 8.2). Thus for a woman who is somewhat irregular—say, with cycles varying from 26 to 33 days in length—a period of abstinence from day 8 to day 22 (a total of 15 days) would be required.

The Basal Body Temperature Method

A somewhat more accurate method for determining ovulation is the **basal body temperature (BBT) method.** The principle behind this was discussed

> **Rhythm method:** A method of birth control that involves abstaining from intercourse around the time the woman ovulates.
> **Calendar method:** A type of rhythm method of birth control in which the woman determines when she ovulates by keeping a calendar record of the length of her menstrual cycles.
> **Basal body temperature method:** A type of rhythm method of birth control in which the woman determines when she ovulates by keeping track of her temperature.

Table 8.2 Determining the Fertile Period Using the Calendar Method*

Shortest Cycle (Days)		Day Fertile Period Begins	Longest Cycle (Days)		Day Fertile Period Ends
22		4	23		12
23		5	24		13
24		6	25		14
25	Minus 18	7	26		15
26	← →	(8)	27		16
27		9	28		17
28		10	29		18
29		11	30		19
30		12	31		20
31		13	32	Minus 11	21
32		14	33	→ →	(22)
			34		23
			35		24

Example: If a woman's cycles vary in length from 26 days to 33 days, she can be fertile any time between days 8 and 22 of the cycle; so to avoid getting pregnant, she must abstain from sexual intercourse from day 8 to day 22.

*See text for further explanation.

in Chapters 6 and 7. The woman takes her temperature every day immediately upon waking. During the preovulatory phase her temperature will be at a fairly constant low level. On the day of ovulation it drops (although this does not always occur), and on the day after ovulation it rises sharply, staying at that high level for the rest of the cycle. Intercourse would be safe beginning about three days after ovulation. Some of the psychological stresses involved in using this method have been noted previously. As a form of contraception, the BBT method has a major disadvantage in that it determines safe days only *after* ovulation; theoretically, according to the method, there are no safe days before ovulation. Thus the BBT method is probably best used in combination with the calendar method, which determines the preovulatory safe period; the BBT method determines the postovulatory safe period.

The Cervical Mucus (Ovulation) Method

Another rhythm method is based on variations over the cycle in the mucus produced by the cervix. This **cervical mucus method** works in the following way:

There are generally a few days just after menstruation during which no mucus is produced and there is a general sensation of vaginal dryness. This is a relatively safe period. Then there are a number of days of mucus discharge around the middle of the cycle. On the first days, the mucus is white or cloudy and tacky. The amount increases, and the mucus becomes clearer, until there are one or two *peak days,* when the mucus is like raw egg white— clear, slippery, and stringy. There is also a sensation of vaginal lubrication. Ovulation occurs within 24 hours after the last peak day. Abstinence is required from the first day of mucus discharge until four days after the peak days. After that the mucus, if present, is cloudy or white, and intercourse is safe.

The Sympto-Thermal Method

The **sympto-thermal method** combines two rhythm methods in order to produce better effectiveness. The woman records changes in her cervical mucus (symptoms) as well as her basal body temperature (thermal). The combination of the two should give a more accurate determination of the time of ovulation.

Cervical mucus method: A type of rhythm method of birth control in which the woman determines when she ovulates by checking her cervical mucus.
Sympto-thermal method: A type of rhythm method of birth control combining both the basal body temperature method and the cervical mucus method.

Home Ovulation Tests

Recently, home tests for the detection of ovulation have been developed, such as the Q Test Ovulation Test Kit. The tests detect hormone levels in urine. They were developed for couples wanting to conceive and, unfortunately, like the temperature method, these tests only determine when ovulation has occurred, and then the safe period after ovulation. They do not warn the woman several days in advance of when she will ovulate, which she also needs to know. Researchers are trying to develop tests that will do both.

Effectiveness

The effectiveness of the rhythm method varies considerably, depending on a number of factors, but basically it is not very effective with typical users (giving rise to its nickname, "Vatican roulette," and a number of old jokes like, "What do they call people who use the rhythm method?" Answer: "Parents"). While the typical-user failure rate is around 25 percent for all methods, ideal-user failure rates vary considerably. They are 5 percent for the calendar method, 2 percent for BBT, 2 percent for the symptothermal method, and 3 percent for the cervical mucus method (Hatcher et al., 1998).

There tend to be fewer failures when the woman's cycle is very regular and when the couple are highly motivated and have been well instructed in the methods. The effectiveness of the rhythm method also depends partly on one's purpose in using it: whether for preventing pregnancy absolutely or for spacing pregnancies. If absolute pregnancy prevention is the goal (as it probably would be, for example, for an unmarried teenager), the method is just not effective enough. But if the couple simply wish to space pregnancies farther apart than would occur naturally, the method will probably accomplish this. Knowing when the woman's fertile times occur can also improve the effectiveness of other methods of contraception.

Advantages and Disadvantages

For many users of the rhythm method, its main advantage is that it is considered an acceptable method of birth control by the Roman Catholic Church.

The method has no side effects except possible psychological stress, and it is cheap. It is easily reversible. It also helps the woman become more aware of her body's functioning. The method requires cooperation from both partners, which may be considered either an advantage or a disadvantage.

Its main disadvantages are its high failure rate and the psychological stress it may cause. Periods of abstinence of at least eight days, and possibly as long as two or three weeks, are necessary, which is an unacceptable requirement for many couples. Actually, the rhythm method would seem best suited to people who do not like sex very much.

A certain amount of time, usually at least six months, is required to collect the data needed to make the method work. Thus one cannot simply begin using it on the spur of the moment.

Sterilization

Sterilization, or voluntary surgical contraception (VSC), is a surgical procedure whereby an individual is made permanently sterile, that is, unable to reproduce. Sterilization is a rather emotion-laden topic for a number of reasons. It conjures up images of government-imposed programs of *involuntary* sterilization in which groups of people—possibly the mentally retarded, criminals, or members of some minority group—are sterilized so that they cannot reproduce. (The following discussion deals only with voluntary sterilization used as a method of contraception for those who want no more children or who want no children at all.) Some people confuse sterilization with castration, though the two are quite different. This is also an emotional topic because sterilization means the end of one's capacity to reproduce, which is very basic to gender roles and gender identity. The ability to impregnate and the ability to bear a child are very important in cultural definitions of manhood and womanhood. We hope that as gender roles become more flexible in our society and as concern about reproduction is replaced by a concern for limiting population size, the word "sterilization" will no longer carry such emotional overtones.

Most physicians are conservative about performing sterilizations; they want to make sure that the patient has made a firm decision on his or her own and

Sterilization: A surgical procedure by which an individual is made sterile, that is, incapable of reproducing.

will not be back a couple of months later wanting to have the procedure reversed. The physician has an obligation to follow the principle of "informed consent." This means explaining the procedures involved, telling the patient about the possible risks and advantages, discussing alternative methods, and answering any questions the patient has.

Despite this conservatism, both male sterilization and female sterilization have become increasingly popular as methods of birth control. Sterilization is the most common method of birth control for married couples in the United States (Marquette et al., 1995). Currently 30 percent of married white women and 20 percent of married white men are sterilized, as are 48 percent of married black women and 2 percent of married black men (Mosher, 1990).

Male Sterilization

The male sterilization operation is called a **vasectomy,** so named for the vas deferens, which is tied or cut. It can be done in a physician's office under local anesthesia and requires only about 20 minutes to perform. In the traditional procedure, the physician makes a small incision on one side of the upper part of the scrotum. The vas is then separated from the surrounding tissues, tied off, and cut. The procedure is then repeated on the other

side, and the incisions are sewn up. For a day or two the man may have to refrain from strenuous activity and be careful not to pull the incision apart. A new *no-scalpel vasectomy* procedure has been developed recently (Hatcher et al., 1998). It involves making just a tiny pierce in the scrotum and has an even lower rate of complications than a standard vasectomy (see Figure 8.12).

Typically, the man can return to having intercourse within a few days. It should not be assumed that he is sterile yet, however. Some stray sperm may still be lurking in his ducts beyond the point of the incision. All sperm are generally gone after 20 ejaculations, and their absence should be confirmed by semen analysis. Until this confirmation is made, an additional method of birth control should be used.

Misunderstandings about the vasectomy abound. In fact, a vasectomy creates no physical changes that interfere with erection. Neither does it interfere in any way with sex hormone production; the testes continue to manufacture testosterone and secrete it into the bloodstream. Neither does a

Vasectomy (vas-EK-tuh-mee): A surgical procedure for male sterilization involving severing of the vas deferens.

Figure 8.12   The no–scalpel vasectomy. (a) The vas (dotted line) is grasped by special ring forceps and the scrotum is pierced by sharp-tipped forceps. (b) The forceps stretch the opening slightly, and (c) The vas is lifted out and then tied off. The other vas is then lifted out through the same small hole and the procedure is repeated.

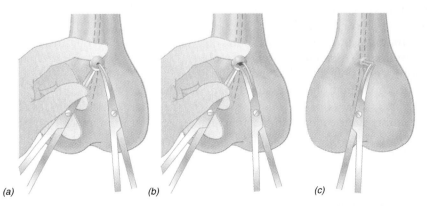

(a)   (b)   (c)

Reprinted with permission from the Population Information Program

vasectomy interfere with the process or sensation of ejaculation. As we noted earlier, virtually all the fluid of the ejaculate is produced by the seminal vesicles and prostate, and the incision is made long before that point in the duct system. Thus the ejaculate is completely normal, except that it does not contain any sperm.

How It Works

The vasectomy makes it impossible for sperm to move beyond the cut in the vas. Thus the vasectomy prevents sperm from being in the ejaculate.

Effectiveness

The vasectomy is essentially 100 percent effective; it has a failure rate of 0.1 percent. Failures occur because stray sperm are still present during the first few ejaculations after surgery, because the physician did not completely sever the vas, or because the ends of the vas have grown back together.

Side Effects

The physical side effects of the vasectomy are minimal. In about 5 percent of cases, there is a minor complication from the surgery, such as infection of the vas (Hatcher et al., 1998).

Some psychologically based sexual problems such as impotence may arise. Thus the man's attitude toward having a vasectomy is extremely important. About 5 to 10 percent regret having had a vasectomy (Hatcher et al., 1994).

Reversibility

Quite a bit of effort has been devoted to developing techniques for reversing vasectomies (the surgical procedure for reversal is termed *vasovasostomy*) and to developing vasectomy techniques that are more reversible. At present, with sophisticated microsurgery techniques, pregnancy rates following reversal are around 50 percent (Hatcher et al., 1998). In making a decision about whether to have a vasectomy, though, a man should assume that it is irreversible.

After a vasectomy some men begin forming antibodies to their own sperm. Because these antibodies destroy sperm, they might contribute further to the irreversibility of the vasectomy.

Advantages and Disadvantages

The major advantages of the vasectomy are its effectiveness and its minimal health risks. Once performed, it requires no further thought or planning on the man's part. As a permanent, long-term method of contraception, it is very cheap. The operation itself is simple—simpler than the female sterilization procedures—and requires no hospitalization or absence from work. Finally, it is one of the few methods that allow the man to assume contraceptive responsibility.

The permanency of the vasectomy may be either an advantage or a disadvantage. If permanent contraception is desired, the method is certainly much better than something like birth control pills, which must be used repeatedly. But if the couple change their minds and decide that they want to have a child, the permanence is a distinct disadvantage. Some men put several samples of their sperm into a frozen-sperm bank so that artificial insemination can be performed if they do decide to have a child after a vasectomy.

Another disadvantage of the vasectomy is the various psychological problems that might result if the man sees sterilization as a threat to his masculinity or virility. However, long-term studies of vasectomized men provide no evidence of such psychological problems (Population Information Program, 1983). In studies done around the world, the majority of vasectomized men say that they have no regrets about having had the sterilization performed, that they would recommend it to others, and that there has been no change or else an improvement in their happiness and sexual satisfaction in marriage. Fewer than 5 percent of vasectomized men report psychological problems such as decreased libido or depression, and this rate is no higher than in control samples of unvasectomized men.

Finally, if a married couple use the vasectomy as a permanent method of birth control, the woman is not protected if she has intercourse with someone other than her husband.

Female Sterilization

Two surgical techniques are used to sterilize a woman: minilaparotomy and laparoscopy (tubal ligation or "having the tubes tied" are terms that are also heard). These techniques differ in terms of the type of procedure used (see Figure 8.13). Both are performed under local or general anesthesia, and both involve blocking the fallopian tubes in some way so that sperm and egg cannot meet.

In a **minilaparotomy** ("minilap"), a small inci-

Figure 8.13 Two methods for performing a tubal ligation or minilaparotomy. In *(a)* the fallopian tubes are tied off and the loop is cut: the cut ends then scar over. In *(b)* the tubes are tied in two places and the section between is cut and removed.

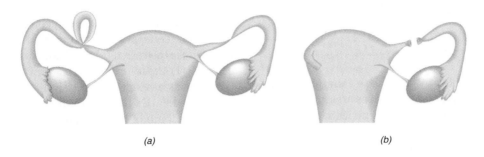

(a)   *(b)*

sion (less than 3 centimeters, or about 1 inch long) is made in the abdomen. Each fallopian tube is in turn gently pulled to the opening. Each tube is blocked either by cutting and tying off the ends or by applying a small clip; the tubes are then allowed to slip back into place. With the *laparoscopy,* a magnifying instrument is inserted into the abdomen. The doctor uses it to identify the fallopian tubes and then blocks them by electrocoagulation or by clips. Either procedure takes only about 10 to 20 minutes and does not require that the woman spend the night in the hospital. Special procedures are also available for women who have just had a baby or an abortion.

The female sterilization procedures do not interfere with the ovaries, and therefore the production of sex hormones continues normally; thus female sterilization does not bring on premature menopause. Some of the misunderstandings arise from confusion of female sterilization procedures with hysterectomy (surgical removal of the uterus) or oophorectomy (surgical removal of the ovaries, which does impair hormonal functioning). These latter two operations do produce sterility, but they are generally performed for purposes other than sterilization, for example, removal of tumors.

How It Works
Female sterilization procedures make it impossible for the egg to move down the fallopian tube toward the uterus. They also prevent sperm from reaching the egg.

Effectiveness
These procedures are essentially 100 percent effective. The failure rate of 0.5 percent is due to an occasional rejoining of the ends of the fallopian tubes, and rare cases in which the woman was pregnant before the sterilization procedure was performed.

Side Effects
Occasionally there are side effects arising from the surgery, such as infections, hemorrhaging, and problems related to the anesthetic. Generally, only 1 percent of women undergoing the surgery experience complications.

Reversibility
Highly refined microsurgery techniques make it possible to reverse female sterilization in some cases. The success rate varies considerably, depending on the method that was used to perform the sterilization. Pregnancy rates range between 45 percent and 80 percent (Hatcher et al., 1998). However, in deciding whether to have sterilization surgery, a woman should assume that it is irreversible.

Advantages and Disadvantages
Female sterilization has some of the same advantages as male sterilization in terms of effectiveness,

Minilaparotomy: A method of female sterilization.

permanence, and cheapness when used for long-term contraception. One disadvantage is that it offers no protection from sexually transmitted diseases.

Psychological Aspects: Attitudes Toward Contraception

It is a favorite old saying among Planned Parenthood workers that contraceptives are only as effective as the people who use them. That is, no contraceptive method is effective if it is not used or if it is used improperly. Thus the user is at least as important as all the technology of contraception.

Each year in the United States, 1 million teenagers, or one out of every eight women aged 15 to 19 (most of them unmarried), become pregnant (Hatcher et al., 1998). It is not an overstatement, then, to say that teenage pregnancy is at epidemic proportions. Approximately 30 percent of these unwanted pregnancies are terminated by abortion (Hatcher et al., 1998), 50 percent result in live births (to single teenagers or to couples joined in "shotgun" matrimony), and the remainder end in miscarriage.

The great majority of these unwanted pregnancies are the result of the failure of sexually active persons to use contraceptives responsibly. According to one study, 27 percent of white teenage women and 35 percent of black teenage women who are sexually active do not use contraceptives (Hofferth, 1990).

If we are to understand this problem and take effective steps to solve it, we must understand the psychology of contraceptive use and nonuse. Many researchers have been investigating this issue.

When adolescents are asked why they do not use contraceptives, the reasons they give tend to fall into the following categories (Morrison, 1985):

1. Beliefs about their own fertility. "I thought I (or my partner) couldn't get pregnant."
2. Wanting to become pregnant, or at least not minding if they did become pregnant.
3. Problems in obtaining contraception. Many factors may be involved, such as not knowing where to get contraceptives, feeling that it is a "hassle" to get contraceptives, or expecting that they would be too expensive to buy.
4. Intercourse is unplanned, and therefore contraception is not planned.

5. Negative attitudes and feelings about contraception. For example, some feel that contraceptives are messy or embarrassing; others have religious objections; and some believe that contraceptives are dangerous.

Social scientists have developed several theories to explain teenagers' use and nonuse of contraceptives. These theories tend to fall into one of three categories: (1) those theories that view contraceptive behavior as a result of a decision-making process; (2) those theories that view contraceptive behavior as the outcome of psychological development; and (3) those theories that focus on personality and emotions, such as Donn Byrne's work on erotophobia. Let us look at each of these three categories.

An example of a decision-making theory has been formulated by Kristin Luker (1975). She argues that teenage women essentially engage in a cost-benefit analysis—perhaps not a very deliberate one—in which they weigh (1) the costs of contraception (e.g., may be difficult to obtain, partner may not like the idea); (2) the benefits of contraception (not getting pregnant); (3) the costs of pregnancy, which may not seem large for a young woman with few bright hopes for the future; and (4) the benefits of pregnancy (feeling like a woman, having a baby who loves you). The woman also makes some estimate of the probability of getting pregnant and generally underestimates it. As a result, she engages in contraceptive risk-taking—much like deciding not to fasten one's seatbelt in a car—because the costs of contraception seem to outweigh the benefits, or perhaps because there seem to be benefits to pregnancy.

Developmental models look at the process of psychological development during adolescence (e.g., Jorgensen, 1980; Morrison, 1985). Teenagers may find their values to be firmly in line with their parents' values, but their behavior increasingly conforms to the norms of their peer group. A conflict between values and behavior results (Zabin et al., 1984). More specifically, teenagers may hold their parents' conservative values about premarital sex, which prize abstinence and therefore nonuse of contraceptives. Meanwhile, their actual behavior conforms to that of the peer group and they engage in premarital intercourse. The hope is that as adolescents mature, their behavior and values will become more consistent with each other.

Social psychologist Donn Byrne (1983; Fisher et al., 1988) focuses on a dimension of personality that he calls erotophobia-erotophilia. According to his analysis, there are five steps in effective contraception:

1. The person must acquire and remember accurate information about contraception.

2. The person must acknowledge that there is a likelihood of engaging in sexual intercourse. Contraceptive preparation, of course, makes sense only if one has some expectations of having intercourse. Gender-role socialization has made it particularly difficult for women to acknowledge such expectations.

3. The person must obtain the contraceptive. This may involve a visit to a doctor or to a drugstore or Planned Parenthood clinic.

4. The person must communicate with his or her partner about contraception. Otherwise, each may assume that the other will take care of it.

5. The person must actually use the method of contraception.

According to Byrne's analysis, a number of psychological factors can intervene in any of these five steps, making the person either more likely or less likely to use contraceptives effectively. These factors include attitudes and emotions, information, expectations, and fantasies.

Attitudes and emotions play an important role. One particular dimension is erotophilia-erotophobia (Byrne, 1983; Fisher et al., 1983; Byrne, 1977). **Erotophobes** don't discuss sex, have sex lives that are influenced by guilt and fear of social disapproval, have intercourse infrequently with few partners, and are shocked by sexually explicit films. **Erotophiles** are just the opposite—they discuss sex, they are relatively uninfluenced by sex guilt, they have intercourse more frequently with more partners, and they find sexually explicit films to be arousing. Research shows that erotophiles are more likely to be consistent, reliable contraceptive users. At every one of the five steps of contraceptive use, the erotophobes are more likely to fail. Research shows that they have less sex information than erotophiles do, and that, when exposed to the same sex information, erotophobes learn less than erotophiles do (Fisher et al., 1983). Because of their fearfulness, erotophobes are less likely to acknowledge that intercourse may occur, which makes contraceptive planning difficult (although extreme erotophobes are likely to abstain from sex com-

pletely, which definitely reduces the risk of unwanted pregnancy). Erotophobes also have more difficulty going to a doctor or a drugstore to obtain contraceptives. Erotophobes don't discuss sex or contraception very much, and therefore effective communication with their partner is unlikely to occur. Finally, erotophobes have trouble with actually using the contraceptive. An erotophobic male isn't going to be thrilled about pulling out a condom. An erotophobic female won't be thrilled with inserting a diaphragm or thinking about sex every day as she takes her pill.

Information is also an important factor in contraceptive use and nonuse. People who lack information about contraceptives and their correct use can scarcely use them effectively.

Expectations play an important role. When thinking about sex and contraception, people have some expectations about how likely it is that intercourse will result in pregnancy. Research shows that many people think that the chance is zero or close to it, expressing the expectation that "It can't happen to me." People with that expectation are unlikely to use contraceptives.

Although it is generally recognized that *fantasy* is an important part of sexual expression, fantasy may also play an important role in contraceptive behavior (Byrne, 1983). Most of us have fantasies about sexual encounters, and we often try to make our real-life sexual encounters turn out like the "scripts" of our fantasies. An important shaper of our fantasies is the mass media. Through movies, television, and romance novels, we learn idealized techniques for kissing, holding, lovemaking. But the media's idealized versions of sex almost never include a portrayal of the use of contraceptives. When Dr. Doug Ross of the hit TV series *E.R.* hops into bed with yet another lovely woman, they never show Doug reaching for a condom or the woman's nightly routine of taking her pill. Thus our fantasy sex, shaped by the media, lacks contraception as part of the script. One exception occurred in the movie *Saturday Night Fever.* John Travolta, lying on top of a willing young woman in the back seat of a car, asks her if she is using a contraceptive. When she replies that she isn't, he zips his pants up and

---

Erotophobia: Feeling guilty and fearful about sex.
Erotophilia: Feeling comfortable with sex, lacking in feelings of guilt and fear about sex.

leaves. If teenagers saw lots of instances of their heroes and heroines behaving responsibly about contraception, it would probably influence their behavior. But right now that is not what the media give them.

What are the solutions? Can this research and theorizing on the social psychology of contraceptive use be applied to reducing the teenage pregnancy problem? The most direct solution would be to have better programs of sex education in the schools. Many districts have no sex education programs, and those that do often skip the important issue of contraception, fearing that it is too controversial. Sex education programs would need to include a number of components that are typically missing (Gross & Bellew-Smith, 1983). These include legitimizing presex communication about sex and contraception (Milan & Kilmann, 1987); legitimizing the purchase and carrying of contraceptives; discussing how one weighs the costs and benefits of pregnancy, contraception, and abortion; legitimizing noncoital kinds of sexual pleasure, such as masturbation and oral-genital sex; and encouraging males to accept equal responsibility for contraception.

Abortion

In the past three decades, **abortion** (the termination of a pregnancy) has been a topic of considerable controversy in North America. Feminist groups talk of the woman's right to control her own body, whereas members of right-to-life groups speak of the fetus's rights. In 1973, the United States Supreme Court made two landmark decisions (*Roe v. Wade* and *Doe v. Bolton*) that essentially decriminalized abortion by denying the states the right to regulate early abortions. The conservative Supreme Court of the 1990s has made some rulings that partly reverse these decisions (see Chapter 22). Nevertheless, 1.3 million legal abortions are performed each year in the United States (Hatcher et al., 1998).

In other countries, policies on abortion vary widely. It is legal and widely practiced in Russia and Japan, parts of eastern and central Europe, and South America. The use of abortion in the developing nations of Africa and Asia is limited be-

Abortion: The termination of a pregnancy.

Table 8.3 Abortion Rates Around the World*

Country	Number of Abortions per Year	Abortion Rate	Abortion Ratio
Australia	63,200	16.6	20.4
Bulgaria	119,900	64.7	50.7
Canada	63,600	10.2	14.7
China	10,394,500	38.8	31.4
India[†]	588,400	3.0	2.2
Israel[†]	15,500	16.2	13.5
Italy[†]	191,500	15.3	25.7
Japan[†]	497,800	18.6	27.0
South Korea	528,000	53	43
Sweden	34,700	19.8	24.9
USSR[†]	6,818,000	111.9	54.9
U.S.	1,588,600	28.0	29.7
Vietnam[†]	170,600	14.6	8.2

*"Abortion rate" is the number of abortions per 1000 women aged 15 to 44. "Abortion ratio" is the number of abortions per 100 known pregnancies.

[†]Data are from 1989 and are of unknown accuracy.

Source: Stanley K. Henshaw. (1990). Induced abortion: A world review, 1990. *Family Planning Perspectives, 22*(2), 76–89.

cause of the scarcity of medical facilities. Table 8.3 gives rates of abortion in various countries.

This section is about methods of abortion and the psychological aspects of abortion; the ethical and legal aspects will be discussed in Chapters 21 and 22.

Abortion Procedures

Several methods of abortion are available; which one is used depends on how far the pregnancy has progressed.

Vacuum Aspiration

The **vacuum aspiration method** (also called *vacuum suction* or *vacuum curettage*) can be performed during the first trimester of pregnancy and up to 14 weeks' gestation. It is done on an outpatient basis with a local anesthetic. The procedure itself takes only about 10 minutes, and the woman stays in the doctor's office, clinic, or hospital for a few hours. It is the most widely used abortion procedure in the United States today, accounting for 97 percent of abortions (Hatcher et al., 1998).

The woman is prepared as she would be for a pelvic exam, and an instrument is inserted into the vagina; the instrument dilates (stretches open) the opening of the cervix. A tube is then inserted into this opening until one end is in the uterus (see Figure 8.14). The other end is attached to a suction-producing apparatus, and the contents of the uterus, including the fetal tissue, are sucked out.

> **Vacuum aspiration:** A method of abortion that is performed during the first trimester and involves suctioning out the contents of the uterus.

Figure 8.14   A vacuum aspiration abortion.

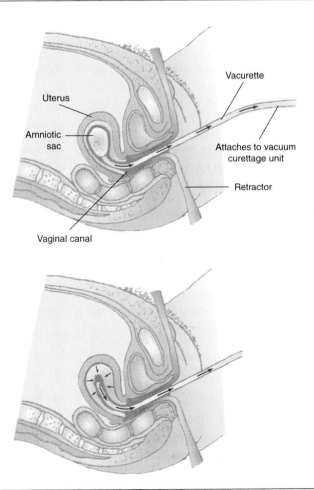

Vacuum aspiration has become the most common method of early (first trimester) abortion because it is simple and entails little risk. There are rare risks of uterine perforation, infection, hemorrhaging, and failure to remove all the fetal material.

Dilation and Evacuation

Dilation and evacuation (D and E) is similar to vacuum aspiration, but it must be done in a hospital under a general anesthetic. It is used especially for later abortions, from 13 to 16 weeks' gestation, and possibly up to 20 weeks. It is somewhat similar to vacuum aspiration, but it is a bit more complicated because the fetus is relatively large by the second trimester.

Induced Labor

During the late part of the second trimester, abortion is usually performed by inducing labor and a miscarriage. The most commonly used version of this method is the **saline-induced abortion.** A fine tube is inserted through the abdomen into the amniotic sac inside the uterus. Some amniotic fluid is removed through the tube, and an equal amount of saline solution is injected into the amniotic sac. Within several hours, the solution causes labor to begin. The cervix dilates, and the fetus is expelled through the contractions of labor. A variation on this technique is the **prostaglandin abortion.** Prostaglandins (hormonelike substances that cause contractions) are injected into the amniotic sac (or intravenously or by means of a vaginal suppository) and cause labor.

Induced labor is used for abortion only if pregnancy has progressed late into the second trimester and accounts for only 1 percent of abortions in the United States (Koonin et al., 1991b). This method is both more hazardous and more costly than the previous methods. The most serious complications of the saline-induced method are shock and possibly death (if the technique is done carelessly and the saline solution gets into a blood vessel) and a bleeding disorder, although these are rare.

Hysterotomy

Hysterotomy is a surgical method of abortion that can be done from 16 to 24 weeks after the woman's last menstrual period. Essentially, a cesarean section is performed, and the fetus is removed. Hysterotomy is more serious and more expensive (more than $1000) than the other methods, and there is greater risk of complications. It is done only rarely, but it may be useful if the pregnancy has progressed to the late second trimester and the woman's health is such that the induction methods should not be used.

A summary of statistics on death rates associated with the various methods of abortion is shown in Table 8.4.

RU-486

In 1986 French researchers announced the development of a new drug called **RU-486** (mifepristone) (Couzinet et al., 1986; Ulmann et al., 1990). It can induce a very early abortion. It has a powerful antiprogesterone effect, causing the endometrium of the uterus to be sloughed off and thus bringing about an abortion. It is administered as a tablet followed two days later by a small dose of prostaglandin (misoprostol), which increases contractions of the uterus, helping to expel the embryo. It is used within 10 days of an expected but missed menstrual period. Research shows that it is effective in 92 to 96 percent of the cases, when combined with prostaglandin (Silvestre et al., 1990; Spitz et al., 1998; Ulmann et al., 1990). It is most effective when the woman has been

---

Saline-induced abortion: A method of abortion that is done in the late second trimester and involves inducing labor by injecting a saline solution into the amniotic sac.
Prostaglandin abortion: A method of abortion that is done in the late second trimester and involves inducing labor by injecting prostaglandins into the amniotic sac.
Hysterotomy: A surgical method of abortion done in the late second trimester.
RU-486: The "abortion pill."

Table 8.4 Summary of Death Rates Associated with Legal Abortion and with Normal Childbirth

Deaths per 100,000 legal abortions*	
Vacuum aspiration	0.5
Induced labor	7.0
Hysterotomy	51.6
Deaths per 100,000 normal childbirths	9.1
Blacks	22.0
Whites	6.5
Others	9.8

*Based on 4,500,000 abortions, Centers for Disease Control, 1985.

Sources: Koonin et al. (1991b); Hatcher et al. (1994).

Focus 8.2
Abortion in Cross-Cultural Perspective

Beliefs about abortion show dramatic variation in different cultures around the world. The following is a sampling from two very different cultures.

Ekiti Yoruba

The Ekiti Yoruba, many of whom have a high school or college education, live in southwest Nigeria. For them, abortion is not a distinct category from contraception but rather is on a continuum with it (Renne, 1996).

Traditionally in the Ekiti Yoruba culture, the ideal was for a woman to have as many children as possible, spaced at 2- to 3-year intervals. The spacing of children is made possible by a period of sexual abstinence for 2 years postpartum. The Ekiti Yoruba believe that sexual intercourse while a woman is still breast-feeding a baby causes illness or death to the child; men whose children have died in infancy have been blamed for breaking the postpartum sex taboo and causing the death. Because of the high value placed on fertility, use of contraceptives and abortion must be kept secret. Even though condoms, foam, and birth control pills are available at a local clinic, few people take advantage of the service because they would not want others to know that they engaged in such practices. Abortion then becomes the chief method of birth control. Estimates are that between 200,000 and 500,000 pregnancies are aborted each year in Nigeria and that about 10,000 women die each year from botched abortions.

If a woman has an unwanted pregnancy, she generally will consult a local divine healer or herbalist first, in order to "keep the pregnancy from staying." They generally provide pills or substances to insert in the vagina. If the treatment does not work, the woman then goes to a clinic and has a dilation and curettage (D&C; a procedure that is similar to D&E) abortion.

Women who abort generally fall into two categories: unmarried high school or college students who want to finish their education, and married women who are pregnant because of an affair. Here is one woman's story:

> In 1991, when awaiting entrance into university, I became pregnant by one boyfriend whom I later decided not to marry in favor of another. Since I did not want my chosen fiancé to know of the pregnancy, I decided to abort it. I first used 3 Beecodeine tablets, Andrew's Liver Salt, and Sprite, mixing them together and then drinking them. When this did not work, I went to a clinic in a neighboring town for D&C. The abortion cost N80 and was paid for by my boyfriend. There were no aftereffects. (Renne, 1996, p. 487)

The ease with which Ekiti Yoruba people rely on abortion is related in part to their understanding of prenatal development. Many believe that the "real child" is not formed until after the fourth month of pregnancy, and before that, the being is lizard-like.

Greece

Birth control for women was legalized in Greece only in 1980, and abortion was legalized in 1986. Yet Greece has had a sharply declining birth rate since World War II, accounted for, in large part, by abortion (Georges, 1996). Among European nations, Greece is unique in its combination of very little use of medical contraception, a low fertility rate, and the highest abortion rate in Europe.

Three powerful institutions—the government, the Greek Orthodox Church, and the medical profession—have exerted a strong pronatalist (in favor of having babies) influence. The Greek Orthodox

Church equates abortion with murder and prohibits all methods of birth control except rhythm and abstinence. The government, for its part, encourages large families by a variety of measures, including paying a monthly subsidy to families with more than three children, making day care centers widely available, and keeping female methods of contraception illegal (until 1980). Despite all this, Greek women achieved a low fertility rate, which is regarded by the government as a threat to the Greek "race," Greek Orthodoxy, and the military strength of Greece in relation to hostile neighbors such as the extremely fertile Turks.

Despite the illegality of abortion in Greece until 1986, abortion was widespread and a very open secret. Abortions were not back-alley affairs but rather were performed by gynecologists in private offices. Physicians, as members of a powerful and prestigious profession, were successful at legal evasion. As a result, Greek women did not have to face the life-threatening risks that occur with illegal abortion in other countries. They had access to safe, illegal abortion.

Why is there so much reliance on abortion and so little access to contraception in this modern European nation? As noted earlier, the Greek Orthodox Church opposes all medical contraception and the Greek government kept contraception illegal until 1980. But even then, contraception did not become widespread. In 1990, only 2 percent of women of reproductive age used the pill. Some blame this on the medical profession, which is thought to block access to contraception in order to continue a thriving abortion practice that is more lucrative. Greek women, too, resist contraception. They distinguish between "contraception" (such as birth control pills) and "being careful" (withdrawal and condoms). They reject contraception, but being careful—especially the use of withdrawal—is widespread. Rhythm is not widely used and would not be very successful if it were, since Greek women commonly believe that they are most fertile for the 4 to 7 days just before and after the menstrual period. The mass media have spread scare messages about the pill, and many women believe that it causes cancer.

How do Greek women, the great majority of whom are Orthodox, deal with the contradiction between their church's teaching—that abortion is murder and that a woman who has had an abortion may not receive communion—and their actual practice of having abortions? First, the Greek Orthodox Church is not as absolutist in its application of doctrines regarding abortion as the Roman Catholic Church is. Some attribute this to the fact that Orthodox priests can be married and are therefore more in touch with the realities of life. In some cases, women do abstain from receiving communion following an abortion but then later make a confession to the priest. Priests typically are forgiving.

In Greece, motherhood is highly esteemed and idealized, yet abortion is not considered contradictory to the high value placed on motherhood. Good motherhood today is thought to require intense investment of time and energy in one's children; by definition, then, the good mother limits family size, and abortion is a means to achieve that goal.

Cross-Cultural Patterns

Several patterns emerge from the study of abortion in these quite distinct cultures and other cultures (e.g., Gursoy, 1996; Johnson et al., 1996; Rigdon, 1996; Rylko-Bauer, 1996). First, no matter how strict the prohibitions against abortion, some women in all cultures choose and manage to obtain abortion. Second, the meaning of abortion is constructed in any particular culture based on factors such as beliefs about prenatal development, when life starts, and how much large families are valued. Third, the legality and morality of abortion in any culture is determined in part by political forces, such as the Greek government's desire to expand the size of the Greek population.

Sources: Georges, 1996; Gursoy (1996); Johnson, Horga, and Andronache (1996); Renne (1996); Rigdon (1996); Rylko-Bauer, 1996.

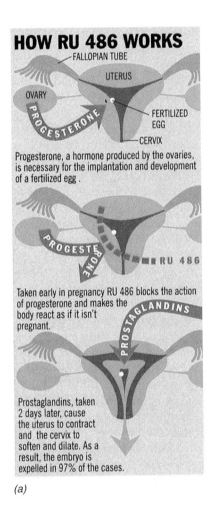

HOW RU 486 WORKS

Progesterone, a hormone produced by the ovaries, is necessary for the implantation and development of a fertilized egg .

Taken early in pregnancy RU 486 blocks the action of progesterone and makes the body react as if it isn't pregnant.

Prostaglandins, taken 2 days later, cause the uterus to contract and the cervix to soften and dilate. As a result, the embryo is expelled in 97% of the cases.

(a)

(b)

Figure 8.15   RU-486. *(a)* How it works. *(b)* The controversy over RU-486: Prolife forces demonstrate against it.

pregnant less than 49 days. Early research has found little evidence of side effects, although the woman experiences some cramping as the uterine contents are expelled.

In France today, more than a quarter of women who decide to terminate an early pregnancy choose RU-486 rather than conventional abortion methods (Ulmann et al., 1990). However, until 1994, the drug was blocked from use in the United States by pressure from antiabortion groups. They fear that the drug, because it can be easily administered in any doctor's office and reduces the use of abortion clinics, will make it more difficult for the antiabortion groups to protest abortions. President Clinton approved research on the drug in the United States, and by the end of 1994 the Population Council was con-

ducting clinical trials. The drug is now widely available.

Scientists who developed RU-486, as well as prochoice groups, prefer not to call it an abortion method, but rather a method for the induction of menstruation or a contragestational drug. It cannot properly be called a contraceptive because it prevents gestation, not conception. RU-486 and the drug discussed below, methotrexate, are referred to as medical methods, compared with the more traditional surgical methods.

Methotrexate
Another recent innovation in drug-induced early abortion has the advantage of avoiding the complex political issues surrounding RU-486. It involves the use of a combination of the drug methotrexate,

which is toxic to the embryo, with misoprostol, which causes uterine contractions that expel the dead embryo (Hausknecht, 1995). Both of these drugs are already widely used for other purposes, methotrexate for the treatment of cancer and misoprostol for ulcers. Therefore, any attempt to block their manufacture would be unlikely to succeed. Like RU-486, they permit the early induction of abortion in a physician's office rather than an abortion clinic, allowing women to avoid sidewalk picketers and potential violence by protestors at abortion clinics.

Psychological Aspects

The discovery of an unwanted pregnancy triggers a complicated set of emotions, as well as a complex decision-making process. Initially women tend to feel anger and some anxiety (Shusterman, 1979). They then embark on a decision-making process studied by psychologist Carol Gilligan (1982). In this process women essentially weigh the need to think of themselves and protect their own welfare against the need to think of the welfare of the fetus. Even focusing only on the welfare of the fetus can lead to conflicting conclusions: Should I complete the pregnancy because the fetus has a right to life, or should I have an abortion because the fetus has a right to be born into a stable family with married parents who have completed their education and can provide good financial support? Many women in Gilligan's study showed considerable psychological growth over the period in which they wrestled with these issues and made their decision.

The best scientific evidence indicates that most women do not experience severe negative psychological responses to abortion (Adler et al., 1992). When women are interviewed a year or so after their abortion, most show good adjustment. Typically they do not feel guilt or sorrow over the decision. Instead, they report feeling relieved, satisfied, and relatively happy, and say that if they had the decision to make over again, they would do the same thing. Nonetheless, some women benefit from talking about their experience, and it is important that postabortion support groups be available (Lodl et al., 1984).

Research in this area raises many interesting questions. Women generally show good adjustment after having an abortion, but good adjustment compared with what? That is, what is the appropriate control or comparison group? One comparison group that could be studied is women who requested an abortion but were denied it. It would be important to know the consequences for their adjustment.

One group that has been studied is children who were born because an abortion request was denied. It is impossible to do this research in the United States now because abortion has been legal since 1973. However, in some other countries access to abortion depends upon obtaining official approval. One such country is the former Czechoslovakia. Researchers followed up 220 children born to women denied abortion (the "study group") and 220 children born to women who had not requested abortion; the children were studied when they were 9 years old and again when they were 14 to 16 years old (David, 1992; David & Matejcek, 1981). By age 14, 43 children from the study group, but only 30 of the controls, had been referred for counseling. Although there were no differences between the groups in tested intelligence, children in the study group did less well in school and were more likely to drop out. Teachers described them as less sociable and more hyperactive compared with the control group. At age 16, the boys (but not the girls) in the study group more frequently rated themselves as feeling neglected or rejected by their mothers, and felt that their mothers were less satisfied with them. By their early twenties, the study group reported less job satisfaction, more conflicts with coworkers and supervisors, and fewer and less satisfying friendships. Several other studies have found results similar to the Czechoslovakian one (David et al., 1988). These results point to the serious long-term consequences for children whose mothers would have preferred to have an abortion.

Men and Abortion

Only women become pregnant, and only women have abortions, but where do men enter the picture? Do they have a right to contribute to the decision to have an abortion? What are their feelings about abortion?

Sociologist Arthur Shostak and his colleagues (1984) surveyed 1000 male "abortion veterans." The most common reaction from the men was a sense of helplessness. Although most men are used to being in control, in this situation they are not, and the feeling of powerlessness is difficult for them. Most of the men also felt isolated, angry at themselves and their partners, and fearful of emotional and physical damage to the woman. Most of them tried to hide their stress and remain unemotional. Nonetheless, 26 percent thought of abortion as murder, and 81 percent said they thought about the child who might have been born. However, few men wanted to be able to overrule the woman's decision; they only wanted to share in it.

Although counseling for women undergoing abortion is a standard procedure, counseling is rarely available for the men who are involved. Given Shostak's findings, it is clear that such counseling is badly needed. On a more political level, some men's activists argue that, just as women should not be forced to carry a pregnancy to term, so men should not be subjected to forced fatherhood and an 18-year financial commitment (Marsiglio & Diekow, 1998).

New Advances in Contraception

According to some, a really good method of contraception is not yet available. The highly effective methods either are permanent (sterilization) or have associated health risks (the pill). Other, safer methods (such as the condom and the diaphragm) have failure rates that cannot be ignored. Most of the methods are for women, not men. Because of the limitations of the currently available methods, contraception research continues. Unfortunately, its pace has been slow in the last few years because pharmaceutical companies are weary of lawsuits. There is little incentive for conducting highly innovative research and much incentive for companies to be cautious. Nonetheless, research and innovation continue, even if at a slow pace. Some of the more promising possibilities for the future are discussed below.

Male Methods
Several possibilities for new or improved male contraception are being explored (Gabelnick, 1998).
Non-latex Condoms.   In efforts to address concerns about lack of sensation, breakage, and slippage with traditional latex condoms, several alternatives are being developed. As noted earlier, a polyurethane condom was introduced in 1994. It is thinner than latex, yet strong, and it should provide more sensation. Other plastics are being tried as well.
Male Hormonal Methods.   Testing has begun on the drug testosterone enanthate. Research shows that it completely inhibited sperm production in 70 percent of men and severely inhibited

sperm production in 98 percent, yielding 99 percent effectiveness (Gabelnick, 1998). The disadvantages are that it must be taken for a fairly long time before it becomes effective and, in its present form, requires weekly injections. Implants that would be active for a year are being explored.
Immunocontraceptives.   Vaccines are being explored that use either FSH or LHRH (luteinizing-hormone releasing hormone). The LHRH version has the effect of shutting down the testes so that they produce neither sperm nor testosterone. Testosterone would then need to be taken in order to maintain sexual desire and the capacity for erection. The FSH version, in tests with monkeys, has eliminated sperm production while not affecting testosterone levels (Gabelnick, 1998).

Female Methods
Lea's Shield.   This is basically a one-size-fits-all diaphragm, modified with a one-way valve that allows air to escape while it is being inserted, achieving a tighter, suctioned fit (see Figure 8.16).[2] The valve also allows cervical secretions to escape, but does not allow sperm to pass through. This device is sold in Canada and a few other countries and, as of this writing, is under review by the FDA in the United States. The major advantage is that Lea's Shield would not require a doctor's appointment for fitting, as the diaphragm does.

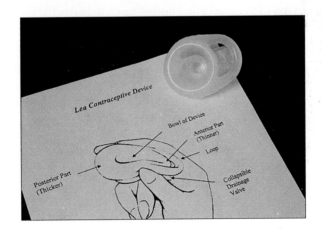

Figure 8.16   Lea's Contraceptive, a new, one-size-fits all diaphragm, is in development and will be sold over the counter without prescription. A similar new diaphragm, Femcap, will be available in 3 sizes.

[2]This reminds one of "burping" the lid on a tupperware container of lettuce, although you probably shouldn't think about this while using it because it isn't very erotic.

Focus 8.3

A Brief Summary of the Development of Sophisticated Methods of Contraception

Late 1700s	Casanova (1725–1798) popularizes and publicizes use of the sheath, or "English riding coat."
1798	Malthus urges "moral restraint" or abstinence.
1840s	Goodyear vulcanizes rubber. Production of rubber condoms soon follows.
1883	Mensinga invents the diaphragm.
1893	Harrison performs the first vasectomy.
1909	Richter uses the intrauterine silkworm gut.
1910–1920	Margaret Sanger pioneers in New York City; the term "birth control" is coined.
1930	Graffenberg publishes information documenting his 21 years of experience with the ring (silver and copper) and catgut as IUDs.
1930–1931	Knaus and Ogino elucidate "safe and unsafe" periods of the woman's menstrual cycle: the rhythm method.
1934	Corner and Beard isolate progesterone.
1937	Makepeace demonstrates that progesterone inhibits ovulation.
1950s	Abortions are utilized extensively in Japan.
1950–1960	Hormonal contraceptive research results in FDA approval of the use of the pill as a contraceptive in 1960.
1960s	Many Western nations liberalize abortion laws. Modern IUDs become available. Contraceptive sterilization becomes more acceptable. The laparoscopic tubal ligation technique is developed.
1973	The United States Supreme Court rules on abortion. The first "minipill," or low-dose progestin pill, wins FDA approval.
1970s	Depo-Provera contraceptive injections become available in over 50 nations (though not in United States until 1992).
1990	Norplant becomes available.
1994	Female condom available over-the-counter.

Source: Robert A. Hatcher et al. (1976). *Contraceptive technology, 1976–1977* (8th ed.). New York: Irvington.

Implants and Injectables.   A new version of Norplant, Norplant 2, has received FDA approval. It is similar to Norplant but uses only 2 rods, making insertion and removal easier and safer.

In regard to Depo-Provera, one of its problems (but also one of the virtues) is that it disrupts normal cycles of menstruation because it contains only progestin. An alternative injectable, Cyclo-Provera, contains both progestin and estrogen and is being tested in the United States.

Vaginal Rings.   This ring-shaped device containing progesterone is inserted into the vagina and delivers progesterone precisely to the organs that need it rather than throughout the body. Another vaginal ring, containing both progesterone and estrogen, is also being developed.

IUDs.   A kinder, gentler IUD is being developed that will not press hard on the uterus and therefore will not create the cramping that some women experience with current IUDs.

SUMMARY

Table 8.5 provides a comparative summary of the various methods of birth control discussed in this chapter.

Table 8.5 Summary of Information on Methods of Contraception and Abortion

Method	Effectiveness Rating	Failure Rate, Perfect Use,%	Failure Rate, Typical Use,%	Death Rate (per 100,000 Women)	Yearly Costs, $*	Advantages	Disadvantages
Norplant	Excellent	0.05	0.05	—	114†	Highly effective; does not require memory; not used at time of coitus	Cost; side effects yet to be documented
Depo-Provera	Excellent	0.3	0.3		232		
Combination birth control pills	Excellent	0.1	5	1.6	198	High effective; not used at time of coitus: improved menstrual cycles	Cost; possible side effects; must be taken daily
Progestin-only pills	Excellent	0.5	5	—	198		
IUD, Copper T	Excellent	0.6	0.8	1.0	75‡	Requires no memory or motivation	Side effects; may be expelled
Progesterone T	Excellent	1.5	2.0				
Condom, male	Very good	3	14	1.7	50	Easy to use; protection from STDs	Used at time of coitus; continual expense
Condom, female	Good	5	21	2.0	300		
Diaphragm with cream or jelly	Good	6	20	2.0§	100	No side effects; inexpensive	Aesthetic objections
Cervical cap with spermicide					100	No side effects; inexpensive	—
Parous women	Fair	26	40	2.0§			
Nulliparous women	Good	9	20	2.0§			
Vaginal foam, cream	Fair	6	26	2.0§	50	Easy to use; availability	Messy; continual expense
Withdrawal	Good	4	19	2.0	None	No cost	Requires high motivation
Rhythm	Poor to fair	2–9	25	2.0§	None	No cost, accepted by Roman Catholic Church	Requires high motivation, prolonged abstinence; not all women can use
Unprotected intercourse	Poor	85	85	9§	None¶		
Legal abortion, first trimester	Excellent	0	0	0.5	300	Available when other methods fail	Expense; moral or psychological unacceptability
Sterilization, male	Excellent	0.10	0.15	0.3	1800**	Permanent; highly effective	Permanence; expense
Sterilization, female	Excellent	0.5	0.5	1.5	1800**	Permanent; highly effective	Permanence; expense

*Based on 150 acts of intercourse. Prices are provided by Wisconsin Planned Parenthood, 1998, for full-paying clients. Prices are reduced for those with low incomes. Prices are higher for private physicians. Prices for sterilization provided by Meriter Hospital.

†Based on a cost of $570 for Norplant and insertion and use for five years.

‡Based on a cost of $120 for the IUD plus the fee for insertion by a physician, and the assumption that the IUD will be used for two years.

§Based on the death rate for pregnancies resulting from the method.

¶But having a baby is expensive.

**These are one-time-only costs.

Source: R. A. Hatcher et al. (1998). Contraceptive technology. 17th ed., New York. Ardent Media.

REVIEW QUESTIONS

1. Combination birth control pills contain the hormones _____ and _____.

2. The birth control pill works mainly by preventing ovulation. True or false?

3. Cancer is the major health risk associated with the use of the birth control pill. True or false?

4. Depo-Provera contains _____ and is effective for _____ (length of time).

5. The typical user failure rate of the IUD is about 25 percent. True or false?

6. In order to be highly effective, the diaphragm must be used with a spermicidal cream or jelly. True or false?

7. _____ is the drug that induces a very early abortion and must be used within 10 days of an expected but missed menstrual period.

8. The calendar method, the basal body temperature method, and the cervical mucus method are all variations on the _____ method of birth control.

9. Two highly effective methods of birth control that can be used by men are _____ and _____.

10. According to the research of Byrne, erotophobes are more likely than erotophiles to be consistent, reliable users of contraceptives. True or false?

(The answers to all review questions are at the end of the book, after the Glossary.)

QUESTIONS FOR THOUGHT, DISCUSSION, AND DEBATE

1. Do you think you are an erotophobe or an erotophile? In what ways do you think your erotophobia or erotophilia has affected or will affect your use of birth control?

2. Debate the following topic. Resolved: The birth control pill is a safe and effective method of birth control for most women.

3. In the United States, few IUDs are available because of lawsuits against companies that make them and concern over possible health risks. In contrast, in the People's Republic of China they are the mainstay of contraception, with 40 million in use. Which country has the better policy?

4. On your campus, as on all campuses, students probably are inconsistent in their use of birth control or use nothing even though they are sexually active. Design a program to improve birth control practices on your campus.

5. Knowing that you are taking a human sexuality course and have gained a lot of expertise, Latoya, your best friend, comes to you in a state of crisis. She and her boyfriend had unprotected intercourse last night and she is terrified that she is pregnant. What options would you explain to her, and which one would you recommend?

SUGGESTIONS FOR FURTHER READING

Gordon, Linda. (1990). *Woman's body, woman's right.* 2nd ed. New York: Penguin Books. This book provides a fascinating and enlightening social history of the development of birth control in America, from the outlawing of contraceptive methods in the early 1800s, to the pioneering efforts of Margaret Sanger, to the influences of the modern women's movement.

Hatcher, Robert A., et al. (1998). *Contraceptive technology.* 17th ed. New York: Ardent Media.

This authoritative book is updated frequently and provides the most recent information on all methods of contraception.

Ulmann, André, Teutsch, Georges, and Philibert, Daniel. (June 1990). RU-486. *Scientific American, 262,* 42–48. An interesting, behind-the-scenes article by three of the French scientists involved in the development of RU-486.

WEB RESOURCES

Contraception

http://plannedparenthood.org
Planned Parenthood: Birth Control pages

http://www.usc.edu/hsc/info/newman/resources/primer.html
A Primer on natural family planning

Emergency Contraception

http://www.opr.princeton.edu/ec/ec.html
Site maintained by Princeton University, Office of Population Research

Abortion

http://www.plannedparenthood.org/
Planned Parenthood; Abortion, Pro-Choice and Political Action pages

http://medicalabortion.com/index.htm
A site with information about medical abortion, maintained by a physician

http://www.naral.org
National Abortion and Reproductive Rights Action League; Information, statistics, and links about reproductive health.

http://www.nrlc.org/
National Right to Life Organization; Information and links about right to life issues

International Women's Health Issues

http://www.iwhc.org/index.html
International Women's Health Center

http://www.qweb.kvinnoforum.se/qwhealth.htm
Swedish Website—women's health issues and issues related to sex and reproduction

chapter 9

THE PHYSIOLOGY OF SEXUAL RESPONSE

CHAPTER HIGHLIGHTS

Here are some colors of different people's orgasms: champagne, all colors and white and gray afterward, red and blue, green, beige and blue, red, blue and gold. Some people never make it because they are trying for plaid.*

This chapter is about the way the body responds during sexual arousal and orgasm and the processes behind these responses. This information is very important in developing good techniques of lovemaking (see Chapter 10) and in analyzing and treating sexual disorders such as premature ejaculation (see Chapter 19).

Masters and Johnson: Four Stages of Sexual Response

Sex researchers William H. Masters and Virginia E. Johnson provided one of the first models of the physiology of human sexual response. Their research culminated in 1966 with the publication of *Human Sexual Response,* which reported data on 382 women and 312 men observed in over 10,000 sexual cycles of arousal and orgasm. (A discussion and critique of the Masters and Johnson research techniques were presented in Chapter 3.)

According to Masters and Johnson, there are four stages of sexual response, which they call *excitement, plateau, orgasm,* and *resolution.* The two basic physiological processes that occur during these stages are vasocongestion and myotonia. **Vasocongestion** occurs when a great deal of blood flows into the blood vessels in a region, in this case the genitals, as a result of dilation of the blood vessels in the region. **Myotonia** occurs when muscles contract, not only in the genitals but also throughout the body. Let us now consider in detail what occurs in each of the stages.

Excitement

The **excitement** phase is the beginning of erotic arousal. The basic physiological process that occurs during excitement is vasocongestion. This produces the obvious arousal response in the male— erection. Erection results when the corpora caver-

nosa and the corpus spongiosum fill (becoming engorged) with blood (see Figure 9.1). Erection may be produced by direct physical stimulation of the genitals, by stimulation of other parts of the body, or by erotic thoughts. It occurs very rapidly, within a few seconds of the stimulation, although it may take place more slowly as a result of a number of factors including age, intake of alcohol, and fatigue.

An important response of females in the excitement phase is lubrication of the vagina. Although this response might seem much different from the male's, actually they both result from the same physiological process: vasocongestion. Masters and Johnson found that vaginal lubrication results when fluids seep through the semipermeable membranes of the vaginal walls, producing lubrication as a result of vasocongestion in the tissues surrounding the vagina. This response to arousal is also rapid, though not quite so fast as the male's; lubrication begins 10 to 30 seconds after the onset of arousing stimuli.[1] Like male sexual response, female responding can be affected by factors such as age, intake of alcohol, and fatigue.

Several other physical changes occur in women during the excitement phase. The glans of the clitoris

---

*Eric Berne. (1970). *Sex in human loving.* New York: Simon & Schuster, p. 238.

[1]Before the Masters and Johnson research, it was thought that the lubrication was due to secretions of the Bartholin glands, but it is now known that these glands contribute little if anything. At this point, you might want to go back to the limerick about the Bartholin glands in Chapter 4 and see whether you can spot the error in it.

Vasocongestion (vay-so-con-JES-tyun): An accumulation of blood in the blood vessels of a region of the body, especially the genitals; a swelling or erection results.
Myotonia (my-oh-TONE-ee-ah): Muscle contraction.
Excitement: The first stage of sexual response, during which erection in the male and vaginal lubrication in the female occur.

Figure 9.1   Changes during the sexual response cycle in the male.

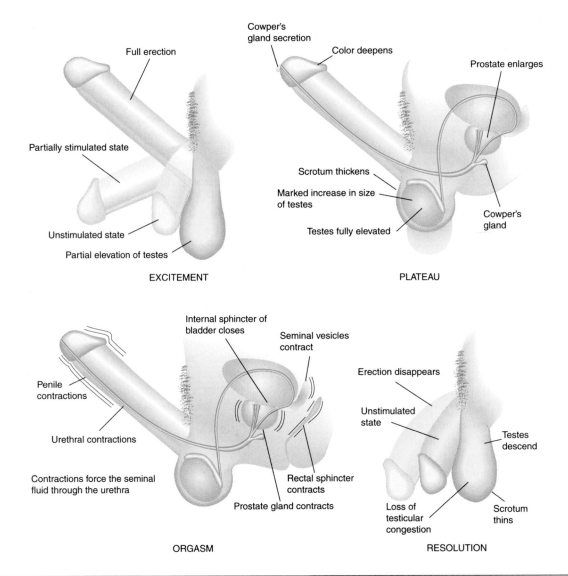

Full erection

Partially stimulated state

Unstimulated state

Partial elevation of testes

EXCITEMENT

Cowper's gland secretion

Color deepens

Prostate enlarges

Scrotum thickens

Marked increase in size of testes

Testes fully elevated

Cowper's gland

PLATEAU

Internal sphincter of bladder closes

Seminal vesicles contract

Penile contractions

Urethral contractions

Contractions force the seminal fluid through the urethra

Rectal sphincter contracts

Prostate gland contracts

ORGASM

Erection disappears

Unstimulated state

Testes descend

Loss of testicular congestion

Scrotum thins

RESOLUTION

(the tip) swells. This results from engorgement of its corpora cavernosa and is similar to erection in the male. The clitoris can be felt as larger and harder than usual.

The nipples become erect; this results from contractions of the muscle fibers (myotonia) surrounding the nipple. The breasts themselves swell and enlarge somewhat in the late part of the excitement phase (a vasocongestion response). Thus the nipples may not actually look erect but may appear somewhat flatter against the breast because the breast has swollen. Many males also have nipple erection during the excitement phase.

In the unaroused state the inner lips are generally folded over, covering the entrance to the vagina, and the outer lips lie close to each other. During excitement the inner lips swell and open up (a vasocongestion response).

Figure 9.2 Changes during the sexual response cycle in the female.

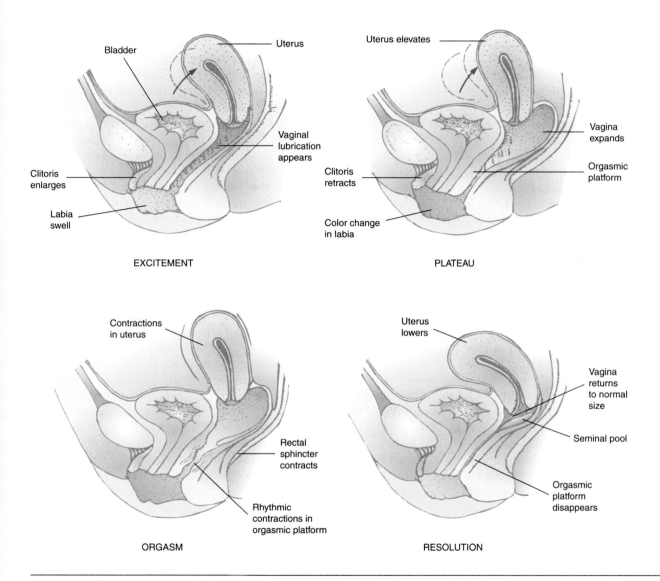

The vagina shows an important change during excitement. Think of the vagina as being divided into two parts, an upper (or inner) two-thirds and a lower (or outer) one-third. In the unaroused state the walls of the vagina lie against each other, much like the sides of an uninflated balloon. During the excitement phase, the upper two-thirds of the vagina expands dramatically in what is often called a "ballooning" response; that is, it becomes more like an inflated balloon (see Figure 9.2). This helps to accommodate the entrance of the penis. As part of the ballooning, the cervix and uterus pull up, creating a "tenting effect" in the vaginal walls (Figure 9.2) and making a larger opening in the cervix, which probably allows sperm to move into the uterus more easily.

During excitement a "sex flush" may appear on the skin of both men and women, though more commonly of women. The sex flush resembles a measles rash; it often begins on the upper abdomen and spreads over the chest. It may also appear later in the sexual response cycle.

Other changes that occur in both men and women include an increase in pulse rate and in blood pressure.

In men, the skin of the scrotum thickens. The scrotal sac tenses, and the scrotum is pulled up and closer to the body (Figure 9.1). The spermatic cords shorten, pulling the testes closer to the body.

Plateau

During the **plateau** phase, vasocongestion reaches its peak. In men, the penis is completely erect, although there may be fluctuations in the firmness of the erection. The glans swells. The testes are pulled up even higher and closer to the body. A few drops of fluid (for some men, quite a few), secreted by the Cowper's gland, appear at the tip of the penis. Although they are not the ejaculate, they may contain active sperm.

In both women and men there is a further increase in the rate of breathing, in pulse rate, and in blood pressure.

In females, the most notable change during the plateau phase is formation of the **orgasmic platform.** This is a tightening of the outer third of the vagina (Figure 9.2). Thus the size of the vaginal entrance actually becomes smaller, and there may be a noticeable increase in gripping of the penis.

Another change is the elevation of the clitoris. The clitoris essentially retracts or draws up into the body.

In short, the processes of the plateau phase are a continuation of the basic processes—vasocongestion and myotonia—of the excitement phase. Both processes continue to build until there is sufficient tension for orgasm.

Orgasm

In the male, **orgasm** consists of a series of rhythmic contractions of the pelvic organs at 0.8-second intervals. Actually, male orgasm occurs in two stages. In the preliminary stage, the vas, seminal vesicles, and prostate contract, forcing the ejaculate into a bulb at the base of the urethra (Figure 9.1). Masters and Johnson call the sensation in this stage one of

"ejaculatory inevitability" ("coming"); that is, there is a sensation that ejaculation is just about to happen and cannot be stopped. And, indeed, it cannot be, once the man has reached this point. In the second stage the urethral bulb and the penis itself contract rhythmically, forcing the semen through the urethra and out the opening at the tip of the penis.

In both males and females, there are sharp increases in pulse rate, blood pressure, and breathing rate during orgasm.[2] Muscles contract throughout the body. The face may be contorted in a grimace; the muscles of the arms, legs, thighs, back, and buttocks may contract; and the muscles of the feet and hands may contract in "carpopedal spasms." Generally, in the passion of the moment, one is not really aware of these occurrences, but an aching back or buttocks may serve as a reminder the next day.

The process of orgasm in females is basically similar to that in males. It is a series of rhythmic muscular contractions of the orgasmic platform. The contractions generally occur at about 0.8-second intervals; there may be three or four in a mild orgasm or as many as a dozen in a very intense, prolonged orgasm. The uterus also contracts rhythmically. Other muscles, such as those around the anus, may also contract.

Female orgasm is a funny thing. As with love, you can almost never get anyone to give you a solid definition of what it is. Instead, people usually fall back on, "You'll know what it is when you have one." This evasiveness is probably related to several factors, most notably that female orgasm leaves no tangible evidence of its occurrence like ejaculation. Also, women often do not reach orgasm as quickly as

---

[2]With all the current attention to aerobics and exercising the heart, we have yet to hear anyone suggest orgasm aerobics. It seems to us that it should work. Jazzercise, watch out. Here comes sexercise!

Plateau: The second stage of sexual response, just before orgasm.
Orgasmic platform: A tightening of the entrance to the vagina caused by contractions of the bulbospongiosus muscle (which covers the vestibular bulbs) that occur during the plateau stage of sexual response.
Orgasm: The third stage of sexual response; an intense sensation that occurs at the peak of sexual arousal and is followed by release of sexual tensions.

men do, a point to be discussed in more detail in Chapter 14. In fact, some women, particularly young women, may think they are having an orgasm when they are not; they have never had an orgasm, and they mistake intense arousal for orgasm.

Just what does orgasm in the female feel like? The main feeling is a spreading sensation that begins around the clitoris and then spreads outward through the whole pelvis. There may also be sensations of falling or opening up. The woman may be able to feel the contraction of the muscles around the vaginal entrance. The sensation is more intense than just a warm glow or a pleasant tingling. In one study, college men and women gave written descriptions of what an orgasm felt like to them (Vance & Wagner, 1976). Interestingly, a panel of experts (medical students, obstetrician-gynecologists, and clinical psychologists) could not reliably figure out which of the descriptions were written by women and which by men. This suggests that the sensations are quite similar for males and females.

Some of the men in our classes have asked how they can tell whether a woman has really had an orgasm. Their question in itself is interesting. In part it reflects a cultural skepticism about female orgasm. There is usually obvious proof of male or-

gasm: ejaculation. But there is no similar proof of female orgasm. The question also reflects the fact that men know that women sometimes "fake" orgasm. Faking orgasm is a complex issue. Basically, it probably is not a very good idea, because it is dishonest. It also leads the woman's partner to think that his technique of stimulation is more effective than it is. On the other hand, one needs to be sympathetic to the variety of reasons women do it. It is often difficult for women to reach orgasm, and our culture currently places a lot of emphasis on everyone's having orgasms. The woman may feel that she is expected to have an orgasm, and realizing that it is unlikely to happen this time, she "fakes" it in order to meet expectations. She may also do it to please her partner.[3] But back to the question: How can one tell? There really is not any very good way. From a scientific point of view, a good method

[3]Indeed, many of the old sex manuals, as well as physicians' textbooks, counseled women to fake orgasm. For example: "It is good advice to recommend to the women the advantage of innocent simulation of sex responsiveness, and as a matter of fact many women in their desire to please their husbands learned the advantage of such innocent deception" (Novak & Novak, 1952, p. 572).

Figure 9.3   Tom Cheney/Reprinted Courtesy of *Penthouse Magazine.*

"Did you come?"

would be to have the woman hooked up to an instrument that registers pulse rate; there is a sudden sharp increase in the pulse rate at orgasm, and that would be a good indicator. We doubt, though, that most men have such equipment available, and we are even more doubtful about whether most women would agree to be so wired up. Probably rather than trying to check up on each other, it would be better for partners to establish good, honest communication and avoid setting up performance goals in sex, points that will be discussed further in later chapters.

Resolution

Following orgasm is the **resolution** phase, during which the body returns physiologically to the unaroused state. Orgasm triggers a massive release of muscular tension and of blood from the engorged blood vessels. Resolution, then, represents a reversal of the processes that build up during the excitement and plateau stages.

The first change in women is a reduction in the swelling of the breasts. As a result, the nipples may appear to become erect, since they seem to stand out more as the surrounding flesh moves back toward the unstimulated size. In women who develop a sex flush during arousal, this disappears rapidly following orgasm. In the 5 to 10 seconds after the end of the orgasm, the clitoris returns to its normal position, although it takes longer for it to shrink to its normal size. The orgasmic platform relaxes and begins to shrink. The ballooning of the vagina diminishes, and the uterus shrinks.

The resolution phase generally takes 15 to 30 minutes, but it may take much longer—as much as an hour—in women who have not had an orgasm. This latter fact helps to account for the chronic pelvic congestion that Masters and Johnson observed in prostitutes (see Chapter 3). The prostitutes frequently experienced arousal without having orgasms. Thus there were repeated buildups of vasocongestion without the discharge of it brought about by orgasm. The result was a chronic vasocongestion in the pelvis. A mild version of this occurs in some women who engage in sex but are not able to have orgasms, and it can be quite uncomfortable.

In both males and females, resolution brings a gradual return of pulse rate, blood pressure, and breathing rate to the unaroused levels.

In men, the most obvious occurrence in the resolution phase is detumescence, the loss of erection in the penis. This happens in two stages, the first occurring rapidly but leaving the penis still enlarged (this first loss of erection results from an emptying of the corpora cavernosa) and the second occurring more slowly, as a result of the slower emptying of the corpus spongiosum and the glans.

During the resolution phase, men enter a **refractory period,** during which they are refractory to further arousal; that is, they are incapable of being aroused again, having an erection, or having an orgasm. The length of this refractory period varies considerably from one man to the next; in some it may last only a few minutes, and in others it may go on for 24 hours. The refractory period tends to become longer as men grow older.

Women do not enter into a refractory period, making possible the phenomenon of multiple orgasm in women, to be discussed in the next section.

Other Findings of the Masters and Johnson Research

A number of other important findings on the nature of sexual response emerged from the Masters and Johnson research, two of which will be discussed here.

Clitoral Orgasm Versus Vaginal Orgasm

Some people believe that women can have two kinds of orgasm: **clitoral orgasm** and **vaginal orgasm.** The words "clitoral" and "vaginal" are not meant to imply that the clitoris has an orgasm or that the vagina has an orgasm. Rather, they refer to the locus of stimulation: an orgasm resulting from clitoral stimulation versus an orgasm resulting

Resolution: The fourth stage of sexual response, in which the body returns to the unaroused state.
Refractory period (ree-FRAK-toh-ree): The period following orgasm during which the male cannot be sexually aroused.
Clitoral orgasm: Freud's term for orgasm in the female resulting from stimulation of the clitoris.
Vaginal orgasm: Freud's term for orgasm in the female resulting from stimulation of the vagina in heterosexual intercourse; Freud considered vaginal orgasm to be more mature than clitoral orgasm.

from vaginal stimulation. The distinction was originated by Sigmund Freud. Freud believed that in childhood little girls masturbate and thus have orgasms by means of clitoral stimulation, or clitoral orgasms. He thought that as women grow older and mature, they ought to shift from having orgasms as a result of masturbation to having them as a result of heterosexual intercourse, that is, by means of vaginal stimulation. (Freud's self-interest as a male in this matter is rather transparent!) Thus the vaginal orgasm was considered "mature" and the clitoral orgasm "immature" or "infantile," and not only did there come to be two kinds of orgasm, but also one was "better" (that is, more mature) than the other.

Freud's formulation is of more than theoretical interest, since it has had an impact on the lives of many women. Many have undertaken psychoanalysis and spent countless hours agonizing over why they were not able to achieve the elusive vaginal orgasm and why they enjoyed the "immature" clitoral one so much. Women who could have orgasms only through clitoral stimulation were called "vaginally frigid" or "fixated" at an infantile stage.

According to the results of Masters and Johnson's research, though, the distinction between clitoral and vaginal orgasms does not make sense. This conclusion is based on two findings. First, their results indicate that all female orgasms are physiologically the same, regardless of the locus of stimulation. That is, an orgasm always consists of contractions of the orgasmic platform and the muscles around the vagina, whether the stimulation is clitoral or vaginal. Indeed, they found a few women who could orgasm purely through breast stimulation, and that orgasm was the same as the other two, consisting of contractions of the orgasmic platform and the muscles around the vagina. Thus physiologically there is only one kind of orgasm. (Of course, this does not mean that psychologically there are not different kinds; the experience of orgasm during intercourse may be quite different from the experience of orgasm during masturbation.) Second, Masters and Johnson found that clitoral stimulation is almost always involved in producing orgasm. Because of the way in which the inner lips connect with the clitoral hood, the movement of the penis in and out of the vagina creates traction on the inner lips, which in turn

pull the clitoral hood so that it moves back and forth, stimulating the clitoris. Thus even the purely "vaginal" orgasm results from quite a bit of clitoral stimulation. It seems that clitoral stimulation is usually the trigger to orgasm and that the orgasm itself occurs in the vagina and surrounding tissues.

Multiple Orgasm

Traditionally it was believed that orgasmically, women behaved like men in that they could have one orgasm and then would enter a refractory period before they could have another. According to the Masters and Johnson research, however, this is not true; rather, women do not enter into a refractory period, and they can have **multiple orgasms** within a short period of time. Actually, women's capacity for multiple orgasms was originally discovered by Kinsey in his interviews with women (Kinsey et al., 1953; see also Terman et al., 1938). The scientific establishment, however, dismissed these reports as another instance of Kinsey's unreliability.

The term "multiple orgasm," then, refers to a series of orgasms occurring within a short period of time. They do not differ physiologically from single orgasms. Each is a "real" orgasm, and they are not minor experiences. One nice thing, though, is that the later ones generally require much less effort than the first one.

How does multiple orgasm work physiologically? Immediately following an orgasm, both males and females move into the resolution phase. In this phase, the male enters into a refractory period, during which he cannot be aroused again. But the female does not enter into a refractory period. That is, if she is stimulated again, she can immediately be aroused and move back into the excitement or plateau phase and have another orgasm.

Multiple orgasm is more likely to result from hand-genital or mouth-genital stimulation than from intercourse, since most men do not have the endurance to continue thrusting for such long periods of time. Regarding capacity, Masters and Johnson found that women in masturbation might have 5 to 20 orgasms. In some cases, they quit only when physically exhausted. When a vibrator is used, less

Multiple orgasm: A series of orgasms occurring within a short period of time.

Focus 9.1

William Masters and Virginia Johnson

William Howell Masters was born in 1915. He attended Hamilton College in Clinton, New York, graduating in 1938 with a B.S. degree. At Hamilton he specialized in science courses, and yet he managed to play on the varsity football, baseball, basketball, and track teams and to participate in the Debate Club. The college yearbook called him "a strange, dark man with a future. . . . Has an easy time carrying three lab courses but a hard time catching up on lost sleep. . . . Bill is a boy with purpose and is bound to get what he is working for." His devotion to athletics persisted, and in 1966 a science writer described him as "a dapper, athletically trim gynecologist who starts his day at 5:30 with a two-mile jog."

He entered the University of Rochester School of Medicine in 1939, planning to train himself to be a researcher rather than a practicing physician. In his first year there he worked in the laboratory of the famous anatomist Dr. George Washington Corner. Corner was engaged in research on the reproductive system in animals and humans, which eventually led to important discoveries about hormones and the reproductive cycle. He had also published *Attaining Manhood: A Doctor Talks to Boys About Sex* and the companion volume, *Attaining Womanhood.*

The first-year research project that Corner assigned to Masters was a study of the changes in the lining of the uterus of the rabbit during the reproductive cycle. Thus his interest was focused early on the reproductive system.

Masters was married in 1942 and received his M.D. in 1943. He and his wife had two children.

After Masters received his degree, he had to make an important decision: To what research area should he devote his life? Apparently his decision to investigate the physiology of sex was based on his shrewd observation that almost no prior research had been done in the area and that he thus would have a good opportunity to make some important scientific discoveries. In arriving at the decision, he consulted with Dr. Corner, who was aware of Kinsey's progress and also of the persecution he had suffered (see Chapter 3). Thus Corner advised Masters not to begin the study of sex until

effort is required, and some women were capable of having 50 orgasms in a row.

It should be noted that some women who are capable of multiple orgasms are completely satisfied with one, particularly in intercourse, and do not wish to continue. We should be careful not to set multiple orgasm as another of the many goals in sexual performance.

In one sample of adult women, all of whom were nurses (and therefore presumably had a good understanding of the anatomy and physiology involved), 43 percent reported that they usually experienced multiple orgasms (Darling et al., 1991). Among these women, 40 percent reported that each successive orgasm was stronger than the previous one, 16 percent said that each successive one

was weaker than the one before, and the remainder said that the orgasms varied or there were no differences in intensity among them. Women who had multiple orgasms were more likely, compared with women who had single orgasms, to have their orgasm before their partner did.

There is evidence that some men are capable of having multiple orgasms (e.g., Zilbergeld, 1992; Hartman & Fithian, 1984). In one study, 21 men were interviewed, all of whom had volunteered for research on multiply orgasmic men (Dunn & Trost, 1989). Some of the men reported having been multiply orgasmic since their sexual debut, whereas others had developed the pattern later in life, and still others had worked actively to develop the capacity after reading about the possibility. The re-he had established himself as a respected researcher in some other area, was somewhat older, and could conduct the research at a major university or medical school.

Masters followed the advice. He completed his internship and residency and then established himself on the faculty of the Washington University School of Medicine in St. Louis. From 1948 to 1954 he published 25 papers on various medical topics, especially on hormone-replacement therapy for postmenopausal women.

In 1954 he began his research on sexual response at Washington University, supported by grants from the U.S. Public Health Service. The first paper based on that research was published in 1959, but the research received little attention until the publication, in 1966, of *Human Sexual Response* and, in 1970, of *Human Sexual Inadequacy* (to be discussed in Chapter 19), both of which received international acclaim.

Virginia Johnson was born Virginia Eshelman in 1925 in the Missouri Ozarks. She was raised with the realistic attitude toward sex that rural children often have, as well as many of the superstitions that have grown up in that area. She began studying music at Drury College but transferred to Missouri University, where she studied psychology and sociology. She was married in 1950 and had two children, one in 1952 and the other in 1955.

Figure 9.4   Virginia Johnson and William H. Masters.

Shortly after that, she and her husband separated, and she went to the Washington University placement office to find a job. Just at that time, Masters had put in a request for a woman to assist him in research interviewing, preferably a married woman with children who was interested in people. Johnson was referred to him, and she became a member of the research and therapy team in 1957.

Following the divorce of Masters and his first wife, he and Virginia Johnson were married. They were divorced in 1992. In 1994, at the age of 79, Masters retired and closed his research institute.

Source: Ruth Brecher & Edward Brecher (Eds.). (1966). *An analysis of* Human Sexual Response. New York: Signet Books, New American Library.

---

spondents reported that multiple orgasm did not occur every time they engaged in sexual activity. For these men, detumescence did not always follow an orgasm, allowing for continued stimulation and an additional orgasm. Some reported that some of the orgasms included ejaculation and others in the sequence did not. This study cannot tell us the incidence of multiply orgasmic men in the general population, but it does provide some evidence that multiply orgasmic men exist.

Cognitive-Physiological Models

Some experts on human sexuality are critical of Masters and Johnson's four-stage model. One im-portant criticism is that the Masters and Johnson model ignores the cognitive and subjective aspects of sexual response (Zilbergeld & Ellison, 1980). That is, Masters and Johnson focused almost entirely on the physiological aspects of the response, ignoring what the person is thinking and feeling emotionally. Desire and passion are not part of the model. This omission would not be a problem except that there can be major discrepancies between physiological response and subjective feelings. Men can have erections without feeling the least bit aroused sexually. People can also feel a high level of sexual desire, yet have no erection or vaginal lubrication.

A second important criticism concerns how research participants were selected and how this process may have created a self-fulfilling prophecy for the outcome (Tiefer, 1991a). To participate in the research, participants were required to have a history of orgasm both through masturbation and through coitus. Essentially, anyone whose pattern of sexual response did not include orgasm—and therefore did not fit Masters and Johnson's model—was excluded from the research. As such, the model cannot be generalized to the entire population. Masters and Johnson themselves commented that every one of their participants was characterized by high and consistent levels of sexual desire. Yet sexual desire is certainly missing among some members of the general population, or it is present sometimes and absent other times. It is not surprising that sexual desire was omitted from Masters and Johnson's model when the participants were preselected to be uniform in their high levels of desire. The research, in short, claims to be objective and universal when it is neither (Tiefer, 1991a).

Once these difficulties with the Masters and Johnson research and model of sexual response were recognized, several alternative models were proposed. We will examine two of them in the following sections. Both of them add a cognitive component to Masters and Johnson's physiological model.

Kaplan's Triphasic Model

On the basis of her work on sex therapy (discussed in Chapter 19), Helen Singer Kaplan (1974; 1979) proposed a **triphasic model** of sexual response. Rather than thinking of the sexual response as having successive stages, she conceptualized it as having three relatively independent phases, or components: *sexual desire, vasocongestion* of the genitals, and the reflex *muscular contractions* of the orgasm phase. Notice that two of the components (vasocongestion and muscular contractions) are physiological, whereas the other (sexual desire) is psychological.

There are a number of justifications for Kaplan's approach. First, the two physiological components are controlled by different parts of the nervous system. Vasocongestion—producing erection in the male and lubrication in the female—is controlled by the parasympathetic division of the autonomic nervous system. In contrast, ejaculation (and presumably orgasm in the female) is controlled by the sympathetic division.

Second, the two components involve different anatomical structures, blood vessels for vasocongestion and muscles for the contractions of orgasm.

Third, vasocongestion and orgasm differ in their susceptibility to being disturbed by injury, drugs, or age. For example, the refractory period following orgasm in the male lengthens with age. Accordingly, there is a decrease in the frequency of orgasm with age. In contrast, the capacity for erection is relatively unimpaired with age, although the erection may be slower to make its appearance. An elderly man may have nonorgasmic sex several times a week, with a firm erection, although he may have an orgasm only once a week.

Fourth, the reflex of ejaculation in the male can be brought under voluntary control by most men, but the erection reflex generally cannot.

Finally, the impairment of the vasocongestion response and impairment of the orgasm response produce different disturbances (sexual disorders). Erection problems in the male are caused by an impairment of the vasocongestion response, whereas premature ejaculation and retarded ejaculation are disturbances of the orgasm response. Similarly, many women show a strong arousal and vasocongestion response, yet have trouble with the orgasm component of their sexual response.

Kaplan's triphasic model is useful both for understanding the nature of sexual response and for understanding and treating disturbances in it. Her writing on the desire phase is particularly useful in understanding disorders of sexual desire, to be discussed in Chapter 19.

Walen and Roth: A Cognitive Model

As noted earlier, an important criticism of the Masters and Johnson model is that it ignores the cognitive and subjective aspects of sexual response. In Chapter 2 we discussed the importance of cognitive approaches in understanding the psychology of human sexuality. Susan R. Walen and David Roth (1987) have applied this approach to understanding the sexual response. Their model is shown in Figure 9.5.

Triphasic model: Kaplan's model of sexual response in which there are three phases: vasocongestion, muscular contractions, and sexual desire.

Figure 9.5 Walen and Roth's cognitive model showing the feedback loop that produces a positive sexual experience.

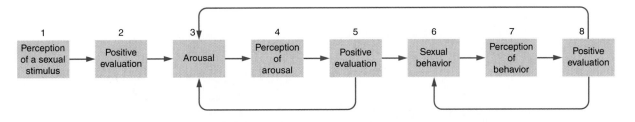

1	2	3	4	5	6	7	8
Perception of a sexual stimulus	Positive evaluation	Arousal	Perception of arousal	Positive evaluation	Sexual behavior	Perception of behavior	Positive evaluation

Recall from Chapter 2 that according to the cognitive approach, how we feel depends tremendously on what we are thinking: how we perceive what is occurring and how we evaluate it. Thus the first step in the cognitive model is *perception:* the perception of a stimulus as sexual. What we perceive to be sexy stimuli (whether they are visual stimuli, touch stimuli, or odors) depends a great deal on the culture in which we've grown up and on our prior learning. If you've just begun a sexual relationship with someone, the very sight of that person may make you feel turned on, whereas looking at 10 other people in the same room produces no turn-on. To a fetishist, the sight of black leather, high-heeled women's boots may produce instant arousal. According to the model, perception is the first step.

The second step in the cognitive model is *evaluation*. If we feel positive about the sexual stimulus, that will lead to the next step, arousal, but if our evaluation of the stimulus is negative, the arousal cycle stops. For example, if you are a married woman and your husband, with whom you normally have a great sex life, begins to kiss you when his breath smells of a cigarette he has just smoked, your evaluation of the sexual stimulus is likely to be negative and you will not feel aroused.

Let's suppose, though, that the evaluation of the sexual stimulus is positive. Physiological *arousal*—as described in the Masters and Johnson model discussed earlier in this chapter—is the next step. But again, the cognitive approach says that it is not so much what happens physically, but how we perceive it, that counts. So the *perception of arousal*, step 4, is critical. For example, some research shows that women, probably because vaginal lu-

brication can be a rather subtle response, are sometimes not aware of their own physical arousal (Heiman, 1975, discussed in Chapter 14). Or a man might pay so much attention to whether his technique is pleasing his partner that he fails to perceive his own arousal and thus does not experience as much sexual pleasure as he might. People can create sexual problems for themselves if they set too high criteria for deciding that they are aroused. How much lubrication, or how firm an erection, is "enough"? You can augment your sexual response by perceiving even a little bit of lubrication, or the first stirrings of an erection, or some other sign (such as increased heart rate) as an indication of arousal.

Again, the cognitive approach argues that it is not only the perception of arousal but also a positive evaluation of the arousal that is important if the sexual response cycle is to continue; thus *evaluation* is the next step. As before, if the evaluation is negative, the response cycle stops. An example would be an adult who feels himself becoming aroused when looking at a child. He realizes that this response is totally inappropriate, evaluates the arousal negatively, and feels anxiety rather than arousal. If the evaluation of the arousal is positive, though—you like the feeling of being turned on in this situation—there is feedback to step 3, so that physical arousal increases further. Essentially, feeling good about being turned on makes you feel even more turned on.

All this propels you to the next step, *sexual behavior*. Again, cognitive psychologists believe that two further steps—*perception of the behavior* and a *positive evaluation* of it—are critical for the arousal cycle to continue. If the evaluation is positive,

Figure 9.6   Can the cognitive model of sexual arousal explain our response to this? In St. Tropez in the south of France, people transact much of their daily business, including banking, in the nude. Most Americans would find a roomful of nude people to be arousing, yet it isn't in St. Tropez. Why not? The context (the bank) leads us not to perceive these people as sexual stimuli.

two kinds of feedback occur: the sexual behavior is likely to continue, and arousal increases.

In sum, the cognitive model of the sexual response cycle stresses the importance of our perception and evaluation of sexual events. It's a testimony to the power of positive thinking, although in a context that Norman Vincent Peale probably didn't intend. As another saying has it, the greatest erogenous zone is the brain.

Hormonal and Neural Control of Sexual Behavior

Up to this point we have focused on the cognitive and genital responses that occur during sexual activity. We have not yet considered the underlying neural and hormonal mechanisms that make this possible; they are the topic of this section.

The Brain, the Spinal Cord, and Sex

The brain and the spinal cord both have important interacting functions in sexual response. First, the relatively simple spinal reflexes involved in sexual response will be discussed; then the more complex brain mechanisms will be considered.

Spinal Reflexes

Several important components of sexual behavior, including erection and ejaculation, are controlled by fairly simple spinal cord reflexes (see the lower part of Figure 9.7). A reflex has three basic components: the *receptors*, which are sense organs that detect stimuli and transmit the message to the spinal cord (or brain); the *transmitters*, which are centers in the spinal cord (or brain) that receive the message, interpret it, and send out a message to produce the appropriate response; and the *effectors*, which are organs that respond to the stimula-

Figure 9.7   Nervous system control of erection. Note both the reflex center in the spinal cord and brain control.

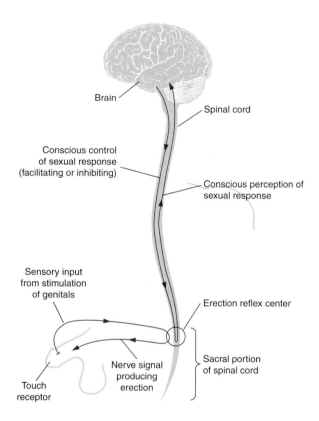

Brain

Spinal cord

Conscious control of sexual response (facilitating or inhibiting)

Conscious perception of sexual response

Sensory input from stimulation of genitals

Erection reflex center

Nerve signal producing erection

Sacral portion of spinal cord

Touch receptor

tion. The jerking away of the hand when it touches a hot object is a good example of a spinal reflex.

Mechanism of Erection

Erection is produced by a spinal reflex with a similar mechanism. Tactile stimulation (stroking or rubbing) of the penis (the receptor) or nearby regions such as the scrotum or inside of the thighs produces a neural signal which is transmitted to an "erection center" in the sacral, or lowest, part of the spinal cord (there may also be another erection center higher in the cord). This center then sends out a message via the parasympathetic division of the autonomic nervous system to the muscles (the effectors) around the walls of the arteries in the penis. In response to the message, the muscles relax; the arteries then expand, permitting a large volume of blood to flow into them, and erection re-

sults. Furthermore, the valves in the veins and the compression of the veins caused by the swelling in the tissue around them reduce the blood flow out of the penis.

The existence of this reflex is confirmed by the responses of men who have had their spinal cords severed, as a result of accidents, at a level above that of the reflex center (see Focus 9.2). They are capable of having erections and ejaculations produced by rubbing their genitals, although it is clear that no brain effects can be operating, since signals from the brain cannot move past the point at which the spinal cord was severed. (In fact, these men cannot "feel" anything because neural signals cannot be transmitted up the spinal cord either.) Thus erection can be produced simply by tactile stimulation of the genitals, which triggers the spinal reflex.

Focus 9.2
Sexuality and Disability

It is commonly believed that a person in a wheelchair is sexless. The physically disabled are thought not to be interested in sex, much less to be capable of engaging in sexual activity. Contrast those stereotypes with the following ideas:

> A stiff penis does not make a solid relationship, nor does a wet vagina. . . .

> Absence of sensation does not mean absence of feelings. . . .

> Inability to move does not mean inability to please.

(Cole & Cole, 1978, p. 118)

About 10 percent of adults in the United States have a physical handicap that imposes a substantial limitation on their activities. Given a chance to express themselves, these people emphasize the importance of their sexuality and sex drive, which are not necessarily altered by their disability.

Space does not permit a complete discussion of all types of disabilities and their consequences for sexuality. Instead, this discussion will concentrate on two illustrative examples: spinal-cord injury and mental retardation.

Spinal Cord Injury

Paraplegia (paralysis of the lower half of the body on both sides) and quadriplegia (paralysis of the body from the neck down) are both caused by injuries to the spinal cord. Many able-bodied people find it difficult to understand what it feels like to be paralyzed. Imagine that your genitals and the region around them have lost all sensation. You would not know they were being touched unless you saw it happen. Further, there is loss of bladder and bowel control, which may produce embarrassing problems if sexual activity is attempted.

The capacity of a spinal-cord-injured man to have erections depends on the level of the spinal cord at which the injury (lesion) occurred, and whether the spinal cord was completely or only partially severed. According to most studies, a majority of spinal-cord-injured men are able to have erections. In some cases only reflex erections are possible, that is, erections produced by direct stimulation of the genitals, even though the man cannot feel the sensation. In a few cases, particularly if the injury was not severe, the man is able to produce an erection by erotic thoughts, but this capacity is typically lost with spinal-cord injury. When the injury is severe, the man is not able to ejaculate, although ejaculation may be possible if the cord was only partially severed. Generally men's fertility is impaired by spinal-cord injury. In sum, many spinal-cord-injured men experience the same sexual responses as able-bodied men—including erection, elevation of the testes, and increases in heart rate—except that they generally cannot ejaculate, nor can they feel the physical stimulation.

The data on women with spinal-cord injuries, interestingly, generally do not focus on their capacity for vaginal lubrication and orgasm, but rather on their menstrual cycles and capacity for pregnancy. Women typically have amenorrhea (absence of menstruation) immediately following a spinal-cord injury, although normal menstrual cycling usually returns within six months. After that time, their ability to conceive a baby is normal. Most pregnancies proceed normally, although there is a higher risk of some complications such as cystitis (bladder infection), anemia, premature labor, and autonomic hyperreflexia (an exaggerated response of the autonomic nervous system, in which the woman experiences a great deal of sweating, abnormally rapid heartbeat, and anxiety). Vaginal deliveries are usually possible, and they can be done without anesthetic because the woman feels no sensation in the pelvic region. The fertility of spinal-cord-injured women means that

(a) (b)

Figure 9.8 In the last twenty years we have become aware of the capacity of people with disabilities for sexual expression. *(a)* There is a need for sex education for children with disabilities. *(b)* Ellen Stohl, the first woman with disabilities to appear in *Playboy* (October, 1987). She is a quadriplegic, having suffered several broken bones in her neck in an automobile accident. She was completely paralyzed initially, but gradually recovered some use of her arms. She has some sensitivity in her legs and is hypersensitive in the genitals. She enjoys sex very much and is orgasmic.

contraception is an important consideration for them.

Spinal-cord-injured women experience many of the same sexual responses as able-bodied women, including engorgement of the clitoris and labia, erection of the nipples, and increases in heart rate. Approximately 50 percent of women with spinal cord injuries are able to have orgasms from stimulation of the genitals. It is interesting to note that some spinal-cord-injured women develop a capacity for orgasm from stimulation of the breasts or lips.

Because sexuality in our culture is so orgasm-oriented, orgasm problems among spinal-cord-injured people may appear to be devastating. But many of them report that they have been able to cultivate a kind of "psychological orgasm" that is as satisfying as the physical one. Fantasy is a perfectly legitimate form of sexual expression that has not been ruled out by their injury.

Mental Retardation

Persons with IQs below 70 are generally classified as mentally retarded. There is a great range in the capacities of retarded individuals, from some who require institutionalization and constant care, to those who function quite normally in the community, who can read and write and hold simple jobs. It is important to recognize that the great majority of retarded persons are only moderately retarded (IQs between 50 and 70) and are in the category of near-normal functioning.

Four issues are especially important when considering the sexuality of mentally retarded persons: their opportunity for sexual expression, the need for sex education, the importance of contraception, and the possibility of sexual abuse.

Retarded persons have normal sexual desires and seek to express them. Unfortunately, because retarded children are often slower to learn the norms of society, they may express themselves sexually in ways that may shock others, such as masturbating in public. For this reason and others, careful sex education for retarded persons is essential. They must be taught about the norms of privacy for sexuality. At the same time, they must be allowed their privacy, a right that institutions often fail to recognize.

It is important that retarded individuals be educated about contraception and that contraceptives be made available to them. Because the retarded have normal sexual desires, they may engage in sexual intercourse but, if they lack sex education, they may not realize that pregnancy can result. An unwanted pregnancy for a retarded woman or couple may be a difficult situation; they may be able to function well when taking care of themselves but not with the added burden of a baby. On the other hand, some retarded persons do function sufficiently well to care for a child. The important thing is that they make as educated a decision as possible and that they have access to contraceptives. Many experts recommend the IUD for retarded women, because it does not require memory and forethought for effective use.

The topic of contraception and the retarded raises the ugly issue of involuntary sterilization. Until the mid-1950s, retarded persons in institutions were routinely sterilized, although certainly not with their informed consent. We now view this as a violation of the rights of retarded persons, especially if they are only mildly retarded. Legally it is now very difficult to gain permission to sterilize a retarded person.

A final concern is that retarded persons may be particularly vulnerable to sexual abuse.

In summary, there are three general points to be made about sexuality and persons with disabilities: (1) They generally do have sexual needs and desires; (2) they are often capable of sexual response quite similar to that of able-bodied people of average intelligence; and (3) there is a real need for more information—and communication—about what people with various disabilities can and cannot do sexually.

Sources: Ames (1991); Baladerian (1991); Bérard (1989); Beretta et al. (1989); Kempton & Kahn (1991); Rawicki & Hill (1991); Siosteen et al. (1990); Sipski & Alexander (1997).

Erection may also be produced by conditions other than tactile stimulation of the genitals; for example, fantasy or other purely psychological factors may produce erection. This points to the importance of the brain in producing erection, a topic that will be discussed in more detail below.

Mechanism of Ejaculation

The ejaculation reflex is much like the erection reflex, except that the ejaculation center is located higher in the spinal cord, the sympathetic division of the nervous system is involved (as opposed to the parasympathetic division in the erection reflex), and the response is muscular, with no involvement of the blood vessels (Rowland & Slob, 1997). In the ejaculation reflex, the penis responds to stimulation by sending a message to the "ejaculation center," which is located in the lumbar portion of the spinal cord. A message is then sent out via the nerves in the sympathetic nervous system, and this message triggers muscle contractions in the internal organs that are involved in ejaculation.

Ejaculation can often be controlled voluntarily.

This fact highlights the importance of brain influences on the ejaculation reflex.

The three main problems of ejaculation are premature ejaculation, male orgasmic disorder (retarded ejaculation), and retrograde ejaculation. Premature ejaculation, which is by far the most common problem, and male orgasmic disorder will be discussed in Chapter 19. **Retrograde ejaculation** occurs when the ejaculate, rather than going out through the tip of the penis, empties into the bladder (Kothari, 1984). A "dry orgasm" results, since no ejaculate is emitted. This problem can be caused by some illnesses, by tranquilizers and drugs used in the treatment of psychoses, and by prostate surgery. The mechanism that causes it is fairly simple (see Figure 9.9). Two sphincters are involved in ejaculation: an internal one, which closes

---

Retrograde ejaculation: A condition in which orgasm in the male is not accompanied by an external ejaculation; instead, the ejaculate goes into the urinary bladder.

Figure 9.9 How a retrograde ejaculation occurs.

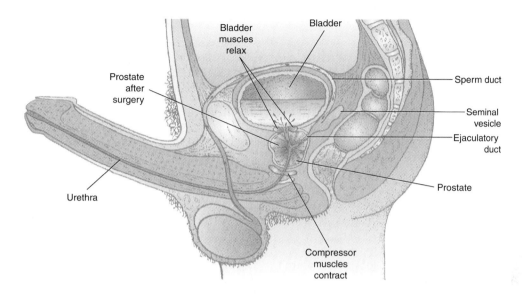

off the entrance to the bladder during a normal ejaculation, and an external one, which opens during a normal ejaculation, allowing the semen to flow out through the penis. In retrograde ejaculation, the action of these two sphincters is reversed; the external one closes, and thus the ejaculate cannot flow out through the penis, and the internal one opens, permitting the ejaculate to go into the bladder. The condition itself is quite harmless, although some men are disturbed by the lack of sensation of emitting semen.

Mechanisms in Women

Unfortunately there is less research on similar reflex mechanisms of arousal in women. Generally it is assumed that since the basic processes of sexual response (vasocongestion and myotonia) are similar in males and females and since their genital organs are derived from the same embryonic tissue (and thus have similar nerve supplies), reflexive mechanisms in women are similar to those in men. That is, since the lubrication response in women results from vasocongestion, it is similar to erection in the male, and thus it might be expected to be controlled by a similar spinal reflex. This is speculative, though.

Some research suggests the possibility that there is such a thing as *female ejaculation* (Perry & Whipple, 1981; Addiego et al., 1981; Belzer, 1981). The research discovered fluid spurting out of the urethra during orgasm; chemically, the fluid in one study was like the seminal fluid of vasectomized men— that is, semen without the sperm. The region responsible seems to be the **Gräfenberg spot** (or **G-spot**), also called the female prostate. It is located on the top side of the vagina (with the woman lying on her back, which is the best position for finding it), about halfway between the pubic bone and the cervix (see Figure 9.10). Stroking it produces an urge to urinate, but if the stroking continues for a few seconds more, it begins to produce sexual pleasure. Perry and Whipple argued that continued stimulation of it produces a *uterine orgasm*, characterized by deeper sensations of uterine contractions than the clitorally induced vulvar orgasm

Gräfenberg spot (GRAY-fen-berg) or **G-spot:** A hypothesized small region on the front wall of the vagina, emptying into the urethra, and responsible for female ejaculation.

Figure 9.10   G-spot: Hypothesized to produce ejaculation in some women.

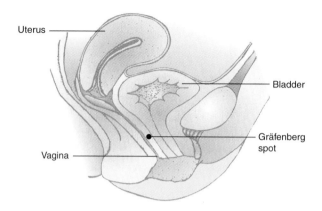

investigated in the Masters and Johnson research. Their ideas raise the whole clitoral-vaginal orgasm controversy again, although from a different research base. Perry and Whipple found that not every woman ejaculates, but it is important to recognize that some women do. Our firm notions of male-female differences are constantly challenged!

But Perry and Whipple's results have also been challenged. In their book, Perry and Whipple claimed to have examined over 400 women and found a G-spot in every one (Ladas, Whipple, & Perry, 1982). Yet in one study, two gynecologists examined 11 women, 6 of whom said they were ejaculators (Goldberg, 1983). The gynecologists found an area fitting the description of the G-spot in only 4 of the women, and 2 of them were ejaculators and 2 were not. Chemical analysis of the fluid ejaculated by the 6 ejaculators did not find it to be chemically like male semen; chemically it was like urine. Thus the study gave some support to the existence of a G-spot, but it gave no support to the hypothesis that some females ejaculate. On the other hand, in one survey of 1289 adult women, 40 percent reported having experienced ejaculation at the time of orgasm at least once, and 66 percent reported having an especially sensitive area on the front wall of the vagina (Darling et al., 1990). (For other mixed evidence, see Alzate & Londono, 1984.) In our judgment, it will take another 10 years of research, by many independent investigators, before we will really be able to sort out whether there is a G-spot and, if so, what it does (Rosen & Beck, 1988; Alzate, 1990).

Brain Control of Sexual Response

Sexual responses are controlled by more than simple spinal reflexes. Sexual responses may be brought under voluntary control, and they may be initiated by purely psychological forces, such as fantasy. Environmental factors, such as having been taught as a child that sex is dirty and sinful, may also affect one's sexual response. All these phenomena point to the critical influence of the brain and its interaction with the spinal reflexes in producing sexual response (see Figure 9.6).

Brain control of sexual response is complex and only partly understood at the present time. It appears that the most important influences come from a set of structures called the **limbic system** (see Figure 9.11). The limbic system forms a border between the central part of the brain and the outer part (the cerebral cortex); it includes the amygdala, the hippocampus, the cingulate gyrus, the fornix, and the septum. The thalamus, the hypothalamus, the pituitary, and the reticular formation are not properly part of the limbic system, but they are closely connected to it.

Several lines of evidence point to the importance of the limbic system in sexual behavior. In experiments with monkeys, an electrode was in-

Limbic system: A set of structures in the interior of the brain, including the amygdala, hippocampus, and fornix; believed to be important for sexual behavior in both animals and humans.

Figure 9.11   The limbic system of the brain, which is important in sexuality.

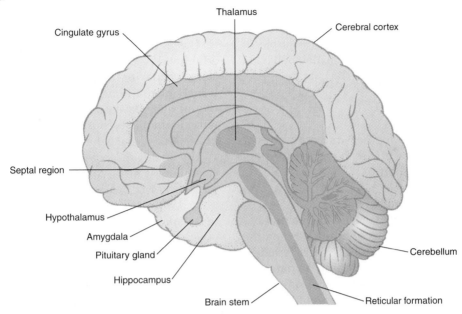

serted into various regions of the brain to deliver electrical stimulation. Stimulation of some areas of the brain produced an erection (MacLean, 1962; Van Dis & Larsson, 1971). In particular, three "erection centers" were found in the limbic system, including one in the septal region. For obvious reasons, little of this research has been done with humans. But in one study, stimulation of the septal region of the limbic system produced orgasm in two human subjects (Heath, 1972).

A particular region of the anterior portion of the hypothalamus (the medical preoptic area, or MPOA/AH) has been especially implicated in male sexual behavior (Paredes & Baum, 1997). If this area of the hypothalamus is given electrical stimulation, male rats increase their sexual behavior; if this region is lesioned (destroyed), they no longer engage in copulation with females (Paredes & Baum, 1997).

The classic work of the physiological psychologist James Olds (Olds, 1956; Olds & Milner, 1954) established the existence of "pleasure centers" in the brain. Electrodes were implanted in various regions of the brains of rats and were wired so that a rat could stimulate its own brain by pressing a lever. When the electrodes were placed in certain regions—in particular, the septal region and the hypothalamus—the rats would press the lever

thousands of times per hour and would forgo food and sleep and endure pain in order to stimulate these regions. The location of these pleasure centers so close to the sex centers may explain why sexual experiences are so intensely pleasurable.

The study of humans noted above also points to an association between the sex centers and the pleasure centers (Heath, 1972). Electrodes were placed in the brains of two humans for therapeutic purposes (one male psychiatric patient and one female epileptic). When the stimulation was to certain areas, they reported it as being very pleasurable; these areas were essentially the same as the pleasure centers found in the brains of animals. When the stimulation was delivered to the septal region or the amygdala (both in the limbic system), the pleasure was sexual in nature. In addition to the subjects' reports of sexual arousal and pleasurable sexual feelings, the male subject stimulated himself almost insatiably (as many as 1500 times per hour) and begged to be allowed to stimulate himself a few more times whenever the apparatus was taken away.

There are also "rage," or "aggression," centers in the brain, the stimulation of which throws the animal into a rage so that it will attack any object in the cage. These centers are located in the hypothalamus and, like the pleasure centers, are close to the sex centers; the closeness of the two may explain the association of sex and aggression in phenomena such as rape and competition over mates (MacLean, 1962).

The brain centers for sex are also close to the olfactory centers. This brings us to the topic of pheromones and their role in sexual behavior, which will be discussed later in the chapter.

Hormones and Sex

The sex hormones are another important physiological force that interacts with the nervous system to influence sexual response.

Organizing Versus Activating Effects

Endocrinologists generally make a distinction between the organizing effects of hormones and the activating effects of hormones. As discussed in Chapter 5, hormones present during prenatal development have important influences on genital anatomy, creating male or female genitals. Hormone effects such as these are called **organizing effects** because they cause a relatively permanent change in the organization of some structure, whether in the nervous system or in the reproductive system. Typically there are "critical periods" during which these hormone effects may occur.

It has also been known for some time that if an adult male mouse or cat is castrated (has the testes removed, which removes the source of testosterone), it will cease engaging in sexual behavior (and will be less aggressive). If that animal is then given injections of testosterone, it will start engaging in sex again. Hormone effects such as these are called **activating effects** because they activate (or deactivate) certain behaviors.

The organizing effects of sex hormones on sexual behavior have been well documented. In a classic experiment, testosterone was administered to pregnant female guinea pigs. The female offspring that had been exposed to testosterone prenatally[4] were, in adulthood, incapable of displaying female sexual behavior (in particular, lordosis, which is a sexual posturing involving arching of the back and raising of the hindquarters so that intromission of the male's penis is possible) (Phoenix et al., 1959). It is thought that this result occurred because the testosterone "organized" the brain tissue (particularly the hypothalamus) in a male fashion. These female offspring were also born with masculinized genitals, and thus their reproductive systems had

also been organized in the male direction. But the important point here is that the prenatal doses of testosterone had masculinized their sexual behavior. Similar results have been obtained in experiments with many other species.

These hormonally masculinized females in adulthood displayed mounting behavior, a male[5] sexual behavior. When they were given testosterone in adulthood, they showed about as much mounting behavior as males did. Thus the testosterone administered in adulthood *activated* male patterns of sexual behavior.

The analogous experiment on males would be castration at birth, followed by administration of ovarian hormones in adulthood. When this was done with rats, female sexual behavior resulted; these males responded to mating attempts by other males essentially in the same way females do (Harris & Levine, 1965). Their brain tissue had been organized in a female direction during an early, critical period when testosterone was absent, and the female behavior patterns were activated in adulthood by administration of ovarian hormones.

It thus seems that males and females initially have capacities for both male and female sexual behaviors; if testosterone is present early in development, the capacity for exhibiting female behaviors is suppressed. Sex hormones in adulthood then activate the behavior patterns that were differentiated early in development.

One question that might be raised is: How relevant is this research to humans? Generally the trend is for the behavior of lower species to be more under hormonal control and for the behavior of higher species to be more under brain (neural) control.

---

[4]Note the similarity of these experiments to John Money's observations of human intersexes (Chapter 5).

[5]The term "male sexual behavior" is being used here to refer to a sexual behavior that is displayed by normal males of the species and either is absent in females of that species or is present at a much lower frequency. Normal females do mount, but they do so less frequently than males do. "Female sexual behavior" is defined similarly.

Organizing effects of hormones: Effects of sex hormones early in development, resulting in a permanent change in the brain or reproductive system.
Activating effects of hormones: Effects of sex hormones in adulthood, resulting in the activation of behaviors, especially sexual behaviors and aggressive behaviors.

Thus human sexual behavior is less under hormonal control than rat sexual behavior; human sexual behavior is more controlled by the brain, and thus learning and past experiences, which are stored in the brain, are more likely to have a profound effect. For example, it is estimated that only about 20 to 50 percent of cases of human sexual disorders (e.g., impotence) are caused by physical factors; the rest are due to psychological factors (see Chapter 19).

Let us now consider in more detail the known activating effects of sex hormones on the sexual behavior of adult humans.

Testosterone and Sexual Desire

Testosterone has well-documented effects on libido, or sexual desire, in humans (Carani et al., 1990; Carter, 1992; Everitt & Bancroft, 1991). In men deprived of their main source of testosterone by castration or by illness, there is a dramatic decrease in sexual behavior in some, but not all, cases (Feder, 1984). Sexual desire is rapidly lost if a man is given an antiandrogen drug. Thus testosterone seems to have an activating effect in maintaining sexual desire in the adult male. However, in cases of castration, sexual behavior may decline very slowly and may be present for several years after the source of testosterone is gone; this points to the importance of experience and brain control of sexual behavior in humans.

It has also been demonstrated that levels of testosterone are correlated with sexual behavior in boys around the time of puberty (Udry et al., 1985). Boys in the eighth, ninth, and tenth grades filled out a questionnaire about their sexual behavior and gave blood samples from which their level of testosterone could be measured. Among those boys whose testosterone level was in the highest quartile (25 percent of the sample), 69 percent had engaged in sexual intercourse, whereas only 16 percent of the boys whose testosterone level was in the lowest quartile had engaged in intercourse. Similarly, of those boys with testosterone levels in the highest quartile, 62 percent had masturbated, compared with 12 percent for the boys in the lowest quartile. These effects were uncorrelated with age, so it wasn't simply a matter of the older boys having more testosterone and more sexual experience. The authors concluded that at puberty, testosterone affects sexual motivation directly.

Research indicates that androgens, not estrogen, are related to sexual desire in women also (Hutchinson, 1995; Sherwin, 1991; Bancroft, 1987).[6] If all sources of androgens in women (the adrenals and the ovaries) are removed, women lose sexual desire.

For example, one study investigated the sexual functioning of women who had marked androgen deficiencies because they had undergone either chemotherapy for cancer or an oophorectomy (surgical removal of the ovaries) (Kaplan & Owett, 1993). These women showed a marked decline in sexual desire and responsiveness. Moreover, androgens are used successfully in the treatment of women who have low sexual desire (Kaplan & Owett, 1993). Some physicians are reluctant to use such treatment because of the masculinizing side effects, such as growth of facial hair, but these side effects can be avoided by the use of lower doses.

Pheromones

There has been a great deal of interest recently in the role that pheromones play in sexual behavior. Pheromones are somewhat like hormones. Recall that hormones are biochemicals that are manufactured in the body and secreted into the bloodstream to be carried to the organs that they affect. **Pheromones,** in contrast, are biochemicals that are secreted outside the body. Thus, through the sense of smell, they are an important means of communication between animals. Often the pheromones are contained in the animal's urine. Thus the dog that does "scent marking" is actually depositing pheromones. Some pheromones appear to be important in sexual communication, and some have even been called "sex attractants."

Much of the research on pheromones has been done with animals and demonstrates the importance of pheromones in sexual and reproductive functioning. For example, pheromones present in female urine have an influence on male sexual behavior. An early study demonstrated that male rats can tell the difference between the odor of estrous females and that of females not in estrus and that the males prefer

---

[6]Lest the reader be distressed by the thought that women's considerably lower levels of testosterone might mean that they have lower sex drives, it should be noted that the sensitivity of cells to hormone levels is critical. Women's cells may be more sensitive to testosterone than men's are. Thus for women, a little testosterone may go a long way.

Pheromones (FARE-oh-mones): Biochemicals secreted outside the body that are important in communication between animals and that may serve as sex attractants.

Focus 9.3

Sentencing Rapists: Castration or Incarceration?

In 1983, a judge delivered a controversial sentence in a rape case: The defendants could choose either 30 years in prison or castration. The three men had been convicted of a brutal gang rape of an 80-pound woman, who had required a transfusion of 4 pints of blood and five days of hospitalization. In 1992, a young man, who was on probation for molesting a 7-year-old and had just been arrested for raping a 13-year-old girl, volunteered to be castrated rather than being jailed.

These cases raise a host of questions, some of them legal, some within the province of the sciences. Legally, the castration sentence could be protested on the grounds that it is cruel and unusual punishment. What purpose did the judge have in mind when he assigned the castration sentence? Was he simply being punitive and letting the punishment fit the crime? Or did he view castration as a solution that would ensure that the men would never commit rape again? The scientific data become pertinent in addressing this last point.

Would castration (surgical removal of the testes, technically known as *bilateral orchidectomy*) prevent a man from committing rape? When the testes are removed, the man is left with little or no natural testosterone in his body. Numerous experiments with lower animals have demonstrated that the effect of this low level of testosterone is a sharply reduced sex drive and the virtual elimination of sexual behavior. However, the effects in humans are not so clear, because we are not as hormone-dependent as other species. There are documented cases of castrated men continuing to engage in sexual intercourse for years after the castration. Thus, castration may reduce sexual behavior in humans, but the effects are not completely predictable. Furthermore, testosterone is available artificially, either by pill or by injection, so that castrated criminals might secretly obtain replacement testosterone.

The discussion up to this point has focused on rape as a form of sexual behavior. Yet many experts believe that rape is better conceptualized as an aggressive or violent crime that happens to be expressed sexually. Thus, the scientific question may be restated from "Does castration eliminate sexual behavior?" to "Does castration eliminate aggressive behavior?" Here, too, there are numerous experiments documenting—in other species—that castration, by lowering testosterone levels, greatly reduces aggressive behavior. But once again, the

estrous females (LeMagnen, 1952). Similar phenomena have been demonstrated in male dogs, stallions, bulls, and rams (Michael & Keverne, 1968). The hormones estrogen and progesterone are critical to the effect of female pheromones on males (Beach & Merari, 1970). If a female's ovaries are removed, thereby removing estrogen and progesterone, her odors are no longer stimulating to males.

The sense of smell (olfaction) seems to be essential in order for pheromone effects to occur. Removal of the olfactory bulbs (a part of the brain important in olfaction), specifically a region called the vomeronasal organ, dramatically reduces sexual behavior of males from species such as mice and guinea pigs (Wysocki & Lepri, 1991).

Some research has been done on pheromones in monkeys. It has been found, for example, that estrogen treatment of a female rhesus monkey sexually excites the males in her cage (Herbert, 1966). If the males' sense of smell is blocked, they no longer are sexually excited.

What relevance does all this have for humans? Humans are not, by and large, "smell animals." Olfaction is much less important for us than for most other species. We tend to rely mostly on vision and, secondarily, hearing. Compare this with a dog's ability to get a wealth of information about who and what has been in a park simply by sniffing around for a few minutes. Does this mean that pheromones have no influence on our sexual behavior?

hormone effects are not as clear or consistent in humans. Thus, castration might be effective in reducing sexual or aggressive behaviors and thus might reduce the chance of the men committing rape again, but such effects cannot be guaranteed.

Other biological treatments for rapists are being explored. One is injections of the drug Depo-Provera, which is also used as a form of birth control for women (see Chapter 8). In male sex offenders, Depo-Provera is a kind of "chemical castration" because its effect is to reduce sharply the levels of testosterone in the body. It seems clear that Depo-Provera should be only part of the treatment, which should also include intensive psychotherapy. In one treatment program combining Depo-Provera and psychotherapy, after three and a half years only 15 percent of the men had repeated their offense, compared with rates as high as 85 percent for men who were only imprisoned. Yet some authorities believe that Depo-Provera is dangerous, possibly increasing the risk of cancer and suicidal depression.

Sources: Michael S. Serrill. (1983, Dec. 12). Castration or incarceration? *Time*, p. 70. Nikolaus Heim. (1981). Sexual behavior of castrated sex offenders. *Archives of Sexual Behavior, 10*, pp. 11–20. Robert T. Rubin, June M. Reinisch, & Roger F. Haskett. (1981). Postnatal gonadal steroid effects on human behavior. *Science, 211*, pp. 1318–1324. Robert Lacayo (1992, Mar. 23). Sentences inscribed on flesh. *Time*, 54.

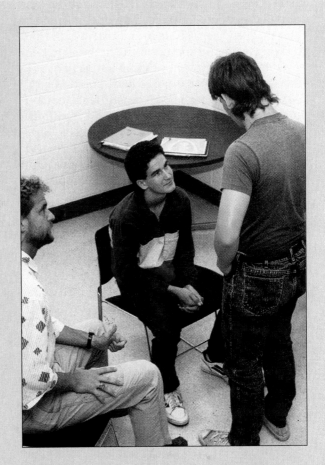

Figure 9.12 Sex offenders in a therapy group. Research shows that treatment of sex offenders, whether with the drug Depo-Provera or not, should always include psychotherapy.

It is now clear that human pheromones exist, and they may play an important role in sexual behavior. Indeed, pheromones may be exactly the "body chemistry" that attracts people to each other. Perfumes with musky scents have become popular and presumably increase sexual attractiveness, perhaps because they smell like pheromones. The perfume industry has eagerly tried to capitalize on pheromone research (Berliner et al., 1991). Indeed, there is a perfume on the market called "Pheromone."

What scientific evidence is there regarding the existence and effects of pheromones in humans? First, the vomeronasal organ—which, as noted earlier, is related to olfaction and sexual behavior in other species, and essentially seems to function as a pheromone sensor—is present in most humans (Garcia-Velasco & Mondragon, 1991). This makes one wonder whether plastic surgery to create a pretty nose might inadvertently harm the person's sex life.

Androstenol, an odorous steroid that is well-documented as a pheromone in pigs, has been isolated in the underarm sweat of humans (Gower & Ruparelia, 1993). Short-chain fatty acids that are known to be sex-attractant pheromones to male rhesus monkeys have been isolated in human vaginal secretions (Cowley & Brooksbank, 1991). It is clear, therefore, that humans do secrete pheromones.

In one clever experiment, female volunteers slept overnight with a plastic tube necklace around their necks (Cowley & Brooksbank, 1991). Inside the tube were pieces of cotton, placed by openings in the tube, that had been soaked in androstenol. The next morning, women exposed to the androstenol engaged in significantly more social interactions with men than did women exposed to no pheromones.

An interesting series of experiments demonstrates the influence of pheromones on the human menstrual cycle (Cutler et al., 1986; Preti et al., 1986). Men and women "donors" wore cotton pads in their armpits. These pads collected axillary (underarm) secretions, including, presumably, pheromones. The secretions were then extracted from the cotton pads, producing "male extract" and "female extract" (it reminds one a bit of vanilla extract). In one of the experiments, a group of volunteer women were exposed to female extract collected at regular intervals across the menstrual cycle. The recipient women showed significant shifts toward having their cycles at the same time. This finding builds on classic research by Martha McClintock (1971), which documented the existence of a phenomenon known as **menstrual synchrony:** the convergence, over several months, of the dates of onset of menstrual periods among women who are in close contact with each other (Graham, 1991; Weller et al., 1995).

In perhaps the most dramatic experiment to date in humans, the results indicated that the timing of ovulation could be experimentally manipulated with human pheromones (Stern & McClintock, 1998). Odorless secretions from women's armpits were collected in the late follicular phase, that is, just before ovulation. Recipient women exposed to these secretions showed an accelerated appearance of the LH surge that triggers ovulation. Underarm secretions from the same donors collected later in the menstrual cycle had the opposite effect: they delayed the LH surge of recipients and lengthened the time to menstruation.

It should be noted that the smell of pheromones would not necessarily have to be consciously perceived in order to have an effect. The olfactory system can respond to odors even when they are not consciously perceived. Thus, pheromones that we are not even aware of may have important influences.

If these speculations about the effects of pheromones on human sexual behavior are correct, our "hyperclean" society may be destroying the scents that attract people to each other. The normal genital secretions (assuming reasonable cleanliness to eliminate bacteria) may contain sex attractants. Ironically, "feminine hygiene" deodorants may destroy precisely the odors that "turn men on."

Future research on human pheromones should be interesting indeed.

> **Menstrual synchrony:** The convergence, over several months, of the dates of onset of menstrual periods among women who are in close contact with each other.

(a)

(b)

Figure 9.13   Pheromones. *(a)* Pheromones are a major means of communication between animals. *(b)* Are there human pheromones that are sex attractants?

SUMMARY

William Masters and Virginia Johnson conducted an important program of research on the physiology of human sexual response. They found that two basic physiological processes occur during arousal and orgasm: vasocongestion and myotonia. They divide the sexual response cycle into four stages: excitement, plateau, orgasm, and resolution.

Their research indicates that there is no physiological distinction between clitoral and vaginal orgasms in women, which refutes an early idea of Freud's. They have also provided convincing evidence of the existence of multiple orgasm in women.

Criticisms of Masters and Johnson's model are that (1) they ignored cognitive factors, and (2) their selection of research participants may have led to a self-fulfilling prophecy in their results.

Two cognitive-physiological models are Kaplan's three-component (desire, vasocongestion, and muscular contraction) model, and Walen and Roth's model, which emphasizes cognitive aspects of sexual response (perception and evaluation).

The nervous system and sex hormones are important in sexual response. The nervous system functions in sexual response by a combination of spinal reflexes (best documented for erection and ejaculation) and brain influences (particularly of the limbic system). There is controversial evidence that some women ejaculate. Hormones are important to sexual behavior, both in their influences on prenatal development (organizing effects) and in their stimulating influence on adult sexual behavior (activating effects). Testosterone seems to be crucial for maintaining sexual desire in both men and women.

Pheromones are biochemicals secreted outside the body and play an important role in sexual communication and attraction. Much of the evidence is based on research with animals, but evidence in humans is accumulating rapidly.

REVIEW QUESTIONS

1. According to Masters and Johnson, the four stages of sexual response are, in order, _____, _____, _____, and _____.

2. According to Masters and Johnson, the two basic physiological processes that occur during sexual arousal are _____ and myotonia.

3. Masters and Johnson's model of sexual response has been criticized because they ignored _____ factors such as sexual desire.

4. Only men are capable of multiple orgasm. True or false?

5. According to Kaplan, the three components of sexual response are _____, _____, and _____.

6. Erection in the male is produced by a simple spinal reflex. True or false?

7. It has been hypothesized that a region called the _____ is responsible for ejaculation in some women.

8. If female guinea pig fetuses are exposed to testosterone prenatally, they are born with masculinized genitals and hypothalamus and show male sexual behaviors; this demonstrates the _____ effects of hormones.

9. _____ are biochemicals secreted outside the body that may serve a role as sex attractants.

10. When a judge offered three convicted rapists a choice of prison or castration, he made a wise decision because research has proven that castration prevents men from committing rape in the future. True or false?

(The answers to all review questions are at the end of the book, after the Glossary.)

QUESTIONS FOR THOUGHT, DISCUSSION, AND DEBATE

1. Debate the following topic. Resolved: Castration is an appropriate and effective treatment for convicted rapists.

2. Do you think that pheromones might play more of a role in human sexual behavior in cultures that do not stress personal hyper-cleanliness as much as we do in the United States? If so, what do you think the effects of pheromones would be in those other cultures?

3. Your little sister, Katie, is a sophomore in college and is in the fourth month of what looks like it will be a wonderful long-term relationship. Katie and her boyfriend now seem to have a full sexual relationship. As a benevolent older brother/sister, you want Katie to enjoy much sexual pleasure in this relationship. What information from this chapter would you tell Katie about, if you hoped to ensure her sexual satisfaction and sexual pleasure?

SUGGESTIONS FOR FURTHER READING

Kroll, Ken et al. (1995). *Enabling romance: A guide to love, sex, and relationships for the disabled (and the people who care for them).* Bethesda, MD: Woodbine House. With the recognition of disabled persons' sexuality comes a need for self-help books, and this one is designed for that purpose.

Tiefer, Leonore. (1995). *Sex is not a natural act and other essays.* Boulder, CO: Westview. Tiefer is both knowledgeable and a witty writer. This book contains some of her insightful criticisms of the Masters and Johnson model of sexual response.

WEB RESOURCES

http://www.pheromones.com
    Website with information about pheromones; James Kohl, Ph.D.

http://sexualhealth.com/
    Information about sexuality for people with disabilities, illness, or other health-related problems.

chapter

10

TECHNIQUES OF AROUSAL AND COMMUNICATION

CHAPTER HIGHLIGHTS

Erogenous Zones

One-Person Sex

Two-Person Sex

Aphrodisiacs

Should Intercourse and Orgasm Be the Goal?

From Inexperience to Boredom

Communication and Sex

S eeking sexual satisfaction is a basic de- sire, and masturbation is our first natural sexual activity. It's the way we discover our eroticism, the way we learn to respond sexually, the way we learn to love ourselves and to build self-esteem. Sexual skill and the ability to respond are not "natural" in our society. "Doing what comes naturally," is to be sexually inhibited. Sex, like any other skill, has to be learned and practiced.*

We live in the era of sex manuals. Books like *The New Joy of Sex, The Good Girl's Guide to Great Sex, How to Make Love All Night, Simultaneous Orgasm,* and *Mindblowing Sex in the Real World,* as well as feature articles and advice columns in many magazines, give us information on how to produce bigger and longer orgasms in ourselves and our partners. The "read-all-about-it" boom has produced not only benefits but also problems. It may turn our attention so much to mechanical techniques that we forget about love and the emotional side of sexual expression. The sex manuals may also set up impossible standards of sexual performance that none of us can meet. On the other hand, we live in a society that has a history of leaving the learning of sexual techniques to nature or to chance, in contrast to some other societies in which adolescents are given explicit instruction in methods for producing sexual pleasure. For human beings, sexual behavior is a lot more than "doin' what comes naturally"; we all need some means for learning about sexual techniques, and the sex manuals may help to fill that need.

The purpose of this chapter is to provide information on techniques of arousal, while attempting to avoid making sex too mechanical or setting up unrealistic performance standards. It also focuses on skills for communicating in a relationship. As you build a sexual relationship with another person, three components are important: (1) building a loving attachment, (2) building the sexual part of the relationship, and (3) developing good communication. This chapter focuses on items 2 and 3. Love in relationships will be discussed in Chapter 13.

*Betty Dodson. (1987). *Sex for one.* New York: Harmony Books (Crown Publishers), p. 36.

Erogenous Zones

While the notion of **erogenous zones** originated in Freud's work, the term is now part of our general vocabulary and refers to parts of the body that are sexually sensitive; stroking them or otherwise stimulating them produces sexual arousal. The genitals and the breasts are good examples. The lips, neck, and thighs are generally also erogenous zones. But even some rather unlikely regions— such as the back, the ears, the stomach, and the feet—can also be quite erogenous. One person's erogenous zones can be quite different from another person's. Thus it is impossible to give a list of sure "turn-ons." The best way to find out is to communicate with your partner, either verbally or nonverbally.

One-Person Sex

It does not necessarily take two to have sex. One can produce one's own sexual stimulation. Sexual self-stimulation is called **autoeroticism.**[1] The best examples are masturbation and fantasy.

[1]For those of you who are interested in the roots of words, "autoeroticism" does not refer to sex in the back seat of a car. The prefix "auto" means "self" (as in "autobiography"); hence self-stimulation is autoeroticism.

Erogenous zones (eh-RAH-jen-us): Areas of the body that are particularly sensitive to sexual stimulation.
Autoeroticism: Sexual self-stimulation; for example, masturbation.

Masturbation

Here the term "hand-genital stimulation" will be reserved for stimulation of another's genitals and the term **masturbation** for self-stimulation, either with the hand or with some object, such as a pillow or a vibrator. Masturbation is a very common sexual behavior; almost all men and the majority of women masturbate to orgasm at least a few times during their lives. In response to questions included in the NHSLS of 3432 Americans, 62 percent of the men and 42 percent of the women reported that they had masturbated in the preceding year (Laumann et al., 1994, Table 3.1). Twenty-seven percent of the men and 8 percent of the women said that they masturbated at least once a week. There are differences in the incidence of masturbation by ethnicity; African American men and women are less likely to report this behavior (see Table 1.2 in Chapter 1). The techniques used by males and females in masturbation are interesting in part because they provide information to their partners concerning the best techniques to use in lovemaking.

Techniques of Female Masturbation

Most commonly women masturbate by manipulating the clitoris and the inner lips (Hite, 1976; Kinsey et al., 1953, pp. 158, 189). They may rub up and down or in a circular motion, sometimes lightly and sometimes applying more pressure to the clitoris. Some prefer to rub at the side of the clitoris, while a few stimulate the glans of the clitoris directly. The inner lips may also be stroked or tugged. One woman described her technique as follows:

> I use the tips of my fingers for actual stimulation, but it's better to start with patting motions or light rubbing motions over the general area. As excitement increases I begin stroking above the clitoris and finally reach a climax with a rapid, jerky circular motion over the clitoral hood. Usually my legs are apart, and occasionally I also stimulate my nipples with the other hand. (Hite, 1976, p. 20)

This finding is in distinct contrast to what many men imagine to be the techniques of female masturbation; the male pictures the woman inserting a finger, a banana, or a similar object into the depths of the vagina (Kinsey et al., 1953). In fact, this is not often done; by far the most common method is clitoral and labial manipulation. Of the women in Kinsey's sample who masturbated, 84 percent used clitoral and labial manipulation; inserting fingers or objects into the vagina was the second most commonly used technique, but it was practiced by only 20 percent of the women.

Other techniques used by women in masturbation include breast stimulation, thigh pressure exerted by crossing the legs and pressing them together rhythmically to stimulate the clitoris, and pressing the genitals against some object, such as a pillow, or massaging them with a stream of water while in the shower. A few women are capable of using fantasy alone to produce orgasm; fantasy-induced orgasms are accompanied by the same physiological changes as orgasms produced by masturbation (Whipple, Ogden, & Komisanak, 1992).

Techniques of Male Masturbation

Almost all males report masturbating by hand stimulation of the penis. For those interested in speed, an orgasm can be reached in only a minute or two (Kinsey et al., 1948, p. 509).

Most men use the technique of circling the hand around the shaft of the penis and using an up-and-down movement to stimulate the shaft and glans. Because the penis produces no natural lubrication of its own, some men like to use a form of lubrication, such as soapsuds while showering. The tightness of the grip, the speed of movement, and the amount of glans stimulation vary from one man to the next. Most increase the speed of stimulation as they approach orgasm, slowing or stopping the stimulation at orgasm because further stimulation would be uncomfortable (Masters & Johnson, 1966). At the time of ejaculation, they often grip the shaft of the penis tightly. Immediately after orgasm, the glans and corona are hypersensitive, and the man generally avoids further stimulation of the penis at that time (Sadock & Sadock, 1976).

> **Masturbation:** Stimulation of one's own genitals with the hand or with some object, such as a pillow or vibrator.

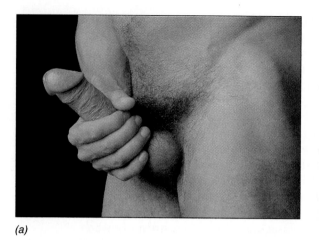

(a)

(b)

Figure 10.1   *(a)* Male masturbation using hand stimulation of the penis. *(b)* Female masturbation using clitoral stimulation.

Fantasy

Sexual fantasy refers to sexual thought or images that alter the person's emotions or physiological state (Maltz & Boss, 1997). Almost all men and women report that they have experienced sexual fantasies (Leitenberg & Henning, 1995).

The Content of Fantasies

The content of sexual fantasies reported by men and women are similar. The most common are touching and kissing sensuously, watching a partner undress, giving or receiving oral sex, and seducing someone or being seduced (Hsu et al., 1994). Other common fantasies are having sex in an unusual location, having sex in an unusual position, and having sex that lasts for hours.

There are some gender differences in the content of fantasies. Men's fantasies focus on sexual activity; they may fantasize having oral sex with a virgin, an experienced older woman, or with two partners at the same time. Women are more likely to focus on playing a role during sexual activity, such as the shy maiden, the victim, the dominatrix, or the voyeur (Maltz & Boss, 1997). For many women, the relationship with the partner is the key, not the activity. One study comparing the fantasies of gays and lesbians with those of heterosexual men and women found that the reported contents were very similar, except that the partner was someone of the same gender (Price et al., 1985). However, heterosexuals may fantasize sexual activity with someone of the same gender, and gay and lesbian fantasies may include persons of the other gender.

The Nature of Fantasies

Where do sexual fantasies come from? The images may come from past experience, dreams, media portrayals you have read or viewed, or stories someone told you. A fantasy may include only one person, either the self or someone else, or it may include other persons of either gender and any sexual orientation. The activity may be dreamlike and sensuous, or explicit and vigorous. Like all fantasy, sexual fantasies represent a fusion of mind, body, and emotion. Sexual fantasies may represent earlier or childhood experiences, pleasant or abusive (Maltz & Boss, 1997).

Sexual fantasy can play a variety of functions for the person doing the fantasizing (Maltz & Boss, 1997). These include enhancing self-esteem and attractiveness, increasing one's own sexual arousal (e.g., during masturbation or partnered sex), and facilitating orgasm. A very important role is enabling the person to mentally rehearse future possibilities. Such rehearsal may enable the person to change behavior, initiate communication with a partner, or change partners.

Sexual fantasy: Sexual thoughts or images that alter the person's emotions or physiological state.

Fantasy During Masturbation

More men than women report ever having had sexual fantasies during masturbation. The results of 13 studies indicate that about 87 percent of men and 69 percent of women fantasized during masturbation. Here is an adolescent male's description of one of his favorite fantasies during masturbation:

> We would be riding in the back seat of the car and I would reach over and fondle her breasts. She would reach into my pants and begin to caress my penis and finally suck me off. (Jensen, 1976a, p. 144)

This is the fantasy of one college woman as she imagines the seduction of her French teacher:

> At exactly 8:00 I knocked on the door. When this guy saw what I was wearing I thought his eyes were going to pop out. Calmly he asked me to come in and sit down . . .
> "Please call me Jim." Now I was getting somewhere. . . . I slipped out of my shoes and loosened my dress. When Jim returned I was ready and waiting. . . . Well the man finally got the hint; he reached around and unzipped my dress. While I was slowly undoing his zipper, he buried his head between my breasts. As his mouth slowly descended down my body I could feel the heat rising from between my legs. . . . To add to my desire he started speaking French to me. You didn't have to be fluent to understand this. As his mouth continued to nibble away, his tongue zeroed in on my clit and sent me to a mind boggling orgasm. As my pleasure subsided I began to return the favor. . . . Although I had enjoyed several orgasms by now the night was far from over. Jim then got on top of me and made love to me for what seemed like an eternity. . . . (Moffatt, 1989, pp. 190–191)

The content of male and female sexual fantasies seems to be influenced by cultural stereotypes of male and female sexuality.

Vibrators, Dildos, and Such

Various sexual devices, such as vibrators and dildos, are used by some people in masturbation or by couples as they have sex together.

Both male and female artificial genitals can be purchased. A **dildo** is a rubber or plastic cylinder, often shaped like a penis; it can be inserted into the vagina or the anus. Dildos are used by women in masturbation (although, as was noted in the previous section, this is not very common), by men, by

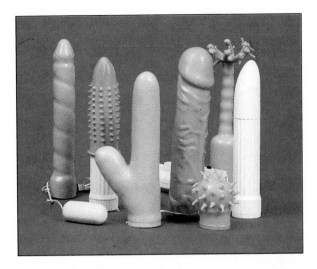

Figure 10.2   Vibrators and dildos, used for sexual stimulation.

lesbians, by gays, and by heterosexual couples. Artificial vaginas, and even inflatable replicas of the entire body, male or female, can also be purchased.

Some *vibrators* are shaped like a penis and others are not; there are models with a cord that plugs into an electric socket and also battery-operated, cordless models. Women may use them to masturbate, stimulating the clitoral and mons area, or they may insert them into the vagina. Males may use them to stimulate the genitals or the anus. They are used either in masturbation or during sex with another person. They can be purchased in "respectable" stores (where they are sometimes euphemistically called "face massagers"), in sex supermarkets, and by mail.

Body oils are also popular for sexual use. In fact, their use has been encouraged by experts in the field; for example, Masters and Johnson and other sex therapists recommend them for the touching or sensate focus exercises that they prescribe for their patients in sex therapy (see Chapter 19). Oils have a sensuous quality that heightens erotic feelings. Furthermore, if you are being stroked or massaged for any extended period of time, the oil helps ensure that the part of your body that is being stimulated will not end up feeling like a piece of wood that has been sandpapered. Oils can be used either while

Dildo: A rubber or plastic cylinder, often shaped like a penis.

masturbating or while having sex with another person. The sex stores sell them in a variety of exotic scents, but plain baby oil will also do nicely.

Two-Person Sex

When many of us think of techniques of two-person sex, the image that flashes across our mind generally reflects several assumptions. One assumption is that one of the people is a male and the other a female—that is, that the sex is heterosexual. This reflects a belief that heterosexual sex is normative. We also tend to assume that the male is supposed to do certain things during the act and that the female is supposed to do certain other things, reflecting the sexual scripts of our culture. He, for example, is supposed to take the initiative in deciding what techniques are used, while she is to follow his lead. Although there is nothing particularly evil in these assumptions, they do tend to impose some limitations on our own sexual expression and to make some people think that their own sexual behavior is "not quite right." Therefore, an attempt to avoid these assumptions will be made in the sections that follow.

Kissing

Kissing (or what we might call, technically, "mouth-to-mouth stimulation") is an activity that virtually everyone in our culture has engaged in (DeLamater & MacCorquodale, 1979). In simple kissing, the partners keep their mouths closed and touch each other's lips. In deep kissing ("French kissing"), both people part their lips slightly and insert their tongues into each other's mouths (somehow these clinical descriptions do not make it sound like as much fun as it is). There are endless variations on these two basic approaches, such as nibbling at the partner's lips or tongue or sucking at the lips; they depend only on your imagination and personal preference. There are also plenty of other regions of the body to kiss: the nose, the forehead, the eyelids, the earlobes, the neck, the breasts, the genitals, and even the feet, to give a few examples.

Touching

Enjoying touching and being touched is essential to sexual pleasure. Caresses or massages, applied to virtually any area of the body, can be exciting. The regions that are exciting vary a great deal from one person to the next and depend on how the person is feeling at the moment; thus it is important to communicate what sort of touching is most pleasurable to you. (For specific exercises on touching and being touched, see Chapter 19.)

As was noted earlier, one of the best ways to find out how to use your hands in stimulating the genitals of another person is to find out how that person masturbates.

Hand Stimulation of the Male Genitals

As a technique of lovemaking, hand stimulation of the male genitals can be used as a pleasurable preliminary to intercourse, as a means of inducing orgasm itself, or as a means of producing an erection after the man has had one orgasm and wants to continue for another round of lovemaking ("rousing the dead").

Alex Comfort, in *The New Joy of Sex*, recommends the following techniques (see Figure 10.3):

> If he isn't circumcised, she will probably need to avoid rubbing the glans itself, except in pursuit of very special effects. Her best grip is just below the groove, with the skin back as far as it will go, and using two hands—one pressing hard near the root, holding the penis steady, or fondling the scrotum, the other making a thumb-and-first-finger ring, or a whole hand grip. She should vary this, and, in prolonged masturbation, change hands often. (1991, p. 85)

As a good technique for producing an erection he also mentions rolling the penis like dough between the palms of the hands. Firm pressure with one finger midway between the base of the penis and the anus is another possibility.

One of the things that make hand stimulation most effective is for the man's partner to have a playful delight in, and appreciation of, the man's penis. Most men think their penis is pretty important. If the partner cannot honestly appreciate it and enjoy massaging it, hand stimulation might as well not be done.

Hand Stimulation of the Female Genitals

The hands can be used to stimulate the woman's genitals to produce orgasm, as a preliminary method of arousing the woman before intercourse, or simply because it is pleasurable.

Figure 10.3 Technique of hand stimulation of the penis.

Generally it is best, particularly if the woman is not already aroused, to begin with gentle, light stroking of the inside of the thighs and the inner and outer lips, moving to light stroking of the clitoris. As she becomes more aroused, the stimulation of the clitoris can become firmer. There are several rules for doing this, though. First, remember that the clitoris is very sensitive and that this sensitivity can be either exquisite or painful. Some care has to be used in stimulating it; it cannot be manipulated like a piece of Silly Putty. Second, the clitoris should never—except, perhaps, for some light stroking—be rubbed while it is dry, for the effect can be more like sandpaper stimulation than sexual stimulation. If the woman is already somewhat aroused, lubrication can be provided by touching the fingers to the vaginal entrance and then spreading the lubrication on the clitoris. If she is not aroused or does not produce much vaginal lubrication, saliva works well too. Moisture makes the stimulation not only more comfortable but also more sensuous. Third, some women find direct stimulation of the clitoral glans to be painful in some states of arousal. These women generally prefer stimulation on either side of the clitoris instead.

With these caveats in mind, the clitoris can be stimulated with circular or back-and-forth movements of the finger. The inner and outer lips can also be stroked, rubbed, or tugged. These techniques, if done with skill and patience, can bring the woman to orgasm. Another technique that can be helpful in producing orgasm is for the partner to place the heel of the hand on the mons, exerting pressure on it while moving the middle finger in and out of the vaginal entrance. Another rule should also be clear by now: The partner needs to have close-trimmed nails that do not have any jagged edges. This is sensitive tissue we are dealing with.

The Other Senses

So far, this chapter has been focused on tactile (touch) sensations in sexual arousal. However, the other senses—vision, smell, and hearing—can also make contributions.

Sights

The things that you see while making love can contribute to arousal. Men seem, in general, to be more turned on by visual stimuli; many men have mild fetishes (see Chapter 16 for more detail on fetishes) and like to see their partners wearing certain types of clothing. Perhaps as female sexuality becomes more liberated, women will become more interested in visual turn-ons. A good rule here, as elsewhere, is to communicate with your partner to find out what he or she would find arousing.

The decor of the room can also contribute to visual stimulation. Furry rugs, placed on the floor or used as a bedspread, are sensuous both visually and tactilely. Large mirrors, hung either behind the bed or on the ceiling above it, can be a visual turn-on, since they allow you to watch yourself make love. Candlelight is soft and contributes more to an erotic atmosphere than an electric light, which is harsh, or complete darkness, in which case there is no visual stimulation at all.

Perhaps the biggest visual turn-on comes simply from looking at your own body and your partner's body. According to the NHSLS, watching a partner undress is one of the most appealing sexual activities (Michael et al., 1994, Table 12).

Smells

Odors can be turn-ons or turn-offs. The scent of a body that is clean, having been washed with soap and water, is a natural turn-on. It does not need to be covered up with an "intimate deodorant."[2] In a sense, the scent of your skin, armpits, or genitals is your "aroma signature," and can be quite arousing (Comfort, 1991).

A body that has not been washed or a mouth that has not been cleaned or has recently been used for smoking cigarettes can be a real turn-off. Breath that reeks of garlic may turn a desire for closeness into a desire for distance. Ideally, the communication between partners is honest and trusting enough so that if one offends, the other can request that the appropriate clean-up be done.

Sounds

Music—whether your preference is for rock or classical music—can contribute to an erotic atmosphere. Another advantage of playing music is that it helps muffle the sounds of sex, which can be important if you live in an apartment with thin walls or if you are worried about your children hearing you. Similarly, music can be a turn-off; a funeral dirge is not conducive to joyous sexual activity.

---

[2]With the increased popularity of mouth-genital sex, some women worry that the scent of their genitals might be offensive. The advertisements for feminine hygiene deodorant sprays prey upon those fears. These sprays should not be used because they may irritate the vagina. Besides, there is nothing offensive about the scent of a vulva that has been washed; some people, in fact, find it arousing.

Fantasy During Two-Person Sex

Fantasies are a means of self-stimulation to heighten the experience of sex with another person. Particularly in a long-term, monogamous relationship, sexual monotony can become a problem; fantasies are one way to introduce some variety and excitement without violating an agreement to be faithful to the other person. It is important to view such fantasies in this way, rather than as a sign of disloyalty to, or dissatisfaction with, one's sexual partner.

Fantasies during two-person sex are generally quite similar to the ones people have while masturbating. In one study, 84 percent of males and 82 percent of females reported that they fantasized at least some of the time during intercourse (Cado & Leitenberg, 1990). This kind of fantasizing is quite common.

Marion, a 32-year-old married woman, reported this fantasy:

> I arrive at a party in a large opulent mansion. A servant [gives me a blindfold]. He tells me that I have to put it on and follow him through to the bedroom. He sits me down on the bed and takes off my shoes. Somebody else unzippers my dress and somebody else unfastens my necklace. I feel fingertips gently pinching my nipples through my bra. Somebody else unfastens my bra. Then I feel two people kissing and sucking my breasts. Then somebody's kissing my inner thighs. . . . She clamps her mouth over the crotch of my panties, . . . starts sucking my panties, and slipping her tongue underneath. After a while she pulls my panties to one side and starts licking my pussy. A man works his finger up into my bottom. The girl slides her finger inside my pussy. I'm getting so turned on I can hardly stand it. The girl's licking my clitoris like a hysterical butterfly. My whole body feels as if it's being swept away. I shake with an orgasm. (Masterton, 1993, pp. 107–109)

Here is a man's fantasy:

> We're in a group of people around a dinner table . . . a group of several couples. We're playing a card game, strip poker, or a type of elaborate strip poker where anybody who wins gets to have the person who lost do anything they want. And the fantasy is always of Colleen becoming very sexual. All she's interested in is sex. Fucking, sucking, exhibiting herself, that's it. It's mostly a seduction fantasy. She might seduce one of the guys at the table by walking up the stairs and stripping off her clothes as she

goes. Then she'll get on her hands and knees and want to fuck there. (Maurer, 1994, pp. 144–145)

Some sexual partners enjoy sharing their fantasies with each other. Describing a sexual fantasy to your partner can be a turn-on for both of you. Some couples act out all or part of a fantasy and find it very gratifying. One man says:

> I fantasized about being a fourteen-year-old girl. This is a long-lived fantasy of mine. The girl is pretty, with a petite body, probably five-one, five-two, cute ass, maybe light blond or dark brown pussy hair. Small to medium size breasts. Kind of sweet-looking, virginal. . . . Anyway, Sue got into it, like we were two girlfriends sleeping over and she was a couple years older and she was going to show me some things. I happen to be very sensitive around my nipples, so she'd say, "Oh, you have nice little nipples, you're going to have very pretty breasts." Then she'd massage my crotch like she was massaging a vagina: "Do you like it when I pet you down there?" And she'd show me what to do to pleasure her, how to masturbate her and go down on her. We played around with that for a few months. (Maurer, 1994, p. 231)

Genital-Genital Stimulation: Positions of Intercourse

One of the commonest heterosexual techniques involves the insertion of the penis into the vagina; this is called **coitus**[3] or *sexual intercourse.* The couple may be in any one of a number of different positions while engaging in this basic sexual activity. Ancient love manuals and other sources illustrate many positions of intercourse.

Some authorities state that there are only four positions of intercourse. Personally, we prefer to believe that there are an infinite number. Consider how many different angles your arms, legs, and torso may be in, in relation to those of your partner, and all the various ways in which you can intertwine your limbs—that's a lot of positions. We trust that given sufficient creativity and time, you can discover them all for yourself. We would agree, though, that there are a few basic positions and a few basic dimensions along which positions can vary. One basic variation depends on whether the

Figure 10.4   Erotic sculptures at the Temple of Kandariya Mahadevo, India, built in A.D. 1000.

couple face each other (face-to-face position) or whether one partner faces the other's back (rear-entry position); if you try the other obvious variation, a back-to-back position, you will quickly find that you cannot accomplish much that way. The other basic variation depends on whether one partner is on top of the other or whether the couple are side by side. Let us consider four basic positions that illustrate these variations. As Julia Child does, we'll give you the basic recipes and let you decide on the embellishments.

Man on Top

The face-to-face, man-on-top position ("missionary" position—see Figure 10.5) is probably the one

---

[3]From the Latin word *coire,* meaning "to go together."

Coitus: Sexual intercourse; insertion of the penis into the vagina.

Figure 10.5   The man-on-top position of intercourse.

used most frequently by couples in the United States. In this position (which used to be called "male-superior," an unacceptable term today) the man and woman stimulate each other until they are aroused, he has an erection, and she is producing vaginal lubrication. Then he moves on top of her as she spreads her legs apart, either he or she spreads the vaginal lips apart, and he inserts his penis into her vagina. He supports himself on his knees and hands or elbows and moves his penis in and out of the vagina (pelvic thrusting). Some men worry that their heavy weight will crush the poor woman under them; however, because the weight is spread out over so great an area, most women do not find this to be a problem at all, and many find the sensation of contact to be pleasurable.

The woman can have her legs in a number of positions that create variations. She may have them straight out horizontally, a position that produces a tight rub on the penis but does not permit it to go deeply into the vagina. She may bend her legs and elevate them to varying degrees, or she may hook them over the man's back or over his shoulders. The last approach permits the penis to move deeply into the vagina. The woman can also move her pelvis, either up and down or side to side, to produce further stimulation.

The man-on-top position has some advantages and some disadvantages. It is the best position for ensuring conception, if that is what you want. It leaves the woman's hands free to stroke the man's body (or her own, for that matter). The couple may feel better able to express their love or to communicate other feelings, since they are facing each other. This position, however, does not work well if the woman is in the advanced stages of pregnancy or if either she or the man is extremely obese. Sex therapists have also found that it is not a very good position if the man wants to control his ejaculation; the woman-on-top position is better for this (see Chapter 19).

Woman on Top

There are a number of ways of getting into the woman-on-top position (Figure 10.6). You can begin by inserting the penis into the vagina while in the man-on-top position and then rolling over. Another possibility is to begin with the man on his back; the woman kneels over him, with one knee on either side of his hips. Then his hand or her hand guides the erect penis into the vagina as she lowers herself onto it. She then moves her hips to produce the stimulation. Beyond that, there are numerous variations, depending on where she puts her legs. She can remain on her knees, or she can straighten out her legs behind her, putting them outside his legs or between them. Or she can even turn around and face toward his feet.

Figure 10.6   The woman-on-top position of intercourse.

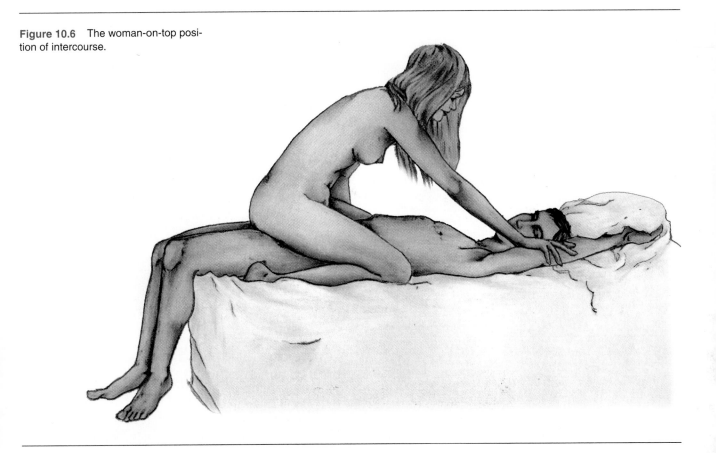

This position has a number of advantages. It provides a lot of clitoral stimulation, and the woman can control the kind of stimulation she gets; thus many women find it the best position for them to have an orgasm. It is also a good position for the man who wants to delay his ejaculation, and for this reason it is used in sex therapy. This position is also a good one if the man is tired and it seems advisable for the woman to supply most of the movement. Furthermore, the couple face each other, facilitating better communication, and each has the hands free to stroke the other.

Rear Entry

In the rear-entry position, the man faces the woman's back. One way to do this is to have the woman kneel with her head down; the man kneels behind her and inserts his penis into her vagina (Figure 10.7). (This is sometimes called the "doggie position," because it is the manner in which dogs and most other animals copulate.) Another possi-

bility is for the woman to lie on her stomach, raising her hips slightly so that the man can insert his penis while kneeling over her. Rear entry can also be accomplished when the couple are in the side-to-side position (see Figure 10.8).

In this position the man's hands are free to stimulate the woman's clitoris or any other part of her body. The couple do not face each other, however, and some couples dislike this aspect of the position. A small amount of air may enter the vagina when this position is used, producing interesting noises when it comes out.

Side to Side

In the side-to-side position, the man and woman lie beside each other, either face to face or in a rear-entry position (Figure 10.8). There are many variations beyond this, depending on where the arms and legs go—so many, in fact, that no attempt will be made to list them. One should be aware, though, that in this position an arm or a leg can get trapped

Figure 10.7 The rear-entry position of intercourse.

under a heavy body and begin to feel numb, and shifting positions is sometimes necessary.

The side-to-side position is good for leisurely or prolonged intercourse or if one or both of the partners are tired. It is also good for the pregnant and the obese. At least some hands are free to stimulate the clitoris, or whatever.

Other Variations

Aside from the variations in these basic positions that can be produced by switching the position of the legs, there are many other possibilities. For example, the man-on-top position can be varied by having the woman lie on the edge of a bed with her feet on the floor while the man kneels on the floor. Or the woman can lie on the edge of a table while

the man stands (don't forget to close the curtains first). Both these positions produce a somewhat tighter vagina and therefore more stimulation for the penis. Or the man can sit on a chair and insert the penis as the woman sits on his lap, using either a face-to-face or a rear-entry approach. Or, with both partners standing, the man can lift the woman onto his erect penis as she wraps her legs around his back, or she can put one leg over his shoulder (you have to be pretty athletic to manage this one, however).

Mouth-Genital Stimulation

One of the most striking features of the sexual revolution of the last few decades is the increased

Figure 10.8   The side-to-side position of intercourse.

popularity of mouth-genital, or oral-genital, techniques (Hunt, 1974). There are two kinds of mouth-genital stimulation ("going down on" one's partner); cunnilingus and fellatio.

Cunnilingus

In **cunnilingus,** or "eating" (from the Latin words *cunnus,* meaning "vulva," and *lingere,* meaning "to lick"), the woman's genitals are stimulated by her partner's mouth. Generally the focus of stimulation is the clitoris, and the tongue stimulates it and the surrounding area with quick darting or thrusting movements, or the mouth can suck at the clitoris. A good prelude to cunnilingus can be kissing of the inner thighs or the belly, gradually moving to the clitoris. The mouth can also suck at the inner lips, or the tongue can stimulate the vaginal entrance or be inserted into the vagina. During cunnilingus, some women also enjoy having a finger inserted into the vagina or the anus for added stimulation. The best way to know what she wants is through communication between partners, either verbal or nonverbal.

Many women are enthusiastic about cunnilingus and say that it is the best way, or perhaps the only way, for them to have an orgasm. Such responses are well within the normal range of female sexuality. As one woman put it:

A tongue offers gentleness and precision and wetness and is the perfect organ for contact. And, besides, it produces sensational orgasms. (Hite, 1976, p. 234)

Cunnilingus (like fellatio) can transmit some sexually transmitted diseases such as gonorrhea. Therefore, you need to be as careful about whom you engage in mouth-genital sex with as about whom you would engage in intercourse with. A small sheet of plastic, called a dental dam, can be placed over the vulva for those wanting to practice safer sex. One other possible problem should be noted, as well. Some women enjoy having their partner blow air forcefully into the vagina. While this technique is not dangerous under normal circumstances, when used on a pregnant woman it has been known to cause death (apparently as the result of air getting into the uterine veins), damage to the placenta, and embolism (Sadock & Sadock, 1976). Thus it should not be used on a pregnant woman.

Cunnilingus (cun-ih-LING-us): Mouth stimulation of the female genitals.

Figure 10.9   Female • Female sexual expression.

Cunnilingus can be performed by either heterosexuals or lesbian couples.

Fellatio

In **fellatio**[4] ("sucking," "a blow job") the man's penis is stimulated by his partner's mouth. The partner licks the glans of the penis, its shaft, and perhaps the testicles. The penis is gently taken into the mouth. If it is not fully erect, an erection can generally be produced by stronger sucking combined with hand stimulation along the penis. After that, the partner can produce an in-and-out motion by moving the lips down toward the base of the penis and then back up, always being careful not to scrape the penis with the teeth. Or the tongue can be flicked back and forth around the tip of the penis or along the corona.

To bring the man to orgasm, the in-and-out motion is continued, moving the penis deeper and deeper into the mouth and perhaps also using the fingers to encircle the base of the penis and give further stimulation. Sometimes when the penis moves deeply toward the throat, it stimulates a gag reflex, which occurs anytime something comes into contact with that part of the throat. To avoid this, the partner should concentrate on relaxing the throat muscles while firming the lips to provide more stimulation to the penis.

When a couple are engaged in fellatio, the big question in their minds may concern ejaculation. The man may, of course, simply withdraw his penis from his partner's mouth and ejaculate outside it. Or he may ejaculate into it, and his partner may even enjoy swallowing the ejaculate. The ejaculate

---

[4]"Fellatio" is from the Latin word *fellare,* meaning "to suck." Women should not take the "sucking" part too literally. The penis, particularly at the tip, is a delicate organ and should not be treated like a straw in an extra-thick milkshake.

Fellatio (feh-LAY-shoh): Mouth stimulation of the male genitals.

Figure 10.10 Male • Male sexual expression.

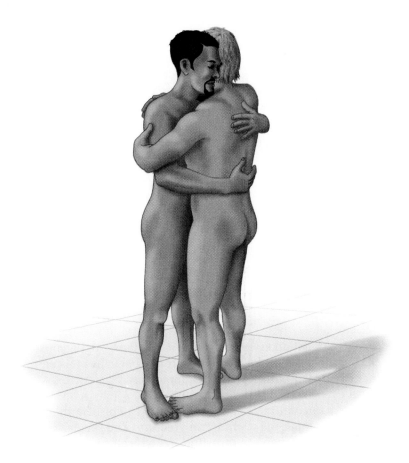

resembles partially cooked egg white in texture; it does not have a very distinctive flavor but often leaves a salty aftertaste. Because some people have mixed feelings about having the semen in their mouths, it is probably a good idea for the couple to discuss ahead of time (or during the activity) what they plan to do, particularly because ejaculation into the mouth is an unsafe practice in the AIDS era (see Chapter 20).

Most men find fellatio to be a highly stimulating experience, which no doubt accounts for the high frequency with which prostitutes are asked to do it. Enjoyment of fellatio is certainly within the normal range of male sexuality.

Fellatio can be performed by either heterosexuals or gay male couples.

Sixty-Nining

Fellatio and cunnilingus can be performed simultaneously by both partners. This is often called **sixty-nining**[5] because the numerals "69" suggest the position of the two bodies during simultaneous mouth-genital sex. Sixty-nining may be done either side to side or with one person on top of the other, each with the mouth on the other's genitals (Figure 10.11).

[5]If you want to be elegant and impress your friends, you can call it *soixante-neuf,* which is just "sixty-nine" in French.

> **Sixty-nining:** Simultaneous mouth-genital stimulation; also called "soixante-neuf."

Figure 10.11 Simultaneous mouth-genital stimulation in the sixty-nine position.

Simultaneous mouth-genital sex allows both people to enjoy the pleasure of that stimulation at the same time. It can give a feeling of total body involvement and total involvement between partners. Some couples, however, feel that this technique requires doing too many things at once and is more complicated than enjoyable. For example, the woman may be distracted from enjoying the marvelous clitoral stimulation she is receiving because she has to concentrate at the same time on using her mouth to give good stimulation to the penis. If sixty-nine is done in the man-on-top position, some women also feel that they have no control over the movement and that they may be choked.

Anal Intercourse

In **anal intercourse** the man inserts his penis into his partner's rectum (Morin, 1981). In legal terminology it is sometimes called sodomy (although this term may also refer to other sexual practices such as intercourse with animals), and it is sometimes referred to as having sex "Greek style." It may be done by either heterosexual couples or gay male couples.

Anal intercourse is somewhat more difficult than penis-in-vagina intercourse because the rectum has no natural lubrication and because it is surrounded by fairly tight muscles. The man should therefore begin by moistening the partner's anus, either with saliva or with a sterile surgical lubricant such as K-Y jelly (*not* Vaseline). He should also lubricate his penis. He then inserts it gently into the rectum and begins controlled pelvic thrusting. It is typically done in the rear-entry position or in the man-on-top position. The more the partner can relax, the less uncomfortable it is; if it is done properly, it need involve no pain. While some heterosexual couples find the idea repulsive, others delight in it. Some women report orgasm during anal intercourse, particularly when it is accompanied by hand stimulation of the clitoris. Men also report orgasms from anal intercourse, primarily due to stimulation of the prostate.

There are some health risks associated with anal intercourse. It can cause damage to the tissue of the rectal lining and anal sphincter. It can lead to infections with various organisms. Of greatest concern, the AIDS virus can be transmitted through

Anal intercourse: Insertion of the penis into the partner's rectum.

anal intercourse. Thus safer sex consists of either refraining from engaging in anal intercourse, or using a condom if one does (or doing it only in a monogamous relationship with an uninfected partner). Furthermore, for heterosexuals the penis should never be inserted into the vagina after anal intercourse unless it has been washed thoroughly. The reason for this is that the rectum contains bacteria that do not belong in the vagina and that can cause a dandy case of vaginitis if they happen to get there.

Another variation is **anilingus** (*feuille de rose* in French, "rimming" in slang), in which the tongue and mouth stimulate the anus. The anus may also be stimulated by the hand, and some people report that having a finger inserted into the rectum near the time of orgasm provides a heightened sexual sensation. Anilingus carries with it some risk of AIDS, hepatitis, or *E. coli* infections.

Techniques of Lesbians and Gays

Some people have difficulty imagining exactly what gays and lesbians do in bed; after all, the important ingredients for sex are one penis and one vagina, aren't they?

The preliminaries consist, as they do for heterosexuals, of kissing, hugging, and petting. Gay men engage in mutual masturbation, oral-genital sex (fellatio), and, less frequently, anal intercourse. Gay men sometimes also engage in **interfemoral intercourse,** in which one man's penis moves between the thighs of the other. Lesbians engage in mutual masturbation, oral-genital sex (cunnilingus), and a practice called **tribadism** ("dry hump"), in which heterosexual intercourse is imitated, with one partner lying on top of the other and making thrusting movements so that both receive genital stimulation. According to *The Gay Report,* this is a common practice (Jay & Young, 1979). Another practice is the use of a dildo by one person to stimulate the other.

An important point to note about these practices is that they are all behaviors in which heterosexuals also engage. That is, homosexuals do the same things sexually that heterosexuals do. The only thing that is distinctive about the homosexual act is that the partners are of the same gender.

Masters and Johnson (1979), in their laboratory studies, have done direct observations of the love-making techniques of gays and lesbians and compared them with those of straights. They found that, in masturbation techniques, there were no differences. However, in couple interactions there were some substantial differences. The major one was that homosexuals "took their time"—that is, they seemed not to have any goal orientation. Heterosexual couples, on the other hand, seemed to be performance-oriented—they seemed to strive toward a goal of orgasm for each partner. In the initial approach to stimulating the female, heterosexuals and lesbians began with holding and kissing, but this lasted only about 30 seconds for the heterosexuals, who quickly moved on to genital stimulation. Lesbians, on the other hand, spent more time in holding and kissing and then went on to a long period of breast stimulation, which sometimes resulted in orgasm in the absence of genital stimulation. This breast stimulation was a major difference between heterosexual and homosexual techniques. Lesbians also appeared to communicate more with each other. In the initial approach to stimulating the male, gays did extensive stimulation of the nipples, generally producing erection; such a technique was rare among heterosexuals (only with 3 of 100 couples). Gay men were also much more likely to stimulate the frenulum (the area of the penis on the lower side, just below the corona). They also used a "teasing technique" in which the man brings his partner near orgasm, then relaxes the stimulation, then increases the stimulation again, and so on, essentially prolonging the pleasure. Among heterosexuals, the husband's most frequent complaint was that the wife did not grasp the shaft of the penis tightly enough. Masters and Johnson argue that heterosexuals can learn from gays and lesbians; the technique of gays and lesbians benefits from stimulating another body like their own.

Anilingus (ay-nih-LING-us): Mouth stimulation of the partner's anus.
Interfemoral intercourse: A sexual technique used by gay men in which one man moves his penis between the thighs of the other.
Tribadism (TRY-bad-izm): A sexual technique used by lesbians in which one woman lies on top of another and moves rhythmically in order to produce sexual pleasure, particularly clitoral stimulation.

Aphrodisiacs

Is There a Good Aphrodisiac?

An **aphrodisiac** is a substance—such as a food, a drug, or a perfume—that excites sexual desire. Throughout history people have searched for the "surefire" aphrodisiac. Before arousing your hopes, we should note that the search has been unsuccessful; there is no known substance that works well as an aphrodisiac.

One popular idea is that oysters are an aphrodisiac. This notion appears to be a reflection of the myth that foods that resemble sexual organs have sexual powers. For example, bananas resemble the penis and have been thought to be aphrodisiacs. Another example is the Asian belief that powdered rhinoceros horn is an aphrodisiac (perhaps this is also the origin of the term "horny") (Taberner, 1985). Perhaps oysters are thought to have such powers because of their resemblance to the testes (MacDougald, 1961). Oysters, however, contain no substances that can in any way influence sexual functioning (Neiger, 1968).

Doubtless some substances gain a continued reputation as aphrodisiacs because simply believing that something will be arousing can itself be arousing. Thus the belief that a bull's testicles ("prairie oysters") or peanuts or clams have special powers may produce a temporary improvement of sexual functioning, not because of the chemicals contained in them but because of a belief in them.

Alcohol also has a reputation as an aphrodisiac. The effects of alcohol on sexual functioning will be discussed in Chapter 19. Briefly, drinking small quantities of alcohol may, for some people, decrease psychological inhibitions and therefore increase sexual desire. Moderate to large quantities, however, rapidly lead to an inability to function sexually.

Users of marijuana report that it acts as a sexual stimulant. Probably this is due, in part, to the fact that marijuana produces the sensation that time is being stretched out, thus prolonging and intensifying sensations, including sexual sensations. There is no scientific documentation of the aphrodisiac effects of marijuana except for the reports of users. Possible negative effects of marijuana on sexual functioning are discussed in Chapter 19.

Unfortunately, some of the substances that are thought to enhance sexual functioning are quite dangerous. For example, cantharides (Spanish fly) has a reputation as an aphrodisiac, but it is poisonous (Kaplan, 1974; Leavitt, 1974; Taberner, 1985).

Amyl nitrite ("poppers") is popular among some people. Because it relaxes the sphincter muscle of the anus, it is used by those engaging in anal intercourse (Taberner, 1985). Users report that it produces heightened sensations during orgasms (Everett, 1975). Probably it acts by dilating the blood vessels in the genitals. It may, however, have side effects, including dizziness, headaches, fainting, and, in rare cases, death; thus it can be dangerous (Taberner, 1985).

Butyl nitrite—sold under such trade names as Rush, Locker Room, and Climax—is a chemical relative of amyl nitrite. It is used to heighten sexual pleasure. Although no deaths have been reported from inhaling it, there are two reported deaths from swallowing it (UPI, 1981).

In contrast to these "street" drugs, some prescription drugs are being tested as potential treatments for some sexual disorders. So far, no single medication has been found to be a safe and effective aphrodisiac (Rosen & Ashton, 1993).

Anaphrodisiacs

Just as people have searched for aphrodisiacs, so they have sought **anaphrodisiacs,** substances or practices that would diminish sexual desire. Cold showers are reputed to have such effects, as is potassium nitrate (saltpeter). The latter contains nothing that decreases sexual drive, but it does act as a diuretic; it makes the person want to urinate frequently, which may be distracting enough so that he or she is not much interested in sex.

There has been some medical interest in finding drugs that would decrease sex drive for use in treating aggressive sexual offenders. One such drug is cyproterone acetate, which is an antiandrogen (Taberner, 1985). Within about two weeks of use, it may cause a reduction of libido; in some men, the sperm count declines substantially and testosterone levels fall to 20 to 30 percent of normal.

Aphrodisiac (ah-froh-DIZ-ih-ak): A substance that increases sexual desire.

Anaphrodisiac (an-ah-froh-DIZ-ih-ak): A substance that decreases sexual desire.

Other drugs that may lead to a loss of sexual functioning are discussed in Chapter 19.

Should Intercourse and Orgasm Be the Goal?

Our culture has traditionally held the belief that a sexual encounter should "climax" with intercourse and orgasm, at least orgasm for the man. In our modern era of multiple orgasms for women and general sexual liberation, the view that intercourse is the important part of sex and that orgasm is the goal toward which both partners must strive is pervasive. This belief system is reflected in the term "foreplay," which implies that activities like hand stimulation of the genitals, kissing, and mouth-genital sex are only preliminaries that take place before intercourse, the latter being "real sex." Similar beliefs are reflected in a commonly used phrase, "achieving orgasm," as if orgasm were something to be achieved like a promotion on the job.[6]

Psychologist Rollo May feels that men particularly, by concentrating on "achieving" orgasm and *satisfying* their desire, miss out on the more important part of the sexual experience: prolonging the feeling of desire and pleasure, building it higher and higher. As he puts it:

> The pleasure in sex is described by Freud and others as the reduction of tension [orgasm]; in eros, on the contrary, we wish not to be released from the excitement but rather to hang on to it, to bask in it, and even to increase it. (1974, pp. 71–72)

Marc Feigen Fasteau, a men's liberation leader, argues that although orgasm is good, a large part of pleasure is the buildup to orgasm:

> What the masculine disdain for feeling makes it hard for men to grasp is that the state of desire . . . is one of the best, perhaps *the* best, part of the experience of love. (1974, p. 31)

Another one of the "goals" of sex that has emerged recently is the simultaneous orgasm. Some people

consider this an event to be worked for rather than a pleasant thing that sometimes happens.

The legacy of the Protestant ethic in our culture is that our achievement drives now seem to be channeled into our sexual behaviors. There is nothing intrinsically wrong with expressing achievement drives in sex, except that any time there is an achievement situation, there is also the potential for a failure. If she does not have an orgasm or if he cannot get an erection, the couple feel as if the whole experience was a disaster. The problem with setting up sexual goals, then, is that the possibility of sexual failures or sexual disorders is also being set up.

The best approach is to enjoy all the various aspects of lovemaking for themselves, rather than as techniques for achieving something, and to concentrate on sex as a feast of the senses, rather than as an achievement competition. We need to broaden our view of sexual expression to recognize that:

> a broad continuum of activities may provide sexual pleasure—a dream, a thought, a conversation, cuddling, kissing, sensual massage, dancing, oral-genital stimulation and intercourse. (Chalker, 1995)

In short, we need to view the entire body as erogenous and shift our focus to *outercourse*.

From Inexperience to Boredom

Some people, after an initial lack of experience with sex, shift rather quickly to becoming bored with it, with perhaps only a brief span of self-confident, pleasurable sexuality in between. Most of us, of course, are sexually inexperienced early in our lives, and most of us feel bored with the way we are having sex at times. How do we deal with these problems?

Sexual Inexperience

In our culture we expect men to be "worldly" about sex—to have had experience with it and to be skillful in the use of sexual techniques. A man or a boy who is sexually inexperienced (perhaps a virgin) or who has had only a few sexual experiences, with little opportunity to practice, may have a real fear about whether he will be able to "perform" (the achievement ethic again) in a sexual encounter.

---

[6]To avoid this whole notion, we never use the phrase "to achieve orgasm" in this book. Instead, we prefer "to have an orgasm" or simply "to orgasm." Why not turn it into a verb so that we will not feel we have to work at achieving it?

With the sexual revolution has come an increasing expectation that women also should have a bag of sexual tricks ready to use, and so they, too, are increasingly expected to be experienced.

How can one deal with this problem of inexperience? First, it is important to question society's assumption that one should be experienced. Everyone has to begin sometime, and there is absolutely nothing wrong with inexperience. Second, there are many good books and articles on sexual techniques that are definitely worth reading, although it is important to be selective, since a few of these may be more harmful than helpful. This chapter should be a good introduction; you also might want to consult *The New Joy of Sex* or any of the self-help manuals listed at the end of Chapter 19. Do not become slavishly attached to the techniques you read about in books, though; they should serve basically as a stimulus to your imagination, not as a series of steps that must be followed. Third, communicate with your partner. Because individual preferences vary so much, no one, no matter how experienced, is ever a sexual expert with a new partner. The best way to please a partner is to find out what that person likes, and communication may accomplish this better than prior experience. A later section in this chapter gives some specific tips on communication. Interestingly, one study found that the best sexual predictor of relationship satisfaction was not frequency of sex or techniques but mutual agreement on sexual issues (Markman, personal communication).

Boredom

The opposite problem to inexperience is the feeling of boredom in a long-term sexual relationship. Boredom, of course, is not always a necessary consequence of having sex with the same person over a long period of time. Certainly there are couples who have been married for 40 or 50 years and who continue to find sexual expression exciting. Unfortunately, the major sex surveys have not inquired about the phenomenon of boredom, so it is not possible to estimate the percentage of people who eventually become bored or who experience occasional fits of boredom. However, such experiences are surely common. As someone once said, a rut is no place to be making love. How can we deal with the problem of boredom?

Communication can help in this situation, as it can in others. Couples sometimes evolve a routine sexual sequence that leads to boredom, and sometimes that sequence is not really what either person wants. By communicating to each other what they really would like to do and then doing it, two people can introduce some variety into their relationship. The various love manuals can also give ideas on new techniques. Finally, a couple's sexual relationship often mirrors the other aspects of their relationship, and sexual boredom may sometimes mean that they are generally bored with each other. Rejuvenating the rest of the relationship—perhaps taking up a hobby or a sport together or going on a good vacation, during which they really try to build their relationship in general—may do wonders for their sexual relationship.

One might also question the meaning of "boredom." Perhaps our expectations for sexual experience are too high. Encouraged by the media, we tend to believe that every time we have intercourse, the earth should move. We do not expect that every meal we eat will be fantastic or that we will always have a huge appetite and enjoy every bite. Yet we do tend to have such expectations with regard to sexuality. Perhaps when boredom seems to be a problem, it is not the real issue; rather, the problem may be unrealistically high expectations.

Communication and Sex

Consider the following situation:

> Sam and Donna have been married for about three years. Donna had had intercourse with only one other person before Sam, and she had never masturbated. Since they've been married, she has had orgasms only twice during intercourse, despite the fact that they make love three or four times per week. She has been reading some magazine articles about female sexuality, and she is beginning to think that she should be experiencing more sexual satisfaction. As far as she knows, Sam is unaware that there is any problem. Donna feels lonely and a bit sad.

What should Donna do? She needs to communicate with Sam. They apparently have not communicated much about sex in the last three years, and they need to begin. The following sections will

Focus 10.1
Gender Differences in Communication

Linguist Deborah Tannen, author of best-selling books such as *You Just Don't Understand: Women and Men in Conversation,* believes that women and men have radically different verbal communication styles, so different that they essentially belong to different linguistic communities. Communication between women and men, according to this point of view, is as difficult as cross-cultural communication. These arguments have captured the imagination of the general public and worked their way into corporate training programs. Does the scientific evidence support Tannen's claims? Are there substantial gender differences in communication styles and if so, what are the implications for sexual interactions?

A number of gender differences in communication have been found repeatedly in research. Women are more skilled at reading nonverbal cues than men are. Women are more likely than men to inquire about upsetting situations that another person is in, and to use comforting messages that acknowledge and legitimize the feelings of others. In same-gender pairs, men are more likely to discuss sports, careers, and politics, whereas women are more likely to talk about feelings and relationships. Men interrupt more than women do.

One research finding is that, in conversation, women are more self-disclosing than men are. That is, women reveal more personal, intimate information about themselves. Yet this pattern is found only with same-gender conversational pairs—that is, men talking with men, and women talking with women. Men, when talking with a woman, disclose far more than when they talk with a man. In one study, women and men were brought to the laboratory to have a conversation with their best friend of the same gender. They were told to discuss something important and to reveal their thoughts and feelings. No gender differences in self-disclosure occurred. Gender differences in self-disclosure, then, are far from universal, and men are capable of being as self-disclosing as women are.

One claim is that women and men have different goals when they speak. Women use speech to establish and maintain relationships, whereas men use speech to exert control, preserve independence, and enhance their status (Wood, 1994). This pattern is consistent with research findings that indicate that women are more concerned than men are with the quality of the relationship in which sex occurs (see Chapter 14) and that some men use sexual assault to exert power and control over women (see Chapter 17).

A review of dozens of studies of gender differences in communication, however, indicates that the differences, overall, are small (Canary & Hause, 1993; Canary & Dindia, 1998). The contention that gender differences in communication are so large that it is as if women and men are from different cultures is simply not supported. Another problem with the "two cultures" approach is that it assumes that patterns of gender differences are the same for all ethnic groups and social classes, when almost all the research has been done with middle-class whites.

What are the implications for sexuality? We should not be led astray by flashy claims that men and women have totally different communication styles, making it difficult at best, impossible at worst, to communicate. Gender differences in communication are small. That is a happy result for sexuality, and particularly for heterosexual interactions. As you have seen in this chapter, good communication is essential for satisfying, mutually pleasurable sex. If men and women could not communicate, it would be a serious problem. Fortunately, the gender differences are small and, with a little effort, heterosexual couples should be able to engage in clear, accurate sexual communication.

Sources: Aries (1996); Canary & Dindia (1998); Canary & Hause (1993); Tannen (1991)

discuss the relationship between sex, communication, and relationships and provide some suggestions on how to communicate effectively.

Communication and Relationships

A good deal of research has looked at differences in communication patterns between nondistressed (happy) married couples and distressed (unhappy, seeking marital counseling) married couples. This research shows, in general, that distressed couples tend to have communication deficits (Markman & Floyd, 1980; Noller, 1984; Gottman, 1994). Research also shows that couples seeking therapy for sex problems have poor communication patterns compared with nondistressed couples (Zimmer, 1983). Of course, there are many other factors that contribute to marital or relationship conflict or sex problems, but poor communication patterns are certainly among them. The problem with this research is that it is correlational (see Chapter 3 for a discussion of this problem in research methods); in particular, we cannot tell whether poor communication causes unhappy marriages or whether unhappy marriages create poor communication patterns.

An elegant longitudinal study designed to meet this problem provides evidence that unrewarding, ineffective communication precedes and predicts later relationship problems (Markman, 1979; 1981). Dating couples who were planning marriage were studied for 5½ years. The more positively couples rated their communication interactions at the beginning of the study, the more satisfaction they reported in their relationship when they were followed up 2½ years later and 5½ years later.

On the basis of this notion that communication deficits cause communication problems, marriage counselors and marital therapists often work on teaching couples communication skills. Recent research suggests that distressed couples do not differ from nondistressed couples in their communication skills or ability; rather, some distressed couples use their skills as weapons, to send negative messages (Burleson & Denton, 1997). These results suggest that therapists should focus on the intent of the partners as they communicate with each other, not just on techniques.

But what are these negative messages? Gottman (1994) used audiotape, videotape, and monitoring of physiological arousal to answer this question.

He identified four destructive patterns of interaction: criticism, contempt, defensiveness, and withdrawal. *Criticism* refers to attacking a partner's personality or character: "You are so selfish; you never think of anyone else." *Contempt* is intentionally insulting or orally abusing the other person: "How did I get hooked up with such a loser?" *Defensiveness* refers to denying responsibility, making excuses, replying with a complaint of one's own, and other self-protective responses, instead of addressing the problem. The meaning of *withdrawal* is probably obvious, whether it is responding to the partner's complaint with silence, turning on the TV, or walking out of the room in anger. You can probably see that these types of communication are likely to lead to an escalation of the hostility rather than a solution to the problem.

It is clear that positive communication is important in developing and maintaining intimate relationships. The sections that follow will describe some of the skills that are involved in positive communication. These ideas arise from extensive research comparing distressed and nondistressed couples, and from the experiences of therapists.

Self-Disclosure

One of the keys to building a good relationship is self-disclosure. **Self-disclosure** involves telling your partner some personal things about yourself. It may range from telling your partner about something embarrassing that happened to you at work today, to disclosing a very meaningful event that happened between you and your parents 15 years ago.

Self-disclosure is closely related to satisfaction with the relationship. Research shows that there is a positive correlation between the extent of a couple's self-disclosure and their satisfaction with the relationship—that is, couples that practice more self-disclosure are more satisfied (Hendrick, 1981). More specifically, self-disclosure of sexual likes and dislikes is associated with sexual satisfaction (Purnine & Carey, 1997).

Research consistently shows that self-disclosure leads to reciprocity (Berg & Derlega, 1987; Hendrick & Hendrick, 1992). That is, if one member of

Self-disclosure: Telling personal things about yourself.

the couple self-discloses, it seems to prompt the other partner to self-disclose also. Self-disclosure by one member of the couple can essentially get the ball rolling. Psychologists have proposed a number of reasons why this occurs (Hendrick & Hendrick, 1992). First, disclosure by our partner may make us like and trust that person more. Second, as social learning theorists would argue, simple modeling and imitation may occur. That is, one partner's self-disclosing serves as a model for the other partner. Norms of equity may also be involved (see Chapter 12 for a discussion of equity theory). After one partner has self-disclosed, the other person may follow suit in order to maintain a sense of balance or equity in the relationship.

Not only does self-disclosure seem to help a relationship progress, but patterns of self-disclosure can actually predict whether a couple stays together or breaks up. Research in which couples are followed for periods ranging from two months to four years shows that the greater the self-disclosure, the greater the likelihood that the relationship will continue, and the less the self-disclosure, the greater the likelihood of breakup (Hendrick et al., 1988; Sprecher, 1987).

Being an Effective Communicator

Back to Donna and Sam: One of the first things to do in a situation like Donna's is to decide to talk to one's partner, admitting that there is a problem. Then the issue is to resolve to communicate, and particularly to be an *effective* communicator. Suppose Donna begins by saying,

> You're not giving me any orgasms when we have sex. (1)

Sam gets angry and walks away. Donna meant to communicate that she wasn't having any orgasms, but Sam thought she meant that he was a lousy lover. It is important to recognize the distinction between **intent** and **impact** in communicating (Gottman et al., 1976; Purnine & Carey, 1997). Intent is what you mean. Impact is what the other person thinks you mean. A good communicator is one whose impact matches her or his intent. Donna wasn't an **effective communicator** in the above example because the impact on Sam was considerably different from her intent. Notice that effectiveness does not depend on the content of the message. One can be as effective at communicating contempt as communicating praise.

Many people value spontaneity in sex, and this attitude may extend to communicating about sex. It is best to recognize that to be an effective communicator, it may be necessary to plan your strategy. It often takes some thinking to figure out how to make sure that your impact matches your intent. Planning also allows you to make sure that the timing is good—that you are not speaking out of anger, or that your partner is not tired or preoccupied with other things.

Finally, we ought to recognize that it is going to be harder for Donna to broach this subject than it would be to ask Sam why he didn't take out the garbage as he had promised. It is hard for most people to talk about sex, particularly sexual problems, with their partners.[7] One woman describes her inability to talk to her partner:

> There was a period when I was furious with him all the time. Every time he touched me I was angry. I didn't tell him about it [her desire for greater gentleness during sex], because I thought it was something I had to deal with. (Maurer, 1994, p. 448)

Ironically, in the last few decades public communication about sex has become relatively open, but private communication remains difficult (Crawford et al., 1994). That doesn't mean that Donna can't communicate. But she shouldn't feel guilty or stupid because it is difficult for her. And she will be better off if she uses some specific communication skills and has some belief that they will work. The sections that follow suggest some skills that are useful in being an effective communicator and how to apply these to sexual relationships.

---

[7]In fact, a survey of students in human sexuality courses at two universities indicated that sexual "pleases" and "displeases" are the most difficult topics to talk about with one's partner. Furthermore, women seem to be more aware of problems in this aspect of the relationship than men are (Markman, personal communication).

Intent: What the speaker means.
Impact: What someone else understands the speaker to mean.
Effective communicator: A communicator whose impact matches his or her intent.

Focus 10.2

A Personal Growth Exercise: Getting to Know Your Own Body

Most experts on sexual communication agree that before you can begin to communicate your sexual needs to your partner, you must get to know your own body and its sexual responsiveness. This exercise is designed to help you do that. Set aside some time for yourself, preferably 30 minutes or more. You'll need privacy and a mirror, preferably a full-length mirror.

1. Undress and stand in front of the mirror. Relax your body completely.

2. Take a good look at your body, top to bottom. Look at the colors, the curves, the textures. Take your time doing this. Try to discover things you haven't noticed before. What pleases you about your body? What don't you like about your body? Can you say these things aloud?

3. Look at your body. What parts of it influence how you feel about yourself sexually?

4. Run your fingers slowly over your body, head to toe. How does it feel to you? Are some parts soft? Are some sensitive? Are you hurrying over some places? Why? How do you feel about doing this?

5. Explore your genitals. *If you're a man,* look at them. Do you like the way they look? Now explore your genitals with your fingers. Gently stroke your penis, scrotum, and the area behind the scrotum. Pay close attention to the various sensations you're producing. Which areas feel particularly good when they're stroked? Try different kinds of touching—light, hard, fast, slow. Which kind feels best? If you get an erection, that's OK. Just take your time and learn as much as you can. Are there differences in sensitivities between the aroused state and the unaroused state? *If you're a woman,* take a hand mirror and look at your genitals. Do you like the way they look? Now explore your genitals with your fingers. Touch your outer lips, inner lips, clitoris, vaginal entrance. Which areas feel particularly good? Try different kinds of touching—light, hard, fast, slow. Which kind feels best? If you get aroused, that's OK. Are there differences in sensitivities between the aroused state and the unaroused state? Just take your time and learn as much as you can.

6. Now you're ready to communicate some new information to your partner!

For more exercises like this, see Zilbergeld's *The New Male Sexuality* (1992) and Heiman, LoPiccolo, and LoPiccolo's *Becoming Orgasmic: A Sexual Growth Program for Women* (1976).

Sources: Myron Brenton. (1972). *Sex talk.* New York: Stein & Day. Julia Heiman, Leslie LoPiccolo, and Joseph LoPiccolo. (1976). *Becoming orgasmic: A sexual growth program for women.* Englewood Cliffs, NJ: Prentice-Hall. Bernie Zilbergeld. (1992). *The new male sexuality.* New York: Bantam Books.

Figure 10.12

Good Messages

Every couple has problems. The best way to voice them is to complain rather than to criticize (Gottman, 1994). Complaining involves the use of **"I" language** (e.g., Brenton, 1972). That is, speak for yourself, not your partner (Miller et al., 1975). By doing this you focus on what you know best—your own thoughts and feelings. "I" language is less likely to make your partner defensive. If Donna were to use this technique, she might say,

> I feel a bit unhappy because I don't have orgasms very often when we make love. (2)

Notice that she focuses specifically on herself. There is less cause for Sam to get angry than there was in message 1.

One of the best things about "I" language is that it avoids mind reading (Gottman et al., 1976). Suppose Donna says,

> I know you think women aren't much interested in sex, but I really wish I had more orgasms (3)

She is engaging in **mind reading.** That is, she is making certain assumptions about what Sam is thinking. She assumes that Sam believes women aren't interested in sex or having orgasms. Research shows that mind reading is more common among distressed couples than among nondistressed couples (Gottman et al., 1977). Worse, Donna doesn't *check out* her assumptions with Sam. The problem is that she may be wrong, and Sam may not think that at all. "I" language helps Donna avoid this by focusing on herself and what

she feels rather than on what Sam is doing or failing to do. Another important way to avoid mind reading is by giving and receiving feedback, a technique to be discussed in a later section.

Documenting is another important component of giving good messages (Brenton, 1972). In documenting you give specific examples of the issue. This is not quite so relevant in Donna's case, because she is talking about a general problem; but even here, specific documenting can be helpful. Once Donna has broached the subject, she might say,

> Last night when we made love, I enjoyed it and felt very aroused, but then I didn't have an orgasm, and I felt disappointed. (4)

Now she has gotten her general complaint down to a specific situation that Sam can remember.

Suppose further that Donna has some idea of what Sam would need to do to bring her to orgasm: he would have to do more hand stimulation of her clitoris. Then she might do specific documenting as follows:

"I" language: Speaking for yourself, using the word "I"; not mind reading.
Mind reading: Making assumptions about what your partner thinks or feels.
Documenting: Giving specific examples of the issue being discussed.

Last night when we made love, I enjoyed it, but I didn't have an orgasm, and then I felt disappointed. I think what I needed was for you to stimulate my clitoris with your hand a bit more. You did it for a while, but it seemed so brief. I think if you had kept doing it for two or three minutes more, I would have had an orgasm.   (5)

Now she has not only documented to Sam exactly what the problem was, but she has given a specific suggestion about what could have been done about it, and therefore what could be done in the future.

Another technique in giving good messages is to offer *limited choices* (Langer & Dweck, 1973). Suppose Donna begins by saying,

I've been having trouble with orgasms. Could we discuss it?   (6)

The trouble with this approach is that a "no" from Sam is not really an acceptable answer to her because she definitely wants to discuss the problem. Yet she set up the question so that he could answer by saying "no." To use the technique of limited choices she might say,

I've been having trouble with orgasms when we make love. Would you like to discuss it now, or would you rather wait until tomorrow night?   (7)

Now, either answer he gives will be acceptable to her; she has offered a set of acceptable limited choices.[8] She has also shown some consideration for him by recognizing that he might not be in the mood for such a discussion now and would rather wait.

Leveling and Editing

Leveling means telling your partner what you are feeling by stating your thoughts clearly, simply, and honestly (Gottman et al., 1976). This is often the hardest step in communication, especially when the topic is sex. It is especially difficult for adults to reach shared understandings about sex, since there is great secrecy about it in our society (Craw-

ford et al., 1994). In leveling, keep in mind that the purposes are

1. To make communication clear
2. To clear up what partners expect of each other
3. To clear up what is pleasant and what is unpleasant
4. To clear up what is relevant and what is irrelevant
5. To notice things that draw you closer or push you apart (Gottman et al., 1976)

When you begin to level with your partner, you also need to do some editing. **Editing** involves censoring (not saying) things that would be deliberately hurtful to your partner or that would be irrelevant. You must take responsibility for making your communication polite and considerate. Leveling, then, should not mean a "no holds barred" approach. Ironically, research indicates that married people are ruder to each other than they are to strangers (Gottman et al., 1976).

Donna may be so disgruntled about her lack of orgasms that she's thinking of having an affair to jolt Sam into recognizing her problem, or perhaps in order to see if another man would stimulate her to orgasm. Donna is probably best advised to edit out this line of thought and concentrate on the specific problem: her lack of orgasms. If she and Sam can solve that, she won't need to have the affair anyway.

The trick is to balance leveling and editing. If you edit too much, you may not level at all, and there will be no communication. If you level too much and don't edit, the communication will fail because your partner will respond negatively, and things may get worse rather than better.

Listening

Up to this point, we have been concentrating on techniques for you to use in sending messages about sexual relationships. But, of course, communication is a two-way street, and you and your partner will exchange responses. It is therefore im-

---

[8]The technique of limited choices is useful in a number of other situations, including dealing with children. For example, when my daughter was a 2-year-old and she had finished watching *Sesame Street* and I wanted the TV turned off, I didn't say, "Would you turn the TV off?" (she might say "no") but, rather, "Do you want to turn off the TV, or would you like me to?" Of course, sometimes she evaded my efforts and said "no" anyway, but most of the time it worked.

> **Leveling:** Telling your partner what you are feeling by stating your thoughts clearly, simply, and honestly.
> **Editing:** Censoring or not saying things that would be deliberately hurtful to your partner or that are irrelevant.

portant for you and your partner to gain some skills in listening and responding constructively to messages. The following discussion will suggest such techniques.

One of the most important things is that you must really *listen*. That means more than just removing the headphones from your ears. It means actively trying to understand what the other person is saying. Often people are so busy trying to think of their next response that they hardly hear what the other person is saying. Good listening also involves positive nonverbal behaviors, such as maintaining eye contact with the speaker and nodding one's head when appropriate. Be a *nondefensive listener;* focus on what your partner is saying and feeling, and don't immediately become defensive, or counterattack with complaints of your own.

The next step, after you have listened carefully and nondefensively, is to give *feedback*. This often involves brief vocalizations—"Uh-huh," "Okay"—nodding your head, or facial movements that indicate you are listening (Gottman et al., 1998). It may involve the technique of **paraphrasing,** that is, repeating in your own words what you think your partner meant. Suppose, in response to Donna's initial statement, "You're not giving me any orgasms when we have sex," Sam hadn't walked away angrily. Instead, he tried to listen and then gave her feedback by paraphrasing. He might have responded,

> I hear you saying that I'm not very skillful at making love to you, and therefore you're not having orgasms. (8)

At that point, Donna would have had a chance to clear up the confusion she had created with her initial message, because Sam had given her feedback by paraphrasing his understanding of what she said. At that point she could have said, "No, I think you're a good lover, but I'm not having any orgasms, and I don't know why. I thought maybe we could figure it out together." Or perhaps she could have said, "No, I think you're a good lover. I just wish you'd do more of some of the things you do, like rubbing my clitoris."

It's also a good idea to *ask for feedback* from your partner, particularly if you're not sure whether you're communicating clearly.

Just as it is important to be a good listener to your partner's verbal messages, so too is it impor-

tant to be good at "reading" your partner's nonverbal messages. There's nothing quite so frustrating as turning your back on another person in order to express anger and having the person miss the signal. In technical language, the ability to comprehend another's nonverbal signals is termed "decoding" nonverbal communication; the opposite process, sending the nonverbal message, is called "encoding." Some research shows that it is the wife's skill in this area that is most important (Sabatelli et al., 1982). Wives who are good encoders (their signals are easy to understand) have husbands with few marital complaints. Other research, though, shows that the husband's ability to decode his wife's nonverbal messages correlates more with the degree of marital happiness than the wife's ability to decode his messages (Gottman & Porterfield, 1981). Probably the best scenario is when both partners are good at encoding and decoding and are committed to such accurate and sensitive nonverbal communication.

Validating

Another good technique in communication is **validation** (Gottman et al., 1976), which means telling your partner that, given his or her point of view, you can see why he or she thinks a certain way. It doesn't mean that you agree with your partner or that you're giving in. It simply means that you recognize your partner's point of view as legitimate, given his or her set of assumptions, which may be different from yours.

Suppose that Donna and Sam have gotten into an argument about cunnilingus. She wants him to do it and thinks it would bring her to orgasm. He doesn't want to do it because he finds the idea repulsive and because he believes no real man would do such a thing. If Donna tried to validate Sam's feelings, she might say,

> I can understand the way you feel about cunnilingus, especially given the way you were brought up to think about sex. (9)

Paraphrasing: Saying, in your own words, what you thought your partner meant.
Validation: Telling your partner that, given his or her point of view, you can see why he or she thinks a certain way.

Sam might validate Donna's feelings by saying,

> I understand how important it is for you to have an
> orgasm.   (10)

Validating hasn't solved their disagreement, but it has left the door open so that they can now make some progress.

Drawing Your Partner Out

Suppose it is Sam who initiates the conversation rather than Donna. Sam has noticed that Donna doesn't seem to get a lot of pleasure out of sex, and he would like to find out why and see what they can do about it. He needs to draw her out. He might begin by saying,

> I've noticed lately that you don't seem to be enjoy-
> ing sex as much as you used to. Am I right about
> that?   (11)

That much is good because he's checking out his assumption. Unfortunately, he's asked a question that leads to a "yes" or "no" answer, and that can stop the communication. So, if Donna replies "yes," Sam had better follow it up with an *open-ended* question like

> Why do you think you aren't enjoying it more?   (12)

If she can give a reasonable answer, good communication should be on the way. One of the standard—and best—questions to ask in a situation like this is

> What can we do to make things better?   (13)

Body Talk: Nonverbal Communication

Often the precise words we use are not so important as our **nonverbal communication**—the way we say them. Tone of voice, expression on the face, position of the body, whether you touch the other person—all are important in conveying the message.

As an example, take the sentence "So you're here." If it is delivered "So *you're* here" in a hostile tone of voice, the message is that the speaker is very unhappy that you're here. If it is delivered "So you're *here*" in a pleased voice, the meaning may be that the speaker is glad and surprised to see you here in Wisconsin, having thought you were in Europe. "So you're here" with a smile and arms outstretched to initiate a hug might mean that the speaker has been waiting for you and is delighted to see you.

Suppose that in Donna and Sam's case, the reason Donna doesn't have more orgasms is that Sam simply doesn't stimulate her vigorously enough. During sex, Donna has adopted a very passive, nearly rigid posture for her body. Sam doesn't stimulate her more vigorously because he is afraid that he might hurt her, and he is sure that no lady like his wife would want such a vigorous approach. The response (or rather nonresponse) of her body confirms his assumptions. Her body is saying "I don't enjoy this. Let's get it over with." And that's exactly what she's getting. To correct this situation, she might adopt a more active, encouraging approach. She might take his hand and guide it to her clitoris, showing him how firmly she likes to have it rubbed. She might place her hands on his hips and press to indicate how deep and forceful she would like the thrusting of his penis in her vagina to be. She might even take the daring approach of using some verbal communication and saying "That's good" when he becomes more vigorous.

The point is that in communicating about sex, we need to be sure that our nonverbal signals help to create the impact we intend rather than one we don't intend. It is also possible that nonverbal signals are confusing communication and need to be straightened out. "Checking out" is a technique for doing this that will be discussed in a later section.

Interestingly, research shows that distressed couples differ from nondistressed couples more in their nonverbal communication than in their verbal communication (Gottman et al., 1977; Vincent et al., 1979). For example, even when a person from a distressed couple is expressing agreement with his or her spouse, that person is more likely to accompany the verbal expressions of agreement with negative nonverbal behavior. Distressed couples are also more likely to be negative listeners—while listening, the individuals are more likely to display frowning, angry, or disgusted facial expressions, or tense or inattentive body postures. Contempt is often expressed nonverbally, by sneering or rolling the eyes, for example. In contrast, harmonious marriages are characterized by closer physical dis-

> **Nonverbal communication:** Communication not through words, but through the body, e.g., eye contact, tone of voice, touching.

(a)

(b)

Figure 10.13   (a) A couple with good body language (good eye contact and body position); (b) A couple with poor body language (poor eye contact and body position).

tances and more relaxed postures than are found in distressed couples (Beier & Sternberg, 1977). Once again, it is not only what we say verbally but how we say it, and how we listen, that makes the difference.

Accentuate the Positive

We have been concentrating on negative communications, that is, communications where some problem or complaint needs to be voiced. It is also important to communicate positive things about sex (Miller et al., 1975). If that was a great episode of lovemaking, or the best kiss you've ever experienced, say so. A learning theorist would say that you're giving your partner some positive reinforcement. Social psychologists' research shows that we tend to like people better who give us positive reinforcements (see Chapter 13). Recognition of the strengths in a relationship offers the potential for enriching it (e.g., Miller et al., 1975; Otto, 1963). And if you make a habit of positive communications about sex, it will be easier to initiate the negative ones, and they will be better received.

Most communication during sex is limited to muffled groans, or "Mm-m's," or an occasional "Higher, José" or "Did you, Latisha?" It might help your partner greatly if you gave frequent verbal and nonverbal feedback such as "That was great" or "Let's do that again." That would make the pos-

itive communications and the negative ones far easier.

Research shows that nondistressed couples make more positive and fewer negative communications than distressed couples (Birchler et al., 1975; Billings, 1979). In fact, Gottman's (1994) research found that there is a *magic ratio* of positive to negative communication. In stable marriages, there is five times as much positive interaction—verbal and nonverbal, including hugs and kisses, as there is negative. Not only do happy couples make more positive communications; they are more likely to respond to a negative communication with something positive (Billings, 1979). Distressed couples, on the other hand, are more likely to respond to negative communication with more negative communication, escalating into conflict. We might all take a cue from the happy couples and make efforts not only to increase our positive communications but even to make them in response to negative comments from our partner.

Fighting Fair

Even if you use all the techniques described above, you may still get into arguments with your partner. Arguments are a natural part of a relationship and are not necessarily bad. Given that there will be arguments in a relationship, it is useful if you and your partner have agreed to a set of rules called

Figure 10.14   Arguments are not necessarily bad for a relationship, but it is important to observe the rules for "fighting fair."

fighting fair (Bach & Wyden, 1969) so that the arguments may help and won't hurt.

Here are some of the basic rules for fighting fair that may be useful to you (Brenton, 1972; Creighton, 1992).

1. Don't make sarcastic or insulting remarks about your partner's sexual adequacy. It generates resentment, opens you to counterattack, and is just a dirty way to fight.

2. Don't bring up the names of former spouses, lovers, boyfriends, or girlfriends to illustrate how all these problems didn't happen with them. Stick to the issue: your relationship with your partner.

3. Don't play amateur psychologist. Don't say things like "The problem is that you're a compulsive personality" or "You acted that way because you never resolved your Oedipus complex." You really don't have the qualifications (even after reading this book) to do that kind of psychologizing. Even if you did, your partner would not be apt to recognize your expertise in the middle of an argument, thinking, quite rightly, that you're probably biased at the moment.

4. Don't threaten to tell your parents or run home. This involves ganging up on your partner or retreating like a child.

5. If you have children, don't bring them into the argument. It is too stressful emotionally to force them to take sides between you and your spouse.

6. Don't engage in dumping. Don't store up gripes for six months and then dump them on your partner all at one time.

7. Don't hit and run. Don't bring up a serious negative issue when there is no opportunity to continue the discussion, such as when you're on the way out the door going to work or when guests are coming for dinner in five minutes.

8. Don't focus on who's to blame. Focus on solutions, not on who's at fault. If you avoid blaming, it lets both you and your partner save face, which helps both of you feel better about the relationship.

Checking Out Sexy Signals

One of the problems with verbal and nonverbal sexual communications is that they are often ambiguous. This problem may occur more often with couples who don't know each other well, but it can cause uncertainty and misunderstanding in married couples as well.

Some messages are very direct. Statements like "I want to have sex with you" are not ambiguous at

---

Fighting fair: A set of rules designed to make arguments constructive rather than destructive.

Focus 10.3

How Solid Is Your Relationship?

Good communication enhances a relationship, and a good relationship facilitates good communication. There are several components of a good relationship. Two of these components are love and respect. The following self-test assesses the degree of love and respect in a relationship. If you are in an intimate relationship, answer yes or no to each of the following statements. If you agree or mostly agree, answer yes. If you disagree or mostly disagree, answer no. You can either ask your partner to take the test too, or you can take it a second time yourself, answering the way you think your partner would answer.

1. My partner seeks out my opinion.
   YOU: Yes   No   YOUR PARTNER: Yes   No

2. My partner cares about my feelings.
   YOU: Yes   No   YOUR PARTNER: Yes   No

3. I don't feel ignored very often.
   YOU: Yes   No   YOUR PARTNER: Yes   No

4. We touch each other a lot.
   YOU: Yes   No   YOUR PARTNER: Yes   No

5. We listen to each other.
   YOU: Yes   No   YOUR PARTNER: Yes   No

6. We respect each other's ideas.
   YOU: Yes   No   YOUR PARTNER: Yes   No

7. We are affectionate toward one another.
   YOU: Yes   No   YOUR PARTNER: Yes   No

8. I feel my partner takes good care of me.
   YOU: Yes   No   YOUR PARTNER: Yes   No

all. Unfortunately, such directness is not common in our society. In a series of studies of tactics people used to promote sexual encounters, college students reported good hygiene, good grooming, and dressing nicely as the actions they most frequently used (Greer & Buss, 1994). These are *very* indirect signals of sexual interest. Consider George, who stands up, stretches, and says "It's time for bed." Does he mean he wants to engage in sexual activity or to go to sleep?

Ambiguous messages can lead to feelings of hurt and rejection, or to unnecessary anger and perhaps complaints to third parties. If George wants to have sex but his partner interprets his behavior as meaning that George is tired, George may go to bed feeling hurt, unattractive, and unloved. A woman who casually puts her arm around the shoulders of a coworker and gives him a hug may find herself explaining to her supervisor that it was a gesture of friendship, not a sexual proposition.

9. What I say counts.
   YOU: Yes   No   YOUR PARTNER: Yes   No
10. I am important in our decisions.
    YOU: Yes   No   YOUR PARTNER: Yes   No
11. There's lots of love in our relationship.
    YOU: Yes   No   YOUR PARTNER: Yes   No
12. We are genuinely interested in one another.
    YOU: Yes   No   YOUR PARTNER: Yes   No
13. I love spending time with my partner.
    YOU: Yes   No   YOUR PARTNER: Yes   No
14. We are very good friends.
    YOU: Yes   No   YOUR PARTNER: Yes   No
15. Even during rough times, we can be empathetic.
    YOU: Yes   No   YOUR PARTNER: Yes   No
16. My partner is considerate of my viewpoint.
    YOU: Yes   No   YOUR PARTNER: Yes   No
17. My partner finds me physically attractive.
    YOU: Yes   No   YOUR PARTNER: Yes   No
18. My partner expresses warmth toward me.
    YOU: Yes   No   YOUR PARTNER: Yes   No
19. I feel included in my partner's life.
    YOU: Yes   No   YOUR PARTNER: Yes   No
20. My partner admires me.
    YOU: Yes   No   YOUR PARTNER: Yes   No

Scoring: If you answered yes to fewer than seven items, it is likely that you are not feeling loved and respected in this relationship. You and your partner need to be more active and creative in adding affection to your relationship.

Source: J. Gottman. (1994). *Why marriages succeed or fail.* New York: Simon & Schuster.

When we confront ambiguous messages, we should check out their meaning. The problem is that most of us are reluctant to do that. Somehow we assume that we ought to know exactly what the other person meant, and that we are dumb or naive if we don't. It is important to recognize that many "sexy signals"—like putting an arm around someone's shoulders, inviting a date to your apartment for coffee, or french-kissing and rubbing your date's (clothed) buttocks—really are ambiguous. Ideally, each of us should be effective communicators, making sure our message clearly matches our intent. As recipients of ambiguous messages, we need to make an effort to clear them up. In response to an invitation to a woman's apartment for coffee, a man might reply, "I would like some coffee, but I'm not interested in sex this time." Or he might draw her out with a question: "I'd like some coffee; is that all you have in mind?" Check out sexy signals. Don't make any assumptions about the meaning of ambiguous messages.

SUMMARY

Sexual pleasure is produced by stimulation of various areas of the body; these are the erogenous zones.

Sexual self-stimulation, or autoeroticism, includes masturbation and sexual fantasies. Women typically masturbate by rubbing the clitoris and surrounding tissue and the inner and outer lips. Men generally masturbate by circling the hand around the penis and using an up-and-down movement to stimulate the shaft. Many people have sexual fantasies while masturbating. Common themes of these fantasies are kissing and touching sensuously, oral sex, and seduction. Similar sexual fantasies are also common while having intercourse.

An important technique in two-person sex is hand stimulation of the partner's genitals. A good guide to technique is to find out how the partner masturbates. Touching other areas of the body and kissing are also important. The other senses—sight, smell, and hearing—can also be used in creating sexual arousal.

While there are infinite varieties in the positions in which one can have intercourse, there are four basic positions: man on top (the "missionary" position), woman on top, rear entry, and side to side.

There are two kinds of mouth-genital stimulation: cunnilingus (mouth stimulation of the female genitals) and fellatio (mouth stimulation of the male genitals). Both are engaged in frequently today and are considered pleasurable by many people. Lesbians and gays use techniques similar to those of straights (e.g., hand-genital stimulation and oral-genital sex). Gays and lesbians, though, seem less goal-oriented, take their time more, and communicate more than heterosexuals do.

Anal intercourse involves inserting the penis into the rectum.

An aphrodisiac is a substance that arouses sexual desire. There is no known reliable aphrodisiac, and some of the substances that are popularly thought to act as aphrodisiacs can be dangerous to one's health.

We have a tendency in our culture, perhaps a legacy of the Protestant ethic, to view sex as work and to turn sex into an achievement situation, as witnessed by expressions such as "achieving orgasm." Such attitudes make sex less pleasurable and may set the stage for sexual failures or sexual disorders.

There are clear differences in communication patterns between happy, nondistressed couples and couples who are unhappy, seeking counseling, or headed for divorce. Destructive patterns of interaction include criticism, contempt, defensiveness, and withdrawal. The key to building a good relationship is reciprocal self-disclosure. The key to maintaining a good relationship is being a good communicator.

Specific tips for being a good communicator include: use "I" language; avoid mind reading; document your points with specific examples; use limited-choice questions; level and edit; be a nondefensive listener; give feedback by paraphrasing; validate the other's viewpoint; draw your partner out; be aware of your nonverbal messages; and engage in positive verbal and nonverbal communication. When you do fight, fight fair. Finally, it is important to check out ambiguous sexy signals to find out what they really mean.

REVIEW QUESTIONS

1. The most common technique of female masturbation involves inserting a dildo or similar object into the vagina. True or false?

2. The majority of both men and women fantasize while they masturbate. True or false?

3. The man-on-top position of intercourse works well for a woman in the late stages of pregnancy. True or false?

4. The scientific term for mouth stimulation of the female genitals is _____.

5. There are no health risks associated with anal intercourse. True or false?

6. Masters and Johnson, in their research on the lovemaking techniques of homosexuals and heterosexuals, found that the gays took their time and were less goal-oriented. True or false?

7. Research shows that couples seeking therapy for sex problems have poor communication patterns compared with nondistressed couples. True or false?

8. _____ is the term for telling your partner what you are feeling by stating your thoughts clearly, simply, and honestly.

9. The use of "I" language is considered to be a poor technique in couple communication. True or false?

10. Communicating to your partner that, given his or her point of view, you can see why he or she thinks a certain way, is termed _____.

(The answers to all review questions are at the end of the book, after the Glossary.)

QUESTIONS FOR THOUGHT, DISCUSSION, AND DEBATE

1. If you are in a long-term relationship, think about the kind of communication pattern you have with your partner. Do you use the methods of communication recommended in this chapter? If not, do you think that there are areas in which you could change and improve? Would your partner cooperate in attempts to improve your communication pattern?

2. What do you think about sexual fantasizing? Is it harmful, or is it a good way to enrich one's sexual expression? Are your ideas consistent with the results of the research discussed in this chapter?

3. You have been in an intimate relationship for two years. You find you are getting bored with your sexual activity. List five things that could make your sexual relationship more satisfying. How would you communicate your desire to do each of these to your partner?

SUGGESTIONS FOR FURTHER READING

Comfort, Alex. (1991). *The new joy of sex: A gourmet guide to lovemaking for the nineties.* New York: Crown. Alex Comfort has rewritten his bestseller to maintain the joy of sex while recognizing the risk of AIDS.

Dodson, Betty. (1987). *Sex for one: The joy of self-loving.* New York: Harmony Books (Crown Publishers). An inspiring ode to masturbation.

Gibbons, Boyd. (1986). The intimate sense of smell. *National Geographic Magazine, 170,* 324–361. A fascinating discussion of the sense of smell, including its role in sexual interactions.

Gottman, John. (1994). *Why marriages succeed or fail.* New York: Simon & Schuster. Summarizes the results of 20 years of research on communication in marriage. The book includes self-assessment questions and specific suggestions to help couples enhance their communication.

Tannen, Deborah. (1986). *That's not what I meant: How conversational style makes or breaks relationships.* New York: Ballantine Books. Tannen, a communications expert and author of very popular books, provides excellent tips for improving communication.

WEB RESOURCES

www.sexuality.org
> The Society for Human Sexuality, a sex-positive organization based at the University of Washington. Extensive on-line library

www.sexhealth.org/infocenter/GuideBS/GuideBS.htm
> The Sexual Health InfoCenter: Guide to Better Sex Page

http://www.umr.edu/~counsel/assert.html
> Assertive communication skills; maintained by the Center for Personal and Professional Development, University of Missouri at Rolla

chapter

11

SEXUALITY AND THE LIFE CYCLE: CHILDHOOD AND ADOLESCENCE

CHAPTER HIGHLIGHTS

My friend said to me, "If you show me yours, I'll show you mine." I said, "All right," but that we should go into the garage where no one would see us. I knew or thought that if someone caught us, we'd both be in real trouble. I don't remember what brought on this fear, but he seemed to have the same idea. So we went into the garage, and that was the first time I ever remember seeing a boy's penis. Many more incidents of sexual interest and exploration took place with this same playmate. . . .*

Stop for a moment and think of the first sexual experience you ever had. Some of you will think of the first time you had sexual intercourse, while others will remember much earlier episodes, like "playing doctor" with the other kids in the neighborhood. Now think of the kind of sex life you had, or expect to have, in your early twenties. Finally, imagine yourself at 65 and think of the kinds of sexual behavior you will be engaging in then.

In recent years, scientists have begun thinking of human sexual development as a process that occurs throughout the lifespan. This process is influenced by biological, psychological, social, and cultural factors. This represents a departure from the Freudian heritage, in which the crucial aspects of development were all thought to occur in childhood. This chapter and Chapter 12 are based on the newer **lifespan,** or "life-cycle," approach to understanding the development of our sexual behavior throughout the course of our lives. The things you were asked to remember and imagine about your own sexual functioning in the preceding paragraph will give you an idea of the sweep of this approach to development.

Data Sources

What kinds of scientific data are available on the sexual behavior of people at various times in their lives? One source we have is the Kinsey report (Kinsey et al., 1948, 1953). The scientific techniques used by Kinsey were discussed and evaluated in Chapter 3. A number of more recent surveys of adults also provide relevant data.

In these surveys, adults are questioned about their childhood sexual behavior, and their responses form some of the data to be discussed in this chapter. These responses may be even more problematic than some of the other kinds of data from those studies, though. For example, a 50- or 60-year-old man is asked to report on his sexual behavior at age 10. How accurately will he remember things that happened 40 or 50 years ago? Surely there will be some forgetting. Thus the data on childhood sexual behavior may be subject to errors that result from adults being asked to recall things that happened a very long time ago.

An alternative would be to interview children about their sexual behavior or perhaps even to observe their sexual behavior. Few researchers have done either, for obvious reasons. Such a study would arouse tremendous opposition from parents, religious leaders, and politicians, who would argue that it is unnecessary, or that it would harm the children who were studied. These reactions reflect in part the widespread beliefs that children are not yet sexual beings and should not be exposed to questions about sex. Such research also raises ethical issues: at what age can a child give truly informed consent to be in such a study?

In a few studies children have been questioned directly about their sexual behavior. Kinsey interviewed 432 children, aged 4 to 14, and the results of the study were published after his death by Elias and Gebhard (1969). A recent innovation is the use of a "talking" computer to interview children (Romer et al., 1997). The computer is programmed to present the questions through headphones, and the child

*Female respondent quoted in Eleanor S. Morrison et al. (1980). *Growing up sexual.* New York: Van Nostrand.

Lifespan development: Development from birth through old age.

enters his or her answers using the keyboard. This process preserves confidentiality, even when others are present, because only the child knows the question. This procedure was used to gather data from samples of high-risk youth ages 9 to 15. More children reported sexual experience to the computer than did children in face-to-face interviews.

Many studies of adolescent sexual behavior have also been done. Particularly notable are the well-done studies by Kantner and Zelnik of young unmarried women (1972, 1973), the survey of adolescent men and women by Zelnik and Kantner (1980), and Coles and Stokes's report, *Sex and the American Teenager* (1985). Sociological studies of premarital sexual behavior have been done by Ira Reiss (1967) and by John DeLamater and Patricia MacCorquodale (1979; see Focus 11.1). Increasingly we have available well-sampled studies of adolescent sexuality, and we can have the most confidence in their statistics.

The studies of child and adolescent sexual behavior have all been surveys, and they have used either questionnaires or interviews. No one has made systematic, direct observations of children's sexual behavior.

Infancy (0 to 2 Years)

A century ago it was thought that sexuality was something that magically appeared at puberty. Historically, we owe the whole notion that children—in fact, infants—have sexual urges and engage in sexual behavior to Sigmund Freud.

The capacity of the human body to show a sexual response is present from birth. Male infants, for example, get erections. Indeed, boy babies are sometimes born with erections. Ultrasound studies indicate that reflex erections occur in the male fetus for several months before birth (Masters et al., 1982). Vaginal lubrication has been found in baby girls in the 24 hours after birth (Masters et al., 1982).

The first intimate relationship most children experience is with their mothers. The mother-infant relationship involves a good deal of physical contact and typically engages the infant's tactile, olfactory, visual, and auditory senses (Frayser, 1994).

Most activities associated with nurturing and hygienic care of babies is intimate and sensuous since it involves contact with sensitive organs—lips, mouth, anus, and genitals—that can produce in the infant a physiological response of a sensuous and sexual nature. These activities include (in addition to breast feeding) toilet training, bathing, cleaning, and diapering. The highly physiological and emotionally charged first encounters of mother and infant play an indispensable part in the development process. (Martinson, 1994, p. 11)

Masturbation

Infants have been observed fondling their own genitals. There is some question as to how conscious they are of what they are doing, but at the least they seem to be engaging in some pleasurable, sexual self-stimulation. The rhythmic manipulation of the genitals associated with adult masturbation does not occur until age 2½ to 3 years (Martinson, 1994). Ford and Beach (1951), on the basis of their survey of sexual behavior in other cultures, noted that if permitted, most boys and girls will progress from absentminded fingering of their genitals to systematic masturbation by ages 6 to 8. In fact, in some cultures adults fondle infants' genitals to keep them quiet, a remarkably effective pacifier.

Orgasms from masturbation are possible even at this early age, although before puberty boys are not capable of ejaculation. Masturbation is a normal, natural form of sexual expression in infancy. It is definitely not a sign of pathology, as some previous generations believed. Indeed, in one study comparing infants who had optimal relationships with their mothers and infants who had problematic relationships with their mothers, it was the infants with the optimal maternal relationships who were more likely to masturbate (Spitz, 1949).

Infant-Infant Sexual Encounters

Infants and young children are very self-centered (what the psychologist Jean Piaget called *egocentric*). Even when they seem to be playing with another child, they may simply be playing alongside the other child, actually in a world all their own. Their sexual development parallels the develop-

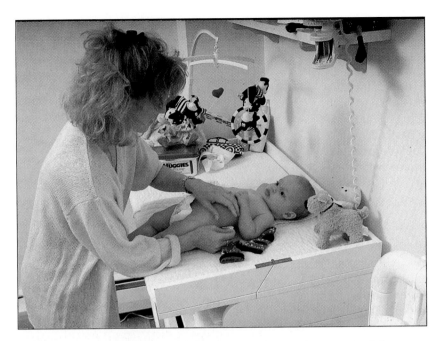

Figure 11.1 Some activities associated with nurturing an infant are potentially sensuous, because they involve pleasant physical contact.

ment of their other behaviors. Thus their earliest sex is typically one-person sex—masturbation. Not until later do they develop social, two-person sex, either heterosexual or homosexual.

Nonetheless, particularly in later infancy, there may be some infant-infant encounters, either affectionate or sexual. In these encounters, children may kiss, hug, pat, stroke, and gaze at each other, behaviors that are part of erotic intimacy later in life.

Nongenital Sensual Experiences

Many of the sensual experiences that infants and young children have are diffuse and not easily classified as masturbation or as heterosexual or homosexual activity. For example, as Freud noted, infants delight in putting things in their mouths. Thus sucking at the mother's breast, or sucking on his or her own fingers, may be a sensuous experience for the infant.

Being cuddled or rocked can also be a warm, sensuous experience. Indeed, the infant's experiences in such early intimate encounters might influence her or his reactions to intimacy and cuddling

Figure 11.2 Infant masturbation.

in adulthood. It seems that some infants are cuddlers and some are noncuddlers (Schaffer & Emerson, 1964). Cuddlers enjoy physical contact, while noncuddlers show displeasure and restlessness when they are handled or held. As soon as they are old enough to do so, they show resistance to such situations or crawl or walk away from them. Cuddling and noncuddling seem to be basically different personality patterns. It would be interesting to know whether these patterns remain consistent into adulthood.

Attachment

The quality of the relationship with the parents at this age can be very important to the child's capacity for later sexual and emotional relationships. In psychological terms, an **attachment** (or bond) forms between the infant and the mother, father, or other caregiver. The bond begins in the hours immediately following birth and continues throughout the period of infancy (Coustau and Augelini, 1995). It is facilitated by cuddling and other forms of physical contact. Later, attachments form to other familiar people. These are the individual's earliest experiences with love and emotional attachment. It seems likely that the quality of these attachments—whether they are stable, secure, and satisfying or unstable, insecure, and frustrating—affects the person's capacity for emotional attachments in adulthood. Recent research with humans (discussed in Chapter 13) suggests that adults' styles of romantic attachment are similar to the kinds of attachment they remember having with their parents in childhood.

Knowing About Boy-Girl Differences

By age 2½ or 3, children know what gender they are (see Chapter 14). They know that they are like the parent of the same gender and different from the parent of the opposite gender and from other children of the opposite gender. At first, infants think that the difference between girls and boys is a matter of clothes or haircuts. But by age 3 there may be some awareness of differences in the genital region and increasing interest in the genitals of other children (Martinson, 1994).

Figure 11.3   Between the ages of 3 and 7 there is a marked increase in sexual interest.

Early Childhood (3 to 7 Years)

Between the ages of 3 and 7, there is a marked increase in sexual interest and activity, just as there is an increase in activity and interest in general.

Masturbation

Children increasingly gain experience with masturbation during childhood. In a study of college students, 15 percent of the males and 20 percent of

Attachment: A psychological bond that forms between an infant and the mother, father, or other caregiver.

the females recalled that their first masturbation experience occurred between ages 5 and 8 (Arafat and Cotton, 1974). In a Norwegian study, kindergarten teachers reported that between one-half and one-third of their pupils engaged in some masturbation (Langfeldt, 1981).

Children also learn during this period that masturbation is something that one does in private.

Heterosexual Behavior

By the age of 4 or 5, children's sexuality has become more social. There is some heterosexual play. Boys and girls may hug each other or hold hands in imitation of adults. "Playing doctor" can be a popular game at this age (Gundersen et al., 1981). It generally involves no more than exhibiting one's own genitals, looking at those of others, and perhaps engaging in a little fondling or touching. As one woman recalled,

> It was at the age of 5 that I, along with my three friends who were sisters and lived next door, first viewed the genitals of a boy. They had a male cousin who came to visit and we all ended up behind the furnace playing doctor. No matter what he would say his symptoms were, we were so fascinated with his penis that it was always the center of our examinations. I remember giggling as I punched it and dunked it in some red food-colored water that we were using for medicine. This seemed to give him great enjoyment. One girl put hand lotion and a bandage on his penis and in the process he had an erection. We asked him to do it again, but their [sic] was no such luck. (Martinson, 1994, p. 37)

By about the age of 5, children have formed a concept of marriage—or at least of its nongenital aspects. They know that a member of the other gender is the appropriate marriage partner, and they are committed to marrying when they get older (Broderick, 1966a, 1966b). They practice marriage roles as they "play house."

Some children first learn about heterosexual behavior by seeing or hearing their parents engaging in sexual intercourse, or the *primal scene experience*. Freud believed that this experience could inhibit the child's subsequent psychosexual development; some contemporary writers share this belief. Limited empirical data suggest that the experience

is not damaging. In surveys, about 20 percent of middle-class parents report that their child observed them when the child was 4 to 6 years of age. Parents report such reactions as curiosity ("Why are you bobbing up and down?"), amusement and giggling, or embarrassment and closing the door (Okami, 1995).

Homosexual Behavior

During late childhood and preadolescence, sexual play with members of one's own gender may be more common than sexual play with members of the other gender (Martinson, 1994). Generally the activity involves no more than touching the other's genitals (Broderick, 1966a). One girl recalled:

> I encountered a sexual experience that was confusing at kindergarten age. . . . Some afternoons we would meet and lock ourselves in a bedroom and take our pants off. We took turns lying on the bed and put pennies, marbles, etc. between our labia. . . . As the ritual became old hat, it passed out of existence. (Martinson, 1994, p. 62)

Sex Knowledge and Interests

At age 3 or 4, children begin to have some notion that there are genital differences between males and females, but their ideas are very vague. By age 7, 30 percent of American children understand what the differences are (Goldman & Goldman, 1982). Children generally react to their discovery of genital differences calmly, though of course there are exceptions.

At age 3, children are very interested in different postures for urinating. Girls attempt to urinate while standing. Children are also very affectionate at this age. They enjoy hugging and kissing their parents and may even propose marriage to the parent of the other gender.

At age 4, children are particularly interested in bathrooms and elimination. Games of "show" are also common at this age. They become less common at age 5, as children become more modest. The development of modesty reflects the child's learning of the restrictions that American society places on sexual expression. Often as early as age 3 the child is taught by parents not to display or touch certain parts of his or her body, at least in public. Children are often taught not to touch the bodies of others. Many parents also restrict conversation about sex. These restrictions come at precisely the time the child is becoming more aware of and curious about sexuality. One young man recalled:

> One of my favorite pastimes was playing doctor with my little sister. During this doctor game we would both be nude and I would sit on her as if we were having intercourse. On one occasion, I was touching my sister's genital area and mother discovered us. We were sternly switched and told it was dirty and to never get caught again or we would be whipped twice as bad. So, we made sure we were never caught again. (Starks & Morrison, 1996)

As a result, children turn to sex play and their peers for information about sex (Martinson, 1994). Cross-cultural data suggest that in less restrictive societies, children continue to show overt interest in sexual activities through childhood and preadolescence (Frayser, 1994). In the rare society that puts no restrictions on childhood sex play, intercourse may occur as young as age 6 or 7. In the United States, first attempts at intercourse may occur 3 to 4 years later.

It is important to remember that children's sex play at this age is motivated largely by curiosity and is part of the general learning experiences of childhood. One man illustrated this well as he recalled:

> At the age of six or seven my friend (a boy) and I had a great curiosity for exploring the anus. It almost seemed *more like scientific research.* (Martinson, 1994, p. 59, italics added)

A longitudinal study of the impact of childhood sex play obtained reports from mothers when the child was 6, and assessed the sexual adjustment of the person at age 17 or 18 (Okami et al., 1997). Forty-seven percent of the mothers reported that their child had engaged in interactive sex play. Looking at a range of outcomes, including social relationships and sexual behavior and "problems," there were no significant differences between males and females whose mothers reported such play, and males and females whose mothers did not report such activity.

Preadolescence (8 to 12 Years)

Preadolescence is a period of transition between the years of childhood and the years of puberty and adolescence. Freud used the term "latency" to refer to the preadolescent period following the resolution of the Oedipus complex. He believed that the sexual urges go "underground" during latency and are not expressed. The evidence indicates, however, that Freud was wrong and that children's interest in and expression of sexuality remain lively throughout this period, perhaps more lively than their parents are willing to believe. For many, "sexual awakening" does not occur until the teens, but for others, it is a very real and poignant part of preadolescence (Martinson, 1994).

At around age 9 or 10, the first bodily changes of puberty begin: the formation of breast buds in girls and the growth of pubic hair. The growth of pubic hair occurs in response to *adrenarche,* the maturation of the adrenal glands, leading to increased levels of androgens. In three recent studies, one of adolescents and two of adults, the average age at which participants reported first experiencing sexual attraction to another person was at age 10 (McClintock & Herdt, 1996). The samples included gays, lesbians, and heterosexual men and women. This experience may reflect the maturation of the adrenal gland and the increase in testosterone and estradiol which result. This research suggests that "adult" sexual development starts as early as age 9 or 10, not at puberty as previously thought.

Masturbation

During preadolescence, more and more children gain experience with masturbation. In a sample of college women, 32 percent recalled masturbating to orgasm by age 13.[1] The comparable figures for males are 49 percent by age 13 (Arafat & Cotton, 1974). These data, as well as those on adolescence, indicate that boys generally start masturbating earlier than girls do.

Interestingly, boys and girls learn about masturbation in different ways. Typically boys are told

---

[1] These are cumulative-incidence figures, to use the terminology introduced in Chapter 3.

about it by their male peers, they see their peers doing it, or they read about it; girls most frequently learn about masturbation through accidental self-discovery (Langfeldt, 1981). One man recalled:

> An older cousin of mine took two of us out to the garage and did it in front of us. I remember thinking that it seemed a very strange thing to do, and that people who were upright wouldn't do it, but it left a powerful impression on me. A couple of years later, when I began to get erections, I wanted to do it, and felt I shouldn't, but I remembered how he had looked when he was doing it, and the memory tempted me strongly. I worried, and held back, and fought it, but finally I gave in. The worry didn't stop me, and doing it didn't stop my worrying. (Hunt, 1974, p. 79)

Heterosexual Behavior

There is generally little heterosexual behavior during the preadolescent period, mainly because of the social division of males and females into separate groups. However, children commonly hear about sexual intercourse for the first time during this period. For example, in a sample of adult women, 61 percent recalled having learned about intercourse by age 12 (Wyatt et al., 1988). Children's reactions to this new information are an amusing combination of shock and disbelief—particularly disbelief that their parents would do such a thing. A college woman recalled:

> One of my girlfriends told me about sexual intercourse. It was one of the biggest shocks of my life. She took me aside one day, and I could tell she was in great distress. I thought she was going to tell me about menstruation, so I said that I already knew, and she said, "No, this is *worse!*" Her description went like this: "A guy puts his thing up a girl's hole, and she has a baby." The hole was, to us, the anus, because we did not even know about the vagina and we knew that the urethra was too small. I pictured the act as a single, violent and painful stabbing at the anus by the penis. Somehow, the idea of a baby was forgotten by me. I was horrified and repulsed, and I thought of that awful penis I had seen years ago. At first I insisted that it wasn't true, and my friend said she didn't know for sure, but that's what her cousin told her. But we looked at each other, and we knew it was true. We held each other and cried. We insisted that "my parents would never do that," and "I'll never let anyone do it to

me." We were frightened, sickened, and threatened by the idea of some lusty male jabbing at us with his horrid penis. (From a student essay)

The age at which youth have their first sexual experience has been declining (see Figure 11.7). Some boys and girls have their first experience during the preadolescent period. The research that used the talking computer studied children ages 9 to 15 who lived in public housing projects. The results indicate that 63 percent of the boys and 14 percent of the girls had "had sex with someone" by age 12 (Romer et al., 1997).

For some preadolescents, heterosexual activity occurs in an incestuous relationship, whether brother-sister or parent-child. This topic is discussed in detail in Chapter 17.

Homosexual Behavior

It is important to understand homosexual activity as a normal part of the sexual development of children. In preadolescence, children have a social organization that is essentially **homosocial.** That is, boys play separately from girls, and thus children socialize mainly with members of their own gender. This separation begins at around age 8. According to one study of children's friendship patterns, the segregation reaches a peak at around 10 to 12 years of age. At ages 12 to 13 children are simultaneously the most segregated by gender and the most interested in members of the opposite gender (Broderick, 1966b).

Observational research suggests that there is greater segregation at school than in neighborhood play groups (Thorne, 1993). Some of the social separation of the genders during preadolescence is actually comical; boys, for example, may have been convinced that girls have "cooties" and that they must be very careful to stay away from them.

Given that children are socializing almost exclusively with other members of their own gender, sexual exploring at this age is likely to be homosex-

---

Homosocial: A general form of social grouping in which males play and associate with other males, and females play and associate with other females; that is, the genders are separate from each other.

ual in nature. These homosexual activities generally involve masturbation, exhibitionism, and the fondling of other's genitals. Boys, for example, may engage in a "circle jerk," in which they masturbate in a group.

Girls do not seem so likely to engage in such group homosexual activities, perhaps because the spectacle of them masturbating is not quite so impressive or perhaps because they already sense the greater cultural restrictions on their sexuality and are hesitant to discuss sexual matters with other girls. In any case, as noted above, boys seem to do their sexual exploring with a gang, while girls do it alone (Kinsey et al., 1953).

A study of psychosexual development among lesbian, gay, and bisexual youth ages 14 to 21 found that the participants reported their first experience of sexual attraction at age 10 or 11 (Rosario et al., 1996). First experience of sexual fantasies occurred several months to one year later. The first sexual activity with another person occurred on the average at ages 12 or 13. All of the young men and women reported experiencing sexual attraction to and fantasies about a person of the same gender, and one-half reported them about a person of the other gender.

Dating

There is some anticipation of adolescent dating in the socializing patterns of preadolescents. Group dating and heterosexual parties emerge first among preadolescents. Boys, particularly, are slow in adjusting to the behavior expected of them, and they may be more likely to roughhouse than to ask a girl to dance.

Kissing games are popular at parties, reaching their peak of popularity among children aged about 10 to 13 (Broderick, 1966b). One 17-year-old man reported:

> In sixth and seventh grade it was the Friday night party, where there was a lot of petting and kissing and fondling. In the eighth grade, double dates were real big, watch a movie and fondle. That was pretty much it. (Maurer, 1994, p. 29)

Paired dating generally begins around the age of 12 or 13. It may consist of walking or riding bikes together, sitting next to each other, or sharing tapes and CDs. At age 12, 48 percent of the boys

Figure 11.4 Group dating, heterosexual parties, and hanging out emerge during preadolescence, and may include making out.

and 57 percent of the girls in one study were dating (Broderick, 1966b). At age 13, the figure was 69 percent for both. Typically, though, boys and girls at this age date only a few times a year. Going steady, or "goin' with," also starts at this age, although often the activity centers more on symbols, such as an exchange of rings or bracelets, than it does on any real dating (Martinson, 1994).

Of course, these patterns are averages for American preadolescents. There is great variability within American culture, and in some cultures boys and girls may already be married by age 13.

Adolescence (13 to 19 Years)

A surge of sexual interest occurs around puberty and continues through adolescence (which is equated here roughly with the teenage years, ages

13 to 19). This heightened sexuality may be caused by a number of factors, including bodily changes and an awareness of them, rises in levels of sex hormones, and increased cultural emphasis on sex and rehearsal for adult gender roles. We can see evidence of this heightened sexuality particularly in the data on masturbation. But before examining those data, let's consider some theoretical ideas about how hormones and social forces might interact as influences on adolescent sexuality.

Udry (1988) has proposed a theoretical model that recognizes that both sociological factors and biological factors are potent in adolescent sexuality. He studied eighth-, ninth-, and tenth-graders (13 to 16 years old), measuring their hormone levels (testosterone, estrogen, and progesterone), and a number of sociological factors (for example, whether they were in an intact family, parents' educational level, the teenager's response to a scale measuring sexually permissive attitudes, and the teenager's attachment to conventional institutions such as involvement in school sports and church attendance). Thirty-five percent of the males had engaged in sexual intercourse, as had 14 percent of the females.

For boys, testosterone levels had a very strong relationship to sexual activity (including coitus, masturbation, and extent of feeling sexually "turned on"). For girls, the relationship between testosterone level and sexual activity was not as strong as it was for boys, but it was a significant relationship, and it was testosterone—not estrogen or progesterone—that was related to sexuality. Among the social variables, sexually permissive attitudes were related to sexuality for boys, although they had a much smaller effect than testosterone did. For girls, pubertal development (developing a "curvy" figure) had an effect, probably by increasing the girl's attractiveness. And the effects of testosterone were accentuated among girls in father-absent families. When girls were asked to rate their plans about sexuality, testosterone level as an important predictor of their ratings, as were the social variables of permissive attitudes and church attendance.

The bottom line on this study is that it shows testosterone level to have a substantial impact on the sexuality of adolescent boys and girls. Social variables (such as permissive attitudes, father absence for girls, and church attendance) then interact with the biological effects, in some cases magnifying them (father absence for girls) and in some cases suppressing them (church attendance).

Masturbation

According to the Kinsey data, there is a sharp increase in the incidence of masturbation for boys between the ages of 13 and 15. This is illustrated in Figure 11.5. Note that the curve is steepest between the ages of 13 and 15, indicating that most boys begin masturbating to orgasm during that period. By age

Figure 11.5 Cumulative incidence of males and females who have masturbated to orgasm, according to Kinsey's data.

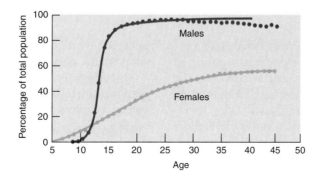

Focus 11.1

The Impact of Mass Media on Adolescent Sexuality

A major developmental task of adolescence is learning how to manage physical and emotional intimacy in relationships with others. It is not surprising, therefore, that young people are curious about sex and about sexual intimacy. An increasingly important source of information is the mass media. In a survey of youth ages 10 to 15, the most frequently named source of information about sexuality and relationships was the mass media, including TV, movies, magazines, and music (Kaiser Family Foundation, 1997). Children who watch the TV program *Dawson's Creek* are introduced to Jen, who lost her virginity at age 12. Youth who watch talk shows learn about impotence, and ways to deal with it. *Kojak* reruns provide information about prostitution and prostitutes. Soap operas deal, sometimes explicitly, with sexual themes like frigidity, menopause, abortion, and infidelity. How much sexual content is there in the mass media? To what extent are children and adolescents exposed to it? And what is the impact of this exposure?

A good deal of research, much of it done in the 1990s, has been devoted to the question of how much sexual content there is in the media. Content analyses of the media define *sexual material* as verbal references to sexual activity, innuendo, implied sexual activity, and explicit presentations. Looking first at prime-time television, analyses in-

dicate that there has been an increase in the number and explicitness of sexual portrayals over the past 20 years. In the fall of 1991, the average was 10 portrayals per hour; these are typically of heterosexual intercourse involving persons not married to each other (Lowry & Shidler, 1993). Consequences such as pregnancy and STDs are rarely discussed or shown (Kunkel et al., 1996). Turning to movies, adolescents see feature films in theatres, on pay TV, or on videotape. Many of these films are rated R, and they contain more frequent and more explicit depictions of sexual behavior, an average of 17.5 per hour, than depictions of sexual behavior on prime-time television (Greenberg et al., 1996). The most frequent portrayals are of unmarried heterosexual intercourse, often in a context of alcohol and drug use; there are no safer sex messages here!

We have already mentioned the variety of sexual situations, most of them problematic, that are included in soap operas. In 1994, each episode of a soap contained an average of 6.6 sexual portrayals, and safe sex was mentioned infrequently (Greenberg & Busselle, 1996). Finally, there are music videos. The visual elements in many MTV videos are implicitly or explicitly sexual (Browne & Steele, 1995), and they frequently combine sexuality with implicit aggression (Sommers-Flanagan et al.,

15, 82 percent of the boys in Kinsey's study had masturbated. Many girls also begin masturbating at around that age, but note that the curve on the graph is flatter for them, indicating that many girls do not begin masturbating until later. Thus the increase in their masturbation behavior is much more gradual than boys' and continues past adolescence.

More recent data suggest that children and adolescents begin to masturbate earlier today, and thus the Kinsey data probably need to be pushed

back about one or two years. However, the general shape of the curves still seems to hold.

One man recalled his adolescent experiences with masturbation and the intense feelings involved as follows:

> When I was fourteen I was like Portnoy—always rushing off to the bathroom when the urge came over me. I did it so much that my dick would get swollen and sore, and even that didn't stop me. By the time I was nineteen I was screwing, but there'd

1993). MTV videos also objectify women, presenting them in revealing clothing and portraying them as receptive to sexual advances.

Clearly, the mass media are providing a great deal of sexual content. But is anybody watching? Children ages 9 to 13 are heavy TV viewers (Comstock, 1991). During middle and late adolescence, TV viewing declines, and time spent watching music videos, reading magazines, and surfing the 'Net increases (for a discussion of the sexual material available on the Internet, see Chapter 18). There are gender differences in exposure to mass media: teenage girls spend more time reading magazines and watching soap operas, while their male counterparts spend more time surfing the 'Net. Media use also varies by social class and ethnicity; children and adolescents from less affluent families watch more television, and African American youth watch more TV than do their European American peers.

What effect do these portrayals have? There has been less research on this issue, partly because it is difficult to isolate the effects of media exposure from other influences on adolescents' sexual attitudes and behavior. Media images can have an immediate effect on the user's emotional state: portrayals may induce arousal, which can influence behavior or activate thoughts or associations (see the discussion of sexual fantasy in Chapter 10). Media portrayals may have long-term effects in that children may learn schemas and scripts that influence their later sexual decision-making and behavior (Kaiser Family Foundation, 1998). One experimental study indicated that teenagers exposed to 15 hours of portrayals of nonmarital sexual relationships had more permissive attitudes toward nonmarital sex than did teens exposed to portrayals of nonsexual relationships (Bryant & Rockwell, 1994). Correlational studies find that teens who watch programs with more sexual content are more likely to have engaged in sexual intercourse. The media may also influence standards of physical attractiveness and contribute to the dissatisfaction with their bodies that many, especially women, feel.

The evidence that mass media portrayals have an important impact on adolescent sexual knowledge, attitudes, and behavior is not conclusive. On the other hand, we noted at the outset that both children and adolescents believe that mass media are the most important source of their knowledge. The problem lies in the fact that these portrayals are unrealistic. In sharp contrast to the high rates of nonmarital sex portrayed in the media, most sexual activity involves persons who are married (see Chapter 12). Many couples, whether married or not, are responsible users of birth control. Many youth and adults use various forms of prophylaxis to prevent STDs. It is unfortunate that these realities are missing from mass media portrayals of sexual behavior. It is also unfortunate that the media have generally not taken advantage of their opportunity to provide positive sexuality education.

be times when I wouldn't be able to get anything, and I'd go back to jacking off—and then I felt really guilty and ashamed of myself, like I was a failure, like I had a secret weakness. (Hunt, 1974, p. 95)

Boys typically masturbate two or three times per week, whereas girls do so about once per month (Hass, 1979). Interestingly, the frequency of masturbation among boys decreases during periods when they are having sexual intercourse; among girls, however, this situation is accompanied by an increased frequency of masturbation (Sorensen, 1973).

Attitudes Toward Masturbation

Attitudes toward masturbation have undergone a dramatic change in this century. As a result, adolescents are now given much different information about masturbation, and this may affect both their behavior and their feelings about masturbation. For example, a popular handbook, *What a Boy*

Should Know, written in 1913 by two doctors, advised its readers:

> Whenever unnatural emissions are produced . . . the body becomes "slack." A boy will not feel so vigorous and springy; he will be more easily tired. . . . He will probably look pale and pasty, and he is lucky if he escapes indigestion and getting his bowels confined, both of which will probably give him spots and pimples on his face. . . .
>
> The results on the mind are the more severe and more easily recognized. . . . A boy who practices this habit can never be the best that Nature intended him to be. His wits are not so sharp. His memory is not so good. His power of fixing his attention on whatever he is doing is lessened. . . . A boy like this is a poor thing to look at. . . .
>
> . . . The effect of self-abuse on a boy's character always tends to weaken it, and in fact, to make him untrustworthy, unreliable, untruthful, and probably even dishonest. (Schofield and Vaughan-Jackson, 1913, pp. 30–42)

Masturbation, in short, was once believed to cause everything from warts to insanity.[2]

---

[2]In case you're wondering why boys' advice books were saying such awful things, there is a rather interesting history that produced those pronouncements (Money, 1986). The Swiss physician Simon André Tissot (1728–1797) wrote an influential book, *Treatise on the Diseases Produced by Onanism,* taking the term from the biblical story of Onan (Genesis 38:9). In this work he articulated a degeneracy theory, in which loss of semen was believed to weaken a man's body; Tissot had some very inventive physiological explanations for his idea. The famous American physician of the 1800s Benjamin Rush was influenced by Tissot and spread degeneracy theory in the United States. The theory became popularized by Sylvester Graham (1794–1851), a religious zealot and health reformer, who was a vegetarian and whose passion for health foods gave us the names for Graham flour and Graham crackers. To be healthy, according to Graham, one needed to follow the Graham diet and practice sexual abstinence. Then John Harvey Kellogg (1852–1943) of—you guessed it—cornflakes fame entered the story. He was an ardent follower of Graham and his doctrines of health food and sexual abstinence. While experimenting with healthful foods, he invented cornflakes. His younger brother, Will Keith Kellogg, thought to add sugar and made a fortune. John Harvey Kellogg contributed further to public fears about masturbation by writing (during his honeymoon, no less) *Plain Facts for Old and Young: Embracing the Natural History and Hygiene of Organic Life,* which provided detailed descriptions of the horrible diseases supposedly caused by masturbation. These ideas then found their way into the advice books for boys of the early 1900s.

Attitudes toward masturbation are now considerably more positive, and few people would now subscribe to notions like those expressed earlier. By the 1970s only about 15 percent of young people believed that masturbation is wrong (Hunt, 1974, p. 74). Indeed, masturbation is now recommended as a remedy in sex therapy. As psychiatrist Thomas Szasz said, the shift in attitudes toward masturbation has been so great that in a generation it has changed from a disease to a form of therapy.

While approval of masturbation is now explicit, people can still have mixed feelings about it. An example of a lingering negative attitude is that of the man quoted previously who likened his adolescent masturbation to Portnoy's,[3] accompanied as it was by feelings of guilt and shame. Among the adolescents interviewed by Sorensen (1973), few felt guilty about masturbation, but many felt defensive or embarrassed about it.

Homosexual Behavior

About 11 percent of adolescent males and 6 percent of adolescent females report having had homosexual experiences, and 25 percent say they have been approached for a homosexual experience. Of those who have had homosexual experiences, 24 percent had their first experience with a younger person, 39 percent with someone of their own age, 29 percent with an older teenager, and 8 percent with an adult (Sorensen, 1973). Thus there is no evidence that adolescent homosexual experiences result from being seduced by adults; most such encounters take place between peers. In many cases the person has only one or a few homosexual experiences, partly out of curiosity, and the behavior is discontinued. Such adolescent homosexual behavior does not seem to be predictive of adult homosexual orientation.

The data indicate that there has been no increase in the incidence of adolescent homosexual behavior (DeLamater & MacCorquodale, 1979). It seems safe to conclude from various studies, taken together, that about 10 percent of adolescents have homosexual experiences, with the percentages being somewhat higher for boys than for girls.

Teenagers can be quite naive about homosexual behavior and societal attitudes toward it. In some

---

[3]Of the novel *Portnoy's Complaint,* which poignantly and humorously describes an adolescent male's obsessive masturbation.

cases, they have been taught that heterosexual sex is "bad"; having been told nothing about homosexual sex, they infer that it is permissible. In some cases, homosexual relationships naively develop from a same-gender friendship of late childhood and adolescence. One woman recalled:

> I did not even know what homosexuality was. I had never heard the term, although I had read extensively. One day at a friend's house, we were listening to music in her bedroom, she came on to me in a very surprising way. We were good friends and spent much time together, but this particular night was different. Her eyes had a new sparkle, she got very close to me, her touch lingered; she was different than she ever had been before. I was 16 and she was 15. I did not understand, but I knew that I was aroused. We were good church going kids who had never heard anything about this. (Starks & Morrison, 1996, p. 97)

Heterosexual Behavior

Toward the middle and end of the adolescent years, more and more young people engage in heterosexual sex, with more and more frequency. Thus heterosexual behavior gains prominence and becomes the major sexual outlet.

In terms of the individual's development, the data indicate that there is a very regular progression from kissing, through French kissing and breast and genital fondling, to intercourse and oral-genital contact; this generally occurs over a period of four or more years (DeLamater & MacCorquodale, 1979). To use terminology introduced in Chapter 2, these behaviors tend to follow a sexual script.

Premarital Sex[4]

One of the most dramatic changes to occur in sexual behavior and attitudes in recent decades is in the area of premarital sexual behavior.

[4]Note that the very term "*pre*marital sex" contains some hidden assumptions, most notably that marriage is normative and that proper sex occurs in marriage. Thus, sex among never-married (young) persons is considered *premarital*—something done before marrying. A more neutral term would be "nonmarital sex," although it fails to distinguish between premarital sex, extramarital sex, and postmarital sex.

Figure 11.6 Sexuality in early adolescence is often playful and unsophisticated.

How Many People Have Premarital Intercourse?

On the basis of the data he collected in the 1940s, Kinsey concluded that about 33 percent of all females and 71 percent of all males have had premarital intercourse by the age of 25. According to the NHSLS (Laumann et al., 1994), 70 percent of women and 78 percent of men interviewed reported engaging in vaginal intercourse before marriage. Thus in the 50 years between these two large-scale surveys, the incidence of premarital intercourse doubled among women, while increasing only slightly among men. Today, about three-fourths of Americans engage in premarital sex.

As premarital intercourse has become common, attention has shifted to "teen sex," the incidence of sexual intercourse among teenagers. The National Survey of Family Growth periodically interviews a national sample of women of childbearing age. The Centers for Disease Control and Prevention conducts a Youth Risk Behavior Survey of high school students every few years. Data from these

Focus 11.2

A Sociological Analysis of Premarital Sexual Behavior

Sociologists John DeLamater and Patricia Mac-Corquodale (1979) conducted a large-scale survey and detailed analysis of patterns of premarital sexuality. They interviewed both students and nonstudents between the ages of 18 and 23 in Madison, Wisconsin. They obtained an initial random sample of 1141 students at the University of Wisconsin. They attempted to interview all these people, obtaining an 82 percent response rate. Nonstudents were contacted by a probability sample of residences in the telephone directory; they had a 63 percent response rate. Data for a total of 1376 respondents were analyzed. The study is notable because it included both students and nonstudents, and because it used excellent sampling techniques and had a good response rate, except that the nonstudent response rate was a bit lower than would be desirable. It is difficult to know to what extent the results are limited to Madison, Wisconsin, and whether they would be generalizable to other areas of the country.

The data on the sexual experience of the respondents are summarized in Table 11.2.

Respondents also answered a number of other questions, and analyses of the answers permitted

the authors to come to some conclusions regarding what factors are most strongly related to premarital sexual expression.

One of the most important factors appeared to be sexual ideology or attitudes. That is, those with the most liberal attitudes had the most premarital sexual experience. DeLamater and MacCorquodale argued that this occurs because ideology forms the basis for self-control. That is, the individual's standards specify the type of relationship, in terms of emotional commitment, that is necessary before particular behaviors are appropriate. Confirming this notion, the variable most closely related to the respondents' current behavior was the emotional quality of their current relationship.

How are the person's attitudes shaped? DeLamater and MacCorquodale found that parents, and sometimes religion, are early shapers of ideology. Later, close friends and dating partners become more important, while the influence of parents wanes. As peers become more important influences, the young person's standards typically become more permissive.

Contrary to what one might expect, the results indicated that a number of psychological variables—

surveys are shown in Table 11.1. These data suggest that rates of adolescent intercourse reached a peak about 1990 and have been declining since then.

Not only are more young women having sex, but young women today are engaging in intercourse for the first time at younger ages, compared to women born 30 years earlier. Figure 11.7 gives data from the National Survey of Family Growth, based on interviews with 8450 women ages 15 to 44 in 1988. The women were grouped by the year they were born, from 1945 to 1972. The figure displays the percentage of women who had premarital intercourse by age 15, 17, and 19 (Trussell & Vaughan, 1991). First, note that few women have intercourse at age 15 or

Table 11.1 Percentage of High School Students Who Have Engaged in Intercourse

Year of Survey	NSFG Females, 15 to 19	CDC Females, 9th–12th grade	CDC Males, 9th–12th grade
1975	36%		
1982	47%		
1990	55%		
1991		51%	57%
1995	50%	52%	54%
1997		48%	49%

Table 11.2 Percentages of Respondents Who Had Ever Engaged in Various Sexual Behaviors

	Male		Female	
	Student	*Nonstudent*	*Student*	*Nonstudent*
Necking	97%	98%	99%	99%
French kissing	93	95	95	95
Breast fondling	92	92	93	93
Male fondling of female genitals	86	87	82	86
Female fondling of male genitals	82	84	78	81
Genitals touching	77	81	72	78
Intercourse	75	79	60	72
Male mouth contact with female genitals	60	68	59	67
Female mouth contact with male genitals	61	70	54	63

self-image, self-esteem, body image, sense of internal or external control, and gender-role definitions —were *un*related to premarital sexual behavior.

Finally, the results indicated increased similarities between women and men and a decline of the double standard, as other recent surveys have found. One of the few differences that remain, though, is that women still require greater emotional commitment before they are accepting of premarital intercourse. This may result in conflicts in some relationships.

DeLamater and MacCorquodale concluded that it is the couple and the nature of their relationship —rather than variables such as social class or religion—that are essential to understanding premarital sexuality.

It is tempting to make causal inferences from these data—to say, for example, that permissive attitudes *cause* increased premarital sexual experience. It is important to remember that these data are correlational in nature and that causality cannot be inferred. Nonetheless, this study provides good evidence of what factors are most closely related to premarital sexual patterns, and these findings have been confirmed by later research.

Source: John DeLamater and Patricia MacCorquodale. (1979). *Premarital sexuality: Attitudes, relationships, behavior.* Madison: University of Wisconsin Press.

---

younger, whether born in 1945 (4 percent of whom had intercourse by age 15) or 1972 (14 percent of whom had intercourse by age 15). Second, among women born in 1945, only 13 percent had premarital intercourse by age 17; this percentage increases steadily to 40 percent for women born in 1972. Finally, there is a similar increase in the percentage reporting first intercourse by age 19. These trends reflect, in part, the impact of the "sexual revolution" of the 1960s and 1970s, which encouraged greater openness about sexuality and acceptance of premarital and other forms of sexual expression. As we saw earlier, the incidence of premarital intercourse

among men has been high since Kinsey collected his data. The impact of the social changes in the 1960s and 1970s was primarily on women.

In the United States, patterns of premarital intercourse vary substantially according to ethnic group. Table 11.3 shows data on this point from a national sample interviewed in 1979 and again in 1983 (Day, 1992). Look first at the differences between men and women of the same race or eth-nicity in mean age of first intercourse. In all four groups, men begin having sex at younger ages than women. Among blacks and Latinos (of Cuban and Puerto Rican origin), males experience first intercourse about 2½ years

Figure 11.7   Young women today are engaging in first intercourse at younger ages than did women born 30 years ago.

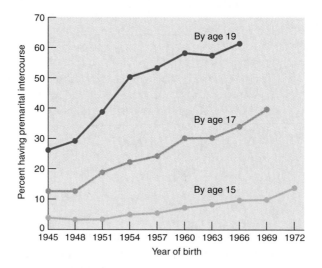

earlier than females; among Mexican Americans and whites of European origin, the difference is only about 1¼ years. There are also differences by race or ethnicity. African Americans have sex for the first time at about 15.5 years of age, Cuban Americans and Puerto Ricans at 16.6, and Mexican Americans and whites at 17. The differences by race or ethnicity reflect differences between these groups in family structure and socioeconomic opportunities. Living in an intact family, having a highly educated mother,

and attending church regularly are all associated with later age of first intercourse (Day, 1992). Living in a neighborhood where average incomes are high and the female unemployment rate is low—that is, where there are good economic opportunities—is also associated with later age of first intercourse (Brewster, 1994). A survey of a representative sample of low-income blacks and Latinos found that males reported the same age of first intercourse as reported by Day (Table 11.3). Females, in contrast, re-

Table 11.3 Ethnicity and Age of First Intercourse in a National Sample of Young Adults

	Subgroup	Number of Respondents	Mean Age of First Intercourse (Years)
Black:	Male	1613	14.3
	Female	1561	16.8
Mexican American:	Male	436	16.3
	Female	385	17.6
Other Latino	Male	220	15.4
	Female	210	17.8
White:	Male	3790	16.3
	Female	3720	17.4

Source: R. D. Day. (1992). The transition to first intercourse among racially and culturally diverse youth. *Journal of Marriage and the Family, 54,* 749–762. Data from the National Longitudinal Survey of Labor Market Experiences of Youth.

ported first intercourse at younger ages; among blacks, the average age was 16.0, and among Latinas it was 17.0 (Norris et al., 1996). In Chapter 1 we discussed variations in the incidence of masturbation and oral sex by ethnicity.

What accounts for these variations among European Americans, Black Americans, and Latinos? Are there ethnic differences in the factors associated with being sexually active as a teenager? A questionnaire study of 15,362 youth ages 12 to 17 in Michigan assessed a variety of these factors (Perkins et al., 1998). In all three ethnic groups, age, alcohol use, lower grade point average, and greater time spent at home unsupervised by an adult were associated with having had sexual intercourse. Religiosity was the only factor whose relationship to sexual activity varied by ethnicity; it was associated with sexual activity among European Americans and Latinos but not among African Americans. This research suggests that differences in cultural norms regarding sexual activity account for the variations noted by ethnicity.

There are substantial variations in patterns of premarital intercourse in different cultures around the world, as the data in Table 11.4 indicate. Most of the data was collected by the Demographic and Health Surveys Program, which interviews women in developing countries. Several interesting points emerge from the data. First, the percentage of unmarried women who have had intercourse is smaller in Latin and South American countries than in African nations; this is partly due to the greater influence of the Catholic Church in the former. Second, looking at the data for 20- to 24-year-old women, the United States has the highest rate of female premarital intercourse. Third, there is much less variation in the average age of first intercourse; it is either 16 or 17 in all but two of the countries.

In many countries around the world, the incidence of premarital intercourse has risen in the last few decades. Around the globe, especially where modernization has been rapid, adolescents are less and less under the influence of family,

Table 11.4 A Global Perspective on Female Premarital Intercourse

Country and Year	Age of Respondents (Years)	Percentage Having Had Intercourse*	Median Age of First Coitus (Years)
Africa			
Cameroon, 1991	20–24	57.0	16.1
Kenya, 1993	20–24	51.5	17.3
Nigeria, 1990	20–24	41.2	16.6
Tanzania, 1991–1992	20–24	44.2	17.3
Zambia, 1992	20–24	52.2	16.6
Latin America			
Mexico, 1985	15–19	13.0	17.0
South America			
Bolivia, 1989	20–24	26.2	16.4
Brazil (NE), 1991	20–24	14.8	20.0
Paraguay, 1990	20–24	35.4	18.9
United States, 1988	15–19	50	17.4[†]
	20–24	76.4	

*These percentages are for all never-married women; typically 75 to 90 percent of these women are 24 years of age or younger.

[†]Mean age of first intercourse.

Source: The data from African and South American countries are from the Demographic and Health Surveys Program. See C. F. Westoff, A. K. Blanc, & L. Nyblade. (1994). *Marriage and entry into parenthood.* Calverton, MD: Marco International; A. J. Gage. (1995). *An assessment of the quality of data on age at first union, first birth, and first sexual intercourse.* Calverton, MD: Marco International. The data from the United States are taken from J. Trussell & B. Vaughan. (1991). *Selected results concerning sexual behavior and contraceptive use from the 1988 National Survey of Family Growth and the 1988 National Survey of Adolescent Males.* Princeton, NJ: Office of Population Research, Working Paper 91–92.

community, and religion and more and more responsive to peers and the mass media (Liskin, 1985). On a trip to Ukraine in 1995, we were struck by the prominence of American programs and films on local television.

To summarize, the trends in premarital intercourse in the last four decades are in the direction of: (1) both in the United States and in most other countries, more adolescents engaging in premarital intercourse; (2) in the United States, a greater increase in incidence for females, thereby narrowing the gap between males and females; (3) first intercourse occurring at somewhat earlier ages; (4) moderately large ethnic-group variations in the United States; and (5) substantial variations from one country to another (Liskin, 1985; Day, 1992).

The increase in premarital sexual intercourse reflects two long-term trends. First, the age of menarche has been falling steadily since the beginning of the twentieth century. The average age is about 12.7 years for whites and 12.5 for blacks (Hofferth, 1990). Second, the age of first marriage has been rising. In 1960, first marriages occurred at age 20.8 for women and 22.8 for men. In 1992, it was 24.4 for women and 26.5 for men (Smith, 1994). The effect is a substantial lengthening of the time between biological readiness and marriage; the gap is typically 12 to 14 years today. Recall that Udry found that hormone levels are a major influence on the initiation of sexual intercourse. Not surprisingly, many more young people are having sex before they get married now than in 1960. Since many of these young people do not consistently use birth control, there has been a corresponding rise in the rate of premarital pregnancy (see Focus 11.3).

First Intercourse

First sexual intercourse is a major transition, with both psychological and social significance. In one study, the researchers analyzed the emotional reactions to first coitus of 1600 college students (Sprecher et al., 1995). Men reported significantly greater pleasure and significantly less guilt than did women. Both men and women who experienced an orgasm rated the experience as more pleasurable than men and women who did not. Men and women who reported a close relationship with the partner reported stronger emotional reactions than those in casual relationships.

The typical female reaction to first intercourse has been described as the **Peggy Lee syndrome** (named for her song "Is That All There Is?"). De-

spite our culture's romanticized high expectations that the first intercourse experience will be like firecrackers popping on the Fourth of July, it turns out to be much less thrilling than that for most females. For example, the women studied by Sprecher and her colleagues on the average gave the experience only a 2.95 on a pleasure scale that ranged from 1 for "not at all" to 7 for "a great deal."

Premarital Sex with a Prostitute

In the 1940s and 1950s, premarital sex with a prostitute was fairly common among males, and many young men received their sexual initiation in this manner. Among the college-educated men under age 35 in one study (Hunt, 1974), 19 percent said they had had their first intercourse with a prostitute. Today, however, having premarital sex with a prostitute is much less common. In the NHSLS of 3300 adults, 3 percent of the men and $1/10$ of 1 percent of the women reported that their first intercourse had involved a paid partner (Laumann et al., 1994).

Techniques in Premarital Sex

Paralleling the increase in the incidence of premarital intercourse is an increase in the variety of techniques that are used in premarital sex. One of the most dramatic changes has been the increased use of oral-genital techniques. In the Kinsey sample, 33 percent of the males had experienced fellatio premaritally and 14 percent had engaged in cunnilingus. In the DeLamater survey, 65 percent of the males had engaged in fellatio premaritally and 64 percent had performed cunnilingus (DeLamater & MacCorquodale, 1979). In our classes, we generally find that about 5 percent of the women students have engaged in fellatio and/or cunnilingus but not in intercourse—perhaps because some have discovered that mouth-genital sex cannot cause pregnancy. Young people today also use a greater variety of positions, not just the traditional man on top.

Doubtless some of this increased variety in techniques is a result of today's "performance ethic" in sexual relations, which was discussed in Chapter 10. Adolescents and young adults may feel

Peggy Lee syndrome: The feelings of disappointment experienced by teenage girls at first intercourse when it is not as thrilling as they expected.

Figure 11.8 In South American nations, the percentage of unmarried young women who engage in premarital intercourse is lower than in the United States owing to the strong influence of the Catholic Church.

pressured to be gold medalists in the sexual Olympics. One man said:

> Sometimes I'm really good; I can make a girl have orgasms until she's about half dead. But if I don't like the girl, or if I'm not feeling confident, it can be hard work—and sometimes I can't even cut the mustard, and that bothers me a lot when that happens. (Hunt, 1974, p. 163)

Attitudes Toward Premarital Intercourse

Attitudes toward premarital intercourse have also undergone marked changes, particularly among young people. Sociologist Ira Reiss (1960) distinguished among four kinds of standards for premarital coitus:

1. **Abstinence** Premarital intercourse is considered wrong for both males and females, regardless of the circumstances.

2. **Permissiveness with affection** Premarital intercourse is permissible for both males and females if it occurs in the context of a stable relationship that involves love, commitment, or being engaged.

3. **Permissiveness without affection** Premarital intercourse is permissible for both males and females, regardless of emotional commitment, simply on the basis of physical attraction.

4. **Double standard** Premarital intercourse is acceptable for males but is not acceptable for females. The double standard may be either "orthodox" or "transitional." In the orthodox case, the double standard holds regardless of the couple's relationship, while in the transitional case, sex is considered acceptable for the woman if she is in love or if she is engaged.

Abstinence: A standard in which premarital intercourse is considered wrong, regardless of the circumstances.
Permissiveness with affection: A standard in which premarital intercourse is considered acceptable if it occurs in the context of a loving, committed relationship.
Permissiveness without affection: A standard in which premarital intercourse is acceptable even if there is no emotional commitment.
Double standard: A standard in which premarital intercourse is considered acceptable for males but not for females.

Focus 11.3
Teen Pregnancy and Parenthood

In the United States, 1 million or more young women under 20 years of age, or 11 percent of all teenage girls, become pregnant each year (Alan Guttmacher Institute, 1994). The rate of teenage pregnancy in the United States is the highest of any Western nation; teenagers in this country are twice as likely to become pregnant as are Canadian or English teens, four times as likely as Swedish teenagers, and nine times as likely as teen women in the Netherlands. About one-half of these women give birth to a child, and the vast majority choose to keep the baby. Teenage birth rates vary by ethnicity. The rate for whites (39.3 per 1,000 15–19-year-olds in 1995) is much lower than the rate for blacks (99.3) or Hispanics (106.7). Of interest is the fact that birth rates for blacks and whites slowly declined from 1992 to 1995. Why is teenage pregnancy considered a major social problem? The reason is that most of these births are *nonmarital;* most of these babies will be raised by single mothers. Nonmarital births have increased dramatically, from 15 percent in 1960 to 75 percent of all teenage births in 1995. They are associated with tremendous social and economic costs.

Several factors have contributed to the incidence of nonmarital pregnancy and childbearing. First, the age at which puberty occurs has declined dramatically in the past century, from about 17 to about 12 years of age. On the other hand, the average age of marriage has increased from 22 to 26 years of age in the past 40 years. Thus, young men and women are at risk for nonmarital pregnancy for up to 14 years! Second, teen childbearing is associated with economic conditions; teenage women who give birth are much more likely to live in a low-income family (Alan Guttmacher Institute, 1994). Poverty and high rates of unemployment in poor neighborhoods lower young people's edu-cational and occupational aspirations (Coley & Chase-Lansdale, 1998). Girls growing up in these circumstances perceive a greater likelihood that they will have a nonmarital birth

(East, 1998). In fact, it has been suggested that for African American adolescent girls with poor employment prospects, motherhood is a career choice (Merrick, 1995).

Sociologist Frank Furstenberg and his colleagues Jeanne Brooks-Gunn and Philip Morgan did an important study of teenage pregnancy that gave essential information on its effects on the mother and her child. The study is particularly impressive because it followed up the women and their children in 1984, 17 years after the women were initially interviewed while pregnant in 1966–1967. There were approximately 400 respondents, most of them black, all of them initially residing in Baltimore.

Furstenberg and his colleagues concluded that although there are many negative consequences to teenage childbearing, they have been exaggerated and there has not been enough attention paid to those women who, despite the odds against them, manage to cope with adversity and succeed.

Let's focus first on the findings for the mothers. When they were first followed up 5 years after the pregnancy, they looked very disadvantaged. For example, 49 percent had not graduated from high school. Approximately one-third of them were on welfare at some point during the 17 years of the study. However, by the time of the 1984 follow-up, an impressive proportion of women had staged a substantial recovery. At that point, an additional 38 percent had graduated from high school, an additional 25 percent had obtained some education beyond high school, and 5 percent had graduated from college. Of those who had been on welfare at some time during the study, two-thirds had managed to get off it by 1984; 67 percent were employed, and fully a quarter had incomes in excess of $25,000 per year.

The study shows clearly that there is great diversity in the outcomes for adolescent mothers. Some remain locked in poverty for the rest of their lives, whereas others manage to succeed despite their circumstances. The most important factor is differential resources. Women with better-educated parents who have more income tend to do better because they have more resources on which to draw. The second most important factor is competence and motivation. Those women who were doing well in school at the time of the pregnancy and had high educational aspirations were more likely to do well following the birth. A third factor is intervention programs such as special schools for pregnant teenagers and hospital intervention programs. When these programs are successful, they help the women complete high school and postpone other births, two factors that are crucial to recovering from the adverse circumstances of a teenage pregnancy. If there are additional births soon after the first, the woman essentially becomes locked out of the job market.

Turning now to the children, the results indicate that they are at risk in many ways. At birth, 11 percent were at low birth weight (2500 grams or less), which puts them at risk for a variety of other problems (see Chapter 7). However, it seems that the excess of low-birth-weight babies is more a function of the adequacy or inadequacy of medical care during pregnancy than it is a function of the mother being a teenager. By 1984 the school record showed evidence of academic failure and behavior problems. Half of the children had had to repeat at least one grade. Thirty-five percent had had to bring their parents to school in the last year because of a behavioral problem, and 44 percent had been suspended or expelled in the past 5 years. The study sample was also more sexually active than randomly chosen nation tends to perpetuate itself.

Teenage pregnancy in the United States is a serious problem, both because of the large numbers of people affected and because the consequences can be so serious. What can be done?

Furstenberg's results enable us to identify factors crucial to successful outcomes and to design social programs to provide these resources to other teen mothers. Two critical factors to success, for example, are finishing high school (and preferably getting even more education) and postponing other births.

Figure 11.9   An important factor in life success for a pregnant teenager is the existence of special programs that allow her to complete high school.

Social programs need to be set up to assist adolescent mothers in finishing high school (including special schools for pregnant teenagers, and child care for mothers while attending school) (Fig 11.9). Information on and access to contraception is essential. Programs such as Head Start that help prepare the children of teen mothers for school are critical, because they are at high risk for academic failure. Marriage to a man with some financial resources was also a route to success for some women in this study. However, the high rate of unemployment among young, black, urban men makes such marriages less likely. This points out the importance of social programs aimed at males as well as females.

In summary, teenage pregnancy is a serious problem, but not an unsolvable one. By studying those women who stage a recovery from the experience, we can gain important insights into how we can break the cycle of poverty and teen pregnancy.

Source: Frank F. Furstenberg, Jr., J. Brooks-Gunn, & S. Philip Morgan. (1987). *Adolescent mothers in later life.* New York: Cambridge University Press.

Historically in the United States, the prevailing standard has been either abstinence or the double standard. However, today, particularly among young people, the standard is one of permissiveness with affection.

We can see evidence of this new standard, and of the shift it represents from previous generations, by comparing the data on current attitudes toward premarital intercourse with those from previous decades, as shown in Table 11.5. Note that in surveys conducted in 1937 and 1959, few people approved of premarital intercourse. By 1996 more people approved than disapproved, representing a real shift in norms.

Motives for Having Premarital Intercourse

Young people mention a variety of reasons for engaging in physical intimacy (Sprecher & McKinney, 1993). The motives include expressing love or affection for the partner, experiencing physical arousal or desire, wanting to please the partner, feeling pressure from peers, or wanting physical pleasure. Women are more likely to mention love and affection, whereas men are more likely to mention physical pleasure. Respondents to the NHSLS gave two additional reasons for having intercourse the first time: it was their wedding night (7 percent of the men, 21 percent of the women) and the person wanted to get pregnant (less than 1 percent of the respondents) (Laumann et al., 1994).

Dating, Going Steady, Getting Engaged

The social forces that have produced changes in premarital sexual behavior and standards are complex. But among them seems to be a change in courtship stages—in the process of dating, going steady, and getting engaged. Dating and going steady occur at much younger ages now than in previous generations. Dating earlier and going steady earlier create both more of a demand for premarital sex and more of a legitimacy for it. For many, sexual intimacy is made respectable by going steady.

Sorensen (1973) found the most common premarital sexual pattern to be **serial monogamy**

Table 11.5 Percentages of People Agreeing That Premarital Intercourse Is Acceptable, 1937, 1959, 1972, 1982, 1990, 1996

Do you think it is all right for either or both parties to a marriage to have had previous sexual intercourse?

	1937	1959
All right for both	22%	22%
All right for men only	8	8
All right for neither	56	54
Don't know or refused to answer	14	16

If a man and a woman have sex relations before marriage, do you think it is . . .

	1972	1982	1990	1996
Always wrong	35%	28%	25%	23%
Almost always wrong	11	9	11	9
Wrong only sometimes	23	21	22	22
Not wrong at all	26	40	39	43
Don't know	4	3	4	3

Source: Morton Hunt. (1974). *Sexual behavior in the 1970s.* Chicago: Playboy Press, pp. 115–116. National Opinion Research Center, *General Social Survey,* 1972, 1982, 1990, 1996.

without marriage. In such a relationship there is an intention of being faithful to the partner, but the relationship is of uncertain duration. Of those in the sample who had premarital intercourse, 40 percent were serial monogamists. Though they averaged about four partners, nearly half of them had had only one partner, and about half of them had been involved in their current relationship for a year or more. A more recent study indicated that about 25 percent of all women have only one partner premaritally (26 percent for white women, 21 percent for black women); but 23 percent have six or more partners (24 percent of white women, 17 percent of black women) (Tanfer & Schoorl, 1992).

Conflicts

We are currently in an era in which there are tensions between a restrictive sexual ethic and a permissive one. In such circumstances, conflicts are bound to arise. One is between parents and children, as parents hold fast to conservative standards while their children adopt permissive ones.

These conflicts within our society are mirrored in the messages of the mass media.

> We are a nation that is deeply ambivalent about sex. On the one hand, sex is so much a part of the landscape that we almost take it for granted, and the message we get is that everybody else seems to be doing it and we are missing the party if we don't get moving. We should liberate our sexual natures, polish up the hot buttons, follow our hormones, and seek fulfillment somewhere across a crowded room. Sex is the ultimate expression of the American dream of freedom, liberation, and mobility. On the other hand, we hear just how frightening sex can be. Some of that comes from the powerful hold of our puritan heritage, but with a uniquely modern twist. AIDS, urban anonymity, sexual abuse and assault, all make sex a dangerous pastime. And this dovetails nicely with the old morality that coexists with our alleged libertine behavior. (Michael et al., 1994, p. 8)

With such conflicting messages so prevalent, it is no wonder that many young people feel conflicts about premarital sex.

Young people may also experience conflicts between their own behaviors and their attitudes or standards. Behaviors generally change faster than attitudes do. As a result, people may engage in premarital sex while still disapproving of it. For example, in one study of inner-city junior high and senior high school students, of those who were sexually active, 83 percent gave an ideal age for first intercourse that was older than when they themselves had first had intercourse (Zabin et al., 1984). And among those who had engaged in premarital intercourse, 25 percent believed that premarital sex is wrong. These inconsistencies between behavior and attitudes can create feelings of conflict.

How Sexuality Aids in Psychological Development

Erik Erikson, whose work represents a major revision of Freudian theory, has postulated a model of psychosocial development according to which we experience crises at each of eight stages of our lives (Erikson, 1950, 1968). Each one of these crises may be resolved in one of two directions. Erikson notes that social influences are particularly important in determining the outcomes of these crises.

The stages postulated by Erikson are listed in Table 11.6. Note that the outcomes of several of them may be closely linked to sexuality. For example, in early childhood there is a crisis between autonomy and shame, and later between initiative and guilt. The child who masturbates at age 5 is showing autonomy and initiative. But if the parents react to this activity by severely punishing the child, their actions may produce shame and guilt. Thus they may be encouraging the child to feel ashamed and consequently to suffer a loss of self-esteem.

In adolescence, the crisis is between identity and role confusion. Gender roles are among the most im-

Serial monogamy: A premarital sexual pattern in which there is an intention of being faithful to the partner, but the relationship may end and the person will then move on to another partner.

Table 11.6 Erikson's Stages of Psychosocial Development

Approximate Stage in the Life Cycle	Crisis
Infancy	Basic trust vs. mistrust
Ages 1½ to 3 years	Autonomy vs. shame and doubt
Ages 3 to 5½ years	Initiative vs. guilt
Ages 5½ to 12 years	Industry vs. inferiority
Adolescence	Identity vs. role confusion
Young adulthood	Intimacy vs. isolation
Adulthood	Generativity vs. stagnation
Maturity	Ego integrity vs. despair

portant; in later adolescence the person may emerge with a stable, self-confident sense of manhood or womanhood or, alternatively, may feel in conflict about gender roles. A choice of career is extremely important in this developing sense of identity, and gender roles influence career choice. A sexual identity also emerges—one's status as, for example, heterosexual or homosexual, popular or unpopular.

In young adulthood, the crisis is between intimacy and isolation. Sexuality, of course, can function in an important way as people develop their capacity for intimacy.

For adolescents particularly, sexuality is related to accomplishing important developmental tasks (A. P. Bell, 1974a). Among these are:

1. Becoming independent of parents. Sexuality is a way of expressing one's autonomy and one's independence from parents. The adolescent boy who masturbates, for example, may be expressing a need to cut the apron strings that tie him to his mother—as you know if you have read *Portnoy's Complaint.*

2. Establishing a viable moral system of one's own. For many adolescents, some of the most critical moral decisions of their lives made independently of parents are in the area of their own sexual conduct. A personal ethical system emerges.

3. Establishing an identity—in particular, a sexual identity.

4. Developing a capacity for establishing an intimate relationship with another person and sustaining it.

Thus we can see that sexuality is an integral part of our psychological development.

SUMMARY

A capacity for sexual response is present from infancy. According to Kinsey's data, about 10 percent of children have masturbated to orgasm by age 10, although recent studies indicate that children begin masturbating at somewhat earlier ages now than a generation or so ago. Children also engage in some heterosexual play, as well as some homosexual activity.

During adolescence there is an increase in sexual activity. According to one theory, this activity is influenced by the interaction of biological factors (increasing testosterone level) and social and psychological factors (for example, sexually permissive attitudes). By age 15, nearly all boys have masturbated. Girls tend to begin masturbating somewhat later than boys, and fewer of them masturbate. Attitudes toward masturbation are considerably more

permissive now than they were a century ago. About 10 percent of adolescents have homosexual experiences to orgasm, with this figure being slightly higher for boys than for girls.

Today the majority of males and the majority of females have premarital sex. This is a considerable increase over the incidence reported in the Kinsey studies, done 50 years ago. Adolescents today are considerably more likely to use a variety of sexual techniques, including mouth-genital sex. There is variation in the incidence of premarital intercourse among various racial and ethnic groups in the United States, and there is even greater variability from one country to another.

The predominant sexual standard today is one of "permissiveness with affection"; that is, sex is seen as acceptable outside marriage, provided there is an emotional commitment between the partners.

Following Erik Erikson's theory, experiences with sexuality can serve important functions in a person's psychological development. They may be important, for example, in the process of becoming independent of parents and in establishing a viable moral system.

REVIEW QUESTIONS

1. The physical capacity for erection in the male and vaginal lubrication in the female are present as early as _____ (age).

2. Children do not begin masturbating until 10 or 11 years of age. True or false?

3. Longitudinal research indicates that children who engage in interactive sex play have poorer sexual adjustment at ages 17 or 18 than children who do not. True or false?

4. The majority of both males and females in the United States engage in premarital intercourse. True or false?

5. One of the few sexual behaviors to show a decline in incidence in the last few decades is sex with a prostitute. True or false?

6. DeLamater and MacCorquodale, in their sociological analysis of premarital sexual behavior, concluded that the particular couple and the nature of their relationship are the most important factors determining whether people engage in premarital intercourse. True or false?

7. According to sociologist Ira Reiss, the current standard under which most adolescents believe that intercourse is acceptable is termed "permissiveness without affection." True or false?

8. According to Udry's theory, the biological factor of _____ interacts with social psychological factors as influences on adolescent sexuality.

9. According to Furstenberg's study, 95 percent of teen mothers remain below the poverty line throughout adulthood. True or false?

10. Sexuality may aid in psychological development, including establishing a moral system of one's own and establishing an intimate relationship with another person. True or false?

(The answers to all review questions are at the end of the book, after the Glossary.)

QUESTIONS FOR THOUGHT, DISCUSSION, AND DEBATE

1. Do you see evidence of conservative trends in sexual attitudes and behaviors, reversing the trends from 1940 to 1990?

2. Does "permissiveness with affection" characterize the standard for premarital intercourse among the 18- to 22-year-olds you know?

3. The mother of a five-year-old child tells you that her son has been masturbating while he watches TV in the family room. She asks you what she should do about his behavior. What would you tell her?

4. In this chapter, we presented data on ethnic differences in adolescent sexual experiences and in teenage pregnancy and parenting (Focus 11.3). In Chapter 1, we presented data on ethnic differences in masturbation and oral sex (Table 1.2). Using these data, create a brief description of adolescent sexuality among black and white youth in the United States. In what ways are they similar? In what ways are they different?

SUGGESTIONS FOR FURTHER READING

Eder, Donna, with Catherine Evans & Stephen Parker. (1995). *School talk: Gender and adolescent culture.* New Brunswick, NJ: Rutgers University Press. A study of adolescent peer groups and relationships in one middle school.

McCormick, Naomi B. (1979). Come-ons and put-offs: Unmarried students' strategies for having and avoiding intercourse. *Psychology of Women Quarterly, 4,* 194–211. An interesting discussion of college students' reported techniques for inviting or avoiding intercourse, and how these techniques relate to gender-role stereotypes.

Starks, Kay, & Eleanor Morrison. (1996). *Growing up sexual* (2nd ed.). New York: HarperCollins. A fascinating view of sexual development with many first-person quotes, based on student autobiographies for a human sexuality course.

WEB RESOURCES

www.plannedparenthood.org
Planned Parenthood: Teen Issues pages

chapter
12

SEXUALITY AND THE LIFE CYCLE: ADULTHOOD

CHAPTER HIGHLIGHTS

Sex and the Single Person

Cohabitation

Marital Sex

Extramarital Sex

Postmarital Sex

Sex and Seniors

Grow old along with me!
The best is yet to be.*

This chapter will continue to trace the development of sexuality across the lifespan. We will look at various aspects of sexuality in adulthood: sex and the single person, cohabitation, marital sexuality, extramarital sexuality, postmarital sexuality, and sex among the elderly. Each of these lifestyles is an option, reflecting the diversity of choices available in the contemporary United States. As we discuss each of these, we will draw on a theory of the development of sexuality in adulthood, formulated by Lorna and Philip Sarrel (1984).

Sex and the Single Person

Sexual Unfolding

Late adolescence and early adulthood are a time of sexual unfolding, as the individual moves toward mature, adult sexuality. First, there is a need to deal with issues of sexual orientation and define one's sexual identity. Heterosexuality is the overwhelming norm in our society, and some people slip into it easily without much thought. Others sense that their orientation is gay or lesbian and must struggle with society's negative messages about lesbians and gays. Others sense that they are attracted to both males and females. Still others feel that their orientation is heterosexual, but wonder why they experience something like homosexual fantasies, thinking that a person's sexual orientation must be perfectly consistent in all areas (research actually shows that heterosexuals sometimes have homosexual fantasies, and vice versa). These struggles over sexual orientation seem to be more difficult for males than for females because hyperheterosexuality is such an important cornerstone of the male role in many societies, including ours (see Chapter 14).

Another step toward maturity is learning our sexual likes and dislikes and learning to communi-

cate them to a partner. Learning what one likes and dislikes may occur naturally, as the individual experiences various behaviors over time. Alternatively, some people intentionally seek opportunities to engage in novel behaviors, or in sexual intimacy with novel partners. Learning to communicate with partners is difficult for many persons, perhaps because there are few role models in our society showing us how to engage in direct, honest communication with sexual partners.

Two more issues are important in achieving sexual maturity: becoming responsible about sex, and developing a capacity for intimacy. Taking responsibility includes being careful about contraception and sexually transmitted diseases, being responsible for yourself and for your partner. Intimacy (see Chapter 13) involves a deep emotional sharing between two people that goes beyond casual sex or manipulative sex.

The Never-Married

The term *never-married* refers to adults who have never been married. This group includes those who intend to marry someday and those who have decided to remain single. The National Survey of Men interviewed a representative sample of men ages 20 to 39 in the United States (Billy et al., 1993). Of these men, 37 percent (most of whom were in their twenties) were never-married.

Most adults in our society do marry. The median age of first marriage in 1992 was 24.4 years for women and 26.5 years for men (Smith, 1994), so the typical person who marries spends several adult years in the never-married category. Some of these men and women spend this entire time in one relationship that eventually leads to marriage. According to the NHSLS, among married persons 20 to 29 years old, 46 percent of the men and 65 percent of the women are in this category (Laumann et al., 1994). Other young adults continue the pattern of *serial monogamy* which (as we saw in Chapter 11) characterizes adolescent intimate re-

*Robert Browning. (1864). *Rabbi Ben Ezra.*

Figure 12.1 Fitness clubs are the current alternative to singles bars for some people who are hoping to meet Mr. or Ms. Right.

lationships; they are involved in two or more sexually intimate relationships prior to marriage. According to the NHSLS, among married persons 20 to 29, 40 percent of the men and 28 percent of the women had two or more sexual partners before they married.

The person who passes age 25 without getting married gradually enters a new world. The social structures that supported dating—such as college—are gone, and most people of the same age are married. Dating and sex are no longer geared to mate selection, and by the time a person is 30, it no longer seems reasonable to call her or his sexual activity "premarital sex."

The attitudes of singles about their status vary widely. Some plan never to get married; they find their lifestyle exciting and enjoy its freedom. Others are desperately searching for a spouse, with the desperation increasing as the years wear on.

At one extreme, there is the *singles scene.* It is institutionalized in such forms as singles' apartment complexes and singles bars. Health clubs, fitness centers, church groups, and Rotary Club lunches provide opportunities for meeting others. The singles group, of course, is composed of the divorced and the widowed as well as the never-married. The singles bar is a visible symbol of the singles scene. Everyone is there for a similar purpose: to meet Mr. or Ms. Right. However, most will settle for a date, and it is fairly well understood that coitus will be a part of the date. The singles bar is somewhat like a meat market; the people there try to display themselves to their best advantage and are judged and chosen on the basis of their physical appearance— and perhaps rejected for too high a percentage of fat.

Many singles, however, do not go to singles bars. Some are turned off by the idea; some feel that they cannot compete, that they are too old, or that they are not attractive enough; and some live in rural areas where they have no access to such places. An alternative way of meeting people is through singles ads, found in most daily newspapers. As one woman said,

> Most of the men I've dated I've met . . . through personals ads. The reason I prefer guys I meet through ads is because I get to know them before meeting them. I get the chance to really get to know them before I actually see them. (Louis, 1997, p. 10)

Singles ads can also be found on the Internet. There are a number of sites, for example Webmatch, where one can post ads or create a personal web page. Persons wishing to place ads are encouraged to indicate the type of partner they are searching for, as well as their age, ethnicity, height, and area of residence. Those seeking partners can search the ads and home pages on these characteristics.

The visibility of singles ads, singles bars, cruises, and other activities geared toward single adults suggests a fun-loving lifestyle with frequent sexual activity. Undoubtedly some single persons live such a life. As Figure 12.2 indicates, 26 percent of the single men and 22 percent of the single women interviewed for the NHSLS reported having sexual intercourse two or more times per week. But the

Figure 12.2   Frequency of sexual activity is closely related to marital status. Note that there is substantial variability in frequency within each status as well (Laumann et al., 1994).

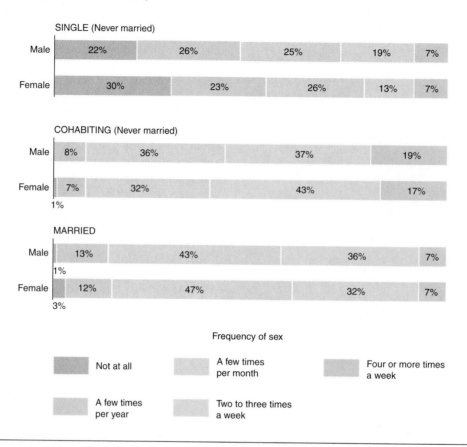

reality is different for other singles; 22 percent of the single men and 30 percent of the single women interviewed did not have sex in the year prior to the interview.

As we noted in Chapter 1, black men and women are more likely to remain single than are their white counterparts (U.S. Bureau of the Census, 1997). In 1996, 47 percent of black family households were headed by a single woman, compared with 14 percent of white (and 26 percent of Hispanic) households. In part, these family arrangements reflect choice. In part, they reflect the fact that there are more adult black women than men (Kiecolt et al., 1995). But they also reflect the structural circumstances of blacks in American society. It is difficult for many black men to find a job that provides the wages and

benefits needed to support a family (Anderson, 1989). As a consequence, some black women are unable to find a suitable black man (Chapman, 1997). When they do, they are more likely than white or Hispanic women to report that they decide whether sex will occur and that they control what behaviors the couple engages in (Quadagno et al., 1998).

Cohabitation

In early adulthood, it is common for couples to experiment with various levels of commitment, such as an exclusive dating relationship or living together. Even when living together, there are different levels of commitment, from "some days and nights" to "all the time." Living together is an important turning point not only because it represents commitment but because it is a public declaration of a sexual relationship. It is rare for a man and woman to live together just because it will save on rent. Cohabiting is an opportunity to try out marriage, at least to some extent.

Cohabitation has become an increasingly common alternative to marriage. In 1996, cohabitors comprised 6.75 percent of all couples in the United States (Bureau of the Census, 1997). Twenty-five percent of people aged 19 to 24 and 42 percent of people aged 25 to 29 have cohabited at least once. These arrangements tend to be short-lived; one-third last less than one year, and only 1 out of 10 lasts 5 years (Bumpass et al., 1991). Almost three-fourths of the men and women who are cohabiting have plans to marry or think they will marry their partner. In fact, 60 percent of these couples do marry. Contrary to what many people think, these marriages are more likely to end in divorce than marriages that are not preceded by cohabitation.

The popular image of cohabitation is that it involves young, never-married couples without children. About one-third of cohabiting couples fit this image, but 40 percent have children. In one-third of these families, the two adults are the biological parents of the children. In the other two-thirds, the children were born into a previous union of one of the adults (Bumpass, Sweet, & Cherlin, 1991). Some formerly married people choose to live with someone instead of remarrying.

With regard to sexual behavior, an analysis of data from a representative sample of more than 7,000 adults found that married persons reported having intercourse 8 to 11 times per month, whereas cohabiting persons reported a frequency of 11 to 13 times per month (Call et al., 1995). The NHSLS found that cohabiting men and women reported more frequent sex than did married men and women (see Figure 12.2) However, notice the wide variation; some cohabitors report having sexual intercourse only a few times per year. It is interesting that on average cohabiting couples have sex more often than married couples. Cohabitors are concerned about the stability of the relationship (Bumpass et al., 1991); they may have sex more frequently in the hope that it will strengthen the relationship (Blumstein & Schwartz, 1983).

Marital Sex

Marriage is a sexual turning point (Sarrel & Sarrel, 1984) for a number of reasons. The decision to get married is a real decision these days, in contrast to previous decades when everyone assumed that they would marry, and the only question was to whom. Today, most couples have had a full sexual relationship, sometimes for years, before they marry. Some psychological pressures seem to intensify with marriage, and these pressures may result in sex problems where there were none previously. Marriage is a tangible statement that one has left the family of origin (the family in which you grew up) and shifted to the family of procreation (in which you become the parents rearing children); for some, this separation from parents is difficult. The pressure for sexual performance may become more intense once married; when just living together, a couple can always say to themselves that if things don't work out in bed, they can simply switch to another partner. And finally, marriage still carries with it an assumption of fidelity or faithfulness, a promise that is hard for some to keep.

In marriage, there is a need to work out issues of gender roles. Who does what? Some of the decisions are as tame as who cooks supper. But who initiates sex is far more sensitive, and who has the right to say "no" to sex is even more so.

This is the era of two-career couples, or at least dual-earner couples. There are issues here of finding time for sex and for just being with each other.

As a marriage progresses, it can't stay forever as blushingly beautiful as it seemed on the day of the wedding. The nature of love changes (see Chapter 13), and for some couples there is a gradual disenchantment with sex. Couples need to take steps to avoid boredom in the bedroom. Sexual disorders (see Chapter 19) occur in many marriages, and couples need to find ways to resolve them.

Frequency of Marital Intercourse

About 90 to 95 percent of all people in the United States marry (Bureau of the Census, 1990). Of those who divorce, a high percentage—80 percent—remarry (Norton, 1987). In our society, marriage is also the context in which sexual expression has the most legitimacy. Therefore, sex in marriage is one of the commonest forms of sexual expression for adults.

The average American married couple have coitus two to three times per week when they are in their twenties, with the frequency gradually declining as they get older. The data on this point from three studies are shown in Table 12.1. Several things can be noted from the table. First, the frequency of marital sex has remained about the same from the 1940s to the 1990s. In each survey, people in their twenties report having intercourse about every 3 days. Second, the frequency of intercourse declines with age; however, in 1993, among couples in their fifties, the frequency was still once per week.

Two general explanations have been suggested for the age-related decline in frequency: biological aging, and habituation to sex with the partner (Call et al., 1995). With regard to aging, there may be physical factors associated with age that affect sexual frequency, such as a decrease in vaginal lubrication in females, or increased likelihood of poor health. The habituation explanation states that we lose interest in sex as the partner becomes more and more familiar. Recent data indicate a sharp decline

in frequency after the first year and a slow, steady decline thereafter. The decline after the first year may reflect habituation (Call et al., 1995). A third factor is the arrival of children, discussed below.

It is important to note that there is wide variability in these frequencies. For example, 2 percent of couples in their twenties report not engaging in intercourse at all; 7 percent of all married couples had not had sex in the 12 months prior to the interview (Smith, 1994). Research on a sample of 6,029 married couples found that sexual inactivity was associated with unhappiness with the marriage, lack of shared activity, presence of children, increased age, and poor health (Donnelly, 1993). In contrast, a married couple in Seattle have claimed the world record, having had intercourse more than 900 times in 700 days! Data from the NHSLS also confirm this wide variability, as shown in Figure 12.2.

Techniques in Marital Sex

The NHSLS (Laumann et al., 1994) included a number of questions about specific aspects of sexual interactions. For example, it asked respondents to es-

(a)

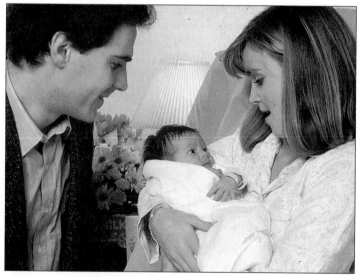

(b)

Figure 12.3 Sexual turning points. *(a)* Marriage and the commitment it represents is a major turning point. *(b)* The birth of a baby is a turning point that can have a negative impact on sexual aspects of the relationship, but couples who are aware of this possibility can work to overcome these problems and keep the romance going.

Table 12.1 Marital Coitus: Frequency per Week (Male and Female Estimates Combined), 1938–1949, 1970, and 1993

1938–1949 (Kinsey)		1970 (Westoff)		1993 (Smith)	
Age	Median Frequency per Week	Age	Mean Frequency per Week	Age	Mean Frequency per Week
16–25	2.45	20–24	2.5	18–29	2.2
26–35	1.95	25–34	2.1	30–39	1.7
36–45	1.40	35–44	1.6	40–49	1.3
46–55	0.85			50–59	1.0
56–60	0.55			60–69	0.6
				70+	0.3

timate the duration of their last sexual interaction. The number of minutes the person reports is probably not precise; Laumann and his colleagues focus on whether the duration reported is 15 minutes or less, or 1 hour or more. Sixteen percent of the married people reported that it lasted 15 minutes or less; about 9 percent reported that it lasted 1 hour or more. Incidentally, one-third of the never-married respondents said it lasted 1 hour or more.

The increased popularity of mouth-genital techniques is one of the most dramatic changes in marital sex to have occurred in the past 50 years. According to Kinsey's data, 54 percent of the married women reported that they had received cunnilingus, and 49 percent reported that they had engaged in fellatio (Kinsey et al., 1948). In the NHSLS data, 74 percent reported that their partners had stimulated their genitals orally, and 70 percent of them had stimulated their partners orally. Women who have attended college are twice as likely to report both techniques as women who did not complete high school.

> Much larger percentages of women under age fifty as compared to those over fifty have given or received oral sex in their lifetime. And those in the oldest age group were less than half as likely to have had or received oral sex the last time they had sex. This is suggestive evidence that oral sex came into vogue in the 1960s. (Michael et al., 1994)

Kinsey did not report data on anal intercourse. According to the NHSLS data, 27 percent of the married men and 21 percent of the married women reported having engaged in this activity (Laumann et al., 1994).

Negotiating Sex

Before any of these techniques are executed, there is typically a "mating dance" between the man and the woman. Sexual scripts are played out in marriages as in other aspects of sex (J. H. Gagnon, 1977, pp. 208–209). Some scripts involve direct verbal statements. One person may say, "I'd love to go to bed with you right now." The partner might reply, "What are we waiting for?", or "Not now. Dinner is almost ready." Some scripts are behavioral. One woman said, "We'd get into bed, and he'd roll over to me and kind of bing me in the back end with his penis, so I'd know, Okay, tonight's the night." (Maurer, 1994, p. 180) For other couples, deciding to have intercourse involves preliminary negotiations, which are phrased in indirect or euphemistic language, in part so that the person's feelings can be salvaged if her or his partner is not interested. For example, the husband may say, "I think I'll go take a shower" or "I think I'll go take a nap" (that means "I want to, do you?"). The wife may respond with, "I think I'll take one too" (that means "yes") or "The kids will be home any moment" or "I have a headache" (that means "no"). Or conversely, she may put on a lot of his favorite perfume and parade around in front of him (that is *her* offer). He may respond with, "I had an exhausting day at work" (his "no") or "I'll meet you upstairs" (his "yes"). To avoid some of the risk of rejection inherent in such negotiations, some couples ritualize sex so they both understand when it will and when it will not occur—Thursday night may be their time, or perhaps Sunday afternoon.

A survey of married couples found that for 33 percent of them the husband and wife are about equally likely to initiate sex; for 51 percent the husband is more likely to be the initiator, and in only 16 percent of the couples is the wife usually the initiator (Blumstein & Schwartz, 1983). Thus there is some evidence of liberation (the couples where both are initiators), but traditional roles persist, with the majority of couples having the male in the initiating role. Women seem to be particularly careful not to initiate sex when they believe their spouse is feeling psychologically vulnerable. The traditional gender-typing of initiation patterns may also be related to how people deal with a refusal. If the man initiates and the woman refuses, he can simply attribute it to her lesser sexual appetite, according to traditional stereotypes. If the woman initiates and the man refuses, she has no stereotype to rescue her, and she is likely to conclude that he is not interested in *her* (Blumstein & Schwartz, 1983). The recent emphasis on a woman's right to sexual fulfillment has lessened this difference and made refusal by the woman threatening to the man's self-esteem (Duncombe & Marsden, 1994). Both men and women in equalitarian relationships may have to work at sustaining (the appearance of) a sexually fulfilling relationship.

Masturbation in Marriage

Many people masturbate; the NHSLS found that 63 percent of the men and 42 percent of the women reported masturbating in the past year. Seventeen percent of the married men and 5 percent of the married women masturbate at least once a week (Laumann et al., 1994).

Many adults continue to masturbate even though they are married and have ready access to heterosexual sex. This behavior is perfectly normal, although it often evokes feelings of guilt and may be done secretly. According to the NHSLS, married people were more likely to report that they masturbated than were single people (Michael et al., 1994). Masturbation can serve very legitimate sexual needs in marriage. It can provide sexual gratification while allowing the partner to remain faithful to a spouse when husband and wife are separated or cannot have sex for some reason such as illness.[1]

Masturbation can also be a pleasant adjunct to marital sex. According to a 49-year-old man:

> One of the other things we do a lot is masturbation. We have it developed to a fine art. We rent a porno movie, take a bath, rub each other down with baby oil. My wife and I not only masturbate each other at the same time, but we get pleasure watching each other masturbate individually as well. It's a terrifically exciting thing. (Janus & Janus, 1993, p. 383)

Satisfaction with Marital Sex

Satisfaction with sex has two components: satisfaction with the sexual activity and emotional satisfaction. In the NHSLS, 51 percent of the married men and 40 percent of the married women said they were "extremely" or "very" physically satisfied by their sexual relationship. Similarly, 48 percent of the husbands and 42 percent of the wives said they were "extremely satisfied" emotionally (Laumann et al., 1994). Analyses of the data indicate that married men and women are significantly more satisfied than are cohabiting or single men and women in a continuing relationship (Waite & Joyner, 1999). The results indicate that this greater satisfaction reflects the stronger emotional commitment and sexual exclusivity associated with marriage.

Another study found that satisfaction with the quality of one's sex life varied considerably as a function of the frequency of sexual intercourse (Blumstein & Schwartz, 1983). For example, among wives who have intercourse three times a week or more, 89 percent are satisfied with their sex life, compared with 32 percent among those who have sex once a month or less. Of course, it is important to remember that these are correlational data, and it is not clear whether the satisfaction is the effect or the cause of the frequency of marital sex.

---

[1]An old Navy saying has it, "If your wife can't be at your right hand, let your right hand be your wife."

In-depth interviews with 52 people, ages 12 to 69, straight, gay, lesbian, and bisexual, provide a different but related picture (Maurer, 1994). Reflecting on what differentiated those who were happy with their sex lives, the researcher identified four factors. First, there is a sense of calm about, an acceptance of their sexuality. Second, happy people are generous; they delight in giving their partners sexual pleasure. Third, these people *listen* to their partners and are aware of the partner's quirks, moods, likes and dislikes. Fourth, they *talk*, both in and out of bed, even though it is often difficult. These interviews remind us that good communication is essential to a satisfying relationship (see Chapter 10).

Sexual Patterns in Marriage

Sexual patterns can change during the course of a marriage. After 10 years of marriage they may be quite different from what they were during the first year. One stereotype is that sex becomes duller as marriage wears on, and certainly there are some marriages in which that happens. As we noted in Chapter 10, a boring sexual relationship can be spiced up by telling each other what you really want to do and then doing it, or by consulting a how-to manual such as *The New Joy of Sex*. One such book changed one 23-year-old married woman's sexual relationship:

> To say that our sex life had been going through the doldrums would be putting it mildly. Sometimes a whole month would go past and I would have another period and realize that during that whole month we hadn't made love once, not even *once,* whereas when we first dated we used to make love six or seven times a week. [I read] *Sex Secrets of the Other Woman,* how "the other woman" takes the trouble to have her hair done well and to look extra good. I went downtown and had my hair highlighted and cut. I bought some really sexy lingerie. When David came home from work that evening, I had a martini ready for him, like I always do. But I wasn't dressed in my usual jeans and sweater. I was wearing my new negligee, and when he sat down I opened it up for him and did a twirl. I was frightened. But he smiled and shook his head and said "Heyyy, that's pretty!" (Masterton, 1993, pp. 86–88)

There are also relationships in which the sex remains very exciting. A 36-year-old engineer said:

> [Though my wife's career] is tremendously important to her, she manages to look attractive, and to dress chicly, and though she is not what many would call a beautiful woman, she is, to me, a handsome woman. In bed, she is the hottest, most exciting woman I have ever known. We have been married seven years. . . . When we get to bed, and she lets herself go, we get wild together. (Janus & Janus, 1993, p. 191)

Having a baby—what researchers call the transition to parenthood—has an impact on a marriage and on the sexual relationship of the couple. According to one 44-year-old mother of three children, "There's the passionate period when you can't get enough of each other, and after a few years it wanes, and after kids it really wanes." (Maurer, 1994, p. 403) Trying to get pregnant and the threat of infertility, which are so much publicized, can be potent forces on one's identity as a sexual being. Pregnancy itself can influence a couple's sexual interactions, particularly in the last few months (see Chapter 7).

For the first few weeks after the baby is born, intercourse is typically uncomfortable for the woman. While estrogen levels are low—which lasts longer when breast-feeding—the vagina does not lubricate well. Then, too, the mother and sometimes the father feel exhausted with 2 A.M. feedings. The first few months after a baby is born are usually not the peak times in a sexual relationship, and so that, too, must be negotiated between partners.

Some people will experience fundamental changes in their sexual experience at least once over the course of the marriage. The change may result from developing a capacity to give as well as receive sexual pleasure. A man may outgrow performance anxiety and enlarge his focus to include his partner. A woman may learn that she can take care of her own sexual needs as well as her partner's. Aging may produce change in sexual experience, a topic we consider later in this chapter. There are changes due to illness, such as breast cancer or testicular cancer, which can lead to disaster or triumph depending upon how the couple copes with it.

Sex and the Two-Career Family

In our busy, achievement-oriented society, is it possible that work commitments—particularly with the increased incidence of wives holding jobs—may interfere with a couple's sex life? One couple, both of whom are professionals,

(a)

(b)

Figure 12.4   Sex and the two-career family. *(a)* Research indicates that marital/sexual relationships do not suffer if the woman works outside the home. *(b)* However, for those working 60 or more hours per week, some experts are concerned because these workaholics literally take their work to bed with them.

commented to us that they actually have to make an appointment with each other to make love.

Research shows that there is little cause for concern. A longitudinal study followed 570 women and 550 of their husbands for one year following the birth of a baby (Hyde, DeLamater, & Hewitt, 1998). The women were categorized according to the number of hours worked per week: homemakers, employed part time (6 to 31 hours/week), full time (32 to 44 hours), and high full time (45 or more hours). There actually were no significant differences among the four groups in frequency of sexual intercourse, sexual satisfaction, or sexual desire. It was not hours of work, but rather the quality of work that was associated with sexual outcomes. Women and men who had satisfying jobs reported that sex was better, compared with people who expressed dissatisfaction with their jobs. For women, fatigue was associated with decreased sexual satisfaction, but that was true for both homemakers and employed women; and homemakers reported the same level of fatigue as employed women.

Problems may occur at the extremes, though. Two-profession couples, when both members are committed to working 60 to 80 hours per week, don't have much time for sex. The issue with such couples is not so much having a career as being workaholics (Sarrel & Sarrel, 1984). An addiction to one's work can spell the death of sex just as readily as addiction to a drug can.

Keeping Your Mate

Most couples who establish a long-term relationship intend to stay together. However, we all know that not all couples succeed. What makes men and women susceptible to infidelity? A study of 107 couples married less than one year asked each partner how likely he or she was to be unfaithful in the next year (Buss & Shackelford, 1997b). Each was asked the likelihood that she or he would: flirt, kiss passionately, and have a romantic date, a one-night stand, a brief affair, or a serious affair with someone of the opposite sex. Thirty-seven percent of the men and 38 percent of the women predicted they would flirt, while 5 percent of men and 7 percent of women said they would kiss. Two percent (of men and of women) predicted a one-night stand, and less than one percent (of men and women) thought they would have a serious affair. In addition, researchers measured a variety of personality, mate value, and relationship characteristics. Among the personality variables, high scorers on narcissism and impulsiveness gave a higher probability of infidelity. Characteristics of the relationship associated with greater likelihood of infidelity included reports of conflict, especially that the partner sexualized others, engaged in sexual withholding, and abused alcohol. Finally, among both men and women, dissatisfaction with the marriage and with marital sex was associated with susceptibility to infidelity.

The role of dissatisfaction with the relationship is obvious in this explanation for a serious affair. "I'm definitely not looking for more sex. The affair I'm having is for emotional reasons. Freddie [her husband] is very self-centered. He's not an emotional support. He's distant, and we have nothing in common." (Maurer, 1994, p. 391)

How would you know if your partner was being emotionally or sexually unfaithful? Researchers asked those two questions of 204 undergraduate students; 82 percent were white, 43 percent were men, and 80 percent reported a past or current committed relationship (Shackelford & Buss, 1997). The participants identified 170 behaviors they thought would be cues of infidelity. In a subsequent study, one-half of a similar sample was asked how likely it is that, if the behavior occurred, the partner was being emotionally/sexually unfaithful. The other half was asked how likely it was that, if the partner was being unfaithful, the behavior would occur. Cues thought to be associated with sexual infidelity included physical signs (partner contracts STD), changes in "normal" sexual behavior with partner, increased or decreased sexual interest, and partner discloses the infidelity. Emotional infidelity was evidenced by expressions of relationship dissatisfaction, emotional disengagement from one's partner, inconsiderateness, being angry and critical toward one's partner, and acting guilty.

Our awareness of the possibility of infidelity sometimes leads us to engage in behaviors designed to preserve the relationship, or *mate retention tactics* (Buss & Shackelford, 1997a). Such tactics may be elicited by our own fear that the partner is losing interest or is dissatisfied, or because we observe some cues to infidelity. Members of the sample of 107 married couples described earlier were given a list of 104 mate-retention behaviors and asked how often they had performed each in the past year. There were marked gender differences in reported actions. Men reported greater use of resources display (giving her money) and more frequent submission to the partner. Women reported more frequent use of enhancing their appearance or attractiveness and use of possessive verbal statements. Evolutionary theory predicts that we make greater effort to retain mates with greater reproductive value. Indeed, men married to young and physically attractive women reported greater use of these tactics, whereas women

married to men with higher incomes and who engaged in resource display and social networking reported greater use of the tactics.

Extramarital Sex

Extramarital sex, or adultery, refers to sexual activity between a married person and someone other than that person's spouse. Extramarital sex can occur under several different circumstances (Pittman, 1993). Sometimes it is *accidental,* unintended and not characteristic of the person; it "just happens." One or both persons may be drunk, or having a bad day, or lonely. More serious is *romantic infidelity,* when the two people fall in love and consider or establish a long-term relationship; this situation can be very destructive to spouses, children, and careers. We noted earlier in this chapter that dissatisfaction with the relationship is especially likely to lead to romantic infidelity. A third type is the *open marriage,* in which the partners agree in advance that each may have sex with other persons. (In contrast, accidental and romantic infidelities begin without the spouse's knowledge, and may remain secret.) Finally, there are *philanderers*—gay, lesbian, or straight—who repeatedly engage in sexual liaisons outside their committed relationship. These men and women are not motivated by the desire for sexual gratification; they are searching for self-affirmation (Pittman, 1993).

How Many People Engage in Extramarital Sex?

Extramarital sexual activity is not as common as many people believe. Laumann and his colleagues, in the NHSLS (1994), found that about 25 percent of the married men and 15 percent of the married women surveyed reported having engaged in extramarital sex at least once.

The incidence of extramarital sex varies from one group to another. According to the General

---

Extramarital sex: Sexual activity between a married person and someone other than that person's spouse; adultery.

Social Survey data, it varies by ethnicity: 27 percent of blacks report extramarital sexual activity, compared to 14 percent of whites (Smith, 1994). Data from the NHSLS suggest that Hispanics have the same rate of extramarital sex as whites (Laumann et al., 1994).

All of these percentages are of persons who had sex with someone other than the spouse while married. At the time of the survey, some of these people were divorced. Others were remarried, perhaps for the second or third time. We can ask a more specific question; how many people engage in extramarital sex during their first marriage? Data from the NHSLS (1994) indicate that from 10 to 23 percent of the men were unfaithful, compared to 6 to 12 percent of the women, depending on the person's age.

For women, at least, there is no indication that extramarital sex is casual or promiscuous. In one survey, of those married women who had had extramarital sex, 43 percent had done so with only one partner (Blumstein & Schwartz, 1983).

Attitudes Toward Extramarital Sex

While, as we have seen, attitudes toward premarital sex have changed substantially during the last several decades, attitudes toward extramarital sex have apparently remained relatively unchanged; most people in the United States disapprove of extramarital sex. According to a well-sampled 1991 survey of adult Americans, 75 percent believe it is always wrong for a married person to have sex with someone other than the marriage partner; this statistic is relatively unchanged from 1973, when it was 69 percent (Davis & Smith, 1991). Some people view unfaithfulness to a partner in any type of committed relationship as the equivalent of adultery. And some don't limit the term to cases of sexual intimacy. "Adultery is absolutely anything," said one young woman.

One of the best predictors of extramarital sexual permissiveness is premarital sexual permissiveness (Thompson, 1983). That is, the person who has a liberal or approving attitude about premarital sex is also likely to hold a liberal or approving attitude toward extramarital sex. On the other hand, attitudes toward extramarital sex are not very good predictors of extramarital sexual behavior (Thompson, 1983). That is, the person who approves of extramarital sex is not more likely

to actually engage in extramarital sex than the person who disapproves of it. Therefore, we have to look to factors other than attitudes in trying to understand why people engage in extramarital sex; some of these other factors are discussed later in this section.

Because our society condemns extramarital sex, the individual who engages in it typically has confused, ambivalent feelings. A young married woman describes her feelings:

> I don't like the illicit part of the affair. Mostly, it's a nuisance, because it's very difficult to find time, and I don't like lying to Freddie and sneaking around. If he wouldn't mind, I'd tell him. I don't think he would go for that. He'd show up with a gun. (Maurer, 1994, p. 393)

Swinging

One form of extramarital sex is **swinging,** in which married couples exchange partners with other couples, or engage in sexual activity with a third person, with the knowledge and consent of all involved.[2]

Swingers may find their partners in several ways. Often they advertise, in tabloid newspapers, in swingers' magazines such as the *Swing Times,* or on specialized bulletin boards and websites on the Internet. The following is an example:

> We are engaged bicouple lookin to meet bim, bif and bicouples for friendship and fun . . . we are into nudism, motorcycling, fishin, volleyball, pool and campin . . . she is 22 5'5 180# 38-d blond blue . . . he is 24 5'10 145# blond green 7' and very thick. email: (www . . . / ~gnkfoxx/ nefriend.htm)

Swingers may also meet potential partners at swingers' clubs, parties, or resorts. Many of these places advertise in swingers' magazines and newsletters and are listed on specialized websites.

Several organizations and many local groups or couples sponsor parties. The date and general

---

[2]Swinging was originally called "wife-swapping." However, because of the sexist connotations of that term and the fact that women were often as eager to swap husbands as men were to swap wives, the more equitable "mate-swapping" or "swinging" was substituted.

Swinging: A form of extramarital sex in which married couples exchange partners with each other.

Focus 12.1

Have Adults Changed Their Sexual Behavior in the AIDS Era?

Time magazine, in a cover story, called it "The Big Chill—How Heterosexuals Are Coping with AIDS." For the past decade, the mass media have published and broadcast a steady stream of features about AIDS, including its risks and recommendations for safer sex practices. Has all this publicity had any impact on people's sexual behavior, or do people continue with high-risk practices? (See Chapter 20 for a discussion of high-risk and safer sex practices.)

If you ask adults whether they have changed their behavior because of the risk of AIDS, some say they have. Moreover, the percent saying yes has increased over time (Smith, 1994). In several surveys conducted between 1986 and 1988, 7 to 13 percent reported changing their behavior. In surveys conducted between 1989 and 1991, 14 to 23 percent said yes.

In order to assess whether behavior has actually changed, we need longitudinal research, where the same people are surveyed or the same questions are asked of comparable samples over time. Unfortunately, there are few such studies.

Research shows that gay men, as a group, have shown the greatest behavior change in the AIDS era (Ehrhardt et al., 1991). Many gay men have reduced the number of sex partners, have fewer anonymous sexual encounters, and engage less in anal intercourse or use condoms consistently. Unfortunately, some gay men have not, especially younger gay men who were not part of the communitywide efforts of the 1980s to change behavior. One recent study surveyed 425 gay and bisexual men aged 17 to 22 in the San Francisco Bay area (Lemp et al., 1994). One-third reported unprotected anal intercourse in the preceding six months; 38 percent of the blacks and 40 percent of the Latinos reported this activity, compared to 27 percent of Asians and 28 percent of whites. Twelve percent reported injecting drugs in the same period.

Among heterosexuals, the number of unmarried adults aged 18 to 39 who report having multiple partners has declined. In 1988, 39 percent reported two or more partners in the prior 12 months; in 1993, only 30 percent reported two or more partners. The 1988 and 1989 General Social Surveys included questions about risky sexual behavior (Smith, 1991). Men were almost four times more likely than women to report risky behavior in the year prior to the survey (11.8 percent compared to 3.1 percent).

One survey assessed condom use in 1988–1989 and again in 1989–1990; the sample consisted of single heterosexuals aged 20 to 45 living in the San Francisco Bay area (Catania et al., 1993). The results showed only a marginal (4 percent) improvement in condom use. Blacks were more likely than Hispanics or whites to increase their use of condoms. Other studies show low levels of condom use by heterosexual men and women. Binson and her colleagues (1993) found that 40 percent of adults with multiple partners never use condoms.

There are other changes that sexually active people can make to reduce their risk. In addition to reducing the number of partners and using condoms more frequently, one can establish a monogamous relationship, select partners more carefully, get to know partners better before initiating sexual intimacy, or abstain from sex. In the NHSLS, 30 percent reported making one or more of these changes (Feinleib & Michael, 1999). Analyses indicated that males were more likely to report change than were females (35 percent vs. 25 percent), blacks were more likely than whites (45 percent vs. 26 percent), and single people were more likely to report at least one change (51 percent) than were those who were cohabiting (40 percent) or married (11 percent).

In conclusion, the data suggest substantial changes toward safer sexual behavior among gay men but at best modest changes among heterosexual men and women. The lack of change among heterosexuals reflects in part their mistaken perception that they are not at risk of HIV infection (Smith, 1991).

location of the party is publicized in magazines and on the Internet. Interested persons call or email a contact person who screens them. If they pass, they are told the exact location of the party, often a private home or a hotel. A fee of $50 or more per couple may be charged for membership or entry to the party. The Lifestyle Organization in southern California sponsors parties and dances at hotels and motels, and it publicizes the exact location in advance. It merely hosts the gathering. Couples who want to swing must connect with others on their own, and they are free to rent a room in the hotel/motel, or travel to some other location to engage in sexual activity.

A man who frequently hosts parties describes what happens:

> A lot of people have the idea that swinger parties are big orgies, where everybody jumps on everybody else. It isn't that way. People are selective, like they are anyplace else. [It starts with the eyes.] So if the interest continues from eye contact to talking, and to desire, there's touching. You just go with it. So you go from talking to touching, and at a swinging party you can go from touching to bed. (Maurer, 1994, p. 120)

Swinging may be closed or open. In *closed swinging*, the couples meet and exchange partners, and each pair goes off separately to a private place to have intercourse, returning to the meeting place at an agreed-upon time. In *open swinging*, the pairs get back together for sex in the same room for at least part of the time. In 75 percent of the cases, this includes the women having sex with each other, although male homosexual sex almost never occurs (Bartell, 1970; Gilmartin, 1975).

What kind of people are swingers? In the non-sexual areas of their lives, they are quite ordinary—perhaps even dull. Although they describe themselves in their ads as exciting people with many interests, in fact they engage in few activities and have few hobbies. As one researcher observed, "These people do nothing other than swing and watch television" (Bartell, 1970, p. 122). Although one might expect them to be politically liberal, in one study the greatest number were Republicans, and only 27 percent described themselves as politically liberal—the rest were moderate or conservative (Jenks, 1985). And in this same sample, 93 percent were white; African-Americans and other minorities are rare among swingers.

Why do people become swingers? In a study of 406 swingers and a control group of 340 non-swingers, it was found that a key factor for swingers was a low degree of jealousy (Jenks, 1985). Swingers also tend to have tolerant attitudes toward various forms of "exotic" sexuality, and they have a strong interest and involvement in sex.

Equity and Extramarital Sex

Equity theory is a social-psychological theory designed to predict and explain many kinds of human relations (Hatfield, Walster, & Berscheid, 1978). In particular, it has been applied to predicting patterns of extramarital sex (Hatfield, 1978).

The basic idea in equity theory is that in a relationship, people mentally tabulate their inputs to it and what they get out of it (benefits or rewards); then they calculate whether these are equitable or not. In an equitable relationship between person A and person B, it would be true that

$$\text{Rewards}_A - \text{Inputs}_A = \text{Rewards}_B - \text{Inputs}_B$$

In a traditional marriage, the wife's inputs might include her beauty, keeping a charming house, cooking good meals, and so on. The husband's inputs might include his income and his pleasant temperament. His rewards from the relationship might include feeling proud when he is accompanied by his beautiful wife, enjoying her cooking, and so on. Notice that this is not an egalitarian relationship in the modern sense; however, it is an equitable relationship (as defined by equity theory) because both partners derive equal benefits from it.

According to equity theory, if individuals perceive a relationship as inequitable (if they feel they are not getting what they deserve), they become distressed. The more inequitable the relationship, the more distressed they feel. In order to relieve the distress, they make attempts to restore equity in the relationship. For example, people who feel they are putting too much into a relationship and not getting enough out of it might let their appearance go, or not work as hard to earn money, or refuse sexual access, or refuse to contribute to conversations. The idea is that such actions will restore equity.

Equity theory: A theory that states that people mentally calculate the benefits and costs for them in a relationship; their behavior is then affected by whether they feel there is equity or inequity, and they will act to restore equity if there is inequity.

Figure 12.5 Equitable sharing of household tasks in a marriage. According to equity theory, if a person perceives that the marital relationship is inequitable and feels underbenefited, he or she is more likely to engage in extramarital sex.

If these equity processes do occur, they might help to explain patterns of extramarital sex. That is, engaging in extramarital sex would be a way of restoring equity in an inequitable relationship. Social psychologist Elaine Hatfield (1978) tested this notion. Her prediction was that people who felt underbenefited in their marriages (that is, they felt that there was an inequity and that they were not getting as much as they deserved) would be the ones to engage in extramarital sex. Confirming this notion, people who felt they were underbenefited began engaging in extramarital sex earlier in their marriages and had more extramarital partners than did people who felt equitably treated or overbenefited. Apparently, feeling that one is not getting all one deserves in a marriage is related to engaging in extramarital sex. (As an aside, equitable marriages were rated as happier than inequitable ones.)

Equity theory includes rewards and costs of all kinds, as indicated by our examples. The *interpersonal exchange model* focuses on the rewards and costs associated with the sexual relationship (Lawrence & Byers, 1995). Research based on this model assesses perceived rewards and costs, perceived rewards and costs relative to what one expects, and perceived rewards and costs relative to one's partner. In a longitudinal study of 244 adults in heterosexual relationships, all six of these measures were related to reported sexual satisfaction with the relationship three months later. The results also indicated that relationship satisfaction was associated with sexual satisfaction. A study of married Chinese men (n=193) and women (n=231) living in Beijing and Shanghai yielded similar results (Renaud & Byers, 1997). In addition to the association of rewards and costs with sexual satisfaction, greater sexual satisfaction was related to higher frequencies of affectionate and sexual behavior, and fewer sexual concerns and problems.

Clearly, our assessments of the rewards and costs in our intimate relationships are associated with both our satisfaction with those relationships and the likelihood that we will become involved in extramarital (or extrarelationship) sexuality.

Evolution and Extramarital Sex

Extramarital sex is not unique to the United States. In fact, it occurs in virtually every society. When sociobiologists observe a behavior that occurs in all societies, they are inclined to explain that behavior in terms of evolutionary processes.

From an evolutionary perspective, those genes that enable their bearers to produce larger numbers of offspring are more likely to survive from one generation to the next. A man who mates with one woman for life could produce a maximum of 6 to 12 offspring, depending on the length of time infants are breast-fed, postpartum sex taboos, and so on. If that same man occasionally has sex with a second woman (or a series of other women), he could produce 12 to 24 offspring. We think you get the picture. Men who seek out the "other woman" will produce more offspring, who in turn will produce more offspring carrying the genetic makeup which leads to extrarelationship liaisons (Fisher, 1992).

What about women? They cannot increase the number of their offspring by increasing the number of their sexual partners. However, there are other ways in which adultery might have been biologically adaptive for women in the past (Fisher, 1992). First, sexual liaisons with other men might have enabled a woman to acquire extra goods and services which enhanced her offspring's chances for survival.

Second, adultery could serve as "insurance"; if her husband died, she would have another man to turn to for food, shelter, and protection. Third, a woman married to a timid, unproductive hunter could "upgrade her genetic line" by mating with another man. Finally, having children with multiple partners increases the genetic diversity of one's offspring, increasing the chances that some of them will survive. Given the lower status of women in most societies, a married woman who had extramarital liaisons would have had to be careful; if caught she risked death.

Genetic research on species that appear to be monogamous reveals that some of the offspring being raised by a male-female pair were fathered by another male. These tests also reveal that females paired with lesser-quality males engage in extrapair mating, whereas females paired with high-quality males do not (Morell, 1998). These results provide solid support for the sociobiological hypothesis about the adaptive value of female infidelity.

According to this sociobiological perspective, then, extramarital sex occurs because some men and women carry in their genetic makeup something that motivates them to be unfaithful. If there are environments in which adultery is adaptive in contemporary society, people with those genes will have a selective advantage.

Postmarital Sex

From the point of view of developmental psychologists, the sexual relationship in a second union, perhaps following a divorce or the death of one's partner, is especially interesting. In what ways is it the same and how does it differ from the sexual relationship in the first marriage? It represents the blending of things that are unique and consistent about the person with those things that are unique to the new situation and new partner. As we develop sexually throughout the lifespan, those two strands continue to be intertwined—the developmental continuities (those things that are us and always will be) and developmental changes (things that differ at various times in our lives, either because we are older or have experienced more, or because our partner or the situation is different).

The Divorced and the Widowed
Divorced and widowed people are in a somewhat unusual situation in that they are used to regular sexual expression and suddenly find themselves in a situation in which the socially acceptable outlet for that expression—marital sex—is no longer available. Partly recognizing this dilemma, our society places few restrictions on postmarital sexual activity, although it is not as approved as marital sex.

Most divorced women, but fewer widowed women, return to having an active sex life. In one well-sampled study, the percentage of formerly married women who had had intercourse in the last 3 months ranged between 50 percent and 67 percent, depending on the age group (Forrest & Singh, 1990). The NHSLS found that 46 percent of the divorced and widowed men and 58 percent of divorced and widowed women had sex a few times per year or not at all (Laumann et al., 1994). In another study, 81 percent of the widowed had been sexually abstinent in the last year, compared with 28 percent of the divorced (Smith, 1994).

The lower incidence of postmarital sex among widows, compared with divorced women, is due in part to the fact that widows are, on the average, older than divorced women; but even when matched for age, widows are still less likely to engage in postmarital sex. There are probably several reasons for this (Gebhard, 1968). Widows are more likely to be financially secure than divorced women and therefore have less motivation for engaging in sex as a prelude to remarriage. They have the continuing social support system of in-laws and friends, and so they are less motivated to seek new friendships. There is also a belief that a widow should be loyal to her dead husband, and having a sexual relationship with another man is viewed as disloyalty. Many widows believe this or tell themselves that they will "never find another one like him."

Divorced women face complex problems of adjustment (Song, 1991). These problems may include reduced income, lower perceived standard of living, and reduced availability of social support. These problems may increase the motivation to establish a relationship with a man.

Widowed and divorced women who have postmarital sex often begin a relationship within one year of the end of the marriage. The evidence suggests that these are long-term relationships. According to a study of a national sample of adults, 74 percent of the divorced men and women reported either zero or one sexual partner in the year prior to the survey (Stack & Gundlach, 1992). Divorced men were more sexually active than were divorced women. The average frequency of intercourse reported was twice a month. A survey of professional women who held teaching and administrative posts in academic institutions found that divorced women had a larger number of sexual partners and more frequent activity than their never-married counterparts (Davidson & Darling, 1988).

For both men and women, there is a higher probability of being sexually active postmaritally for those who are younger (under 35) and those who have no children in the home (Stack & Gundlach, 1992). Single parents involved in sexual relationships report that they and their partners worry about the impact of the sexual relationship on their children (Darling et al., 1989).

Sex and Seniors

When Freud suggested that young children, even infants, have sexual thoughts and feelings, his ideas met with considerable resistance. When, 50 years later, researchers began to suggest that elderly men and women also have sexual thoughts and feelings, there was similar resistance (Pfeiffer et al., 1968). This section deals with the sexual behavior of elderly men and women, the physical changes they undergo, and the attitudes that influence them.

Physical Changes

Changes in the Female
There is a gradual decline in the functioning of the ovaries around menopause, and with this comes a

Figure 12.6   Affection, romance, and sex are not just for the young.

gradual decline in the production of estrogen (see Figure 12.7). Because of the decline in estrogen, several changes take place in the sexual organs. The walls of the vagina, which are thick and elastic during the reproductive years, become thin and inelastic. Because the walls of the vagina are thinner, they cannot absorb the pressures from the thrusting penis as they once did, and thus nearby structures—such as the bladder and the urethra—may be irritated. As a result, elderly women may have an urgent need to urinate immediately after intercourse. Furthermore, the vagina shrinks in both width and length, and the labia majora also shrink; thus there is a constriction

Figure 12.7   Levels of estrogen production in women across the lifespan.

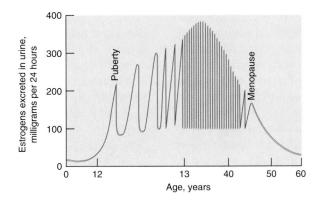

at the entrance to the vagina, which may make penile insertion somewhat more difficult, and the vagina may be less able to accommodate the size of a fully erect penis. By about five years after menopause, the amount of vaginal lubrication has decreased noticeably. Intercourse can then become somewhat more difficult and painful.

Because of hormonal imbalance, the contractions of the uterus that occur during orgasm can become painful, to the point where the woman avoids intercourse. Nonetheless, the woman has the same physical capacity for orgasm at 80 that she did at 30.

Lest these changes sound discouraging, it is important to realize that there are a number of ways to deal with them successfully. Some women's physicians prescribe hormone-replacement therapy for them after menopause (see Chapter 6); with these added doses of estrogen, the changes described above are minimized or may not occur at all. Also, research indicates that adding testosterone to the therapy will enhance sexual desire and enjoyment (Sherwin, 1991). A simple alternative is to use a sterile lubricant as a substitute for vaginal lubrication.

It also appears that these changes are related in part to sexual activity—or, rather, to the lack of it. A study of 59 healthy women between 60 and 70 years of age found consistent differences between those who were sexually active ($n = 39$) and those who were not ($n = 20$) (Bachmann & Leiblum, 1991). All of the women had a male partner. Those who were sexually active reported intercourse an average of five times per month. They reported higher levels of sexual desire, greater comfort about expressing their sexual needs, and greater sexual satisfaction. A pelvic exam by a physician who was unaware of which women were sexually active found less genital atrophy in the sexually active women. Sexual problems reported by the women included reduced sexual desire in both self and partner, decreased vaginal lubrication, and erectile difficulties. These are correlational data, so we cannot be sure what is cause and what is effect.

It appears that the continuing of sexual activity depends more on the opportunity for such activity and on various psychological factors than it does on any physical changes, a point that holds true for both men and women.

Some people believe that a **hysterectomy** means the end of a woman's sex life. In fact, sex hormone production is not affected as long as the ovaries are not removed (surgical removal of the ovaries is called **oophorectomy** or ovariectomy). The major-

ity of women report that hysterectomy has no effect on their sex lives. However, approximately one-third of women who have had hysterectomies report problems with sexual response (Zussman et al., 1981). There are two possible physiological causes for these problems. If the ovaries have been removed, hormonal changes may be responsible; specifically, the ovaries produce androgens, and they may play a role in sexual response. The other possibility is that the removal of the cervix, and possibly the rest of the uterus, is an anatomical problem if the cervix serves as a trigger for orgasm.

Changes in the Male

Testosterone production declines gradually over the years (Schiavi, 1990) (see Figure 12.8). Vascular diseases such as hardening of the arteries are increasingly common with age in men, but good circulation is essential to erection (Riportella-Muller, 1989). A major change is that erections occur more slowly. It is important for men to know that this is a perfectly natural slowdown so that they will not jump to the conclusion that they are developing an erection problem. It is also important for partners to know about this so that they will use effective techniques of stimulating the man and so that they will not mistake slowness for lack of interest.

The refractory period lengthens with age; thus for an elderly man, there may be a period of 24 hours after an orgasm during which he cannot get an erection. (Note that women do not undergo a similar change; most women do not enter into a refractory period and are still capable of multiple orgasm at age 80.) Other signs of sexual excitement—the sex flush and muscle tension—diminish with age.

The volume of the ejaculate gradually decreases, and the force of ejaculation lessens. The testes become somewhat smaller, but viable sperm are produced by even very old men. Ninety-year-old men have been known to father children.

One advantage is that middle-aged and elderly men may have better control over orgasm than young men; thus they can prolong coitus and may be better sexual partners.

A study of healthy men, ages 45 to 74, all married, assessed their biological, psychological, and behav-

> **Hysterectomy (hiss-tur-EK-tuh-mee):** Surgical removal of the uterus.
> **Oophorectomy (OH-uh-fuh-REK-tuh-mee):** Surgical removal of the ovaries.

Figure 12.8   Levels of testosterone production in men across the lifespan.

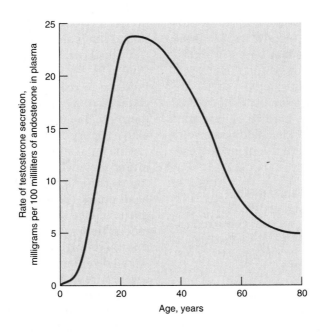

ioral functioning (Schiavi et al., 1994). Satisfaction with sexual functioning was significantly related to whether the man had erectile difficulties. General satisfaction was related to erectile problems, sexual information, and marital adjustment. Accurate information is important because it may result in more realistic expectations for sexual performance.

Some people believe that prostate surgery or removal of the prostate (**prostatectomy**) means the end of a man's sex life. It is true that the volume of the ejaculate will decrease. Prostatectomy can cause damage to the nerves supplying the penis, creating erectile problems. In other cases, retrograde ejaculation may result. Whether there are such problems depends on which of several available methods of surgery is used.

In sum, the evidence suggests that there need be no time limit on sexual expression for either men or women.

A 73 year-old man reported:

I can't begin to tell you how happy I am. I am married to a wonderful woman who loves me as much as I love her. My children gave me a hard time of it at first, especially because she is a bit younger than me. [My son] was telling me that marrying again

and trying to have a lot of sex—imagine that, saying to me *trying* to have sex—could be dangerous to the marriage. So, I said to him with a straight face, "Do you think she'll survive it?" He was so shocked, he laughed. (Janus & Janus, 1993, p. 8)

Attitudes About Sex and the Elderly

Our society has a negative attitude toward sexual expression among the elderly. Somehow it seems indecent for two 70-year-old people to have sex with each other, and even more indecent for a 70-year-old to masturbate. These negative attitudes become particularly obvious in nursing homes, where staff members may frown on sexual activity among the residents. Somehow what is "virility" at 25 becomes "lechery" at 75.

Cross-cultural research indicates that the sexual behavior of the elderly is related to these cultural expectations (Winn & Newton, 1982). The elderly

Prostatectomy (pros-tuh-TEK-tuh-mee): Surgical removal of the prostate.

continue to be sexually active in 70 percent of societies and in precisely those societies where they are expected to be sexually active. Indeed, in 22 percent of societies, women are expected to become more uninhibited about sexuality when they become old.

Why does our society have such negative attitudes toward sex among the elderly? In part, these attitudes are due to the fact that ours is a youth-oriented culture. We value youth, and the physical characteristics that are considered "sexy" are youthful ones, such as a trim, firm body and smooth skin. It is therefore hard to believe that someone with old, wrinkled skin could be sexually active. According to one youthful-looking 59-year-old woman:

> No one looks at me today with any kind of sexual interest. I mean, I don't even get the time of day from any man. I'm an older woman. I want men to be aggressive now. I'm not going to approach a man, because I'm afraid to be rejected. So I've just given up. (Maurer, 1994, p. 475)

Our negative attitudes may be a holdover from the belief that sex was for reproductive purposes only—those past the age of reproduction should therefore not engage in it (Pfeiffer, 1975). The incest taboo may also be involved in our negative attitudes. We tend to identify old people with our parents or grandparents, and we find it hard to think of them as sexual beings. This attitude is encouraged by the fact that many parents take great pains to hide their sexual activity from their children.

These attitudes affect the way elderly people are treated, and the elderly may even hold such attitudes themselves. One remedy that has been proposed for these negative attitudes is a "coming out of the closet"; as one 67-year-old commented:

> The common view that the aging and aged are nonsexual, I believe, can only be corrected by a dramatic and courageous process—the *coming-out-of-the-closet* of sexually active older women and men, so that people can see for themselves what the later years are really like. (Brecher, 1984, p. 21)

Various specific misunderstandings may influence sexuality. For example, a man may believe that sex will precipitate a heart attack or, if he has already had a heart attack, that it will bring on another one. Although Masters and Johnson found that the heart rate accelerated during sexual intercourse, another study showed that the mean heart rate during orgasm was only 117 beats per minute,

which is about that attained during many common forms of daily exercise (Hellerstein & Friedman, 1969). It is about the equivalent of climbing two flights of stairs at a moderate pace. Thus the demands on the heart are not unreasonable. A study of 1,775 patients who had a heart attack questioned them about activities immediately prior to the attack and in the year prior to the attack (Muller et al., 1996). The results indicate that the increase in risk caused by sexual activity is one chance in one million for a healthy individual. Furthermore, the relative risk is no greater in patients with a history of cardiac disease.

One interesting study of male coronary patients found that two-thirds had a sexual dysfunction (most commonly erectile problems) *before* the heart attack (Wabrek & Burchell, 1980). This puts the incidence of sexual difficulties after a heart attack in better perspective. The authors suggested that there might be complex interactions among sexual disorders, stress, and heart attacks. That is, severe stress might precipitate both a sexual disorder and a heart attack. Or a sexual difficulty might be the stress that precipitated the heart attack.

A study of 29 men (mean age 56) who had been diagnosed with cardiovascular disease found that 41 percent reduced their frequency of sexual activity after diagnosis (Quadagno et al., 1995). Interestingly, men who received counseling about resuming sexual activity were no more likely to maintain or resume their predisease frequency. However, we do not know what specifically these men were told. Older men and men who reported that their wives were concerned about resuming sex were more likely to report a decrease in frequency.

Some men also mistakenly believe that sexual activity saps their "vital strength." In some cases men may believe that they can have only a fixed number of orgasms during their lifetimes (as a woman is born with a fixed number of ova) and therefore adopt a strategy of saving them now so that they will have some left later on. One woman wrote:

> My husband has reached the age of sixty-five. He has decided that, in order to ensure a longer life and health, he will no longer engage in sex activity. He is convinced that intercourse and the emission of semen are quite debilitating, particularly in his years. (I. Rubin, 1966, p. 258)

Ideas such as this, as well as factors such as illness or hospitalization, may lead to a period of sexual inactivity. But being sexually inactive is one of the most effective ways of diminishing sexuality. Masters and Johnson emphasized that two factors are critical in maintaining sexual capacity in old age.

1. *Good physical and mental health* An excellent study confirms this notion (Persson, 1980). A representative sample of 70-year-olds in one town in Sweden were selected, and 85 percent agreed to participate in the detailed interviews. In the entire sample, 46 percent of the men and 16 percent of the women still had sexual intercourse; when only those who were currently married were considered, the figures rose to 52 percent for men and 36 percent for women. For both men and women, those who continued to have sexual intercourse had better mental health as rated by a psychiatrist and more positive attitudes toward sexual activity among the aged.

2. *Regularity of sexual expression* As was noted earlier, evidence exists that some physical changes of the sex organs in old age are related to sexual inactivity. As the saying goes, "If you don't use it, you lose it." In fact, a longitudinal study suggests that, for men, frequency of orgasm is positively associated with longevity. The study involved 910 men aged 45 to 59. At the beginning of the study, the men completed a standard medical history and a questionnaire that assessed sexual behavior. Ten years later, the researchers found out who had died, and compared their questionnaire answers with those of the survivors. Men who reported less than one orgasm per month at the beginning of the study were more than twice as likely to die as the men who reported two orgasms per month. (Smith et al., 1997)

Apparently some elderly people have caught on to this fact. As one 80-year-old husband said of his relationship with his 75-year-old wife,

> My wife and I both believe that keeping active sexually delays the aging process . . . if we are troubled with an erection or lubrication, we turn to oral methods or masturbation of each other. We keep our interest alive by a great deal of caressing and fondling of each other's genitals. We feel it is much better to wear out than to rust out. (Brecher, 1984, p. 33)

Women's beliefs about menopause may influence their sexuality in the later years. Some believe that menopause means the end of sex. Certainly such attitudes could lead to a diminishing, or even a cessation, of sexual activity. Probably in a few cases, women who never really enjoyed sex use menopause as an excuse to stop having intercourse. On the other hand, some women who spent their younger years worrying about getting pregnant find menopause to be a liberating experience; their sexual activity may actually increase.

Reformers urge us to change our attitudes about sex and the senior citizen. Nursing homes particularly need to revise their practices; even such simple changes as knocking before entering a resident's room would help (people masturbate, you know). Other reforms would include making provisions for spouses to stay overnight and allowing couples—married or unmarried—to share a bedroom. Indeed, some experts even advocate sex as a form of therapy for persons in nursing homes (Rice, 1974):

> Sex relations can provide a much needed and highly effective resource in the later years of life, when so often men face the loss of their customary prestige and self-confidence and begin to feel old, sometimes long before they have begun to age significantly. The premature cessation of sexual functioning may accelerate physiological and psychological aging since disuse of any function usually leads to concomitant changes in other capacities. After menopause, women may find that continuation of sexual relations provides a much needed psychological reinforcement, a feeling of being needed and of being capable of receiving love and affection and renewing the intimacy they earlier found desirable and reassuring. (Frank, 1961, pp. 177–178)

Sexual Behavior of the Elderly

While sexual behavior and sexual interest do decline somewhat with age, there are still substantial numbers of elderly men and women who have active sex lives, even when in their eighties. In a sample of healthy 80- to 102-year-olds, 62 percent of the men and 30 percent of the women reported that they still engaged in sexual intercourse (Bretschneider & McCoy, 1988). There does not seem to be any age beyond which all people are sexually inactive.

Some elderly people do, for various reasons, stop having intercourse after a certain age. For women, this occurs most often in the late fifties and early sixties, while it occurs somewhat later for

Figure 12.9   Romance is important even for nursing-home residents.

men (Pfeiffer et al., 1968). Contrary to what one might expect, though, when a couple stop having intercourse, the husband is most frequently the cause; both wives and husbands agree that this is true. In some cases, the husband has died (and we can hardly blame him for that), but even excluding those cases, the husband is still most frequently the cause. Thus the decline in female sexual expression with age may be directly related to the male's decline. Death of the spouse is more likely to put an end to intercourse for women than it is for men, in part because women are less likely to remarry (Riportella-Muller, 1989).

One of the most important influences on sexuality in the elderly, then, is that there are far more elderly women than elderly men. Because of both men's earlier mortality and their preference for younger women, elderly women are more likely to be living alone and to have less access to sexual partners. For example, in 1990, among those 85 and over, 47 percent of the men were married and living with their spouse, compared with only 10 percent of women (Bureau of the Census, 1990). Some innovative solutions have been proposed, such as elderly women forming lesbian relationships.

The largest survey to date on sexuality and the elderly has provided rich detail on sexual patterns among the elderly (Brecher et al., 1984). The study was sponsored by Consumers Union, the nonprofit organization that publishes the magazine *Con-sumer Reports.* In the November 1977 issue of *Con-sumer Reports,* a notice appeared requesting cooperation from men and women born before 1928, and thus over 50 years of age at the time. Readers were asked to write in to obtain a questionnaire on personal relationships—family, social, and sexual—during the later years of life. Over 10,000 questionnaires were mailed out in response, and 4246 were returned. Although in Chapter 3 we concluded that magazine surveys are of little value because of severe problems of sampling bias, we view this study as an exception. It is one of the few important sources of information concerning sex and the elderly, a topic about which there is a real shortage of knowledge. The problem of a nonrandom and volunteer-biased sample remains. Specifically, elderly people who are sick or in nursing homes, or whose sight has failed so that they cannot read, are highly unlikely to have responded to the survey. Thus we must regard this as a survey of elderly people who are above average in health, activity, and intelligence and who are doubtless more sexually active than sick or disabled persons. In a sense, the survey gives a view of the richest potential of sexuality in the later years.

Some statistics from the survey are summarized in Table 12.3. Notice that even among respondents over 70 years of age, 33 percent of the women and 43 percent of the men still masturbate. And 65 percent of the married women and 59 percent of the married men over 70 years old reported that they continue to have sex with their spouse.

The questionnaire also contained questions about patterns of extramarital sexuality and homosexuality. The results indicated that 8 percent of the wives and 24 percent of the husbands had engaged in extramarital sex at least once after age 50. For the entire sample, 13 percent of the men and 8 percent of the women had had a homosexual experience at some time in their lives, and 4 percent of the men and 2 percent of the women had engaged in homosexual activity after the age of 50.

What we see from this survey, then, is that among older people who are healthy and active and have regular opportunities for sexual expression, sexual activity in all forms—including masturbation and same-gender behavior—continues past 70 years of age. The sexuality of the elderly has indeed come out of the closet.

Table 12.3 Sexual Activity in a Sample of Elderly Persons

	In Their 50s	In Their 60s	70 and Over
Women			
Orgasms when asleep or while waking up	26%	24%	17%
Women who masturbate	47%	37%	33%
Frequency of masturbation among women who masturbate	0.7/week	0.6/week	0.7/week
Wives having sex with their husbands	88%	76%	65%
Frequency of sex with their husbands	1.3/week	1.0/week	0.7/week
Men			
Orgasms when asleep or while waking up	25%	21%	17%
Men who masturbate	66%	50%	43%
Frequency of masturbation among men who masturbate	1.2/week	0.8/week	0.7/week
Husbands having sex with their wives	87%	78%	59%
Frequency of sex with their wives	1.3/week	1.0/week	0.6/week

Source: Edward M. Brecher. (1984). Love, sex, and aging. Mount Vernon, NY: Consumers Union, p. 316.

SUMMARY

Sexuality continues to develop throughout the lifespan. It may be expressed in singlehood, cohabitation, marriage, extramarital relationships, relationships following divorce, or in a variety of contexts as the individual ages.

Young adults grow toward sexual maturity. Many do so in the context of a single relationship which results in marriage. Others are involved in two or more relationships before they begin to live with or marry someone. Never-married people over 25 may find themselves part of the "singles scene." Blacks are more likely to remain single than whites.

Cohabitation is a stage that up to 40 percent of people experience. The time couples spend living together varies from a few months to several years. Sixty percent of cohabiting couples marry. Some cohabiting couples have children, either together or with previous partners. Men and women who are living together engage in sexual activity more often, on the average, than those who are married or dating.

Marriage represents a major turning point, as couples face new responsibilities and problems, and try to find time for each other. Married couples in their twenties engage in sexual intercourse two or three times per week, on the average, with the frequency declining to two or three times per month among couples over 60. Perhaps the most dramatic change in marital sex practices in recent decades is the increased popularity of oral-genital sex. Many people continue to masturbate even though they are married. Most people today—both women and men—express general satisfaction with their marital sex life. Sexual patterns in marriage, however, show great variability.

About 25 percent of all married men and 15 percent of all married women engage in extramarital sex at some time. Extramarital sex is disapproved of in our society and is generally carried on in secrecy. In a few cases, it is agreed that both husband and wife can have extramarital sex, as in open marriage and swinging. Equity theory and the sociobiological perspective may be helpful in understanding patterns of extramarital sex.

Virtually all widowed and divorced men return to an active sex life, as do most divorced women and about half of widowed women. A particular set of sexual norms characterizes the divorce subculture.

Research indicates that gay men have modified their sexual practices somewhat in the AIDS era. They have reduced their number of partners and have shifted away from risky sex practices. Among heterosexuals, there is some evidence of a reduction in number of partners, and slight improvements in condom use. Blacks and single persons are more likely to have changed. Most heterosexuals do not consider themselves at risk for HIV infection, however, so they are less concerned with safer sex practices.

While sexual activity declines somewhat with age, it is perfectly possible to remain sexually active into one's eighties or nineties. Problems with sex or the cessation of intercourse may be related to physical factors. In women, declining estrogen levels result in a thinner, less elastic vagina and less lubrication; in men, there is lowered testosterone production and increased vascular disease, combined with slower erections and longer refractory periods. Psychological factors can also be involved, such as the belief that the elderly cannot or should not have sex. Masters and Johnson emphasized that two factors are critical to maintaining sexual capacity in old age: good physical and mental health, and regularity of sexual expression. The Consumers Union survey indicates that all sexual behaviors—including heterosexual intercourse, masturbation, and homosexual behaviors—may continue past age 70.

REVIEW QUESTIONS

1. The birth of a baby has an impact on the sexual relationship of a couple. True or false?

2. In general, the frequency of marital intercourse declines with age. True or false?

3. One of the greatest changes in marital sex during the last few decades is the increased popularity of mouth-genital sex. True or false?

4. Masturbation is rare among married adults—only about 10 percent of husbands and wives report that they still engage in masturbation. True or false?

5. People's attitudes toward extramarital sex have changed in the last several decades, so that today a majority of Americans approve of extramarital sex. True or false?

6. _____ refers to married couples exchanging sexual partners with other married couples, with the knowledge and consent of all involved.

7. According to _____, people who feel they are in inequitable marriages are more likely to engage in extramarital sex than those who are in equitable marriages.

8. Most divorced women, but fewer widowed women, return to having an active sex life. True or false?

9. According to cross-cultural research, in societies in which the elderly are expected to be sexually active, they are. True or false?

10. According to Masters and Johnson, the most important factors in maintaining sexual capacity in old age are good physical and mental health and _____.

(The answers to all review questions are at the end of the book, after the Glossary.)

QUESTIONS FOR THOUGHT, DISCUSSION, AND DEBATE

1. What is your response when you see an elderly couple expressing affection physically with each other, perhaps kissing or holding hands? Why do you think you respond that way?

2. What is your opinion about extramarital sex? Is it ethical or moral? What are its effects on a marriage—does it destroy marriage or improve it, or perhaps have no effect?

3. If you are currently in a relationship, apply equity theory to your relationship. Do you feel that it is equitable or inequitable? If you view it as inequitable, what effects does that have on your behavior?

4. You and your partner are talking about establishing a long-term relationship. Your partner says that his main concern is that living to-

gether will lead to a decline in how often you have sex. Is that a realistic concern? If so, what could the two of you do to prevent that from occurring?

SUGGESTIONS FOR FURTHER READING

Brecher, Edward M. (1984). *Love, sex, and aging.* Mount Vernon, NY: Consumers Union. This large-scale survey offers a liberated view of sexuality in the elderly.

Fisher, Helen. (1992). *Anatomy of love: The mysteries of mating, marriage, and why we stray.* New York: Fawcett Columbine. This book presents a provocative, sociobiological account of human sexual behavior, including extramarital sex.

Maurer, Harry. (1994). *Sex: Real people talk about what they really do.* New York: Penguin Books.

Maurer interviewed 52 people of diverse ages, preferences, and orientations about their sexual experiences. He presents lengthy excerpts organized around themes, such as Awakenings, Wild Oats, and the Long Haul.

Sarrel, Lorna, & Sarrel, Philip. (1984). *Sexual turning points: The seven stages of adult sexuality.* New York: Macmillan. This book, written for a general audience, provides an interesting theory of adult sexual development.

WEB RESOURCES

http://members.aol.com/acmefla
   Association for Couples in Marriage Enrichment; resources for couples who want to strengthen their relationship.

http://www.wwme.org/
   Worldwide Marriage Encounter Homepage; resources for couples who want "GREAT marriages."

http://mail.bris.ac.uk/~plmlp/celibate.html
   A discussion of issues related to celibacy.

http://www.sexhealth.org/infocenter/SexAging/sexaging.htm
   The Sexual Health InfoCenter: Sexuality & Aging Page.

chapter
13

ATTRACTION, INTIMACY, AND LOVE

CHAPTER HIGHLIGHTS

To be in love is merely to be in a state of perceptual anesthesia—to mistake an ordinary young man for a Greek god or an ordinary young woman for a goddess.*

We made love, Your Honor. He didn't have any and neither did I. So we made some. It was good.†

The lyrics to the old song, "Love and marriage, love and marriage, go together like a horse and carriage," might today be rewritten as "Love and sex, love and sex, go together. . . ." The standard of today for many is that sex is appropriate if one loves the other person (see Chapter 11), and sex seems to be the logical outcome of a loving relationship. Therefore, it is important in a text on sexuality to spend some time considering the emotion we link so closely to sex: love.

This chapter is organized in terms of the way relationships usually progress—if they progress. That is, we begin by talking about attraction, what brings people together in the first place. Then we consider intimacy, which develops as relationships develop. Next, we look at four different views of what love is. Finally we discuss some of the research on love, including cross-cultural research.

Attraction

What causes you to be attracted to another person? Social psychologists have done extensive research on interpersonal attraction. The major results of this research are discussed in this section.

The Girl Next Door

Our opportunities to meet people are limited by geography and time. You may meet that attractive person sitting two rows in front of you in "human sex," as the course is referred to at the University of Wisconsin, but you will never meet the wealthy, brilliant engineering student who sits in your seat two classes later. You are much more likely to meet and be attracted to the boy next door than the one who lives across town. The NHSLS (introduced in Chapter 3) asked participants where they met their current dating partner, cohabitant, or spouse. More than half met at school, work, church, or a party (Michael et al., 1994, Fig. 2).

Among those who work in the same place or take the same class, we tend to be more attracted to people with whom we have had contact several times than we are to people with whom we have had little contact (Harrison, 1977). This has been demonstrated in laboratory studies in which the amount of contact between participants was systematically varied. At the end of the session, people gave higher "liking" ratings to those with whom they had had much contact and lower ratings to those with whom they had had little contact (Saegert et al., 1973). This is the **mere-exposure effect;** repeated exposure to any stimulus, including a person, leads to greater liking for that stimulus (Bornstein, 1989).

Birds of a Feather

We tend to like people who are similar to us. We are attracted to people who are approximately the same as we are in age, race or ethnicity, and economic and social status. Similarity on these social characteristics is referred to as **homophily,** the tendency to have contact with people equal in social status. Data on homophily from the NHSLS are displayed in Table 13.1. Note that the greatest homophily is by race, followed by education and age. Couples are least likely to be the same on religion. It is interesting that short-term partnerships are as homophilous as marriages.

Mere-exposure effect: The tendency to like a person more if we have been exposed to him or her repeatedly.
Homophily: The tendency to have contact with people who are equal in social status.

---

*H. L. Mencken. (1919). *Prejudices.* (First Series). New York: Knopf, p. 16.
†Julie, in Lois Gould. (1988). *Such good friends.* New York: Farrar, Straus, Giroux, p. 161.

Table 13.1 Percentage of Relationships That Are Homophilous, by Type of Relationship

Type of Homophily	Type of Relationship			
	Marriages*	Cohabitations*	Long-Term Partnerships	Short-Term Partnerships
Racial/ethnic	93%	88%	89%	91%
Age[†]	78	75	76	83
Educational[‡]	82	87	83	87
Religious[§]	72	53	56	60

*Percentages of marriages and cohabitational relationships that began in the ten years prior to the survey.

[†]Age homophily is defined as a difference of no more than five years in partners' ages.

[‡]Educational homophily is defined as a difference of no more than one educational category. The educational categories used were less than high school, high school graduate, vocational training, four-year college, and graduate degree.

[§]Cases in which either partner was reported as "other" or had missing data are omitted.

Source: Laumann et al., 1994, Table 6.4.

Social psychologist Donn Byrne (1971) has done numerous experiments demonstrating that we are attracted to people whose attitudes and opinions are similar to ours. In these experiments, Byrne typically has people fill out an opinion questionnaire. They are then shown a questionnaire that was supposedly filled out by another person and are asked to rate how much they think they would like that person. In fact, the questionnaire was filled out to show either high or low agreement with the subject's responses. Subjects report more liking for a person whose responses are similar to theirs than for one whose responses are quite different.

There are a number of reasons why we are attracted to a person who is similar to us in, say, attitudes (Huston & Levinger, 1978). We get positive reinforcement from that person agreeing with us. The other person's agreement bolsters our sense of rightness. And we anticipate positive interactions with that person.

Folk sayings are sometimes wise and sometimes foolish. The interpersonal-attraction research indicates that the saying "Birds of a feather flock together" contains some truth. This tendency for men and women to choose as partners people who match them on social and personal characteristics is called the **matching phenomenon** (Feingold, 1988). At the same time, "opposites attract" may be more accurate with regard to interpersonal style. In one study, dominant people paired with submissive people reported greater satisfaction with their relationship than dominant or submissive people paired with a similar partner (Dryer & Horowitz, 1997).

"Hey, Good-Lookin' "

We also tend to be attracted to people who are physically attractive, or "good-looking" (Berscheid & Walster, 1974b; Hatfield & Sprecher, 1986b). For example, in one study snapshots were taken of college men and women (Berscheid et al., 1971). A dating history of each person was also obtained. Judges then rated the attractiveness of the men and women in the photographs. For the women there was a fairly strong relationship between attractiveness and popularity; the women judged attractive had had more dates in the last year than the less attractive women. There was some relationship between appearance and popularity for men, too, but it was not so marked as it was for women. This phenomenon has even been found in children as young as 3 to 6 years of age, who are more attracted to children with attractive faces (Dion, 1977; 1973).

Matching phenomenon: The tendency for men and women to choose as partners people who match them, i.e., who are similar in attitudes, intelligence, and attractiveness.

In general, then, we are most attracted to good-looking people. However, this effect depends on gender to some extent; physical attractiveness is more important to males evaluating females than it is to females evaluating males (Feingold, 1990). And this phenomenon is somewhat modified by our own feelings of personal worth, as will be seen in the next section.

The Interpersonal Marketplace

Although this may sound somewhat callous, whom we are attracted to and pair off with depends a lot on how much we think we have to offer and how much we think we can "buy" with it. Generally, the principle seems to be that women's worth is based on their physical beauty, whereas men's worth is based on their success. There is a tendency, then, for beautiful women to be paired with wealthy, successful men.

Data from many studies document this phenomenon. In one study, high school yearbook pictures of 601 males and 745 females were rated for attractiveness (Udry & Eckland, 1984). These people were followed up 15 years after graduation, and measures of education, occupational status, and income were obtained. Females who were rated the most attractive in high school were significantly more likely to have husbands who had high incomes and were highly educated (see also Elder, 1969).

In another study, women students were rated on their physical attractiveness (Rubin, 1973, p. 68). They were then asked to complete a questionnaire about what kinds of men they would consider desirable dates. A man's occupation had a big effect on his desirability as a date. Men in high-status occupations—physician, lawyer, chemist—were considered highly desirable dates by virtually all the women. Men in low-status occupations—janitor, bartender—were judged hardly acceptable by most of the women. A difference emerged between attractive and unattractive women, however, when rating men in middle-status occupations—electrician, bookkeeper, plumber. The attractive women did not feel that these men would be acceptable dates, whereas the unattractive women felt that they would be at least moderately acceptable. Here we see the interpersonal marketplace in action. Men with more status are more desirable. But how desirable a man is judged to be depends on the woman's sense of her own worth. Attractive women are not much interested in middle-status men because they apparently think they are "worth more." Unattractive women find middle-status men more attractive, presumably because they think such men are reasonably within their "price range."

From the Laboratory to Real Life

The phenomena discussed so far—feelings of attraction to people who are similar to us and who are good-looking—have been demonstrated mainly in psychologists' laboratories. Do these phenomena occur in the real world?

Donn Byrne and his colleagues (1970) did a study to find out whether these results would be obtained in a real-life situation. They administered an attitude and personality questionnaire to 420 college students. Then they formed 44 "couples." For half of the couples, both people had made very similar responses on the questionnaire; for the other half of the couples, the two people had made very different responses. The two people were then introduced and sent to the student union on a brief date. When they returned from the date, an unobtrusive measure of attraction was taken—how close they stood to each other in front of the experimenter's desk. The participants also evaluated their dates on several scales.

The results of the study confirmed those from previous experimental work. The couples who had been matched for similar attitudes were most attracted to each other, and those with dissimilar attitudes were not so attracted to each other. The students had also been rated as to their physical attractiveness both by the experimenter and by their dates, and greater attraction to the better-looking dates was reported. In a follow-up at the end of the semester, those whose dates were similar to them and were physically attractive were more likely to remember the date's name and to express a desire to date the person again in the future. Thus in an experiment that was closer to real life and real dating situations, the importance of similarity and physical appearance was again demonstrated.

Attraction On-Line

Technology has created a new way to meet potential partners—on-line (Elias, 1997). Some websites have tens of thousands of "personals ads," and one site claims 500,000 hits per day. Surveys suggest that the people seeking partners on-line are educated, affluent, 20- to 40-year-olds who don't have the time or the taste for "singles bars." Users say that one advantage of meeting on the Net is that the technology forces you to focus on the person's interests and values. This facilitates finding a person with whom you have a lot in common. Also, in many instances you cannot see the person and so you are not influenced by his/her physical attractiveness (or lack thereof). A major disadvantage is the risk that the other person may not be honest about his or her interests, occupation, or marital status. Chat rooms with names like "Women with Other Men" attract married people. Some of the relationships established in these rooms lead to "Divorce, Internet Style" (Quittner, 1997).

Playing Hard to Get

The traditional advice that has been given to girls—by Ann Landers and others—is that boys will be more attracted to them if they play hard to get. Is there any scientific evidence that this is true?

In fact, two experiments provide no support for this kind of strategy; according to these experiments, playing hard to get does not work (Walster et al., 1973). In one of these experiments, college men were recruited for a computer-dating program. They were given the phone number of their assigned date and were told to phone her and arrange a date from the experimenters' laboratory; after they phoned her, the researchers assessed the men's initial impression of her. In fact, all the men were given the same telephone number—that of an accomplice of the experimenters. For half of the men, she played easy to get; she was delighted to receive the phone call and to be asked out. With the other half of the men, she played hard to get; she accepted the date with reluctance and indicated that she had many other dates. The results failed to support the hard-to-get strategy: the men had equally high opinions of the hard-to-get and easy-to-get woman.

The same experimenters reported an ingenious field experiment in which a prostitute played either hard to get or easy to get with her clients and then recorded the clients' responses, such as how much they paid her. Once again, the hard-to-get

hypothesis was not supported; the men seemed to like the easy-to-get and the hard-to-get prostitute equally well.

The experimenters were faced with the bald fact that a piece of folk wisdom just did not seem to be true. They decided that they needed a somewhat more complex hypothesis. They hypothesized that it is not the woman who is generally hard to get or generally easy to get who is attractive to men but rather the one who is *selectively hard to get*. That is, she is easy to get for you, but she is hard to get for other men or unavailable to them. A computer-dating experiment supported this notion; the selectively hard-to-get woman was the most popular with men.

Recent research looked at the amount of effort needed to establish a relationship, to determine its effect on attractiveness (Roberson & Wright, 1994). Male students were led to believe that it would be easy, difficult, or impossible for them to work with an apparently warm, friendly woman. They were then asked to rate her on two dimensions: how nice she was and how appealing she was as a coworker. Men who thought she would be difficult to work with rated her as more appealing than did men who thought she would be easy or impossible.

In practical terms, this means that if a woman is going to use hard-to-get strategies, she had better use them in a skillful way. It seems that the optimal strategy would be to give the impression that she has many offers for dates with others but refuses them, while indicating that she is willing to date the young man in question, although it will take some effort on his part to persuade her.

Interestingly, in all the research discussed, it is always the woman who is playing hard to get and the man who is rating her. This reflects cultural gender-role stereotypes, in which it is the woman's role to do things like play hard to get. What we do not know is how men's use of hard-to-get strategies would affect the way women perceive them.

Explaining Our Preferences

The research data are quite consistent in showing that we select as potential partners people who are similar to us in social characteristics—age, race, education—and who share our attitudes and beliefs. Moreover, both men and women prefer physically attractive people, although women place greater emphasis on a man's social status or earning potential (Sprecher, Sullivan, & Hatfield, 1994). The obvious question is, Why? Two answers are suggested, one drawing on reinforcement theory and one drawing on sociobiology (see Chapter 2 for discussions of these theories).

Reinforcement Theory: Byrne's Law of Attraction

It is a rather commonsense idea—and one that psychologists agree with—that we tend to like people who give us reinforcements or rewards and to dislike people who give us punishments. Social psychologist Donn Byrne (1997) has formulated the law of attraction. It says that our attraction to another person is proportionate to the number of reinforcements that person gives us relative to the total number of reinforcements plus punishments the person gives us. Or, simplified even more, we like people who are frequently nice to us and seldom nasty.

According to this explanation, we prefer people who are similar because interaction with them is rewarding. People who are similar in age, race, and education are likely to have similar outlooks on life, prefer similar activities, and like the same kinds of people. These shared values and beliefs provide the basis for smooth and rewarding interaction. It will be easy to agree about such things as how important schoolwork is, what TV programs to watch, and what to do on Friday night. Disagreement about such things would cause conflict and hostility, which are definitely not rewards (for most people, anyway). We prefer pretty or handsome partners because we are aware of the high value placed on physical attractiveness in American society; we believe others will have a higher opinion of us if we have a good-looking partner. Finally, we prefer someone with high social status or earning potential because all the material things that people find rewarding cost money.

These findings have some practical implications (Hatfield & Walster, 1978). If you are trying to get a new relationship going well, make sure you give the other person some positive reinforcement. Also, make sure that you have some good times together, so that you *associate* each other with rewards. Do not spend all your time stripping paint off old furniture or cleaning out the garage. And do not forget to keep the positive reinforcements (or "strokes," if you like that jargon better) going in an old, stable relationship.

Figure 13.1 According to Byrne's law of attraction, our liking for a person is influenced by the reinforcements we receive from interacting with them. Shared activities provide the basis for smooth and rewarding interaction.

Sociobiology: Sexual Strategies Theory

Sociobiologists view sexual behavior within an evolutionary perspective. Historically, the function of mating has been reproduction. Men and women who selected mates according to some preferences were more successful than those who chose based on other preferences (Allgeier & Wiederman, 1994). The successful ones produced more offspring, who in turn produced more offspring, carrying their mating preferences to the present.

Men and women face different adaptive problems in their efforts to reproduce (Buss & Schmitt, 1993). Since women bear the offspring, men need to identify reproductively valuable women. Other things being equal, younger women are more likely to be fertile than older women; hence a preference for youth, which results in young men choosing young women (homophily). Also, sociobiologists assert that men want to be certain about the paternity of offspring; hence, they want a woman who will be sexually faithful—that is, one who is hard to get.

Other things being equal, a physically attractive person is more likely to be healthy and fertile—thus the preference for good-looking partners. If attractiveness is an indicator of health, we would expect it to be more important in societies where chronic diseases are more prevalent. Gangestad and Buss (1993) measured the prevalence of seven pathogens, including those which cause malaria and leprosy, in 29 cultures, and also obtained ratings of the importance of 18 attributes of mates. They found that physical attractiveness was considered more important by residents in those societies with greater prevalence of pathogens. However, one study found that there was no relationship between rated facial attractiveness (based on a photograph) and a clinical assessment of health in a sample of adolescents (169 females, 164 males). At the same time, raters rated more attractive persons as healthier (Kalick et al., 1998).

Women must make a much greater investment than men in order to reproduce. They will be pregnant for 9 months, and after the birth they must care for the infant and young child for several years. Thus, women want to select as mates men who are reproductively valuable; hence the preference for good-looking mates. They also want mates who are able and willing to invest resources in them and in their children. Obviously, men must have resources in order to invest them; hence, the preference for men with higher incomes and status. Among young people, women will prefer men with greater earning potential; thus, the preference for men with greater education and higher occupational aspirations. This problem of resources is more important than the problem of identifying a reproductively valuable male, so women rate income and earning potential as more important than good looks.

Research provides evidence that is consistent with this theory. For example, researchers presented a list of 31 tactics to a sample of undergraduate students and asked them to rate how effective each would be in attracting a long-term mate (Schmitt & Buss, 1996). Tactics communicating sexual exclusivity or faithfulness were judged highly effective in attracting a mate for women. Tactics that displayed resource potential were judged most effective for men.

These two explanations—reinforcement theory and sociobiology—are not inconsistent. We can think about reinforcement in more general terms. Reproduction is a major goal for most adults in every society. Successful reproduction—that is,

having a healthy child who develops normally—is very reinforcing. Following the sexual strategies that we have inherited is likely to lead to such reinforcement.

Intimacy

Intimacy is a major component of any close or romantic relationship. Today many people are seeking to increase the intimacy in their relationships. And so, in this section, we will explore intimacy in more detail to try to gain a better understanding of it.

Defining Intimacy

Psychologists have offered a number of definitions of intimacy, including the following (Perlman & Fehr, 1987, p. 17):

1. Intimacy's defining features include: "openness, honesty, mutual self-disclosure; caring, warmth, protecting, helping; being devoted to each other, mutually attentive, mutually committed; surrendering control, dropping defenses; becoming emotional, feeling distressed when separation occurs."

2. "Emotional intimacy is defined in behavioral terms as mutual self-disclosure and other kinds of verbal sharing, as declarations of liking and loving the other, and as demonstrations of affection."

Notice that the first definition focuses on intimacy as a characteristic of a person and the second as a characteristic of a relationship. One way to think about intimacy is that certain persons have more of a capacity for intimacy or engage in more intimacy-promoting behaviors than others. But we can also think of some relationships as being more intimate than others.

A definition of **intimacy** in romantic relationships is: "the level of commitment and positive af-

> **Intimacy:** A quality of relationships characterized by commitment, feelings of closeness and trust, and self-disclosure.

fective, cognitive and physical closeness one experiences with a partner in a reciprocal (although not necessarily symmetrical) relationship" (Moss & Schwebel, 1993, p. 33). The emphasis in this definition is on closeness or sharing, which has three dimensions—affective (emotional), cognitive, and physical. Note, too, that while intimacy must be reciprocal, it need not be equal. Many people have had the experience of feeling closer to another person than that person seems to feel toward them. Finally, note that while intimacy has a physical dimension, it need not be sexual.

In one study, college students were asked to respond to an open-ended question asking what they thought made a relationship one of intimacy (Roscoe et al., 1987). The qualities that emerged, with great agreement, were sharing, sexual interaction, trust in the partner, and openness. Notice that these qualities are quite similar to the ones listed in the definitions given above.

Intimacy and Self-Disclosure

One of the key characteristics of intimacy, appearing in psychologists' and college students' definitions, is self-disclosure, that is, telling personal information about oneself to the other person (Derlega, 1984). Consider the following scenario:

> John is 41 years old and a professor at a state university. He has been dating Susan, a 35-year-old junior faculty member, for 3 months. He has felt depressed in the past year. His father died recently after a chronic illness. John has also felt unhappy about his teaching and research efforts. One evening he and Susan talked for several hours about a wide range of topics. They talked most about John's uncertainties and pessimism about taking control over his life. Susan listened patiently as John talked in an open way about himself. She wished to understand him and to be supportive. Susan said she cared about John regardless of the personal problems he was having. John felt grateful and perceived that she understood him. He was much closer to her after their talk. (Derlega, 1984, p. 1)

This story illustrates how self-disclosure promotes intimacy in a relationship and makes us feel close to the other person. It also illustrates how important it is for the partner to be accepting in respond-

Figure 13.2 Intimacy occurs in a relationship when there is warmth and mutual self-disclosure.

ing to self-disclosure. If the acceptance is missing, we can feel betrayed or threatened, and we certainly will not feel on more intimate terms with the partner.

A study of naturally occurring interactions examined the relationships between self-disclosure, perceived partner disclosure, and the degree of intimacy experienced (Laurenceau et al., 1998). Young people recorded data about every interaction lasting more than 10 minutes, for 7 or 14 days. Data were analyzed for more than 4,000 dyadic (that is, two-person) interactions recorded by 158 participants. Both self-disclosure and partner disclosure were associated with the participant's rating of the intimacy of the interaction. In addition, self-disclosure of emotion was more closely related to intimacy than was self-disclosure of facts.

Self-disclosure and intimacy, then, mutually build on each other. Self-disclosure promotes our feeling that the relationship is intimate, and when we feel that the relationship is intimate, we feel comfortable engaging in further self-disclosure. However, self-disclosure and intimacy don't necessarily consistently increase over time. In some relationships, the pattern may be that an increase in intimacy is followed by a plateau or even a pulling back (Collins & Miller, 1994).

We noted earlier the role of physical attractiveness in initial attraction. Research suggests that physical attractiveness is positively related to self-disclosure. Men and women were randomly paired to create 38 previously unacquainted couples. Each couple was left alone in a room for six minutes. Attractive people self-disclosed more, and people disclosed more to attractive partners (Stiles et al., 1996). Thus, attractiveness may facilitate the development of intimacy.

Measuring Intimacy

Psychologists have developed some scales for measuring intimacy, and these scales can give us further insights. One such scale is the Personal Assessment of Intimacy in Relationships (PAIR) Inventory (Schaefer & Olson, 1981). It measures emotional intimacy in a relationship with items such as the following:

1. My partner listens to me when I need someone to talk to.
2. My partner really understands my hurts and joys.

Another scale measuring intimacy in a relationship includes items such as these (Miller & Lefcourt, 1982):

1. How often do you confide very personal information to him or her?
2. How often are you able to understand his or her feelings?
3. How often do you feel close to him or her?
4. How important is your relationship with him or her in your life?

If you are currently in a relationship, answer those questions for yourself and consider what the quality of the intimacy is in your relationship.

In summary, an intimate relationship is characterized by commitment, feelings of closeness and trust, and self-disclosure. We can promote intimacy in our relationships by engaging in self-disclosure (provided, of course, that we trust the person, but it is quite difficult to develop intimacy when there is lack of trust) and being accepting of the other person's self-disclosures.

Figure 13.3 Communicating about love is often difficult.

PEANUTS reprinted by permission of United Feature Syndicate, Inc.

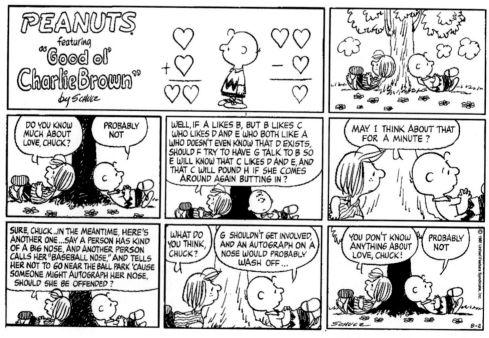

Theories of Love

At the beginning of this chapter, we noted the close connection between love and sex in our society. In this section, we will consider four different views of love: the triangular theory, attachment theory, love as a story, and a biochemical perspective.

Triangular Theory of Love

Robert Sternberg (1986) has formulated a triangular theory of the nature of love. According to his theory, love has three fundamental components: intimacy, passion, and commitment.

Three Components of Love

Intimacy.   Intimacy is the emotional component of love. It includes our feelings of closeness or bondedness to the other person. The feeling of intimacy usually involves a sense of mutual understanding with the loved one; a sense of sharing one's self; intimate communication with the loved one, involving a sense of having the loved one hear and accept what is shared; and giving and receiving emotional support to and from the loved one.

Intimacy, of course, is present in many relationships besides romantic ones. Intimacy here is definitely *not* a euphemism for sex (as when someone asks "Have you been intimate with him?"). The kind of emotional closeness involved in intimacy may be found between best friends and between parents and children, just as it is between lovers.

Passion.   Passion is the motivational component of love. It includes physical attraction and the drive for sexual expression. Physiological arousal is an important part of passion. Passion is the component that differentiates romantic love from other kinds of love, such as the love of best friends or the love between parents and children. Passion is generally the component of love that is faster to arouse, but, in the course of a long-term relationship, it is also the component that fades most quickly.

Intimacy and passion are often closely intertwined. In some cases passion comes first, when a couple experience an initial, powerful physical attraction to each other; emotional intimacy may then follow. In other cases, people know each other only casually, but as emotional intimacy develops, passion follows. Of course, there are also cases where intimacy and passion are completely separate. For example, in cases of casual sex, passion is present but intimacy is not.

Decision or Commitment.   The third component is the cognitive component, decision or commitment. This component actually has two aspects. The short-term aspect is the decision that one loves the other person. The long-term aspect is the commitment to maintain that relationship. Commitment is what makes relationships last. Passion comes and goes. All relationships have their better times and their worse times, their ups and their downs. When the words of the traditional marriage service ask whether you promise to love your spouse "for better or for worse," the answer "I do" is the promise of commitment.

Figure 13.4   The triangle in Sternberg's triangular theory of love.

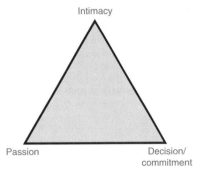

Intimacy

Passion   Decision/ commitment

Figure 13.5   Partners can be well-matched or mismatched, depending on whether their levels of intimacy, passion, and decision/commitment match.

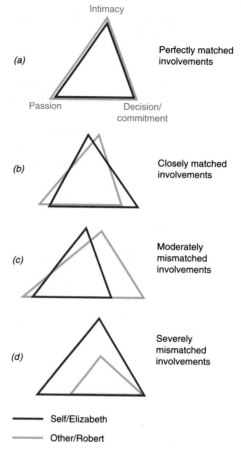

The Triangular Theory

Sternberg (1986) calls his theory a triangular theory of love. Figure 13.4 shows Sternberg's love triangle.[1] The top point of the triangle is intimacy, the left point is passion, and the right point is decision or commitment.

This triangle metaphor allows us to show how the two people in a couple can be well matched or mismatched in the love they feel toward each other. In Figure 13.5(a), Elizabeth feels as much intimacy toward Robert as he does toward her, they both feel equal levels of passion, and they both have the same level of commitment. According to the theory, that is a perfect match. Figure 13.5(b) shows a situation in which the couple are slightly mismatched, but not seriously. Figure 13.5(d) shows a situation in which there is a severe mismatch. Both partners are equally committed, but Elizabeth feels significantly more intimacy and passion than Robert.

Sternberg's research indicates that when there is a good "match" [as shown in Figure 13.5(a) or (b)] between the two partners' love, the partners tend to feel satisfaction with the relationship. When there is a mismatch in the triangles, there is dissatisfaction with the relationship.

[1]This terminology should not be confused with the popular use of the term "love triangle," which refers to a situation in which three people are involved in love, but the love is not reciprocated and so things don't work out quite right. For example, A loves B, B loves C, and C loves A, but A doesn't love C and B doesn't love A. Alas.

Thinking about practical applications of the theory, if a relationship seems to be in trouble, it may be because there is a mismatch of the triangles. One could analyze the love in the relationship in terms of the three components (intimacy, passion, and commitment) to see where the partners are mismatched. It could be that they are well matched for passion, but that one feels and wants more intimacy or commitment than the other person does.

Love in Action
Sternberg also argues that each of the three components of love must be translated into action. The intimacy component is expressed in actions such as communicating personal feelings and information, offering emotional (and perhaps financial) support, and expressing empathy for the other. The passion component is expressed in actions such as kissing, touching, and making love. The decision or commitment component is demonstrated by actions such as saying "I love you," getting married, and sticking with a relationship through times when it isn't particularly convenient.

As the great psychoanalyst Erich Fromm wrote in his book *The Art of Loving* (1956), love is something one *does,* not a state one is *in.* Fromm believed that loving is an art, something that one must learn about and practice. And as Sternberg says, "Without expression, even the greatest of loves can die" (1986, p. 132).

Evidence for Sternberg's Triangular Theory of Love
Sternberg has developed a questionnaire, the Sternberg Triangular Love Scale (STLS), to measure the three components in his theory. Several studies have been done on the characteristics of the scale itself (e.g., Sternberg, 1987, 1997; Whitley, 1993). The scale provides good measures of the components, especially of passion and commitment. Scores for the same relationship are stable for up to two months.

Sternberg makes several predictions about how scores ought to change over time. Acker and Davis (1992) recruited 204 adults, ages 18 to 68; 65 percent were married. The average length of the relationship was 9.5 years. As predicted, commitment scores increased as relationships progressed from dating to marriage. Sternberg expects intimacy to decrease over time as familiarity with the partner increases; sure enough, behavioral intimacy (sharing inner feelings, trying to understand the partner) decreased as predicted. However, contrary to prediction, two other measures of intimacy, including Sternberg's, increased. A study of a sample of German adults assessed the relationship between the three components and sexual activity and satisfaction (Grau & Kimpf, 1993). The theory predicts that the amount of passion should be most closely related to sexual activity, but the results indicated that intimacy was most closely related to sexual behavior and sexual satisfaction.

On the whole, these early results are promising. More refined measures and further research are needed in order to more thoroughly assess the theory.

Attachment Theory of Love
In Chapter 11 we discussed the earliest attachment that humans experience, the attachment between infant and parent. One hypothesis is that the quality of this early attachment—whether secure and pleasant or insecure and unpleasant—profoundly affects us for the rest of our lives, and particularly affects our capacity to form loving attachments to others when we are adults.

The *attachment theory of love* is based on these ideas (Hazan & Shaver, 1987; Simpson, 1990). According to the attachment theory, adults are characterized, in their romantic relationships, by one of three styles. *Secure lovers* are people who find it easy to get close to others and are comfortable having others feel close to them. Mutual dependency in a relationship (depending on the partner and the partner depending on you) feels right to them. Secure lovers do not fear abandonment. In contrast, *avoidant lovers* are uncomfortable feeling close to another person or having that person feel close to them. It is difficult for them to trust or depend on a partner. The third type, *anxious-ambivalent lovers,* want desperately to get close to a partner, but often find the partner does not reciprocate the feeling, perhaps because anxious-ambivalent lovers scare away others. They are insecure in the relationship, worrying that the partner does not really love them. Research shows that about 53 percent of adults are secure, 26 percent are avoidant, and 20 percent are anxious-ambivalent (Hazan & Shaver, 1987). This research also shows that separation from a parent in

Focus 13.1

Jealousy

Jealousy—the green-eyed monster—is an unpleasant emotion that is often associated with love and sexual relationships. Several perspectives contribute to our understanding of this emotion.

Sociologist Ira Reiss (1986) defines jealousy as a boundary-setting mechanism established by the group or society. Every society defines certain relationships as important, and jealousy helps to preserve these relationships. Jealousy in marriage seems to be a cultural universal. Reiss believes that marriage is important because it is the relationship in which we engage in physical intimacy and self-disclosure. In fact, any relationship in which physical and emotional intimacy occur will be subject to jealousy. Not surprisingly, a number of researchers have observed that threats to sexual relationships can invoke intense jealousy.

Bringle and Buunk define jealousy as "any aversive emotional reaction that occurs as a result of a partner's extradyadic [outside the partnership] relationship that is real, imagined, or considered likely to occur." (1991, p. 135). They suggest that jealousy is most intense when the extradyadic liaison threatens the continued existence of the primary relationship. There is evidence of a gender difference in reactions; men are more likely to become jealous in response to their partner's sexual activity with a third person, whereas women are more likely to become jealous in response to their partner's emotional involvement with a third person. This difference is found in research conducted in Germany, the Netherlands, and the United States (Buunk, et al., 1996).

According to sexual strategies theory, males are especially concerned with guaranteeing the paternity of offspring (Buss, et al., 1996). Extradyadic sexual relationships create the possibility of offspring that are not the man's. Hence, he reacts very strongly to sexual infidelity by his mate. Females are especially concerned with having adequate resources to support themselves and their offspring. If the male becomes romantically involved with another woman, he may give her some of his resources; of course, according to this view, the other woman is interested in him precisely because she wants some of his resources. And why is jealousy universal? Because jealous men kept rivals away, thus helping to ensure the passing on of their genes, and jealous women got greater protection and resources from their mate, thus helping to ensure the survival of their offspring.

Psychologists Gregory White and Paul Mullen (1989) see jealousy as a constellation including thoughts, emotions, and actions. Two situations, according to their research, activate jealousy. One is a situation in which there is a threat to our self-esteem. For example, in a good relationship, our romantic partner helps us feel good about ourselves—makes us feel attractive or fun to be with, for example. If a rival appears and our partner shows interest, we may think things like "He finds her more attractive than me" or "She finds him more fun to be with than me." We then feel less attractive or less fun to be with—that is, our self-esteem is threatened.

The second situation that activates jealousy is a threat to the relationship. If a rival appears on the

childhood—perhaps because of divorce or death—is not related to adult attachment styles. That is, children of divorced parents are no more or less likely to be secure lovers than children from intact marriages (a finding that is probably fortunate, given the high divorce rate in the United States).

What did predict adult attachment style was the person's perception of the *quality* of the relationship with each parent.

This research has some important implications. First, it helps us understand that adults bring to any particular romantic relationship their own personal

Figure 13.6   One situation that activates jealousy is a perceived threat to the relationship.

scene, we may fear that our partner will separate from us and form a new relationship with the rival. Jealousy is activated because of our negative thoughts and feelings about the loss of a relationship that has been good for us and the loss of all the pleasant things that go along with that relationship, such as companionship and sex.

According to White and Mullen, we go through several stages in the jealousy response, sometimes very quickly. The first is cognitive, in which we make an initial appraisal of the situation and find that there is a threat to self-esteem or to the relationship. Next we experience an emotional reaction which in itself has two phases. The first is a rapid stress response, the "jealous flash." To use the terminology of the two-component theory of love discussed later in this chapter, this stress response is the physiological component of the jealous emotion. The second phase of emotional response occurs as we reappraise the situation and decide how to cope with it. In the reappraisal stage, we may shift from seeing the situation as a threat to seeing

it as a challenge, for example. The intense initial emotions quiet down and may be replaced by feelings of moodiness.

Attempts to cope with jealousy lead to a variety of behaviors. Some of these are constructive, such as effective communication with one's partner (see Chapter 10 for a discussion of techniques of effective communicators). Such discussion may lead to an evaluation of the relationship, and attempts to change problematic aspects of it. If the problems seem sufficiently serious, a couple may seek advice from a mediator or therapist. Other behavioral responses to jealousy are destructive. The threat to one's self-esteem may lead to depression, substance abuse, or suicide. Aggression may be directed at the partner, the third person, or both, and may result in physical or sexual abuse or even murder.

Research suggests that one's attachment style may be an important influence on how one responds to jealousy (Sharpstein & Kirkpatrick, 1997). Undergraduates were asked how they had reacted in the past to jealousy. Those with a secure attachment style reported that they had expressed their anger to the partner and maintained the relationship. Those with an anxious style reported the most intense anger, but they were most likely to say they did not express their anger. People with an avoidant style were more likely to direct their anger toward the third person.

Sources: Robert Bringle and Bram Buunk. (1991). Extra-dyadic relationships and sexual jealousy. In Kathleen McKinney and Susan Sprecher (Eds.). *Sexuality in close relationships.* Hillsdale, NJ: Lawrence Erlbaum, 135–153; Helen Fisher. (1992). *Anatomy of love.* New York: Fawcett Columbine; Ira Reiss (1986). *Journey into sexuality.* Englewood Cliffs, NJ: Prentice Hall; Gregory White and Paul Mullen. (1989). *Jealousy: Theory, research, and clinical strategies.* New York: Guilford Press.

history of love and attachment. The forces of that personal history can be strong, and one good and loving partner may not be able to change an avoidant lover into a secure lover. Second, it helps us understand that conflict in some relationships may be caused by a mismatch of attachment styles.

A secure lover, who wants a close, intimate relationship, is likely to feel frustrated and dissatisfied with an avoidant lover, who is uncomfortable with feeling close. Attachment theory suggests that the important form of similarity is similarity in attachment style (Latty-Mann & Davis, 1996). Finally, this theory provides some explanation for jealousy, which is most common among anxious-ambivalent lovers (although present among the others) because of their early experience of feeling anxious about their attachment to their parents.

A study of 354 heterosexual couples in serious dating relationships looked at the dynamics of adult attachment styles (Kirkpatrick & Davis, 1994). In over half the couples, both partners had a secure attachment style. About 10 percent consisted of one person with a secure style and one with an avoidant style, and 10 percent consisted of a secure-anxious pairing. As one might expect, there was not a single anxious-anxious or avoidant-avoidant couple; such couples would be very incompatible. Partners with a secure style reported the greatest commitment to and satisfaction with their relationships. Relationships in which the woman had an anxious style were rated more negatively by both partners. Avoidant men gave the most negative ratings, not surprisingly, since they are uncomfortable with emotional closeness. These results lend strong support to the attachment theory.

Attachment style affects relationships by affecting the way the partners interact. A study of 128 established couples (average length of relationship 47 months) assessed attachment style, patterns of accommodation, and satisfaction with the relationship (Scharfe & Bartholomew, 1995). Individuals with a secure attachment reported responding constructively to potentially destructive behavior by the partner, for example, with efforts to discuss and resolve the problem. People who were fearful of attachment to another responded with avoidance or withdrawal.

Love as a Story

When we think of love, our thoughts often turn to the great love stories: Romeo and Juliet, Cinderella and the Prince (Julia Roberts and Richard Gere), King Edward VIII and Wallis Simpson, and *Pygmalion/My Fair Lady*. According to Sternberg's latest work (1998), these stories are much more than entertainment. They shape our beliefs about love and relationships, and our beliefs in turn influence our behavior.

> Zach and Tammy have been married 28 years. Their friends have been predicting divorce since the day they were married. They fight almost constantly. Tammy threatens to leave Zach; he tells her that nothing would make him happier. They lived happily ever after.

> Valerie and Leonard had a perfect marriage. They told each other and all of their friends that they did. Their children say they never fought. Leonard met someone at his office, and left Valerie. They are divorced. (Adapted from Sternberg, 1998)

Wait a minute! Aren't those endings reversed? Zach and Tammy should be divorced, and Valerie and Leonard should be living happily ever after. If love is merely the interaction between two people, how they communicate and behave, you're right; the stories have the wrong endings. But there is more to love than interaction; what matters is how each partner *interprets* the interaction. To make sense out of what happens in our relationships, we rely on our love stories.

A **love story** is a story about what love should be like; it has *characters,* a *plot,* and a *theme.* There are two central characters in every love story, and they play roles that complement each other. The plot details the kinds of events that occur in the relationship. The theme is central; it provides the meaning of the events that make up the plot and gives direction to the behavior of the principals. The story guiding Zach and Tammy's relationship is the War story. Each views love as war; a good relationship involves constant fighting. The two central characters are warriors, doing battle, fighting for what they believe. The plot consists of arguments, fights, threats to leave—in other words, battles. The theme is that love is war. One may win or lose particular battles, but the war continues. Zach and Tammy's relationship endures because they share this view, and because it fits their temperaments. Can you imagine how long a wimp would last in a relationship with either of them?

According to this view, *falling in love* occurs when you meet someone with whom you can create a relationship that fits your love story. Further, we are satisfied with relationships in which we and our partner match the characters in our story (Beall & Sternberg, 1995). Valerie and Leonard's marriage looked great on the surface but it didn't fit Leonard's love story. He left when he met his "true

---

Love story: A story about what love should be like, including characters, a plot, and a theme.

love," that is, a woman who could play the complementary role in his primary love story.

Where do our stories come from? Many of them have their origins in the culture, in folk tales, literature, theater, films, and television programs. The cultural context interacts with our personal experience and characteristics to create the stories that each of us has (Sternberg, 1996). As we experience relationships, our stories evolve, taking into account unexpected events. Each person has more than one story; the stories often form a hierarchy. One of Leonard's stories was "House and Home," where home was the center of the relationship, and he (in his role of Caretaker) showered attention on the house and kids (not on Valerie). But when he met Sharon, with her aloof air, ambiguous past, and dark glasses, he was hooked—she elicited the "Love Is a Mystery" story, which was more salient to Leonard. He could not explain why he left Valerie and the kids; like most of us he was not consciously aware of his love stories.

It should be obvious from the examples that love stories derive their power from the fact that they are self-fulfilling. We create in our relationships events according to the plot and then interpret those events according to the theme. Our love relationships are literally social constructions. Because our love stories are self-confirming, they can be very difficult to change.

Sternberg and his colleagues have identified five categories of love stories found in United States culture, and several specific stories within each category. They have also developed a series of statements that reflect the themes in each story. People who agree with the statements "I think fights actually make a relationship more vital," and "I actually like to fight with my partner" are likely to believe in the War story. Sternberg and Hojjat studied samples of 43 and 55 couples (Sternberg, 1998). They found that couples generally believed in similar stories. The more discrepant the stories of the partners, the less happy the couple was. Some stories were associated with high satisfaction, for example, the Garden story, in which love is a garden that needs ongoing cultivation. Two stories associated with low satisfaction were the Business story (especially the version in which the roles are Employer and Employee?), and the Horror story, in which the roles are Terrorizer and Victim.

The Chemistry of Love

The three theories considered so far define love as a single phenomenon. A fourth perspective differentiates between two kinds of love: passionate love and companionate love (Berscheid & Hatfield, 1978). **Passionate love** is a state of intense longing for union with the other person and of intense physiological arousal. It has three components: cognitive, emotional, and behavioral (Hatfield & Sprecher, 1986a). The cognitive component includes preoccupation with the loved one and idealization of the person or of the relationship. The emotional component includes physiological arousal, sexual attraction, and desire for union. Behavioral elements include taking care of the other and maintaining physical closeness. Passionate love can be overwhelming, obsessive, all-consuming.

By contrast, **companionate love** is a feeling of deep attachment and commitment to a person with whom one has an intimate relationship (Hatfield & Rapson, 1993b). Passionate love is hot; companionate love is warm. Passionate love is often the first stage of a romantic relationship. Two people meet, fall wildly in love, and make a commitment to each other. But as the relationship progresses, a gradual shift to companionate love takes place (Driscoll et al., 1972; Cimbalo et al., 1976). The transformation tends to occur when the relationship is between 6 and 30 months old (Hatfield & Walster, 1978).

Some may find this a rather pessimistic commentary on romantic love. But it may actually be a good way for a relationship to develop. Passionate love may be necessary to hold a relationship together in the early stages, while conflicts are being resolved. But past that point, most of us find that what we really need is a friend—someone who shares our interests, who is happy when we succeed, and who sympathizes when we fail—and that is just what we get with companionate love.

What causes the complex phenomena of passionate and companionate love? Where does

Passionate love: A state of intense longing for union with the other person and of intense physiological arousal.

Companionate love: A feeling of deep attachment and commitment to a person with whom one has an intimate relationship.

the rush of love at first sight come from? Liebowitz (1983) suggests that it is caused by bodily chemistry (see also Fisher, 1992). Passionate love is like an amphetamine high; both activate certain neurochemical circuits, and it is this activation that causes high energy, euphoria, elation, and idealization (or should we say bad judgment?). Liebowitz suggests that phenylethylamine (PEA) is the chemical responsible. Like all chemically induced highs, this one must end. Either the neural circuits become accustomed to PEA, so it has less effect than before, or levels of PEA fall.

According to this theory, the frequent presence of the loved one, produced by passionate love, triggers the production of other chemicals, called *endorphins*. These chemicals produce a sense of calm, peace, and tranquility—in other words, the warm feelings of companionate love. A third chemical, oxytocin, may contribute to long-term relationships. It has been shown to play an important role in pair bonding in some animals (McEwen, 1997). In humans, it is stimulated by touch, including sexual touching, and produces feelings of pleasure and satisfaction.

There is little research that directly tests these hypotheses about the neurochemical basis of romance. The theory is consistent with many observations about love made by poets, novelists, and social scientists.

Research on Love

Measuring Love

So far the discussion has focused on theoretical definitions of various kinds of love. You can see that Sternberg, Hazan and Shaver, and Berscheid and Hatfield mean different things when they use the word "love." One of the ways psychologists and sociologists define terms is by using an operational definition. In an **operational definition,** a concept is defined by the way it is measured. Thus, for example, *IQ* is sometimes defined as the kinds of abilities that are measured by IQ tests. *Job satisfaction* can be defined as a score on a questionnaire that measures a person's attitudes toward his or her job. Operational definitions are very useful because they are precise and because they help clarify exactly what a scientist means by a complex term such as *love*.

We introduced the concept of *passionate love* in the preceding section. Hatfield and Sprecher (1986a) decided to develop a paper-and-pencil measure of this concept. They wrote statements intended to measure the cognitive, emotional, and behavior components of passionate love. The respondent rates each statement on a scale from 1 (not true at all) to 9 (definitely true of him or her). If you feel that you are in love with someone, think about whether you would agree with each of the following statements, keeping that person in mind.

1. *Cognitive component:*
   Sometimes I feel I can't control my thoughts; they are obsessively on _____.
   For me, _____ is the perfect romantic partner.

2. *Emotional component:*
   I possess a powerful attraction for _____.
   I will love _____ forever.

3. *Behavioral component:*
   I eagerly look for signs indicating _____'s desire for me.
   I feel happy when I am doing things to make _____ happy.

Hatfield and Sprecher administered their questionnaire to students at the University of Wisconsin who were in relationships ranging from casually dating to engaged and living together. The results indicated that scores on the Passionate Love Scale (PLS) were positively correlated with other measures of love and with measures of commitment to and satisfaction with the relationship. These correlations give us evidence that the PLS is valid. Students who got high scores on the PLS reported a stronger desire to be with, held by, and kissed by the partner, and said that they were sexually excited just thinking about their partner. These findings confirm that the scale is measuring passion. Finally, passionate love scores increased as the nature of the relationship moved from dating to dating exclusively. Hatfield and Sprecher's research is a good example of how to study an important but complex topic—such as love—scientifically.

Operational definition: Defining some concept or term by how it is measured, for example, defining intelligence as those abilities that are measured by IQ tests.

Gender Differences

The stereotype is that women are the romantics—they yearn for love, fall in love more easily, cling to love. Do the data support this idea?

In fact, research measuring love in relationships indicates that just the opposite is true. Men hold a more romantic view of male-female relations than women do (Hobart, 1958). They fall in love earlier in a relationship (Kanin et al., 1970; Rubin et al., 1981). Men also cling longer to a dying love affair (Hill et al., 1976; Rubin et al., 1981). Indeed, three times as many men as women commit suicide after a disastrous love affair (Hatfield & Walster, 1978). In a word, it seems that men are the real romantics.

Three decades ago, men and women held very different views of the importance of romantic love in marriage (Simpson et al., 1986). In response to the question, "If (someone) had all the other qualities you desired, would you marry this person if you were not in love?" about 30 percent of women said they would refuse to marry, compared with over 60 percent of men. That is, men were more likely to view love as an essential requirement for marriage. However, those patterns have changed. This survey was repeated in 1976 and again in 1984 (Simpson et al., 1986). The gender differences had disappeared, and love was considered more essential to marriage in 1984 than it was in the 1960s. Over 80 percent of both men and women would refuse to marry under the conditions stated in the question. The researchers interpreted this dramatic shift as being a result of great changes in the social roles of men and women. In particular, women are more likely to hold paid employment and to be more economically independent of men. Therefore, they feel less need to be in a marriage—whether in love or not—in order to be financially supported. Consequently, love can be a necessary requirement for them, too. It is ironic that the sexual revolution—which stressed the right to liberated, even casual, sex—was accompanied by an increased, not a decreased, emphasis on love.

Love and Adrenaline

Two-Component Theory of Love

Social psychologists Ellen Berscheid and Elaine Walster (1974a) have proposed a **two-component theory of love.** According to their theory, passionate love occurs when two conditions exist simultaneously: (1) the person is in a state of intense *phys-iological arousal,* and (2) the situation is such that the person applies a particular *label*—"love"—to the sensations being experienced. Their theory is derived from an important theory developed by Stanley Schachter (1964).

Suppose that your heart is pounding, your palms are sweating, and your body is tense. What emotion are you experiencing? Is it love—has reading about passionate love led to obsessive thoughts of another person? Is it fear—are you frantically reading this text because you have an exam tomorrow morning? Is it sexual arousal—are you thinking about physical intimacy later tonight? It could be any of these, or it could be anger or embarrassment. A wide variety of emotions are accompanied by the same physiological states: increased blood pressure, increased heart rate, increased myotonia (muscular tension), sweating palms. What differentiates these emotions is the way we interpret or label what we are experiencing.

Schachter's (1964) two-component theory of emotion says just this; an emotion consists of a physiological arousal state plus the label the person assigns to it (for a critical evaluation of this theory, see Reisenzein, 1983). Berscheid and Walster have applied this to the emotion of "love." They suggest that we feel passionate love when we are aroused and when conditions are such that we identify what we are feeling as love.

Evidence on the Two-Component Theory

Several experiments provide evidence for Berscheid and Walster's two-component theory of love. In one study, male research participants exercised vigorously by running in place; that produced the physiological arousal response of pounding heart and sweaty palms (White et al., 1981). Afterward they rated their liking for an attractive woman, who actually was a confederate of the experimenters. Men in the running group said they liked the woman significantly more than did men who were in a control condition and had not exercised. This result is consistent with Berscheid and Walster's theory. The effect is called

> **Two-component theory of love:** Berscheid and Walster's theory that two conditions must exist simultaneously for passionate love to occur: physiological arousal and attaching a cognitive label ("love") to the feeling.

Figure 13.7 The misattribution of arousal. If people are physically aroused (e.g., by jogging), they may misattribute this arousal to love or sexual attraction, provided the situation suggests such an interpretation.

the **misattribution of arousal;** that is, in a situation like this, the men misattribute their arousal—which is actually due to exercise—to their liking for the attractive woman.

Another study suggests that fear can increase a man's attraction to a woman (Dutton & Aron, 1974; see also Brehm et al., cited in Berscheid & Walster, 1974a). An attractive female interviewer approached male passersby either on a fear-arousing suspension bridge or on a non-fear-arousing bridge. The fear-arousing bridge was constructed of boards, attached to cables, and had a tendency to tilt, sway, and wobble; the handrails were low, and there was a 230-foot drop to rocks and shallow rapids below. The "control" bridge was made of solid cedar; it was firm, and there was only a 10-foot drop to a shallow rivulet below. The interviewer asked subjects to fill out questionnaires that included projective test items. These items were then scored for sexual imagery.

The men in the suspension-bridge group should have been in a state of physiological arousal, while those in the control-bridge group should not have been. In fact, there was more sexual imagery in the questionnaires filled out by the men in the suspension-bridge group, and these men made more attempts to contact the attractive interviewer after the experiment than the men on the control

bridge. Intuitively, this might seem to be a peculiar result: that men who are in a state of fear are more attracted to a woman than men who are relaxed. But in terms of the Berscheid and Walster two-component theory, it makes perfect sense. The fearful men were physiologically aroused, while the men in the control group were not. And according to this theory, arousal is an important component of love or attraction.[2]

Now, of course, if the men (most of them heterosexuals) had been approached by an elderly man or a child, probably their responses would have been different. In fact, when the interviewer in the experiment was male, the effects discussed above did not occur. Society tells us what the appropriate objects of our love, attraction, or liking are. That is, we know for what kinds of people it is appropriate to have feelings of love or liking. For these men, feelings toward an attractive woman could reasonably be labeled "love" or "attraction," while such labels would probably not be attached to feelings for an elderly man.

The physical arousal that is important for love need not always be produced by unpleasant or frightening situations. Pleasant stimuli, such as sexual arousal or praise from the other person, may produce arousal and feelings of love. Indeed, Berscheid and Walster's theory does an excellent job of explaining why we seem to have such a strong tendency to associate love and sex. Sexual arousal is one method of producing a state of physiological arousal, and it is one that our culture has taught us to label as "love." Thus both components necessary to feel love are present: arousal and a label. On the other hand, this phenomenon may lead us to confuse love with lust, an all-too-common error.

Cross-Cultural Research

In the past decade, researchers have studied people from other ethnic or cultural groups, to see whether attraction, intimacy, and love are experi-

[2]According to the terminology of Chapter 3, note that the Dutton and Aron study is an example of *experimental* research.

Misattribution of arousal: When one is in a stage of physiological arousal (e.g., from exercising or being in a frightening situation), attributing these feelings to love or attraction to the person present.

enced in the same way outside the United States. Three topics that have been studied are the impact of culture on how people view love, on whom people fall in love with, and on the importance of love in decisions to marry.

Cultural Values and the Meaning of Love

Cross-cultural psychologists have identified two dimensions on which cultures vary (Hatfield & Rapson, 1993a). The first is individualism–collectivism. *Individualistic cultures,* like those of the United States, Canada, and western European countries, tend to emphasize individual goals over group and societal goals and interests. *Collectivist cultures,* like those of China, Africa, and southeast Asian countries, emphasize group and collective goals over personal ones. Several specific traits have been identified which differentiate these two types of societies (Triandis et al., 1990). In individualistic cultures, behavior is regulated by individual attitudes and cost-benefit considerations; emotional detachment from the group is accepted. In collectivist cultures, the self is defined by group membership; behavior is regulated by group norms; attachment to and harmony within the group are valued.

There are different conceptions of love in the two types of cultures. American society, for example, emphasizes passionate love as the basis for marriage (Dion & Dion, 1993b). Individuals select mates on the basis of such characteristics as physical attractiveness, similarity (compatibility), and wealth or resources. We look for intimacy in the relationship with our mate. In Chinese society, by contrast, marriages are arranged; the primary criterion is that the two families be of similar status. The person finds intimacy in relationships with other family members.

The second dimension on which cultures differ is independence–interdependence. Many Western cultures view each person as independent, and value individuality and uniqueness. Many other cultures view the person as interdependent with those around him or her. The self is defined in relation to others. Americans value standing up for one's beliefs. The people of India value conformity and harmony within the group.

Dion and Dion (1993a) studied university students in Toronto, representing four ethnocultural groups. Students from Asian backgrounds were more likely to view love as companionate, as friendship, in contrast to those from English and Irish backgrounds. This tendency is consistent with the collectivist orientation of Asian cultures. Castaneda (1993) studied Mexican American students. They were similar to American students of European background in the emphasis placed on trust and communication/sharing as components of romantic love, but they placed greater emphasis on mutual respect. One student wrote, "[In a love relationship] we must respect each other's feelings as we would expect them to show us respect" (p. 265). Such respect allows each partner to express his or her needs to the other.

Figure 13.8   Whether a culture is individualistic or collectivistic determines its views on love and marriage. In the United States, which is an individualistic culture, individuals choose each other and marry for love. In India, a collectivistic culture, marriages are arranged by family members to serve family interests.

Cultural Influences on Mate Selection

Buss (1989) conducted a large-scale survey of 10,000 men and women from 37 societies. The sample included people from 4 cultures in Africa, 8 in Asia, and 4 in eastern Europe, in addition to 12 western European and 4 North American ones. Each respondent was given a list of 18 characteristics one might value in a potential mate and was asked to rate how important each was to him or her personally. Regardless of which society they lived in, most respondents, male and female, rated intelligence, kindness, and understanding at the top of the list; note that these are characteristics of companionate love. Men worldwide placed more weight on cues of reproductive capacity, such as physical attractiveness; women rated cues of resources as more important. The results clearly support the sociobiological perspective and suggest that there are not large cultural differences.

Many people prefer mates who are physically attractive. We often hear that "beauty is in the eye of the beholder." This saying suggests that standards of beauty might vary across cultures. In one study, researchers had students from varying cultural backgrounds rate 45 photographs of women on a scale ranging from very attractive to very unattractive (Cunningham et al., 1995). The photographs portrayed women from many different societies. Overall, Asian, Hispanic, and white students did not differ in their ratings of the individual photographs. However, Asian students' ratings were less influenced by indicators of sexual maturity (such as facial narrowness) and expressivity (such as the vertical distance between the lips when the person smiled). In a separate study, black men and white men gave similar ratings to most aspects of female faces, but black men preferred women with heavier bodies than did white men. Again, the results indicate more similarity than difference across cultures, in this case in standards of physical attractiveness.

Love and Marriage

We noted above that individualistic cultures place a high value on romantic love, while collectivist cultures emphasize the group. The importance of romantic love in American society was illustrated earlier when we discussed responses to the question, "If a man (woman) had all the other qualities you desired, would you marry this person if you were not in love with him (her)?" Over time, increasing percentages of American men and women answer no. Levine and his colleagues (1995) asked this question of men and women in 11 different cultures. We would predict that members of individualistic cultures would answer no, whereas those in collectivist cultures would answer yes. The results are displayed in Table 13.2. Note that, as predicted, many Indians and Pakistanis would marry even though they didn't love the person. In Thailand, which is also collectivist, a much smaller percentage said yes. In the individualistic cultures

Table 13.2 "Would you marry someone you didn't love?"

Cultural Group	Responses (Percent)		
	Yes	Undecided	No
Australia	4.8%	15.2%	80.0%
Brazil	4.3	10.0	85.7
England	7.3	9.1	83.6
Hong Kong	5.8	16.7	77.6
India	49.0	26.9	24.0
Japan	2.3	35.7	62.0
Mexico	10.2	9.3	80.5
Pakistan	50.4	10.4	39.1
Philippines	11.4	25.0	63.6
Thailand	18.8	47.5	33.8
United States	3.5	10.6	85.9

Source: Hatfield, Elaine. (1994). *Passionate love and sexual desire: A cross-cultural perspective.* Paper presented at the Annual Meeting, Society for the Scientific Study of Sexuality, Miami.

of Australia, England, and the United States, few would marry someone they did not love.

The Pattern of the Cross-Cultural Findings

When we look at the findings of the cross-cultural research on love, attraction, and marriage, the pattern that emerges is one of *cross-cultural similarities and cross-cultural differences,* a theme we introduced in Chapter 1. That is, some phenomena are similar across cultures, for example, valuing intelligence, kindness, and understanding in a mate. Other phenomena differ substantially across cultures, for example, whether love is a prerequisite for marriage.

SUMMARY

Research indicates that mere repeated exposure to another person facilitates attraction. We tend to be attracted to people who are similar to us socially (age, race or ethnicity, economic status) and psychologically (attitudes, interests). In first impressions, we are most attracted to people who are physically attractive; we also tend to be attracted to people whom we believe to be "within reach" of us, depending on our sense of our own attractiveness or desirability. Playing hard to get seems to work only if the person does so selectively.

According to reinforcement theory, we are attracted to those who give us many reinforcements. Interaction with people who are similar to us is smooth and rewarding; they have similar outlooks and like the same things that we like. According to sexual strategies theory, we prefer young, attractive people because they are likely to be healthy and fertile. Men prefer women who are sexually faithful (hard to get); women prefer men with resources who will invest in them and their children.

Intimacy is a major component of a romantic relationship. It is defined as a quality of a relationship characterized by commitment, feelings of closeness and trust, and self-disclosure.

According to the triangular theory, there are three components to love: intimacy, passion, and decision or commitment. Love is a triangle, with each of these components as one of the points. Partners whose love triangles are substantially different are mismatched and are likely to be dissatisfied with their relationship.

According to the attachment theory of love, adults vary in their capacity for love as a result of their love or attachment experiences in infancy. This theory says that there are three types of lovers: secure lovers, avoidant lovers, and anxious-ambivalent lovers.

Love can also be viewed as a story, with characters, a plot, and a theme. People use their love stories to interpret experiences in relationships. Falling in love happens when a person meets someone who can play a compatible role in his or her story.

Love may have a neurochemical component. Passionate love, a state of intense longing and arousal, may be produced by phenylethylamine. Like all chemically induced "highs," passionate love eventually comes to an end. It may be replaced by companionate love, a feeling of deep attachment and commitment to the partner. This type of love may be accompanied by elevated levels of endorphins and oxytocin, which may be produced by physical closeness and touch.

Hatfield and Sprecher have constructed a scale to measure passionate love. Such scales make it possible to do scientific research on complex phenomena like love. Scores on this scale were correlated with measures of commitment to and satisfaction with romantic relationships. Research indicates that, in general, men are more romantic than women and fall in love earlier in a relationship.

Berscheid and Walster have hypothesized that there are two basic components of romantic love: being in a state of physiological arousal and attaching the label "love" to the feeling. Several studies report evidence consistent with the hypothesis.

Cross-cultural research indicates that individualistic cultures like that of the United States emphasize love as the basis for marriage and encourage intimacy between partners. Collectivist cultures emphasize intergroup bonds as the basis for marriage, and discourage intimacy between partners. Culture influences the importance of various characteristics in choosing a mate; it also affects our standards of beauty and the likelihood that we would marry someone we didn't love.

REVIEW QUESTIONS

1. The mere-exposure effect refers to the tendency of women to be secretly attracted to exhibitionists. True or false?

2. Research indicates that, in the dating marketplace, women's worth is based on their physical beauty, whereas men's worth is based on their success. True or false?

3. According to Byrne's law of attraction, we tend to be attracted to people who give us many _____ and few punishments.

4. According to sexual strategies theory, men want mates who have resources and women want mates who are fertile. True or false?

5. An _____ relationship is characterized by commitment, feelings of closeness and trust, and mutual self-disclosure.

6. According to the triangular theory, love has three components: _____ , _____ , and decision or commitment.

7. According to the attachment theory of love, conflict in some relationships is caused by a mismatch of attachment styles. True or false?

8. Research indicates that women are romantics: they yearn for love and fall in love more easily than men. True or false?

9. According to Berscheid and Walster's two-component theory, passionate love occurs when two conditions exist simultaneously: _____ and attaching the cognitive label "love" to the experience.

10. Cross-cultural psychologists tell us that in _____ cultures, like China, people emphasize group and societal goals in selecting mates.

(The answers to all review questions are at the end of the book, after the Glossary.)

QUESTIONS FOR THOUGHT, DISCUSSION, AND DEBATE

1. If you are currently in love with someone, how would you describe the kind of love you feel, using the various concepts and theories of love discussed in this chapter?

2. Resolved: Selecting mates on the basis of individualistic considerations, such as whether you love the person, contributes to the high rates of divorce, and single-parent families.

3. Your best friend has been dating another person exclusively for the past year. One day you ask how the relationship is going. Your friend replies, "I don't know. We get along really well. We like to do the same things, and we can tell each other everything. But I feel like something is missing. How do you know if you are in love?" How would you answer her question?

SUGGESTIONS FOR FURTHER READING

Fisher, Helen. (1992). *Anatomy of love.* New York: Fawcett Columbine. Fisher explains sexual anatomy, sexual emotions, mate selection, adultery and the sexual double standard, among others, using evolutionary perspectives. A provocative book.

Hendrick, Susan, and Hendrick, Clyde. (1992). *Liking, loving and relating.* 2nd ed. Pacific Grove, CA: Books/Cole. This textbook explains psychologists' research on interpersonal attraction, love, and the formation and maintenance of relationships.

Lerner, Harriet G. (1989). *The dance of intimacy.* New York: Harper and Row. Lerner, a prominent psychotherapist, gives tips of how to promote intimacy in our relationships.

Sternberg, Robert. (1998). *Love is a story: A new theory of relationships.* Sternberg describes his theory and the twenty-seven love stories he has identified. The book includes items from a scale designed to identify which stories a person holds.

chapter
14

GENDER ROLES, FEMALE SEXUALITY, AND MALE SEXUALITY

CHAPTER HIGHLIGHTS

Gender Roles and Stereotypes

Male-Female Psychological Differences

Male-Female Differences in Sexuality

Why the Differences?

Beyond the Young Adults

Transsexualism

> "The majority of women (happily for them) are not very much troubled with sexual feelings of any kind. What men are habitually, women are only exceptionally."*

> Hoggity higgamous,
>   men are polygamous,
> Higgity hoggamous,
>   women monogamous.†

When a baby is born, what is the first statement made about it? "It's a boy" or "It's a girl," of course. Sociologists tell us that gender is one of the most basic of status characteristics. That is, in terms of both our individual interactions with people and the position we hold in society, gender is exceptionally important. We experience consternation in those rare cases when we are uncertain of a person's gender. We do not know how to interact with the person, and we feel quite flustered, not to mention curious, until we can ferret out some clue as to whether the person is a male or a female. We seem to need to know a person's gender before we can figure out how to interact with the person. In this chapter we explore gender roles and the impact they may have on sexuality, and transsexualism (a disturbance of gender identity).

Gender Roles and Stereotypes

One of the basic ways in which societies codify this emphasis on gender is through gender roles.[1] A **gender role** is a set of norms, or culturally defined expectations, that define how people of one gender ought to behave. A closely related phenomenon is a **stereotype,** which is a generalization about a group of people (for example, men) that distinguishes those people from others (for example, women). Research shows that even in modern American society, and even among college students, there is a belief that males and females do differ psychologically in many ways, and these

stereotypes have not changed much since 1972 (Bergen & Williams, 1991).

Heterosexuality is an important part of gender roles. The "feminine" woman is expected to be sexually attractive to men and in turn to be attracted to them. Women who violate any part of this role—for example, lesbians—are viewed as violators of gender roles and are considered masculine (Storms, 1980). Heterosexuality is equally important in the male role.

Gender Roles and Ethnicity

Gender-role stereotypes vary somewhat among the various ethnic groups of the United States. In one recent study, data were collected about this very issue (Niemann et al., 1994). College students at the University of Houston—51 percent of whom were European American and the rest of whom were, in order of frequency, Latino, African American, Asian American, and Native American—were asked to list ten adjectives that came to mind when they thought of members of the following groups: Anglo-American males, Anglo-American females, African American males, African American females, Asian American males, Asian American females, Mexican American males, and Mexican American females. The most frequently listed adjectives are shown in Table 14.1.

Two important patterns can be seen in Table 14.1: (1) Within an ethnic group, males and females have some stereotyped traits in common but are also seen as having some traits that differ. For example, both Mexican American males and Mexican American

---

*Dr. William Acton. (1857). *The functions and disorders of the reproductive organs.*

†Dorothy Parker.

[1]The distinction between sex and gender will be maintained in this chapter. Male-female roles—and thus gender roles—are being discussed here.

> **Gender role:** A set of norms, or culturally defined expectations, that define how people of one gender ought to behave.
> **Stereotype:** A generalization about a group of people (e.g., men) that distinguishes them from others (e.g., women).

Table 14.1 The Interaction of Gender and Ethnicity: Stereotypes of Males and Females from Different Ethnic Groups

Anglo-American Males	Anglo-American Females
Intelligent	Attractive
Egotistical	Intelligent
Upper class	Egotistical
Pleasant/friendly	Pleasant/friendly
Racist	Blond/light hair
Achievement oriented	Sociable

African American Males	African American Females
Athletic	Speak loudly
Antagonistic	Dark skin
Dark skin	Antagonistic
Muscular appearance	Athletic
Criminal activities	Pleasant/friendly
Speak loudly	Unmannerly
	Sociable

Asian American Males	Asian American Females
Intelligent	Intelligent
Short	Speak softly
Achievement oriented	Pleasant/friendly
Speak softly	Short
Hard workers	

Mexican American Males	Mexican American Females
Lower class	Black/brown/dark hair
Hard workers	Attractive
Antagonistic	Pleasant/friendly
Dark skin	Dark skin
Noncollege education	Lower class
Pleasant/friendly	Overweight
Black/brown/dark hair	Baby makers
Ambitionless	Family oriented

Source: Yolanda F. Niemann, Leilani Jennings, Richard M. Rozell, James C. Baxter, and Elroy Sullivan (1994). Use of free responses and cluster analysis to determine the stereotypes of eight groups. *Personality and Social Psychology Bulletin, 20,* 379–390.

American females are stereotyped as pleasant and friendly, but only Mexican American females are stereotyped as overweight. (2) Within a gender, some stereotyped traits are common across ethnic groups, but others differ. For example, females from all ethnic groups are stereotyped as pleasant and friendly. However, Anglo-American and Asian American females are stereotyped as intelligent, whereas African American and Mexican American females are not.

As we consider variations in gender roles across various ethnic groups, it is crucial to understand how these gender roles are a product of *culture.* In the sections that follow, we consider some aspects of the cultures of four ethnic groups and their relevance to gender roles and sexuality.

African Americans

Two factors are especially significant in the cultural heritage of African Americans: the heritage of African culture and the experience of slavery in America and subsequent racial oppression (Sudarkasa, 1997). Two characteristics of African women have been maintained to the present: an important economic function, and a strong bond between mother and child (Ladner, 1971; Dobert, 1975). African women have traditionally been economically independent, functioning in the marketplace and as traders. African American women in the United States continue to assume this crucial economic function in the family to the present day. Mother-child bonds also continue to be extremely important in the structure of African American society.

Some say that the central theme for African American men today is pain (Doyle, 1989). Locked in chains during slavery, they are now locked behind bars in prisons. Just walking down the street, black men observe white women holding their handbags tighter. As one comedian said, "I was born a suspect . . . I asked this white guy for the time and he gave me his watch" (Doyle, 1989, p. 282). In the context of these overwhelmingly negative forces, the courage of the black men who have not given in and have gone on to forge successful lives for themselves and their families has to be admired.

The provider role is difficult for some African American men because of their high unemployment rate. For example, in 1997 the unemployment rate was 4.7 percent for adult white males; for adult black males, it was 11.1 percent (U.S. Bureau of the Census, 1997), more than double the rate for whites. Much of this discrepancy is accounted for by the disappearance of industrial jobs, which had provided good earnings for black men. The high unemployment rate creates a gender-role problem because the role of breadwinner or good provider is an important part of the male role in the United States. The inability to fulfill this part of the male role may be expressed in a number of ways. It may

Figure 14.1   Fathers and sons at the Million Man March in Washington, D.C., in 1995. Leaders of the march wanted to encourage African American men to take more responsibility for their families and community, and about 1 million men seemed to agree.

turn into antisocial behavior, violence, and crime, accounting for the high crime rate among male African American teenagers. It has been suggested that being in the army becomes an alternative means of fulfilling the male role. Twenty-seven percent of men in the army are African American (Bureau of the Census, 1997).

The role of husband is closely tied to the breadwinner role. African American men are understandably reluctant to take on the responsibility of marriage when unemployment is such a justified fear, and the rules of the welfare system essentially force low-income men to be absent from the home. In this context, it is not surprising that African American women expect to hold paying jobs. And, compared with white men, African American men hold more liberal (positive) attitudes about women working, although they are more conservative than white men on many other gender-role issues (Blee & Tickamyer, 1995).

Latinos

Hispanic Americans are currently the nation's second largest minority and are expected to be the largest in the near future (Vasquez & Barron, 1988). When we speak of the cultural heritage of Latinos, we must first understand the concept of *acculturation*, which is the process of incorporating the beliefs and customs of a new culture. The culture of Chicanos (Americans of Mexican heritage) is different from both the culture of Mexico and the dominant Anglo culture of the United States. Chicano culture is based on the Mexican heritage, modified through acculturation to incorporate Anglo components.

The family is the central focus of Hispanic life. Traditional Latinos place a high value on family loyalty and on warm, mutually supportive relationships, so that family and community are highly valued. This emphasis on family places special stresses on employed Latino women, who are expected

Figure 14.2   Asian American women have often been stereotyped as exotic sex toys. In the film *The World of Suzie Wong*, Nancy Kwan portrayed an alluring prostitute.

to be the preservers of family and culture and to do so by staying in the home.

As noted in Chapter 1, in traditional Latin American cultures, gender roles are rigidly defined (Comas-Diaz, 1987). Such roles are emphasized early in the socialization process for children. Boys are given greater freedom, are encouraged in sexual exploits, and are not expected to share in household work. Girls are expected to be passive, obedient, and virginal, and to stay in the home. These rigid roles are epitomized in the concepts of machismo and marianismo, discussed in Chapter 1.

Asian Americans

Chinese—almost all of them men—were recruited first in the 1840s to come to America as laborers in the West and later in the 1860s to work on the transcontinental railroad (for an excellent summary of the cultural heritage of Asian Americans, see Tsai & Uemura, 1988). Racist sentiment against

the Chinese grew, however, and there was a shift to recruiting first Japanese and Koreans and then Filipinos. Then, in the late 1960s and the 1970s, there was a mass exodus to the United States of refugees from war-torn Southeast Asia.

The cultural values of Asian Americans are in some ways consistent with white middle-class American values but in other ways contradict them. Asian Americans share with the white middle class an emphasis on achievement and on the importance of education. For example, Asian American women have a higher level of education, on the average, than white American women (Hyde, 1996). On the other hand, Asian Americans place far more value on family and group interdependence, compared with the white American emphasis on individualism and self-sufficiency. For Asian Americans, the family is a great source of emotional nurturance. One has an obligation to the family, and the needs of the family must take

Figure 14.3 Among some Native American tribes there are three gender roles, the third being known as a "manly hearted woman," or "warrior woman." Chiricahua Tah-des-te was a messenger and warrior in Geronimo's band. She participated in negotiations with several U.S. military leaders and surrendered with Geronimo in 1886.

precedence over the needs of the individual. For Asian American women, there can be a conflict in cultural values, between the traditional gender roles of Asian culture and those of modern Anglo culture, which increasingly prizes independence and assertiveness in women.

Just as the sexuality of African Americans has been stereotyped, so too has the sexuality of Asian Americans. The Asian American man has been stereotyped as asexual (lacking in sexuality), whereas the Asian American woman has been stereotyped as an exotic sex toy (Kim, 1990).

American Indians

At least some Indian tribes, including the Cherokee, Navajo, Iroquois, Hopi, and Zuñi, traditionally had relatively egalitarian gender roles (LaFromboise et al., 1990). That is, their roles were more egalitarian than that of white culture of the same period. The process of acculturation and adaptation to Anglo

society seems to have resulted in increased male dominance among American Indians.

Among the more than 200 Native languages spoken in North America, at least two-thirds have a term that refers to a third (or more) gender beyond male and female (Tafoya & Wirth, 1996). Anglo anthropologists termed this additional category "berdache," a term rejected by Native peoples, who prefer the term "Two-Spirit" (Jacobs et al., 1997). These same anthropologists concluded that the people were homosexuals, transsexuals, or transvestites, none of which are accurate from a Native point of view. A man might be married to a Two-Spirit male, but the marriage would not be considered homosexual because the two are of different genders (Tafoya & Wirth, 1996).

There was also a role of the "manly hearted woman," a role that a woman who was exceptionally independent and aggressive could take on. There was a "warrior woman" role among the Apache, Crow, Cheyenne, Blackfoot, Pawnee, and Navajo tribes (e.g., Buchanan, 1986; House, 1997). In both cases, women could express masculine traits or participate in male-stereotyped activities while continuing to live and dress as women.

In summary, research indicates that gender roles in the United States are not uniform. Different ethnic groups define gender roles differently. Let us turn now to some of the processes that create gender stereotypes.

Gender Schema Theory

In Chapter 2 we discussed psychologist Sandra Bem's gender schema theory, her cognitive approach to understanding gender stereotypes. Recall that according to that theory, a gender schema is the set of ideas (about behaviors, personality, appearance, and so on) that we associate with males and females (Bem, 1981). The gender schema influences how we process information. It causes us to tend to dichotomize information on the basis of gender. It also leads us to distort or fail to remember information that is stereotype-inconsistent. Recall that children who were shown pictures of children engaging in stereotype-inconsistent activities such as boys baking cookies recalled, when tested a week later, that they had seen girls baking cookies.

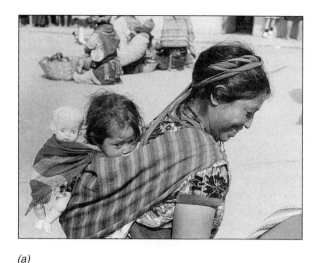

(a)

(b)

Figure 14.4 Children are very interested in achieving adult gender roles.

Gender schema theory points out the extent to which gender stereotypes or gender schemas lead us to gender-dichotomized thinking. It also provides evidence that it is relatively difficult to change people's stereotyped notions because we tend to filter out information that contradicts stereotypes.

Socialization

Many adult women and men do behave as gender roles say they should. Why does this happen? Psychologists and sociologists believe that it is a result of gender-role socialization. **Socialization** refers to the ways in which society conveys to the individual its norms or expectations for his or her behavior. Socialization occurs especially in childhood, as children are taught to behave as they will be expected to in adulthood. Socialization may involve several processes. Children may be rewarded for behavior that is appropriate for their gender ("What a brave little man he is"), or they may be punished for behavior that is not appropriate to their gender ("Nice young ladies don't do that"). The adult models they imitate—whether these are parents of the same gender, teachers, or women and men on television—also contribute to their socialization. In some cases, simply telling children what is expected of males and females may be sufficient for role learning to take place. Socialization continues in adulthood, as society conveys its norms of appropriate behavior for adult women and men. These norms extend from appropriate jobs to who initiates sexual activity.

Who (or what) are society's agents in accomplishing this socialization? Certainly parents have an early, important influence, from buying dolls for girls and footballs or baseball bats for boys to giving boys more freedom to explore. Research indicates that parents treat girls and boys similarly in many ways, with the exception that parents strongly encourage gender-typed activities (Lytton & Romney, 1991).

Parents are not the only socializing agents, though. The peer group can have a big impact in socializing for gender roles, particularly in adolescence. Other teenagers can be extremely effective in enforcing gender-role standards; for example, they may ridicule or shun a boy whose behavior is effeminate. Thus peers can exert great pressure for gender-role conformity.

The media are also important socializing agents. Many people assume that things have changed a lot in the last 20 years and that gender stereotypes are a thing of the past. On the contrary, various media—from television to teen magazines—continue

Socialization: The ways in which society conveys to the individual its norms or expectations for his or her behavior.

to show females and males in stereotyped roles. For example, an analysis of articles in *Seventeen* magazine over time showed that, in 1955, 43 percent of these articles intended for a female audience focused on personal appearance; in 1995 it was 45 percent (Schlenker et al., 1998). But surely the teen magazines are modernizing and providing more stories about today's relevant matters such as career development? The same study of *Seventeen* showed that in 1955 only 10 percent of the articles were about career development; by 1995 the figure had risen to a staggering 14 percent! In short, the distribution of articles conveys to girls that their appearance is important and a career is not.

An analysis of popular television situation comedies (sitcoms) from the 1950s to the 1990s indicated that there were small trends toward more egalitarian gender roles, but traditional stereotyping was still common (Olson & Douglas, 1997). *The Cosby Show* of the late 1980s earned the highest ratings for equality of gender roles of spouses and equality of gender roles of children (Olson & Douglas, 1997). But the 1990s series *Home Improvement* earned the lowest scores on equality of gender roles—lower even than the *Father Knows Best* series of the 1950s. Traditional gender roles are still alive on prime time.

Does the gender stereotyping in the media have an effect on people? In one study, preschool children were individually read a picture book featuring a character of their own gender engaged in play with either a stereotypic or a nonstereotypic toy (doll or dump truck) (Ashton, 1983). Afterward, the children were given the opportunity to play in a playroom with six experimental toys: two female-stereotypic toys (doll and china set), two male-stereotypic toys (truck and gun), and two neutral toys (ball and pegboard). The girls who had been exposed to the stereotyped book played significantly more with the female-stereotypic toys than did the girls who had been exposed to the nonstereotyped book. And boys who had been exposed to the nonstereotyped books played significantly more with the female-stereotyped toys than did boys exposed to the stereotyped book. Gender stereotypes in the media, then, can affect behavior.

Dozens of studies show that gender stereotypes shown on television affect children's stereotyped ideas as well (reviewed by Signorielli, 1990). For example, 3- to 6-year-olds who view more TV have more stereotyped ideas about gender roles than do children who view less. In a naturalistic experiment, children in a town with little availability of television showed less gender-stereotyped attitudes than children in a town with great availability of television. Television then became more available in the first town; two years later, the children in that town were as stereotyped in their attitudes as the children in the town that had had great availability of television all along.

Although gender roles themselves are universal (Rosaldo, 1974)—that is, all societies have gender roles—the exact content of these roles varies from one culture to the next, from one ethnic group to another, and from one social class to another. For example, Margaret Mead (1935) studied several cultures in which gender roles are considerably different from those in the United States. One such group is the Mundugumor of New Guinea. In that culture both females and males are extremely aggressive.

Male-Female Psychological Differences

Gender differences in personality and behavior have been studied extensively by psychologists (e.g., Hyde, 1996). Here we will focus on gender differences in two areas that are particularly relevant to gender and sexuality: aggressiveness and communication styles.

Males and females differ in *aggressiveness*. Males are generally more aggressive than females. This is true for virtually all indicators of aggression (physical aggression such as fighting, verbal aggression, and fantasy aggression) (Hyde, 1984). It is also true at all ages; as soon as children are old enough to perform aggressive behaviors, boys are more aggressive, and males dominate the statistics on violent crimes. The gender difference in aggression tends to be largest among preschoolers, but it gets smaller with age, so that gender differences in adults' aggression are small (Hyde, 1984).

Researchers have found that men and women differ in their style of communicating, both verbally and nonverbally (Wood, 1994). This research was reviewed in Chapter 10, Focus 10.1. Of particular relevance to sexuality, social psychologists have found gender differences in studies of **self-disclosure.** In these studies, people are brought into a laboratory

Self-disclosure: Telling personal information to another person.

and are asked to disclose personal information either to friends or to strangers. Women are more willing to disclose information than men are, at least in situations like these (Dindia & Allen, 1992).

Norms about self-disclosure are changing, though. Traditional gender roles favored emotional expressiveness for females, but emotional repressiveness and avoidance of self-disclosure for males. There is a contemporary ethic, though, of good communication and openness which demands equal self-disclosure from males and females (Rubin et al., 1980). Research with college students who are dating couples confirms the existence of this norm; the majority of both males and females reported that they had disclosed their thoughts and feelings fully to their partners (Rubin et al., 1980). However, women revealed more in some specific areas, particularly their greatest fears. And couples with egalitarian attitudes disclosed more than couples with traditional gender-role attitudes. Thus the traditional expectation that men should not express their feelings seems to be shifting toward an expectation that they be open and communicative.

There are gender differences in people's ability to understand the nonverbal behaviors of others. The technical phrase for this is "decoding nonverbal cues"—that is, the ability to read others' body language correctly. It might be measured, for example, by accuracy of interpreting facial expressions. Research shows that women are better than men at decoding such nonverbal cues and discerning others' emotions (Wood, 1994). Certainly this is consistent with the gender-related expectation that women will show greater interpersonal sensitivity.

What are the implications of these gender differences in communication styles for sexuality? For example, if men are unwilling to disclose personal information about themselves, consider whether this might not hamper their ability to communicate their sexual needs to their partners.

Male-Female Differences in Sexuality

In this section, the discussion will focus on areas of sexuality in which there is some evidence of male-female differences. As we will point out, differences do exist, but they are in a rather small number of areas—masturbation, attitudes about casual sex, and consistency of orgasm during sex. There is

a danger in focusing on these differences to the point of forgetting about gender similarities. You should keep in mind that males and females are in many ways quite similar in their sexuality—for example, in the physiology of their sexual response (Chapter 9)—as you consider the evidence on male-female differences that follows.

Masturbation

In a review of 177 studies of gender differences in sexuality the authors found that the largest gender difference was the incidence of masturbation (Oliver & Hyde, 1993).

Recall that in the Kinsey data, 92 percent of the males had masturbated to orgasm at least once in their lives, as compared with 58 percent of females. Not only did fewer women masturbate, but, in general, those who did masturbate had begun at a later age than the males. Virtually all males said they had masturbated before age 20 (most began between ages 13 and 15), but substantial numbers of women reported masturbating for the first time at age 25, 30, or 35. This gender difference shows no evidence of diminishing, according to more recent studies. The NHSLS, although it did not collect data on lifetime incidence of masturbation, did ask about masturbation in the last year; 63 percent of the men, compared with 42 percent of the women, reported that they had masturbated (Laumann et al., 1994). The data suggest, then, that there is a substantial gender difference in the incidence of masturbation; men are considerably more likely to have masturbated than women are.

Attitudes About Casual Sex

In the review mentioned above, the second largest gender difference noted was in attitudes toward casual sex—that is, premarital (or nonmarital) intercourse in a situation, such as a "one night stand," in which there is no emotionally committed relationship between the partners (Oliver & Hyde, 1993). Men are considerably more approving of such interactions, and women tend to be disapproving. Many women feel that premarital intercourse is ethical or acceptable only in the context of an emotionally committed relationship. For many men, that is a nice context for sex, but it isn't absolutely necessary.

In the NHSLS sample, 76 percent of white women, but only 53 percent of white men, said that they would have sex with someone only if they were in love (Mahay et al., 1999). That gender difference is consistent across other U.S. ethnic groups; the comparable statistics were 77 percent for African American women, 43 percent for African American men, 78 percent for Mexican American women, and 57 percent for Mexican American men.

In one study, 249 undergraduates were surveyed about their motives for having sex (Carroll, Volk, & Hyde, 1985). The results illustrate the importance for women of relationship and emotional connectedness as prerequisites for sex. Men are less concerned about these prerequisites and focus more on the physical pleasure of sex. Consistent with stereotypes, males and females gave considerably different responses to the question "What are your motives for having sexual intercourse?" Females emphasized love and emotional commitment, as in the following examples:

Emotional feelings that were shared, wonderful way to express LOVE!!

My motives for sexual intercourse would all be due to the love and commitment I feel for my partner.

Contrast those responses with the following typical responses from males:

Need it.

To gratify myself.

When I'm tired of masturbation. (Carroll, Volk, & Hyde, 1985, p. 137)

Clearly males—at least in the college years—emphasize physical needs and pleasure as their motives for intercourse, whereas females emphasize love, relationships, and emotional commitment. No wonder there is some conflict in relationships between women and men!

Arousal to Erotica

Traditionally in our society most erotic material—sexually arousing pictures, movies, or stories—has been produced for a male audience. The corresponding assumption presumably has been that women are not interested in such things. Does the scientific evidence bear out this notion?

Laboratory research shows that men are more aroused by erotic materials, but the gender difference is not large (Murnen & Stockton, 1997). A clas-sic study by psychologist Julia Heiman (1975, for a similar study with similar results, see Steinman et al., 1981) provides a good deal of insight into the responses of males and females to erotic materials. The participants were sexually experienced university students, and Heiman studied their responses as they listened to tape recordings of erotic stories. Not only did she obtain the participants' self-ratings of their arousal, as other investigators have done, but she also got objective measures of their physiological levels of arousal. To do this, she used two instruments: a penile strain gauge and a photo-plethysmograph (Fig. 14.5). The **penile strain gauge** (which our students have dubbed the "peter meter") is used to get a physiological measure of arousal in the male; it is a flexible loop that fits around the base of the penis. The **photoplethys-mograph** measures physiological arousal in the female; it is an acrylic cylinder, about the size of a tampon, that is placed just inside the entrance to the vagina. Both instruments measure vasocongestion in the genitals, which is the major physiological response during sexual arousal (see Chapter 9).

Research participants heard one of four kinds of tapes. There is a stereotype that women are more turned on by romance, whereas men are more aroused by "raw sex." The tapes varied according to which of these kinds of content they contained. The first group of tapes was *erotic;* they included excerpts from popular novels giving explicit descriptions of heterosexual sex. The second group of tapes was *romantic;* a couple were heard expressing affection and tenderness for each other, but they did not actually engage in sex. The third group of tapes was *erotic-romantic;* they included erotic elements of explicit sex and also romantic elements. Finally, the fourth group of tapes served as a control; a couple were heard engaging in conversation but nothing else. The plots of the tapes also varied according to whether the man or the woman initiated the activity and whether the description

Penile strain gauge: A device used to measure physiological sexual arousal in the male; it is a flexible loop that fits around the base of the penis.

Photoplethysmograph (foh-toh-pleth-ISS-moh-graf): An acrylic cylinder that is placed inside the vagina in order to measure physiological sexual arousal in the female. Also called a photometer.

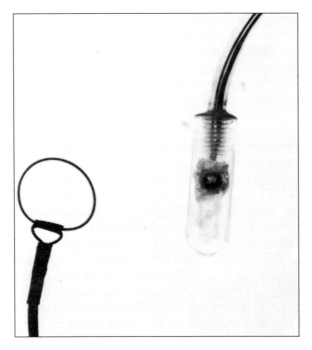

Figure 14.5 Two devices used to measure physiological sexual response in males and females. The penile strain gauge (left) consists of a flexible band that fits around the base of the penis. The photoplethysmograph (right) is an acrylic cylinder containing a photocell and light source, which is placed just inside the vagina.

centered on the woman's physical and psychological responses or on the man's. Thus the tapes were either male-initiated or female-initiated and either female-centered or male-centered. Three important results emerged from the study:

1. Explicit heterosexual sex (the erotic and erotic-romantic tapes) was most arousing, both for women and for men. The great majority of both males and females responded most strongly, both physiologically and in self-ratings, to the erotic and erotic-romantic tapes. Women, in fact, rated the erotic tapes as more arousing than men did. Neither men nor women responded—either physiologically or in self-reports—to the romantic tapes or to the control tapes (except for a couple of men who were aroused by a discussion of the relative merits of an anthropology major versus premed—ah, well).

2. Both males and females found the female-initiated, female-centered tape to be most arousing. Perhaps the female-initiated plot was most arousing because of its somewhat forbidden or taboo nature.

3. Women were sometimes not aware of their own physiological arousal. Generally there was a high correlation between self-ratings of arousal and objective physiological measures of arousal, both for men and for women. When men were physically aroused, they never made an error in reporting this in their self-ratings—it is pretty hard to miss an erection. But when the women were physically aroused, about half of them did not report arousal in their self-ratings (see also Laan et al., 1994). (One might assume that women who were sophisticated enough to volunteer for an experiment of this nature and who were willing to insert a photo-plethysmograph into their vagina would not suddenly become bashful about reporting their arousal; that is, it seems likely that these women honestly did not perceive themselves to be aroused.)

In sum, then, Heiman's study indicates that males and females are quite similar in their responses to erotic materials but that women can sometimes be unaware of their own physical arousal.

In statistical terms, Heiman found a low correlation between women's self-reports of arousal and physiological measures of their arousal. In an interesting follow-up study, one experimental group of women was instructed to attend to their genital signs of sexual arousal ("While rating these slides, I would like you to attend to various changes that may occur in your genital area such as vaginal lubrication, pelvic warmth, and muscular tension"), and a second group was told to attend to nongenital signs of arousal ("While rating these slides, I would like you to attend to various changes that may occur in your body. These are heart rate increase, nipple erection, breast swelling, and muscular tension"), while a control group was given no instructions (Korff & Geer, 1983). Both experimental groups showed high correlations between self-reports and physiological measures of arousal, while the control group showed the same low correlation that Heiman found. This shows that women can be quite accurate in realizing their physical arousal if they are simply told to focus their attention on it. The broader culture, of course, does not give women such instructions, but rather tells them to focus on the environment outside themselves—the love, romance, partner—so that many women have not learned to focus on their body. But the experiment described here shows quite clearly that they can.

Orgasm Consistency

Men are more consistent than women at having orgasms during sex. For example, according to the NHSLS, 75 percent of men, but only 29 percent of women, always have an orgasm during sex with their partner (Laumann et al., 1994, p. 116). The gap is narrower for orgasm consistency during masturbation, but even here men seem to be more effective: 80 percent of men, compared with 60 percent of women, report that they usually or always have an orgasm when masturbating (Laumann et al., 1994, p. 84).

Why the Differences?

The previous section reviewed the evidence on differences in male and female sexuality. Three differences—the lower percentage of females, compared with males, who masturbate, women's more disapproving attitudes toward casual sex and women's lesser orgasm consistency—are fairly well documented and in need of explanation. What factors lead some women not to masturbate, to be disapproving of casual sex, and to be inconsistent at having orgasms? Many possible explanations have been suggested by a wide variety of scholars; these will be considered below.

Biological Factors

There has been some speculation that gender differences in sexuality are created by two biological factors: anatomy and hormones.

Anatomy

The male sexual anatomy is external and visible and has a very obvious response: erection. While the male is nude, he can easily see his sexual organs, either by looking down or by looking in a mirror. The female sexual organs, in contrast, are hidden. The nude female looks down and sees nothing except pubic hair (which really is not very informative); she looks in a full-length mirror and sees the same thing. Only by doing the mirror exercise described in Chapter 4 can she get a good view of her own genitals. Furthermore, the female's genitals do not have an obvious arousal response like the male's erection. As a result, she may be less aware of her own arousal, a notion that is supported by Heiman's research.

The anatomical explanation, then, is that because the woman's genitals are not in plain view and because their arousal response is less obvious than that of the man's genitals, she is less likely to masturbate and less likely to develop her full sexual potential (Baldwin & Baldwin, 1997). If this explanation is correct, or at least part of the answer, could steps be taken to help women develop their sexuality? Perhaps parents could tell their daughters about the mirror exercise at an early age and encourage them to become more aware of their own sexual organs. And parents might want to discuss the idea of masturbation with their daughters.

Hormones

The hormonal explanation rests on the finding that testosterone is related to sexual behavior. This evidence was reviewed in Chapter 9. Basically, the evidence comes from studies in which male animals are castrated (and thus lose their natural source of testosterone), with the result that their sexual behavior disappears, presumably reflecting a decrease in sex drive. If replacement injections of testosterone are given, the sexual behavior returns.

Females generally have lower levels of testosterone in their tissues than males do. Human females, for example, have about one-tenth the level of testosterone in their blood that human males do (Hoyenga & Hoyenga, 1993).

The hormonal explanation, then, is that if testosterone is important in activating sexual behavior and if females have only one-tenth as much of it as males have, this might result in a lower level of sexual behavior such as masturbation in women, or a lower "sex drive."

There are several problems with this logic. First, it may be that cells in the hypothalamus or the genitals of women are more sensitive to testosterone than the comparable cells in men; thus a little testosterone may go a long way in women's bodies.

Second, we must be cautious about making inferences to human males and females from studies done on animals. Although some recent studies demonstrate the effects of testosterone on sexual interest and behavior in humans, the effects are less consistent and more complex than in other species (Chapter 9).

Cultural Factors

Our culture has traditionally placed tighter restrictions on women's sexuality than it has on men's, and vestiges of these restrictions linger today. It seems likely that these restrictions have acted as a damper on female sexuality, and thus they may help to explain why some women do not masturbate, why some women have difficulty having orgasms, and why some women are wary about casual sex.

One of the clearest reflections of the differences in restrictions on male and female sexuality is the double standard. As we saw in Chapter 11, the double standard says that the same sexual behavior is evaluated differently, depending on whether a male or a female engages in it. The sexual double standard gives men more sexual freedom than women (Muehlenhard & McCoy, 1991). An example is premarital sex. Traditionally in our culture, premarital intercourse has been more acceptable for males than for females. Indeed, premarital sexual activity might be a status symbol for a male but a sign of cheapness for a female.

Generally there seems to be less of a double standard today than there was in former times. For example, as the data in Chapter 11 indicate, people now approve of premarital intercourse for females about as much as they do for males. This change in attitudes is reflected in behavior. A much higher percentage of women report having engaged in premarital intercourse now than in the 1940s. In one sample, 80 percent of the males and 63 percent of the females had had premarital intercourse (Robinson et al., 1991). Thus there is much less of a difference between males and females now than there was a generation ago; the vast majority of both males and females are now engaging in premarital intercourse. Premarital intercourse still remains somewhat more common among males, though.

The decline of the double standard may help to explain why some of the gender differences found in older studies of sex behavior have disappeared in more recent studies. When cultural forces do not make such a distinction between male and female, males and females become more similar in their sexual behavior. The last vestige remaining of the double standard today is in regard to casual sex, which is approved more for men than women. In the United States, men, but not women, agree with this double standard (Sprecher & Hatfield, 1996).

Gender roles are another cultural force that may contribute to differences in male and female sexuality, as was discussed earlier in this chapter. Gender roles dictate proper behavior for females and males in sexual interactions—that is, they specify the script. For example, there is a stereotype of the male as the initiator and the female as the passive object of his advances; surely this does not encourage the woman to take active steps to bring about her own orgasms. As a result of such stereotypes, the man has borne the whole weight of responsibility, both for his response and for the woman's response, and women have not been encouraged to take responsibility for producing their own pleasure.

Marital and family roles may play a part. When children are born, they can act as a damper on the parents' sexual relationship. The couple lose their privacy when they gain children. They may worry about their children bursting through an unlocked door and witnessing what Freud called "the primal scene" of their parents making love. Or they may be concerned that their children will hear the sounds of lovemaking. Generally, though, the woman is assigned the primary responsibility for child rearing, and so she may be more aware of the presence of the children in the house and more concerned about possible harmful effects on them of witnessing their parents engaging in sex. Once again, her worry and anxiety do not contribute to her having a satisfying sexual experience.

Other Factors

A number of other factors, not easily classified as biological or cultural, may also contribute to differences between male and female sexuality.[2]

---

[2]Other possible causes of orgasm problems in women are discussed in Chapter 19.

Women get pregnant and men do not. Particularly in the days before highly effective contraceptives were available, pregnancy might be a highly undesirable consequence of sexuality for a woman. Thinking that an episode of lovemaking might result in a nine-month pregnancy and another mouth to feed could put a damper on anyone's sexuality. Even today, pregnancy fears can be a force. For example, according to one study, 27 percent of white teenage women and 35 percent of black teenage women who are sexually active do not use contraceptives (Hofferth, 1990). A woman who is worried about whether she will become pregnant —and, if she is not married, about whether others will find out that she has been engaging in sexual activity—is not in a state conducive to the enjoyment of sex, much less the experience of orgasm (although this scarcely explains why more women than men do not masturbate).

Ineffective techniques of stimulating the woman may also be a factor. The commonest techniques of intercourse, with the penis moving in and out of the vagina, may provide good stimulation for the male but not the female, since she may not be getting sufficient clitoral stimulation. Perhaps the problem, then, is that women are expected to orgasm as a result of intercourse, when that technique is not very effective for producing orgasms in women.

A relationship probably exists between the evidence that fewer women masturbate than men and gender differences in orgasm consistency. Childhood and adolescent experiences with masturbation are important early sources of learning about sexuality. Through these experiences we learn how our bodies respond to sexual stimulation and what the most effective techniques for stimulating our own bodies are. This learning is important to our experience of adult, two-person sex. For example, Kinsey's data suggested that women who masturbate to orgasm before marriage are more likely to orgasm in intercourse with their husbands;[3] 31 percent of the women who had never masturbated to orgasm before marriage had not had an orgasm by the end of their first year of marriage, while only 13 to 16 percent of the women who had masturbated had not had orgasms in their first year of marriage (Kinsey et al., 1953, p. 407). One woman spoke of how she discovered masturbation late and how this may be related to her orgasm capacity in heterosexual sex:

> I thought I was frigid, even after three years of marriage, until I read this book and learned how to turn myself on. After I gave myself my first orgasm, I cried for half an hour, I was so relieved. Afterwards, I did it a lot, for many months, and I talked to my doctor and to my husband, and finally I began to make it in intercourse. (Hunt, 1974, pp. 96–97)

Not only may women's relative inexperience with masturbation lead to a lack of sexual learning, but it also may create a kind of "erotic dependency" on men. Typically, boys' earliest sexual experiences are with masturbation, which they learn how to do from other boys. More important, they learn that they can produce their own sexual pleasure. Girls typically have their earliest sexual experiences in heterosexual petting. They therefore learn about sex from boys, and they learn that their sexual pleasure is produced by the male. As sex researcher John Gagnon commented:

> Young women may know of masturbation, but not know *how* to masturbate—how to produce pleasure, or even what the pleasures of orgasm might be. . . . Some young women report that they learned how to masturbate after they had orgasm from intercourse and petting, and decided they could do it for themselves. (1977, p. 152)

Once again, such ideas might lead to a recommendation that girls be given information about masturbation.

Numerous factors that may contribute to shaping male and female sexuality have been discussed. Our feeling is that a combination of several of these factors produces the differences that do exist. The early differences in experiences with masturbation are very important. Although these differences may result from differences in anatomy, they could be eliminated by giving girls information on masturbation. Women may enter into adult sexual relationships with a lack of experience in the bodily sensations of arousal and orgasm, and they may be unaware of the best techniques for stimulating their own bodies. Put this lack of experience together with various cultural forces, such as the double

---

[3]Note that this is in direct contradiction to the old-fashioned advice given in manuals that suggested that "getting hooked" on masturbation might impair later marital sexuality; if anything, just the reverse is true.

standard and ineffective techniques of stimulation, and it is not too surprising that there are some gender differences in sexuality.

Beyond the Young Adults

One of the problems with our understanding of gender differences in sexuality is that so much of the research has concentrated on college students or other groups of young adults (as is true of much behavioral research). For example, Heiman (1975) used college students in her studies of male-female differences in arousal to erotic materials. Using this population may provide a very narrow view of male-female differences; they are considered during only a very small part of the lifespan. In reality, female sexuality and male sexuality change in their nature and focus across the lifespan. For example, it is a common belief in our culture that men reach their sexual "peak" at around age 19, whereas women do not reach theirs until they are 35 or 40. There is some scientific evidence supporting this view. Kinsey (1953) found, for example, that women generally had orgasms more consistently at 40 than they did at 25.

Psychiatrist Helen Singer Kaplan, a specialist in therapy for sexual disorders, advanced an interesting view of differences between male sexuality and female sexuality across the lifespan (Kaplan & Sager, 1971). According to her analysis, the teenage male's sexuality is very intense and almost exclusively genital in its focus. As the man approaches age 30, he is still highly interested in sex, but not so urgently. He is also satisfied with fewer orgasms, as opposed to the adolescent male, who may have four to eight orgasms per day through masturbation. With age the man's refractory period becomes longer. By the age of 50, he is typically satisfied with two orgasms a week, and the focus of his sexuality is not so completely genital; sex becomes a more sensuously diffuse experience and has a greater emotional component.

In women, the process is often quite different. Their sexual awakening may occur much later; they may, for example, not begin masturbating until age 30 or 35. While they are in their teens and twenties, their orgasmic response is slow and inconsistent. However, by the time they reach their mid-thirties, their sexual response has become quicker and more intense, and they orgasm more consistently than they did during their teens and twenties. They initiate sex more frequently than they did in the past. Also, the greatest incidence of extramarital sex for women occurs among those in their late thirties. Vaginal lubrication takes place almost instantaneously in women in this age group.

Men, then, seem to begin with an intense, genitally focused sexuality and only later develop an appreciation for the sensuous and emotional aspects of sex. Women have an early awareness of the sensuous and emotional aspects of sex and develop the capacity for intense genital response later. To express this in another way, we might use the terminology suggested by Ira Reiss: **person-centered sex** and **body-centered sex.** Adolescent male sexuality is body-centered, and the person-centered aspect is not added until later. Adolescent female sexuality is person-centered, and body-centered sex comes later.

It is important to remember, though, that these patterns may be culturally, rather than biologically, produced. In some other cultures—for example, Mangaia in the South Pacific (see Chapter 1)—females have orgasms 100 percent of the time during coitus, even when they are adolescents.

Transsexualism

Many texts cover transsexualism in the chapter on sexual variations or deviations. However, we have included it in the chapter on gender because it is fundamentally a problem of gender and, more specifically, a problem of gender identity.

A **transsexual** is a person who believes that he or she is trapped in the body of the other gender. This condition is also known as **gender dysphoria,**

Person-centered sex: Sexual expression in which the emphasis is on the relationship and emotions between the two people.
Body-centered sex: Sexual expression in which the emphasis is on the body and physical pleasure.
Transsexual: A person who believes he or she is trapped in the body of the other gender. Also known as a *transgender person.*
Gender dysphoria (dis-FOR-ee-uh): Unhappiness with one's gender; another term for transsexualism.

Focus 14.1
Male Sexuality

Bernie Zilbergeld wrote *Male Sexuality*, published in 1978, on the basis of his experience as a sex therapist and psychotherapist. The book was much respected and had a large readership over more than a decade. In 1992 Zilbergeld wrote an updated version, *The New Male Sexuality*, to reflect trends in the 1990s.

He argues that the media have taught us a Fantasy Model of Sex, which is ultimately detrimental to men, and to women as well. He captures this idea in his chapter title "It's Two Feet Long, Hard as Steel, Always Ready, and Will Knock Your Socks Off," describing the Fantasy Model of the erect penis and its power over women. The Fantasy Model of Sex creates unrealistic expectations and performance pressures on men.

Zilbergeld discusses a number of cultural myths based on the Fantasy Model. Here are some of them.

Myth 1. We're liberated folks who are very comfortable with sex. The media of the 1990s taught us that we have completely shed our Victorian heritage and everyone is totally comfortable with sex. The men and women in the movies and on TV never have any concerns or problems with sex. The women don't worry about their ability to have orgasms. The men don't worry about the size or hardness of their penises. But if all of that is true, why do we have such poor sex education in the United States? Why do parents have such difficulty talking about sex with their children? The truth is that, although public manifestations (like the movies) are very open about sex, in our private lives we have all kinds of discomforts and uncertainties about sex.

Myth 2. A real man isn't into sissy stuff like feelings and communicating. Boys are trained into the male role, which discourages the expression of emotions such as tenderness. Communicating about personal feelings becomes difficult if not impossible. As one man said, "What it really comes down to is that I guess I'm not very comfortable with expressing my emotions—I don't think many men are—but I am pretty comfortable with sex, so I just sort of let sex speak for me" (Zilbergeld, 1992, p. 45). Men are crippled in forming emotional relationships, as a result, and sexual interactions are less satisfying than they might be if there were more communication.

Myth 3. All touching is sexual or should lead to sex. For men, touching is a means to an end: sex. For women, touch more often is a goal in itself, as when women hug each other. Men need to learn that sometimes they just need to be held or stroked, and that that can provide more emotional satisfaction than sexual intercourse.

Myth 6. Sex is centered on a hard penis and what's done with it. Adolescent boys have a fixation on their penis and its erections, and this fascination persists throughout life. It creates heavy performance pressure for an erection—and not just any old erection, but a really big one. As Zilbergeld puts it, "Penises in fantasyland come in only three sizes: large, extra large, and so big you can't get through the door." Men need to learn that the penis is not the only sexual part of their bodies, and that many very enjoyable forms of sexual behavior require no erection at all. That relieves a lot of performance pressure.

Zilbergeld's books are not based on a survey or laboratory research, but rather on his experiences as a sex therapist. His work with people having problems and seeking therapy may bias his views. But his observations are tremendously insightful, and many people not seeking therapy have benefited from his books.

Sources: Bernie Zilbergeld. (1978). *Male sexuality.* New York: Bantam Books. Bernie Zilbergeld. (1992). *The new male sexuality.* New York: Bantam Books.

meaning unhappiness or dissatisfaction with one's gender. Transsexuals are the candidates for the **sex-change operations** (see below) that have received so much publicity. The term "transsexual" is used to refer to the person both before and after the operation. There are two kinds of transsexuals: those with male bodies who think they are females (called **male-to-female transsexuals**) and those with female bodies who think they are males (called *female-to-male transsexuals*). Male-to-female transsexuals have been more likely to seek help at clinics and have more often been given sex-reassignment surgery (Abramowitz, 1986), in part because the surgery required in such cases is easier. Accordingly, most of the discussion that follows will focus on male-to-female transsexuals.[4]

Keeping in mind the distinction between sex and gender, it is important to note that transsexualism is a problem not of sexual behavior but of gender and gender identity. The term *transgender* has become popular for this condition because it better reflects the idea that the issue is gender, not sexuality. That is, the transsexual is preoccupied not with some special kind of sexual behavior but rather with wanting to be a female when her body is male. Sex is involved insofar as being sexually attracted to a member of the other gender (which is expected in our society) is concerned. But we know of one transsexual who has never engaged in any sexual activity beyond kissing since she had surgery to make her a female. She is delighted with the results of the surgery and loves being a woman, but she is not particularly interested in sex.

References to transsexuals are found in much of recorded history, although of course they are not referred to by that modern, scientific term (Devor, 1997). Philo, the Jewish philosopher of Alexandria, described them as follows: "Expending every possible care on their outward adornment, they are not ashamed even to employ every device to change artificially their nature as men into women. . . . Some of them . . . craving a complete transformation into women . . . have amputated their generative members." Transsexualism is therefore by no means a phenomenon of modern, industrialized cultures.

In addition to the distinction between male-to-female (MTF) and female-to-male (FTM) gender dysphorics, we can also make a distinction between homosexual and nonhomosexual (also termed "autogynephilic") gender dysphorics (see Blanchard, Dickey, & Jones, 1995). The distinction refers to the person's sexual orientation *before* sex change. For example, if we think of a female-to-male transsexual, she would be classified as homosexual if she preferred female partners and nonhomosexual if she preferred male partners. Homosexual MTFs tend to be shorter and lighter in weight compared with nonhomosexual MTFs and compared with men in the general population (Blanchard, Dickey, & Jones, 1995). This may be one reason why homosexual MTFs are more successful in their new gender—that is, they are more convincing women. Nonhomosexual MTFs, in contrast, are often heterosexually married and have children; they are masculine in appearance and often make the transition to the new gender after they are 40.

Psychologically the transgender individual is, to put it mildly, in an extreme conflict situation. The body says, "I'm a man," but the mind says, "I'm a woman." The person may understandably react with fright and confusion. Believing herself truly to be a female, she may try desperate means to change her body accordingly. Particularly in the days before the sex-change operation was performed, or among people who were unaware of it, self-castrations have been reported. The woman we mentioned above ate large quantities of women's face cream containing estrogens to try to bring about the desired changes in her body.

The Sex-Change Operation

Gender reassignment is rather complex and proceeds in several stages (Peterson & Dickey, 1995). The first step in the process is very careful counsel-

---

[4]Because this kind of transgender individual thinks of himself as a female, he prefers to be called "she," and to simplify matters in this discussion, "she" will be used to refer to the transgender individual.

Sex-change operation: The surgery done on transsexuals to change their anatomy to the other gender.
Male-to-female transsexual: A person who is born with a male body but who has a female identity and wishes to become a female biologically in order to match her identity.

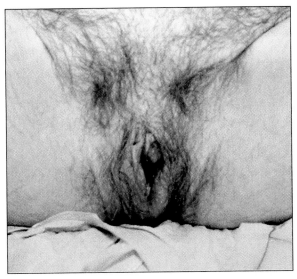

(a)

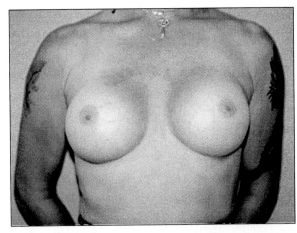

(b)

Figure 14.6 (*a*) The appearance of the genitals following male-to-female transsexual surgery. (*b*) Breast augmentation for the FTM Transexual. Photos courtesy of Dr. Daniel Greenwald

ing and psychiatric evaluation. It is important to establish that the person is a true transsexual, that is, someone whose gender identity does not match her body type. Some people mistakenly seek gender reassignment; for example, a man who is simply poorly adjusted, unhappy, and not very successful might think that things would go better for him if he were a woman. Sometimes schizophrenics display such confused gender identity that they might be mistaken for transsexuals. It is important to establish that the person is a true transgender before going ahead with a procedure that is fairly drastic.

The next step is hormone therapy. The male-to-female transsexual is given estrogen and must remain on it for the rest of her life. The estrogen gradually produces some feminization. The breasts enlarge. The pattern of fat deposits becomes feminine; in particular, the hips become rounded. Balding, if it has begun, stops. Secretions by the prostate diminish, and eventually there is no ejaculate. Erections become less and less frequent, a phenomenon that pleases the transsexual, since they were an unpleasant reminder of the unwanted penis. The female-to-male transsexual is given androgens, which bring about a gradual masculinization. A beard may then develop, to varying degrees. The voice deepens. The pattern of fat deposits becomes more masculine. The clitoris enlarges, although not nearly to the size of a penis,

and becomes more erectile. The pelvic bone structure cannot be reshaped, and breasts do not disappear except with surgery.

Next comes the "real life test," which is the requirement that the person live as a member of the new gender for a period of one or two years. This is done to ensure that the person will be able to adjust to the role of the new gender; once again, the idea is to be as certain as possible that the person will not regret having had the operation. Some transsexuals, even before consulting a physician, spontaneously enter this cross-dressing stage in their efforts to become women. Problems may arise, though. Cross-dressing is illegal in many cities, and they may be arrested.

The final step is the surgery itself. For the male-to-female transsexual, the penis and testes are removed, but without severing the sensory nerves of the penis. The external genitalia are then reconstructed to look as much as possible like a woman's. Next, an artificial vagina—a pouch 15 to 20 centimeters (6 to 8 inches) deep—is constructed. It is lined with the skin of the penis so that it will have sensory nerve endings that can respond to sexual stimulation. For about six months afterward, the vagina must be dilated with a plastic device so that it does not reclose. Other cosmetic surgery may also be done, such as reducing the size of the Adam's apple.

The female-to-male change is more complex and generally less successful. A penis and scrotum

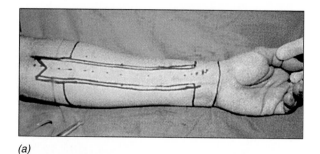

(a)

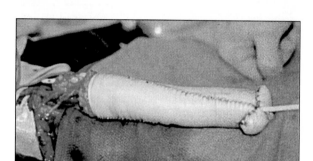

(b)

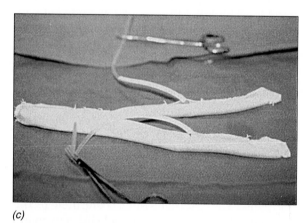

(c)

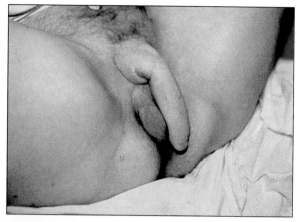

(d)

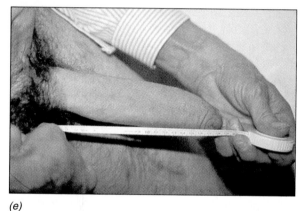

(e)

Figure 14.7   Female-to-male transsexual surgery. *(a)* Skin on the forearm marked before transfer to the groin. *(b)* The penis is constructed (blood vessels and nerves shown on the left). *(c)* An inflatable prosthesis, wrapped in Goretex and ready for insertion. *(d)* Penis before insertion of the implant. *(e)* Erect penis. Photos courtesy of Dr. Daniel Greenwald

are constructed from tissues in the genital area and the forearm (see Figure 14.7). The new penis does not have erectile capacity; in some cases a rigid silicone tube is implanted in the penis so that it can be inserted into a vagina, making coitus possible. Some female-to-male transsexuals choose not to have genital surgery and just go through breast removal and possibly hysterectomy.

It is important to note that transsexualism as a diagnosis, and the means of treating it (hormones and sex-change surgery), are products of modern European and American culture. As discussed earlier, some other societies simply consider these individuals to be members of a third gender and they live comfortably in that category. Examples are the two-spirit people among Native Americans and the Hijras of India (Jacobs et al., 1997; Nanda, 1997).

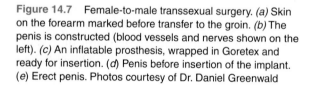

What Causes Transsexualism?

Scientists have not found a definite cause of transsexualism. One likely reason is that there may be more than one path to it. As usual, both biological and environmental theories have been proposed.

On the biological side, John Money (1986) has argued that the issue is a critical period during prenatal development. Some event, not yet known, may lead to atypical development of some brain structure—possibly the hypothalamus, corpus callosum, or anterior commissure (Devor, 1997). Other theorists, following this same line of thinking, have proposed that during prenatal development, the fetus, if it is to become a male, must be both defeminized and masculinized (Pillard & Weinrich, 1987). A failure or mistiming of either or both of those processes could produce a person with a male body but a female identity. To understand this distinction between defeminization and masculinization, at least for anatomy, recall the discussion in Chapter 5 about the Wolffian and Mullerian ducts. Both are present in all fetuses early in prenatal development. In a normally developing male, the Mullerian ducts degenerate (defeminization) and the Wolffian ducts thrive (masculinization). If some developmental process failed, though, both might thrive. The same might be true of cells in the hypothalamus or other brain regions having to do with gender identity.

On the environmental side, noted sex researcher Richard Green (1987) has extensively studied the phenomenon of extreme femininity in boys, which might be a forerunner of either homosexuality or transgenderism. He has found that the parents of these boys basically treat them as if they were girls; for example, they dress them in girls' clothing and then tell them how cute they look. In the case of female-to-male transsexuals, he believes the origins lie in parental practices that include giving the child a gender-ambiguous name and encouraging rough play, and a family constellation in which the mother is unpleasant and emotionally distant (thereby discouraging identification with her) and the father is pleasant and warm (thereby encouraging identification with him).

Research is sparse regarding either the biological or environmental causes, so the theories for now are speculative.

Male-to-female transsexuals account for the great majority of cases, outnumbering female-to-male transsexuals by a ratio of 3:1 (Abramowitz, 1986). Several explanations for this lopsided ratio have been offered. Perhaps male prenatal development is more complex and error-prone, or perhaps the problem is that preschool boys spend so much more time with their mothers than with their fathers.

Other Issues

The phenomenon of transsexualism raises a number of interesting psychological, legal, and ethical questions for our contemporary society.

One case that attracted attention was that of Dr. Renée Richards, formerly Richard Raskind, a physician who had her gender reassigned to that of a woman. When she was a man, she was a successful tennis player. In 1976 she attempted to enter a women's tennis tournament. The women players protested that she was not a woman, and she protested that she was. Officials subsequently decided to use the **buccal smear** test for gender, which is also the one used in the Olympics. In this test a sample of cells is scraped from the inside of the mouth and is stained. If the sex chromosomes are XX, a **Barr body** should be present and will show up under the stain; if the chromosomes are XY, the Barr body should be absent. The test is therefore one of genetic gender. Richards protested that this was not the appropriate test to be used on her. Psychologically she is a female, she has female genitals, and she functions socially as a female, and she feels that these are the appropriate criteria. She does, though, have a male pelvic bone structure and other bone structures that are masculine, and these may have important consequences for athletic performance. The important question raised by this case is: What should the

---

Buccal smear: A test of genetic sex, in which a small scraping of cells is taken from the inside of the mouth, stained, and examined under a microscope.
Barr body: A small, black dot appearing in the cells of genetic females; it represents an inactivated X chromosome.

(a)   (b)

Figure 14.8   Transsexual Renee Richards before operation *(a)* and after *(b)* her sex-change operation.

criteria be for determining a person's gender? Should it be chromosomal gender (XX or XY) as tested by the buccal smear? Should it be the gender indicated by the external genitals? Should it be psychological gender identity?

Another question that might be raised concerns religious groups that do not permit women to become members of the clergy. Is a male-to-female transsexual, for example, qualified to be a priest before "the operation" but not after? Is a female-to-male transsexual qualified to be a priest by virtue of having had a sex-change operation?

The transsexual also encounters a number of practical problems when undergoing gender reassignment. Official records, such as the social security card, must be changed to show not only the new name but also the new gender. Sometimes a new birth certificate is issued and the old one is sealed away. If the person was married before the sex change, often—though not always—the spouse must be divorced. Changing one's gender is, to say

the least, a complicated process.

Transsexuals should be able to give us, through their personal accounts, new insights into the nature of sex and gender. For example, most of us have wondered, at some time, how members of the other gender feel during sexual intercourse. The transgender individuals are in a unique position for giving us information on this question.

Criticisms of Sex-Change Surgery

A number of criticisms of sex-change surgery for transsexuals have been raised. One of these came from a study by Johns Hopkins researcher Jon Meyer (1979). He did a follow-up study of the adjustment of 50 transsexuals, 29 of whom received surgery and 21 of whom did not. His conclusion, much publicized, was that there were no significant differences in the adjustment of the two groups. If that is the case, he claimed, then transsexual surgery is unnecessary and should not be done.

Then criticisms of Meyer's study appeared (e.g., Fleming et al., 1980). Meyer's adjustment scale was somewhat peculiar and involved debatable values. After the Meyer study and the criticisms of it, some clinics ceased doing transsexual surgery, but most continue to do it. Almost surely we will see more attempts in the future to treat transsexualism with psychotherapy rather than surgery, but these methods remain to be worked out. Unfortunately, attempts to use psychotherapy, such as psycho-analysis, as an alternative to surgery have generally been unsuccessful (Roberto, 1983). That is, trying to change the gender identity to match the anatomy—rather than the reverse as in the sex-change operation—does not seem to work very well. In contrast, adjustment of transsexuals has been shown to be significantly better following surgery (Bodlund & Kullgren, 1996; Blanchard et al., 1983; Green & Fleming, 1990). Experts have concluded that approximately two-thirds of those who have sex reassignment surgery are improved by it in terms of adjustment indicators such as re-duction in depression (Abramowitz, 1986). In one study, 86 percent were satisfied with the outcome of their surgery (Lief & Hubschman, 1993). On the other hand, about 7 percent of the cases result in tragic outcomes such as suicide or a request for a reversal (Abramowitz, 1986).

SUMMARY

A gender role is a set of norms, or culturally defined expectations, that specify how people of one gen-der ought to behave. Children are socialized into gender roles first by parents and later by other forces such as peers and the media.

Gender roles are not uniform in the United States. They vary according to ethnic group and other factors. African American women, for exam-ple, have traditionally played an important eco-nomic role in their families. Among Latinos, gen-der roles tend to be more rigidly defined than they are among Anglos. The sexuality of Asian Ameri-cans has been stereotyped, with Asian American men seen as being sexless and Asian American women viewed as exotic sex toys. Some American Indian tribes traditionally had egalitarian gender roles compared with white culture.

Psychological gender differences have been documented in aggressiveness and communica-tion styles.

The two largest male-female differences in sexual-ity are in the incidence of masturbation (males hav-ing the higher incidence) and attitudes toward casual sex (females being more disapproving). Heiman's study of arousal to erotic materials illustrates how males and females are in some ways similar and in some ways different in their responses. Males are more consistent at having orgasms, especially during heterosexual intercourse, than females are.

Three sets of factors have been proposed to ex-plain gender differences in sexuality: biological factors (anatomy, hormones); cultural factors (gender roles, the double standard); and other fac-tors (fear of pregnancy, differences in masturba-tion patterns creating other gender differences).

Most research on gender and sexuality has been done with college-age samples. There is reason to believe that patterns of gender differences in sexu-ality change in middle age and beyond.

Transsexuals—people who seek sex-reassign-ment surgery—represent an interesting variation in which gender identity does not match anatomy. Generally, their adjustment is good following the sex-change operation.

REVIEW QUESTIONS

1. A _____ is a set of norms or culturally defined expectations that define how people of one gender ought to behave.

2. Gender stereotypes vary according to ethnic group in the United States. For example, Asian American men have been stereotyped as

asexual and Asian American women have been stereotyped as exotic sex objects. True or false?

3. The data indicate that the largest gender differences in sexuality are in the areas of _____ and _____ .

4. According to the NHSLS, _____ percent of males and _____ percent of females report having masturbated in the last year.

5. The _____ is a device that measures physiological sexual arousal in the male.

6. Data from the 1990s indicate that today, women are as consistent as men are at having orgasms during heterosexual intercourse. True or false?

7. Zilbergeld uses the phrase "It's two feet long, hard as steel, always ready, and will knock your socks off" to describe the ideal penis according to the Fantasy Model of Sex. True or false?

8. _____ is the term for a person who feels trapped in a body of the wrong gender.

9. An alternative term for transsexualism is _____ .

10. A homosexual female-to-male transsexual would prefer women as sex partners. True or false?

(The answers to all review questions are at the end of the book, after the Glossary.)

QUESTIONS FOR THOUGHT, DISCUSSION, AND DEBATE

1. Do you think that transsexual surgery is the appropriate treatment for transsexuals? Why or why not?

2. Recalling from your childhood, do you think you were socialized in a stereotyped masculine or feminine way? What impact do you think those socialization experiences have had on your current sexual attitudes and behaviors?

3. Do you think there is still a double standard for male and female sexuality today? Explain your answer.

4. Dominic is a strong advocate of equality between men and women, yet his 6-year-old daughter's favorite television program shows both the parents and the children in very stereotyped roles. The mother is a secretary and the father is a physician. The teenage daughter is a cheerleader and thinks of nothing else, while the son plays football. What should Dominic do? Why?

SUGGESTIONS FOR FURTHER READING

Hyde, Janet S. (1996). *Half the human experience: The psychology of women* (5th ed.). Boston, MA: Houghton-Mifflin. We are not in a very good position to give an objective appraisal of this book, but for what it's worth, we think it is an interesting, comprehensive summary of what is known about the psychology of women and gender roles.

Devor, Holly (1997). *FTM: Female-to-male transsexuals in society.* Bloomington: Indiana University Press. FTMs are the understudied group of gender dysphorics, and Devor's book fills the gap in an exceptional and fascinating way.

Zilbergeld, Bernie. (1992). *The new male sexuality.* New York: Bantam Books. Zilbergeld's original *Male Sexuality* was a great success, and this updated version is every bit as insightful.

WEB RESOURCES

http://www.symposion.com/ijt/ijtc0405.htm
This site presents the Standards of Care for Gender Identity Disorders (Transsexuals) adopted by the Harry Benjamin International Gender Dysphoria Association in 1998.

http://www.savina.com/confluence/hormone/
FAQ: Hormone therapy for transsexuals.

http://www.zzapp.org/tgea/home.htm
The Transgender Education Association website; extensive library of information, links.

For information about specific ethnic groups, see Chapter 1.

chapter

15

SEXUAL ORIENTATION: GAY, STRAIGHT, OR BI?

CHAPTER HIGHLIGHTS

Attitudes Toward Gays and Lesbians

Life Experiences of LGBs
Coming Out
Lesbian, Gay, and Bisexual Communities
Gay and Lesbian Relationships
Lesbian and Gay Families

How Many People Are Gay, Straight, or Bi?

Sexual Orientation and Mental Health
Sin and the Medical Model
Research Results
Can Sexual Orientation Be Changed by
   Psychotherapy?

**Why Do People Become Homosexual or
   Heterosexual?**
Biological Theories
Psychoanalytic Theory
Learning Theory
Interactionist Theory
Sociological Theory
Empirical Data

**Sexual Orientation in Multicultural
   Perspective**

Bisexuality

The Western social norm *proscribes* homosexual activity and *prescribes* only heterosexual (fantasy and) intercourse, depending upon one's class, ethnic group, social situation, and historical period. However much individuals conform to this norm it is not universally valid. . . . In Melanesia, our Western norm does not apply; males who engage in ritual homosexual activities are not "homosexuals," nor have they ever heard the concept "gay."*

One night in June 1969, in response to police harassment, gay men and lesbians rioted in the Stonewall, a gay bar in New York City's Greenwich Village. This may have been the first open group rebellion of homosexual persons in history. Gay liberation was born. Since then, the public has been forced into an awareness of an issue—sexual orientation—that it had previously preferred to ignore. Gay liberationists proclaim that gay is good. Meanwhile, many Americans charitably maintain that homosexuals are sick (but can be cured).

Most of us want to know more about sexual orientation. The purpose of this chapter is to try to provide a better understanding of people's sexual orientations, whether homosexual, heterosexual, or bisexual, as well as an understanding of homophobia (the fear and hatred of homosexuals).

Sexual orientation is defined by whom we are sexually attracted to and also have the potential for loving. Thus a **homosexual** is a person whose sexual orientation is toward members of her or his own gender; a heterosexual is a person whose sexual orientation is toward members of the other gender; and a **bisexual** is a person whose sexual orientation is toward both genders. The word homosexual is derived from the Greek root *homo,* meaning "same" (not the Latin word *homo,* meaning "man"). The term "homosexual" may be applied in a general way to homosexuals of both genders or specifically to male homosexuals. The term **lesbian,** which is used to refer to female homosexuals, can be traced to the great Greek poet Sappho, who lived on the island of Lesbos (hence "lesbian") around 600 B.C. She is famous for the love poetry that she wrote to other women. Sappho was actually married, apparently happily, and had one daughter, but her lesbian feelings were the focus of her life.

*Gilbert H. Herdt. (1984). *Ritualized Homosexuality in Melanesia,* pp. ix–x. Berkeley: University of California Press.

Several other terms are also used in conjunction with homosexuality. Gay activists prefer the term **gay** to "homosexual" because the latter emphasizes the sexual aspects of the lifestyle and can be used as a derogatory label, since there are so many negative connotations to homosexuality. A heterosexual is then referred to as **straight.** The term "gay" is generally used for male homosexuals, and "lesbian" for female homosexuals. There are, of course, a number of slang terms for gays and lesbians, such as "queer," "fairy," "dyke," and "faggot" or "fag," which are derogatory when used by straight persons to belittle homosexuals.

In this chapter, we will use the abbreviation "LGB" for lesbians, gays, and bisexuals, because it is awkward to repeat the phrase "gays and lesbians," and even that phrase omits bisexuals.

Attitudes Toward Gays and Lesbians

Your sexual orientation has implications for the attitudes people have toward you. First, there is the belief that all people are heterosexual, that heterosexuality is the norm. Furthermore, just as there are

Sexual orientation: A person's erotic and emotional orientation toward members of his or her own gender or members of the other gender.
Homosexual: A person whose sexual orientation is toward members of the same gender.
Bisexual: A person whose sexual orientation is toward both men and women.
Lesbian: A woman whose sexual orientation is toward other women.
Gay: Homosexual; especially male homosexuals.
Straight: Heterosexual; that is, a person whose sexual orientation is toward members of the opposite gender.

stereotypes about other minority groups—for example, the stereotype that all Asian American men are asexual—so there are stereotypes about homosexuals. These stereotypes and negative attitudes lead to discrimination and hate crimes against gays and lesbians. In this section we will examine some of the scientific data on these negative attitudes.

Attitudes

Many Americans disapprove of homosexuality. For example, as Table 15.1 shows, in a well-sampled 1996 survey of adult Americans, 57 percent expressed the opinion that sexual relations between two adults of the same sex are always wrong. In addition, 22 percent believe that a homosexual man should not be allowed to teach in a college or university.

Has the gay liberation movement succeeded in changing the negative attitudes of Americans? The answer seems to be, yes, but slowly. Table 15.1 shows that the percentage of people who believe that homosexual behavior is always wrong showed a substantial change from 1973 to 1996.

Some experts believe that many Americans' attitudes toward homosexuals can best be described as homophobic (Fyfe, 1983; Hudson & Ricketts, 1980). **Homophobia** may be defined as a strong, irrational fear of homosexuals and, more generally, as fixed negative attitudes and reactions to homosexuals (Fyfe, 1983). Some scholars dislike the term "homophobia" because, although certainly some people have antigay feelings so strong that they

could be called a phobia, what is more common is negative attitudes and prejudiced behaviors. Therefore, some prefer the term **antigay prejudice.**

The most extreme expressions of antigay prejudice occur in *hate crimes* against LGBs. Perhaps the most horrifying recent case occurred in Wyoming. Matthew Shepard, a University of Wyoming freshman, was found tied to a fence, savagely beaten and comatose, on the outskirts of Laramie. He died five days later. Two men, both 21 and high school dropouts, were charged with the murder. Apparently they had led Shepard to believe that they, too, were gay and lured him from a bar to ride in their pickup truck. In the truck, they began beating him with a revolver, then got out and tied him to a fence, beat him more, and left him for dead.

Averaged across 24 different studies of LGBs, the results indicated that 9 percent had been assaulted with a weapon, 17 percent had been physically assaulted, 19 percent had experienced vandalism or other property crimes, 44 percent had been threatened with violence, 13 percent had been spat upon, and 80 percent had been verbally harassed on account of their sexual orientation (Berrill, 1992; Cogan, 1996). A survey of LGB students at

Homophobia: A strong, irrational fear of homosexuals; negative attitudes and reactions to homosexuals.
Antigay prejudice: Negative attitudes and behaviors toward gays and lesbians.

Table 15.1 Attitudes of Adult Americans Toward Homosexuality, 1973 and 1996

Question and Responses	Percentage of Sample	
	1973	*1996*
1. Are sexual relations between adults of the same sex:		
Always wrong	74	57
Almost always wrong	7	5
Wrong only sometimes	8	6
Not wrong at all	11	26
2. Should an admitted homosexual man be allowed to teach in a college or university?		
Yes	49	75
No	51	22

Source: National Opinion Research Center (NORC) Poll, 1973, 1996.

Penn State University showed a similar pattern of harassment on campus (D'Augelli, 1992). For example, 17 percent of the lesbians and 26 percent of the gay men had been threatened with physical violence. These studies show that hate crimes against and harassment of gays and lesbians are not rare, isolated incidents; instead, they are common, even on supposedly peaceful university campuses.

In 1990, Congress passed the Hate Crimes Statistics Act in which gays and lesbians were included with ethnic minority persons as having a special status necessitating legal protection from hate-motivated crimes (Morin & Rothblum, 1991). Although this may seem like cold comfort to a person who has already been the victim of such a crime, it is a step in the direction of providing some legal protection.

But we should also recognize the other side of the coin. As we can see from the statistics in Table 15.1, some Americans are tolerant of or supportive of homosexuals. For example, more than half of Americans approve of an overt homosexual teaching in a college or university. Thus Americans are a strange mixture of bigots and supporters on the issue of homosexuality. As one woman said,

> I really don't feel that I've ever been oppressed as a lesbian or suffered any abuse. I've been careful who I've told, but those people have been really accepting. (Jay & Young, 1979, p. 716)

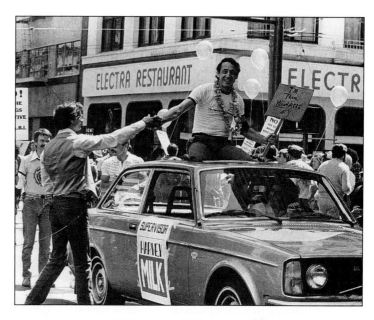

Figure 15.1   Harvey Milk, a gay activist, was an elected member of San Francisco's Board of Supervisors in 1978, representing a district including many gays. Milk fought for gay rights throughout the state of California and was supported by San Francisco's mayor, George Moscone. On November 17, 1978, Dan White, himself a former police officer and supervisor, entered City Hall and fired shots that killed both Milk and Moscone. White confessed within hours. In May 1979, a jury declined to convict White of first-degree murder, instead finding him guilty of voluntary manslaughter, a lesser offense carrying a reduced jail sentence. The gay community, as well as many supporters of gay rights, were shocked and furious. A protest march and the White Night Riot ensued. The entire incident symbolizes the ambivalent progress achieved by gay liberation. In 1986 White was released from prison and later committed suicide. An observance of these events is held in San Francisco every year.

Gays and Lesbians as a Minority Group

From the foregoing, it is clear that LGB people are the subject of many negative attitudes, just as other minorities are. Like members of other minority groups, they also suffer from job discrimination. Just as blacks and women have been denied access to certain jobs, so too have homosexuals. Wage discrimination occurs as well. According to research on a national, random sample, gay men earn 11 to 27 percent less than heterosexual men matched for relevant characteristics (Badgett, 1995). Homosexuality has often been grounds for a dishonorable discharge from the armed forces, a fact that became an issue when President Bill Clinton took office in 1993. The result of the controversy was a "Don't ask, don't tell" policy in which gays and lesbians could be members of the armed services as long as they kept their sexual orientation to themselves. Homosexuality has also been grounds for firing a person from federal employment and for denial of a security clearance. The reason commonly given for the latter two rules is that homosexuals are susceptible to being blackmailed and therefore should not be put in situations in which they have access to sensitive information. Actually, many heterosexuals might also be susceptible to blackmail because of their sexual activities—for example, the married man who is having an affair—although such factors are not grounds for denial of security clearance. In fact, in one sample 90 percent of lesbians and 86 percent of gay men had never been the object of blackmail or threats of blackmail (Jay & Young, 1979). In addition, as LGBs feel free to live more openly, they will be less subject to blackmail.

Discrimination goes hand in hand with stereotypes. One such stereotype is that gay men are child molesters. As with many stereotypes, this one is false. Research shows that only 2 to 3 percent of those who sexually abuse children are homosexual (Jenny et al., 1994).

In a spirit of reform in the 1980s, a number of states and cities passed laws prohibiting discrimination on the basis of sexual orientation. For example, in the state of Wisconsin it is illegal to discriminate against gays and lesbians in matters such as employment and housing. Massachusetts, Hawaii, and 17 other states have similar laws (Epstein, 1995).

There is an important way in which homosexuals differ from other minorities, though. In the case of most other minorities, appearance is a fairly good indicator of minority-group status. It is easy to recognize an African American or a woman, for example, but one cannot tell simply by looking at a person what his or her sexual orientation is. Thus LGBs, unlike other minorities, can hide their status. There are certain advantages to this. It makes it fairly easy to get along in the heterosexual world—to "pass." However, it has the disadvantage of encouraging the person to live a lie and to deny her or his true identity; not only is this dishonest, but it may also be psychologically stressful (Meyer, 1995). A study of gay men (all of whom were uninfected with HIV) indicated that those who concealed their identity had a significantly higher incidence of cancer and infectious diseases than those who did not conceal their identity (Cole et al., 1996).

We shouldn't leave this discussion of discrimination and prejudice against LGBs without asking a crucial question (Rothblum & Bond, 1996): What can be done to prevent or end this prejudice? Change must occur at many levels: the individual, the interpersonal, and the organizational levels (e.g., corporations, educational institutions), as well as society as a whole and its institutions (e.g., the federal government). At the individual level, all of us must examine our own attitudes toward gays and lesbians to see if they are consistent with basic values we hold, such as a commitment to equality and justice. Some people may need to educate themselves or attend anti-homophobia workshops to examine their attitudes. These attitudes, though, were formed as we grew up, influenced by our parents, our peers, and the media. Parents must consider the messages they convey to their children about homosexuals. The adolescent peer group is strongly homophobic. What could be done to change it? How can the media change (see Focus 15.5) in order not to promote antigay prejudice? At the interpersonal level, people must recognize that LGBs are often a hidden minority. Eric, for example, just told a joke that ridiculed gay men. What he didn't know was that one of his three listeners, his friend, is gay—just not "out" with him (for obvious reasons). We must examine our interactions with other people, recognizing the extent to which many of us assume that everyone is heterosexual until proven otherwise. At the institutional level, how can education be changed in order to reduce anti-gay discrimination? A strong program of sex education across the grades, with open discussion of sexual orientation, would be a good start (see Chapter 23). At the level of our federal government, despite the fact that numerous states have passed laws banning discrimination on the basis of sexual orientation, the U.S. government has failed to do so. Such a law would be an important first step.

Life Experiences of LGBs

In understanding lesbian and gay lifestyles, it is important to recognize that there is a wide variety of experiences. One of the most important aspects of this variability is whether the person is **covert** (in the closet) or **overt** (out of the closet) about his or her homosexuality. The covert homosexual may be heterosexually married, have children, and be a respected professional in the community, spending only a few hours a month engaging in secret same-gender sexual behavior. The overt homosexual, on the other hand, may live almost entirely within an LGB community, particularly if he or she lives in a large city like New York or San Francisco where there is a large gay subculture, and may have relatively few contacts with heterosexuals. There are also various degrees of overtness (being "out") and

Covert homosexual: A homosexual who is "in the closet," who keeps his or her sexual orientation a secret.
Overt homosexual: A homosexual who is "out of the closet," who is open about his or her sexual orientation.

Figure 15.2 Colonel Margarethe Cammermeyer, 50, was notified in July 1992 that her 26-year military career was over because she acknowledged that she is lesbian. A Vietnam veteran, she is the highest ranking soldier to be removed from service for being gay or lesbian.

covertness. Many lesbians and gays are out with trusted friends but not with casual acquaintances. The lifestyle of gay men differs from that of lesbians, as a result of the different roles assigned to males and females in our society and the different ways that males and females are reared. In addition, there is more discrimination against gay men than there is against lesbians. For example, it is considered quite natural for two women to share an apartment, but if two men do so, eyebrows are raised.

The lifestyles of LGBs are thus far from uniform. They vary according to whether one is male or female and overt or covert about the homosexuality and also according to social class, occupation, personality, and a variety of other factors.

Coming Out

As we noted earlier, there are significant variations in the gay experience, depending on whether or not one is out of the closet. The process of coming out of the closet, or **coming out,** involves acknowledging to oneself, and then to others, that one is gay or lesbian (Coleman, 1982). The person is psychologically vulnerable during this stage. Whether the person experiences acceptance or rejection from friends and others to whom he or she comes out can be critical to self-esteem.

Following the period of coming out, there is a stage of exploration, in which the person experiments with the new sexual identity; the person makes contact with the lesbian and gay community and practices new interpersonal skills. Typically, next comes a stage of forming first relationships. These relationships are often short-lived and characterized by jealousy and turbulence, much like many heterosexual dating relationships. Finally, there is the integration stage, in which the person becomes a fully functioning member of society and is capable of maintaining a long-term, committed relationship (Coleman, 1982).

Of course, before the coming out process can occur, the person must have arrived at a homosexual identity. This identity development seems to proceed in six stages (Cass, 1979):

1. *Identity confusion*   The person most likely began assuming a heterosexual identity because heterosexuality is so normative in our society. As same-gender attractions or behaviors occur, there is confusion. Who am I?

2. *Identity comparison*   The person now thinks "I *may* be homosexual." There may be feelings of alienation because the comfortable heterosexual identity has been lost.

3. *Identity tolerance*   At this stage the person thinks "I probably am homosexual." The person now seeks out homosexuals and makes

> **Coming out:** The process of acknowledging to oneself, and then to others, that one is gay or lesbian.

Figure 15.3   Today, many LGB communities exist in neighborhoods in large cities, with restaurants (such as this one in Greenwich Village in New York) and other organizations that are an integral part of the community.

contact with the gay subculture, hoping for affirmation. The quality of these initial contacts is critical.

4. *Identity acceptance*   The person now can say "I am homosexual," and accepts rather than tolerates this identity.

5. *Identity pride*   The person dichotomizes the world into homosexuals (who are good and important people) and heterosexuals (who are not). There is a strong identification with the gay group, and an increased coming out of the closet.

6. *Identity synthesis*   The person no longer holds an "us versus them" view of homosexuals and heterosexuals, recognizing that there are some good and supportive heterosexuals. In this final stage, the person is able to synthesize public and private sexual identities.

Lesbian, Gay, and Bisexual Communities

There is a loose network of lesbian, gay, and bisexual communities around the world (Esterberg, 1996). As one woman put it,

I have seen lesbian communities all over the world (e.g., Zimbabwe) where the lesbians of that nation have more in common with me (i.e., they play the same lesbian records, have read the same books, wear the same lesbian jewelry) than the heterosexual women of that nation have in common with heterosexual women in the U.S. (Rothblum, 1994)

Gay and lesbian communities began flourishing in the United States after World War II (D'Augelli & Garnets, 1995). Ironically, in the gender-segregated military, gay men were able to find each other and lesbians were able to find each other in a way that had not been possible previously. Activist groups slowly formed in the 1950s and 1960s, energized particularly by the Stonewall rebellion discussed at the beginning of the chapter. The HIV/AIDS crisis of the 1980s cemented together the gay community as it had never been before. Support networks and activist groups formed rapidly in response to the epidemic.

Today, many LGB communities exist in neighborhoods in large cities, with bookstores, restaurants, theaters, and social organizations that are an integral part of the community (D'Augelli & Garnets, 1995). The lesbian community in particular has been involved in creating a lesbian culture, expressed in music and literature and celebrated at festivals and conferences.

Symbols and rituals are important in defining the LGB community. The pink triangle, which the Nazis used to label gay men, has been adopted as a symbol of pride. The Greek letter lambda is another. Lesbian/gay pride marches held in June each year commemorate the Stonewall uprising. The use of slang is another sign of solidarity among LGBs (see Table 15.2).

Gay bars are one aspect of the LGB social life. These are cocktail lounges that cater exclusively to LGBs. Drinking, perhaps dancing, socializing, and the possibility of finding a sexual partner or a lover are the important elements. Some gay bars look just like any other bar from the outside, while others may have names—for example, The Open Closet—that indicate to the alert who the clientele are. Bars are typically gender-segregated—that is, they are for either gay men or lesbians—although a few are mixed. There are far more bars for gay men than for lesbian women. Typically, the atmo-sphere is different in the two, the male bars being more for finding sexual partners and the female bars more for talking and socializing. Lest the reader be shocked at the none-too-subtle nature of pickup bars, it is well to remember that there are many bars—singles' bars—that serve precisely the same purpose for heterosexuals.

The **gay baths** are another aspect of some gay men's social and sexual lives. The baths are clubs with many rooms in them, generally including a swimming pool or whirlpool, as well as rooms for dancing, watching television, and socializing; most areas are dimly lit. Once a man has found a sexual partner, they go to one of a number of small rooms furnished with beds, where they can engage in sexual activity. The baths feature casual, impersonal sex, since a partner can be found and the act completed without the two even exchanging names, much less making any emotional commitment to each other.

The majority of bath houses were closed in the 1980s because public health officials feared that they encouraged risky sexual practices and the spread of HIV. Bath houses were resurrected in the 1990s, however, and have created a controversy within the gay community. Some see the baths as an aspect of gay culture that spreads HIV and will

> **Gay bar:** A cocktail lounge catering to lesbians or gays.
> **Gay baths:** Clubs where gay men can socialize; features include a swimming pool or whirlpool and access to casual sex.

Table 15.2 Some Slang Terms from LGB Culture

In the closet	Keeping one's homosexuality hidden, not being open or public
Coming out	Coming out of the closet, or becoming open about one's homosexuality
Queen	An effeminate gay man
Nellie	An effeminate gay man
Closet queen	A homosexual who is covert or in the closet
Drag queen	A gay man who dresses in women's clothing ("drag")
Butch	A masculine gay man or a masculine lesbian
Dyke	A masculine lesbian
Femme	A feminine lesbian
Straight	A heterosexual
Trick	A casual sexual partner
Cruising	Looking for a sexual partner
Tearoom	A public rest room where gay men engage in casual sex

continue to do so, killing thousands; they believe the baths must be closed and the destructive practices they encourage should stop (Rotello, 1997; Signorile, 1997). Others celebrate the liberated sexuality fostered by the baths and see it as an essential part of gay men's lifestyle.

Casual sex with many partners is a preference for some gay men. In a pre-AIDS study conducted in San Francisco, approximately half the gay men reported 250 or more sexual partners in their lifetime, whereas only 1 percent of lesbians reported such a pattern (Bell & Weinberg, 1978). These statistics, though, may say more about male sexuality than about gay sexuality. As we saw in Chapter 14, men have more of a preference for casual sex than women do. Gay men may simply have more opportunities for casual sex, because they are negotiating with another man rather than a woman.

Certainly in the last three decades the *gay liberation movement* has had a tremendous impact on the gay lifestyle and community. In particular, it has encouraged homosexuals to be more overt and to feel less guilty about their behavior. LGB liberation meetings and activities provide a social situation in which gay people can meet and discuss important issues rather than simply play games as a prelude to sex, which tends to be the pattern in the bars. In addition, they provide a political organization that can attempt to bring about legal change, combat police harassment, and fight cases of job discrimination, as well as do public relations work. The National Gay and Lesbian Task Force[1] is the central clearinghouse for all these groups; it can provide information on local organizations.

There are thus many places for LGBs to socialize besides bars, including the Metropolitan Community Church (a gay and lesbian church), gay athletic organizations, and gay political organizations.

Among other accomplishments, members of the gay liberation movement have founded numerous gay newspapers and magazines. These have many of the same features as other newspapers: forums for political opinions, human-interest stories, and fashion news. In addition, the want-ads feature advertisements for sexual partners;

similar ads for homosexual and heterosexual partners can be found in underground newspapers in most cities. Probably the best-known LGB magazine is *The Advocate,* published in Los Angeles and circulated throughout the United States. *Lambda Rising News,* published in Washington, D.C., is an important newspaper. There are also several publications that list all the gay bars and baths by city in the United States, which is handy for the traveler or for those newly arrived on the LGB scene.

Gay and Lesbian Relationships

When you hear the term "couple," what do you think of? Most likely a heterosexual married couple, or perhaps a heterosexual dating couple. Sociologists Philip Blumstein and Pepper Schwartz (1983) took an innovative approach in their major study, *American Couples.* They defined "couples" to include heterosexual married couples, heterosexual cohabiting couples, gay male couples, and lesbian couples. They found that many of the characteristics and problems of gay couples are similar to those of heterosexual couples. For example, the frequency of sex declines when the couple have been in the relationship for more than 10 years; this occurs for gay male couples, lesbian couples, and heterosexual married couples. For gay and lesbian couples, there are sometimes problems involved in who initiates sex, just as there are for straight couples. Among heterosexual couples, it is typically the man who initiates sex; among gay couples, the more emotionally expressive partner usually does so.

Various surveys indicate that between 45 and 80 percent of lesbians and between 40 and 60 percent of gay men report currently being in a steady romantic relationship (Kurdek, 1995b). Contrary to stereotypes, a substantial number of lesbians and gay men form long-term, cohabiting relationships. One such relationship is described in Focus 15.1. In a study of 706 lesbian couples and 560 gay couples, 14 percent of the lesbian couples had been together for 10 or more years, as had 25 percent of the gay male couples (Bryant & Demian, 1990, cited in Kurdek, 1995b).

Gay and lesbian couples—like heterosexual couples—must struggle to find a balance that suits both persons, in three aspects of the relationship: attachment, autonomy, and equality (Cochran & Peplau, 1985; Kurdek, 1995a). *Attachment* refers to the qualities of closeness and secure love in the re-

---

[1] The National Gay and Lesbian Task Force, 1734 14th Street N.W., Washington, DC 20009; (202) 332–6483. See the Appendix at the back of this book for a list of other organizations dealing with various aspects of sexuality.

Focus 15.1
A Gay Couple: Tom and Brian

Tom and Brian have been living together as lovers for three years. Tom is 29, and Brian is 22; they live in a medium-sized midwestern city.

Tom grew up in a Roman Catholic family and attended parochial schools. His father was killed in an automobile accident when Tom was 4 days old. His mother remarried when he was 8 years old; although he gets along well with his stepfather, he feels that he has never had a real father. When asked why he thought he was gay, he said he felt it was because of the absence of a father in his early years.

After graduating from high school, Tom joined the army. His first sexual experience was with a Japanese prostitute. Though he enjoyed the physical aspects of sex with her, he felt that something was missing emotionally. When he was 23, after returning to the United States, he had his first homosexual experience with someone he met at a bar.

Tom had sex with about 20 different men before meeting Brian, including one with whom he had a long-term relationship. He is currently a student at a technical college and is working toward a career in electronics.

Brian also grew up in a Roman Catholic family and attended parochial schools. He has always gotten along well with both his parents and his four siblings, and he recalls his childhood as being uneventful. He realized he was gay when he was in his early teens, but he chose to ignore his feelings and instead tried to conform to what society expected of him. He had a girlfriend when he was in high school. After he graduated from high school, he took a job; he had his first homosexual experience when he was 18 with a man he met at work. He currently works as a salesperson at J C Penney. He had sex with eight different men before meeting Tom.

He has never had intercourse with a woman.

Brian and Tom have not discussed their sexual orientation directly with their families. Both think their parents may have some vague ideas, but the parents apparently prefer not to have to confront the situation directly and instead prefer to think of their sons as just being "roommates." Both Brian and Tom wish they could be more honest about their gayness with their parents. However, they also do not want to hurt them, and so they have not pressed the matter.

Their relationship with each other is exclusive; that is, they have an agreement that they will be faithful to each other and that neither will have sex with anyone else. Both of them said they would feel hurt if they found out that the other had been seeing someone else on the sly. They both consider casual sex unfulfilling and want to have a real relationship with their sexual partner.

Household chores are allocated depending on who has the time and inclination to do them. Both men like to cook, so they share that duty, and the laundry, about equally. Tom is good at carpentry and tends to do tasks of that sort.

Tom and Brian feel that the greatest problem in their relationship is lack of communication, which sometimes creates misunderstandings and arguments. Brian feels that the greatest joy in their relationship is the security of knowing that they love each other and can count on each other. Tom agrees, and he also says that it is important to him to know that there is someone who really cares about him and loves him.

Source: Based on an interview conducted by the authors.

Figure 15.4   A gay male couple. A large percentage of lesbians and gay men report currently being in a steady romantic relationship.

lationship. *Autonomy* refers to the quality of independence and individuality for each person. *Equality* refers to the balance of power between the two partners in matters ranging from decisions about finances to the division of household chores. Lesbian couples place a special value on equality in their relationships (Kurdek, 1995a). Attachment and autonomy are, to some extent, in opposition to each other. Some people want a great deal of attachment and little autonomy in their relationship, whereas others want just the opposite. The task is to find a balance that seems reasonable to both people. This task, clearly, is easier if the two partners hold similar values and ideals going into the relationship.

Lesbian and gay couples must negotiate conflicts and issues of power. For both lesbian and gay couples, the areas of conflict, in order of frequency, are finances, driving style, affection and sex, being overly critical, and household tasks (Kurdek, 1995b). This list probably sounds familiar to most heterosexual couples.

Long-term gay and lesbian relationships pass through stages as the years go by and as issues and challenges change. In one study, 156 gay couples were interviewed (McWhirter & Mattison, 1984). The couples had been together, on average, for 9 years, and eight of the couples had been together for 30 years or longer. The researchers concluded that long-term gay relationships pass through six stages. Stage 1, blending, occurs during the first year, in which there is a real sense of forming a couple and being in love. Stage 2, nesting, occurs during the second and third years, and involves creating a home together. In later stages, couples work at maintaining the relationship, once the initial thrills of falling in love and nesting have passed. Stage 6, which occurs in relationships that have lasted 20 years or more, is characterized by a sense that security has been achieved and by shared remembering of the many good times the couple have had together.

What is striking about all the research on gay and lesbian relationships is how similar they are—in their satisfactions, loves, joys, and conflicts—to heterosexual relationships (Peplau et al., 1996).

Lesbian and Gay Families

Increasingly, gay couples and lesbian couples are creating families that include children. This is a controversial concept to many heterosexual people in the United States, who view a lesbian family or gay family as a damaging setting for children to grow up in. The courts have often assumed that lesbians and gay men are unfit parents, and the same-gender sexual orientation of a parent has been grounds for the other, heterosexual parent to gain custody of children following a divorce. What does the research say about these families and the effects on children in them?

It is important to recognize that these families are diverse along dimensions of race, social class, and gender (Allen & Demo, 1995). In some, the children were born to one of the partners in a previous heterosexual relationship. In others, the children were adopted or, in the case of lesbian couples, born by means of artificial insemination.

Some have even said that a "lesbian baby boom" is under way (Patterson, 1995). Some are single-parent families, with, for example, a lesbian mother rearing her children from a previous heterosexual marriage.

How do the children fare in these families? Three concerns have been raised about these children (Patterson, 1992). First, will they show "disturbances" in gender identity or sexual identity? Will they become gay or lesbian? Second, will they be less psychologically healthy than children who grow up with two heterosexual parents? Third, will they have difficulties in relationships with their peers, perhaps being stigmatized or teased because of their unusual family situation?

Research on children growing up in lesbian or gay families, compared with those growing up in heterosexual families, dismisses these fears. For example, the overwhelming number of children growing up in lesbian or gay households have a heterosexual orientation (Bailey et al., 1995; Patterson, 1992).

The adjustment and mental health of children in lesbian and gay families are no different from those of children in heterosexual families (Kirkpatrick, 1996; Patterson, 1996, 1992; Tasker & Golombok, 1997). For example, one study compared adult daughters of lesbian and heterosexual mothers on a personality inventory (Gottman, 1990). There were no significant differences on 17 of the 18 personality scales. The only difference was on the scale measuring psychological well-being, and daughters of lesbians scored more favorably!

As for the third concern, about peer relationships, research indicates that children in lesbian or gay families fare about as well in terms of social skills and popularity as children in heterosexual families (Patterson, 1992).

In conclusion, although concerns have been raised about children growing up in lesbian and gay families, research consistently shows no difference between these children and those in heterosexual families (Patterson, 1992). As one expert in clinical psychology concluded, "it appears that traditional family structures, including father presence and heterosexuality, are not essential for healthy child development. Well-adjusted children of both sexes can be reared in families of varying configurations with the most crucial ingredient appearing to be the presence of at least one supportive, accepting caregiver" (Strickland, 1995).

How Many People Are Gay, Straight, or Bi?

Most people believe that homosexuality is rare. What percentages of people in the United States are gay and lesbian? As it turns out, the answer to this question is complex. Basically, it depends on how one defines "a homosexual" and "a heterosexual."

One source of information we have on this question is Kinsey's research (see Chapter 3 for an evaluation of the Kinsey data). Kinsey found that about 37 percent of all males had had at least one homosexual experience to orgasm in adulthood. This is a large percentage. Indeed, it was this statistic, combined with some of the findings on premarital sex, that led to the furor over the Kinsey report. The comparable figure for females was 13 percent. However, experts agree that, because of problems with sampling, Kinsey's statistics on homosexuality were probably inflated (Pomeroy, 1972).

Today, several well-sampled surveys of the U.S. population have given us improved estimates. One of those is the NHSLS (discussed in Chapter 3). Data from that study are shown in Table 15.3 (Laumann et al., 1994). The statistics are complex because much depends on how "homosexual" is defined. Does the definition require someone to have had exclusively same-gender sexual experiences, or just some same-gender experiences, or perhaps just to have experienced sexual attraction to members of his or her own gender without ever acting upon it? We will return to this point below. What we can say here is that, according to the NHSLS, about 2 percent of men and 1 percent of women are exclusively homosexual in their sexual behavior and in their identity. About 4 percent of both men and women have had at least one same-gender sexual experience in adulthood, and about 4 percent of men and 2 percent of women have experienced sexual attraction to members of their own gender.

These percentages are considerably smaller than Kinsey's. What accounts for the difference? The NHSLS was better sampled; it is generally agreed that Kinsey's unsystematic sampling methods led to overestimates of the incidence of homosexuality. But the NHSLS may not be per-

Table 15.3 The NHSLS Statistics on Same-Gender Sex, Identity, and Attraction

	Men	Women
Partners:		
Same-gender partners only, last year	2.0%	1.0%
Both male and female partners, last year	0.7	0.3
Same-gender partners only, since age 18	0.9	0.4
Both male and female partners, since age 18	4.0	3.7
Sexual Identity:		
Heterosexual	96.9	98.6
Bisexual	0.8	0.5
Homosexual	2.0	0.9
Sexual Attraction:		
Only opposite gender	93.8	95.6
Mostly opposite gender	2.6	2.7
Both	0.6	0.8
Mostly same gender	0.7	0.6
Only same gender	2.4	0.3

Source: E. O. Laumann, J. H. Gagnon, R. T. Michael, & S. Michaels. (1994). *The social organization of sexuality.* Chicago: University of Chicago Press, p. 311, Tables 8.3A and 8.3B.

fectly accurate, either. The data were collected using face-to-face interviews, which may have created problems of concealment, particularly when the questions were about a socially disapproved behavior such as homosexuality. Therefore, the NHSLS statistics are best seen as minimum estimates.

The NHSLS statistics are comparable to those found in a well-sampled international survey. The results indicated that 6.2, 4.5, and 10.7 percent of males in the United States, United Kingdom, and France, respectively, had engaged in sexual behavior with someone of their own gender in the last 5 years (Sell et al., 1995). The comparable statistics for women were 3.6, 2.1, and 3.3 percent. These and other surveys confirm that the incidence of homosexuality among men is higher than the incidence among women. Probably twice as many men as women have a homosexual experience to orgasm in adulthood, and the same ratio is probably true for exclusive homosexuality.

After reading these statistics, though, you may still be left wondering how many people are homosexuals. As Kinsey soon realized in trying to answer this question, it depends on how you count. A prevalent notion is that like black and white, homosexual and heterosexual are two quite separate and distinct categories. This is what might be called a *typological conceptualization* (see Figure 15.5). Kinsey made an important scientific breakthrough when he decided to conceptualize homosexuality and heterosexuality not as two separate categories but rather as variations on a continuum (Figure 15.5, section 2). The black and white extremes of heterosexuality and homosexuality have a lot of shades of gray in between: people who have had both some heterosexual and some homosexual experience, in various mixtures. To accommodate all this variety, Kinsey constructed a scale running from 0 (exclusively heterosexual) to 6 (exclusively homosexual), with the midpoint of 3 indicating equal amounts of heterosexual and homosexual experience.

Many sex researchers continue to use Kinsey's scale today, but the question remains, when is a person a homosexual? If you have had one homosexual experience, does that make you a homosexual, or do you have to have had substantial homosexual experience (say, a rating of 2 or 3 or higher)? Or do you have to be exclusively homosexual to be a homosexual? Kinsey dealt with this problem in part by devising the scale, but he also made another important point. He argued that we should not talk about "homosexuality" but rather about "homosexual behavior." "Homosexuality," as we have seen, is exceedingly difficult to define. "Ho-

Figure 15.5 Three ways of conceptualizing homosexuality and heterosexuality.

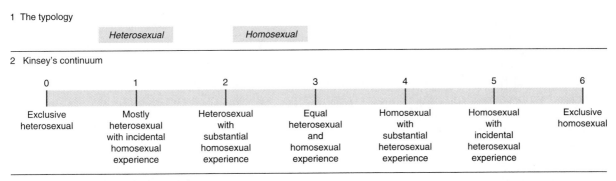

1 The typology

Heterosexual   Homosexual

2 Kinsey's continuum

0	1	2	3	4	5	6
Exclusive heterosexual	Mostly heterosexual with incidental homosexual experience	Heterosexual with substantial homosexual experience	Equal heterosexual and homosexual experience	Homosexual with substantial heterosexual experience	Homosexual with incidental heterosexual experience	Exclusive homosexual

3 Two-dimensional scheme (Storms, 1980)

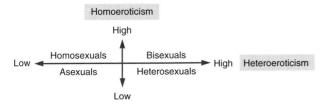

mosexual behavior," on the other hand, can be scientifically defined as a sexual act between two people of the same gender. Therefore, we can talk more precisely about people who have engaged in varying amounts of homosexual behavior or who have had varying amounts of homosexual experience, thus avoiding the problem of deciding exactly when a person is a homosexual, which is difficult, if not impossible, to do.

Other theorists have suggested that Kinsey's one-dimensional scale is too simple (Sell, 1997; Storms, 1980). The alternative is to form a two-dimensional scheme. The idea here is to have one scale for heteroeroticism (the extent of one's arousal to members of the opposite gender), ranging from low to high, and another for homoeroticism (extent of arousal to members of one's own gender), ranging from low to high (see Figure 15.5, section 3). Thus if one is high on both heteroeroticism and homoeroticism, one is a bisexual; the person high on heteroeroticism and low on homoeroticism is heterosexual; the person high on homoeroticism and low on heteroeroticism is homosexual; and finally, the person who is low on both

scales is asexual. This scheme allows even more complexity in describing homosexuality and heterosexuality than Kinsey's scale does.

The answer to the original question—How many people are homosexual and how many are heterosexual?—is complex. Probably about 92 percent of men and 95 percent of women are exclusively heterosexual. About 7 to 8 percent of men and 4 to 5 percent of women have had at least one same-gender sexual experience in adulthood. About 2 percent of men and 1 percent of women are exclusively homosexual. These figures are based on the NHSLS, but adjusted somewhat to allow for concealment by some respondents.

One other statistic that has been popularly cited for decades for the incidence of homosexuality is 10 percent. In fact, one gay and lesbian organization is called the Ten Percent Society. This statistic comes from no single study, but rather from an amalgamation of many. It represents those people whose orientation is predominantly homosexual, although they may have had some heterosexual experience. In view of the most recent surveys, the 10 percent figure may be a bit high.

Focus 15.2

The Ethics of Sex Research: The Tearoom Trade

Sociologist Laud Humphreys's study entitled *Tearoom Trade: Impersonal Sex in Public Places* (1970) is a classic in the field of sex research. In light of concerns on the part of both scientists and the general public about ethical standards in research, however, his methods of data collection are questionable from a contemporary perspective. Important issues are raised about the difficulty of doing good sex research within ethical bounds.

As the title of the book implies, the term "tearoom trade" refers to impersonal sexual acts in places like public rest rooms. Typically, a man enters the rest room and conveys to another man who is already there an interest in having sex. He may do this by making tapping sounds while in one of the stalls, for example. The men generally perform the sexual act in a stall and may not even exchange a word. The activity is typically fellatio, which can be done rapidly and with a minimum of encumbrance.

In the tearoom situation, a third person generally serves as a "lookout" who watches for police or other intruders while the other two engage in sex. To obtain his data, Humphreys became a lookout. Not only did he observe the behaviors involved in the tearoom trade, but he also wrote down the license-plate numbers of the participants. He traced the numbers through state records and thus was able to get the addresses of the persons involved. He then went to the homes of the people and administered a questionnaire (which included questions on sexual behavior) to them under the pretense of conducting a general survey.

The research provided some important findings, particularly that a large proportion of the men who engaged in the tearoom trade were respectable, heterosexually married men, and many were leaders in their community. This finding provoked quite a controversy over the book; the notion that "heterosexual" men could engage in homosexual behavior was shocking to many. Indeed, many gays find the tearoom trade to be shocking.

In his report of the research, Humphreys maintained the complete anonymity of the participants. However, his work still entails numerous ethical problems. There was no informed consent procedure (this study was carried out before scientific societies and universities instituted such standards). Participants were deceived—a problem made worse by the fact that they were never debriefed and told the true purpose of the research. But these considerations in turn raise the question, "Could Humphreys have obtained good data within the bounds of research ethics?" The clearly negative aspects of the study have to be weighed against the benefits that knowing more about this form of sexual behavior offers to society.

Source: Laud Humphreys. (1970). *Tearoom trade: Impersonal sex in public places.* Chicago: Aldine.

Sexual Orientation and Mental Health

Many Americans believe that homosexuality is a kind of mental illness. Is this really true? Do psychologists and psychiatrists agree that homosexuals are poorly adjusted or deviant? What are the implications of sexual orientation for a person's adjustment?

Sin and the Medical Model

Actually, the belief that homosexuality is a form of mental illness is something of an improvement over previous beliefs about homosexuality. Before this century, the dominant belief in Europe and the United States was that homosexuality was a sin or a heresy. During the Inquisition, people who were accused of being heretics were also frequently accused of being homosexuals and were burned at the stake. Indeed, in those times, all mental illness was regarded as a sin. In the twentieth century, this view has been replaced by the **medical model,** in which mental disturbance, and homosexuality in particular, is viewed as a sickness or illness (Bullough & Bullough, 1997).[2] This view is widely held now by the general public.

Psychiatrist Thomas Szasz and others are critical of the medical model. In his well-known writing on "the myth of mental illness," Szasz argues that the medical model is now obsolete and that we need to develop a more humane and realistic way of dealing with mental disorders and variations from the norm. He has argued the case particularly for homosexuality (Szasz, 1965). LGB activists have joined in, saying that they do not like being called "sick" and that this is just another form of persecution of gays and lesbians.

Research Results

What do the scientific data say? Once again, the answers provided by the data are complex and depend on the assumptions of the particular investigator and on the research design used. Basically, three kinds of research designs have been used, representing progressive sophistication and changing assumptions about the nature of homosexuality.

Clinical Studies

The first, and earliest, approach was clinical; homosexuals who were in psychotherapy were studied by the investigator (usually the therapist). He or she looked for disturbances in their current adjustment or in past experiences or home life. The data were then reported in the form of a case history of a single individual or a report of common factors that seemed to emerge in studying a group of homosexuals (for example, Freud, 1920; see also the review by Rosen, 1974). These clinical studies provided evidence that the homosexual was sick or abnormal; she or he typically was found to be

poorly adjusted and neurotic. But the reasoning behind this research was clearly circular. The homosexual was assumed to be mentally ill, and then evidence was found supporting this view.

Studies with Control Groups

The second group of studies made significant improvements over the previous ones by introducing control groups. The question under investigation was rephrased. Rather than "Do homosexuals have psychological disturbances?" (after all, most of us have some problems) it became "Do homosexuals have more psychological disturbances than heterosexuals?" The research design involved comparing a group of homosexuals in therapy with a group of randomly chosen heterosexuals not in therapy. These studies tended to agree with the earlier ones in finding more problems of adjustment among the homosexual group than among the heterosexual group (Rosen, 1974). The homosexuals tended to make more suicide attempts, to be more neurotic, and to have more disturbed family relationships.

Once again, though, it became apparent that there were some problems with this research design. It compared a group of people in therapy with a group of people not in therapy and found, not surprisingly, that the people in therapy had more problems. It, too, was circular in assuming that homosexuals were abnormal (in therapy) and that heterosexuals were normal (not in therapy) and then finding exactly that. Of course, the investigator used the homosexuals in therapy as research participants at least in part because they were convenient, but this does not help to make the results any more accurate.

Nonpatient Research

A major breakthrough came with the third group of studies, which involved nonpatient research. In these studies, a group of homosexuals not in therapy (nonpatients) were compared with a group of heterosexuals not in therapy. The nonpatient homosexuals were generally recruited through LGB

---

[2]As one gay comedian quipped, "If homosexuality is an illness, hey, I'm going to call in queer to work tomorrow."

> **Medical model:** A theoretical model in psychology and psychiatry in which mental problems are thought of as sickness or mental illness; the problems in turn are often thought to be due to biological factors.

organizations, advertisements, or word of mouth. Such nonpatient research has found no differences between the groups (Ross et al., 1988; Rothblum, 1994). That is, gays and lesbians seem to be as well adjusted as heterosexuals. This finding is quite remarkable in view of the very negative attitudes that members of the general public tend to hold toward LGBs (Gonsiorek, 1996).

On the basis of these studies, it must be concluded that the evidence does not support the notion that the homosexual is by definition "sick" or poorly adjusted. This position has received official professional recognition by the American Psychiatric Association. Prior to 1973, the APA had listed homosexuality as a disorder under Section V, "Personality Disorders and Certain Other Nonpsychotic Mental Disorders," in its authoritative *Diagnostic and Statistical Manual of Mental Disorders.* In 1973, the APA voted to remove homosexuality from that listing; thus it is no longer considered a psychiatric disorder.

Despite these overall positive findings about the mental health of LGBs, there are some causes for concern (Gonsiorek, 1996). Earlier in this chapter we noted the high incidence of victimization—for example, verbal abuse and threats of attack—directed at LGBs. Such victimization may be associated with depression (Otis & Skinner, 1996). Research indicates, though, that positive forces such as a supportive family can help to counteract the negative effects of victimization (Hershberger & D'Augelli, 1995).

Can Sexual Orientation Be Changed by Psychotherapy?

Masters and Johnson (1979) used the term **sexual dissatisfaction** to refer to lesbians and gays who seek therapy to become heterosexual. They treated 54 males and 13 females for this (using behavioral techniques similar to those we will describe in Chapter 19). Traditionally, psychologists have achieved a change rate of only about 10 percent in "converting" gays to straights. Masters and Johnson, however, reported a failure rate of 21 percent immediately after therapy and 28 percent five years after therapy, translating to a 72 percent success rate. It was this high success rate that created the controversy over their book on homosexuality.

This research is subject to some criticism. For

example, Masters and Johnson did not define clearly what they meant by a "success" or "failure." Low motivation has often been pinpointed as the cause of the traditionally low success rate for therapy for sexual dissatisfaction; Masters and Johnson carefully screened applicants for therapy and probably kept only those with very high motivation. And many of their "homosexuals" might more accurately have been described as bisexuals at the beginning of therapy.

Masters and Johnson's treatment program can actually be seen as one of the latest versions of **conversion therapy,** which has been around for over 100 years (Haldeman, 1994; Murphy, 1992). Many earlier techniques for conversion were downright inhumane. They included crude behavior therapy that involved giving gay men electrical shock while they viewed slides of nude men, as well as surgeries ranging from castration to brain surgery. All these treatments rested on the assumption that homosexuality is an illness that should be cured. None were successful.

Given the evidence discussed earlier in this section—that LGBs are not mentally ill or less well adjusted than heterosexuals—such therapies make no sense. Ethical issues are raised as well: Should a person be changed from gay to straight against his or her will? Only in rare cases would conversion therapy be justified, and then the treatment should be chosen carefully and should not be abusive to the individual receiving it.

In sum, it is probably about as easy to change a homosexual person into a happy heterosexual as it is to change a heterosexual person into a happy homosexual—that is, not very.

Why Do People Become Homosexual or Heterosexual?

A fascinating psychological question is: Why do people become homosexual or heterosexual? Sev-

---

Sexual dissatisfaction: Masters and Johnson's term for the discontent of gays and lesbians who seek therapy to become heterosexual.

Conversion therapy: Any one of a number of treatments designed to turn LGBs into heterosexuals.

(a)

(b)

Figure 15.6 Gay and lesbian political issues: *(a)* The custody issue—lesbian mothers want the right to keep their children after a divorce; *(b)* the right to adoption—a gay couple with their two adopted children.

eral theoretical answers to this question, as well as the results of empirical research, are discussed in this section. You will notice that the older theorists and researchers considered it their task to explain homosexuality; more recent investigators, realizing that heterosexuality needs to be explained as well, are more likely to consider it their task to explain sexual orientation.

Biological Theories

A number of scientists have speculated that homosexuality might be caused by biological factors. The likeliest candidates for these biological causes are genetic factors, prenatal factors, differences in brain structure, and endocrine imbalance.

Genetic Factors

The most recent carefully done study on this question recruited gay and bisexual men who had a twin brother or an adopted brother (Bailey & Pillard, 1991). Among the 56 gay men who had an identical twin brother, 52 percent of the cotwins were themselves gay (in the terminology of geneticists, this is a 52 percent concordance rate). Among the 54 gay men who had a nonidentical twin

brother, 22 percent of the cotwins were themselves gay. Of the adoptive brothers of gay men, 11 percent were gay. The same research team later repeated the study with lesbians (Bailey et al., 1993). Among the 71 lesbians who had an identical twin sister, 48 percent of the cotwins were also lesbian. Among the 37 lesbians who had a nonidentical twin sister, 16 percent of the cotwins were lesbian. Of the adoptive sisters of lesbians, 6 percent were lesbian. The statistics for women are therefore quite similar to those for men. (For another study with similar results, see Whitam et al., 1993. For a critique of these studies, see Byne, 1995; Byne & Parsons, 1993; McGuire, 1995.)

The fact that the rate of concordance is substantially higher for identical twins than for nonidentical twins argues in favor of a genetic contribution to sexual orientation. If genetic factors absolutely *determined* sexual orientation, however, there would be a 100 percent concordance rate for the identical twin pairs, and the rate of 52 percent is far from that. This rate indicates that factors other than genetics also play a role in influencing sexual orientation.

One research group believes that they have discovered a gene, located on the X chromosome, for homosexuality; this research is highly controversial (Hamer et al., 1993; Marshall, 1995). One study, by Hamer, has replicated the finding but another has not (Bailey & Pillard, 1995).

Prenatal Factors

Another speculation about a possible biological cause is that homosexuality develops as a result of factors during the prenatal period. As was seen in Chapter 5, exposure to inappropriate hormones during fetal development can lead a genetic female to have male genitals, or a genetic male to have female genitals. It has been suggested that a similar process might account for homosexuality (and also for transsexualism—see Chapter 14).

According to the most recent theory, homosexuality is caused by a variation in development. There is a critical time from the middle of the second month to the middle of the fifth month of fetal development, during which the hypothalamus differentiates and sexual orientation is determined (Ellis & Ames, 1987). Any of several biological variations during this period will produce homosexuality.

One line of research that supports this theory has found evidence that severe stress to a mother during pregnancy tends to produce homosexual offspring. For example, exposing pregnant female rats to stress produces male offspring that assume the female mating posture, although their ejaculatory behavior is normal (Ward & Stehm, 1991). The stress to the mother reduces the amount of testosterone in the fetus, which is thought to produce homosexual rats.

Research with humans designed to test the prenatal stress hypothesis reports mixed results. Perhaps the best study found no evidence of prenatal stress effects on sexual orientation for either males or females (Bailey et al., 1991).

Another research group has suggested that prenatal exposure to abnormally high levels of estrogen produces human female offspring who are more likely to be lesbian (Meyer-Bahlburg, 1997; Meyer-Bahlburg et al., 1995). To test this hypothesis they studied adult women who had been exposed to DES. DES, or diethylstilbestrol, is a powerful estrogen, which was used until 1971 to prevent miscarriage, when its use was discontinued because of harmful side effects. More DES-exposed women than controls were rated as homosexual or bisexual (Kinsey ratings of 2 through 6).

Another research group studied the birth order of gay men. Their research shows consistently, across many samples, that, compared with heterosexual men, gay men are more likely to have a late birth order and to have more older brothers but not more older sisters (Blanchard & Bogaert, 1996a, b; Blanchard & Klassen, 1997; Blanchard, Zucker, Bradley, & Hume, 1995; Blanchard, 1997). The researchers find no birth order or sibling effects for lesbians compared with heterosexual women. They believe that they have uncovered a prenatal effect, hypothesizing that, with each successive pregnancy with a male fetus, the mother forms more antibodies against an antigen (H-Y antigen) produced by a gene on the Y chromosome (Blanchard & Klassen 1997). H-Y antigen is known to influence prenatal sexual differentiation, and the hypothesis is that the mother's antibodies to H-Y antigen may affect sexual differentiation in the developing fetal brain.

The theories of prenatal influence are intriguing, but we need far stronger research with humans before they can be accepted. John Money (1987), for example, reviewing current research findings, argues that although there are prenatal hormone effects on sexual orientation, postnatal socialization influences are also important.

Brain Factors

Another line of theorizing has been that there are anatomical differences between the brains of gays and straights that produce the differences in sexual orientation. There have been a number of studies deriving from this point of view, all looking at somewhat different regions of the brain (Swaab et al., 1995). A highly publicized study by neuroscientist Simon LeVay (1991) is an example. LeVay found significant differences between gay men and straight men in certain cells in the anterior portion of the hypothalamus. Anatomically, the hypothalamic cells of the gay men were more similar to those of women than to those of straight men, according to LeVay. However, the study had a number of flaws: (1) The sample size was very small: only 19 gay men, 16 straight men, and 6 straight women were included. This small sample size was necessitated by the fact that the brains had to be dissected in order to examine the hypothalamus, so that the brains of living persons could not be studied. (2) All of the gay men in the sample, but only six of the straight men and one of the straight women, had died of AIDS. The groups are not comparable, then. Perhaps the brain differences were caused by the neurological effects of AIDS. (3) Lesbian women were omitted from the study, making them invisible in the research—as they often have been in psychological and biological research. (4) The gay men were known to have been gay based on records at the time of death; the others, however, were just "presumed" to be heterosexual—if there was no record of sexual orientation, the assumption was that the person had been heterosexual, scarcely a sophisticated method of measurement. Finally, this study has not been replicated; it is difficult to know how much confidence to place in it.

Hormonal Imbalance

Investigating the possibility that endocrine imbalance is the cause of homosexuality, many researchers have tried to determine whether the testosterone ("male" hormone) levels of male homosexuals differ from those of male heterosexuals. These studies have not found any hormonal differences between male homosexuals and male heterosexuals (Banks & Gartrell, 1995; Gooren et al., 1990).

Despite these results, some clinicians have attempted to cure male homosexuality by administering testosterone therapy (Glass & Johnson, 1944). This therapy fails; indeed, it seems to result in even more homosexual behavior than usual. This is not an unexpected result, since, as was seen in Chapter 9, androgen levels seem to be related to sexual responsiveness. As a clinician friend of ours replied to an undergraduate male who was seeking testosterone therapy for his homosexual behavior, "It won't make you heterosexual; it will only make you horny."

Other, more complex kinds of hormonal differences between heterosexuals and homosexuals are also being explored (e.g., Gladue et al., 1984); however, it is too early to be able to conclude much from these studies. In addition, virtually all these hormone studies have been done on male homosexuals.

In conclusion, of the biological theories, the genetic theory and the prenatal theory have new evidence, and the genetic evidence seems especially strong; but much more research is needed.

Psychoanalytic Theory

Freudian Theory

Since Freud believed sex to be the primary motivating force in human behavior, it is not surprising that he concerned himself with sexual orientation and its development (his classic work on the subject is *Three Essays on the Theory of Sexuality,* published in 1910).

According to Freud, the infant is **polymorphous perverse;** that is, the infant's sexuality is totally undifferentiated and is therefore directed at all sorts of objects, both appropriate and inappropriate. As the child grows up and matures into an adult, sexuality is increasingly directed toward "appropriate" objects (members of the other gender), while the desire for "inappropriate" objects (for example, members of the same gender) is increasingly repressed. Therefore, according to Freud, the homosexual is fixated at an immature stage of development.

According to Freud, homosexuality also stems from the **negative Oedipus complex.** In the (positive) Oedipus complex, discussed in Chapter 2, the child loves the parent of the other gender but eventually gives this up and comes to identify with the parent of the same gender, thereby acquiring a sense of gender identity. In the negative Oedipus complex, things are just the opposite: The child loves the parent of the same gender and identifies with the parent of the other gender. For example, in the negative Oedipus complex, a little boy would love his father and identify with his mother. In the process of maturation, once again, the child is supposed to repress this negative Oedipus complex. The homosexual person, however, fails to repress it and remains fixated on it. Thus according to Freud, for example, a woman becomes a homosexual because of a continuing love for her mother and identification with her father. In Freud's view, homosexuality is a continuation of love and lust for the parent of the same gender.

Polymorphous perverse: Freud's term for the infant's indiscriminate, undifferentiated sexuality.
Negative Oedipus complex: Freud's term for the opposite of the Oedipus complex; in the negative Oedipus complex the child loves and sexually desires the parent of the same gender and identifies with the parent of the other gender.

Consistent with Freud's belief that the infant is polymorphous perverse was his belief that all humans are inherently bisexual; that is, he believed that all people have the capacity for both heterosexual and homosexual behavior. Thus he viewed homosexuality as very possible, if not desirable. This notion of inherent bisexuality also led to his concept of the *latent homosexual,* the person with a repressed homosexual component of the personality.

Bieber's Research

Because Freud had such a great influence on psychiatric thought, he inspired a great deal of theorizing and research, including research on homosexuality. Irving Bieber and his colleagues (1962) did one of the more important of these psychoanalytically inspired studies. They compared 106 male homosexuals with 100 male heterosexuals; all the subjects were in psychoanalysis, which makes the results somewhat questionable. The family pattern that Bieber and his colleagues tended to find among the homosexuals was that of a dominant mother and a weak or passive father. The mother was both overprotective and overly intimate. Bieber thus originated the concept of the **homoseductive mother** as an explanation of male homosexuality. This family pattern, according to Bieber, has a double effect: The man later fears heterosexual relations both because of his mother's jealous possessiveness and because her seductiveness has produced anxiety in him. Bieber thus suggested that homosexuality results, in part, from fears of heterosexuality. While Bieber's findings on the homoseductive mother have received a great deal of attention, a more striking finding in his research was of a seriously *disturbed relationship between the homosexual male and his father* (Bieber, 1976). The fathers were described as detached or openly hostile or both; thus the homosexual son emerged into adulthood hating and fearing his father and yet deeply wanting the father's love and affection.

Evaluation of Psychoanalytic Theories

Psychoanalytic theories of the genesis of homosexuality clearly operate under the assumption that homosexuality is deviant or abnormal. As Irving Bieber wrote:

> We consider homosexuality to be a pathologic, bio-social, psycho-sexual adaptation to pervasive fears surrounding the expression of heterosexual impulses. (Bieber et al., 1962, p. 22)

> All psychoanalytic theories assume that adult homosexuality is psycho-pathologic. (Bieber et al., 1962, p. 18)

Actually, in his later writings Freud came to consider homosexuality to be within the normal range of variation of sexual behavior; this view, however, appears not to have had much of an impact, compared with that of his earlier writings on the subject. Psychoanalytic theory could thus be criticized for the abnormality assumption, since, as we have seen, there is no evidence for the notion that the homosexual is poorly adjusted.

The psychoanalytic approach can also be criticized for its confusion of the concepts of "gender identity" and "sexual orientation." As was previously noted, the homosexual differs from the heterosexual in sexual orientation but not in gender identification: The male homosexual generally has a masculine identification, and the lesbian has a feminine identification. Psychoanalytic theory, however, assumes that the homosexual not only makes an inappropriate object choice but also has an abnormal gender identification—that the gay man has not identified with his father and has therefore not acquired a masculine identity and that the lesbian has not identified with her mother and has therefore not acquired a feminine identity. This basic assumption of abnormal gender identification is not supported by the data and is therefore another basis for criticizing psychoanalytic theories of homosexuality.

Learning Theory

Behaviorists emphasize the importance of learning in the development of sexual orientation. They note the prevalence of bisexual behavior both in other species and in young humans, and they argue that rewards and punishments shape the individual's behavior into predominant homosexuality or predominant heterosexuality. The assumption, then, is that humans have a relatively amorphous, undifferentiated pool of sex drive which, depending on circumstances (rewards and punishments), may be channeled in any of several directions. In short, people are born sexual, not heterosexual or homo-

---

Homoseductive mother: Bieber's term for the mother who is seductive toward her son, traumatizing the boy and turning him into a homosexual.

sexual. Only through learning does one of these behaviors become more likely than the other. For example, a person who has early heterosexual experiences that are very unpleasant might develop toward homosexuality. Heterosexuality has essentially been punished and therefore becomes less likely. This might occur, for instance, in the case of a girl who is raped at an early age; her first experience with heterosexual sex was extremely unpleasant, so she avoids it and turns to homosexuality. Parents who become upset about their teenagers' heterosexual activities might do well to remember this notion; punishing a young person for engaging in heterosexual behavior may not eliminate the behavior but rather rechannel it in a homosexual direction.

Another possibility, according to a learning-theory approach, is that if early sexual experiences are homosexual and pleasant, the person may become homosexual. Homosexual behavior has essentially been rewarded and therefore becomes more likely.

The learning-theory approach treats homosexuality as a normal form of behavior and recognizes that heterosexuality is not necessarily inborn but must also, like homosexuality, be learned. There is a problem with the learning-theory approach, however (Whitam, 1977). The rewards in our society go overwhelmingly to heterosexuality. Society gives few rewards to homosexuality, and often punishes it. Why, then, does anyone become homosexual? While human sexual behavior is no doubt determined in part by reinforcement contingencies, it probably also has much more complex determinants which are not yet understood.

Interestingly, though, research indicates that children who grow up with a homosexual parent are not themselves likely to become gay (Bailey et al., 1995; Golombok & Tasker, 1996; Patterson, 1992). In this sense, then, homosexuality is not "learned" from parents.

Interactionist Theory

Bem: The Exotic Becomes Erotic
Psychologist Daryl Bem (1995) has proposed a theory of the development of sexual orientation that encompasses the interaction of biological factors and experiences with the environment. Bem's theory is diagramed in Figure 15.7.

The theory begins with biological influences, relying on the evidence discussed earlier about biological contributions to sexual orientation (box A

in Figure 15.7). However, Bem does not believe that genes and other biological factors directly and magically determine a person's sexual orientation. Rather, he believes that biological factors exert their influence on sexual orientation through their influence on temperament in childhood (box A to box B). Psychologists have found abundant evidence that two aspects of temperament have a biological basis: aggression and activity level. Moreover, these two aspects of temperament show reliable gender differences. According to Bem, most children show levels of aggression and activity level that are typical of their gender; boys are generally more aggressive and more active than girls. These tendencies lead children to engage in gender-conforming activities (B to C). Most boys play active, aggressive sports, and most girls prefer quieter play activities. These play patterns also lead children to associate almost exclusively with members of their own gender. The boy playing tackle football is playing in a group that consists either entirely or almost entirely of boys. This chain of events will eventually lead to a heterosexual orientation in adulthood.

A minority of children, however, have temperamental characteristics that are not typical of their gender: some boys are not particularly active or aggressive and some girls are. These children are then gender-nonconforming in their play patterns. The boys prefer quieter, less active play and have more girls as friends, and the gender-nonconforming girls prefer active sports and have more boys as friends.

These experiences with childhood play and playmates create a feeling in children that certain other children are different from them and therefore are exotic (box C to D). For the boy who spends most of his time playing active, aggressive sports with other boys, girls are different, mysterious, and exotic. For the gender-nonconforming girl who plays active sports with boys, most girls, too, seem different and exotic to her.

The presence of an exotic other causes a person to feel generalized arousal, whether in childhood, adolescence, or adulthood (box D to E). Those of you who are heterosexual certainly remember many instances in your past when you felt ill at ease and nervous in the presence of a member of the other gender.

In the final link of the model, this generalized arousal is transformed into erotic/romantic attraction. Essentially, the exotic becomes erotic. This

Figure 15.7   Bem's theory of the development of sexual orientation: The exotic becomes erotic.

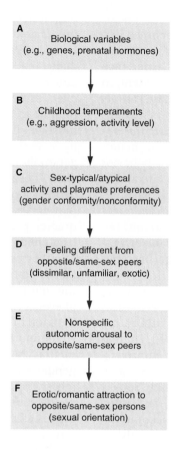

transformation may be a result of processes described in Berscheid and Walster's (1974) two-component theory of arousal, discussed in Chapter 13. Generalized arousal can easily be transformed into sexual arousal and attraction if the conditions are right. For heterosexual persons, the exotic people are members of the other gender, with whom they had less contact in childhood and who have become eroticized. For gays and lesbians, the exotic people are members of their own gender, from whom they felt different in childhood and who have become eroticized.

One well-documented phenomenon is that there are considerably more heterosexuals than homosexuals. The theory deals with this fact by recognizing that American society, like most other societies around the world, is gender polarized.

Gender distinctions are emphasized, and gender roles are strong. For most individuals, then, members of the other gender are exotic. It is only in the minority of cases, those whose childhood temperament leads them to be gender-nonconforming, that members of their own gender are exotic.

How good is this theory? It is too new to have had much direct testing, so we will have to wait for research evidence to accumulate. Bem cites evidence from many earlier studies in support of the links in his model. Certainly one virtue of the theory is that it is designed to explain sexual orientation (homosexual or heterosexual) rather than homosexuality. That is, the essential process is that the exotic becomes erotic, for both heterosexuals and homosexuals. The difference is that, because of childhood temperament and play activities, members of the other gender seem exotic to heterosexuals, whereas members of their own gender seem exotic to gays and lesbians.

One body of evidence that is consistent with Bem's theory indicates that gay men and lesbians, on average, are more likely than heterosexuals to have had a childhood history of gender nonconformity (Bell et al., 1981; Bailey & Zucker, 1995). This pattern has been found cross-culturally as well. For example, in one study of lesbians and heterosexual women in Brazil, Peru, the Philippines, and the United States, the lesbians were, as children, significantly more gender-nonconforming than were the heterosexual women (Whitam & Mathy, 1991). The lesbians were significantly more rejecting of gender-typical activities, such as playing with girls' toys and paying attention to women's fashions, and more involved with gender-atypical activities, such as playing with boys' toys and being a tomboy. The consistency of these results is particularly striking given how vastly different these four cultures are. Although this evidence is consistent with Bem's theory, it is also consistent with the biological theories.

At the same time, Bem's theory and evidence have been thoroughly criticized (Peplau et al., 1998; for Bem's response, see Bem, 1998). Two criticisms have been raised: (1) evidence not discussed by Bem contradicts some central propositions of the theory; and (2) the theory reflects male experience and neglects female experience. Regarding the evidence, Bem noted that, in one major study, lesbians (70 percent) were significantly more likely than were heterosexual women (51 percent) to recall feeling somewhat or very different from other girls their age (Bell et al., 1981). The difference is significant, but perhaps more important is the finding that a majority of heterosexual women felt different from other girls. Other girls seemed different or exotic to them. Why didn't they become lesbian, then?

Sociological Theory

Sociologists emphasize the effects of *labeling* in explaining homosexuality. The label "homosexual" has a big impact in our society. If you are heterosexual, suppose that someone said to you, "I think you are homosexual." How would you react? Your immediate reaction would probably be negative: anger, anxiety, and embarrassment. The label "homosexual" has derogatory connotations and may

even be used as an insult, reflecting our society's predominantly negative attitudes toward homosexuality.

A clever experiment demonstrated the effects of labeling someone homosexual. Half the participants (all men) were led to believe that a particular member of their group was homosexual. For the other half of the subjects (the control group), the man was not labeled. People in the experimental group, in which the man was labeled, rated him as being significantly less clean, softer, more womanly, more tense, more yielding, more impulsive, less rugged, more passive, and quieter (Karr, 1978). Thus labeling a person homosexual does influence people's perception of that person.

But the label "homosexual" may also act as a self-fulfilling prophecy. Suppose that a young boy—possibly because he is slightly effeminate, because he is poor in sports, or for no reason at all—is called a "homosexual." He reacts strongly and becomes more and more anxious and worried about his problem. He becomes painfully aware of the slightest homosexual tendency in himself. Finally he convinces himself that he is homosexual. He begins engaging in homosexual behavior and associates with a gay group. In short, a homosexual has been created through labeling.

Recall that in Chapter 2 we discussed Reiss's sociological theory of human sexuality. In his theorizing he has addressed the issue of sexual orientation, focusing particularly on gay men. Recognizing the need to explain cross-cultural differences in sexual patterns, he contends that it is male-dominant societies with great rigidity of gender roles that produce the highest incidence of homosexuality. In such societies, there is a rigid male role that must be learned and conformed to, but young boys have little opportunity to learn it from adult men precisely because the gender roles are rigid, so that women take care of children and men have little contact with them. It is therefore difficult to learn the heterosexual component of the male role. In addition, because the male role is rigid, there will be a certain number of males who dislike it and reject its heterosexual component. Cross-cultural studies support his observations (Reiss, 1986). Societies that have a great maternal involvement with infants and low father involvement with infants and that have rigid gender roles are precisely those that have the highest incidence of same-gender sexual behavior in males.

This pattern describes the negative pathway to homosexuality. Reiss argues that there is also a positive pathway. It exists in less gender-rigid societies with more permissiveness about sexuality. In such societies, individuals feel freer to experiment with same-gender behavior and may find it satisfying. Examples are provided by several American Indian tribes in which three gender roles have been recognized (see Chapter 14).

Empirical Data

Theoretical positions on the origins of sexual orientation are many and varied, as the previous sections indicate. Do any of these proposed explanations have a basis in fact?

The most comprehensive study of the causes of sexual orientation was done by Alan Bell, Martin Weinberg, and Sue Hammersmith, of the Kinsey Institute (1981). They interviewed 979 gay men and lesbian women and a comparison sample of 477 heterosexual men and women, all in San Francisco. The interviews, which took 3 to 5 hours each, included approximately 200 questions about the person's childhood and adolescence; the questions were designed specifically to allow the researchers to test all the major theories that have been proposed as explanations for the development of sexual orientation. In general, their results indicated that all the environmental explanations are inadequate and are not supported by the data. Specifically, the researchers concluded that:

1. The influence of disturbed relationships with one's parents—as proposed in psychoanalytic theory—is grossly exaggerated. Parental relationships seem to make little or no difference in whether one becomes heterosexual or homosexual.

2. The sociologists' notion that homosexuality results from labeling by others received no support from the data.

3. The notion—proposed by learning theory—that homosexuality results from early unpleasant heterosexual experience received no support. For example, lesbians were no more likely to have been raped than heterosexual women were.

4. The idea—proposed by learning theory—that homosexuality might result from a boy or girl being seduced by an older member of his or her own gender (an early positive homosexual experience) also was not supported.

Having shot down all the standard theories, Bell, Weinberg, and Hammersmith did reach some positive conclusions. Two are especially important.

1. Sexual orientation seems to be determined before adolescence. This would be important if, for example, it was discovered that the high school football coach was gay. According to Bell and his colleagues, parents and the principal should not worry that he will be a "bad influence" on the team members; their sexual orientation is already determined.

2. It is likely that there is a biological basis for homosexuality. The researchers reached this conclusion because none of the standard environmental explanations was supported by the data. They actually collected no biological data (e.g., measurements of hormone levels). Thus this conclusion amounts to no more than speculation, although it received a great deal of publicity.

In summary, when all the studies in this area are surveyed, no single factor emerges consistently as a cause of male homosexuality or of lesbianism (the two might have different causes). To be blunt, we do not know what causes sexual orientation. But there may be a good theoretical lesson to be learned from this somewhat frustrating statement. It has generally been assumed not only that gays form a distinct category (which, we have already seen, is not very accurate) but also that they form a homogeneous category, that is, that all gays are fairly similar. Not so. Probably there are many different kinds or "types" of homosexuals. Indeed, one psychologist, expressing this notion, has suggested that we should refer not to "homosexuality" but rather to "the homosexualities" (A. P. Bell, 1974b; Bell & Weinberg, 1978). If this is the case, then one would not expect a single "cause" of homosexuality but rather many causes, each corresponding to its type. The next step in research, then, should be to identify the various types of homosexuals—not to mention the various types of heterosexuals—and the different kinds of development that lead to each.

Sexual Orientation in Multicultural Perspective

Just as different cultures around the world hold different views of same-gender sexual behavior (see Focus 15.3), so do various American ethnic minority groups have different cultural definitions for same-gender behaviors.

Generally it is thought that there is less tolerance for homosexuality in the African American community (Icard, 1996). A survey of the attitudes of 2006 state employees confirmed this view (Ernst et al., 1991). In response to the statement "AIDS will help society by reducing the number of homosexuals (gay people)," African Americans were significantly more likely to agree, that is, to express a negative view of gays. Black women were especially negative about homosexuals, compared with white women. The researchers suggested that black women's more negative attitudes may be due to the perception, by some black women, that homosexuality is one of many forces draining the pool of eligible, marriageable black men (other factors include a higher rate of premature death among black males and a higher rate of unemployment).

We should not overemphasize ethnic differences, though. According to one large-scale study of African American lesbians and gay men, respondents typically had partners who were similar in age, education, and income (Peplau et al., 1997). This pattern has been found in numerous studies of white gays, lesbians, and heterosexuals (see Chapter 13).

It is also true that black and Latino men are more likely than white men to engage extensively in homosexual behavior while still considering themselves to be heterosexual (Peterson & Marin, 1988; Peterson et al., 1992). An interesting example of these different cultural definitions comes from a study of Mexican and Mexican American men and their same-gender sexual behavior (Magana & Carrier, 1991). In Mexico, there is a dichotomizing of same-gender sexual behaviors that parallels the rigidly defined gender roles. Anal intercourse, because it most resembles penis-in-vagina intercourse, is the preferred behavior, and fellatio is practiced relatively little. A man adopts the role either of receptive partner or inserting partner and does this exclusively. Those who take the receptive role are considered unmanly, feminine, and "homosexual." Those who take the inserting role are considered masculine, are not labeled "homosexual," and are not stigmatized. This approach differs substantially from that in Anglo culture, where men commonly switch roles. Among Mexican American men who have migrated to the United States, their views and sexual practices seem to be determined by whether their sexual socialization in adolescence occurred with Anglo peers or with Mexican American peers.

Such different definitions of homosexuality are not limited to Mexican and Mexican American culture. One researcher described the scene in Egypt as follows:

> In Egypt, because there was so little sense of homosexuality as an identity, what position you took in bed defined all. Between men, the only sex that counted was anal sex. . . . In the minds of most Egyptians, "gay," if it meant anything at all, signified taking the receptive position in anal sex. On the other hand, a person who took the insertive role—and that seemed to include virtually all Egyptian men, to judge by what my acquaintances told me—was not considered gay. . . . Many of the insults in the Arabic language concern being penetrated anally by another man. (Miller, 1992, p. 76)

As for lesbians, Latinas experience conflicts in the complexities of ethnicity and sexual orientation (Espin, 1987). Although in Latin cultures emotional and physical closeness among women is considered acceptable and desirable, attitudes toward lesbianism are even more restrictive than in Anglo culture. The special emphasis on family—defined as mother, father, children, and grandparents—in Latin cultures makes the lesbian even more of an outsider. As a result, Latina lesbians often become part of an Anglo lesbian community while remaining in the closet with their family and among Latinos, creating difficult choices among identities. As one Cuban woman responded to a questionnaire, "I identify myself as a lesbian more intensely than as a Cuban/Latin. But it is a very

Focus 15.3

Ritualized Homosexuality in Melanesia

Melanesia is an area of the southwest Pacific that includes the islands of New Guinea and Fiji as well as many others. Anthropologists' research on homosexual behavior in those cultures provides great insight into the ways in which sexual behaviors are the products of the scripts of a culture. This research is rooted in sociological and anthropological theory (see Chapter 2). As such, the analysis focuses on the norms of the society and the symbolic meaning that is attached to sexual behaviors.

Among Melanesians, homosexual behavior has a very different symbolic meaning from the one it has in Western culture. It is viewed as natural, normal, and indeed necessary. The culture actually prescribes the behavior, in contrast to Western cultures, in which it is forbidden or proscribed.

Sociologists and anthropologists believe that most cultures are organized around the dimensions of social class, race, gender, and age. Among Melanesians, age organizes the homosexual behavior. It is not to occur among two men of the same age. Instead, it occurs between an adolescent and a preadolescent, or between an adult man and a pubertal boy. The older partner is always the inserter for the acts of anal intercourse, and the younger partner is the insertee.

Ritualized homosexual behavior serves several social purposes in these cultures. It is viewed as a means by which a boy at puberty is incorporated into the adult society of men. It is also thought to encourage a boy's growth, so that it helps to "finish off" his growth in puberty. In these societies, semen is viewed as a scarce and valuable commodity. Therefore the homosexual behaviors are viewed as helpful and honorable, a means of passing on strength to younger men and boys. As one anthropologist observed:

> Semen is also necessary for young boys to attain full growth to manhood. . . .They need a boost, as it were. When a boy is eleven or twelve years old, he is engaged for several months in homosexual intercourse with a healthy older man chosen by his father. (This is always an in-law or unrelated person, since the same notions of incestuous relations apply to little boys as to marriageable women.) Men point to the rapid growth of adolescent youths, the appearance of peach fuzz beards, and so on, as the favorable results of this child-rearing practice. (Schieffelin, 1976, p. 124)

In all cases, these men are expected later to marry and father children. This points up the contrast between sexual identity and sexual behavior. The sexual behaviors are ones that we would surely term "homosexual," yet these cultures are so structured that the boys and men who engage in homosexual behaviors do not form a homosexual identity.

Ritualized homosexual behaviors are declining as these cultures are colonized by Westerners. It is fortunate that anthropologists were able to make their observations over the last several decades to document these interesting and meaningful practices before they disappear.

Source: Gilbert H. Herdt (Ed.). (1984). *Ritualized homosexuality in Melanesia.* Berkeley: University of California Press.

painful question because I feel that I am both, and I don't want to have to choose" (Espin, 1987, p. 47).

Among Chinese Americans, two features of Asian American culture shape attitudes toward homosexuality and its expression: (1) a strong distinction between what may be expressed publicly and what should be kept private; and (2) a stronger value placed on loyalty to one's family and on the performance of family roles than on the expression of one's own desires (Chan, 1995). Sexuality is something that must be expressed only privately, not publicly. And having an identity, much less a sexual identity or gay lifestyle, apart from one's family is almost incomprehensible to traditional Chinese Americans. As a result, a relatively small proportion of Chinese American LGBs seem to be "out" compared with non-Asians. Those Chinese American LGBs who are out tend to be more acculturated—that is, influenced by American culture. They echo the sentiments of the Latina lesbians mentioned earlier, saying that they would prefer not to have to choose between their ethnic identity and their sexual identity but that when forced to make the choice, they are more closely tied to the LGB part of their identities (Chan, 1995; Liu & Chan, 1996).

In sum, when we consider sexual orientation from a multicultural perspective, two main points emerge: (1) The very definition of homosexuality is set by culture. In the United States, we would say that a man who is the inserting partner in anal intercourse with another man is engaging in homosexual behavior, but other cultures (such as Mexico and Egypt) would not agree. (2) Some ethnic groups are even more disapproving of homosexuality than are U.S. whites. In those cases, LGBs feel conflicts between their sexual identity and loyalty to their ethnic group.

Bisexuality

Here is a riddle: What is like a bridge that touches both shores but doesn't meet in the middle? The answer: research and theories on sexual orientation (MacDonald, 1982; you'll have to admit that this is classier than the average riddle). The point of the riddle is that scientists, as well as lay people,

Figure 15.8   Ethnicity and sexual orientation. Among Latina women, warmth and physical closeness are very acceptable, but there are strong taboos against female–female sexual relationships.

focus on heterosexuals and homosexuals, ignoring all the bisexuals in between.

A bisexual is a person whose sexual orientation is toward both women and men, that is, toward members of the same gender as well as members of the other gender. A slang term is "ac–dc" (alternating current–direct current). Some scientists use the term **ambisexual.**

Bisexuality is not rare; in fact, it is more common than exclusive homosexuality (if a "bisexual" is defined as a person who has had at least one sexual experience with a male and at least one with a

Ambisexual: Another term for bisexual.

female). Earlier in this chapter we saw that about 4 percent of men and women have had both male and female partners since age 18. About 1 percent of men and 0.5 percent of women claim a bisexual identity (Laumann et al., 1994), although there is probably some underreporting.

The proponents of bisexuality argue that it has some strong advantages. It allows more variety in one's sexual and human relationships than does either exclusive heterosexuality or exclusive homosexuality. The bisexual does not rule out any possibilities and is open to the widest variety of experiences.

On the other hand, the bisexual may be viewed with suspicion or downright hostility by the gay community (Hutchins & Kaahumanu, 1991). Radical lesbians refer to bisexual women as "fence sitters," saying that they betray the lesbian cause because they can act straight when it is convenient and act lesbian when it is convenient. Some gays even argue that there is no such thing as a true bisexual.

Sexual Identity and Sexual Behavior

A consideration of the phenomenon of bisexuality will illuminate several theoretical points and will provide some insights into homosexuality and heterosexuality. First, though, several concepts need to be clarified. A distinction has already been made between sex (sexual behavior) and gender (being male or female) and between gender identity (the male's association with the male role and the female's association with the female role) and sexual orientation (heterosexual, homosexual, or bisexual). To this, the concept of **sexual identity** should be added; this refers to one's self-label or self-identification as heterosexual, homosexual, or bisexual.[3] Sexual identity, then, is the person's concept of herself or himself as a heterosexual, homosexual, or bisexual.

[3]Clearly, the term "sexual identity" is not being used here the way it is generally used, which is to refer to one's sense of maleness or femaleness (for which the term "gender identity" is used in this book). However, it seems important to be able to talk about a person's sense that he or she is straight, gay, or bisexual, and "sexual identity" seems the most useful term for that purpose.

There may be contradictions between people's sexual identity (which is subjective) and their actual choice of sexual partners viewed objectively (Lever et al., 1992). For example, a woman might identify herself as a lesbian, and yet occasionally sleep with men. Objectively, her choice of sexual partners is bisexual, but her identity is lesbian. More common are persons who think of themselves as heterosexuals but who engage in both heterosexual and homosexual sex. A good example of this is the tearoom trade—the successful, heterosexually married men in gray flannel suits who occasionally stop off at a public rest room to have another male perform fellatio on them. Once again, the behavior is objectively bisexual, in contradiction to the heterosexual identity. Another example is the group of women who claim to have bisexual identities but who have experienced only heterosexual sex. These women, often as a result of feminist beliefs, claim bisexuality as an ideal which they are capable of attaining at some later time. Once again, identity contradicts behavior.

Bisexuals tend to be stereotyped as nonmonogamous (Spalding & Peplau, 1997). Some of those who have a bisexual identity, too, believe that they have to have regular sexual relations with both women and men (Weinberg et al., 1994). If you have been monogamously heterosexually married for the past 5 years, how can you think of yourself as bisexual? Monogamy is therefore an issue for bisexuals.

Some bisexuals are heterosexually married. One study examined 26 married couples in which the husband was bisexual (Wolf, 1985; see also Matteson, 1985). The couples had, on the average, been married 13 years, and there had been open disclosure of the man's homosexuality for an average of 5.5 years. By and large, the marriages were happy. When asked to rate the quality of the marriage, 42 percent of the men and 32 percent of the women said it was outstanding. The majority of both husbands and wives said that they remained in the

Sexual identity: One's self-identity as homosexual, heterosexual, or bisexual.

Focus 15.4

Bisexual Tendencies

I am a lesbian with bisexual tendencies. I find myself admitting that I am a bisexual, and then quickly saying, "But I'm involved with a woman." My brief experiences identifying as a single, available bisexual were thrilling, but difficult. During that time I dated a lesbian who decided to stop seeing me partly because she was put off by "that bisexual thing." Similarly, I don't think that the man I saw soon after believed that I wouldn't end up leaving him for a woman. Both of these people also thought that being bisexual means being nonmonogamous. Not for me. One lover is all I can handle.

In my late teens and early twenties I was often in a group of men, one of whom was my lover, arguing about feminism. I wanted a partnership of equals and I just never felt that I could be equal with a man in the eyes of the world.

I began to identify as a lesbian in the late seventies in Ann Arbor. I wanted to be surrounded by women who had the same stake in feminism and leftist politics that I did. Coming out was like dropping backwards into a pillow. After several years and a number of relationships with women, I began to admit that I wasn't just interested in feminism or politics, but that I had—and have—strong sexual feelings for women.

Eventually I admitted that I was still attracted to men and had several flirtations. When I did begin to go out with a man again after eight years of love relationships with women, I encountered some surprises. I appreciated the level of understanding of feminist issues I now found in some men.

Then there were my friends' reactions. One said that it was fine if I wanted to see the guy I was dating, but that she didn't ever want to hang out with us together. This is all too familiar to me, having faced the pain of not having my relationships with women accepted, much less welcomed and celebrated, by my family and co-workers. Most of my lesbian friends were supportive, although some appeared concerned. Others revealed that they too had been attracted to men and seemed to take a vicarious pleasure in watching my progress.

At least two otherwise pleasant dinner parties were spoiled by my casual mention of my bisexual tendencies. I brought them up innocent of the inferno of reactions the *bi* word produces. One lesbian attorney argued quite convincingly that I had no right to call myself a lesbian if I ever have sex with men. At a safe distance, I'm not so sure. I think "lesbian with bisexual tendencies" is fairly descriptive.

Sex, after all, is something to be contained. Sex unties all of the nice packages we use to keep ourselves—and to try to keep one another—from slipping off the edge of the cosmos. Our sexual identities tell us how to live. But I need to remember that I can define my own sexual identity. I take courage from some of the bisexual activists I see around me. As a feminist, I want to see more viable alternatives for women. When I acknowledged my attraction to women *and* men, the world became exponentially larger. It's a big, big world, full of interesting and attractive people.

Source: Condensed from Lisa Yost. (1991). Bisexual tendencies. In L. Hutchins & L. Kaahumanu (Eds.), *Bi any other name: Bisexual people speak out.* Boston: Alyson.

Focus 15.5

Bisexual Characters in Film

If you were asked to name some gay or lesbian characters in film or on TV, you could probably come up with several names quickly. But what if you had to identify bisexual characters? You would find the task harder and perhaps wouldn't be able to name a single one. As in the rest of society, bisexuals have been the hidden category in the media. Here we will focus on their intriguing presence in movies.

How can we know whether a film character is bisexual? With no opportunity to interview the character, the most practical approach is to use a behavioral definition, in which the character is classified as bisexual if he or she shows evidence of engaging in sex with at least one man and one woman, or of feeling erotic attraction to both men and women. The identification of bisexual characters is still difficult, though, because often films leave the viewer wondering in frustration whether a particular couple did or will engage in sex.

As early as the era of silent films, just before World War I, bisexual characters can be found. The best example is *A Florida Enchantment*, in which a young heiress discovers some seeds from an African tree of sexual change. She swallows them and turns into a man, yet continues to dress as a woman. She makes overtures to several women, who seem to return the feeling. *The Wild Party* (1929), directed by film pioneer Dorothy Arzner, explored situational bisexuality in a girls' boarding school. The star, Clara Bow, is in love with her professor (Frederic March), yet there is an undercurrent of female-female attraction, demonstrated by some significant embraces. A number of silents displaying bisexual behavior were produced in Germany about the same time.

Scandals began to rock Hollywood in the 1920s. Olive Thomas, whom Selznick Pictures had billed as the Ideal American Girl, was found dead in a Paris hotel room, apparently having committed suicide after not being able to buy enough heroin for her husband, actor Jack Pickford (actress Mary Pickford's brother). Paramount executive William Desmond Taylor was found shot to death; police investigations unearthed a sizeable collection of pornographic photographs of him with well-known actresses. It then was revealed that he was having simultaneous affairs with three women and visited "queer meeting places." Amidst these and other scandals, the film industry decided that it had better regulate itself before censorship was imposed. The result was the Motion Picture Production Code (The Code) of 1930. Written by several Catholic clergy, it formulated a self-imposed code for the industry that would eliminate anything offensive to the Catholic Church. As a result, producers reduced the sexual content in their films considerably. Strict rules always lead some to want to subvert them, however. Marlene Dietrich, in *Morocco* (1930), performs in top hat and tails and kisses a woman, but behaves heterosexually throughout the rest of the film.

Under pressure from the Catholic Church, The Code was strengthened in 1934 to eliminate even hints of homosexuality. Only negative or comic portrayals of gays and lesbians were allowed, and then only if they were not explicit. Every movie had to pass through The Code office. Even Walt Disney's *Fantasia* was censored.

Figure 15.9   Same-gender attraction in recent movies. Sharon Stone and friend in *Basic Instinct*.

The Bride of Frankenstein (1935) somehow slipped through. Recall that Frankenstein is not the monster, but rather the scientist who created him. Frankenstein marries Elizabeth, yet spends his wedding night with his research partner, Dr. Pretorius, who is in all likelihood gay. Who, then, is Frankenstein's bride? Some film critics have argued that this movie is an allegory on the persecution of homosexuals, with the monster hunted down by townspeople. Otherwise, the ban on homosexual or bisexual characters was completely successful, and they did not return to film until World War II.

American culture changed much as a result of World War II. During the war, a large percentage of the population spent long periods of time isolated with only members of their own gender. Men in the service came to depend on other men for emotional support, and in more cases than one might think, for sexual pleasure as well. Moreover, The Code suffered a serious blow in the Supreme Court decision in *United States vs. Paramount Pictures*. The Court ruled that the major studios were in violation of antitrust laws. This decision opened up opportunities for independent filmmakers, who had not agreed to any voluntary conformity to The Code. In addition, in 1952 the Supreme Court overturned an earlier decision that films were not protected under the First Amendment. There was now some guarantee of freedom of expression for filmmakers.

Implications of lesbianism and bisexuality increasingly appeared in film. In *Screaming Mimi* (1958), Gypsy Rose Lee has an affair with a stripper. Otto Preminger filmed *Advise and Consent* in 1961 and did not remove the homosexual theme from the story. The Code continued to be revised in response to such challenges, and various restrictions gradually disappeared. In 1961 the English film *Darling* showed bisexuality directly; a sexy young Italian waiter sleeps with both Julie Christie and a gay photographer. Perhaps even more daring, the waiter is happy and well adjusted.

How are bisexuals portrayed in film today? Sometimes they are ordinary, run-of-the-mill murderers, as in *Basic Instinct,* the screenplay for which was snapped up for $3 million. Meanwhile, Paul Newman was unable to find financial backing for a love story about two men. Sometimes bisexuals are psychos. *Lilith* is a schizophrenic bisexual. Married bisexual men are wife batterers in numerous films, including *American Gigolo* (1980) and *Total Eclipse* (1995). Another theme is the bisexual female vampire, as shown, for example, in *The Velvet Vampire* (1971) and *The Hunger* (1983), featuring Catherine Deneuve and Susan Sarandon, respectively. Bisexuals in film are often promiscuous and sex-crazed, as is the character Frank in *Blue Velvet* (1986).

The general public's stereotypes about bisexuals and film portrayals of bisexuals feed each other. The daring filmmaker will create a well-adjusted, monogamous bisexual.

Source: Bryant, Wayne M. (1997). *Bisexual characters in film: From Anais to Zee.* New York: Haworth.

marriage because they valued the friendship of the spouse. Nonetheless, there were some conflicts, as there are in all marriages. Trust was one issue. One woman said, "I feel more suspicious and jealous at times since he lied in the past" (p. 142). Others, however, commented that their trust had deepened. One factor that seemed most related to positive adjustment in these marriages was communication. The couples expressing the most satisfaction with the marriage were the most likely to have very direct styles of communication and to have communicated about the homosexuality early in the marriage or from the beginning of the marriage. In the AIDS era, a different meaning may be attached to a husband's bisexuality; now it is not just an alternative sexual pattern, but also a potentially dangerous pattern that could introduce HIV infection into the marriage.

Bisexual Development

The available data on bisexual development suggest several important points (see Focus 15.4, about a bisexual woman). Bisexual men and women generally begin to think of themselves as bisexual in their early to mid-twenties (Fox, 1995; Weinberg, 1994). There are some gender differences in the sequence of behaviors, though. Bisexual women typically have their first heterosexual attractions and sexual experiences before their first homosexual ones. Bisexual men, in contrast, are more likely to have homosexual experiences first, followed by heterosexual ones. The timing and flexibility of these sequences argue for the importance of late-occurring experiences in the shaping of one's sexual behavior and identity. As we have seen in this chapter, most of the theory and research rest on the assumption that homosexuality is determined by pathological conditions in childhood, or even prenatally or by genetic factors. Yet some people have their first heterosexual and then their first homosexual experience in their twenties. It is difficult to believe that these behaviors were determined by some pathological condition at age 5. **Deprivation homosexuality,** or situational homosexuality, is also a good example of the influence of late-occurring experience. A heterosexual man may engage in homosexual behavior

while in prison but return to heterosexuality after his release. Once again, it seems likelier that such a man's homosexual behavior was determined by his circumstances (being in prison) than by some problem with his Oedipus complex 20 years before. Unlike gender identity, which seems to be fixed in the preschool years, sexual identity continues to evolve throughout one's lifetime (Riddle, 1978). This contradicts Bell, Weinberg, and Hammersmith's assertion that sexual orientation is determined before adolescence. We think that when sexual orientation is determined is still an open question. For some, it may be determined by genetic factors or experiences early in life, but for others it may be determined in adulthood.

Second, a question is raised as to whether heterosexuality is really the "natural" state. The pattern in some theories has been to try to discover those pathological conditions which cause homosexuality (for example, a father who is an inadequate role model or a homoseductive mother)—all on the basis of the assumption that heterosexuality is the natural state and that homosexuality must be explained as a deviation from it. As we have seen, this approach has failed; there appear to be multiple causes of homosexuality, just as there may be multiple causes of heterosexuality. The important alternative to consider is that bisexuality is the natural state, a point acknowledged by Freud, the learning theorists, and sociological theorists (Weinberg et al., 1994). This chapter will close, then, with some questions. Psychologically, the real question should concern not the pathological conditions that lead to homosexuality but rather the causes of exclusive homosexuality and exclusive heterosexuality. Why do we eliminate some people as potential sex partners simply on the basis of their gender? Why isn't everyone bisexual?

Deprivation homosexuality: Homosexual activity that occurs in certain situations, such as prisons, when people are deprived of their regular heterosexual activity.

SUMMARY

Sexual orientation is defined as a person's erotic and emotional attraction toward members of his or her own gender, toward members of the other gender, or both.

The majority of Americans believe that homosexuality is wrong. This belief is the basis for much antigay prejudice. In some cases this prejudice is so strong that it results in hate crimes and harassment directed at gays and lesbians.

There are lesbian, gay, and bisexual communities around the world. These communities are defined by a common culture and social life and by rituals such as pride marches.

In surveys, the majority of gay men and lesbians report being in a steady romantic relationship. In these relationships, the couples must find a balance on the dimensions of attachment, autonomy, and equality. Although concerns have been voiced about the sexual orientation and psychological well-being of children who grow up in lesbian and gay families, these concerns seem to be unfounded, according to the available studies.

The most recent well-sampled surveys (when corrected for some underreporting) indicate that about 92 percent of men and 95 percent of women are exclusively heterosexual. About 7 to 8 percent of men and 4 to 5 percent of women have at least one same-gender sexual experience in adulthood, and about 2 percent of men and 1 percent of women are exclusively homosexual. Kinsey devised a scale ranging from 0 (exclusively heterosexual) to 6 (exclusively homosexual) to measure this diversity of experience.

Well-conducted nonpatient research indicates that there are no differences in adjustment between homosexuals and heterosexuals. Masters and Johnson developed a program of therapy for sexual dissatisfaction—that is, for people who wanted to change their sexual orientation from homosexual to heterosexual. Although they reported a fairly high success rate, their data have been criticized. Most therapists believe that it is extremely difficult to change a person's sexual orientation.

In regard to the causes of sexual orientation, biological explanations include hormone imbalance, prenatal factors, brain factors, and genetic factors. The genetic explanation has some support from the data, and there is some new, tentative evidence of prenatal factors. According to the psychoanalytic view, homosexuality results from a fixation at an immature stage of development and a persisting negative Oedipus complex. Learning theorists stress that the sex drive is undifferentiated and is channeled, through experience, into heterosexuality or homosexuality. Bem's interactionist theory proposes that homosexuality results from the influence of biological factors on temperament, which in turn influences whether a child plays with boys or girls; the less familiar (exotic) gender may become associated with sexual arousal. Sociologists emphasize the importance of roles and labeling in understanding homosexuality. They also note that gender-rigid, male-dominant societies are likely to produce a higher incidence of gay men. Available data do not point to any single factor as a cause of homosexuality but rather suggest that there may be many types of homosexuality ("homosexualities") with corresponding multiple causes.

Different ethnic groups in the United States, as well as different cultures around the world, hold diverse views of same-gender sexual behaviors.

Bisexuality has been overlooked both by researchers and by the general public. A person's sexual identity may be discordant with his or her actual behavior. Bisexuality may be more "natural" than either exclusive heterosexuality or exclusive homosexuality.

REVIEW QUESTIONS

1. "LGB" stands for _____.
2. Attitudes toward homosexuality have become more liberal in recent years, so that now a majority of Americans are accepting of sexual relations between two adults of the same gender. True or false?

3. About 5 to 10 percent of LGBs have been the victims of hate crimes or harassment. True or false?

4. The process of "coming out" involves acknowledging to oneself and then to others that one is gay or lesbian. True or false?

5. There are LGB communities in large cities in the United States, but typically such communities are not found outside the United States and European nations. True or false?

6. The three major aspects of their relationships that gay and lesbian couples must negotiate are _____, _____, and _____.

7. According to the NHSLS, about _____ percent of men and _____ percent of women are exclusively homosexual.

8. According to Kinsey's scale, someone who is a "3" is exclusively heterosexual. True or false?

9. According to Reiss's sociological theory, the highest incidence of male homosexuality should occur in male-dominant cultures with rigid gender roles. True or false?

10. Studies of Mexican Americans and Chinese Americans indicate that the definition of homosexuality is a product of culture. True or false?

(The answers to all review questions are at the end of the book, after the Glossary.)

QUESTIONS FOR THOUGHT, DISCUSSION, AND DEBATE

1. Debate the following topic. Resolved: Homosexuals should not be discriminated against in employment, including such occupations as high school teaching.

2. Do you feel that you are homophobic, or do you feel that your attitude toward gays is positive? Why do you think your attitudes are the way they are? Are you satisfied with your attitudes or do you want to change them?

3. Does your college or university, in addition to prohibiting discrimination on the basis of race and sex, also prohibit discrimination on the basis of sexual orientation? Do you think it should?

4. Imagine that you are a gay man employed at a managerial level in an advertising agency in Minneapolis. You and your partner have been together for 11 years and intend to stay that way. It is becoming increasingly awkward for you to pretend that you have no partner and that you are straight when you attend parties for the staff or when people ask you how your weekend was. Should you come out to your colleagues at work? Why or why not?

SUGGESTIONS FOR FURTHER READING

Herek, Gregory M., Kimmel, Douglas C., Amaro, Hortensia, & Melton, Gary B. (1991). Avoiding heterosexist bias in psychological research. *American Psychologist, 46,* 957–963. This stimulating article on research methodology points out ways in which heterosexist bias can enter research and suggests ways to avoid this bias.

Hutchins, Loraine, & Kaahumanu, Lani (Eds.) (1991). *Bi any other name: Bisexual people speak out.* Boston: Alyson Publications. This is a wonderful collection of first-person essays by bisexual people who represent great diversity in age, race, class, gender, ablebodiedness, and marital status.

LeVay, Simon (1996). *Queer science: The use and abuse of research into homosexuality.* Cambridge, MA: MIT Press. Neuroscientist Simon LeVay, whose research was discussed in this chapter, has put together a comprehensive book on scientists' research on homosexuality, focusing particularly on the causes of sexual orientation.

Miller, Neil. (1992). *Out in the world: Gay and lesbian life from Buenos Aires to Bangkok.* New York: Random House. The author traveled around the world, observing gay and lesbian communities. The accounts are facinating.

WEB RESOURCES

http://www.sexhealth.org/infocenter/LBGIssue/LBGIssue.htm
   The Sexual Health InfoCenter: LGBT Issues Page.

http://www.mtholyoke.edu/~maclayto/rainbow.html
   The Rainbow page: gay, lesbian, bisexual.

http://www.biresource.org/
   Bisexual Resource Center: access to the Center's publications and other resources.

http://comp9.psych.cornell.edu/dbem/ebe_theory.html
   Exotic Becomes Erotic (EBE) theory, Daryl Bem's original paper.

chapter

16

VARIATIONS IN SEXUAL BEHAVIOR

CHAPTER HIGHLIGHTS

S ome men love women, some love other men, some love dogs and horses, and occasionally you find one who loves his rain-coat.*

Most laypeople, as well as most scientists, have a tendency to classify behavior as normal or abnormal. There seems to be a particular tendency to do this with regard to sexual behavior. Many terms are used for abnormal sexual behavior, including "sexual deviance," "perversion," "sexual variance," and paraphilias. The term "sexual variations" will be used in this chapter because it is currently favored in scientific circles and because defining exactly what is "deviant" or what is a "perversion" is rather difficult.

In Chapter 15 we argued that homosexuality per se is not an abnormal form of sexual behavior. This chapter will deal with some behaviors that more people might consider to be abnormal, so it seems advisable at this point to consider exactly when a sexual behavior is abnormal. That is, what is a reasonable set of criteria for deciding what kinds of sexual behavior are abnormal?

When Is Sexual Behavior Abnormal?

As we saw in Chapter 1, sexual behavior varies a great deal from one culture to the next. There is a corresponding variation across cultures in what is considered to be "abnormal" sexual behavior. Given this great variability, how can one come up with a reasonable set of criteria for what is abnormal? Perhaps it is best to begin by considering the way others have defined "abnormal" sexual behavior.

One approach is to use a *statistical definition.* According to this approach, an abnormal sexual behavior is one that is rare, or not practiced by many people. Following this definition, then, standing on one's hands while having intercourse would be considered abnormal because it is rarely done, although it does not seem very abnormal in other ways. This definition, unfortunately, does not give us much insight into the psychological or social functioning of the person who engages in the behavior.

In the *sociological approach,* the problem of culture dependence is explicitly acknowledged. A sociologist might define a deviant sexual behavior as a sexual behavior that violates the norms of society. Thus if a society says that a particular sexual behavior is deviant, it is—at least in that society. This approach recognizes the importance of the individual's interaction with society and of the problems that people must face if their behavior is labeled "deviant" in the culture in which they live.

A *psychological approach* was stated by Arnold Buss in his text entitled *Psychopathology* (1966). He says, "The three criteria of abnormality are discomfort, inefficiency, and bizarreness." The last of these criteria, bizarreness, has the problem of being culturally defined; what seems bizarre in one culture may seem normal in another. However, the first two criteria are good in that they focus on the discomfort and unhappiness felt by the person with a truly abnormal pattern of sexual behavior and also on inefficiency. For example, a male clerk in a Minneapolis supermarket was having intercourse with willing shoppers in their cars several times a day. This apparently compulsive behavior led to his being fired. This is an example of inefficient functioning, and the behavior can reasonably be considered abnormal.

The *medical approach* is exemplified by the definitions included in the *Diagnostic and Statistical Manual of Mental Disorders (DSM-IV)* (American Psychiatric Association, 1994). It recognizes eight paraphilias: fetishism, transvestism, masochism, sadism, voyeurism, exhibitionism, pedophilia, and other conditions. **Paraphilia** is defined as:

*Max Schulman. *I was a teen-age dwarf.*

Paraphilia (par-uh-FILL-ee-uh): Recurring, unconventional sexual behavior that is obsessive and compulsive.

recurrent, intense sexually arousing fantasies, sexual urges, or behaviors involving 1) nonhuman objects, 2) the suffering or humiliation of oneself or one's partner, or 3) children or other nonconsenting persons . . . The behavior, sexual urges or fantasies cause clinically significant distress in social, occupational, or other important areas of functioning. (*DSM-IV*, pp. 522–523)

Persons with one (or more) of these conditions are considered eligible for or in need of treatment.

In this chapter, we will discuss six of the eight paraphilias: fetishism, transvestism, masochism, sadism, voyeurism, and exhibitionism. We will also consider several other atypical behaviors, including hypersexuality and asphyxiophilia. The most serious paraphilia, pedophilia or sexual abuse of children, is discussed in Chapter 17.

Fetishism

One sexual variation is **fetishism.** In fetishism, a person becomes sexually fixated on some object other than another human being and attaches great erotic significance to that object. In extreme cases the person is incapable of becoming aroused and having an orgasm unless the fetish object is present. Typically, the fetish item is something closely associated with the body, such as clothing. Inanimate-object fetishes can be roughly divided into two subcategories: media fetishes and form fetishes.

Media Fetishes and Form Fetishes

In a **media fetish,** the material out of which an object is made is the source of arousal. An example would be a leather fetish, in which any leather item is arousing to the person. Media fetishes can be subdivided into hard media fetishes and soft media fetishes. In a hard media fetish, the fetish is for a hard substance, such as leather or rubber. Hard media fetishes may often be associated with sadomasochism (discussed later in the chapter). In a soft media fetish, the substance is soft, such as fur or silk.

In a **form fetish,** it is the object and its shape that are important. An example would be a shoe fetish, in which shoes are highly arousing (see Focus 16.1). Some shoe fetishes require that the shoes be high-heeled; this fetish may be associated with sadomasochism, in which the fetishist derives sexual satisfaction from being walked on by a woman in high heels. Other shoe fetishes require that the shoes be leather boots that have been worn. Other examples of form fetishes are fetishes for leather clothing (which also may be associated with sadomasochism), garters, and lingerie.

The Normal-Abnormal Continuum

Perhaps reading that being aroused by lingerie is deviant has made you rather uncomfortable because you yourself find lingerie arousing. Fetishes are discussed first in this chapter because they provide an excellent example of the continuum from normal to abnormal sexual behavior. That is, normal sexual behavior and abnormal sexual behavior (like other normal and abnormal behaviors) are not two separate categories but rather gradations on a continuum. Many people have mild fetishes—they find things such as silk underwear arousing—and that is well within the range of normal behavior; only when the fetish becomes extreme is it abnormal. Indeed, in one sample of college men, 42 percent reported that they had engaged in voyeurism and 35 percent had engaged in frottage (sexual rubbing against a woman in a crowd) (Templeman & Stinnett, 1991). Unfortunately, the researchers did not ask about fetishes. But the point is that many of these behaviors are common even in normal populations.

This continuum from normal to abnormal behavior might be conceptualized using the scheme

Fetishism: A person's sexual fixation on some object other than another human being and attachment of great erotic significance to that object.
Media fetish: A fetish whose object is anything made of a particular substance, such as leather.
Form fetish: A fetish whose object is a particular shape, such as high-heeled shoes.

Figure 16.1 A common fetish is for leather, often in association with sexual sadism and masochism. This store caters to clientele interested or involved in those activities.

shown in Figure 16.2. A mild preference, or even a strong preference, for the fetish object (say, silk panties) is within the normal range of sexual behavior. When the silk panties become a necessity—when the man cannot become aroused and have intercourse unless they are present—we have crossed the boundary into abnormal behavior. When the man becomes obsessed with white silk panties and shoplifts them at every opportunity, so that he will always have them available, the fetish has become a paraphilia. In extreme forms, the silk panties may become a substitute for a human partner, and the man's sexual behavior consists of masturbating with the silk panties present. In these extreme forms, the man may commit burglary or even assault to get the desired fetish object, and this would certainly fit our definition of abnormal sexual behavior.

The continuum from normal to abnormal behavior holds for many of the sexual variations discussed in this chapter, such as voyeurism, exhibitionism, and sadism.

Why Do People Become Fetishists?

Psychologists are not completely sure what causes fetishes to develop. Here we will consider three theoretical explanations: learning theory, cognitive theory, and the sexual addiction model. These theories can be applied equally well to explaining many of the other sexual variations in this chapter.

According to learning theory (for example, McGuire et al., 1965), fetishes result from classical conditioning, in which a learned association is built between the fetish object and sexual arousal and orgasm. In some cases a single learning trial

Focus 16.1

A Case History of a Shoe Fetishist

The following case history is taken directly from the 1886 book *Psychopathia Sexualis*, by Richard von Krafft-Ebing, the great early investigator of sexual deviance. It should give you the flavor of his work.

Case 114. X., aged twenty-four, from a badly tainted family (mother's brother and grandfather insane, one sister epileptic, another sister subject to migraine, parents of excitable temperament). During dentition [teething] he had convulsions. At the age of seven he was taught to masturbate by a servant girl. X. first experienced pleasure in these manipulations when the girl happened to touch his member [penis] with her shoe-clad foot. Thus, in the predisposed boy, an association was established, as a result of which, from that time on, merely the sight of a woman's shoe, and finally, merely the idea of them, sufficed to induce sexual excitement and erection. He now masturbated while looking at women's shoes or while calling them up in imagination. The shoes of the school mistress excited him intensely, and in general he was affected by shoes that were partly concealed by female garments. One day he could not keep from grasping the teacher's shoes—an act that caused him great sexual excitement. In spite of punishment he could not keep from performing this act repeatedly. Finally, it was recognized that there must be an abnormal motive in play, and he was sent to a male teacher. He then revelled in the memory of the shoe scenes with his former school mistress and thus had erections, orgasms, and, after his fourteenth year, ejaculation. At the same time, he masturbated while thinking of a woman's shoes. One day the thought came to him to increase his pleasure by using such a shoe for masturbation. Thereafter he frequently took shoes secretly and used them for that purpose.

Nothing else in a woman could excite him; the thought of coitus filled him with horror. Men did not interest him in any way. At the age of eighteen he opened a shop and, among other things, dealt in ladies' shoes. He was excited sexually by fitting shoes for his female patrons or by manipulating shoes that came for mending. One day while doing this he had an epileptic attack, and, soon after, another while practicing onanism in his customary way. Then he recognized for the first time the injury to health caused by his sexual practices. He tried to overcome his onanism, sold no more shoes, and strove to free himself from the abnormal association between women's shoes and the sexual function. Then frequent pollutions, with erotic dreams about shoes occurred, and the epileptic attacks continued. Though devoid of the slightest feeling for the female sex, he determined on marriage, which seemed to him to be the only remedy.

He married a pretty young lady. In spite of lively erections when he thought of his wife's shoes, in attempts at cohabitation he was absolutely impotent because his distaste for coitus and for close intercourse in general was far more powerful than the influence of the shoe-idea, which induced sexual excitement. On account of his impotence the patient applied to Dr. Hammond, who treated his epilepsy with bromides and advised him to hang a shoe up over his bed and look at it fixedly during coitus, at the same time imagining his wife to be a shoe. The patient became free from epileptic attacks and potent so that he could have coitus about once a week. His sexual excitation by women's shoes also grew less and less.

Source: Richard von Krafft-Ebing. (1886). *Psychopathia sexualis*, p. 288. (Reprinted by Putnam, New York, 1965.)

Figure 16.2 The continuum from normal to abnormal behavior in the case of fetishes.

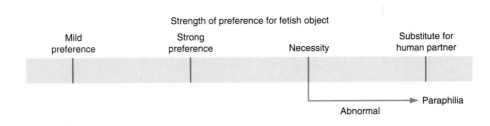

Strength of preference for fetish object

| Mild preference | Strong preference | Necessity | Substitute for human partner |

Paraphilia

Abnormal

might serve to cement the association. For example, one adult male recalled:

> I was home alone and saw my uncle's new penny loafers. I went over and started smelling the fresh new leather scent and kissing and licking them. It turned me on so much that I actually ejaculated my first load into my pants and have been turned on ever since. (Weinberg, Williams, & Calhan, 1995, p. 22)

In this case, shoes were associated with sexual arousal as the result of an early learning experience. Another example appears to be the shoe fetishist described in Focus 16.1. There was even an experiment that demonstrated that males could, in the laboratory, be conditioned to become sexually aroused when viewing pictures of shoes (Rachman, 1966).

A second possible theoretical explanation comes from cognitive psychology, discussed in Chapter 2 (Walen & Roth, 1987). According to cognitive theorists, fetishists (or other paraphiliacs) have a serious cognitive distortion in that they perceive a nonconventional stimulus—such as black leather boots—as erotic. Further, their perception of arousal (link 4 in the model; see Figure 9.4) is distorted. They feel "driven" to the sexual behavior when aroused, but the arousal may actually be caused by feelings of guilt and self-loathing. Thus there is a chain in which there are initial feelings of guilt at thoughts of the unconventional behavior, which produces arousal, which is misinterpreted as sexual arousal, which leads to a feeling that the fetish ritual must be carried out; it is, there is orgasm and temporary feelings of relief, but the evaluation of the event is negative, leading to further feelings of guilt and self-loathing, which perpetuates the chain.

A third theory that has been advanced to explain some paraphilias, especially those that seem com-

pulsive, is the theory of sexual addiction, discussed in Focus 16.2.

Whatever the cause, fetishism typically develops early in life. In one sample of foot/shoe fetishists, the mean age at which respondents reported first being sexually aroused by feet/shoes was 12 years (Weinberg, Williams, & Calhan, 1995).

Transvestism

Transvestism ("trans" = "cross"; "vest" = "dressing") refers to dressing as a member of the other gender. Cross-dressing may be done by a variety of people for a variety of reasons. As has been noted, male transsexuals may go through a stage of cross-dressing in the process of becoming women. Some male homosexuals—**drag queens**—dress up as women, and some lesbians dress in masculine clothes ("butch"); these practices, though, are basically caricatures of masculinity and femininity. **Female impersonators** are men who dress as women, often as part of their jobs as entertainers. For example, Jack Lemmon and Tony Curtis were female impersonators in the classic film with Marilyn Monroe *Some Like It Hot*. Robin Williams as *Mrs. Doubtfire* and Dustin Hoffman as *Tootsie* won praise from critics and big box office profits for their impersonations of women. Finally, some—perhaps many—adolescent

Transvestism: The practice of deriving sexual gratification from dressing as a member of the other gender.
Drag queen: A male homosexual who dresses in women's clothing.
Female impersonator: A man who dresses up as a woman as part of a job in entertainment.

Focus 16.2

Sexual Addictions?

Patrick Carnes, in his book *The Sexual Addiction,* has advanced the theory that some cases of abnormal sexual behavior are actually a result of an addictive process much like alcoholism. One definition of alcoholism or other drug dependency is that the person has a pathological relationship with the mood-altering substance. In the case of the sexual addiction, the person has a pathological relationship to a sexual event or process, substituting it for a healthy relationship with others.

One common characteristic of alcoholics and sexual addicts is that they have a faulty belief system in which there is denial and distortion of reality. For example, the sex addict may deny the possibility of sexually transmitted disease. Sex addicts also engage in self-justification, such as "If I don't have it every few days, the pressure builds up." Like alcoholism, the addiction leads to many self-destructive behaviors. And, like alcoholism, the chief distinguishing feature of the sexual addiction is that the person has lost control of the behavior. Consider these examples:

- George has to take out a secret loan for $2700 to cover his frequent payments to prostitutes.
- Jeffrey, a respected lawyer, is arrested for the third time for flashing.
- In an example from film, Theresa (played by Diane Keaton) in *Looking for Mr. Goodbar* is murdered by one of her sexual partners as a consequence of her compulsive search for anonymous sex with bar pickups.

According to Carnes's analysis, each episode of the sexually compulsive behavior proceeds through a four-step cycle, and the cycle intensifies each time it is repeated.

1. *Preoccupation*   The person can think of nothing other than the sexual act to which he or she is addicted.
2. *Rituals*   The person enacts certain rituals that have become a prelude to the addictive act.
3. *Compulsive sexual behavior*   The sexual behavior is enacted and the person feels that he or she has no control over it.
4. *Despair*   Rather than feeling good after the sexual act is completed, the addict falls into a feeling of hopelessness and despair.

According to Carnes, it is important to understand that not all sexual variations are addictions, and therefore the addiction model will not explain all paraphilias. There are also addictions to some behaviors, such as masturbation, that in and of themselves are perfectly normal. Thus, for example, the man who masturbates while looking at pornographic magazines two times per week is probably not an addict, and the behavior is well within the normal range. However, the man who buys 20 porn magazines a week, masturbates four or five times a day while looking at them, for a total of perhaps two or three hours, and can think of nothing else but where he can buy the next porn magazine and find the next private place to masturbate—that person is addicted. The key is the compulsiveness, the lack of control, the obsession (constant thoughts of the sexual scenario), and the obliviousness to danger or harmful consequences.

Interestingly, there is often a history of alcoholism in the families of sex addicts. Sometimes, too, there is a history of the same sexual addiction, such as the father who sexually abuses his son, who in turn abuses his son, so that the cycle perpetuates itself. Sex addicts often report that as children they felt a sense of being abandoned and not loved.

Just as our culture celebrates alcohol yet despises the alcoholic, so the culture celebrates sex yet abhors the sex addict. The media glorify alcohol. "You only go around once, so do it with Michelob." Similarly, advertising is filled with sexual messages (see Chapter 18). These sexual sells are difficult reminders for the sex addict who is trying to recover.

According to Carnes, the most effective therapy for the sex addict is the Alcoholics Anonymous program applied to sexual addictions. Several groups have adapted the AA program to sexual addiction, among them Sexaholics Anonymous (SA), Sex Addicts Anonymous (SAA), and Sex and Love Addicts Anonymous (SLAA). These groups can usually be found by calling the local phone number for Alcoholics Anonymous. The first step in the process of recovery is admitting that one is sexually addicted, that the behavior is out of control, and that one's life has become unmanageable. These are hard admissions to make for someone who has spent years denying the existence of a problem. There are frequent meetings with a support group, and there is a strong emphasis on building feelings of self worth.

There has generally been much praise for Carnes's analysis of sexual addictions. Criticism has come, however, from some gay activists, who feel that the approach is homophobic in its assertion that gay men are addicted to sex. However, it is important to recognize that there are many possible addictions, many of them heterosexual (such as men addicted to anonymous sex with prostitutes and women addicted to anonymous sex with bar pickups). Surely there are some gay men who are addicted to behaviors such as having impersonal tearoom sex, but the majority of gay men are not sex addicts. The key feature, again, is that the behavior must be out of control and compulsive.

Other criticisms have come from therapists and researchers in the field. The term "addiction," such as addiction to alcohol or heroin, has a very specific definition among professionals, and sexual addictions do not meet the definition in some ways. For example, if one is addicted to alcohol and suddenly stops using it, there is a withdrawal phenomenon that involves striking physical symptoms. If a person abstains from an addictive sexual behavior, there are no physiological withdrawal symptoms. A second criticism is that "addiction" may become an excuse for illegal, destructive behavior. For example, a rapist might say "I'm sexually addicted to raping and therefore can't stop myself."

There has also been criticism of the application of the AA model to sexual behavior. Applied to alcohol or drugs, the AA model demands that the addict abstain from contact with the substance. This "abstinence model" cannot be applied to sexuality, because sexual expression is a basic human need.

In order to resolve this debate, some experts have recommended that we use the term *compulsive sexual behavior* instead of "sexual addiction" (e.g., Coleman, 1991). A behavior is compulsive if the person is obsessed with it, feels compelled to engage in it, and has experienced personal, relationship, medical, or legal difficulties as a result of it.

The criticisms notwithstanding, Carnes' work has opened up a new way to think about out-of-control sexual behaviors and how to treat them.

Source: Patrick Carnes. (1983). *The sexual addiction.* Minneapolis, MN: CompCare Publications.

boys cross-dress, usually only once or a few times (Green, 1975). This behavior does not necessarily mean a life of transvestism; it may simply reflect the sexual drives, confusions, and frustrations of adolescence.

In contrast to people who engage in cross-dressing for the reasons discussed above, the transvestite[1] is a person who derives sexual gratification from cross-dressing. Thus transvestism is probably basically a fetish (Pomeroy, 1975) and seems to be quite similar to a clothing fetish. The cross-dressing may often be done in private, perhaps by a married man without his wife's knowledge.

As a result of such secrecy, no one has accurate data on the incidence of transvestism. One authority, though, estimates that more than a million males in the United States are involved in it, if all instances of men getting at least a temporary erotic reward from wearing female clothing are counted (Pomeroy, 1975).

Transvestism is almost exclusively a male sexual variation; it is essentially unknown among women. There may be a number of reasons for this difference, including our culture's tolerance of women who wear masculine clothing and intolerance of men who wear feminine clothing. The phenomenon illustrates a more general point, namely, that many sexual variations are defined for, or practiced almost exclusively by, members of one gender; the parallel practice by members of the other gender is often not considered deviant. Most sexual variations are practiced mainly by males.[2]

A common means of studying persons with atypical sexual behavior patterns is to place ads in specialty newsletters and magazines, and to solicit participants at meetings and conventions attended by such persons. Using these procedures, researchers gathered survey data from 1,032 cross-dressers (Docter & Prince, 1997). The sample did not include drag queens or female impersonators. The vast majority of the men (87 percent) were heterosexual, 60 percent were married, 65 percent had a college education, and 76 percent reported being raised by both parents through age 18. Sixty-six percent reported that their first cross-dressing experience occurred before age 10. Sexual excitement and orgasm were reported by 40 percent as often or almost always associated with cross-dressing. Almost all (93 percent) preferred complete cross-dressing, but only 14 percent frequently went out in public dressed as a woman.

Another survey was conducted by mailing questionnaires to 1,200 members of a national cross-dressing organization (Bullough & Bullough, 1997). There were 372 questionnaires returned. The median age at which the men began to cross-dress was 8.5; 32 percent reported that they first dressed as a female before they were 6. Most of them reported cross-dressing as children, and 56 percent said they were never caught.

Research indicates that there are four basic motivations for men to engage in transvestite activities (Talamini, 1982):

1. *Sexual arousal*   As we noted earlier, transvestism appears to result from a conditioned association between sexual arousal and women's clothing. As one male transvestite commented,

   My older sister had beautiful clothes. When I was thirteen I wanted to see what I looked like when I dressed as a woman. We're about the same height. I put some of her lingerie on and I never had such powerful sexual feelings. It's twenty years later and I still dress up. (Talamini, 1982, p. 20)

2. *Relaxation*   Transvestites report that they periodically need a break from the confining, pressured male role. Dressing up as women allows them to express emotionality and grace, traits that are taboo for men in our society.

3. *Role playing*   Just as many actors derive great satisfaction from playing roles in the theater, so transvestites get a great sense of achievement from being able to pass as women when in public.

---

[1]A common abbreviation for transvestism, used both by scientists and by members of the transvestite subculture, is TV. Therefore, if you see an ad in an underground newspaper placed by a person who is a TV, this is *not* someone who has delusions of broadcasting the six-o'clock news but rather a transvestite.

[2]Several theories have been proposed to explain why there are so many more male than female paraphiliacs (Finkelhor & Russell, 1984). Sociobiologists believe that the difference lies in the evolutionary selection of males to inseminate many partners and to be aroused by sexual stimuli devoid of emotional content (Wilson, 1987). Alternatively, sociologists point to gender role socialization, which teaches males to be instrumental and to initiate sexual interactions. Females, on the other hand, are taught to be nurturing and to empathize with others, for example, with the vulnerability of children (Traven et al., 1990). Psychoanalytic theory (see Chapter 2) suggests that paraphilias result from *castration anxiety,* which is relieved by a forceful sexual act; since women do not fear castration, they are not subject to paraphilias.

(a)

(b)

Figure 16.3   Two examples of cross-dressing: *(a)* The lead actor in several films has portrayed a cross-dresser. (e.g. the three leads from To Wang Foo, With Love, Julie Newmar. *(b)* Men in "drag".

4. *Adornment*   Men's clothes are relatively drab, whereas women's are more colorful and beautiful. Transvestites enjoy this sense of being beautiful.

How do the wife and children of the transvestite react to his unusual behavior? In one sample of 50 heterosexual transvestites, 60 percent of the wives were accepting of their husband's cross-dressing (Talamini, 1982). Most of these women commented that otherwise he was a good husband. Some of the wives felt fulfilled being supportive of the husband, and some even helped him in dressing and applying makeup. In the same sample, 13 of the couples had told their children about the cross-dressing. They claimed that the relationship with the children was undamaged and that the children were tolerant and understanding.

Transvestism is one of the harmless, victimless sexual variations, particularly when it is done in private. Like other forms of fetishism, it is a problem only when it becomes so extreme that it is the person's only source of erotic gratification, or when it becomes a compulsion which the person cannot control.

"sadism" derives from the name of the historical character the Marquis de Sade, who lived around the time of the French Revolution. Not only did he practice sadism—several women apparently died from his attentions (Bullough, 1976)—but he also wrote novels about these practices (the best known is *Justine*), thus assuring his place in history.

A **masochist** is a person who is sexually aroused by experiencing pain. This variation is named after Leopold von Sacher-Masoch (1836–1895), who was himself a masochist and who wrote novels expressing masochistic fantasies. Notice that the definitions of these variations make specific their *sexual* nature; the terms are often loosely used to refer to people who are cruel or to people who seem to bring misfortune on themselves, but these are not the meanings used here. Sadism and masochism are considered paraphilias in *DSM-IV* (American Psychiatric Association, 1994). These two are often referred to as a pair since the two behaviors or roles (giving and receiving pain) are complementary.

There are two other styles of interaction that are related to but distinct from sadism-masochism

Sadism and Masochism

Definitions

A **sadist** is a person who derives sexual satisfaction from inflicting pain on another person. The term

Sadist: A person who derives sexual satisfaction from inflicting pain on another person.
Masochist: A person who derives sexual satisfaction from experiencing pain.

(S-M). These are bondage and discipline (B-D) and dominance and submission (D-S) (Ernulf & Innala, 1995). **Bondage and discipline** refers to the use of physically restraining devices or psychologically restraining commands as a central aspect of sexual interactions. These devices or commands may enforce obedience and servitude without inducing any physical pain. **Dominance and submission** refers to interaction that involves a consensual exchange of power; the dominant partner uses his/her power to control and sexually stimulate the submissive partner. Both B-D and D-S encompass a variety of specific interactions that range from atypical to paraphilic.

Sadomasochistic Behavior

Sadomasochism is often accompanied by elaborate rituals and gadgetry, such as tight black leather clothing, pins and needles, ropes, whips, and hot wax.

Interestingly, sadists and masochists do not consistently find experiencing pain and giving pain to be sexually satisfying. For example, the masochist who smashes a finger in a car door will yell and be unhappy just like anyone else. Pain is arousing for such people only when it is part of a carefully scripted ritual. As one woman put it,

> Of course, he doesn't *really* hurt me. I mean quite recently he tied me down ready to receive "punishment," and then by mistake he kicked my heel with his toe as he walked by. I gave a yelp, and he said, "Sorry love—did I hurt you?" (Gosselin & Wilson, 1980, p. 55)

Sadomasochism (S-M) is a rare form of sexual behavior, although in its milder forms it is probably more common than many people think. Kinsey found that 26 percent of females and the same percentage of males had experienced definite or frequent erotic responses as a result of being bitten during sexual activity (Kinsey et al., 1953, pp. 677–678). Sadistic or masochistic fantasies appear to be considerably more common than real-life sadomasochistic behavior.

In one study, 178 S-M men who responded to an ad in an S-M magazine or belonged to an S-M support group filled out a questionnaire (Moser & Levitt, 1987). The majority were heterosexual, well-educated, and interested in both dominant and submissive roles (switchable). The following are behaviors that the majority of them had both tried and enjoyed: humiliation, bondage, spanking, whipping, fetish behavior, tying up with ropes, and master and slave role playing.

Another study administered questionnaires to 130 males and 52 females who responded to ads placed in S-M magazines; the study focused particularly on the women respondents, who are often not recognized as part of the S-M subculture (Breslow et al., 1985). Thirty-three percent of the males and 28 percent of the females preferred the dominant role; 41 percent of the males and 40 percent of the females preferred the submissive role; and 26 percent of males and 32 percent of females were versatile. The majority of these S-M respondents were heterosexual. Men involved in S-M frequently report having been interested in such activity since childhood; women are more likely to report having been introduced to the subculture by someone else (Weinberg, 1994). Women prefer bondage, spanking, oral sex, and master-slave role playing (Levitt, Moser, & Jamison, 1994).

Causes of Sadomasochism

The causes of sadism and masochism are not precisely known. The theories discussed in the section on fetishes can be applied here as well. For example, learning theory points to conditioning as an explanation. A little boy is being spanked over his mother's knee; in the process, his penis rubs against her knee, and he gets an erection. Or a little girl is caught masturbating and is spanked. In both cases, the child has learned to associate pain or spanking with sexual arousal, possibly setting up a lifelong career as a masochist. On the other hand, in one sample of sadomasochists, over 80 percent did *not* recall receiving erotic enjoyment from being punished as a child (Moser, 1979, cited in Weinberg, 1987). Thus forces besides conditioning must be at work.

Bondage and discipline: The use of physical or psychological restraint to enforce servitude, from which both participants derive sensual pleasure.
Dominance and submission: The use of power consensually given to control the sexual stimulation and behavior of the other person.

Another psychological theory has been proposed to explain masochism specifically, although not sadism (Baumeister, 1988a, 1988b). According to the theory, the masochist is motivated by a desire to escape from self-awareness. That is, the masochistic behavior helps the individual escape from being conscious of the self in the same way that drunkenness and some forms of meditation do. In an era dominated by individualism and self-interest, why would anyone want to escape from the self? Probably because high levels of self-awareness can lead to anxiety as a result of a focus on pressures on the self, responsibilities, the need to keep up a good image in front of others, and so on. Masochistic activity allows the person to escape from being an autonomous, separate individual. Masochism may be an unusually powerful form of escape because of its link to sexual pleasure. This theory can also explain why patterns of masochism seem to be so gender-linked (Baumeister, 1988b). According to the theory, the male role is especially burdensome because of the heavy pressures for autonomy, separateness, and individual achievement. Masochism accomplishes an escape from these aspects of the male role, explaining why masochism is more common among males than among females.

Bondage and Discipine

Sexual bondage, the use in sexual behavior of restraining devices which have sexual significance, has been a staple of erotic fiction and art for centuries. Current mainstream and adult films and videos portray this activity. In some communities, individuals interested in B-D have formed clubs.

We noted earlier the difficulty of gathering data on participation in variant forms of sexual expression. One innovative study downloaded all messages about bondage mailed to an international computer discussion group (Ernulf & Innala, 1995). Of the messages in which senders indicated their gender, 75 percent were male. Of those indicating a sexual orientation, most were heterosexual; 18 percent said they were gay, and 11 percent said they were lesbian. The messages were coded for discussion of what the person found sexually arousing about B-D. Most frequently mentioned (12%) was play: "sex is funny, and sex is lovely, and sex is PLAY." Next was the exchange of power (4%): "It is a power trip because the active is

responsible for the submissive's pleasure." The next most common themes were intensified sexual pleasure, tactile stimulation associated with the use of ropes and cuffs, and the visual enjoyment experienced by the dominant person.

There is a marked imbalance in preferences for the active ("top") and passive ("bottom") roles. Most men and women, regardless of sexual orientation prefer to be "bottom." This may be the reason why there are an estimated 2,500 professional dominatrices in the United States.

Dominance and Submission

Sociologists emphasize that the key to S-M is not pain, but rather dominance and submission (D-S) (Weinberg, 1987). Thus it is not an individual phenomenon, but rather a social phenomenon embedded in a subculture and controlled by elaborate scripts.

Sociologists feel that to understand D-S, one must understand the social processes that create and sustain it (Weinberg, 1987). There is a distinct D-S subculture, involving magazines (such as *Corporal*), clubs, and bars. It creates culturally defined meanings for D-S acts. Thus a D-S act is not a wild outbreak of violence, but rather a carefully controlled performance with a script (recall the concept of scripts, Chapter 2). One woman reported:

> we got into dominance and submission. Like him giving me orders. Being very rough and pushing me around and giving me orders, calling me a slut, calling me a cunt. Making me crawl around . . . on all fours and beg to suck his cock. Dominance-submission is more important than the pain. I've done lots and lots of scenes that involve no pain. Just a lot of taking orders, being humiliated. (Maurer, 1994, pp. 253, 257)

Within the play, people take on roles such as master, slave, or naughty child. Thus American men can play the submissive role in D-S culture, even though it contradicts the American male role, because it is really not they who are the naughty child, just as an actor can play the part of a murderer and know that he is not a murderer.

One interesting phenomenon, from a sociological point of view, is the social control over risk taking in the D-S subculture (Weinberg, 1987). That is, having allowed oneself to be tied up or restrained and then whipped, one could be seriously injured or even murdered, yet such outcomes are rare. Why? Research shows that complex social arrangements are made in order to reduce the risk (Lee, 1979). First, initial contacts are usually made in protected territories such as bars or meetings, which are inhabited by other D-Sers who play by the same rules. Second, the basic scripts are widely shared, so that everyone understands what will and will not occur. When the participants are strangers, the scenario may be negotiated before it is enacted. Third, as the activity unfolds, very subtle nonverbal signals are used to control the interaction (Weinberg, 1994). By using these signals, the person playing the submissive role can influence what occurs. Thus, as two people enact the master and slave script, the master is not in complete control and the slave is not powerless. So it is the *illusion* of control, not actual control, that is central to D-S activity for both the master and the slave.

Voyeurism

There are two types of **voyeur** ("Peeping Tom").[3] In **scoptophilia,** sexual pleasure is derived from observing sexual acts and the genitals; in *voyeurism,* technically, the sexual pleasure comes from viewing nudes, often while the voyeur is masturbating.

Voyeurism appears to be much more common among males than among females. According to FBI reports, nine men to one woman are arrested on charges of "peeping."

Voyeurism provides another good illustration of the continuum from normal to abnormal behavior. For example, many men find it arousing to watch a nude woman—otherwise there would be no stripper bars—and this is certainly well within the normal range of behavior. Some women are "crotch watchers," much as men are breast watchers (Friday, 1973, 1975). The voyeurism is abnormal when it replaces sexual intercourse or when the person commits a crime such as breaking and entering to observe others, thereby risking arrest.

Figure 16.4   Bondage.

Peepers typically want the woman they view to be a stranger and do not want her to know what they are doing (Yalom, 1960). The element of risk is also important; while one might think that a nudist camp would be heaven to a peeper, it is not, because the elements of risk and forbiddenness are missing (Sagarin, 1973).

Voyeurs may or may not be dangerous (Tutuer, 1984). Potentially dangerous voyeurs can be identified by the following characteristics: (1) They enter the confines of a building or other structure in order to view their subject, and (2) they draw the attention of their subject to the fact that they are watching (Yalom, 1960).

A study of 561 males who sought treatment for paraphilias included 62 voyeurs (Abel & Rouleau, 1990). One third reported that their first experience occurred before they were 12 years old. One half said they recognized their interest in peeping prior to age 15. These men estimated that, on the average, they had peeped at 470 persons.

In one study of arrested peepers, it was found that they were likely to be the youngest child in their family and to have good relationships with their parents but poor relationships with their peers (Gebhard et al., 1965). They had few sisters

---

[3]"Voyeur" comes from the French word *voir,* meaning "to see." "Peeping Tom" comes from the story of Lady Godiva; when she rode through town nude to protest the fact that her husband was raising his tenants' taxes, none of the townspeople looked except one, Tom of Coventry.

Voyeur: A person who becomes sexually aroused from secretly viewing nudes.
Scoptophilia: A sexual variation in which the person becomes sexually aroused by observing others' sexual acts and genitals.

and few female friends. Few were married. These studies, however, point out one of the major problems with the research on sexual variations: Much of it has been done only on people who have been arrested for their behavior or sought treatment. The "respectable paraphiliac" who has the behavior under somewhat better control or who is skilled enough or can pull enough strings not to get caught is not studied in such research. Thus the picture that research provides for us of these variations may be very biased.

Exhibitionism

The complement to voyeurism is **exhibitionism** ("flashing"), in which the person derives sexual pleasure from exposing his genitals to others in situations where this is clearly inappropriate.[4] The pronoun "his" is used advisedly, since exhibitionism is defined exclusively for men. The woman who wears a dress that reveals most of her bosom is likely to be thought of as attractive rather than abnormal. When the male exposes himself, however, his behavior is considered offensive. Here again, whether a sexual behavior is considered abnormal depends greatly on whether the person doing it is a male or a female. Homosexual exhibitionism is also quite rare; so the prototype we have for exhibitionism is a man exposing himself to a woman. About 30 percent of all arrests for sexual offenses are for exhibitionism (Cox, 1988). According to one survey, 33 percent of college women have been the objects of indecent exposure (Cox, 1988).

According to the study of males seeking treatment for paraphilias (Abel & Rouleau, 1990), 15 percent of the exhibitionists had exposed themselves at least once by age 12; one half had done so by age 15. According to other research (Blair & Lanyon, 1981), exhibitionists generally recall their childhoods as being characterized by inconsistent discipline, lack of affection, and little training in appropriate forms of social behavior.

[4]Here is a classic limerick on exhibitionism:

There was a young lady of Exeter
So pretty, men craned their necks at her.
  One was even so brave
  As to take out and wave
The distinguishing mark of his sex at her.

Figure 16.5 Exhibitionism.

In adulthood, exhibitionists do not seem to be psychiatrically disturbed (Blair & Lanyon, 1981). However, they generally are timid and unassertive and lacking in social skills. They also seem to have trouble recognizing and handling their own feelings of hostility. Many are married, but they do not seem to be gratified with heterosexual sex.

The exact causes of exhibitionism are not known, but a social learning-theory explanation offers some possibilities (Blair & Lanyon, 1981). According to this view, the parents might have subtly (or perhaps obviously) modeled such behavior to the man when he was a child. In adulthood, there may be reinforcements for the exhibitionistic behavior because the man gets attention when he performs it. In addition, the man may lack the social skills to form an adult relationship, or the sex in his marriage may not be very good, so he receives little reinforcement for normal sex.

The learning-theory approach has been used to devise some programs of therapy that have been successful in treating exhibitionists. For example, in one therapy program exhibitionists were shown photos of scenes in which they typically engaged in exhibitionism; simultaneously, an unpleasant-smelling substance was placed at their nostrils (Maletzky, 1974, 1977, 1980). After 11 to 19 twice-weekly sessions of this conditioning and some self-administered home sessions, all but one of the men passed a temptation test in which they were placed in a naturalistic situation with a volunteer female and managed not to flash at her.

Exhibitionist: A person who derives sexual gratification from exposing his genitals to others in situations in which this is inappropriate.

Many women, understandably, are alarmed by exhibitionists. But since the exhibitionist's goal is to produce shock or some other strong emotional response, the woman who becomes extremely upset is gratifying him. Probably the best strategy for a woman to use in this situation is to remain calm and make some remark indicating her coolness, such as suggesting that he should seek professional help for his problem.

A study of 62 female sex offenders in Great Britain identified five women who had exhibited themselves (O'Connor, 1987). One 21-year-old woman stripped off her clothes and masturbated in public on several occasions. A 25-year-old single woman exposed her genitals and invited passersby to have sex with her. A 40-year-old woman entered private residences, took off her clothes, and invited any male present (including one child) to have sex with her. Two women were arrested while urinating in public. All five women had histories of unusual behavior and had been diagnosed with alcohol or psychiatric problems. The atypical sexual behaviors appear to reflect these problems, rather than sexual motivations.

Notice that both voyeurism and exhibitionism are considered problematic behavior when the other person involved is an unwilling participant. A man who derives erotic pleasure from watching his partner undress, or a woman who is aroused by exhibiting her body in new lingerie to her husband, are not engaging in criminal or paraphilic behavior.

Hypersexuality

Hypersexuality includes nymphomania and satyriasis, conditions in which there is an extraordinarily high level of sexual activity and sex drive, to the point where the person is apparently insatiable and where sexuality overshadows all other concerns and interests. When it occurs in women, it is called **nymphomania;** in men it is called **satyriasis** (or Don Juanism).[5] While this definition seems fairly simple, in practice it is difficult to say when a person has an abnormally high sex drive. As was seen in Chapters 11 and 12, there is a wide range in the frequencies with which people engage in coitus; therefore, the range we define as "normal" should also be broad. In real life, the "nymphomania" or "satyriasis" is often defined by the spouse. Some men, for example, might think that it was unreasonable for a wife to want intercourse once a day or even twice a week, and they would consider such a woman a nymphomaniac.[6] Other men might think it would be wonderful to be married to a woman who wanted to make love every day.

Because these two terms are imprecise, some researchers prefer the term hypersexuality. **Hypersexuality** refers to an excessive, insatiable sex drive in either a man or a woman. It leads to compulsive sexual behavior, in the sense that the person feels driven to it even when there may be very negative consequences (Goldberg, 1987). The person is also never satisfied by the activity, and she or he may not be having orgasms, despite all the sexual activity. Such cases meet the criteria for abnormal behavior discussed at the beginning of this chapter: The compulsiveness of the behavior leads it to become extremely inefficient, with the result that it impairs functioning in other areas of the person's life.

A study of 100 male patients with paraphilias or related disorders focused on creating an operational definition of hypersexuality (Kafka, 1997b). The results supported the use of the criterion of seven or more orgasms per week for a minimum duration of six months. The men reported an average of 7.4/8.0 orgasms per week in the preceding six months; the modal time per day the men spent in unconventional sexual activity was one to two hours. They reported that their hypersexual activity began between the ages of 19 and 21. The most common unconventional behaviors were compul-

---

[5]Satyriasis is named for the satyrs, who were part-human, part-animal beasts in Greek mythology. A part of the entourage of Dionysus, the god of wine and fertility, they were jovial and lusty and have become a symbol of the sexually active male.

[6]Someone once defined a nymphomaniac as a woman whom a man can't keep up with.

Nymphomania (nim-foh-MANE-ee-uh): An excessive, insatiable sex drive in a woman.
Satyriasis (sat-ur-EYE-uh-sis): An excessive, insatiable sex drive in a man.
Hypersexuality: An excessive, insatiable sex drive in either men or women.

Figure 16.6   Historical painting of a satyr, which gives the name to satyriasis, a sexual variation in which a man has an excessive, insatiable sex drive.

sive masturbation (67 percent of the sample), protracted promiscuity (56 percent), and dependence upon pornography (41 percent). The most common paraphilias were exhibitionism (35 percent of those with a paraphilia), voyeurism (27 percent), and pedophilia (25 percent).

While this research provides a useful operational definition for men, note that the suggested criterion should not be applied to women. The criterion is stated as number of orgasms per week. Some women rarely or never experience orgasms; in fact, their anorgasmia might cause them to engage in compulsive sexual behavior. Another problem is that women who are orgasmic are capable of multiple orgasms during a single session of activity (see Chapter 9). A woman who engages in sexual activity three times a week could experience seven or eight orgasms, and this would not be atypical or abnormal. Once again, we see that a person's gender is very important in defining abnormality. A valid criterion for hypersexuality in men may not be valid for women.

(Zaviačič, 1994). A variety of techniques are used, including temporary strangulation by a rope around the neck, a pillow against the face, or a plastic bag over the head or upper body. Obviously, this is very dangerous behavior; a miscalculation can lead to death. In fact, it is estimated that it causes between 250 and 1000 deaths per year in the United States (Innala & Ernulf, 1989). The average age of males who die during this activity is 26, leading investigators to suggest that it may be novices who die, due to their inexperience (Lowery & Wetli, 1982).

Little is known about asphyxiophilia. Most of the deaths attributed to the practice involve men. Such cases are often obvious to the trained investigator. Characteristics that distinguish these deaths from intentional suicides include a male who is nude, cross-dressed, or dressed with genitals exposed, and evidence of sexual activity at the time of death (Hucker & Blanchard, 1992). Pornography or other props such as mirrors are often present (Zaviačič, 1994).

Recently, some cases have been identified involving women (Byard, Hucker, & Hazelwood,

Asphyxiophilia

Asphyxiophilia is the desire to induce in oneself a state of oxygen deficiency in order to create sexual arousal or to enhance sexual excitement and orgasm

Asphyxiophilia: The desire to induce in oneself a state of oxygen deficiency in order to create sexual arousal or to enhance excitement and orgasm.

Figure 16.7   Advertising for partners. Many newspapers and magazines carry such "personal ads." © Tribune Media Services, Inc. All Rights Reserved. Reprinted with permission.

1993). A review of eight fatal cases among women found that only one involved unusual clothing, and none involved pornography or props. Two of the cases were initially ruled homicide, one suicide, and five accidental death. The investigators suggest that death due to asphyxiophilia may be much more common among women than we realize, because these deaths are less often recognized for what they are by investigators.

Men and women engage in asphyxiophilia in the belief that arousal and orgasm are intensified by reduced oxygen. There is no way to determine whether this is true. If the experience is more intense, it may be due to heightened arousal created by the risk rather than by reduced oxygen. Some believe that some women may experience an orgasm accompanied by urethral ejaculation; this belief has been identified as one reason women engage in asphyxiophilia. Again, there is no evidence.

There is probably a range from those who try this activity once out of curiosity to those who engage in it repeatedly and compulsively. Obviously, practitioners don't want to kill themselves; most include a self-release mechanism, but these safeguards sometimes fail.

Other Sexual Variations

The sexual variations discussed below are too rare to have had much research devoted to them; they are nonetheless interesting because of their bizarreness.

Troilism, or *triolism,* refers to three people having sex together.

Saliromania is a disorder found mainly in men; there is a desire to damage or soil a woman or her clothes or the image of a woman, such as a painting or statue. The man becomes sexually excited and may ejaculate during the act.

Coprophilia and **urophilia** are both variations having to do with excretion. In coprophilia the

> **Troilism (TROY-uhl-ism):** Three people having sex together.
> **Saliromania:** A desire to damage or soil a woman or her clothes.
> **Coprophilia (cop-roh-FILL-ee-uh):** Deriving sexual satisfaction from contact with feces.
> **Urophilia (YUR-oh-fill-ee-uh):** Deriving sexual satisfaction from contact with urine.

feces are important to sexual satisfaction. In urophilia it is the urine that is important. The urophiliac may want to be urinated on as part of the sexual act.

Frotteurism is inappropriate sexual touching or rubbing. A man may approach a woman from the rear and covertly press his penis against her buttocks; sometimes she is not even aware of it, in a place like a crowded subway or elevator.

Necrophilia is sexual contact with a dead person. It is a very rare form of behavior and is considered by experts to be psychotic and extremely deviant. Necrophiliacs derive sexual gratification from viewing a corpse or actually having intercourse with it; the corpse may be mutilated afterward (Thorpe et al., 1961). The highly publicized murder and mutilation case involving Jeffrey Dahmer in Milwaukee may well fall into this category. Experts believe that he was motivated by necrophilia.

Zoophilia is sexual contact with an animal; this behavior is also called *bestiality* or *sodomy*, although the latter term is also used to refer to anal intercourse or even mouth-genital sex between humans. About 8 percent of the males in Kinsey's sample reported sexual experiences with animals. Most of this activity was concentrated in adolescence and probably reflected the experimentation and diffuse sexual urges of that period. Not too surprisingly, the percentage was considerably higher among boys on farms; 17 percent of boys raised on farms had had animal contacts resulting in orgasm. Kinsey found that only about 3 to 4 percent of all females have had some sexual contact with animals. Contemporary therapists report a few cases of men and women engaging in sexual activity with household pets. Activities include masturbating the animal, oral genital contact, and intercourse.

Prevention of Sexual Variations

For many of the variations discussed in this chapter, there is a continuum from normal to abnormal. People whose behavior falls at the normal end enjoy these activities at no expense to self or others. People whose behavior falls at the abnormal end are cause for concern.

The misery that many people—e.g., the sexually addicted S-Mer—suffer, not to mention the harm

they may do to others (e.g., the child molester), is good reason to want to develop programs for preventing sexual variations (Qualls et al., 1978). In preventive medicine, a distinction is made between primary prevention and secondary prevention. Applied to the sexual variations, primary prevention would mean intervening in home life or in other factors in childhood to help prevent problems from developing or trying to teach people how to cope with crises or stress so that problems do not develop. In secondary prevention, the idea is to diagnose and treat the problem as early as possible, so that difficulties are minimized.

It would be highly advantageous to do primary prevention of sexual variations—that is, to head them off before they even develop. Unfortunately, this is proving to be difficult, for a number of reasons. One problem is in diagnostic categories. The categories for the diagnosis of sexual variations are not nearly so clear-cut in real life as they may seem in this chapter, and multiple diagnoses for one person are not uncommon. That is, a given man might have engaged in incest, pedophilia, and exhibitionism. If it is unclear how to diagnose sexual variations, it is going to be rather difficult to figure out how to prevent them. If one is not sure if there is a difference between chicken pox and measles, it is rather difficult to start giving inoculations.

An alternative approach that seems promising—rather than figuring out ways to prevent each separate variation—is to analyze the *components of sexual development*. Disturbance in one or more of these components in development might lead to different sexual variations. One proposal for these components is as follows (Bancroft, 1978):

1. *Gender identity* The sense of maleness or femaleness developed in early childhood
2. *Sexual responsiveness* Arousal to appropriate stimuli
3. *Formation of relationships with others*

It seems clear that different developmental components are disturbed in different variations.

Necrophilia: Deriving sexual satisfaction from contact with a dead person.
Zoophilia: Sexual contact with an animal; also called bestiality or sodomy.

For example, in transsexualism, it is the first component, gender identity, that is disturbed. In the case of the fetishist, it is the second component, sexual responsiveness to appropriate stimuli, that is disturbed. And in the case of the exhibitionist, it may be that it is the last component, the ability to form relationships, that is disturbed.

The idea would then be to try to ensure that as children grow up, their development in each of these three components is healthy. Ideally, sexual variations should not occur then.

Space does not permit us to consider what prevention programs might look like for all the different variations (see Bancroft, 1978, for further discussion), so let us consider one example, transsexualism, in some detail (Green, 1978b; see Chapter 14 for a discussion of transsexualism).

Suppose that we have a typical case of a very feminine boy, Billy, whose parents bring him in for treatment. Billy prefers to dress in girls' clothes, wants to play with dolls and play house, and dislikes playing with boys because they are too rough. He might be considered a high risk for becoming a transsexual, because virtually all transsexuals recall a sense of being trapped in the wrong body from earliest childhood.

What kind of therapy can be used? Some efforts are made at simple education—making sure Billy understands the anatomical differences between boys and girls, and comprehends that one cannot change gender magically. Positive aspects of maleness are emphasized. Male playmates are found who are not rough. Parents are encouraged not to engage in behavior that may reinforce his conflict—for example, commenting that he is cute when dressed up as a girl. The father-son relationship is encouraged, and a male therapist is used so that the boy can identify with him. Finally, intervention may simply involve helping the child accept his own atypical behavior.

Such therapy raises a host of ethical issues. Is it right to make a traditional, stereotyped male out of a boy who might simply be androgynous? Is it right to intervene when one is not sure that he will otherwise become a transsexual? Indeed, in longitudinal follow-up studies of 26 feminine boys, 14 did become transsexuals, transvestites, or homosexuals, but 12 became heterosexuals. Therefore, one cannot be sure about the eventual development of a feminine 5-year-old. And even if superficial masculine behavior is successfully encouraged, what if a host of conflicts continues to simmer below the surface, creating a more seriously disturbed individual? Richard Green summarizes the complex problems this way:

> It may be argued that to induce intervention (which may be prevention) reinforces societal sexism. Regrettably, to a degree it does. But while we have a responsibility to reduce sexism, we have a responsibility to an individual child caught in the cross fire between sex role idealism and the real world in which he is embedded. (1978b, p. 88)

We have a long way to go in preventing sexual variations, to say the least.

Treatment of Sexual Variations

Some of the sexual variations discussed in this chapter—such as the mild fetishes—are well within the normal range of sexual expression. There is no need for treatment. Others, however, fall into the abnormal range, causing personal anguish to the individual and possibly causing harm to unwilling victims. Treatments are needed for this category of variations. Many different treatments have been tried, each based on a different theoretical understanding of the causes of sexual variations. We will look at four categories of treatments: medical treatments, cognitive-behavioral therapies, skills training, and AA-type 12-step programs.

Medical Treatments

Inspired by the notion that sexual variations are caused by biological factors, various medical treatments for sexual variations have been tried over the last century. Some of them look today like nothing other than cruel and unusual punishment. Nonetheless, people would love to have a "pill" that would cure some of these complex and painful or dangerous paraphilias, so the search for such treatments continues.

Surgical castration was used fairly commonly in the United States in the 1800s as a treatment for various kinds of uncontrollable sexual urges (Bullough, 1976). The idea resurfaced recently in some court cases in which castration was proposed as a treatment for rapists, as discussed in Focus 9.3 in Chapter 9. Such treatments are based on the notion that removing a man's testosterone by removing the testes will lead to a drastic reduction in sex drive, which will in turn erase urges to commit sex offenses. However, as we saw in Chapter 9, reduction in testosterone levels in humans does not always lead to a reduction in sexual behavior. Surgical castration cannot be recommended as a treatment for sex offenders either on humanitarian grounds or on grounds of effectiveness.

Hormonal treatment involves the use of drugs to reduce sexual desire. Sexual arousability is heavily dependent on maintaining the level of androgen in the bloodstream above a threshold. Two ways to reduce this level are to administer 1) drugs that reduce the production of androgen in the testes ("chemical castration"), or 2) antiandrogens which bind to androgen receptors in the brain and genitals, blocking the effects of androgen. Use of either should produce a sharp decline in sexual desire. Several drugs have been tried in the past 40 years. The most commonly used in the United States is *medroxyprogesterone acetate* (MPA), which binds to androgen receptors. The drug is given by injection, often weekly. A review of the research on its effectiveness found that most studies report decreases in sexual desire and in paraphilic behaviors in men who take MPA (Gijs & Gooren, 1996).

An alternative, *psychopharmacological treatment*, has increased dramatically in frequency in the 1990s. Here, psychotropic medications, i.e., anti-depressants such as Prozac, are administered to offenders. These medications influence psychological functioning and behavior by their action on the central nervous system. Due to the newness of the drugs being used, there is little research on the effectiveness of this technique. Antidepressants are being used with paraphilics who are also diagnosed with obsessive-compulsive disorder or depression. These drugs appear to change the obsessive-compulsive behavior rather than sexual desire (Gijs & Gooren, 1996).

Both hormonal and psychopharmacological treatment should be used as only one element in a complete program of therapy, which includes counseling and treatment for other emotional and social deficits. The best results are obtained with men who are highly motivated to change their behavior and therefore comply with the prescribed treatment regimen. If the paraphilic stops taking the drug or participating in other aspects of treatment, the program will fail. Unfortunately, one of the limitations of research on the effectiveness of these treatments is the drop-out rate, which was 46 percent in one study.

Cognitive-Behavioral Therapies

Some treatment programs are based on cognitive-behavioral therapies. Comprehensive programs include (Abel et al., 1992):

1. Behavior therapy to reduce inappropriate sexual arousal and enhance appropriate arousal.
2. Social skills training.
3. Modification of distorted thinking; challenging the rationalizations that the person uses to justify the undesirable behavior.
4. Relapse prevention; helping the person identify and control or avoid triggers of the behavior.

Although the media sometimes carry flashy stories about applications of behavior therapy that involve administering electric shocks to a sex offender if he becomes aroused at, say, a picture of a nude child, in fact much milder techniques have been used and are effective. *Covert sensitization* is one such therapy. It involves pairing aversive imagery (thoughts) with fantasies of the target behavior. In the treatment of an exhibitionist, for example, he repeatedly practices a vivid fantasy in which, just as he imagines getting ready to expose himself, he experiences waves of nausea and vomiting. The details, of course, are individualized to the person and his particular problem. Case reports indicate that this approach has been effective in cases such as sadism, exhibitionism, and pedophilia (Walen & Roth, 1987; Barlow et al., 1969).

Another approach is *orgasmic reconditioning* (Walen & Roth, 1987; Marquis, 1970). The patient is told to masturbate to his usual paraphiliac fantasies. Then, just at the moment of orgasm, he switches to an acceptable fantasy. After practicing this for some time, he becomes able to have an orgasm regularly while having an acceptable fantasy. He then is told to move the fantasy progressively to an earlier phase of masturbation. Gradually he becomes conditioned to experiencing sexual arousal in the context of acceptable behavior.

One program designed for female offenders combined cognitive-behavioral with psychodynamic techniques, and relied on one-on-one rather than group therapy (Traven et al., 1990). Male offenders typically deny responsibility for their behavior, so the initial stage of treatment may focus on acknowledging one's behavior and its consequences. Women in this program readily acknowledged what they had done, and were

Figure 16.8   The centerpiece of 12-step programs such as Sex Addicts Anonymous are group meetings in which participants confront their addiction with the support of other group members.

overwhelmed by guilt and shame, so the initial stage focused on self-esteem. Thus, different treatment programs may be needed for male and female paraphilics.

Skills Training

According to yet another theoretical understanding, paraphiliacs engage in their behavior because they have great difficulty forming relationships, and so do not have access to appropriate forms of sexual gratification. Many of these people do not have the skills to initiate and maintain conversation. They may find it difficult to develop intimacy (see Chapter 13). Such people may benefit from a treatment program that includes social skills training. Such training may include how to carry on a conversation, how to develop intimacy, how to be appropriately assertive, and identifying irrational fears which are inhibiting the person (Abel et al., 1992). These programs may also include basic sex education.

If a person needs to learn and practice sexual interaction skills, one approach would be to have him interact with a trained partner. This is the basis for a very controversial practice, the use of *sex surrogates* as part of a treatment program. The surrogate works with the therapist, interacting socially and sexually with the client to provide opportunities for using the newly acquired information and skills. Some therapists believe that the use of surrogates is ethical, but others see it as a type of prostitution. Just as the definition of "abnormal" depends on one's point of view, so does the definition of "sex therapy."

AA-Type 12-Step Programs

As we saw in Focus 16.2, sexual addiction theory argues that many people who engage in uncontrollable, inappropriate sexual patterns are addicted to their particular sexual practice. The appropriate treatment, according to this approach, is one of the 12-step programs modeled on Alcoholics Anonymous.

Treatment programs based on this approach have become very common in recent years. Some of the programs are run by group members, whereas others are affiliated with professional health care facilities. Twelve-step programs combine cognitive restructuring, obtaining support from other members who have the same or similar problem behaviors, and enhancing spirituality. This last aspect involves increasing one's awareness of a "higher power" who can be relied on to help one recover. AA-based groups are generally unwilling to cooperate with researchers, believing that to do so would prevent group members from concentrating on recovery. As a result, little research data exist on these programs.

What Works?

What is needed is carefully controlled research on the effectiveness of various approaches to the treatment of sexual variations. Research to date has tended to apply one method to a heterogeneous group of people. It is likely, however, that each method will be more effective with some paraphilias than with others. Research involving drugs should systematically assess side effects; some drugs have severe side effects on some people, especially if used more than six months. Research should consider the relative likelihood of relapse; some paraphiliacs pose grave danger to themselves (e.g., asphyxiophiliacs) or to others if they resume their problematic behavior.

SUMMARY

It seems reasonable to define "abnormal sexual behavior" as behavior that is uncomfortable for the person, inefficient, bizarre, or physically or psychologically harmful to the person or others. The American Psychiatric Association defines paraphilias as recurrent, unconventional sexual behaviors that are obsessive and compulsive.

Four theoretical approaches have been used in understanding the paraphilias: learning theory, cognitive theory, the sexual addiction model, and sociological theory. Several explanations have been proposed for the fact that there are many more male than female paraphilics.

A fetishist is a person who becomes erotically attached to some object other than another human being. Most likely, fetishism arises from conditioning, and it provides a good example of the continuum from normal to abnormal behavior.

The transvestite derives sexual satisfaction from dressing as a member of the other gender. Like many other sexual variations, transvestism is much more common among men than among women. Survey data suggest that many men begin cross-dressing in childhood.

Three styles of sexual interaction involve differences in control over sexual interactions. Dominance and submission involve a consensual exchange of power, and the enacting of scripted performances. Bondage and discipline involve the use of physical restraints or verbal commands by one person to control the other. Both D-S and B-D may occur without genital contact or orgasm. Sadism and masochism involve deriving sexual gratification from giving and receiving pain. Both are recognized as paraphilias if they become compulsive.

The voyeur is sexually aroused by looking at nudes. The exhibitionist displays his or her sex organs to others. Both are generally harmless.

Nymphomania and satyriasis are terms used to describe women and men with extraordinarily high sex drive. Both terms are ambiguous and subject to misuse. The term hypersexuality is potentially more precise, particularly if it is defined behaviorally.

Other sexual variations include asphyxiophilia, zoophilia, and necrophilia.

The possibility of programs to prevent sexual variations is being explored, although the topic is complex.

Four types of programs to treat sexual variations are: medical treatments including hormonal and psychopharmacological interventions, cognitive-behavioral therapies, social skills training, and AA-type 12-step programs. We need careful research to determine which program works best in the treatment of which behaviors.

REVIEW QUESTIONS

1. According to a psychological definition, sexual behavior can be considered abnormal if it causes discomfort to the person, if it is bizarre, or if it is inefficient. True or false?

2. _____ is the term for becoming sexually fixated on some object other than another human being and attaching great erotic significance to that object.

3. Transvestism is considered to be a fetish, because the man becomes sexually aroused from dressing in women's clothes. True or false?

4. A _____ is a person who becomes sexually aroused when experiencing pain.

5. A voyeur is a person who derives sexual gratification from exposing his or her genitals to others. True or false?

6. When a woman has an insatiable sex drive it is termed nymphomania, but the preferred scientific term is _____.

7. According to Patrick Carnes, if a person's sexual ritual has become an obsession and can't be controlled, the person has a _____.

8. It is rare to find a transvestite who is heterosexually married. True or false?

9. Bondage and discipline refers to the use of physical or psychological restraint to enforce servitude, from which participants derive sexual pleasure. True or false?

10. Covert sensitization is a treatment for sex offenders based on sexual addiction theory. True or false?

(The answers to all review questions are at the end of the book, after the Glossary.)

QUESTIONS FOR THOUGHT, DISCUSSION, AND DEBATE

1. What do you think of the idea about preventing sexual variations presented in this chapter? Do you think the schools or some other agency should institute a program to screen children, trying to detect those with characteristics that might indicate they would develop a sexual variation later in life, and then give therapy to those children?

2. Of the sexual variations in this chapter, which seems to you to be the most abnormal? Why? Does the one you have chosen fit the criteria for abnormality discussed at the beginning of the chapter?

3. A common phenomenon in medical school is for medical students to think they have contracted one of the diseases they are studying. The analogous phenomenon would be for students who read this chapter to think that they have one or more of the paraphilias. Did you notice yourself becoming sexually aroused as you read about one (or more) of the variations? If so, do you think you are abnormal? Why, or why not?

4. Most persons who seek treatment or are arrested for atypical or paraphilic behaviors are white men. Do you think that this is an accurate picture of our society, or are we just unaware of nonwhites who engage in these behaviors? Based on the data we presented and the causes we discussed, would you expect black, Asian, or Hispanic men to engage in these behaviors? Why, or why not?

SUGGESTIONS FOR FURTHER READING

Griffin-Shelley, Eric. (1991). *Sex and Love: Addiction, treatment, and recovery.* Westport, CT: Praeger. This book describes the addiction model as it is applied to sex and love. It also describes treatment programs.

Wilson, Glenn D. (Ed.). 1987. *Varient sexuality: Research and theory.* Baltimore: Johns Hopkins University Press. The chapters in this book review different theoretical explanations for sexual variations, from genetics and sociobiology to cross-cultural perspectives.

WEB RESOURCES

http://www.avitale.com/ReadingList.html
  A bibliography of books on transvestism and related topics.

http://www.ohsu.edu/cliniweb/F3/F3.800.800.600.html
  A site with links to the scientific literature on diagnosis and treatment of each of the paraphilias.

http://www.sexualrecovery.com
  Diagnosis and treatment of sexual addiction; the Sexual Recovery Institute.

chapter
17

SEXUAL COERCION

CHAPTER HIGHLIGHTS

Rape
Incidence Statistics
The Impact of Rape
Date Rape
Marital Rape
Causes of Rape
Rapists
Men as Victims of Rape
Prison Rape
Ethnicity and Rape
Preventing Rape

Child Sexual Abuse
Patterns of Child Sexual Abuse
Patterns of Incest
Psychological Impact on the Victim
The Offenders

Sexual Harassment
Sexual Harassment at Work
Sexual Harassment in Education: An A for a Lay
Psychotherapist-Client Sex

Rape Poem

ere is no difference between being raped
d being pushed down a flight of cement steps
cept that the wounds also bleed inside.

ere is no difference between being raped
d being run over by a truck
cept that afterward men ask if you enjoyed it.

There is no difference between being raped
and being bit on the ankle by a rattlesnake
except that people ask if your skirt was short
and why you were out alone anyhow.*

This chapter is about sexual activity that involves coercion and is not between consenting adults; specifically, we will consider rape, child sexual abuse, and sexual harassment at work and in education. All these topics have been highly publicized in the last 10 or 15 years, and some good scientific research on them has appeared.[1]

Rape

Rape is typically defined, following current laws in many states, as "nonconsensual oral, anal, or vaginal penetration, obtained by force, by threat of bodily harm or when the victim is incapable of giving consent" (Koss, 1993, p. 1062). Notice that the definition includes not only forced vaginal intercourse, but forced oral sex and anal sex as well. The crucial point is that the activity is nonconsensual—that is, the victim did not consent to it. One type of nonconsent occurs when the victim is incapable of giving consent, perhaps because of being drunk, unconscious, or high on drugs.

Incidence Statistics
In 1996 95,800 rapes (completed or attempted) were reported in the United States; that means there were 71 reported rapes for every 100,000 women (FBI, 1997). However, according to the FBI, forcible rape is one of the most underreported crimes. One study found that only about 1 in 5 (21 percent) of stranger rapes had been reported to the police, and only 2 percent of acquaintance rapes had been reported (Koss et al., 1988). A well-sampled national study of women college students found that 15 percent had experienced an act that met the legal definition of rape (Koss et al., 1987). Similar results have been found in research with Australian college students (Patton & Mannison, 1995). The NHSLS found that 22 percent of women had been involved in an incident in which they were forced to have sex (Laumann et al., 1994). Statistics vary somewhat from one study to another, but most find that a woman's lifetime risk of being raped is between 14 and 25 percent (Koss, 1993).

The Impact of Rape
A large number of studies have investigated the psychological reactions of women following rape (e.g., Burgess & Holmstrom, 1974a, 1974b; Calhoun et al., 1982; Resick, 1983; see review by Koss, 1993). This research shows that rape is a time of crisis for a woman and that the effects on her adjustment may persist for a year or more. The term **rape trauma syndrome** has been used to refer to the emotional and physical effects a woman undergoes following a rape or an attempted rape (Burgess & Holmstrom, 1974a).

---

*Excerpted from Marge Piercy. (1976). Rape poem. In *Living in the open*. New York: Knopf. Reprinted by permission of Wallace & Sheil Agency, Inc. Copyright 1974, 1976 by Marge Piercy.

[1]Portions of this chapter are excerpted from J. S. Hyde. (1996). *Half the human experience: The psychology of women*, 5th ed. Boston: Houghton Mifflin.

Rape: Nonconsenting oral, anal, or vaginal penetration obtained by force, by threat of bodily harm, or when the victim is incapable of giving consent.
Rape trauma syndrome: The emotional and physical effects a woman undergoes following a rape or attempted rape.

Focus 17.1
A Date Rape Victim Tells Her Story

During my sophomore year at Northwestern University, I realized I wanted to socialize more. I broke up with my hometown honey, began attending college parties, started drinking, and dated other guys at my school. I was a virgin and didn't want to be anymore. I met my second boyfriend in physics. "G" was a football player and a big, handsome man, the best-looking man I'd ever met. We began to date and at first it was wonderful. He even carried me home once from a party, and I thought, "This is the one." Our first attempt at intercourse was difficult and I began to beg him to stop, but he just kept trying until he was successful and there was blood everywhere. It was awful.

I continued to date G, but he began to act very differently. When he drank, he became extremely violent, and on different occasions I watched him break a vending machine, put his foot through the soft top of a Porsche, and pull a toilet out of a wall at a fraternity. He was unhappy with the way he was treated on the football team, and I tried to console him. He became angrier and paranoid. He demanded to know where I was at all times and accused me of cheating on him. I wanted to break up with him, especially since sex was rough and not always consensual, but I was scared of him. I tried avoiding him but he always found me.

One night, G arrived very drunk at a party I was attending. He threatened a guy who was there talking to me. I tried to sneak out of the party. He noticed I left and ran out after me. Another guy slowed me down to try to persuade me to let him drive me home since I had been drinking. G accused me of trying to pick this guy up and threatened the guy, who even volunteered to drive both of us home. But G wouldn't let him and said I would take him home. I didn't want trouble so I drove him to his dorm. He claimed he was too drunk to walk to his dorm, so I tried to help him into his room. When I turned around to leave, he sprang up and locked me in. He attacked me. I tried fighting him, but he wouldn't listen or stop. He hit my head into the wall several times and tried to force me to perform oral sex. I bit him, and that made him more angry. He then tried to force anal sex, and I fought as hard as I could. I finally started crying, and he stopped when he lost his erection. The rest of the night, I felt completely trapped in his dorm room. I lay awake all night and tried to leave, but he would wake up and stop me. I've never forgotten how scared I felt that entire night. He got up that morning and showered and acted as if nothing had happened. He was in all my classes for the rest of my college career.

Emotional reactions immediately after a rape (the acute phase) are often severe. The high levels of distress generally reach a peak 3 weeks after the assault and continue at a high level for the next month. There is then gradual improvement beginning 2 or 3 months after the assault (Koss, 1993; Rothbaum et al., 1992). Many differences between raped and nonvictimized women disappear after 3 months, except that raped women continue to report more fear, anxiety, self-esteem problems, and sexual disorders. These effects may persist for 18 months or longer (Koss, 1993).

See Focus 17.1 for one woman's account of these effects in her life.

Some women experience self-blame. A woman may spend hours agonizing over what she did to bring on the rape or what she might have done to prevent it: "If I hadn't worn that tight sweater . . . "; "If hadn't worn that short skirt . . . "; "If I hadn't been dumb enough to walk on that dark street . . . "; "If I hadn't been stupid enough to trust that guy . . . " This is an example of a tendency on the part of both the victim and others to *blame the victim*.

I began to drink very heavily afterwards. I told my roommate, J, and another woman on my floor, D. I never thought to report it, since he was my "boyfriend." A week later, D was assaulted by another student. She went to the campus police afterwards. They took her to the head of student housing. He told her it was not their problem or the police's but her drinking problem.

A year and a half later, I began to hear voices. It was a male voice calling me a fucking bitch, whore, and other names. I thought I was going crazy and became very depressed. I decided I must be schizophrenic and decided to kill myself. My attempt was unsuccessful.

Soon after, I was shopping in a feminist bookstore. I saw this title staring right at me, "I Never Called It Rape." I started reading it right there in the store, and I began crying and thinking, "This is what happened to me." I spoke with a faculty member, who arranged for immediate counseling.

The first time I went to see the counselor, I couldn't even speak. I sat in her office and cried for the entire hour. She kept saying, "It's not your fault, it's not your fault." I couldn't believe it. Later we discussed how most of the times G and I had sex had actually been rape, including the first and the last. I participated in a "Take Back the Night March." One fraternity threw bottles at us.

I went through medical school and residency. During my first year of residency, I was assaulted again, this time by a man in a stairwell during a New Year's Eve party at a hotel. I started screaming "You're raping me, you're raping me!" He stopped and I got away! But I didn't go to the ER, I just went home and crawled in bed. My old shame came back. I began to drink heavily again. One night I drank all night and never showed up to work that next morning. I finally rolled into my director's office, depressed, hung over, and still smelling of alcohol, and my boss said I had to stop and straighten up right away or he wouldn't let me back for the next year. So I stopped drinking. I also made two very good friends around the same time. Through their support, I really turned my life around.

Three years later, I moved to Madison, Wisconsin. I was living alone for the first time and had a great deal of anxiety. I joined a Sexual Assault Survivors Support Group. Then in the spring, I was invited to speak at my old university for Career Day for high school students, so I went back there 10 years after the incidents. I was finally successful in my career, had strong, loving relationships with my friends and parents, and was happy. I look back at what happened now and think that I really survived a lot. I feel it helps me to be a better physician because I can empathize with how bad life can be for people.

Source: Based on an interview conducted by Janet Hyde.

Researchers are finding increased evidence of the damage to women's *physical health* that may result from rape (Heise, 1993; Koss et al., 1991; Koss & Heslet, 1992). The woman may have physical injuries, such as cuts and bruises, and vaginal pain and bleeding. Women who have been forced to have oral sex may suffer irritation or damage to the throat; rectal bleeding and pain are reported by women forced to have anal intercourse. A raped woman may contract a sexually transmitted disease such as HIV/AIDS or herpes. In about 5 percent of rape cases, pregnancy results (Koss et al., 1991).[2] Women who have been sexually or physically assaulted at some time in the past visit their physician twice as often per year as nonvictimized women (Koss et al., 1991).

Recently it has been suggested that using the term "rape trauma syndrome" is not the best way to label and understand the effects of rape; rather, some experts argue, we should recognize that rape

[2]Tests for sexually transmitted diseases should routinely be done as part of the hospital treatment of rape victims. Pregnancy tests can be done if the woman's period is late. If pregnancy seems likely, emergency contraception can be used (see Chapter 8).

Figure 17.1   Rape crisis counseling. Many women experience severe emotional distress after a rape, and it is important that crisis counseling be available to them.

victims are experiencing **post-traumatic stress disorder** (e.g., Koss, 1993). Post-traumatic stress disorder (PTSD) is an official diagnosis that was originally developed to describe the long-term psychological distress suffered by war veterans, most of whom are men. Symptoms can include persistently re-experiencing the traumatic event (flashbacks, nightmares), avoiding stimuli associated with it (avoiding certain locations or activities), and hyperarousal (sleep difficulties, difficulty concentrating, irritability). According to a cognitive-behavioral view of PTSD, people who have experienced a terrifying event form a memory schema that involves information about the situation and their responses to it (Foa et al., 1989). Because the schema is large, many cues can trigger it and therefore evoke the feelings of terror that occurred at the time; the schema is probably activated at some level all the time.

Schemas also affect how we interpret new events, so that the consequences are far-reaching and long-lasting.

It is important to recognize that rape affects many people besides the victim. Most women routinely do a number of things that stem from rape fears. For example, a single woman is not supposed to list her full first name, but rather her first initial or a man's name, in the telephone book, so as not to reveal that she lives alone. Many women, when getting into their car at night, almost reflexively check the backseat to make sure no one is hiding there. Most college women avoid walking alone through dark parts of the campus at night. At least once in their lives, most women have been afraid of spending the night alone. If you are a woman, you can probably extend this list from your own experience. The point is that most women experience the fear of rape, if not rape itself (Burt & Estep, 1981; Warr, 1985), and this fear restricts their activities.

Spouses or partners of victims may also be profoundly affected. At the same time, they can provide important support for the woman as she recovers (see Focus 17.2).

Date Rape

In Mary Koss's national study of college women, among those who had experienced an act that met the legal definition of rape, 57 percent of the rapes involved a date, often a steady dating partner (Koss et al., 1988; Koss & Cook, 1994). Date rape is one of the most common forms of rape, especially on college campuses.

In some cases, date rape seems to result from male-female miscommunication. Men's traditional view in dating relationships has been that a woman who says "no" really means "yes." Men need to learn that "no" means "no." Consider this example of miscommunication and different perceptions in a case of date rape:

> *Bob:* Patty and I were in the same statistics class together. She usually sat near me and was always

Post-traumatic stress disorder (PTSD): Long-term psychological distress suffered by someone who has experienced a terrifying event.

Focus 17.2
How Can Friends Help a Rape Victim?

What the Victim Needs to Do

Obtain medical assistance.

Feel safe. Rape is a traumatic violation of the person. Especially in the beginning, it is often difficult for victims to be alone.

Be believed. With date rape especially, victims need to be believed that what occurred was, in fact, a rape.

Know it was not her fault. Most rape victims feel guilty and feel that the attack was somehow their fault.

Take control of her life. When a person is raped, she may feel completely out of control of what is happening to her. A significant step on the road to recovery is to regain a sense of control in little, as well as big, things.

Things You Can Do to Help

Listen, do not judge. Accept her version of the facts and be supportive.

Offer shelter. If possible, stay with her at her place or let her spend at least one night at your place. This is not the time for her to be alone.

Be available. She may need to talk at odd hours, or a great deal in the beginning. Also encourage her to call a hotline or go for counseling. Be available even months later.

Give comfort. She needs to be nurtured.

Let her know she is not to blame.

Encourage action. For example, suggest she call a hotline, go to a hospital, and/or call the police. Respect her decision if she decides not to file charges. Do not make her decisions for her, because she needs to regain a sense of control of her life.

Put aside your feelings and deal with them somewhere else. Although it is supportive for a rape survivor to know that others are equally upset with what happened, it does her no good if she also has to deal with, for example, your feelings of rage. If you have strong feelings, talk to another friend or to a local hotline.

Source: Condensed from Jean O. Hughes & Bernice R. Sandler. (1987). *"Friends" raping friends: Could it happen to you?* Washington, DC: Association of American Colleges. Used with permission.

---

very friendly. I liked her and thought maybe she liked me, too. Last Thursday I decided to find out. After class I suggested that she come to my place to study for midterms together. She agreed immediately, which was a good sign. That night everything seemed to go perfectly. We studied for a while and then took a break. I could tell that she liked me, and I was attracted to her. I was getting excited. I started kissing her. I could tell that she really liked it. We started touching each other and it felt really good. All of a sudden she pulled away and said "Stop." I figured she didn't want me to think that she was "easy" or "loose." A lot of girls think they have to say "no" at first. I knew once I showed her what a good time she could have, and that I would respect her in the morning, it would be OK. I just ignored her protests and eventually she stopped struggling. I think she liked it but afterwards she acted bummed out and cold. Who knows what her problem was?

Patty: I knew Bob from my statistics class. He's cute and we are both good at statistics, so when a tough midterm was scheduled, I was glad that he suggested we study together. It never occurred to me that it was anything except a study date. That night everything went fine at first, we got a lot of studying done in a short amount of time, so when he suggested we take a break I thought we deserved it. Well, all of a sudden he started acting really romantic and starting kissing me. I liked the kissing but then he started touching me below the waist. I pulled away and tried to stop him but he didn't listen. After a while I stopped struggling; he was hurting me and I was scared. He was so much bigger and stronger than me. I couldn't believe it was happening to me. I didn't know what to do. He actually forced me to have sex with him. I guess looking back on it I should have screamed or done something besides trying to reason with him but it was so unexpected. I couldn't believe it was happening. I still can't believe it did. (Hughes & Sandler, 1987, p. 1)

Several explanations have been proposed for why sexually aggressive men misperceive women's communications. The first proposes that aggressors lack competence in "reading" women's negative emotions; they just don't get it when she's "bummed out." The second suggests that sexually aggressive men fail to make subtle distinctions between women's friendliness and seductiveness. The third proposes that they have a "suspicious schema" and automatically doubt that women are communicating truthfully and accurately. Research testing these explanations supports the third explanation—sexually aggressive men generally believe that women do not communicate honestly, particularly when the woman communicates clearly and assertively that she is rejecting an advance (Malamuth & Brown, 1994). These findings have important implications for prevention and treatment programs for sexual aggressors. They suggest that simple skills training—to read women's emotions or distinguish between friendly and seductive behavior—may not be the key. Cognitive therapy using cognitive restructuring may be the most effective, the goal being to get the man to change his suspicious schema. Such programs might be used with incarcerated rapists, but they might also be used in prevention programs with high school or college men who have been identified as rape-prone.

Kanin (1985) studied 71 unmarried college men who were self-disclosed date rapists, comparing them with a control group of unmarried college men. The date rapists tended to be sexually predatory. For example, when asked how frequently they attempt to seduce a new date, 62 percent of the rapists said "most of the time," compared with 19 percent of the controls. The rapists were also much more likely to report using a variety of manipulative techniques with their dates, including getting them high on alcohol or marijuana, falsely professing love, and falsely promising "pinning," engagement, or marriage.

One of the most frightening problems today is the emergence of the so-called "date-rape drug," rohypnol (row-Hip-nawl, "roofie," drug name flunitrozepam). Numerous cases have been reported of men who slipped the drug into a woman's drink. The drug causes drowsiness or sleep and the man rapes the woman while she is asleep. The drug also causes the woman not to remember the event the next day. Several strategies for avoiding this situation have been suggested, including, especially, not accepting a drink from a stranger.

Marital Rape

How common is **marital rape?** In a random sample of San Francisco women, 14 percent of those who had ever been married had been raped by a husband or ex-husband (Russell, 1983). Other studies generally find that the prevalence of marital rape in the general population is between 7 and 14 percent (Whatley, 1993).

One phenomenon that emerges from the research is an association between marital violence and marital rape—that is, the man who batters his wife is also likely to rape her (Hanneke et al., 1986). For example, in a study of 137 women who had reported beatings from their husbands, 34 percent reported being raped by their husbands (Frieze, 1983). Reflecting the fact that some women are unwilling to define certain acts as marital rape, 43 percent of that sample said that sex was unpleasant because their husband forced them to have sex, a higher percentage than those admitting being raped. The response of the majority of the women was anger toward the husband. However, women who had been raped frequently began to experience self-blame. Marital rape also appeared to have consequences for the marriage: The raped women were more likely to say that their marriages had been getting worse over time.

A man might rape his wife for many motives, including anger, power and domination, sadism, or a

---

Marital rape: The rape of a person by her or his spouse.

desire for sex regardless of whether his wife is willing (Russell, 1990). In some cases the husband is extremely angry, perhaps in the middle of a family argument, and he expresses his anger toward his wife by raping her. In other cases, power and domination of the wife seem to be the motive; for example, the wife may be threatening to leave him, and he forces or dominates her into staying by raping her. Finally, some rapes appear to occur because the husband is sadistic—enjoys inflicting pain—and is psychiatrically disturbed.

Causes of Rape

To provide a perspective for the discussion that follows, we can distinguish among four major theoretical views of the nature of rape (Albin, 1977; Baron & Straus, 1989):

1. *Victim-precipitated* This view holds that a rape is always caused by a woman "asking for it." Rape, then, is basically the woman's fault. This view represents the tendency to "blame the victim."

2. *Psychopathology of rapists* This theoretical view holds that rape is an act committed by a psychologically disturbed man. His deviance is responsible for the crime occurring.

3. *Feminist* Feminist theorists view rapists as the product of gender-role socialization in our culture. They have theorized about the complex links between sex and power: In some rapes, men use sex to demonstrate their power over women; in other rapes, men use their power over women to get sex. Feminists also point to the eroticization of violence in our society. Gender inequality is both the cause and the result of rape.

4. *Social disorganization* Sociologists believe that crime rates, including rape rates, increase when the social organization of a community is disrupted. Under such conditions the community cannot enforce its norms against crime.

You personally may subscribe to one or more of these views. It is also true that researchers in this area have generally based their work on one of these theoretical models, and this may influence their research. You should keep these models in mind as you read the rest of this chapter.

What do the data say? Research indicates that a number of factors contribute to rape, ranging from forces at the cultural level to forces at the individual level. These factors include the following: cultural values; sexual scripts; early family influences; peer-group influences; characteristics of the situation; miscommunication; sex and power motives; and masculinity norms and men's attitudes. The data on each of these factors are considered below.

Cultural values can serve to support rape. Cross-culturally, in preliterate societies, rape is significantly more common in cultures that are characterized by male dominance, a high degree of general violence, and an ideology of male toughness (Sanday, 1981).

Sociologists Larry Baron and Murray Straus (1989), experts in violence research, did an extensive study to test the last two theories listed earlier, feminist theory and social disorganization theory. Both theories deal with rape as a result of cultural context. Baron and Straus collected extensive data on each of the 50 states in the United States, seeing them as representing variations in cultural context (think, for example, of the different cultures of Louisiana, New York, and North Dakota). To test feminist theory, they collected data on the extent of gender inequality in each state (for example, the gap between men's and women's wages); they also examined the feminist hypothesis that use of pornography encourages rape, by collecting data on the circulation of pornographic magazines in each state. They also obtained measures of social disorganization, such as the number of people moving into or out of the state, the divorce rate, and even the number of tourists flowing into the state. Their data gave strong support to three conclusions: (1) Gender inequality is related to rape (the states with the greatest gender inequality had the highest rape rates); (2) pornography provides ideological support for rape (the states with the highest circulation of pornographic magazines tended to have the highest rape rates); and (3) social disorganization contributes to rape (those states with the

Victim-precipitated: The view that rape is a result of a woman "asking for it."

greatest social disorganization tended to have the highest rape rates). This research emphasizes how important cultural context is in creating a social climate that encourages or discourages rape.

Sexual scripts play a role in rape as well (Byers, 1996). Adolescents quickly learn society's expectations about dating and sex through culturally transmitted sexual scripts. These scripts support rape when they convey the message that the man is supposed to be oversexed and the sexual "aggressor." By adolescence, both girls and boys endorse scripts that justify rape (Koss et al., 1994). A study of 1700 middle school students revealed that approximately 25 percent of the boys said that it was acceptable for a man to force sex on a woman if he had spent money on her (Koss et al., 1994). These findings have been replicated in a number of studies of high school and college students (e.g., Goodchilds & Zellman, 1984; Muehlenhard, 1988).

Early family influences may play a role in shaping a man into becoming a sexual aggressor. Specifically, young men who are sexual aggressors are likely to have been sexually abused themselves in childhood (Friedrich et al., 1988; Koss et al., 1994).

The peer group can have a powerful influence, encouraging men to rape. For an example, see Focus 17.3, which describes the ways in which the peer group in a fraternity created a climate that encouraged its members to rape.

Characteristics of the situation play a role. Secluded places foster rape, as do parties in which excessive alcohol use is involved (Koss et al., 1994). Another situational factor is social disorganization, as noted earlier. An extreme example is war, in which rape of women is common (Brownmiller, 1975). In the 1990s we have seen graphic examples of this in the war in the former Yugoslavia. Bosnian women—Croats and Muslims—were frequently raped by the attacking Serbs.

Miscommunication between women and men is a factor. In the section on date rape we saw a case in which the man and the woman had totally different understandings of what had occurred. Because many people in the United States are reluctant to discuss sex directly, they try to infer sexual interest from subtle nonverbal cues, a process that is highly prone to errors (Abbey, 1991). Specifically, men are likely to interpret a woman's friendly behavior as carrying a sexual message that she did not intend (Abbey, 1991).

Sex and power motives are involved in rape. Feminists have stressed that rape is an expression of power and dominance by men over women (Brownmiller, 1975). Current theory emphasizes that sexual motives and power motives are both involved and interact with each other. A number of processes may be involved (Barbaree & Marshall, 1991). For example, rapists may differ from nonrapists in their ability to suppress sexual arousal when it occurs under inappropriate circumstances. Rapists may be capable of experiencing sexual arousal and hostile aggression simultaneously, whereas other men find that hostile aggression inhibits sexual arousal.

Finally, *masculinity norms and men's attitudes* are another factor (Koss et al., 1994), as we will see in the next section.

Rapists

What is the profile of the typical rapist? The basic answer to that question is that there is no typical rapist. Rapists vary tremendously in occupation, education, marital status, previous criminal record, and motivation for committing rape.

A massive program of research by Neil Malamuth, Mary Koss, and their colleagues identified four factors that predispose men to engage in sexual coercion of women (Malamuth, 1998; Malamuth et al., 1991):

1. *Violent home environment* A boy who grows up in a hostile home environment has a higher likelihood of engaging in sexual aggression against women. Factors that create a hostile home environment include violence between the parents or abuse directed toward the child, whether battering or sexual abuse.

2. *Delinquency* Being involved in delinquency is itself made more likely by coming from a hostile home. But the delinquency in turn increases the likelihood of engaging in sexual coercion—the boy associates with delinquent peers who, for example, encourage hostile attitudes and rationalizations for committing illegal acts and reward a tough, aggressive image.

3. *Sexual promiscuity* The male, often in the context of the delinquent peer group, develops a heavy emphasis on sexual conquests to bring him self-esteem and status with the peer group.

Focus 17.3

Fraternity Gang Rape

Anthropologist Peggy Sanday (1990) has analyzed a widely publicized case of gang rape in a fraternity at a particular university, as well as many other similar cases documented at other universities.

Men join fraternities for many possible reasons. Some may anticipate establishing networks of friendships that will help them in their future careers. But often freshmen, insecure in a complex new environment, join the fraternity to find security. According to Sanday's analysis, the initiation rituals of many fraternities follow a sequence of creating high levels of anxiety in the new members, followed by a male bonding ritual that makes them "brothers." Essentially the young man's identity as an individual is undermined while loyalty to the group is prized, indeed enforced.

In the case investigated by Sanday, the XYZ fraternity (she used this name to guard the anonymity of the population being studied, as required by the ethical standards for anthropologists) had a practice called the "XYZ express," referring to an express train. It involved a gang rape in which a woman, typically drunk or surreptitiously drugged so that she was barely conscious, was raped successively by a series of brothers who stood in line to take their turn, just as cars in a train are in a line. Often this occurred toward the end of a party, when the brothers themselves were drunk.

Sanday points out how this practice has two consequences: it establishes dominance over a woman, and it promotes strong bonds among the fraternity brothers. The practice, of course, fits the definition of rape and is illegal. Yet many of the brothers, when the case was brought to court, said that they had no idea that their activities were wrong or illegal. The culture of the fraternity had dulled their capacity to make a rational judgment. The judge who heard the case was astounded that universities would tolerate, indeed support, institutions that created an environment in which such acts could occur.

Sanday notes anthropologists' findings that, cross-culturally, some societies are free of sexual assault while others are rape-prone. She concludes, "Social ideologies, not human nature, prepare men to abuse women" (Sanday, 1990, p. 192). The XYZ fraternity and others like it are essentially a subculture that socializes men to have sexist attitudes toward women and creates an environment in which gang rape is likely to occur.

Source: Peggy R. Sanday. (1990). *Fraternity gang rape.* New York: New York University Press.

Coercion may seem to him a reasonable way of making conquests.

4. *A hostile masculine personality* This personality constellation involves deep-seated hostility toward women together with negatively defined, exaggerated masculinity—masculinity defined as rejecting anything feminine such as nurturance, and emphasizing power, control, and "macho" characteristics.

Surprisingly, this research was not based on convicted rapists but rather on a national representative sample of male college students. The factors that contribute to sexual aggression against women can be present even in such apparently benevolent populations.

One factor seems to attenuate or reduce a man's likelihood of raping: empathy (Dean & Malamuth, 1997). That is, a man who has several of the risk factors listed above, but who also is sensitive to others' feelings and needs and is not self-centered, is not likely to rape, compared with a man who has the risk factors and lacks empathy and is self-centered. These research findings have important implications for programs of therapy for convicted rapists. Empathy training should be emphasized, as it is in the most modern programs (Marshall, 1993; Pithers, 1993).

Men as Victims of Rape

Women are far more likely than men to be the victims of rape; according to the NHSLS, 22 percent of women had been the objects of forced sex with a man, whereas only about 1 percent of men had been the objects of forced sex with a woman (Laumann et al., 1994). In fact, it is more common for a man to have been forced to have sex by another man (1.9 percent of men) than by a woman (1.3 percent of men).

Nonetheless, it is possible for a woman to rape a man; research shows that men may respond with an erection in emotional states such as anger and terror (Sarrel & Masters, 1982). In a study of 115 men who had been sexually assaulted, 7 percent had been assaulted by a woman or group of women and an additional 6 percent by both a man and a woman (King & Woollett, 1997). Forced vaginal intercourse occurred in only 2 of the cases. Research shows that men who have been raped experience symptoms of PTSD, as women do (Sarrel & Masters, 1982). It is important for counselors and others in helping professions to recognize this possibility of male rape victims.

In a study of date rape in a sample of college students, 13 percent of the women and 9 percent of the men reported that they had experienced forced sex at some time while at the university (Struckman-Johnson, 1988). The type of coercion, however, was considerably different for men and women. The majority of men reported that they were coerced by psychological pressure, whereas the majority of women said they were coerced by physical force. One man reported,

> I was invited over for a party, unaware that it was a date. As the evening wore on, I got the message that the girl was my date. I didn't have to make a move on her because she was all over me. She wouldn't take no for an answer. Usually I like to get

to know the person. I felt I was forced into sex. After, I felt terrible and used. (Struckman-Johnson, 1988, p. 238)

Another study investigated a broader category: unwanted sexual activity; that is, engaging in sexual activities when you really don't want to (Muehlenhard & Cook, 1988). The researchers actually found that more men (63 percent) than women (46 percent) had experienced unwanted intercourse. The most common reasons men gave for engaging in the unwanted activity were enticement or seduction, altruism (wanting to satisfy the other person or not wanting the other person to feel rejected), inexperience (wanting to get experience or to have something to talk about with peers), and intoxication. Most of these incidents would not fit the legal definition of rape, but they do indicate the ways in which men, too, can be coerced into sexual activity.

Having recognized the possibility of men being raped by women, it is important to note that the great majority of male rape victims are raped by men, not women, often in prison (Calderwood, 1987).

Prison Rape

According to a study of 516 men and women prisoners in a state prison system, 22 percent of the men and 7 percent of the women had been the objects of sexual coercion (Struckman-Johnson et al., 1996). Prison staff were the perpetrators in 18 percent of the cases, and fellow prisoners were the perpetrators in the remainder. Among the male victims, 53 percent had been forced to have receptive anal sex, sometimes with multiple male perpetrators, and 8 percent were forced to have receptive oral sex. The men reported severe emotional consequences. Inmates offered a number of suggestions for ending prison sexual violence. The most frequent was to segregate those who are most vulnerable: those who are young, nonviolent, new in prison, white. Many also favored allowing conjugal visits.

Prison rape is a particularly clear example of the way in which rape is an expression of power and aggression; prisoners use it as a means of establishing a dominance hierarchy.

Ethnicity and Rape

We have seen how cultural context can promote or inhibit rape and affect the meaning people attach to rape. The different cultural heritages of the various ethnic groups in the United States provide different cultural contexts for people of those groups, so it is important to consider patterns of rape in U.S. ethnic groups.

Rape has a highly charged meaning in the history of African Americans (Wyatt, 1992). In the period following the Civil War, an African American man convicted of rape or attempted rape of a white woman was typically castrated or lynched. In sharp contrast, there was no penalty for a white man who raped a black woman. Moreover, stereotypes originating at that time and continuing to the present portray both African American men and African American women as being highly sexual. Black women are so highly sexual, the reasoning goes, that they cannot be raped. The result is that African American women have a long history of nondisclosure of rape, a pattern that exceeds even white women's. Many African American women think that no one will believe they can be raped and that they will have no credibility as rape victims.

Research on a random sample of women in Los Angeles indicates that the rate of attempted or completed rape incidents was nearly the same for the two groups—25 percent for African American women and 20 percent for white women (Wyatt, 1992). However, only 23 percent of the black women reported the incident to the police or a rape crisis center, compared with 31 percent of the white women. Black women and white women were similar in their experience of the effects of the rape, such as its negative impact on their later sexual functioning.

Another survey of a random sample of Los Angeles women compared the rape experiences of Anglo and Latin women (Sorenson & Siegel, 1992). The results indicated that the Latinas were considerably less likely to have been the victims of a sexual assault—8.1 percent of Latinas compared with 19.9 percent of Anglos. The researchers interpreted this difference as being due to values in Latino culture, particularly among those born in Mexico, that place strong emphasis on the family and uphold patriarchal attitudes that insist that men should protect women.

Preventing Rape

Strategies for preventing rape fall into three categories: (1) avoiding situations in which there is a high risk of rape; (2) if the first strategy has failed, knowing some self-defense techniques if a rape attempt is actually made; and (3) changing attitudes that contribute to rape.

The first strategy, of course, is to be alert to situations in which there is a high risk of rape and to avoid them. The Association of American Colleges, for example, recommends the following to avoid date rape situations (Hughes & Sandler, 1987, p. 3):

Set sexual limits.   No one has a right to force you to do something with your body that you don't want to do. If you don't want to be touched, for example, you have a right to say "Don't touch me," and to leave if your wishes are not respected.

Decide early if you would like to have intercourse. The sooner you communicate your intentions firmly and clearly, the easier it will be for your partner to understand and accept your decision.

Do not give mixed messages; be clear.   Say "yes" when you mean "yes" and "no" only when you mean "no."

Be forceful and firm.   Do not worry about being polite if your wishes are being ignored.

Do not do anything you do not want to do just to avoid a scene or unpleasantness.   Do not be raped because you are too polite to get out of a dangerous situation or because you are worried about hurting your date's feelings. If things get out of hand, be loud in protesting, leave, and go for help.

Be aware that alcohol and drugs are often related to date rape.   They compromise your ability (and that of your date) to make responsible decisions.

Trust your gut-level feelings.   If the situation feels risky to you, or if you feel you are being pressured, trust your feelings. Leave the situation or confront the person immediately.

Be careful when you invite someone to your home or you are invited to your date's home.   These are the most likely places for date rapes to occur.

If this first set of strategies—avoiding rape situations—does not work, self-defense strategies are needed. Always remember that the goal is to get away from the attacker and run for help.

Many universities, YWCAs, and other organizations offer self-defense classes for women, and we believe that every woman should take at least one such course. Many techniques are available. Judo

Figure 17.2   Ethnicity and rape. Rape has a highly charged meaning in the history of African Americans. In the time of slavery, although there was no penalty for a white man who raped a black woman, a black man convicted of raping a white woman was typically castrated or put to death.

(and aikido, which is similar) emphasizes throwing and wrestling. Tae kwon do (Korean karate) emphasizes kicking. Jujitsu uses combinations of these strategies. The exact method the woman chooses is probably not too important, as long as she does know some techniques. Related to this is the importance of getting exercise and keeping in shape; this gives a woman the strength to fight back and the speed to run fast. Research shows that fighting back—fighting, yelling, fleeing—increases a woman's likelihood of thwarting a rape attempt (Ullman & Knight, 1993; Zoucha-Jensen & Coyne, 1993).

Self-defense, though, is useful to the woman only in defending herself once an attack is made. It would be better if rape could be rooted out at a far earlier stage so that attacks never occurred. To do this, our society would need to make a radical change in the way it socializes males (Hall & Barongan, 1997). If little boys were not so pressed to be aggressive and tough, perhaps rapists would never

develop. If adolescent boys did not have to demonstrate that they are hypersexual, perhaps there would be no rapists. As we noted earlier, rape is unheard of in some societies where males are socialized to be nurturant rather than aggressive.

Changes would also need to be made in the way females are socialized, particularly if women are to become good at assertiveness and self-defense. Weakness is not considered a desirable human characteristic, so it should not be considered a desirable feminine characteristic, especially because it makes women vulnerable to rape. Mothers particularly need to think of the kinds of role models they are providing for their daughters and should consider whether they are providing models of weakness. The stereotype of female weakness and passivity is so pervasive in our society that a complex set of strategies will be required to change it. Girls also need increased experience with athletics as they grow up. This change should have a number of beneficial effects: building strength, speed,

Figure 17.3 Self-defense classes for women. Many experts believe that all women should take such classes to gain the skills necessary to defend themselves in the case of an attempted rape.

and agility in women; building confidence in their ability to use their own bodies; and decreasing their fear of rough body contact. All these developments would help them defend themselves against rape. While some people think that it is silly for the federal government to rule that girls must have athletic teams equal to boys' teams, it seems quite possible that the absence of athletic training for girls has contributed to making them rape victims.

Finally, for both males and females, we need a radical restructuring of our ideas concerning sexuality. As long as females are expected to pretend to be uninterested in sex and as long as males and females continue to play games on dates, rape will persist.

Child Sexual Abuse

In this section we discuss the sexual coercion of children, including the broad category of child sexual abuse and one specific subcategory, incest, when the sexual abuse occurs within the family.

Patterns of Child Sexual Abuse

How common is child sexual abuse? According to the NHSLS, 17 percent of women and 12 percent of men had had sexual contact, as a child, with an adolescent (aged 14 to 17) or an adult (Laumann et al., 1994). Another survey found rates of 15 percent for females and 6 percent for males (Finkelhor, 1984). It seems clear that child sexual abuse is not rare and that females are more often its victims than males are.

Most cases are never reported. In the NHSLS, only 22 percent of victims said that they had told anyone.

The great majority of perpetrators of child sexual abuse are men. According to the NHSLS, for girls almost all the cases involved sexual contact with men; for boys, some cases involved men and

Figure 17.4 Child sexual abuse has become a major concern, as exemplified by this educational poster.

some women, although cases involving men were considerably more common. In another study, 94 percent of all perpetrators were men (Finkelhor, 1984). A number of factors probably account for this great imbalance. Men in our culture are socialized more toward seeing sexuality as focused on sexual acts rather than as part of an emotional relationship. The sexual script for men involves partners who are smaller and younger than themselves, whereas women's sexual script involves partners who are larger and older than they are.

In the great majority of cases, both for boys and girls, the sexual activity involved only touching of the genitals (Laumann et al., 1994). However, for girls, 10 percent of cases involved forced oral sex, 14 percent of cases involved forced vaginal intercourse, and 1 percent involved forced anal sex. For boys, 30 percent of cases involved forced oral sex and 18 percent involved forced anal sex.

Sexual abuse may occur at astoundingly young ages. For example, for girls, 33 percent of cases occurred when they were under 7 years of age, and an additional 40 percent occurred among girls between 7 and 10 years of age (Laumann et al., 1994).

Table 17.1 shows the relationship between adults who committed child sexual abuse and their victims, according to the NHSLS. Notice that sexual abuse by strangers is not common. Most abusers are family friends and relatives.

Patterns of Incest

Incest is typically defined as sexual contact between blood relatives, although the definition is often extended to include sex between non-blood relatives—for example, between stepfather and stepdaughter.

Fifty years ago, incest was widely believed to be a rare and bizarre occurrence. Early research confirmed this notion, indicating that the incidence of incest cases prosecuted by the police was only about one or two people per million per year in the United States (Weinberg, 1955). The catch, though, is that the overwhelming majority of cases go unreported to authorities and unprosecuted. The NHSLS data (Table 17.1) show what a large percentage of child sexual abuse cases are perpetrated by adults within the family. However, because it specified that the sexual contact had to be with an adult or an adolescent aged 14 or older, the NHSLS missed one category of incest, namely sibling incest. In a general survey of undergraduates, 15 percent of the females and 10 percent of the males said that they had had a sexual experience with a sibling (Finkelhor, 1980). It is likely that sibling incest is the most common form of incest.

Psychological Impact on the Victim

Many therapists who are experienced with cases of child sexual abuse feel that the effects on the victim are serious and long-lasting, despite the fact that the incidents were not reported and seem to have been forgotten (Herman, 1981). Consider the following case:

Incest: Sexual contact between relatives.

A 25-year-old office worker was seen in the emergency room with an acute anxiety attack. She was pacing, agitated, unable to eat or sleep, and had a feeling of impending doom. She related a vivid fantasy of being pursued by a man with a knife. The previous day she had been cornered in the office by her boss, who aggressively propositioned her. She needed the job badly and did not want to lose it, but she dreaded the thought of returning to work. It later emerged in psychotherapy that this episode of sexual harassment had reawakened previously repressed memories of sexual assaults by her father. From the age of 6 until midadolescence, her father had repeatedly exhibited himself to her and insisted that she masturbate him. The experience of being entrapped at work had recalled her childhood feelings of helplessness and fear. (Herman, 1981, p. 8)

One study reached different conclusions, though. In a general survey of 526 undergraduates, 17 percent of the students reported having had a sibling sexual encounter in childhood (Greenwald & Leitenberg, 1989; see also Finkelhor, 1980). There were no differences between this group and those who had had no such encounters, on a variety of measures of sexual behavior and adjustment, including incidence of premarital intercourse, age at first intercourse, number of sexual partners, sexual satisfaction, and sexual disorders. The researchers concluded that childhood sexual experiences with a sibling close in age have no effect—positive or negative—on adult sexual adjustment.

In a major review of studies of children who were sexually abused (either by family members or by nonrelatives), the researchers concluded that there is strong evidence of a number of negative effects on these children, compared with control groups of nonabused children (Kendall-Tackett et al., 1993; see also Spaccarelli, 1994). Sexually abused children were significantly more likely to have symptoms of anxiety, post-traumatic stress disorder (PTSD), depression, poor self-esteem, health complaints, aggressive and antisocial behavior, inappropriate sexual behavior, school problems, and behavior problems such as hyperactivity (Cosentino et al., 1995; Rodriguez et al., 1997). Victims had more severe symptoms when (1) the perpetrator was a member of the family; (2) the sexual contact was frequent or occurred over a long period of time; and (3) the sexual activity involved penetration (vaginal, oral, or anal). The child's gender did not seem to be a factor; that is, there were no dif-

Table 17.1 Categories of People Who Sexually Abuse Children, as Reported by Adults Recalling Incidents of Sexual Abuse in Their Childhood

Perpetrators	Percentage of Adults Abused as Children*	
	Women	Men
Stranger	7%	4%
Teacher	3	4
Family friend	29	40
Older friend of respondent	1	4
Older brother	9	4
Stepfather or mother's boyfriend	9	2
Father	7	1
Other relative	29	13
Other	19	17

*Percentages do not total 100 because some respondents reported on multiple categories of abuse.

Source: Edward O. Laumann, John H. Gagnon, Robert T. Michael, & Stuart Michaels. (1994). *The social organization of sexuality.* Chicago: University of Chicago Press, adapted from Table 9.14, p. 343.

ferences in symptoms between boys and girls. However, the researchers noted that gender was not investigated in many studies, probably because so few boys appeared in most samples.

What, then, are the psychological consequences of incest or other childhood sexual abuse for the victim? Incest may not be damaging to the victim in some cases, particularly if it is brother-sister incest when the two are close in age and it is consensual. However, in most cases childhood sexual abuse is psychologically damaging, and may lead to symptoms such as anxiety and PTSD. Several factors affect how severe the psychological consequences are; they seem to be more severe when the perpetrator is a close family member who is an adult, and when there is extensive sexual contact that involves penetration.

In contrast, some women who were the objects of child sexual abuse perceive some benefit from this adverse life experience. They believe that it made them better at protecting their own children from abuse and made them stronger people (McMillen et al., 1995).

The Offenders

In 1994, Leroy Hendricks was released from prison in Kansas, having served his 10-year term for

Focus 17.4

False Memory Syndrome? Recovered Memory?

One of the nastiest professional controversies today concerns the issue of what some call recovered memory, and what others call false memory syndrome. The issue is sexual abuse or other severe trauma in childhood and whether the child victim can forget (repress) the memory of the event and later recover that memory.

On one side of the argument, the *recovered memory* side, psychotherapists see adult clients who display serious symptoms of prior trauma, such as severe depression and anxiety. Sometimes these clients have clear memories of being sexually abused in childhood and have always remembered the events but never told anyone until they told the therapist. In other cases, the client doesn't remember that any abuse occurred but, during the process of therapy, or sometimes spontaneously before therapy, something triggers the memory and the client now recalls the sexual abuse. Psychotherapists are understandably outraged about the psychological trauma that results from childhood sexual abuse.

On the other side, some psychologists believe that these memories for events that had been for-gotten and then are remembered are actually *false memories.* They argue that unscrupulous or overzealous therapists may induce these memories by hypnotizing clients or strongly suggesting to clients that they had been abused in childhood.

What do the data say? First, there is evidence from laboratory studies that information associated with unpleasant emotions is more likely to be forgotten (e.g., Bootzin & Natzoulas, 1965). Research directly on the issue of child sexual abuse also provides support for the idea that forgetting does occur in some cases. In one study, 129 women who were known to have been sexually abused as children—they had been brought to a hospital for treatment at the time and the abuse had been medically verified—were interviewed 17 years later; 38 percent did not remember their prior abuse (Williams, 1994). The possible flaw in this study is that some of the respondents may have remembered but not have reported it to the interviewer. However, they were reporting many other intimate sexual experiences, so it seems likely that they would report accurately about this one, too. In a study of adult women who reported to a

molesting two 13-year-old boys (Collins, 1997). Rather than walking free, however, he was immediately transferred to a correctional mental health facility where he could remain for the rest of his life. A 1994 Kansas law, the Sexually Violent Predator Act, allowed him to be put away for life, on the grounds that his mental problems made him likely to attack again. In fact, the 1994 conviction was his fifth over a span of about 30 years. Hendricks challenged the constitutionality of the law, but the Supreme Court, in 1997, upheld the law and his treatment under it.

What do the data say about child sexual abusers? Are they likely to repeat the offense? Are there effective treatments for them?

A review of studies of recidivism (repeat offending) by sex offenders indicated the recidivism rate was only about 13 percent for child molesters, within 4 to 5 years after the offense (Hanson & Bussiere, 1998). Although this is an optimistic conclusion, there is still cause for concern because certain subgroups of child molesters had predictably much higher rates of recidivism. For example, recidivism was higher among those who had committed previous sexual offenses, had begun sexual offending at an early age, and had targeted male victims. The strongest predictor of recidivism was phallometric measures of sexual deviance (recall the penile strain gauge discussed in Chapter 14). Subjects are shown slides of children and their arousal (erection) is measured. Those who show the strongest sexual arousal to pictures of children have the highest recidivism

researcher that they had been victims of child sexual abuse, 30 percent said that they had completely blocked out any memory of the abuse for a full year or more (Gold et al., 1994). In another, similar study, 19 percent of adult women who reported child sexual abuse said that they forgot the abuse for a period of time (Loftus et al., 1994). In a national survey of a sample of psychologists, 24 percent reported being the victims of child sexual abuse; of those, 40 percent reported a period of forgetting (Feldman-Summers & Pope, 1994).

Abuse was more likely to have been forgotten when it was more severe. Therefore, the evidence seems to indicate that in from 19 to 40 percent of cases memories of childhood sexual abuse are forgotten for a period of time and then remembered again. In fact, it has been theorized that amnesia for traumatic events of this kind, particularly when the child has been betrayed by someone like a parent, is an adaptive response that helps the child survive in a terribly distressed family situation (Freyd, 1996).

The other question is, can memories of events that did not occur be "implanted" in someone? In one study, the researcher was able to create false memories of childhood events in 25 percent of the adults given the treatment (Loftus, 1993). Certain conditions seem to increase the chances that people will think they remember things that did not actually occur, including suggestion by an authority figure and suggestion under hypnosis.

What is the bottom line? There is evidence that some people do forget childhood sexual abuse and remember it later. There is also evidence that some people can form false memories based on suggestions made by another person. It seems likely that most of the cases of recovered memory of child sexual abuse are true, but that some are false and the product of suggestion. To put the matter in perspective, each year thousands of children are sexually abused; the vast majority of these cases go unreported and the perpetrators go unpunished. It is also probably true that false accusations of past child abuse are made, often by a well-intentioned "victim" who is highly suggestible to press reports of other cases or who has been misled by an overly zealous therapist. As a result, an accused person who is actually innocent may be arrested. There are errors of justice on both sides. Nonetheless, there are many more cases of unreported and unpunished perpetrators than of falsely convicted persons.

Sources: Bootzin & Natzoulas (1965); Feldman-Summers & Pope (1994); Freyd (1996); Gold et al. (1994); Loftus (1993); Loftus et al. (1994); Williams (1994).

---

rates. Therefore, depending on the particular case, there may be a very low or very high risk of repeating the offense, and we know some of the factors that predict which category a particular offender might belong in. It is also true that, in the studies that were reviewed, many of the offenders had been given some treatment. The low recidivism rate might therefore tell us more about the success of the treatment programs than about natural rates of recidivism among those who have not been given some rehabilitation.

A number of treatments for child sexual abusers are in use: surgical castration, antiandrogen drugs, hormones, SSRIs, and cognitive-behavioral therapy (Bradford & Greenberg, 1996; Hall, 1995; Marshall & Pithers, 1994). The idea behind surgical castration is that removal of a man's testes sharply reduces his levels of testosterone, with the hope that his sexual and aggressive behavior will also be reduced sharply. Cyproterone acetate (CPA) is an antiandrogen drug—that is, it reduces the action of testosterone in the body and is therefore a kind of chemical castration—that has been used in the treatment of child sexual abusers. Research indicates that CPA greatly reduces pedophiles' sexual arousal to children (Bradford & Greenberg, 1996). Medroxyprogesterone acetate (MPA) is a hormone that has a similar antiandrogen effect and is used in the treatment of sex offenders. Its effects are similar to those with CPA but, for a variety of technical reasons, CPA seems to be the preferred drug. Results with a new anti-GnRH drug are quite promising (Rosler & Witztum, 1998). By blocking the action of GnRH, pituitary and gonadal functioning is inhibited, again reducing levels of testosterone. A relatively new class of antidepressants (the SSRIs, which include Prozac and Zoloft) are also proving to be successful in the treatment of sex offenders (Bradford & Greenberg, 1996). Their use is based on the assumption that sex offending can be a particular kind of obsessive-compulsive disorder, and such disorders generally respond well to these antidepressants. Cognitive-behavioral therapy makes use of a number of techniques, including cognitive restructuring, masturbation reconditioning [the man learns to experience arousal to appropriate persons (adults) rather than inappropriate persons (children)], role-playing, desensitization, and stress management (Marshall & Pithers, 1994).

A review of studies assessing the effectiveness of the various treatments found that overall, the recidivism rate was 27 percent for untreated sex offenders and 19 percent for treated sex offenders, a significant improvement (Hall, 1995). Hormonal treatments and cognitive-behavioral therapy were equally effective. Cognitive-behavioral therapy has advantages, though, because a large proportion of offenders refuse hormone treatment or discontinue it.

Sexual Harassment

The issue of sexual harassment exploded into public awareness during the dramatic and widely televised hearings involving Anita Hill and Clarence Thomas during Thomas's confirmation to the Supreme Court in 1992. The issue is a powerful one—it can force a victim out of a job, but it also might force a perpetrator from a job.

The official definition of sexual harassment, given by the U.S. Equal Employment Opportunity Commission (EEOC), is as follows:

Unwelcome sexual advances, requests for sexual favors, and other verbal or physical conduct of a sexual nature constitute sexual harassment when

A. Submission to such conduct is made either explicitly or implicitly a term or condition of an individual's employment or academic advancement,

B. Submission to or rejection of such conduct by an individual is used as the basis for academic or employment decisions affecting that individual, or

C. Such conduct has the purpose or effect of unreasonably interfering with an individual's work or academic performance or creating an intimidating, hostile, or offensive working or educational environment.

The key ingredients for sexual harassment, then, are that the sexual advances are unwelcome and that they are coercive in the sense that the victim's job or grade is at stake. This is termed "quid pro quo harassment" (*quid pro quo* meaning "I'll do something for you if you'll do something for me"). Point C of the definition specifies that a "hostile environment" also constitutes harassment—that is, if an employee is in a work environment that is so hostile (constant lewd innuendoes, verbal intimidation, and so on) that he or she cannot work effectively, then that fits the definition of harassment, even if there has been no explicit sexual proposition directed to the employee.

The EEOC definition addresses sexual harassment at work and in education. Sexual harassment may occur in other contexts as well, such as in psychotherapy or on the street.

Sexual Harassment at Work

Sexual harassment at work may take a number of different forms. A prospective employer may make it clear that sexual activity is a prerequisite to being hired. Stories of such incidents are rampant among actresses. Once on the job, sexual activity may be made a condition for continued employment, for a promotion, or for other benefits such as a raise. Here is one case:

A woman who works at a factory in Monroe, Wisconsin, has been subjected to an onslaught of outrageous conduct during the seven years she has been there. For her birthday, her male coworkers constructed and presented her with an inflatable dildo. "I just laughed it off because, if I let them know it bothered me, they'd (harass me) more," she said.

She said the men call the women in the plant "office c---" in front of them and one woman is always being called "a life support system for a pr---."

One man came up to her and said that he was to be married in two weeks. "Last chance to f--- me." he said.

(a)   *(b)*

Figure 17.5   In 1992, Anita Hill's testimony *(a)* in the televised confirmation hearings of Supreme Court nominee Clarence Thomas *(b)* captured the nation's attention and sparked much debate over sexual harassment.

When she was pregnant, she was constantly asked who at the factory was the father.

Things got worse when she divorced six months ago; a coworker called her a whore when she declined his invitation to go out. Her managers ordered the man not to talk to her—but now everyone in the office ignores her. "I'm looked down on for turning him in," she said. "When you work with men and go up against them, you can't walk the halls without being talked about." Even the female coworkers think she lied about harassment, she said.

She said she feels like an "outsider" now. "It's very hard to show up at work every day." She can't quit, though, because she supports two children. (*Source:* Told to newspaper reporter, 1992)

It is clear in such incidents how damaging a hostile work environment can be.

Surveys indicate that sexual harassment at work is far more common than many people realize. In one well-sampled study, 21 percent of women workers reported that they had been the object of incidents that an expert definitely classified as sexual harassment, compared with 9 percent of the men workers (Gutek, 1985). In a well-sampled survey of over 20,000 members of the federal work force, 42 percent of the women and 15 percent of the men reported having been sexually harassed at work within the preceding two years (Tangri et al., 1982; see also Levinson et al., 1988, for similar results). The preponderance of harassers (78 percent) were male.

Both women and men victims report that harassment has negative effects on their emotional and physical condition, their ability to work with others on the job, and their feelings about work (Tangri et al., 1982). However, men are more likely to feel that the overtures from women ended up being reciprocal and mutually enjoyable. Women, on the other hand, are more likely to report damaging consequences, including being fired or quitting their job (Gutek, 1985). There is evidence linking the experience of sexual harassment to depression and PTSD (Fitzgerald, 1993).

Figure 17.6 This man has positioned himself so that the woman cannot avoid contact, and, if he is her supervisor, she may be hesitant to protest.

Why does sexual harassment at work occur? According to a theory proposed by psychologists Susan Fiske and Peter Glick (1995), it results from a combination of gender stereotyping and men's ambivalent motives. Stereotypes of women in American culture are complex and include three distinct clusters: "sexy," "nontraditional" (e.g., feminist), and "traditional" (e.g., mother). Many men have ambivalent motives in their interactions with women because they desire both dominance and intimacy. Fiske and Glick argue that there are four types of harassment. With the first, *earnest harassment,* the man is truly motivated by a desire for sexual intimacy, but he won't take "no" for an answer and persists with unwelcome sexual advances. He stereotypes women as sexy. With the second type, *hostile harassment,* the man's motivation is domination of the woman, often because he perceives her as being competitive with him in the workplace. He holds the stereotype of women as nontraditional and therefore competitive with him. His response to rejection by a woman is

increased harassment. The third and fourth types of harassment involve ambivalent combinations of the two basic motives, dominance and a desire for intimacy. In the third type, *paternalistic-ambivalent harassment,* the man is motivated by a desire for sexual intimacy but also by a paternalistic desire to be like a father to the woman. This type of harassment may be particularly insidious because the man thinks of himself as acting benevolently toward the woman. Finally, the fourth type, *competitive-ambivalent harassment,* mixes real sexual attraction and a stereotype of women as sexy with the man's hostile desire to dominate the woman, which is based on his belief that she is nontraditional and competitive with him. This theory gives us an excellent view of the complex motives that underlie men's sexual harassment of women.

Sexual harassment at work is more than just an annoyance. Particularly for women, because they are more likely to be harassed by supervisors, it can make a critical difference in career advancement. For the working-class woman who supports her family, being fired for sexual noncompliance is a catastrophe. The power of coercion is enormous.

Sexual Harassment in Education: An A for a Lay

Sexual harassment in education was brought to public attention when, in 1977, women students sued Yale University, complaining of sexual harassment, in the important case *Alexander v. Yale.* The case recognized that sexual harassment of women in education was a possible violation of Title IX of the Civil Rights act.

The data indicate that about 50 percent of female students have been harassed by professors, with acts ranging from insults and come-ons to sexual assault (Fitzgerald, 1993). Women report dropping courses, changing majors, or dropping out of higher education as a result of sexual harassment (Fitzgerald, 1993).

In the wake of the Yale case and others, many universities have set up grievance procedures for sexual harassment cases.

Psychotherapist-Client Sex

Legal definitions of sexual harassment focus on these problems in the workplace or in education.

However, there is another category of coercive and potentially damaging sexual encounters—those between a psychotherapist and client, or between other professionals, such as physicians, and patients. Professional societies such as the American Psychological Association state clearly in their codes that such behaviors are unethical. Nonetheless, they occur, and can be damaging.

One survey of a sample of licensed Ph.D. psychologists found that 5.5 percent of the male and 0.6 of the female psychologists admitted having engaged in sexual intercourse with a client during the time the patient was in therapy, and an additional 2.6 percent of male and 0.3 percent of female therapists had intercourse with clients within three months of termination of therapy (Holroyd & Brodsky, 1977). These are probably best regarded as minimum figures because they are based on the self-reports of the therapists and some might not be willing to admit such activity even though the questionnaire was anonymous. Of those therapists who had intercourse with clients, 80 percent repeated the activity with other clients.

Experts regard this kind of situation as having the potential for serious emotional damage to the client (Williams, 1992). Like the cases of sexual harassment discussed earlier, it is a situation of unequal power, in which the more powerful person, the therapist, imposes sexual activity on the less powerful person, the client. The situation is regarded as particularly serious because people in psychotherapy have opened themselves up emotionally to the therapist and therefore are extremely vulnerable emotionally.

SUMMARY

Rape is defined as nonconsensual oral, anal, or vaginal penetration obtained by force, by threat of bodily harm, or when the victim is incapable of giving consent. A woman's lifetime risk of being raped is between 14 and 25 percent. Victims may experience post-traumatic stress disorder (PTSD) as a result of the assault. Date rape and marital rape are more common than many people realize. There are four major theoretical views of rape: victim-precipitated, psychopathology of rapists, feminist, and social disorganization. Rape has particularly charged meanings for some ethnic groups within the United States.

Approximately 17 percent of women and 12 percent of men report that, when they were children, they had sexual contact with an adult or adolescent over 14 years of age. Most sexual abuse of children is committed by a relative or by a family friend. Sexually abused children are more likely than other children to have symptoms such as anxiety, PTSD, depression, and health complaints. More severe psychological consequences are likely to occur when the perpetrator is a close family member who is an adult (sibling incest seems less harmful) and when the sexual contact is extensive and involves penetration. Some types of child molesters have a low rate of recidivism, but certain types are highly likely to repeat their offense. Drugs such as MPA and CPA are effective treatments for sex offenders, as is cognitive-behavioral therapy.

There is a controversy among professionals over whether adults can recover memories of child sexual abuse that they had forgotten (recovered memory), or whether these are cases of false memory syndrome, in which the supposedly remembered incidents never actually occurred.

Sexual harassment, whether on the job or in education, involves unwelcome sexual advances when there is some coercion involved, such as making the sexual contact a condition of being hired or receiving an A grade in a course. In another form of sexual harassment, the work or educational environment is made so hostile, on a sexual and gender basis, that the employee cannot work effectively. Surveys show that sexual harassment at work is fairly common. In severe cases it can lead to damaging psychological consequences for the victim, such as PTSD. In education, the data indicate that about 50 percent of female students have been harassed by professors. This abuse can lead to negative consequences for the student, such as being forced to change majors or drop out of school. Sex between psychotherapist and client also carries the potential for psychological damage to the client.

REVIEW QUESTIONS

1. Averaged over many studies, the lifetime risk of a woman being raped is about 3 percent. True or false?

2. A person who believes that a raped woman is really a slut who got what she was asking for holds the _____ view of rape.

3. Currently, experts believe that _____, rather than rape trauma syndrome, is the best label for the responses of rape victims.

4. Research shows that a hostile masculine personality predisposes a man to rape. True or false?

5. Forced sexual intercourse on a date does not meet the legal definition of rape. True or false?

6. In the 1800s, there were no penalties for a white man who raped an African American woman. True or false?

7. Research indicates that the recidivism rate for child molesters is approximately 90 percent. True or False?

8. Antiandrogen drugs are an effective treatment for sex offenders, but cognitive-behavioral therapy is not effective. True or false?

9. Surveys of the general population indicate that the most common form of incest is father-daughter incest. True or false?

10. A job interviewer promising a woman that she will get a job if she has intercourse with him is an example of sexual harassment. True or false?

(The answers to all review questions are at the end of the book, after the Glossary.)

QUESTIONS FOR THOUGHT, DISCUSSION, AND DEBATE

1. On your campus, what services are available for rape victims? Do these services seem adequate, given what you have read in this chapter about victim responses to rape? What could be done to improve the services?

2. Find out what procedures are available on your campus to deal with incidents of sexual harassment of a student by a professor.

3. Apply the four theoretical views of rape to child sexual abuse.

4. Angie was raped last night at a party, by a man she had met previously in one of her classes. Should she report it to the police?

5. In this chapter we have discussed sexual harassment, which involves repeated unwelcome sexual advances, or a requirement of sex in return for something like being hired or getting a raise, or an environment that is so sexually hostile that the person has difficulty working. Do you think there is a racial parallel to sexual harassment? That is, do you think that racial harassment exists? If yes, how would you define it, and do you think it should be illegal? If not, why not?

SUGGESTIONS FOR FURTHER READING

Brady, Katherine. (1979). *Father's days.* New York: Dell paperback. This autobiography of an incest victim is both moving and insightful.

Koss, Mary B., et al. (Eds.). (1994). *No safe haven: Male violence against women at home, at work, and in the community.* Washington, DC: American Psychological Association. An excellent collection, with chapters by experts on rape and sexual harassment.

Raine, Nancy V. (1998). *After silence: Rape and my journey back.* New York: Crown. Raine, a professional writer, provides an intense account of the aftermath of being raped.

White, Jacquelyn W., & Sorenson, Susan B. (1992). Adult sexual assault. *Journal of Social Issues, 48*(1). The whole special issue of this journal is on sexual assault of adults and is packed with interesting articles.

WEB RESOURCES

http://www.mincava.umn.edu/
University of Minnesota Center Against Violence and Abuse; information, resources, and links about sexual assault and sexual harassment.

http://www.dpscs.state.md.us/pct/ccpi/sexi.htm
Sexual Assault Prevention Resources

http://www.cs.utk.edu/~bartley/saInfoPage.html
Sexual Assault Information Page, developed by University of Tennessee student Chris Bartley.

chapter
18

Sex for Sale

CHAPTER HIGHLIGHTS

> **P**ornography is an expression not of human erotic feeling and desire, and not a love of the life of the body, but of a fear of bodily knowledge, and a desire to silence eros.*

The exchange of sexual gratification for money is a prominent feature of many contemporary societies. It is estimated that it involves at least $20 billion a year in economic activity (The Sex Industry, 1998). In this chapter, we consider two ways in which sex can be bought and sold: prostitution and pornography. Both involve complex legal issues and public controversy, but also attract a steady stream of eager customers.

Prostitution

Prostitutes ("hookers," commercial sex workers) engage in sexual acts in return for money or some other payment such as drugs. As some social critics have pointed out, though, some dating, living arrangements, and marriages also fall into this category. Thus one should probably add the conditions that prostitutes are promiscuous and fairly nondiscriminating in their choice of partners. Female prostitutes will be discussed first.

Kinds of Prostitutes

There are many kinds of prostitutes; they vary in terms of social class, status within prostitution, and lifestyle. In turn, the social status of their customers and the risks they face vary accordingly.

The **call girl** (notice the diminutive "girl") is the elite of prostitutes. She is typically from a middle-class background and is better educated than many prostitutes; some are college graduates. She dresses expensively and lives in an attractive apartment in an upscale neighborhood. A call girl in a medium-size city may charge a minimum of $100 per hour (more if she is classier or engages in atyp-

ical sexual activity); call girls in major metropolitan areas charge $200 or more per hour. A call girl can earn a great deal of money, tax free. But she does have large business expenses: an expensive residence, an extensive wardrobe, high medical bills for maintaining good health (prostitutes rarely have medical insurance), expenses for makeup and hairdressers, large tips for doormen and landlords, and the cost of a reliable answering service.

A call girl may obtain her clients through personal referrals or through a madam (see p. 471). She may have many regular customers. She makes dates only by telephone, which enables her to exercise close control over whom she services. She usually sees clients in her residence, which allows her to control the setting in which she works. In addition to sexual gratification she may provide other services, such as accompanying clients to business or social gatherings.

Below call girls on the status ladder are prostitutes who work out of **brothels.** In the 1800s and early 1900s there were many successful brothels in the United States. They varied from storefront clipjoints, where the customer's money was stolen while he was sexually occupied, to elegant mansions where the customer was treated like a distinguished dinner guest. Brothels declined in number after World War II. A few remain in places like Nevada, where prostitution is legal in 5 counties. In the past twenty years, they have been replaced by **in-call services,** which employ women working

---

*Susan Griffin. *Pornography and silence.* (1981). New York: Harper & Row.

Prostitute: A person who engages in sexual acts in return for money or drugs and does so in a promiscuous, fairly nondiscriminating fashion.

Call girl: The most expensive and exclusive category of prostitutes.

Brothel (BRAH-thul): A house of prostitution where prostitutes and customers meet for sexual activity.

In-call service: A residence in which prostitutes work regular shifts, selling sexual services on an hourly basis.

Figure 18.1   Sex for sale.

regular shifts in an apartment or condo, servicing clients who come to the apartment. These services provide sexual gratification on an hourly basis. In major cities the charge is $150 or $200 per hour; in exchange, the client can participate in standard sexual activity including fellatio, cunnilingus, and vaginal intercourse. Many in-call services also require initial contact by telephone, although others advertise their location in specialized media or even telephone books.

Another contemporary setting for commercial sex is the **massage parlor.** Some massage parlors provide legitimate massage therapy. In others, the employees sell sexual services (Perkins & Bennett, 1985); these often advertise "sensual massage" or "stripassage," making it pretty clear which type of parlor they are. Some parlors offer a standard list of services and prices; other parlors allow the masseuse/masseur to decide what she or he will do with a particular client, and possibly how much of a "tip" is required for that activity. Massage parlors vary greatly in decor and price. Some are located in "professional" buildings, expensively decorated, and provide food and drinks in addition to sexual gratification. Charges may range from $100 to $300 or more. Such parlors may accept charge cards, with the business listed on the monthly statement as a restaurant. At the other end of the scale, storefront parlors, often located in "commercial sex districts," offer no amenities and charge rates of $40 to $100.

Another way in which a veneer of respectability is added to a sex business is by calling it an *escort service.* These services have revealing names such as Alternative Lifestyle Services, First Affair, All Yours, Versatile Entertainment, and Hubbies for Hire. Most escort services employ both men and women who will engage in sexual activity; like massage parlors, the service may

Massage parlor: A place where massages, as well as sexual services, can generally be purchased.

have a standard menu, or the escort may have the autonomy to decide what activities he or she will do with a client. Prostitution in this setting is referred to as **out-call service,** since the escorts go to the clients. This is obviously a more risky business, since the escort cannot control the setting in which the services are provided. Escorts are usually required to telephone the service when they arrive and when they leave the client's location. This not only contributes to their safety, but allows the service to monitor how long the escort spent with the clients, and therefore the amount owed the service.

At the lower end of the status hierarchy is the **streetwalker.** She sells her wares on the streets of cities. She is generally less attractive and less fashionably dressed than the call girl, and she charges correspondingly less for services, perhaps as little as $20 for a "quickie." She is more likely to impose strict time constraints on the customer. Because her mode of operation is obvious, she is likely to be arrested, and she is also more likely to be exposed to dangerous clients. A California streetwalker, Divine Brown, achieved notoriety in 1995 when she and British actor Hugh Grant were arrested as she performed oral sex in his sports car. Grant pleaded guilty to charges of "lewd conduct in public" and was fined.

A **baby pro** is generally younger than 16 years old and works conventions, resorts, and hotels (Shoemaker, 1977). There has been an increasing demand in recent years for this kind of prostitute, especially because it is believed, often erroneously, that girls this young are less likely to be infected with STDs.

Pimps and Madams

The **pimp** ("The Man") is the companion-master of the prostitute. She supports him with her earnings, and in return he gives her companionship and sex, bails her out of jail, and provides her with food, clothing, shelter, and drugs, especially cocaine. He also provides protection against theft and other crimes, since the prostitute is scarcely in a position to go to the police if she is robbed by a customer. The *procurer,* or **panderer,** helps the prostitute and client find each other. Pimps sometimes, though not always, function as panderers.

A **madam** is a woman who manages or owns an in-call service, out-call service, a brothel or an escort service. Typically, she is quite intelligent and has excellent financial and social skills. The sensation of the 1980s was Sydney Biddle Barrows, who operated an escort service in New York City (Barrows, 1986). She was called the "Mayflower Madam" because her ancestors came to the United States on the *Mayflower.* The best-known madam of the 1990s was Heidi Fleiss, dubbed the "Hollywood Madam" by the media. She was convicted of pandering for supplying call girls for rich and famous clients associated with the entertainment industry. Reportedly her customers paid from $200 for a "quick hit" to $1500 for an entire night with one of the attractive, stylish young women who worked for her (Fleming & Ingrassia, 1993). In 1995 she was fined $1500 and sentenced to 3 years in prison.

The Career of a Prostitute

The first step in a prostitute's career is entry. Women enter prostitution for a variety of reasons (Vanwesenbeeck, 1994). The most important is economics. Some women are motivated by a desire for money, material goods, and an exciting lifestyle. These women are attracted by the image of the call girl, and some of them are fortunate enough to attain that status. For some women—for example, a poor but attractive woman—prostitution can be a means of upward economic mobility. Other women enter out of economic necessity, in order to survive. A poorly educated single mother may have no alternative means of earning a living. Some women become prostitutes in order to support a drug addiction.

Force or coercion is another factor. Some women report being coerced physically or psychologically

---

Out-call service: A service that sends a prostitute to a location specified by the client to provide sexual services.
Streetwalker: A lower-status prostitute who walks the streets selling sexual services.
Baby pro: A young prostitute, generally 16 years of age or younger.
Pimp: A prostitute's companion, protector, and master.
Panderer: The person who helps the prostitute and client find each other.
Madam: A woman who manages a brothel, in-call, out-call, or escort service.

Focus 18.1
An Australian Prostitute Tells Her Story

Zoe is an ex-prostitute who has worked in a parlor and as an escort girl. A member of the Australian Prostitutes Collective, she is anxious to help establish a movement in Australia similar to those in other countries. Zoe is a social worker and active feminist, both areas of interest which she wants to incorporate with her concern for prostitute women. She says,

My mother divorced my father when I was 6 years old, so there was my mother, myself, my twin sister, and my little sister in an all-female household. In the early stages we did miss our father and there used to be a lot of fights. We always got on well with our mother, who was very family-oriented. When I started going out with boys at age 14 she put me on the pill. She encouraged us to bring our boyfriends home and that way we didn't have to go screwing in the bicycle shed. At school I was always very much a tomboy, loved sport, and was always a leader of a group.

My first sexual experience was at 15, with my mother's lover. He was living in our house since I was 11, and he used to give us cuddles. In looking back now it was sexual molestation I suppose, but

my twin sister and I competed for his favors. Mum was out a lot so she didn't know what was going on.

In adolescence I had no stable relationships. I think I was fairly promiscuous and I know I was searching for something. Sex was a very important part of my adolescent identity. I experimented a lot with different boys, and I had my first sexual experience with a girl when I was 15 as well.

In 1979 I came from England to Australia to meet my father. I hadn't seen him in 12 years. I came over by myself, did some painting and decorating for about 8 months, but became very bored with it as I wasn't meeting anybody so I did some bar work, receptionist work and waitressing, and then I was unemployed for 5 months. I was feeling really miserable at the time and wasn't getting on very well with my father. He had become an alcoholic.

I was still unemployed, isolated, and lonely. I went for five interviews one day and the last one said "escort." I wondered if it were prostitution or not and when I went for the interview I knew it was a massage parlor. They told me the prices and what the job was, asked me if I wanted it. I said "Yes," and I walked out with $150 that night.

A friend told my father. He offered me money to get out of it, but I refused because there were other

by a husband or lover into selling sex for money. Coercion is reportedly a major factor in prostitution in some countries. Young women in Slavic and Asian countries answer ads in local papers promising wealth and glamour; they are taken to countries such as Israel and Thailand expecting to work in restaurants or hotels but instead are forced into prostitution (Specter, 1998). The "employer" may destroy the woman's passport and threaten her with deportation if she does not comply with his demands. Another category of reasons involves gaining power. For example, a woman who serves as a call girl to famous politicians may think of herself as having access to real

political power. Some women become involved in prostitution through a family member or friend who is already a hustler and can teach them the ropes (Miller, 1986).

On entering prostitution, most women go through an apprenticeship in which they learn the skills of the profession. The apprentice learns sexual techniques, especially fellatio since many customers want oral sex. She learns how to hustle, to successfully negotiate services and price with potential customers. She learns how to maintain control over the interaction so that she can protect herself from being hurt or being robbed by clients. She learns values, like "the customer is always right," and fairness to

reasons for remaining in prostitution than money, such as experimenting, being sexually anonymous, being independent, earning my own money rather than ask him for support.

We had clients coming in for mild, medium, and heavy bondage, but I couldn't do heavy. They were mostly businessmen, but we had army and navy guys as well asking for these. We had one navy guy, about 19, who used to come in for medium bondage. He wanted to be dressed up in a pink nightie and have a dildo shoved up his ass, while there were two of us who used to dress up in black leather raincoats with black underwear, high black shoes and whips. He would grovel around the room in the nightie and we would beat him on the bum while shouting abuse at him. That was $80 for half an hour.

When you're screwing the whole day it can be pretty heavy going, but with bondage it is good money where you don't have to have sex and it's without beating yourself around too much. Also it has the advantage of being able to abuse a guy rather than having the emotional pull of lying down and being fucked by a guy.

When the police closed the parlor, I applied for work at an escort agency. I would then phone up and tell them I would be on call that night. Then I got dressed ready to go out, and sat home waiting for the phone to ring. They might ring up and say there's a client at the WOL Hotel, a businessman with a bankcard, for two hours at $125 an hour. I would catch a cab to the hotel, meet the client in the bar, fill in the bankcard or take the money,

phone through to the office to tell them I'd arrived, have a drink with the client, and go out or up to his room. Most of the work was fairly chatty, chat about his business or silly small talk, do the job in his room and phone through after it to let them know I'd finished.

The job risk is much higher than in parlors. You were very vulnerable in the client's room and have no control over the situation, which can be pretty frightening if things get nasty. You always let each client know that you have to phone the office before and after the job so that he is aware that you are being guarded. Nevertheless, it does not reduce the danger from clients who decide to mug you in the room.

I left the escort work when I fell in love with the guy I was going out with and I decided to set up a stable relationship with him. The relationship went on for 14 months, but I was totally frigid for six months after I gave up prostitution. I had lost myself in prostitution and had become so well established in my identity and role as a prostitute that once I had stopped I couldn't then relate to my lover or myself.

My immediate future plans are basically to finish my studies in humanities at university. I'd like to see all laws against prostitutes repealed, all of them.

Source: Roberta Perkins & Garry Bennett. (1985). *Being a Prostitute.* London: Allen & Unwin, pp. 106–113.

---

other "working girls."[1] Women who are recruited into the life by a pimp may be trained by one of his more experienced "wives." Some women are trained by an experienced madam, in exchange for a large percentage of their fees (Heyl, 1979).

One problem with prostitution is that it is a short-lived career, in this respect resembling the career of a professional basketball or football player. Even a woman who starts as a high-priced call girl may find herself drifting down in status, either as she ages and begins to show wrinkles or if

she gets hooked on alcohol or drugs. In prostitution, seniority is not rewarded.

"Squaring up" or "leaving the life" refers to giving up prostitution. Financially it is a difficult thing to do, particularly for the woman with no job skills; recognizing this, some rehabilitation programs provide job training, as well as a halfway house, to integrate the woman back into society (Winick & Kinsie, 1971).

The married prostitute may simply go back to being a housewife. The unmarried woman may escape through marriage, since she may get proposals from her regular customers.

Other reasons for leaving include arrest and the threat of a long-term jail sentence, government

[1] A 1986 film entitled *Working Girls* provides a realistic look at a contemporary in-call service.

agencies' insistence that she give up her children, and the knowledge that a friend was the victim of violence while she worked as a prostitute (Bess & Janus, 1976). The major hazards associated with being a prostitute are violence (including rape, assault, and murder), drug addiction, sexually transmitted diseases, and arrest by the police (Perkins & Bennett, 1985).

Prostitutes' Well-Being

There are a variety of images of the contemporary prostitute: young, attractive, autonomous, healthy, "the happy hooker"; young, brazen, aggressive; not-so-young, bruised emotionally and physically, a victim. Which one is valid?

According to a landmark study (Vanwesenbeeck, 1994), all these images are accurate. In the first phase, researchers recruited 90 prostitutes and former prostitutes and conducted extended interviews with them about their daily lives. Two years later, 100 women who had been working for at least one year were recruited for a study of "sex and health"; these women were interviewed and completed several measures of coping style and well-being. The samples included women who worked on the street, in windows (in the Netherlands), in clubs and brothels, for escort services, and in their own homes. The results indicated that one-fourth of the women were doing well. They had few physical or psychosocial complaints, used problem-focussed coping strategies, and were satisfied with their lives. Another quarter were at the opposite end. They complained of headaches, backaches, anxiety and depression; their coping strategies involved dissociation (seeing problems as unrelated to the self) and denial, and they were dissatisfied with prostitution. The remaining women were in the middle.

What variables are associated with poor well-being? The women with poor well-being reported a history of victimization and trauma as children or adolescents, before they entered prostitution. They entered prostitution out of economic necessity. They worked in riskier settings, the windows or the streets; they worked faster, had more clients, and earned less per client. These results suggest strong associations between childhood and adolescent experiences, circumstances of entry into prostitution, risk experienced in working as a prostitute, and well-being.

Customers

At the time of the Kinsey research, about 69 percent of all white males had had some experience with prostitutes (Kinsey et al., 1948). The NHSLS in 1992 asked all respondents whether they had had sex with someone they paid or who paid them (Laumann et al., 1994). Only 17 percent of men and 2 percent of women reported that they had had sex with such a partner since age 18. Thus, the use of prostitutes has declined dramatically in the past 50 years. This reflects the increased frequency of nonmarital and casual sexual activity during this same period (see Chapters 11 and 12).

Prostitutes refer to their customers as "johns." About 50 percent of the clients are occasional johns; they may be respectable businessmen who seek only occasional contacts with prostitutes, perhaps while on business trips. Nearly 50 percent are repeat clients who seek a regular relationship with a prostitute or a small group of prostitutes (Freund et al., 1991). The remainder are compulsive johns, who use prostitutes for their major sexual outlet. They are driven to them and cannot stay away (see Chapter 16). Some of these men are able to function sexually only with prostitutes (Bess & Janus, 1976). About 40 percent of the men are married (Freund et al., 1991).

A study of the clients of male and female prostitutes in Atlanta found that the two groups of men were very similar. On the average, they were 35 years old and had completed 1 or 2 years of college. Fifty-four percent were white and the rest were black. The men reported having had an average of four sexual partners during the preceding month of whom 2.5 were paid. The clients of male prostitutes had been using prostitutes for 6 years, while the clients of female prostitutes had been doing so for 7½ years (Boles & Ellifson, 1994). The participants were recruited through newspaper ads, so they are probably not a representative sample.

Men use the services of prostitutes for a variety of reasons. Some are married but want sex more frequently than their wives do or want to engage in practices—such as fellatio—that they feel their wives would not be willing to do (McKeganey, 1994). Some use prostitutes to satisfy their exotic sexual needs, such as being whipped or having sex with a woman who pretends to be a corpse. The motivation for the unmarried man or the one who is away from home for a long period of time (for example, during a war) may simply be release of sexual tension. Others, particularly adolescents, may have sex with prostitutes to prove their manhood or to get sexual experience. Finally, some men enjoy sex with prostitutes because it is "forbidden."

Male Prostitutes

Some male prostitutes serve a heterosexual clientele, selling their services to women. These prostitutes work in three settings. Some of them work for escort services and provide companionship and sexual gratification on an out-call basis. Working in this setting is much less risky for male prostitutes than it is for female prostitutes. Some men work in massage parlors, under the same conditions as female employees. These male prostitutes virtually never work the street, in contrast to female streetwalkers and male *hustlers* (see below). This reflects gender-role socialization; female clients are unlikely to cruise the streets and pick up a prostitute, because they have been taught to let the male take the initiative.

A third type is the **gigolo** (in French, "one who dances"), a male who provides companionship and sexual gratification on a continuing basis to a woman in exchange for money. The gigolo often, though not always, has only one client at a time. There are several types, including the "Golden Boy," the pampered playboy kept by a very wealthy woman; the "Lap Dog," who enters a series of marriages of convenience; and the "Toy Boy," or stud, who works as a companion on a limited term basis (Nelson & Robinson, 1994). The demand for gigolos reflects the fact that women, like men, desire sexual gratification on a continuing basis and will pay for it when circumstances require or allow them to do so. On the other hand, women, unlike men, often prefer their sexual activity to be part of an ongoing relationship.

Hustlers are male prostitutes who cater to a homosexual clientele. Interestingly, some of them consider themselves to be heterosexual, not homosexual (Coombs, 1974). They may have strict rules for their customers to follow, such as only permitting the customer to perform fellatio on them. To indicate their masculinity, they may wear leather jackets and tight jeans. There is some market for "chickens" (young boys) as prostitutes.

Parallel to patterns of heterosexual prostitution,

there are male homosexual escort services for a more upscale clientele (Salamon, 1989).

Male hustlers seem to fall into four categories (Allen, 1980). First are the full-time street and bar hustlers, who operate much as female streetwalkers do. Second are full-time call boys or kept boys. They tend to have a more exclusive clientele and to be more attractive and more sexually versatile than the streetwalkers. Surprisingly, by far the largest group is the third: part-time hustlers, who are typically students or individuals employed in another occupation. They generally work at prostitution only when they need money. The part-time hustlers are notable because unlike those in the other groups, they are less likely to come from inadequate families. They also have the best long-term chance for getting an education and a stable job and achieving a good social adjustment. Finally, a fourth group is made up of delinquents; they use prostitution as an extension of other criminal activities, such as assault and robbery. They are taught by older gang members how to pick up homosexuals and then threaten, blackmail, or assault them.

In one study of adolescent male prostitutes in San Francisco, the main reason for engaging in prostitution, stated by 87 percent, was money (Weisberg, 1985). Most often they had left home because of conflict in the family, typically leaving when they were 15 or 16 years old, although some had done so when they were 11 or 12. The majority (72 percent) reported using drugs while engaging in acts of prostitution. The reason most often cited was the enjoyment of being high. Drugs were also used to reduce feelings of anxiety or fear stemming from the scary nature of the work.

Pornography

A debate over pornography has been raging for more than three decades. Religious fundamentalists and some feminists (strange bedfellows, indeed!)

Gigolo (JIG-uh-loh): A male who provides companionship and sexual gratification on a continuing basis to a woman in exchange for money.
Hustler: A male prostitute who sells his services to men.

agree that some kinds of pornography should be made illegal, while civil liberties groups and some other feminists argue that freedom of expression, guaranteed in the Constitution, must be preserved and therefore pornography should not be restricted by law. Meanwhile, Joe Brown stops at his local video store, buys *Wall To Wall Sex, v. 16,* and drives home for a pleasurable evening's entertainment. Here we will examine what the issues are, paying particular attention to social scientists' research on the effects of pornography on people who are exposed to it. First, we need to clarify some terminology.

Terms

We can distinguish between pornography, obscenity, and erotica. Pornography comes from the Greek work "porneia," which means, quite simply, "prostitution," and "graphos" which means "writing." In general usage today, **pornography** refers to literature, films, and so on, that are intended to be sexually arousing (Malamuth, 1998).

In legal terminology the word used is "obscenity," not pornography. **Obscenity** refers to that which is foul, disgusting, or lewd, and it is used as a legal term for that which is offensive to the authorities or to society (Wilson, 1973). The U.S. Supreme Court has had a rather hard time defining exactly what is obscene and what can be regulated legally, a point to be discussed in more detail in Chapter 22.

In the current debate over pornography, some make the distinction between pornography (which is unacceptable to them) and erotica (which is acceptable). For example, sociologist Diana E. H. Russell defines *pornography* as "explicit representations of sexual behavior, verbal or pictorial, that have as a distinguishing characteristic the degrading or demeaning portrayal of human beings, especially women" (Russell, 1980, p. 218). In contrast, **erotica** is defined as differing from pornography "by virtue of not degrading or demeaning women, men, or children" (Russell, 1980, p. 218). According to this distinction, a movie of a woman being raped would be pornography, whereas a movie of two mutually consenting adults who are both enjoying having sexual intercourse together would be considered erotica.

Beyond these definitions given by scholars in the area, it is interesting to see how typical Americans

define pornography. Research shows that there is great diversity in what people consider pornography. *Time* magazine's well-sampled survey on this topic (July 21, 1986) showed that 56 percent of Americans consider books describing sex acts to be pornographic; that means that about half of Americans consider such books to be pornographic, but half do not. There are also substantial gender differences. Fifty-two percent of the women polled consider nude photos in magazines to be pornographic, whereas only 39 percent of men hold this view. There is not a clear consensus on what is pornographic, a difficult problem when the topic under debate is what laws we should or should not have on this issue. This same poll also showed that the majority of Americans have been exposed to various kinds of sexually explicit materials.

Types of Pornography

Pornography is a multibillion-dollar business in the United States. Included in that business are a number of products: magazines directed to various audiences, films, X-rated videocassettes, live sex shows, telephone sex, the latest technological advance, computer porn, and kiddie porn. Some of this activity is legal (e.g., publishing *Penthouse*); some of it is illegal (e.g., producing films featuring children engaged in sex); and the legality of the rest is hotly debated. It is impossible to obtain precise data on the economics of pornography; one estimate is that Americans spent $8 billion on pornography in 1994.

Magazines

A large chunk of the pornography market consists of magazines, ranging from *Playboy* and *Penthouse*—soft-core—to *Hustler* and hundreds of other hard-core magazines with less well-known names. The soft-core magazines mushroomed in the 1970s. In the 1990s the market was large and included both general magazines and those catering

---

Pornography: Sexually arousing art, literature, or films.
Obscenity: Something that is offensive according to accepted standards of decency; the legal term for pornography.
Erotica: Sexually arousing material that is not degrading or demeaning to women, men, or children.

Figure 18.2   Porn magazines designed for a male audience. They run the gamut from the relatively tame *Playboy* to the raunchier and more sexually explicit *Hustler.*

Figure 18.3   The sexual content of many music videos played on MTV is unmistakable.

to specialized tastes. *Playboy* reports a circulation of 3.4 million copies per issue, *Penthouse* a circulation of 1.3 million, and *Hustler* a circulation of 1 million (*Ulrich's,* 1995).

Much pornography is designed for the heterosexual male reader. In the 1970s, *Playgirl* appeared, featuring "beefcake" in an attempt to attract heterosexual women. The magazine now has a circulation of 575,000 copies per month. There is also a large variety of printed material designed for gay men, lesbians, people interested in interracial sex, swingers, and other groups.

Hard-core magazines have a no-holds-barred approach to what they present. Photographs may include everything from vaginal intercourse to anal intercourse, sadomasochism, bondage, and sex with animals. A study of the titles of magazines and books found in "adult" bookstores revealed that 17 percent were about a paraphilia or sexual variation (Lebeque, 1991). Of those, 50 percent featured sadomasochism. An additional 21 percent were about incest.

Here, again, the profit is great. The markups may be as high as 600 percent, and it is estimated that there are approximately 20,000 stores in the United States selling hard-core magazines. If run properly, such a business will have as much as $200,000 per year in gross sales (Serrin, 1981). Yet more profit comes from customers who have regular subscriptions for pornographic magazines by mail, and repeat, frequent customers account for a large part of the market.

Films and Videos

Although sexually explicit movies were made as early as 1915, only in the last three decades have these films been slick and well produced. The *hard-core film* industry began to emerge in a big way around 1970. Two films were especially important in this breakthrough. *I Am Curious, Yellow,* appearing in 1970, showed sexual intercourse explicitly. In part because it was a foreign film with an intellectual tone, it became fashionable for people, including married couples, to see it. The other important early film was *Deep Throat,* appearing in 1973. With its humor and creative plot, it was respectable and popular among the middle class. Linda Lovelace, the female star, gained national recognition and later appeared on the cover of *Esquire.*

After the success of *Deep Throat,* many more full-length, technically well-done hard-core films soon appeared. *Deep Throat* had made it clear that there were big profits to be made. It cost $24,000 to make, yet by 1982 it had yielded $25 million in profits ("Video turns big profit," 1982).

Loops are short (10-minute) hard-core films. They are set up in coin-operated projectors in private booths, usually in adult bookstores. The patron can enter and view the film in privacy and perhaps masturbate while doing so (Jewksbury, 1990).

In the early 1980s, X-rated *videocassettes* for home viewing began to replace porn theaters. For example, *Deep Throat* became available on cassette in 1977, and, by 1982, 300,000 copies had

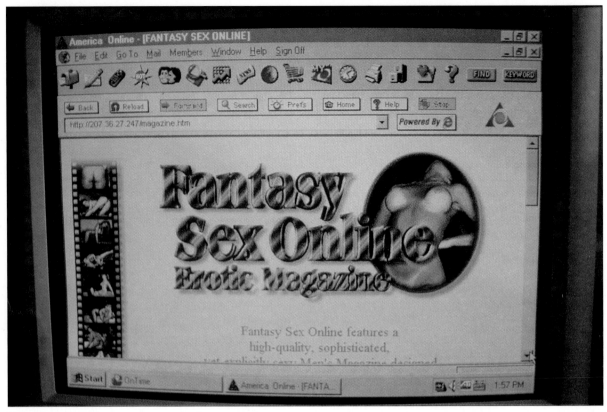

Figure 18.4   The newest innovation in the porn business is computer porn.

been sold ("Video turns big profit," 1982; Cohn, 1983). Cable television has also entered the arena, with porn channels thriving in some areas of the country.

Many hard-core films and X-rated videocassettes are made for a heterosexual audience. Nationally, "adult videos" earn 2.5 billion dollars per year and account for more than 25 percent of all video sales and rentals (The Sex Industry, 1998). They portray couples engaging in both male-active and female-active oral sex, and vaginal intercourse in various settings and bodily positions. A content analysis of 50 videos produced between 1979 and 1988 identified several changes over that 10-year period (Brosius, Weaver, & Staab, 1993). There were increases in portrayals of sexual activity within casual relationships, sexual activity involving a male and a female superordinate (e.g., his boss), encounters in which the female persuades the male to engage in sexual activity, and fellatio. Portrayals of males having sex with a coworker or prostitute, however, declined in frequency. Less often, films and videos show sexual

activity involving three or more people, or two women (Davis & Bauserman, 1993).

A rapidly expanding part of the porn industry is the "amateur" video. The development of the home-video camera has enabled anybody with a willing partner, friends, or neighbors to produce home-made porn. Such videos cost virtually nothing to make, and distributors are eager to purchase them. These films account for at least 20 percent of all adult videos made in the United States (The Sex Industry, 1998).

In the 1990s, a number of companies began marketing videos designed to educate people about various aspects of human sexuality. With names like the "Better Sex" video series, these include explicit portrayals of a wide variety of consenting heterosexual activities. As such, these are erotica, not pornography. They often include commentary by a psychologist or sex therapist reassuring viewers that the activities portrayed are normal. These series are advertised in national magazines and some daily newspapers.

The continuum noted in magazines from the subtle to the explicit also exists in video. The subtle end is found in *music videos*. The sexual content of many videos shown on MTV is unmistakable. One team of researchers analyzed 40 MTV videos, coding the content at 30-second intervals. They found that in these videos, men engaged in significantly more aggressive behavior, women engaged in significantly more implicitly sexual behavior, and women were the object of implicit, explicit, and aggressive sexual advances (Sommers-Flanagan, Sommers-Flanagan, & Davis, 1993).

Live Sex Shows

Live sex shows are yet another part of the sex industry. Strip shows, of course, have a long tradition in our culture. They have declined in popularity recently, but male strippers, catering to a female audience, have become common. In the sex districts of many cities, there are also live sex shows featuring couples or groups engaging in sexual acts onstage.

Telephone Sex

Telephone sex provides another example of enlisting technology to sell sexual titillation. Initially, "dial-a-porn" was available through 900-number services, but after the Federal Communications Commission tightened the regulation of such services in 1991, most telephone sex moved to 800-number or regular long-distance services (Glascock & LaRose, 1993).

A study of a sample of prerecorded messages in 1991 identified several patterns (Glascock & LaRose, 1993). The typical recording was of a female voice describing a series of sexual activities in which the caller was a participant. Callers to 900 numbers generally heard descriptions of hugging and kissing. Callers to 800 and long-distance numbers were much more likely to hear fantasies involving masturbation, vaginal intercourse, and oral sex. Few of the descriptions included violence or rape. More frequent were descriptions of activities in which the woman dominated and degraded the man.

There are also phone sex services that provide live conversation. In some cases, large numbers of workers may be in the same location talking to callers. In other cases, the phone sex worker takes calls at his or her home. A male described one conversation:

I say, "When you give a guy a blow job, what's your favorite way to do it?" She says, "Well, I love being on my knees, 'cause being on my knees with him standing is so submissive." I said, "Do you like looking at his cock in his pants, does that really turn you on?" She says, "I love that." I say, "I would love to be standing in front of you." She says, "Oh, that would really turn me on." "Do you like having your breasts played with when you're giving a blow job?" "Yes, I like it very much. I also like to be fingered." "How many fingers do you like inside you?" "I love two fingers." "And why do you like blowing a guy so much?" "Because I'm getting him excited and I know I can't wait for him to fuck me." "Do you like to fuck for a long time?" "Yes, I get lost in it." So then I said, "Well, I've really been thinking about you on top of me, and while you're riding me, I'll be spanking you." She said, "Oh God, I love that." I was masturbating and I came. It was great. (Maurer, 1994, 349–350)

Electronic Porn

In the past decade the World Wide Web and Internet have made available a wide variety of pornographic services for every computer attached to a modem, whether at work, at school, or at home. These services include: on-line chats or conversations, exchanging messages with like-minded persons via discussion groups, access to sexually arousing stories and photographs via Usenet groups, and access to a broad array of goods and services via specialized websites.

Chat Rooms

Chat rooms, or Internet relay chat groups, provide a location where individuals can meet and carry on conversations electronically. These rooms are often oriented toward persons with particular sexual interests, often captured by their names. An Internet resource guide had links to 155 sex-oriented chat rooms in April 1998. The conversations often involve graphic descriptions of sexual activities or fantasies. The phone sex conversation reproduced above could have taken place electronically, with the words displayed on a computer screen instead of spoken over a phone line. Some people have left relationships, including marriages, for someone they knew only through electronic conversation. In this context, an interesting feature of these chats is that the other person can not see you. This allows you to present yourself in any way you desire, to rehearse or try out a broad range of identities.

> Doug is a midwestern college junior. He plays four characters . . . One is a seductive woman. One is a macho, cowboy type whose self-description stresses that he is a 'Marlboros-rolled-in-the-T-shirt sleeve kind of guy.' The third is a rabbit of unspecified gender . . . a character he calls Carrot. (Turkle, 1995, p. 13)

Doug would not describe the fourth character, beyond saying it was furry. Remember that the character you are chatting with may not be who she or he seems to be.

News Groups

People can also log on to sex-oriented news or discussion groups, read messages posted by others, and post messages themselves. The messages may include personal information, or they may be a story or a file containing pornographic pictures in digital format. Stories can be printed by the user; picture files can be downloaded and viewed through a "plug-in." Often, the messages are simply advertisements for or links to sex-oriented websites that sell pornographic material. The April 1998 resource guide mentioned previously had links to 620 sex-oriented news groups. A research team at Carnegie-Mellon University studied Usenet activity in 1994 (Rimm, 1995). Only 3 percent of all groups included sexually explicit content, but they were accessed much more frequently than were nonsexual groups.

Commercial Bulletin Boards

There are numerous commercial bulletin boards that contain sexually explicit photographs in digital form. The aforementioned April 1998 Internet resource guide listed 43 sex-oriented bulletin boards. Users may log on to these bulletin boards and download images for a fee. Each image is listed in an electronic catalog with a short description. The Carnegie-Mellon researchers studied the content of these images by classifying 5.5 million image descriptions. Table 18.1 displays the results. Hard-core images (sample description: "Girl with big tits gets fucked by guy") made up 38 percent of the downloaded images. Another 33 percent were images of paraphilias ("Girl with big tits gets

Table 18.1 Total Surveyed "Adult" Usenet Files and Downloads by Classification

Classification	Total Files	Total Downloads
Hard-core	133,180 (45.6%)	2,102,329 (37.9%)
Soft-core	75,659 (25.9%)	760,009 (13.7%)
Paraphilia	63,232 (21.6%)	1,821,444 (32.8%)
Pedophilia	20,043 (6.9%)	864,333 (15.6%)

Source: Rimm, M. (1995). Marketing pornography on the information superhighway. Georgetown Law Journal, 83, Table 5, p. 1891.

fucked by horse"). Of particular concern is the popularity of pedophilic images, which accounted for 15.6 percent of the downloaded images. Such images are illegal and not readily available elsewhere. The researchers concluded: "The 'adult' BBS market is driven largely by the demand for paraphilic and pedophilic imagery. The availability of, and the demand for, vaginal sex imagery is relatively small" (Rimm, 1995, p. 1890). The U.S. Congress and various governmental agencies are debating what, if anything, can be done to regulate the flow of these materials on computer networks.

Adult Websites

Adult websites sell a variety of pornographic services and sexual materials. An April 1998 Internet resource guide had links to 660 such sites, with names like Bizar, Honey's Hot Horny Spot, House of Sin, IIISome, and Smutland. Each site typically includes thousands of pornographic photos organized by content, pornographic videos that can be viewed on your computer screen, stories, links to live sex shows, and links to live video cameras in places such as men's and women's locker rooms. Some also sell videos, CD-ROMS, sex aids such as dildos, and other sexual devices and costumes. Some also include "interactive" sex shows, where the viewer can request that the actors perform specific acts. Many of these sites specialize, featuring "teenagers" (if the actors are under 18, the material violates the law), black, Asian, or Hispanic women, gays, lesbians, pregnant women, and on and on. Each site charges a daily, weekly or monthly "membership fee" for access; the fee can usually be paid by supplying a valid credit card number.

Figure 18.5   While there are differences between men and women in their response to pornography, some women enjoy watching a stripper as much as some men do. The final scenes of *The Full Monty* portray the pleasure both the male performers and the female audience experience.

Electronic porn is cause for concern, for several reasons. One is that the large and ever-increasing number of chat rooms, news groups, and websites facilitate a person becoming dependent on or addicted to (see Chapter 16) this type of sexual content. None of these involve face-to-face social interaction, which is central to most sexual relationships; the risk is that they become a substitute. There is also a great deal of concern that children will access these materials. All sites include a printed notice that one must be over 18 to access the site, and that one who is offended by sexually explicit material should not enter the site. This "honor system" is not an effective control. It is true that to access a website or download photographs from bulletin boards, the user must supply a valid credit card number, but this is not a major deterrent for many adolescents. We will discuss the regulation of Internet access by the law in Chapter 22.

Kiddie Porn

Kiddie porn features photographs or films of sexual acts involving children. It is viewed as the most reprehensible part of the porn industry because it produces such an obvious victim, the child model. Children, by virtue of their developmental level, cannot give truly informed consent to participation in such activities, and the potential for doing psychological and physical damage to them is great. Many states have moved to outlaw kiddie porn, making it illegal to photograph or sell such material; as of 1984, 49 states prohibited the production of child pornography and 36 prohibited its distribution (Burgess, 1984, p. 202).

Kiddie porn: Pictures or films of sexual acts involving children.

Focus 18.2
Ernie: A Pedophile and Child Pornographer

In a study of child sexual abuse and child pornography, the investigators reported the following description of one offender:

Contact was made with Ernie, a northern Indiana man, and arrangements were made to meet with him to share child pornography collections. Ernie arrived at a motel room carrying a small suitcase containing approximately 75 magazines and a metal file box containing twelve super-8-mm movies. The metal box was also filled with numerous photographs. Ernie then began to discuss his collection. He described himself as a pedophile and showed a series of instant photographs he had taken of his 7-year-old niece while she slept. The pictures revealed Ernie's middle finger inserted into the young girl's uncovered genitals, and he described how he had "worked" his finger up into her. Other photographs featured the girl being molested in various ways by Ernie while she remained asleep.

Despite the fact that he had engaged in numerous incidents of child molestation, Ernie had not been discovered because his victims remained asleep. He had molested and exploited both males and females, his own children, grandchildren, and neighborhood children. In order to photograph the uncovered genitals of his sleeping victims, Ernie had devised a string and hook mechanism. The hook was attached to the crotch of the panties, and he would uncover his sleeping victims' genitals as he photographed them with an instant-developing camera.

Ernie displayed his collection with the pride of a hobbyist. He exhibited photographs he had reproduced from magazines; he had reproduced these same photographs repeatedly and had engaged in a child pornography business from his residence.

While displaying his magazines and films, Ernie was arrested. A search warrant was obtained for his residence, and material seized from his one-bedroom apartment filled two pickup trucks. Numerous sexually explicit films, photographs, magazines, advertisements, and children's soiled underwear were confiscated. The panties had been encased in plastic, the child's school photograph featured with the panties. Also confiscated were nine cameras and a projector.

Several months passed before investigators found proof that Ernie had processed, through a central Indiana photographic lab, approximately 1500 photographs per week. It is believed that Ernie sold these pictures at $2 each, grossing an estimated $3000 per week.

Source: Ann W. Burgess. (1984). *Child pornography and sex rings.* Lexington, MA: Lexington Books of D. C. Heath, pp. 26–27.

Again, the profit motives are strong. An advertisement in the magazine *Screw* offered $200 for little girls to appear in porn films, and dozens of parents responded. A reporter covering the scene said:

Some parents appeared in the movie with their children; others merely allowed their children to have sex. One little girl, age 11, who ran crying from the bedroom after being told to have sex with a man of 40, protested, "Mommy I can't do it." "You have to do it," her mother answered. "We need the money." And of course the little girl did. (Anson, 1977)

Some major, well-known films could easily be classified as kiddie porn. *Taxi Driver* featured Jodie Foster as a 12-year-old prostitute. And *Pretty Baby* launched the career of Brooke Shields, playing the role of a 12-year-old brothel prostitute in New Orleans. Shields herself was 12 years old when the film was made.

A major study of offenders (the people who produce child pornography) and victims profiled the offender as follows: All of the 69 offenders studied were male. They ranged in age from 20 to 70, with an average age of 43. And 38 percent had an already established relationship with the child before the illicit activity began—they were family friends or relatives, neighbors, teachers, or counselors (Burgess, 1984). For a profile of one child pornographer, see Focus 18.2.

Advertising

Let's close our discussion of pornography by considering a mating of sex and money that all of us encounter every day—*sex in advertising*. Both subtle and obvious sexual promises are used to sell a wide variety of products. A muscular young man wearing low-slung jeans and no shirt sells Calvin Kleins. A 1995 Klein campaign caused an uproar because it used models who appeared to be minors in sexual poses. Perfumes promise that they will make women instantly sexually attractive. One brand of coffee seems to guarantee a warm, romantic, sensuous evening for the couple who drink it. An important part of the presence of sexuality in the media is advertising. One study analyzed the sexual content of magazine advertising in 1964 and 1984 (Soley & Kurzbard, 1986). The percentage of ads with sexual content had not increased over that 20-year period, although the illustrations had become more explicit. In 1984, 23 percent of ads contained sexual content. And in those ads, female models were considerably more likely (41 percent) than male models (15 percent) to be shown partially clad or nude.

The Customers

What is known about the consumer of pornography? Studies consistently find that the average customer in a pornographic bookstore is an educated, middle-class male between the ages of 22 and 34 (Mahoney, 1983). That is, the use of the materials sold in such stores is "typical" or "normal" (in the statistical sense) among males. The manager of one store said,

> We get everyone in here from millionaires to scum of the earth. The blue-collar and white-collar men come for the tapes. Married couples come in for

things to help their sex life. The gay crowd cruises the booths in back. Groups of women come in for gag items. ("Porn shop," 1994)

Surveys also suggest that many students use sexually explicit materials. In one survey, 59 percent of white male college students and 36 percent of white female students said they went to X-rated movies or read pornographic books (Houston, 1981). Over 5 percent of the females and 9 percent of the males said they did so frequently or very frequently.

Repeat customers are a crucial part of the success of the porn business. "There are people who come in here three to five times a day for their fix. Before work. At lunch. After work. Late at night. Their addiction isn't to the perversion of it, just the pornography" ("Porn Shop," 1994).

Computer porn attracts a more varied clientele. Chat rooms and news groups attract both men and women of diverse ages (assuming that those who describe themselves are doing so accurately). Some of these people are married. There have been reports of a man or woman leaving a spouse/partner in order to live with someone he or she met via the Internet. Depending on the focus of the room or group, participants may be from diverse racial or ethnic backgrounds. Bulletin boards and adult websites probably attract the same types of clients as adult bookstores—middle-aged, middle-class, white men, although some emphasize materials oriented toward other clienteles.

Feminist Objections to Pornography

Some—though not all—feminists are very critical of pornography (e.g., Lederer, 1980; Griffin, 1981; Morgan, 1978). Why would feminists, who prize sexual liberation, be opposed to pornography?

There are three basic reasons some feminists object to pornography. First, they argue that pornography debases women. In the milder, soft-core versions it portrays women as sex objects whose breasts, legs, and buttocks can be purchased and then ogled. In the hard-core versions women may be shown being urinated upon or being chained. This scarcely represents a respectful attitude toward women.

Second, pornography associates sex with

Figure 18.6 Sex in advertising. A series of ads by Calvin Klein (one shown here on a bus) stirred public protest, particularly because it sexualized children.

violence toward women. As such, it contributes to rape and other forms of violence against women and girls. Robin Morgan put it bluntly: ". . . pornography is the theory and rape is the practice" (Morgan, 1980, p. 139). This is a point that can be tested with scientific data, and this evidence will be covered in the next section.

Third, pornography shows, indeed glamorizes, unequal power relationships between women and men. A common theme in pornography is men forcing women to have sex, so the power of men and subordination of women are emphasized. Consistent with this point, feminists do not object to sexual materials that portray women and men in equal, humanized relationships—what we have termed *erotica*.

Feminists also note the intimate relationship between pornography and traditional gender roles. Pornography is enmeshed as both cause and effect. That is, pornography in part results from traditional gender roles that make pornography socially acceptable for men to use and require hypersexuality and aggressiveness as part of the male role. One study of androgyny and the use of pornography among college students found that the likeliest users of pornography are traditionally gender-typed, masculine males and androgynous females (Kenrick et al., 1980). In turn, pornography may serve to perpetuate traditional gender roles. By seeing or reading about dominant males and submissive, dehumanized females, each new generation of adolescent boys is socialized to accept these roles.

The Effects of Pornography

Some of the assertions summarized above—for example, that using violent pornography may predispose men to committing violent crimes against women—can be tested using the methods of social

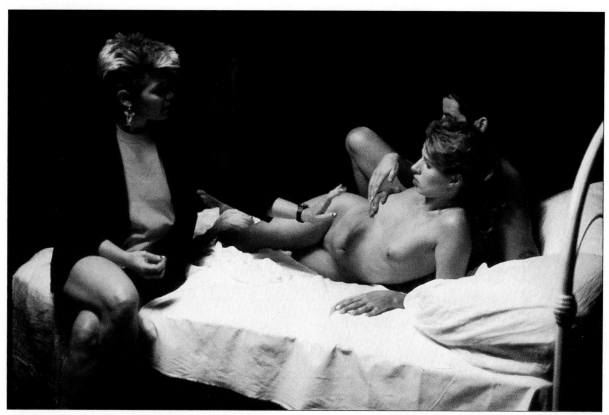

Figure 18.7   Candida Royale (left), a former porn star, now produces and directs soft-core films geared toward a female audience.

science. A number of social psychologists have been collecting data for 20 years to test such assertions.

Four questions can be asked about the effects of using pornography. First, does it produce sexual arousal? Second, does it affect users' attitudes, particularly attitudes about aggression toward women and rape? Third, does it affect the sexual behavior of users? Fourth, does it affect the aggressive or criminal behavior of users, particularly aggressive behavior toward women?

More than 40 studies have examined the effect of sexually explicit material on sexual arousal. This research consistently finds that exposure to *material that the viewer finds acceptable* does produce arousal (Davis & Bauserman, 1993). Exposure to portrayals that the viewer finds objectionable produces a negative reaction. Most people disapprove of paraphilic behaviors (see Chapter 16), rape, and sexual activity involving children, so they react

negatively to hard-core and kiddie pornography. The one exception is that men who report, before they view pornography, that they would commit rape under some circumstances are aroused by portrayals of rape.

There are gender differences in self-reports of response to sexually explicit materials. Men report higher levels of arousal to such portrayals than do women (Malamuth, 1998). The differences are larger in response to pornography than to erotica, and the difference is much larger among college students than among older persons (Murnen & Stockton, 1997). This difference between men and women is often attributed to the fact that most erotica/pornography is male-oriented. The focus is almost exclusively on sexual behavior, with little character development or concern for relationships. There is limited foreplay and afterplay; the male typically ejaculates on some part of the woman's body (the "cum shot") rather than inside her. Former porn film star Candida Royale produces videos made for women. An experiment found that male college students responded positively to and were aroused by videos made for men and for women; females reported negative responses to the videos intended for males and positive responses and sexual arousal to the videos designed for women (Mosher & MacIan, 1994).

What about the effect of pornography on attitudes? The research indicates that a single exposure to stories, photographs, or videos has little or no effect. Massive exposure (such as viewing videos for 5 hours) does lead to more permissive attitudes. In this situation, viewers become more tolerant of the behavior observed and less in favor of restrictions on it (Davis & Bauserman, 1993). What about attitudes toward aggression against women? Some

(a)

Figure 18.8 The controversy over pornography has put some members of the political left wing on the same side as some from the right wing. *(a)* Feminist Gloria Steinem protests pornography. *(b)* Conservatives have encouraged citizens to picket "adult" businesses located in or near residential neighborhoods.

(b)

studies show that exposure to portrayals of rape lead men to be more tolerant of sexual assault, but other studies do not find a relationship between exposure and attitudes (Fisher & Grenier, 1994). Men exposed to portrayals of sexual aggression against women do not report a greater willingness to rape a woman (Davis & Bauserman, 1993).

With regard to sexual behavior, the research shows that, in response to erotic portrayals of consenting heterosexual activity, both men and women may report an increase in sexual thoughts and fantasies and in behaviors such as masturbation and intercourse. Exposure to portrayals of behavior the person has not personally engaged in does *not* lead to an increase in those behaviors (Davis & Bauserman, 1993).

Finally, there has been great interest in whether exposure to portrayals of sexual aggression (which almost always involve men behaving aggressively toward women) increase aggressive behavior. In laboratory studies, males who are insulted or provoked by a woman will respond aggressively toward her if given the opportunity. Males who have been exposed to violent pornography are significantly more aggressive toward a woman in this situation, compared with men exposed to sexually explicit but nonviolent material. If the comparison group is men exposed to nonsexual violent films, many studies find no differences between the two, but some find the exposure to sexual violence increases aggression toward a woman more than exposure just to violence (Davis & Bauserman, 1993).

In sum, then, we can conclude that exposure to sexually explicit material that the viewer finds acceptable is arousing to both men and women. Exposure to aggressive pornography does increase males' aggression toward women, and may affect males' attitudes, making them more accepting of violence against women. (For an excellent review, see Linz, 1989).

What Is the Solution?

These are disturbing conclusions. What is the solution? Should pornography be censored or made illegal? Or would that only make it forbidden and therefore more attractive, and still available on a black market? Or should all forms of pornography be legal and readily available, and should we rely on other methods—such as education of parents and students through the school system—to abolish its use? Or should we adopt some in-between strategy, making some forms of pornography—kiddie porn and violent porn—illegal, while allowing free access to erotica?

Our own opinion is that legal restrictions—known less politely as censorship—are probably not the solution. We agree with the view of Donnerstein and his colleagues that a better solution is education (Linz et al., 1987; Donnerstein et al., 1987). In their experiments, they have debriefed male participants at the conclusion of the procedures. They convey to the participants that media depictions are unreal and that the portrayal of women enjoying forced sex is fictitious. They dispel common rape myths, especially any that were shown in the film used in the experiment. Participants who have been debriefed in this way show less acceptance of rape myths and more sensitivity to rape victims than participants shown a neutral film (Donnerstein et al., 1987). It is also possible that *pre*briefing participants—telling them about the possible harmful effects even before they view the films—would be effective. More research is needed on the effectiveness of these educational approaches. In the meantime, we are optimistic that education will turn out to be a considerably better solution than censorship to the problem of pornography.

SUMMARY

Commercial sex is a major industry in the United States. Two prominent aspects of it are prostitution and pornography.

There are several kinds of female prostitutes, ranging in status and income from the call girl to the streetwalker. They work in a variety of settings, including their own homes, in-call services, escort services, and massage parlors. Women enter prostitution for several reasons, the most important being economic needs or desires. Many newcomers serve an apprenticeship, during which a more experienced prostitute or a madam teaches them the "tricks of the trade." The career is usually short, because of the negative effects of aging and the many risks that prostitution entails. Research suggests that a prostitute's well-being depends upon the risk level of the setting in which she works, the reasons she entered prostitution, and whether she experienced victimization as a child or adolescent.

Data indicate that the use of prostitutes has declined substantially in the United States in the past 50 years. About one-half of the clients of female prostitutes are occasional johns; the other 50 percent are repeat clients. Some men rely on prostitutes for their sexual outlet.

Some male prostitutes serve a female clientele. They may work as escorts, employees of massage parlors, or gigolos. Hustlers cater to a male homosexual clientele.

Distinctions are made among pornography (sexually arousing art, literature, or film), obscenity (material offensive to authorities or society), and erotica (sexual material that shows men and women in equal, humane relationships). Pornographic magazines, films, and videocassettes, both soft-core (erotica) and hard-core, are a multi-billion-dollar business. Electronic porn has mushroomed in the past 10 years; people can discuss explicit sexual activity online, read sexually arousing stories, download sexually explicit images, or purchase a variety of goods and services at adult websites. Children—often runaways—are the star-victims in kiddie porn.

Some feminists object to pornography on the grounds that it debases women, encourages violence against women, and portrays unequal relationships between men and women.

Social-psychological research indicates that exposure to portrayals that the viewer finds acceptable is arousing to both men and women. Men are more likely to report arousal than women. Massive exposure leads to more favorable attitudes toward the behavior observed. Some studies find that exposure to violent pornography creates more tolerant attitudes toward violence against women but others find no such effect. Exposure to portrayals of consenting heterosexual activity leads to an increase in sexual thoughts and behavior. Exposure to portrayals of sexual or nonsexual violence toward women increases men's aggression against women. Education about the effects of pornography is probably the best solution to the problems created by pornography.

REVIEW QUESTIONS

1. The call girl is the most expensive and exclusive of prostitutes. True or false?
2. The _____ is a modern version of a brothel.
3. The _____ is the companion-master of the prostitute.
4. _____ are male prostitutes who sell companionship and sexual gratification to a woman on a continuing basis.
5. The most important reason why women enter prostitution is economics. True or false?

6. Some experts distinguish between pornography and erotica, erotica being sexually explicit but not degrading to women, men, or children. True or false?

7. Kiddie porn is considered to be the worst part of the porn industry; it is illegal to produce kiddie porn in all the states but one. True or false?

8. The typical customer of pornography is a male in his forties or fifties who is psychiatrically disturbed, usually a psychopath. True or false?

9. Men exposed to portrayals of sexual aggression against women are more likely to behave aggressively toward a woman in a laboratory situation. True or false?

10. The best solution to the problems created by pornography is to increase censorship of it. True or false?

(The answers to all review questions are at the end of the book, after the Glossary.)

QUESTIONS FOR THOUGHT, DISCUSSION, AND DEBATE

1. What is your position on the issue of censoring pornography? Do you think that all pornography should be illegal? Or should all pornography be legal? Or should some kinds—such as kiddie porn and violent porn—be illegal, but not other kinds? What reasoning led you to your position?

2. Prostitution is becoming more visible in your community. Some people are demanding stricter laws and longer sentences for women who exchange sex for money. They claim this will deter women from entering prostitution. Based on what you know about why women enter prostitution, their varied lifestyles, and the risks these women face, is this claim plausible? If not, what alternative approach would you suggest?

3. Many of the streetwalkers in major cities are black or Asian women. Adult websites often prominently display photographs, stories, and videos about "interracial sex." Yet most of the customers of both prostitutes and adult websites are white males. Why are white men attracted by materials featuring women from other ethnic groups? Why aren't more black men among the customers for sexual services?

SUGGESTIONS FOR FURTHER READING

Bullough, Vern, & Bullough, Bonnie. (1987). *Women and prostitution: A social history.* Buffalo, NY: Prometheus Books. A fascinating history of the oldest profession, from ancient Greece and Rome, through medieval times, India, and China, to the present.

Rimm, Marty. (1995). Marketing pornography on the information superhighway. *Georgetown Law Journal, 83,* 1849–1925. This article reports the findings of the first systematic study of computer pornography.

Vanwesenbeeck, Ine. (1994). *Prostitutes' well-being and risk.* Amsterdam: VU University Press. An excellent empirical study of the determinants of psychological well-being among prostitutes.

WEB RESOURCES

http://www.bayswan.org
    Website of Bay Area Sex Worker Advocacy Network; information about sex worker rights and issues.

http://www.worldsexguide.org
    Website with information about prostitution throughout the world.

http://www.spectacle.org/freespch/faq.html
    Website with information about censorship and the Internet.

http://www.eff.org/CAF/cafuiuc.html
    Website with information about attempts by universities to censor/limit access to Internet.

http://www2000.ogsm.vanderbilt.edu/cyberporn.debate.html
    The Cyberporn Debate Homepage; articles, references and links about sex on the Internet.

chapter
19

SEXUAL DISORDERS AND SEX THERAPY

CHAPTER HIGHLIGHTS

Kinds of Sexual Disorders

What Causes Sexual Disorders?

Therapies for Sexual Disorders

Critiques of Sex Therapy

Sex Therapy in the AIDS Era

Some Practical Advice

T he only thing we have to fear is fear itself.*

Sexual disorders—such as premature ejaculation in men and inability to have orgasms in women—cause a great deal of psychological distress to the individuals troubled by them, not to mention to their partners. Until the 1960s, the only available treatment was long-term psychoanalysis, which is costly and inaccessible for most people. A new era in understanding and treatment was ushered in with the publication, in 1970, of *Human Sexual Inadequacy* by Masters and Johnson. This book reported on the team's research on sexual disorders, as well as on their rapid-treatment program of behavioral therapy. Since then, many additional developments in the field have taken place, including cognitive-behavior therapy and medical (drug) treatments. Sex disorders and treatments for them are the topics of this chapter.

We begin by defining the term "sexual disorder." A **sexual disorder** is a problem with sexual response that causes a person mental distress. The term *sexual dysfunction* is also used. Examples are a man's inability to get an erection and a woman's inability to have an orgasm. This definition seems fairly simple; as we will see in the following sections, however, in practice it can be difficult to determine exactly when something is a sexual disorder. In addition, there is a tendency to think in terms of only two categories, people with a sexual disorder and "normal" people. In fact, though, there is a continuum, much like the Kinsey scale for gradations in sexual orientation that we discussed in Chapter 15. Most of us have had, at one time or another, a sexual disorder that went away in a day or a few months without treatment. These cases represent the shades of gray that lie between absolutely great sexual functioning and long-term difficulties that require sex therapy.

First we will consider the kinds of sexual disorders. Following that, we will review the causes of these disorders, and then the treatments for them.

*Franklin Delano Roosevelt, First Inaugural Address, March 4, 1933.

Kinds of Sexual Disorders

In this section we will consider first male sexual disorders and then female sexual disorders, followed by sexual desire disorders, which affect both women and men.

Erectile Disorder

Erectile disorder is the inability to have an erection or maintain one. Other terms for it are *erectile dysfunction, inhibited sexual excitement,* and *impotence,* the term used by laypersons. One result of erectile disorder is that the man cannot engage in sexual intercourse. A case of erectile disorder may be either primary or secondary. In **primary erectile disorder,** the man has never been able to have an erection that is satisfactory for intercourse. In **secondary erectile disorder,** the man has difficulty getting or maintaining an erection but has had erections sufficient for intercourse at other times.

According to the NHSLS, about 10 percent of men have experienced an erection problem within the last 12 months (Laumann et al., 1994). This statistic varies a great deal by age: it is only 5.6 percent for 18- to 24-year-olds, but 20 percent for 55- to 59-year-olds. Erectile disorder is currently the most common of the disorders among men who seek sex therapy (Spector & Carey, 1990).

Sexual disorder: A problem with sexual response that causes a person mental distress.
Erectile (eh-REK-tile) disorder: The inability to have or maintain an erection.
Primary erectile disorder: Cases of erectile disorder in which the man has never had an erection sufficient to have intercourse.
Secondary erectile disorder: Cases of erectile disorder in which the man at one time was able to have satisfactory erections but now no longer is.

Psychological reactions to erectile disorder may be severe. For many men, it is one of the most embarrassing things they can imagine. Depression may follow from repeated episodes. It may also cause embarrassment or worry to the man's partner.

The causes of erectile disorder and its treatment will be discussed later in this chapter.

Premature Ejaculation

Premature ejaculation (or PE, if you want to be on familiar terms with it) occurs when a man has an orgasm and ejaculates too soon. In extreme cases, ejaculation may take place so soon after erection that it occurs before intercourse can even begin. In other cases, the man is able to delay the orgasm to some extent, but not as long as he would like or not long enough to meet his partner's preferences. Some experts prefer the terms *early ejaculation* or *rapid ejaculation* as having fewer negative connotations (McCarthy, 1989; Grenier & Byers, 1995).

While the definition given above—having an orgasm and ejaculating too soon—seems simple enough, in practice it is difficult to specify when a man is a premature ejaculator (Grenier & Byers, 1995; Metz et al., 1997). What should the precise criterion for "too soon" be? Should the man be required to last for at least 30 seconds after erection? For 12 minutes? For 2 minutes after insertion of the penis into the vagina? The definitions used by authorities in the field vary widely. One source defines "prematurity" as the occurrence of orgasm less than 30 seconds after the penis has been inserted into the vagina. Another group has extended this to 1½ minutes; for a third, the criterion is ejaculation before there have been 10 pelvic thrusts. Masters and Johnson defined premature ejaculation as the inability to delay ejaculation long enough for the woman to have an orgasm at least 50 percent of the time. This last definition has merit because it stresses the importance of the interaction between the two partners; however, it carries with it the question of how easily the woman is stimulated to orgasm. Psychiatrist and sex therapist Helen Singer Kaplan (1974; see also McCarthy, 1989) believed that the key to defining the premature ejaculator is the absence of voluntary control of orgasm; that is, the real problem is that the premature ejaculator has little or no control over when he orgasms. Another good defini-

tion is self-definition: if a man finds that he has become greatly concerned about his lack of ejaculatory control or that it is interfering with his ability to form intimate relationships, or if a couple agree that it is a problem in their relationship, then it may reasonably be called premature ejaculation.

Premature ejaculation is a common problem in the general male population. In the NHSLS, 29 percent of the men reported having problems in the last 12 months with climaxing too early (Laumann et al., 1994). The great majority of men probably never seek therapy for the problem, either because it goes away by itself or because they are too embarrassed.

Like erectile disorders, premature ejaculation may create a web of related psychological problems. Because the ability to postpone ejaculation and "satisfy" a partner is so important in our concept of a man who is a competent lover, early ejaculation can cause a man to become anxious about his sexual competence. Furthermore, the partner may become frustrated because she or he is not having a satisfying sexual experience either. So the condition may create friction in the relationship.

The negative psychological effects of early ejaculation are illustrated by a young man in one of our sexuality classes who handed in an anonymous question. He described himself as a premature ejaculator and said that after several humiliating experiences during intercourse with dates, he was now convinced that no woman would want him in that condition. He no longer had the courage to ask for dates, so he had stopped dating entirely. He wanted to know how the women in the class would react to a man with such a problem. The question was discussed in class, and most of the women agreed that their reaction to his problem would depend a great deal on the quality of the relationship they had with him. If they cared deeply for him, they would be sympathetic and patient and help him overcome the difficulty. The point is, though, that the early ejaculation had created problems so severe that the young man not only had stopped having sex but also had stopped dating.

Premature ejaculation: A sexual disorder in which the man ejaculates too soon and he feels he cannot control when he ejaculates; early ejaculation.

Men, of course, have developed home remedies for dealing with premature ejaculation. Perhaps the most common is to think of something else. In one study college men reported the thoughts they used to delay ejaculation (Grenier & Byers, 1997). These thoughts fell into 5 categories: sex negative (thinking of an ugly monster); sex positive (thinking "we're in no hurry" or visualizing a past episode of prolonged intercourse); nonsexual and negative (thinking of a sad event, unpaid debts); sex neutral (counting backwards from 100); and sexually incongruous (thinking of your grandmother, reciting the Lord's Prayer). We recommend the sex positive alternative.

Male Orgasmic Disorder

Male orgasmic disorder (also sometimes called *retarded ejaculation*) is the opposite of premature ejaculation. The man is unable to have an orgasm, even though he has a solid erection and has had more than adequate stimulation. The severity of the problem may range from only occasional problems with orgasming to a history of never having experienced an orgasm. In the most common version, the man is incapable of orgasm during intercourse but may be able to orgasm as a result of hand or mouth stimulation.

Male orgasmic disorder is far less common than premature ejaculation. In the NHSLS, 8 percent of the male respondents had had a problem in the last 12 months with being unable to orgasm (Laumann et al., 1994). The incidence varies as a function of age: it was 4.6 percent for 18- to 24-year-olds and 14 percent for 50- to 54-year-olds. It also varies somewhat as a function of ethnicity, being more common among Asian Americans (19 percent) than whites or blacks. Among men seeking sex therapy, it constitutes about 3 to 8 percent of the cases (Rosen & Leiblum, 1995b).

Male orgasmic disorder is, to say the least, a frustrating experience for the man. One would think that any woman would be delighted to have intercourse with a man who has a long-lasting erection that is not terminated by orgasm. In fact, though, some women react negatively to this condition, seeing their partner's inability to have an orgasm as a personal rejection. Some men, anticipating these negative reactions, have adopted the practice of "faking" orgasm. In some cases, too, the

man's orgasmic disorder can create painful intercourse in the woman because intercourse simply goes on too long.

We turn now to sexual disorders in women.

Female Orgasmic Disorder

Female orgasmic disorder is the inability to have an orgasm. This condition goes by a variety of other terms, including *orgasmic dysfunction, anorgasmia,* and *inhibited female orgasm.* Laypersons may call it "frigidity," but sex therapists reject this term because it has derogatory connotations and because it is imprecise. "Frigidity" may refer to a variety of conditions ranging from total lack of sexual arousal to arousal without orgasm. Therefore, the term "female orgasmic disorder" is preferred.

Like some other sexual disorders, cases of female orgasmic disorder may be classified into primary and secondary. **Primary orgasmic disorder** refers to cases in which the woman has never in her life experienced an orgasm (Andersen, 1983). **Secondary orgasmic disorder** refers to cases in which the woman had orgasms at some time in her life but no longer does so. A common pattern is **situational orgasmic disorder,** in which the woman has orgasms in some situations but not others. For example, she may be able to have orgasms while masturbating, but not while having sexual intercourse.

Orgasmic disorders are common among women. According to the NHSLS, 24 percent of the female respondents reported difficulty in the last 12 months with having orgasms (Laumann et al.,

---

Male orgasmic disorder: A sexual disorder in which the male cannot have an orgasm, even though he is highly aroused and has had a great deal of sexual stimulation.
Female orgasmic disorder: A sexual disorder in which the woman is unable to have an orgasm.
Primary orgasmic disorder: A case of female orgasmic disorder in which the woman has never in her life had an orgasm.
Secondary orgasmic disorder: A case of female orgasmic disorder in which the woman had orgasms at some time in her life but no longer does so.
Situtational orgasmic disorder: A case of orgasmic disorder in which the woman is able to have an orgasm in some situations (e.g., while masturbating) but not in others (e.g., while having sexual intercourse).

1994). Female orgasmic disorder accounts for 25 to 35 percent of the cases of women seeking sex therapy (Spector & Carey, 1990).

Once again, though, these definitions become more complicated in practice than they are in theory. Consider the case of the woman who has orgasms as a result of masturbation or hand or mouth stimulation by a partner but who does not have orgasms in vaginal intercourse. Is this really a sexual disorder? The notion that it is a disorder can be traced to sexual scripts and beliefs that there is a "right" way to have sex—with the penis inside the vagina—and a corresponding "right" way to have orgasms. Because this pattern of situational orgasmic disorder is so common, some experts consider it to be well within the normal range of female sexual response (Wincze & Carey, 1991). Perhaps the woman who orgasms as a result of hand or mouth stimulation, but not penile thrusting, is simply having orgasms when she is adequately stimulated and is not having them when she is inadequately stimulated.

On the other hand, there should be room for self-definition of disorders. If a woman has situational orgasmic disorder, is truly distressed that she is not able to have orgasms during vaginal intercourse, and wants therapy, then it may be appropriate to classify her condition as a disorder and provide therapy. The therapist, however, should be careful to explain to such a woman the problems of definition raised above, in order to be sure that her request for therapy stems from her own dissatisfaction with her sexual responding rather than from an overly idealistic sexual script. Therapy in such cases probably is best viewed as an attempt to enrich the client's experience rather than to fix a problem.

Female Sexual Arousal Disorder

Female sexual arousal disorder refers to a lack of response to sexual stimulation (Rosen & Leiblum, 1995b). The disorder involves both a subjective, psychological component and a physiological component (Morokoff, 1993). It is defined partly by the woman's own subjective sense that she does not feel aroused and partly by difficulties with vaginal lubrication.

Difficulties with lubrication are common; they were reported by 19 percent of the women in the NHSLS (Laumann et al., 1994). These problems become particularly frequent among women during and after menopause: as estrogen levels decline, vaginal lubrication decreases. The use of sterile lubricants is an easy way to deal with this problem. The absence of subjective feelings of arousal is more complex to treat.

Painful Intercourse

Painful intercourse, or **dyspareunia,** refers simply to pain experienced during intercourse (Meana & Binik, 1994). It is usually thought of as a female sexual disorder, but males occasionally experience it as well. In the NHSLS, 14 percent of the women reported pain during sex, compared with 3 percent of the men (Laumann et al., 1994). While complaints of occasional pain during intercourse are fairly common among women, persistent dyspareunia is not very common. In women, the pain may be felt in the vagina, around the vaginal entrance and clitoris, or deep in the pelvis. In men, the pain is felt in the penis or testes. To put it mildly, dyspareunia decreases one's enjoyment of the sexual experience and may even lead one to abstain from sexual activity.

Painful intercourse may be related to a variety of physical factors, to be discussed below.

Vaginismus

Vaginismus (the suffix "-ismus" means "spasm") is a spastic contraction of the outer third of the vagina; in some cases it is so severe that the entrance to the vagina is closed, and the woman cannot have intercourse (Leiblum & Rosen, 1989). Vaginismus and dyspareunia may be associated. That is, if intercourse is painful, one result may be spasms that close off the entrance to the vagina.

Female sexual arousal disorder: A sexual disorder in which there is a lack of response to sexual stimulation.

Dyspareunia (dis-pah-ROO-nee-uh): Painful intercourse.

Vaginismus (Vaj-in-IS-mus): A sexual disorder in which there is a spastic contraction of the muscles surrounding the entrance to the vagina, in some cases so severe that intercourse is impossible.

Figure 19.1   Vaginismus. *(a)* A normal vagina and other pelvic organs, viewed from the side, and *(b)* vaginismus, or involuntary constriction of the outer third of the vagina.

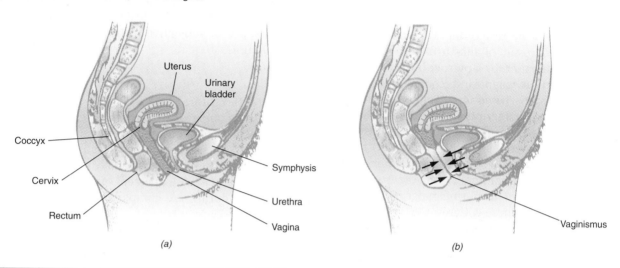

(a)   (b)

Vaginismus is not a very common sexual disorder in the general population. It is, however, relatively common among women seeking sex therapy, accounting for 12 to 17 percent of the cases (Spector & Carey, 1990). Women may be more likely to seek treatment for it than for other disorders because sometimes it makes intercourse impossible, creating difficulties in the couple's relationship.

Disorders of Sexual Desire

Sexual desire, or *libido*, refers to an interest in sexual activity, leading the individual to seek out sexual activity or to be pleasurably receptive to it (Kaplan, 1995). When sexual desire is inhibited, so that the individual is not interested in sexual activity, this is a disorder termed **hypoactive sexual desire** (HSD; the prefix "hypo-" means "low") (Beck, 1995; Rosen & Leiblum, 1995a). It is also sometimes termed *inhibited sexual desire* or *low sexual desire*. This disorder is found in both women and men.

People with HSD typically avoid situations that will evoke sexual feelings. If, despite their best efforts, they find themselves in an arousing situation, they experience a rapid "turn-off" so that they feel nothing. The turn-off may be so intense that they report negative, unpleasant feelings; they may

even report *sexual anesthesia*, that is, no feeling at all, even though they may respond to the point of having an orgasm.

The identification of low sexual desire as a sexual disorder began only about two decades ago. It arose from the study of persons whom traditional sex therapy had failed to help. Typically, these patients had been misdiagnosed into one of the categories discussed earlier. Therapists came to realize that they were seeing a new, and increasingly frequent, disorder of desire rather than of excitement or orgasm (e.g., Kaplan, 1979; LoPiccolo, 1980). Surveys of the general population indicate that lack of interest in sex is common. In the NHSLS, 16 percent of men and 33 percent of women reported experiencing it (Laumann et al., 1994).

Like other sexual disorders, HSD poses complex problems of definition. There are many circumstances when it is perfectly normal for a person's desire to be inhibited. For example, one cannot be expected to find every potential partner a turn-on.

Hypoactive sexual desire (HSD): A sexual disorder in which there is a lack of interest in sexual activity; also termed "inhibited sexual desire" or "low sexual desire."

It is also often true that the problem is not the individual's absolute level of sexual desire, but a discrepancy between the partners' levels (Zilbergeld & Ellison, 1980). That is, if one partner wants sex considerably less frequently than the other partner does, there is a conflict. This problem is termed a **discrepancy of sexual desire.**

Among those seeking therapy for HSD, some gender differences appear (Donahey & Carroll, 1993). Women usually report HSD in their early thirties, whereas men usually do in their mid to late forties. There are several possible reasons for this age difference. It may be that women are willing to report this problem earlier. Another possibility is that their male partners, who are themselves young, may be more likely to see the low desire as a problem and insist that something be done about it. Women with HSD were also more likely than men with HSD to be dissatisfied with the quality of their relationship, and particularly with the expression of affection. As we saw in Chapter 14, expressions of love and affection are particularly important to women. It may be that when women's emotional needs are not met by their partners, their sexual desire decreases.

What Causes Sexual Disorders?

There are many causes of sexual disorders, varying from person to person and from one disorder to another. Three general categories of factors may be related to sexual disorders: physical factors (organic factors and drugs), individual psychological factors, and interpersonal factors. Each of these categories is discussed below.

Physical Causes

Physical factors that cause sexual disorders include **organic factors** (physical factors such as diseases) and drugs. Organic factors that have been implicated in various disorders are discussed below, followed by a discussion of the effects of drugs.

Erectile Disorder

Perhaps 50 percent or more of cases of erectile disorder may be due to organic factors or to a combination of organic factors and other factors (Richardson, 1991; Buvat et al., 1990).

Diseases associated with the heart and the circulatory system are particularly likely to be associated with the condition, since erection itself depends on the circulatory system. Any kind of vascular pathology (problems in the blood vessels supplying the penis) can produce erection problems. Erection depends on a great deal of blood flowing into the penis via the arteries, with simultaneous constricting of the veins so that the blood cannot flow out as rapidly as it is coming in. Thus damage to either these arteries or veins may produce erectile disorder.

There is an association between erectile disorder and diabetes mellitus, although the exact mechanism by which diabetes or a prediabetic condition may cause erectile disorder is not known (Bancroft & Gutierrez, 1996). In fact, erectile disorder may in some cases be the earliest symptom of a developing case of diabetes. Of course, not all diabetic men have erectile disorders; indeed, the majority do not. One estimate is that 35 percent of men with diabetes have an erectile disorder (Weinhardt & Carey, 1996).

Any disease or injury that damages the lower part of the spinal cord may cause erectile disorder, since that is the location of the erection reflex center (see Chapter 9). Erectile disorder may also result from severe stress or fatigue. Finally, some—though not all—kinds of prostate surgery may cause the condition.

With erectile disorders, as with most sexual disorders, it is important to recognize that the distinction between organic causes and psychological causes is too simple (Buvat et al., 1990). Many sexual disorders result from a complex interplay of the two causes. For example, a man who has circulatory problems that initially cause him to have erection problems is likely to develop anxieties about erection, which in turn create further difficulties. This notion of dual causes has important implications

Discrepancy of sexual desire: A sexual disorder in which the partners have considerably different levels of sexual desire.
Organic causes of sexual disorders: Physical factors, such as disease or injury, that cause sexual disorders.

for therapy. Many people with such disorders require both medical treatment and psychotherapy.

Premature Ejaculation

Early ejaculation is more often caused by psychological rather than physical factors. In cases of secondary disorder, though, in which the man at one time had ejaculatory control but later lost it, physical factors may be involved. A local infection such as prostatitis may be the cause, as may degeneration in the related parts of the nervous system, which may occur in neural disorders such as multiple sclerosis.

An intriguing explanation for early ejaculation comes from the sociobiologists (Hong, 1984). Their idea is that rapid ejaculation has been selected for in the process of evolution—what we might call "survival of the fastest." In monkeys and apes, the argument goes, copulating and ejaculating rapidly would be advantageous in that the female would be less likely to get away and the male would be less likely to be attacked by other sexually aroused males while he was copulating. Thus males who ejaculated quickly were more likely to survive and to reproduce. Interestingly, among chimpanzees, which some see as our nearest evolutionary relatives, the average time from intromission (insertion of the penis into the vagina) to ejaculation is 7 seconds (Tutin & McGinnis, 1981). In modern U.S. society, of course, rapid ejaculation is not particularly advantageous and might even lead a man to have difficulty finding partners. Nonetheless, according to the sociobiologists, plenty of genes for rapid ejaculation are still hanging around from natural selection that occurred thousands of years ago. (For a critique, see Bixler, 1986.)

Male Orgasmic Disorder

Male orgasmic disorder, or retarded ejaculation, may be associated with a variety of medical or surgical conditions, such as multiple sclerosis, spinal cord injury, and prostate surgery (Rosen & Leiblum, 1995). Most commonly, though, it is associated with psychological factors.

Female Orgasmic Disorder

Orgasmic disorder in women may be caused by severe illness, by general ill health, or by extreme fatigue. However, most cases are caused by psychological factors.

Painful Intercourse

Dyspareunia in women is often caused by organic factors. These include:

1. *Disorders of the vaginal entrance* Irritated remnants of the hymen; painful scars, perhaps from an episiotomy or sexual assault; or infection of the Bartholin glands.
2. *Disorders of the vagina* Vaginal infections; allergic reactions to spermicidal creams or the latex in condoms or diaphragms; a thinning of the vaginal walls, which occurs naturally with age; or scarring of the roof of the vagina, which occurs after hysterectomy.
3. *Pelvic disorders* Pelvic infection such as pelvic inflammatory disease, endometriosis, tumors, cysts, or a tearing of the ligaments supporting the uterus.

Painful intercourse in men can also be caused by a variety of organic factors. For an uncircumcised man, poor hygiene may be a cause; if the penis is not washed thoroughly with the foreskin retracted, material may collect under the foreskin, causing infection. Phimosis, a condition in which the foreskin cannot be pulled back, can also cause painful intercourse. An allergic reaction to spermicidal creams or to the latex in condoms may also be involved. Finally, various prostate problems may cause pain on ejaculation.

Vaginismus

Vaginismus is sometimes caused by painful intercourse, and therefore by organic factors that cause that condition. More frequently, though, it is caused by individual psychological factors or interpersonal factors (Rosen & Leiblum, 1995).

Drugs

Some drugs may have side effects that cause sexual disorders (Rosen, 1991; Segraves, 1988, 1989). For example, some drugs used to treat high blood pressure increase problems with erection in men and with decreased sexual desire in both men and women. Although it would be impossible to list every drug effect on every aspect of sexual functioning, a list of some of the major drugs that may cause sexual disorders is provided in Table 19.1. Here we will consider the effects of alcohol, illicit drugs, and prescription drugs.

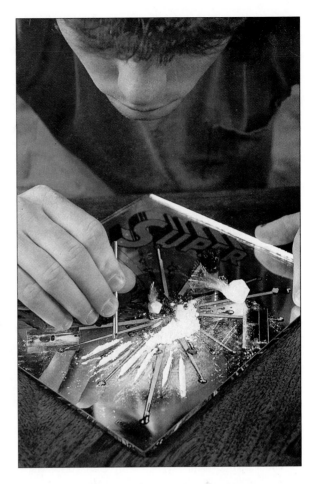

Figure 19.2   Alcohol and cocaine are popular recreational drugs which many people believe enhance sexual experience. Research shows, though, that high levels of alcohol suppress sexual arousal, and repeated use of cocaine is associated with loss of sexual desire, orgasm disorders, and erection problems.

Alcohol

The effects of alcohol on sexual responding vary considerably. We can think of these effects as falling into three categories: (1) short-term pharmacological effects; (2) expectancy effects; and (3) long-term effects of chronic alcohol abuse. In regard to the last category, alcoholics, particularly in the later stages of alcoholism, frequently have sexual disorders, typically including erectile disorder, orgasmic disorder, and loss of desire (Rosen, 1991). These sex problems may be the result of any of a number of organic effects of long-term alcoholism. For example, chronic alcoholism in men may cause disturbances in sex hormone production because of atrophy of the testes. Chronic alcohol abuse, too, generally has negative effects on the person's interpersonal relationships, which may contribute to sexual disorders.

What about the person who is not an alcoholic, but rather has had one or many drinks on a partic-ular evening and then proceeds to a sexual interaction? As noted above, there is an interplay of two effects: expectancy effects and actual pharmacological effects (Crowe & George, 1989). Many people have the expectation that alcohol will loosen them up, making them more sociable and sexually uninhibited. These expectancy effects in themselves produce increased physiological arousal and subjective feelings of arousal. Expectancy effects, though, interact with the pharmacological effects and work mainly at low doses, that is, when only a little alcohol has been consumed. At high dosage levels, alcohol acts as a depressant and sexual arousal is markedly suppressed, both in men and women.

Illicit or Recreational Drugs

There is a widespread belief that *marijuana* has aphrodisiac properties. Little scientific research has been done on its actual effects, and most of what has been done is old and based on small samples (Rosen, 1991). Therefore, we can provide only tentative ideas about the effects of marijuana on sexual functioning. In surveys of users, many respondents report that it increases sexual desire and makes sexual interactions more pleasurable. In regard to potential negative effects, one earlier study indicated that long-term marijuana use lowers testosterone levels and sperm count in males (Kolodny et al., 1974). Later studies indicated that marijuana has no effect on sex hormones (Crenshaw & Goldberg, 1996).

Among drug users, *cocaine* is reported to be the drug of choice for enhancing sexual experiences (Rosen, 1991). It is said to increase sexual desire, enhance sensuality, and delay orgasm. Newer evidence, though, indicates that chronic use of cocaine is associated with loss of sexual desire, orgasmic disorders, and erectile disorders (MacDonald et al., 1988). The effects also depend on the means of administration—whether the cocaine is inhaled, smoked, or injected. The most negative effects on sexual functioning occur among those who regularly inject the drug (MacDonald et al., 1988).

Stimulant drugs, notably *amphetamines,* are associated with increased sexual desire and better control of orgasm in some studies (Rosen, 1991).

Table 19.1 Drugs That May Impair (or Improve) Sexual Response

Drug	How It Affects Sexual Functioning	Common Medical Uses
1. Psychoactive Drugs		
Antianxiety drugs/tranquilizers		Anxiety, panic disorders
Buspirone	Enhanced desire, orgasm	
Benzodiazepines (Librium, Valium, Ativan)	Decreases hypoactive desire, improves premature ejaculation	
Antidepressants I:	Desire disorders, erection problems, orgasm problems, ejaculation problems	Depression
Tricyclics and MAO inhibitors	May treat hypersexuality, premature ejaculation	
Antidepressants II:	Desire disorders, erection problems, orgasm problems	Depression, obsessive-compulsive disorder, panic disorders
Serotonin Reuptake Inhibitors (Prozac, Zoloft)		
Lithium	Desire disorders, erection problems	Bipolar disorder
Antipsychotics	Desire disorders, erection problems, orgasm problems, ejaculation problems	Schizophrenia
2. Antihypertensives		High blood pressure
Reserpine, Methyldopa	Desire disorders, erection difficulties, orgasm delayed or blocked	
Ace inhibitors (Vasotec)	Erection difficulties	
3. Substance Use and Abuse		
Alcohol	At low doses, increases desire	
	At high doses, decreases erection, arousal, orgasm	
	Alcoholism creates many disorders and atrophied testicles, infertility	
Nicotine	Decreases blood flow to penis, creates erectile disorder	
Opiods		
Endogenous: Endorphins	Sense of well-being and relaxation	
Heroin	Decrease in desire, orgasm, ejaculation, replaces sex	
Marijuana	Enhances sexual pleasure but not actual "performance"	

Source: Theresa L. Crenshaw & James P. Goldberg (1996). *Sexual Pharmacology.* New York: Norton.

Injection of amphetamines itself causes a physical sensation that is described by some as a total-body orgasm. In some cases, though, orgasm becomes difficult or impossible when using amphetamines.

Central nervous system depressants, or *sedatives* (e.g., barbiturates), are preferred by some female drug users because of their disinhibiting effects (Rosen, 1991). The effects seem to be much like those of alcohol: disinhibition with low doses, and sexual disorders with high doses.

The *opiates* or narcotics, such as morphine, heroin, and methadone, have strong suppression effects on sexual desire and response (Rosen, 1991). Long-term use of heroin, in particular, leads to decreased testosterone levels in males.

Prescription Drugs

Table 19.1 provides a partial list of prescription drugs that can affect sexual responding.

Some *psychiatric drugs*—that is, drugs used in the treatment of psychological disorders—may affect sexual functioning (Rosen, 1991; Crenshaw & Goldberg, 1996). In general, these drugs have their beneficial psychological effects because they alter the functioning of the central nervous system. But these CNS alterations in turn affect sexual functioning. For example, the drugs used to treat schizophrenia may cause delayed orgasm or "dry orgasm" in men—that is, orgasm with no ejaculation. Tranquilizers and antidepressants often improve sexual responding as a result of improvement of the person's mental state. However, there may also be negative effects. Some of the antidepressants, for example, are associated with both arousal and delayed orgasm problems in both men and women.

The list of other prescription drugs that can affect sexual functioning is long, so we will mention just two examples. Some of the antihypertensive drugs (used to treat high blood pressure) can cause erection problems in men (Rosen, 1991). Most of the research on antihypertensive drug effects has been done with men, so we have less knowledge of these effects on women, although sexual problems among women using antihypertensive drugs have been reported. Some of the drugs used to treat epilepsy appear to cause erection problems and decreased sexual desire, although epilepsy by itself also seems to be associated with sexual disorders.

Psychological Causes

The psychological sources of sexual disorders have been categorized into immediate causes and prior learning (Kaplan, 1974). **Prior learning** refers to the things that people have learned earlier—for example, in childhood—which now inhibit their sexual response. **Immediate causes** are various things that happen in the act of lovemaking itself that inhibit the sexual response.

Immediate Psychological Causes

The following four factors have been identified as immediate psychological causes of sexual disorder: (1) anxieties such as fear of failure, (2) cognitive interference, (3) failure of the partners to communicate, and (4) failure to engage in effective, sexually stimulating behavior.

Masters and Johnson theorized that *anxiety* during intercourse can be a source of sexual disorders. Anxiety may be caused by fear of failure—that is, fear of being unable to perform. But anxiety itself can block sexual response in some people. Often anxiety can create a vicious circle of self-fulfilling prophecy in which fear of failure produces a failure, which produces more fear, which produces another failure, and so on. For example, a man may have one episode of erectile dysfunction, perhaps after drinking too much at a party. The next time he has sex, he anxiously wonders whether he will "fail" again. His anxiety is so great that he cannot get an erection. At this point he is convinced that the condition is permanent, and all future sexual activity is marked by such intense fear of failure that erectile disorder results. The prophecy is fulfilled.

Cognitive interference is a second immediate cause of sexual disorders. "Cognitive interference" refers to thoughts that distract the person from focusing on the erotic experience. The problem is basically one of attention and of whether the person is focusing attention on erotic thoughts or on

Prior learning: Things that people have learned earlier—for example, in childhood—that now affect their sexual response.
Immediate causes: Various factors that occur in the act of lovemaking that inhibit sexual response.
Cognitive interference: Negative thoughts that distract a person from focusing on the erotic experience.

distracting thoughts (Will my technique be good enough to please her? Will my body be beautiful enough to arouse him?). **Spectatoring,** a term coined by Masters and Johnson, is one kind of cognitive interference. The person behaves like a spectator or judge of his or her own sexual "performance." People who do this are constantly (mentally) stepping outside the sexual act in which they are engaged, to evaluate how they are doing, mentally commenting, "Good job," or "Lousy," or "Could stand improvement." These ideas on the importance of cognition in sexual disorder derive from the cognitive theories of sexual responding discussed in Chapters 2 and 9.

Sex researcher David Barlow (1986) has done an elegant series of experiments to test the ways in which anxiety and cognitive interference affect sexual functioning. He has studied men who are functioning well sexually and men with sexual disorders, particularly erectile disorder. We will call these two groups the "functionals" and the "dysfunctionals." He finds that functionals and dysfunctionals respond very differently to stimuli in sexual situations. For example, anxiety (induced by the threat of being shocked) *increases* the arousal of functional men, but *decreases* the arousal of dysfunctional men while watching erotic films. Demands for performance (e.g., the experimenter says the research participant must have an erection or he will be shocked) increase the arousal of functionals but are distracting to (create cognitive interference in) and decrease the arousal of dysfunctionals. When both self-reports of arousal and physiological measures of arousal (the penile strain gauge) are used, dysfunctional men consistently underestimate their physical arousal, whereas functional men are accurate in their reporting.

From these laboratory findings, Barlow has constructed a model that describes how anxiety and cognitive interference act together to produce sexual disorders such as erectile disorder (see Figure 19.3). When dysfunctionals are in a sexual situation, there is a performance demand. This causes them to feel negative emotions such as anxiety. They then experience cognitive interference and focus their attention on nonerotic thoughts, such as thinking about how awful it will be when they don't have an erection. This increases arousal of their autonomic nervous system. To them, that feels like anxiety, whereas a functional would expe-rience it as sexual arousal. For the dysfunctionals, the anxiety creates further cognitive interference, and eventually the sexual performance is dysfunctional—they don't manage to get an erection. This leads them to avoid future sexual encounters or, when they are in one, to experience negative feelings, and the vicious cycle repeats itself.

This analysis is insightful and is backed by numerous well-controlled experiments. It fails to tell us, however, how the dysfunctionals got into this pattern in the first place. That explanation probably has to do with prior learning, to be discussed in the next section.

It is also important to note that anxiety produces sex problems only in some men—the dysfunctionals. For the majority of men, who function well sexually, anxiety does not impair sexual responding. The same is true for women (Elliott & O'Donohue, 1997).

Third, *failure to communicate* is one of the most important immediate causes of sexual disorders. Many people expect their partners to have ESP concerning their own sexual needs. You are the leading expert in the field of what feels good to you, and your partner will never know what turns you on unless you make this known, either verbally or nonverbally. But many people do not communicate their sexual desires. For example, a woman who needs a great deal of clitoral stimulation to have an orgasm may never tell her partner this; as a result, she does not get the stimulation she needs and consequently does not have an orgasm.

A fourth immediate cause of sexual disorders is a *failure to engage in effective sexually stimulating behavior.* Often this is a result of simple ignorance. For example, some couples seek sex therapy because of the wife's failure to have orgasms; the therapist soon discovers that neither the husband nor wife is aware of the location of the clitoris, much less of its fantastic erotic potential. Often such cases can be cleared up by simple educational techniques.

Prior Learning

Besides immediate causes, the other major category of psychological sources of sexual disorders is

---

Spectatoring: Masters and Johnson's term for acting as an observer or judge of one's own sexual performance; hypothesized to contribute to sexual disorders.

Figure 19.3 This model shows how anxiety and cognitive interference can produce erectile dysfunction and other sexual disorders (Barlow, 1986).

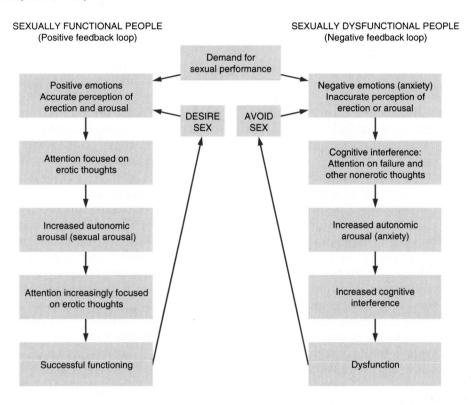

prior learning. This category includes various things that were learned or experienced in childhood, adolescence, or even adulthood.

In some cases of sexual disorders, the person's first sexual act was traumatic. An example would be a young man who could not get an erection the first time he attempted intercourse and was laughed at by his partner. Such an experience sets the stage for future erectile disorder.

Seductive behavior by parents and child sexual abuse by parents or other adults are probably the more serious of the traumatic early experiences that lead to later sexual disorders. A history of sexual abuse is frequently reported by women seeking therapy for problems with sexual desire, arousal, or vaginismus (Becker, 1989). The findings are similar for men with desire or arousal problems (McCarthy, 1990).

In some other cases of sexual disorders, the person grew up in a very strict, religious family and was taught that sex is dirty and sinful. Such a person may have grown up thinking that sex is not pleasurable, that it should be gotten over with as quickly as possible, and that it is for purposes of procreation only. Such learning inhibits the enjoyment of a full sexual response; in fact, to use Byrne's terminology (Chapter 8), it may create an erotophobic personality.

Another source of disorders originating in the family occurs when parents punish children severely for sexual activity such as masturbation. An example is the little girl who is caught masturbating, is punished severely, and is told never to "touch herself" again; in adulthood she finds that she cannot have an orgasm through masturbation or as a result of hand stimulation by her partner.

Parents who teach their children the double standard may contribute to sexual disorders, particularly in their daughters. Women whose sexual response is inhibited in adulthood often were taught as children that no nice lady is interested in sex or enjoys it.

Combined Cognitive and Physiological Factors

In Chapter 13 we discussed the two-component theory of love, which holds that we experience love when two conditions are present: physiological arousal and a cognitive label of "love" attached to it (Berscheid & Walster, 1974a). An analogous cognitive-physiological model of sexual functioning and dysfunction has been proposed (Palace, 1995a, 1995b). According to this model, we function well sexually when we are physiologically aroused and we interpret it as sexual arousal (rather than something else, like nervousness). As Barlow's research shows, people with sexual disorders tend to interpret that arousal as anxiety. In addition, the physiological processes and cognitive interpretations form a feedback loop (see discussion of cognitive theories in Chapter 2). That is, interpreting arousal as sexual arousal increases one's arousal further.

In a clever experiment based on this model, women with sexual disorders were exposed, in a laboratory setting, to a frightening movie, which increased their general autonomic arousal (Palace, 1995b). The women were then shown a brief erotic video and given feedback (actually false) that their genitals had shown a strong arousal response to it. This feedback created a cognitive interpretation for the way they were feeling. The combination of general autonomic arousal and the belief that they were responding with strong sexual arousal led these women, compared with controls, to greater vaginal arousal responses and subjective reports of arousal in subsequent sessions. This demonstration of the effectiveness of combined physiological and cognitive factors is particularly striking because the women began with problems in sexual responding.

Interpersonal Factors

Disturbances in a couple's relationship are another leading cause of sexual disorders. Anger or resentment toward one's partner does not create an optimal environment for sexual enjoyment. Sex can also be used as a weapon to hurt a partner; for example, a woman can hurt her husband by refusing to engage in a sexual behavior that he wants. Conflicts over power may contribute to sex problems.

Intimacy problems in the relationship can be a factor in sexual disorders. These problems typically represent a combination of individual psychological factors and relationship problems. Some individuals have a fear of intimacy—that is, of deep emotional closeness to another person (Kaplan, 1979). Indeed, some people seem to like sex but fear intimacy. They would prefer to watch TV or talk about the weather or have sex than to engage in a truly intimate, emotionally vulnerable, and trusting conversation with another person. They typically progress in a dating relationship to a certain degree of closeness and then lose interest. This pattern is repeated with successive partners. The fear of intimacy may be a result of negative or disappointing intimate relationships—particularly with parents—in early childhood. The fear of intimacy causes a person to draw back from a sexual relationship before it becomes truly fulfilling.

Therapies for Sexual Disorders

A variety of therapies for sexual disorders are available, each relying on a different theoretical understanding of what causes sexual disorders. Here we examine four major categories of therapies: behavior therapy, cognitive-behavioral therapy, couple therapy, and biomedical therapies.

Behavior Therapy

Behavior therapy has its roots in behaviorism and learning theory. The basic assumption is that sex problems are the result of prior learning (as discussed above) and that they are maintained by on-

Behavior therapy: A system of therapy based on learning theory, in which the focus is on the problem behavior and how it can be modified or changed.

going reinforcements and punishments (immediate causes). It follows that these problem behaviors can be unlearned by new conditioning. One of the key techniques is systematic desensitization, in which the client is gradually led through exercises that reduce anxiety.

In 1970, Masters and Johnson reported on their development of a set of techniques for sex therapy and ushered in a new era of sex therapy. They operated from a behavior therapy model because they saw sexual disorders as learned behaviors rather than as symptoms of psychiatric illness. If sexual disorders are the result of learning, they can be unlearned. Masters and Johnson used a rapid two-week program of intensive therapy that consisted mainly of discussions and specific behavioral exercises, or "homework assignments." They used a male-female therapy team to treat heterosexual couples so that each member of the couple would have a therapist of his or her own gender and would not feel outnumbered.

One of the basic goals of Masters and Johnson's therapy is to eliminate goal-oriented sexual performance. Many clients believe that in sex they must perform and achieve certain things. If sex is an achievement situation, it also can become the scene of failure, and it is perceived failures that lead people to believe that they have a sexual problem. The form of cognitive interference known as *spectatoring* (discussed earlier) contributes to this problem because it generates anxiety and other unpleasant feelings. The idea is to use therapy techniques to reduce anxiety.

In one technique Masters and Johnson devised to eliminate a goal-oriented attitude toward sex, the couple is forbidden to have sexual intercourse until they are specifically permitted to by the therapists. They are assigned **sensate focus exercises** that reduce demands on them. As patients successfully complete each of these exercises, the sexual component of subsequent exercises is gradually increased. The couple chalk up a series of successes until eventually they are having intercourse and the disorder has disappeared.

Sensate focus exercises are based on the notion that touching and being touched are important forms of sexual expression and that touching is also an important form of communication; for example, a touch can express affection, desire, understanding, or lack of caring. In the exercises, one member of the couple plays the "giving" role (touches and strokes the other), while the other person plays the "getting" role (is touched by the other). The giving partner is instructed to massage or fondle the other, while the getting partner is instructed to communicate to the giver what is most pleasurable. Thus the exercise fosters communication. The partners switch roles after a certain period of time. In the first exercises, the giver is not to stroke the genitals or breasts but may touch any other area. As the couple progress through the exercises, they are instructed to begin touching the genitals and breasts. These exercises also encourage the partners to focus their attention or concentrate on the sensuous pleasures they are receiving. Many people's sexual response is dulled because they are distracted; they are thinking about how to solve a family financial problem, or they are spectatoring their own performance. To use Barlow's (1986) terminology, they are victims of cognitive interference. The sensate focus exercises train people to concentrate only on their sexual experience, thereby increasing pleasure.

In addition to these exercises, Masters and Johnson supplied simple education. The couple is given thorough instruction in the anatomy and physiology of the male and female sexual organs. Some couples, for example, have no idea of what or where the clitoris is. These instructions may also clear up misunderstandings that either member of the couple may have had since childhood. For example, a man with an erectile disorder may have learned as a child that men can have only a fixed number of orgasms in their lifetime. As he approaches middle age, he starts to worry about whether he may have used up almost all of his orgasms, and this creates the erectile disorder. It is important for such men to learn that nature has imposed no such quota on them.

Masters and Johnson collected data on the success and failure rates of their therapy. In their book *Human Sexual Inadequacy*, they reported on the treatment of 790 persons. Of these, 142 still had a

Sensate focus exercise: A part of the sex therapy developed by Masters and Johnson in which one partner caresses the other, the other communicates what is pleasurable, and there are no performance demands.

Focus 19.1
A Case of Low Sexual Desire

Mr. and Mrs. Brown were in their early thirties, had middle-class backgrounds, and were college-educated; they had been married for 4 years when they entered therapy. The presenting complaint was this: Mr. Brown never initiated sexual contact and rarely appeared interested in sex. Sexual intercourse had occurred only once during the past seven months. He never experienced erectile difficulty. He professed love for his wife and denied interest in any other women.

Mrs. Brown expressed a similarly positive picture of her marriage during her separate interview. She loved her husband very much, was interested in and enjoyed sex, had no interests in other men, and could not identify any problems in their relationship other than her husband's lack of sexual interest. She felt his lack of sexual interest was the likely result of his strict religious background. Whenever she tried to act sexy, such as wearing erotic undergarments, Mr. Brown would laugh and discourage her. Currently, she said that she had just shut down and didn't bother thinking of sex any more. She was worried about this because she wanted to become pregnant.

The medical assessment by a urologist revealed no medical factors that could explain Mr. Brown's low desire. The interview with Mr. Brown revealed some factors that might be contributors to his lack of sexual initiative and low arousal. Although Mr. Brown's religious upbringing did not affect his overall interest in sex, it did seem to dichotomize in his mind sex with "good girls" versus "bad girls." He had had some very arousing sexual experiences with a number of women before marriage. In his words, if the women were very "slutty" and sexually demanding, he became highly aroused. On the other hand, if a woman was "proper" and deserving of respect, he found it very difficult to become aroused. His wife was very attractive but very wholesome; this wholesome image seemed

---

disorder at the end of the 2-week therapy program. This translates to a failure rate of 18 percent, or a *success rate of 82 percent.* While the failure rate ran around 18 percent for most disorders, there were two exceptions: Therapy for premature ejaculation had a very low failure rate (2.2 percent), and therapy for primary erectile disorder had a very high failure rate (40.6 percent). That is, premature ejaculation was quite easy to cure, whereas primary erectile disorder was very difficult. Other research shows that couples' communication skills increase significantly during treatment (Tullman et al., 1981). Masters and Johnson's success rate is impressive, although their results have been called into question, as we shall see later in this chapter.

In Masters and Johnson's initial development of their therapy techniques, all of the couples were heterosexual. They later used the same techniques in treating sexual disorders in gay and lesbian couples, with a comparable success rate (Masters & Johnson, 1979; see also McWhirter & Mattison, 1980).

Cognitive-Behavioral Therapy

As we discussed in Chapter 2, cognitive theories are increasingly important in psychology. Paralleling this increased importance in theory, cognitive approaches in psychotherapy are also becoming important. Today, many sex therapists use a combination of the behavioral exercises pioneered by

to contribute to a restrained approach to sex on his part.

Other possible factors emerged. Mrs. Brown used to initiate sex, but during the past two years she had left the responsibility up to Mr. Brown. Furthermore, Mr. Brown never used sexual fantasies to enhance his arousal and did not expose himself to erotic materials. Finally, Mr. Brown said that he would approach his wife for sex only if he felt fully aroused. It did not occur to him that arousal would develop as a by-product of sexual activity.

All of these possible etiological factors were discussed with Mr. and Mrs. Brown. The Browns were encouraged to try sensate focus, with Mr. Brown taking the initiative. He was reminded that he did *not* have to be sexually aroused to start. He had a great deal of trouble initiating any kind of physical activity; in fact, after failing to initiate during home practice, he disclosed in therapy that he had *never* initiated contact with any female. By not initiating, he never took the chance of being rejected. He also had difficulty in expressing his emotions and found it unmanly to do so. These issues were discussed at length, with a focus on his cognitions, and he was encouraged to continue with his initiation of the sensate focus sessions.

Mrs. Brown was encouraged to be expressive of her sexual feelings in dress and in action, whereas Mr. Brown was encouraged to work on accepting her sexuality. Cognitive restructuring with Mr. Brown focused on his long-standing beliefs regarding women and sexual expression, especially as these related to Mrs. Brown's sexual behavior.

Mr. Brown was also encouraged to practice bringing erotic thoughts into his life. He was given assignments to read erotic passages and to enjoy sexual material he viewed on TV rather than turning away from it.

Fifteen therapy sessions were held over 10 months, and the couple experienced a very positive change in their sexual relationship. Mr. Brown became much more emotionally expressive and felt comfortable initiating sexual contacts. At the end of therapy, the couple was participating in sexual intercourse approximately once a week, and both partners were satisfied with this rate. Both took equal responsibility for initiating sex.

Source: John P. Wincze & Michael P. Carey. (1991). *Sexual dysfunction: A guide for assessment and treatment.* New York: Guilford, pp. 174–175.

Masters and Johnson and cognitive therapy (Bancroft, 1997a; McCarthy, 1989). This is termed **cognitive-behavioral therapy.**

Cognitive restructuring is an important technique in a cognitive approach to sex therapy (Wincze & Carey, 1991). In cognitive restructuring, the therapist essentially helps the client restructure his or her thought patterns, helping them to become more positive (for an example, see Focus 19.1). In one form of cognitive restructuring, the therapist challenges the client's negative attitudes. These attitudes may be as general as a woman's negative, distrusting attitudes toward all men, or as specific as a man's negative attitudes toward masturbation. The client is helped to reshape these attitudes into more positive ones.

Earlier in this chapter we noted that cognitive interference is one of the immediate causes of sexual disorders. That is exactly the kind of issue that a cognitive-behavioral therapist likes to address. The general idea is to reduce the presence of interfering thoughts during sex. First the therapist must help the client identify the presence of such thoughts. The therapist then suggests techniques for reducing these thoughts, generally by replacing them with erotic thoughts—perhaps focusing attention

Cognitive-behavioral therapy: A form of therapy that combines behavior therapy and restructuring of negative thought patterns.

on a particular part of one's body and how it is responding with arousal, or perhaps having an erotic fantasy. Out go the bad thoughts, in come the good thoughts.

Couple Therapy

As we noted earlier, a significant cause of sexual disorders is interpersonal difficulties. Accordingly, some sex therapists use couple therapy as part of the treatment. This approach rests on the assumption that there is a reciprocal relationship between interpersonal conflict and sex problems. Sex problems can cause conflicts, and conflicts can cause sex problems. In couple therapy, the relationship itself is treated, with the goal of reducing antagonisms and tensions between the partners. As the relationship improves, the sex problem should be reduced.

For certain disorders and certain couples, therapists may use a combination of cognitive-behavioral and couple therapy. For example, sex therapists Raymond Rosen, Sandra Leiblum, and Ilana Spector use a five-part model in treating men with erectile disorder (Rosen, Leiblum, & Spector, 1994):

1. *Sexual and performance anxiety reduction* Men with erectile disorder often have a great deal of performance anxiety. This can be treated using such techniques as the sensate focus exercises discussed earlier in this chapter.

2. *Education and cognitive intervention* Men with erectile disorder often lack sexual information and have unrealistic expectations about sexual performance and satisfaction. For example, older men may not be aware of the natural effects of aging on male sexual response. Cognitive interventions may help the man to overcome "all or nothing" thinking— that is, the belief that if any aspect of his sexual performance is not perfect, the whole interaction is a disaster. An example is the belief "I failed sexually because my erection was not 100 percent rigid."

3. *Script assessment and modification* The man with erectile disorder and his partner have a sexual script that they enact together. People with sexual disorders typically have a restricted, repetitive, and inflexible script, using a small number of techniques that they never

change. Novelty is one of the greatest turn-ons, so therapy is designed to help the couple break out of their restricted script.

4. *Conflict resolution and relationship enhancement* As we have discussed, conflicts in a couple's relationship can lead to sexual disorders. In therapy, these conflicts are identified and the couple can work to resolve them.

5. *Relapse prevention training* Sometimes a relapse—a return of the disorder—occurs following therapy. Therapists have developed techniques to help couples avoid or deal with such relapses. For example, they are told to engage in sensate focus sessions at least once a month.

Notice that part 1 represents the behavior therapy techniques pioneered by Masters and Johnson; parts 2 and 3 are cognitive therapy techniques; and part 4 is couple therapy. Most skilled sex therapists today use combined or integrated techniques such as these, tailored to the specific disorder and situation of the couple.

Specific Treatments for Specific Problems

Some very specific techniques have been developed for the treatment of certain sexual disorders.

The Stop-Start Technique

The stop-start technique is used in the treatment of premature ejaculation (see Figure 19.4). The woman uses her hand to stimulate the man to erection. Then she stops the stimulation. Gradually he loses his erection. She resumes stimulation, he gets another erection, she stops, and so on. The man learns that he can have an erection and be highly aroused without having an orgasm. Using this technique, the couple may extend their sex play to 15 or 20 minutes, and the man gains control over his orgasm. Another version of this method is the squeeze technique, in which the woman adds a squeeze around the coronal ridge, which also stops orgasm.

Masturbation

The most effective form of therapy for women with primary orgasmic disorder is a program of directed masturbation (LoPiccolo & Stock, 1986; Rosen & Leiblum, 1995). The data indicate that

Figure 19.4 The stop-start technique for treating premature ejaculation and the position of the couple while using the stop-start technique.

masturbation is the technique most likely to produce orgasm in women; it is therefore a logical treatment for women who have problems with having orgasms, many of whom have never masturbated. Masturbation is sometimes recommended as therapy for men as well (Zilbergeld, 1992).

Kegel Exercises

One technique that is used with women is the **Kegel exercises,** named for the physician who devised them (Kegel, 1952). They are designed to exercise and strengthen the *pubococcygeal muscle,* or PC muscle, which runs along the sides of the entrance of the vagina (see Figure 4.8 in Chapter 4). The exercises are particularly helpful

for women who have had this muscle stretched in childbirth and for those who simply have poor tone in the muscle. The woman is instructed first to find her PC muscle by sitting on a toilet with her legs spread apart, beginning to urinate, and stopping the flow of urine voluntarily. The muscle that stops the flow is the PC muscle. After that, the woman is told to contract the muscle 10 times during each of six sessions per day.

Kegel (KAY-gul) exercises: A part of sex therapy for women with orgasmic disorder, in which the woman exercises the muscles surrounding the vagina; also called pubococcygeal or PC muscle exercises.

Figure 19.5   A position used in treating female sexual disorders.

Gradually she can work up to more.[1] The most important effect of these exercises is that they seem to increase women's sexual pleasure by increasing the sensitivity of the vaginal area (Messe & Geer, 1985). They also permit the woman to stimulate her partner more because her vagina can grip his penis more tightly, and they are a cure for women who have problems with involuntarily urinating as they orgasm. Kegel exercises are sometimes also used in treating men.

[1]Students should recognize the exciting possibilities for doing these exercises. For example, they are a good way to amuse yourself in the middle of a boring lecture, and no one will ever know you are doing them.

Bibliotherapy

Bibliotherapy refers simply to the use of a self-help book to treat a disorder. Research shows that bibliotherapy is effective for orgasmic disorders in women (van Lankveld, 1998). Julia Heiman's *Becoming Orgasmic: A Sexual Growth Program for Women* (1976) has been used extensively for this purpose. Unfortunately, few self-help books are

Bibliotherapy: The use of a self-help book to treat a disorder.

available for other sexual disorders; therefore, almost no research has been done to test whether bibliotherapy is effective with other disorders.

Biomedical Therapies

In the last decade, there has been increased recognition of the biological bases of some sexual disorders. Consistent with that emphasis, many developments in medical/drug treatments and even surgical treatment have occurred.

Drug Treatments

Many promising advances have been made in the identification of drugs that cure sexual disorders or work well when used together with cognitive-behavioral therapy or other psychological forms of sex therapy (Rosen & Leiblum, 1995b). Some are drugs that have direct sexual effects, whereas others are psychotherapeutic drugs (such as antidepressants) that work by improving the person's mood.

Certainly the most widely publicized breakthrough among these treatments was the release, in 1998, of **Viagra** (sildenafil) for the treatment of erectile disorder. Earlier biomedical treatments were unsatisfactory for various reasons. For example, intracavernosal injections (discussed in the following section) are not exactly romantic. Viagra is taken by mouth approximately one hour before anticipated sexual activity. It does not, by itself, produce an erection. Rather, when the man is stimulated sexually after taking Viagra, the drug facilitates the physiological processes that produce erection. Specifically, it relaxes the smooth muscles in the corpora cavernosa, allowing blood to flow in and create an erection. A study of the effectiveness of Viagra in the treatment of 532 men with erectile disorder indicated that 69 percent of attempts at intercourse were successful for those on Viagra, compared with 22 percent for those who took a placebo (Goldstein et al., 1998). The men were generally quite satisfied with Viagra. Side effects were not common; they included headache, flushing, and vision disturbances.

Within a few months of the release of Viagra, 16 deaths had been associated with its use (Associated Press, 1998). Viagra should never be used by men taking nitroglycerin, a medication for

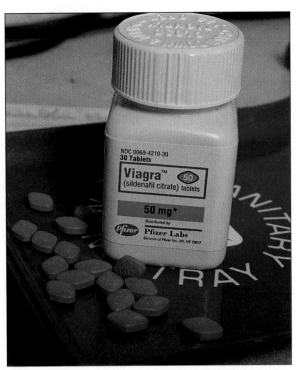

Figure 19.6   One of the new anti-impotence drugs.

heart patients. The combination of the two accounted for several of the deaths. It is also true that the drug would be taken mainly by older men, many of whom have difficulties with their circulatory system, creating both a risk of heart attack and erection problems. That is, the population of men who take Viagra are probably rather high in risk for heart attack with or without Viagra. In a testimony to the crucial importance of sexuality in people's lives, many men have indicated that they are willing to assume serious risks in order to take the drug and experience erections. One physician was quoted as saying "I've had a lot of patients say 'If I have to go, that's the way I want to go out'" (Associated Press, 1998).

On balance, Viagra seems to be quite safe (Morales et al., 1998). It does not seem to cause priapism (an erection that just won't go away). Yet the very ease of its use may lead physicians to overprescribe it and men to demand it in inappropriate

Viagra: A drug used in the treatment of erectile disorder. Sildenafil.

Figure 19.7   At the time Viagra was released, the media were full of jokes about this "delicate" subject.

circumstances. If the erection difficulties are due to a relationship problem, Viagra will provide at most a temporary solution. It is not helpful for sexual disorders other than erectile disorder. And there is no evidence that it enhances sexual performance in men who function sexually within the normal range (Rosen, 1998).

Intracavernosal Injection

Intracavernosal injection (ICI) is a treatment for erectile disorder (Gregoire, 1992; Levitt & Mulcahy, 1995; Wagner & Kaplan, 1993). This treatment involves injecting a drug into the corpora cavernosa of the penis. The drugs used are vasodilators—that is, they dilate the blood vessels in the penis so that much more blood can accumulate there, producing an erection. In one study of a sample of men treated with ICI, the erections lasted on the average 39 minutes (reported in Levitt & Mulcahy, 1995).

ICI is used in cases in which the erectile problem has an organic cause. It is also used in conjunction with cognitive-behavioral therapy in the treatment of cases that are psychologically based or have combined organic and psychological causes. ICI can have positive psychological effects because it restores the man's confidence in his ability to get erections, and it reduces his performance anxiety because he is able to engage in intercourse successfully. Some men experience penile pain from this treatment; switching to a different combination of drugs may solve the problem (Light, 1997). There are also potential abuses. Men who have normal erections should not use ICI in an attempt to produce a "super erection."

Figure 19.8   A treatment for erectile disorder. An external tube, with a rubberband around it, is placed over the lubricated penis. Suction applied to the tube produces erection, which is maintained by the constricting action of the rubberband once the plastic tube has been removed.

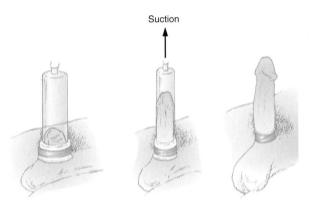

Suction

Suction Devices

Suction devices are another treatment for erectile disorders (Gregoire, 1992). Essentially, they pump you up! A tube is placed over the penis (see Figure 19.8). With some devices, the mouth can produce enough suction; with others, a small hand pump is used. Once a reasonably firm erection is present, the tube is removed and a rubber ring is placed around the base of the penis to maintain the penis's engorgement with blood. These devices have been used successfully with, for example, diabetic men.

Surgical Therapy: The Inflatable Penis

For severe cases of erectile disorder, surgical therapy is possible. The surgery involves implanting a **prosthesis** into the penis (see Figure 19.9) (Kabalin & Kuo, 1997). A sac or bladder of water is implanted in the lower abdomen, connected to two inflatable tubes running the length of the corpus spongiosum, with a pump in the scrotum. Thus the man can literally pump up or inflate his penis so that he has a full erection.

The surgery takes approximately one and a half hours and requires only one incision, where the penis and scrotum meet. The total cost is about $10,000.

It should be emphasized that this is a radical treatment and that it should be reserved only for those cases that have not been cured by sex therapy or drug therapy. Typically it should be a case of primary erectile disorder that is due to organic factors such as diabetes. The patient must understand that the surgery itself destroys some portions of the penis, so that a natural erection will never again be possible. Research shows that about one-fourth of men who have had this treatment are dissatisfied afterward. Reasons for dissatisfaction include the penis being smaller when erect than it was presurgery, different sensations during arousal, and different sensations during ejaculation (Steege et al., 1986). There is concern that unnecessary implants are being performed when less drastic therapy could be used (Shaw, 1989). Although the treatment is radical and should be used conservatively, it is a godsend for some men who have been incapable of erection because of organic difficulties. Indeed, more than a dozen children have been born as a result of this surgery, to women whose partners have previously been incapable of intercourse.

Penile prosthesis (prahs-THEE-sis): A surgical treatment for erectile dysfunction, in which inflatable tubes are inserted into the penis.

Figure 19.9 A surgically implanted prosthesis can be used in treating erectile dysfunction, although it should be regarded as a treatment of last resort.

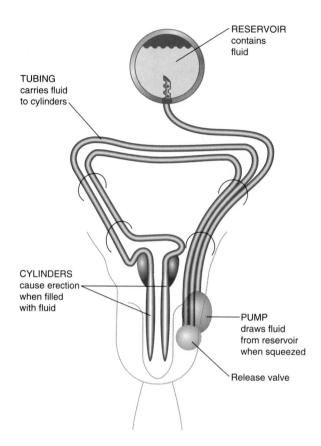

RESERVOIR
contains
fluid

TUBING
carries fluid
to cylinders

CYLINDERS
cause erection
when filled
with fluid

PUMP
draws fluid
from reservoir
when squeezed

Release valve

In another version of a surgical approach, a semirigid, siliconelike rod is implanted into the penis (Melman & Tiefer, 1992; Shandera & Thompson, 1994). This noninflatable device is less costly than the inflatable version and has a lower rate of complications (Rosen & Leiblum, 1995b).

Critiques of Sex Therapy

One of the most basic questions we must ask about sex therapy is, Is it effective?

Psychologists Bernie Zilbergeld and Michael Evans (1980) did an extensive critique of the research methods used by Masters and Johnson in evaluating the success of their sex therapy. Zilbergeld and Evans concluded that there are a number of substantial problems. In brief, the criticisms suggest that we really do not know what the success rate of Masters and Johnson's therapy is, and it is almost surely lower than 80 percent. A discussion of the specific criticisms follows. These points are actually important to bear in mind when evaluating any program of sex therapy.

First, Masters and Johnson never actually reported a *success* rate for the therapy. Instead, they reported a *failure* rate of about 20 percent. Thus, most people have concluded, as we did earlier in the chapter (and you probably thought it was logical as you read it), that this implies a success rate of about 80 percent. But Masters and Johnson say that this is not the case. That is, the 80 percent apparently includes a mixture of clear successes and cases that are ambiguous as to whether they are successes or failures—in short, they have 80 percent nonfailures, but that does not mean 80 percent successes.

Furthermore, Masters and Johnson never defined what they meant by a "success" in therapy. This is an important issue. How improved does a person have to be to be counted as a success? Suppose a woman seeks help for anorgasmia; she has never had an orgasm. By the end of therapy, she is able to have orgasms from a vibrator, but not from hand stimulation or mouth stimulation by her partner, nor by intercourse. Is that a success? How would Masters and Johnson have classified her? We cannot tell from their book.

Zilbergeld and Evans initiated their critical appraisal after finding that they and other sex therapists were unable to get success rates as dramatic as the ones Masters and Johnson reported. The obvious possibility is that other sex therapists have been using definitions of therapy success that are much stricter and more precise than the definition used by Masters and Johnson.

Masters and Johnson did not report clearly how the initial population of clients for therapy was chosen. They said that they rejected some people from therapy, but they did not specify who made the decision, how the decision was made, and how many people were thus rejected. It seems quite possible that Masters and Johnson weeded out the most difficult cases, leaving themselves with the easier ones and consequently a high success rate.

With their 5-year follow-up of patients, Masters and Johnson reported an amazingly low relapse rate of 7 percent, but other sex therapists find much higher relapse rates. Once again, Masters and Johnson did not specify their exact criterion for relapse, so it is hard to evaluate or replicate the 7 percent figure.

Masters and Johnson were somewhat misleading about the duration of their therapy. They described it as two weeks of rapid treatment. Other therapists typically find that patients need more sessions than that. What Masters and Johnson failed to highlight is the fact that patients were instructed to call Masters and Johnson any time they ran into problems; in addition, there were regularly scheduled telephone calls between the couple and the cotherapists. Essentially, a couple could get a great deal more than two weeks of therapy.

Finally, Masters and Johnson never discussed the possible harmful effects of their therapy. They made fleeting references to a couple of cases in which the therapy apparently ended in divorce, but they made no systematic attempt to assess problems of this sort. Their 5-year follow-up was of "successes" only, not failures. It seems likely that the failures were precisely those who might have been harmed, yet no information on them is provided.

One recent critique pointed out that a major shortcoming of the field of sex therapy is the lack of carefully controlled studies that (1) investigate the success of various therapies compared with other therapies and with untreated controls; and (2) examine what aspect of a particular therapy or combination of therapies seems to have the beneficial effect (Rosen & Leiblum, 1995b).

A recent authoritative review concluded that there was sufficient evidence evaluating certain treatments for certain disorders, to reach the following conclusions (Heiman & Meston, 1997):

- Primary orgasmic dysfunction is successfully treated with directed masturbation, and the treatment can be enhanced with sensate focus exercises.

- Treatments for secondary orgasmic dysfunction are somewhat less successful. Therapy that combines some or all of the following components seems to be most effective: sex education, sexual skills training, communication skills training, and body image therapy. The problem here, most likely, is that there are many different patterns of secondary anorgasmia, with a need to match treatment to the pattern of the disorder, something that research has not been able to untangle.

- Vaginismus is successfully treated with progressive vaginal dilators; relaxation and Kegel exercises may also be helpful, but the evidence is not as strong.

- Several of the biomedical treatments for erectile disorder (intracavernosal injection, insertion of a drug capsule into the urethra, suction method) are successful in producing erections (this review was published before Viagra became available).

The evidence also suggests that systematic desensitization is effective.

- The squeeze technique is effective for treating premature ejaculation. Drugs, specifically some antidepressants (serotonin reuptake inhibitors), may be effective but there is less data.

- For some disorders—sexual desire disorders, dyspareunia, and delayed orgasm in men—research is insufficient to conclude that there is an effective treatment.

Another recent critique pointed out the *medicalization* of sexual disorders, particularly male erectile problems (Tiefer, 1994). Research has increasingly identified organic sources of sexual disorders, and with these advances have come attempts to identify drugs and surgeries, rather than psychotherapies, to treat problems. In part, political issues are involved, as physicians try to seize the treatment of sexual disorders from psychologists. But there is also a cost to the patient, as the disorder is given a quick fix with drugs while the patient's anxieties and relationship problems are ignored.

In contrast to these scientific criticisms is the criticism by psychiatrist Thomas Szasz. In his book *Sex by Prescription* (1980) he criticized the philosophical basis of sex therapy. Szasz has long been an outspoken critic of psychotherapy. He is particularly critical of the medical model in dealing with psychological problems (see, for example, his classic *The Myth of Mental Illness*). His essential argument is that psychologists and psychiatrists take people who have problems in living, or perhaps have freely chosen lifestyles, and classify them as "sick" or "mentally ill" (the medical model) and in need of therapy. Although the professionals may think they are being helpful, they may do more harm than good. For example, once persons are classified as "sick," the implication is that they need a psychologist or physician to fix them up; whereas in fact, Szasz maintains, it might be better for them to make active efforts to solve their own problems.

Applying this thinking to the sex therapy field, Szasz argues that the sex therapists have essentially created a lot of illnesses by creating the (somewhat arbitrary) diagnostic categories of the sexual disorders. For example, the man who cannot have intercourse because he cannot manage an erection is said to have "erectile dysfunction," yet the man who cannot bring himself to perform cunnilingus is not regarded as having any dysfunction. Why should the first problem be an "illness" and the second one not? A man who ejaculates rapidly is termed a "premature ejaculator" and considered in need of therapy, but what exactly is wrong with ejaculating rapidly? Szasz believes that instead of regarding the so-called sexual disorders as diseases, it would be better to see them as individuals' solutions to various life situations.

Szasz criticizes Masters and Johnson for presenting their work as medical and scientific; he believes that it is moral and political and laden with values. For example, Masters and Johnson claim that homosexuality is not a disease but that nonetheless they can cure it in two weeks.

Szasz summarizes his arguments as follows:

I do not deny that sexual problems exist or are real. . . . I maintain only that such problems—including sexual problems—are integral parts of people's lives. . . .

As some of the examples cited in the book illustrate, one medical epoch's or person's sexual problem may be another epoch's or person's sexual remedy. Today, it is dogmatically asserted—by the medical profession and the official opinion-makers of our society—that it is healthy or normal for people to enjoy sex, that the lack of such enjoyment is the symptom of a sexual disorder, that such disorders can be relieved by appropriate medical (sex-therapeutic) interventions, and that they ought, whenever possible, to be so treated. This view, though it pretends to be scientific, is, in fact, moral or religious: it is an expression of the medical ideology we have substituted for traditional religious creeds. (1980, pp. 164–165)

In summary, a number of criticisms have been raised about the field of sex therapy. The research methods used by Masters and Johnson to evaluate the success of their therapy had a number of problems and, as a result, their implied success rate of 80 percent is probably unrealistically high. Despite a decade of rapid advances in both psychological and medical treatments of sexual disorders, adequate research has not yet been done on the effectiveness of many of these treatments. There is a trend to medicalize sexual disorders, particularly

Focus 19.2

Sex Therapy On-line?

Dr. Patti Britton is a trained sex therapist who runs a successful sex therapy website. It is difficult to find such legitimate websites because searching for terms such as "sex therapy" leads to a deluge of porn sites (as of this writing, Dr. Britton's address is yoursexcoach.com). Her website includes a chat room where people can talk with others about their sexual experiences and problems. And, for $45/hr, you can have a half-hour individual session with Dr. Britton, during which you type your back-and-forth conversation with her. In addition, Dr. Britton is one of the professional experts at ivillage.com, a "club" geared for women, with a membership fee. Dr. Britton hosts a weekly one-hour "chat" at that site.

Is sex therapy on-line the wave of the future? What are its advantages and disadvantages? Proponents argue that it is more affordable than traditional in-person therapy and that its anonymity is a major advantage. People who are too shy or embarrassed to tell their story to a sex therapist can log on and type their questions anonymously. A person could easily obtain help without even their partner knowing about it. The advice columns at these sites can provide accurate, explicit, and non-judgmental information. Interactions with a thera-

pist online can break the wall of isolation surrounding a person with a sexual disorder. Specialized message boards and chat rooms for people who share a common theme (e.g., bisexuals, persons with disabilities) can help to create a sense of community, especially for those who are geographically isolated. Because the web is international, people in countries in which sex therapy is unknown can obtain helpful information that would otherwise be unavailable.

There are disadvantages, too. Currently there is no system for licensing on-line sex therapists, so unqualified and perhaps unethical persons could easily present themselves as therapists. Moreover, on-line sex therapists probably will not be able to give true intensive therapy of the kind one would get in multiple in-person sessions with a therapist. What on-line therapists can do is provide permission and positive encouragement as well as accurate information, and that is enough to solve many people's problems.

Source: Presentations by Patti Britton, Al Cooper, Sandor Gardos, and others at the 1998 meetings of the Society for the Scientific Study of Sexuality.

---

erectile disorder, which may lead to the neglect of patients' needs for psychological treatment. Finally, Szasz questions the whole notion of sexual disorders as such.

Where do these criticisms leave us? In our opinion, they do not invalidate the work of sex therapists. Rather, they urge us to be cautious. Most sexual disorders will not have cure rates of 80 percent, but they may well turn out to have cure rates of 60 percent or more (Kaplan, 1979). A single method of therapy, such as Masters and Johnson's behavior therapy, will not be effective with every disorder.

Finally, we must be sensitive to the values expressed in labeling something as being, or someone as having, a "disorder."

Sex Therapy in the AIDS Era

The threat of AIDS—not to mention herpes and genital warts (HPV)—puts modern sex therapy into a new cultural context (Leiblum & Rosen, 1989). Here are a few examples of some of these new factors that are changing sex therapy.

People are now more interested in maintaining a long-term, monogamous relationship and less interested in giving up on a partner and finding a new one just because the sex has lost some excitement. Sex therapists are seeing more couples who want help in rejuvenating their sex lives.

Communication skills training (see Chapter 10), which is a standard part of sex therapy, has become increasingly important. Being able to communicate directly, openly, and effectively with a partner about condom use can literally be a matter of life or death today. The demand for communication skills training will surely increase as the AIDS epidemic continues.

Sex therapists will need to do more work with people on how to have pleasurable sex using condoms. Therapists have documented case histories of men who have such negative feelings about using a condom that they literally lose their erection if they try to put one on (Leiblum & Rosen, 1989). As condoms become an increasing necessity, this might even become a sexual disorder—condom-induced erectile disorder! Sex therapists can work with such people to diminish uncomfortable feelings and substitute comfortable—perhaps even erotic—feelings about condoms.

Some Practical Advice

Avoiding Sexual Disorders

The National Safety Council's motto is: "Prevention is better than cure." This principle could well be applied not only to accidents but also to sexual disorders. That is, people could use some of the principles that emerge from sex therapists' work to avoid having sexual disorders in the first place. The following are some principles of good sexual mental health:

1. *Communicate with your partner* Don't expect him or her to be a mind reader concerning what is pleasurable to you. One way to do this is to make it a habit to talk to your partner while you are having sex; verbal communication then does not come as a shock. Some people, though, feel uncomfortable talking at such times; nonverbal communication, such as

placing your hand on top of your partner's and moving it where you want it, works well too (see Chapter 10 for more detail).

2. *Don't be a spectator* Don't feel as if you are putting on a sexual performance that you constantly need to evaluate. Concentrate as much as possible on the giving and receiving of sensual pleasures, not on how well you are doing.

3. *Don't set up goals of sexual performance* If you have a goal, you can fail, and failure can produce disorders. Don't set your heart on having simultaneous orgasms or, if you are a woman, on having five orgasms before your partner has one. Just relax and enjoy yourself.

4. *Be choosy about the situations in which you have sex* Don't have sex when you are in a terrific hurry or are afraid you will be disturbed. Also be choosy about who your partner is. Trusting your partner is essential to good sexual functioning; similarly, a partner who really cares for you will be understanding if things don't go well and will not laugh or be sarcastic.

5. *"Failures" will occur* They do in any sexual relationship. What is important is how you deal with them. Don't let them ruin the relationship. Instead, try to think, "How can we make this turn out well anyhow?"

Choosing a Sex Therapist

Unfortunately, most states do not have licensing requirements for sex therapists (most states do have requirements for marriage counselors and psychologists). Particularly with the popularizing of Masters and Johnson's work, quite a few quacks have hung out shingles saying "Sex Therapist," and many states have made no attempt to regulate this. Some of these "therapists" have no more qualifications than having had a few orgasms themselves.

How do you go about finding a good, qualified sex therapist? Your local medical association or psychological association can provide a list of psychiatrists or psychologists and may be able to tell you which ones have special training in sex therapy. There are also professional organizations of sex therapists. The American Association of Sex Educators, Counselors, and Therapists certifies sex therapists (see Appendix A for complete information on this organization and many other useful ones). Choose a therapist or clinic that offers an *integrated approach* that recognizes the potential biological, cognitive-behavioral, and relationship influences on any sexual disorder and is prepared to address all of these.

SUMMARY

In men, the major kinds of sexual disorders are erectile disorder, premature ejaculation, and male orgasmic disorder. In women, the major disorders are female orgasmic disorder, female sexual arousal disorder, painful intercourse, and vaginismus. Disorders of sexual desire may occur in men or women.

Sexual disorders may be caused by physical factors, individual psychological factors, and interpersonal factors. Organic causes include some illnesses, infections, and damage to the spinal cord. Certain drugs may also create problems with sexual functioning. Individual psychological causes are categorized into immediate causes, such as anxiety or cognitive interference, and prior learning. Interpersonal factors include conflict in the couple's relationship and intimacy problems.

Therapies for sexual disorders include behavior therapy (pioneered by Masters and Johnson) based on learning theory, cognitive-behavioral therapy, couple therapy, specific treatments for specific problems (e.g., stop-start for premature ejaculation), and a variety of biomedical treatments, which include drug treatments and surgery.

A number of criticisms of sex therapy have been raised, including concerns about the research methods used to evaluate the success of Masters and Johnson's therapy, the lack of evaluation research on recently developed therapies, the medicalization of sexual disorders, and the entire enterprise of identifying and labeling sexual disorders.

REVIEW QUESTIONS

1. _____ is the proper term for the inability to have an erection or maintain one.

2. The term that describes the disorder in which a woman has never in her life had an orgasm is _____.

3. One common cause of dyspareunia is a vaginal infection. True or false?

4. Sexual disorders such as erectile disorder and hypoactive sexual desire are common among long-term alcoholics. True or false?

5. Negative sexual experiences in childhood are termed an immediate cause of sexual disorders. True or false?

6. When negative thoughts distract a person from having an enjoyable sexual experience, this is termed _____.

7. Masters and Johnson pioneered the use of sensate focus exercises to reduce performance demands. True or false?

8. Cognitive restructuring is used in _____ therapy.

9. Intracavernosal injections are used in the treatment of _____.

10. Antidepressants are the best therapy for women with primary orgasmic disorder. True or false?

(The answers to all review questions are at the end of the book, after the Glossary.)

QUESTIONS FOR THOUGHT, DISCUSSION, AND DEBATE

1. When (or if) you engage in sexual activity with a partner, do you feel that you are under pressure to perform and that you engage in spectatoring? If so, what might you do to change this pattern?

2. Considering the prior learning causes of sexual disorders, what are the implications for parents who want to raise sexually healthy children? Could parents do certain things that would avoid or prevent sexual disorders in their children?

3. Your best friend, Steve, who is 22, discloses to you a long history of premature ejaculation, which has been very embarrassing and frustrating for him. He has heard about Viagra and, knowing that you are taking a human sexuality course, comes to you for advice about whether he should go to a doctor to get a prescription for it. What advice would you give him?

4. Hypoactive sexual desire is a common sexual disorder in the United States today. Given what you know about this disorder and its causes, do you think it is also common in other cultures? Think about Britain, China, India, and Mexico as examples. (For further information, see Bhurga & de Silva, 1993.)

SUGGESTIONS FOR FURTHER READING

Barbach, Lonnie G. (1975). *For yourself: The fulfillment of female sexuality.* Garden City, NY: Doubleday. Provides good information for women with orgasmic disorders, based on the author's experiences as a sex therapist. Still the classic in the field.

Barbach, Lonnie G. (1983). *For each other: Sharing sexual intimacy.* Garden City, NY: Anchor Books. Barbach's sequel to *For Yourself* (see above); this volume is designed for couples.

Tiefer, Leonore. (1995). *Sex is not a natural act, and other essays.* Boulder, CO: Westview. Tiefer is a brilliant and entertaining writer, and her criticisms of sex therapy are insightful.

Zoldbrod, Aline P. (1998). *Sex smart: How your childhood shaped your sexual life and what to do about it.* Oakland, CA: New Harbinger. Zoldbrod, a sex therapist, provides a detailed guide to uncovering the sources of sex problems in early childhood learning.

WEB RESOURCES

http://www.sexhealth.org/infocenter/SexualDy/sexualdy.htm
The Sexual Health InfoCenter: Sexual Disorders Page.

chapter 20

SEXUALLY TRANSMITTED DISEASES

CHAPTER HIGHLIGHTS

Chlamydia

HPV

Genital Herpes

HIV Infection and AIDS

Gonorrhea

Syphilis

Viral Hepatitis

Pubic Lice

Preventing STDs

Other Genital Infections

Maria and Luis get home after an evening on the town and enter the house hungry for passion. The two embrace, clinging to each other, longing for each other. Luis slowly undresses Maria, hungry for the silky flesh he feels beneath him. As their passion grows, she begins reaching for him, ripping his clothes off as she explores his body with her tongue. As Luis gets more and more excited, Maria rips a condom package open with her teeth and slowly slides the condom over Luis's erect penis. After an hour of incredible lovemaking, Luis, exhausted with pleasure, turns to Maria and says, "You were right, the BEST sex is SAFE sex with LATEX!"*

The sexual scene is not the same as it was 30 years ago. Herpes and AIDS pose real threats. We need to do many things to combat these dangers. One is that we must rewrite our sexual scripts, as the quotation above illustrates. We also need to inform ourselves, and the goal of this chapter is to provide you with the important information you need to make decisions about your sexual activity.[1]

Your health is very important, and a good way to ruin it or cause yourself a lot of suffering is to have an untreated case of a sexually transmitted disease (STD). Consequently, it is very important to know the symptoms of the various kinds of STDs so that you can seek treatment if you develop any of them. Also, there are some ways to prevent STDs or at least reduce your chances of getting them, and these are certainly worth knowing about. Finally, after you have read some of the statistics on how many people contract STDs every year and on your chances of getting one, you may want to modify your sexual behavior somewhat. If you love, love wisely.

The STDs are presented in this chapter in the following order. First we look at a group of three diseases—chlamydia, HPV (genital warts), and herpes—that are all quite frequent among college students. Following that is a discussion of HIV infection and AIDS, which is less common among college students but is one of the world's major public-health problems and is generating an enormous amount of research. Next we discuss gonorrhea, syphilis, and viral hepatitis. Next comes not an infection but a bug, the pubic louse. After a practical section for you on preventing STDs, the chapter ends with a section about various other genital infections that, for the most part, are not sexually transmitted.

Many statistics throughout this chapter are taken from the Centers for Disease Control (CDC) website:

http://www.cdc.gov/nchstp/dstd

The CDC is in Atlanta, GA. It is the federal government's agency that monitors diseases in the United States, and conducts research and prevention programs. Data from the CDC are used so frequently throughout the chapter that we do not provide a citation every time. Information on STDs changes quickly so, to get the most up-to-date information, check the website.

One final note before we proceed: There are a lot of illustrations in this chapter showing the symptoms of various STDs, and some of the photos may make you say "Nasty!" These illustrations are in the chapter not to scare you, but rather to help you recognize the symptoms of STDs. You should know what herpes blisters, for example, look like, in case you spot them on a prospective sexual partner and in case they appear on you.

*From a student essay.

[1]The diagnosis and treatment of sexually transmitted diseases is an area of furiously active research. New discoveries are announced almost monthly. Therefore, by the time you read this book, some of the statements in this chapter may have been superseded by newer research.

Chlamydia

Chlamydia trachomatis is a bacterium that is spread by sexual contact and infects the genital organs of both males and females. The female is said to have a **chlamydia** infection. Men with a chlamydia infection in the urethra are said to have chlamydia or **nongonococcal urethritis,** or **NGU.** NGU is any inflammation of the male's urethra that is not caused by a gonorrhea infection. (For a review, see Stamm & Holmes, 1990.) *Chlamydia* is one organism known to cause NGU, but it may be caused by several other organisms.

Statistics indicate that chlamydia has become one of the major sexually transmitted diseases. The Centers for Disease Control estimates that there are approximately 1.3 million new cases of chlamydia in women annually and 1.8 million in men, compared with 1 million cases of gonorrhea (Schachter & Barnes, 1996). Chlamydia is more prevalent in members of higher socioeconomic groups and among university students. Adolescent girls have a particularly high rate of infection. When a man consults a physician because of a urethral discharge, his chances of having chlamydia are greater than his chances of having gonorrhea. It is important that the correct diagnosis be made since chlamydia does not respond to some of the drugs used to cure gonorrhea.

Symptoms

The main symptoms in men are a thin, usually clear discharge and mild discomfort on urination appearing 7 to 14 days after infection. The symptoms are somewhat similar to the symptoms of gonorrhea in the male. However, gonorrhea tends to produce more painful urination and a more profuse, puslike discharge. Diagnosis is made from a sampling of cells from the genitals (the urethra in males, the cervix in females). Tests are then used to detect the bacterium. Unfortunately, 75 percent of the cases of chlamydia infection are **asymptomatic** in women. This means that the woman never goes to a clinic for treatment, and she goes undiagnosed and untreated. The consequences of untreated chlamydia in women are discussed in the next section. Even among men, 50 percent of the cases are asymptomatic.

Treatment

Chlamydia is quite curable. It is treated with azithromycin or doxycycline; it does not respond to penicillin. Poorly treated or undiagnosed cases may lead to a number of complications: urethral damage, epididymitis (infection of the epididymis), Reiter's syndrome,[2] and proctitis in men who had anal intercourse. Women with untreated or undiagnosed chlamydia may experience the following serious complications if not treated: damage to the cervix, *salpingitis* (infection of the fallopian tubes[3]), **pelvic inflammatory disease (PID),** and possibly infertility due to scarring of the fallopian tubes (Weinstock et al., 1994). Chlamydia infection is associated with an increased risk of ectopic pregnancy and with increased rates of prematurity and low-birth-weight babies (Cohen et al., 1990; Chow et al., 1990; Sherman et al., 1990). A baby born to an infected mother may develop pneumonia or an eye infection.

Prevention?

Scientists doing research on chlamydia have a major goal of developing a vaccine that would prevent infection (Rasmussen, 1998). Vaccines have been developed that are effective in mice, but certain technical obstacles prevent their use with humans. An effective vaccine for humans should be available in the next decade.

Until a vaccine is available, one of the most effective tools for prevention is screening. The problem with chlamydia is that so many infected people are asymptomatic and spread the disease

---

[2]Reiter's syndrome involves the following symptoms: urethritis, eye inflammations, and arthritis.

[3]"Salpinges" (singular, "salpinx") is another name for the fallopian tubes.

Chlamydia (klah-MIH-dee-uh): An organism causing a sexually transmitted disease; the symptoms in males are a thin, clear discharge and mild pain on urination; females are frequently asymptomatic.
Nongonococcal urethritis (NGU): (non-gon-oh-COK-ul yur-ith-RITE-is): An infection of the male's urethra usually caused by chlamydia.
Asymptomatic (ay-simp-toh-MAT-ik): Having no symptoms.
Pelvic inflammatory disease (PID): An infection and inflammation of the pelvic organs, such as the fallopian tubes and the uterus, in the female.

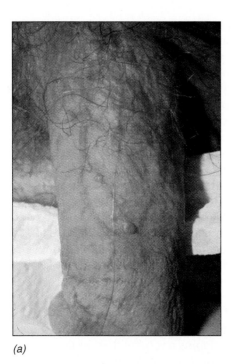

(a)

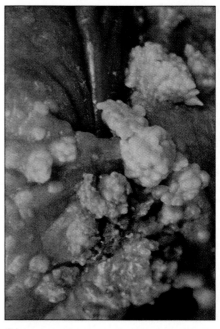

(b)

Figure 20.1   Genital warts *(a)* on the penis, and *(b)* on the vulva.

unknowingly. In screening programs, asymptomatic carriers are identified, treated, and cured so that they do not continue to spread the disease. The CDC has conducted several demonstration projects that have yielded impressive results. In one, screening was conducted in family planning clinics in Alaska, Idaho, Oregon, and Washington. In the first 8 years of the program, from 1988 to 1995, a 65 percent decline occurred in the rate of infection.

On an individual level, the best method of prevention is the consistent use of a condom.

HPV

HPV stands for human papilloma virus, which causes genital warts. **Genital warts** are cauliflower-like warts appearing on the genitals, usually around the urethral opening of the penis, the shaft of the penis, or the scrotum in the male, and on the vulva, the walls of the vagina, or the cervix in the fe-

male; warts may also occur on the anus (Moreland et al., 1996). Typically they appear 3 to 8 months after intercourse with an infected person.

Experts say that, if AIDS were not around, the big news story of the decade would be genital warts. New data show that infection with HPV is widespread. In the United States, approximately 3 million new cases are diagnosed each year (Hatcher et al., 1994). The disease is highly infectious. About two-thirds of the sex partners of persons with genital warts develop the disease (Oriel, 1990).

In one study, researchers randomly selected 454 women attending an STD clinic and 545 college women undergoing a routine annual examination (Kiviat et al., 1989). HPV infection was found in 11 percent of the STD clinic patients and 2 percent of

HPV: Human papilloma virus, the organism that causes genital warts.
Genital warts: A sexually transmitted disease causing warts on the genitals.

the college women. There were cervical abnormalities in 40 to 50 percent of the women infected with HPV.

The evidence also makes it clear that HPV increases the risk of cervical cancer (McCance, 1994). Because of this increased risk, any woman with recurrent genital warts should have a Pap smear every six months. Additional evidence indicates that HPV infection is associated not only with cancer of the cervix but also with cancer of the vagina, vulva, penis, and anus (Kiviat et al., 1989).

Diagnosis

Diagnosis can sometimes be made simply by inspecting the warts, because their appearance is distinctive. However, some strains of warts are flat and less obvious. Also, the warts may grow inside the vagina and may not be detected there. A new test involves analysis of the DNA from a sample of cells taken from the cervix, vagina, or other area of suspected infection and tests directly for the presence of the HPV virus.

Treatment

Several treatments for genital warts are available (Moreland et al., 1996). Podophyllin or trichloroacetic acid (TCA) can be applied directly to the warts. Typically these treatments have to be repeated several times, and the warts then fall off. With cryotherapy (often using liquid nitrogen), the warts are frozen off; again, it is typically necessary to apply more than one treatment. Laser therapy can also be used to destroy the warts. In one study of college women, HPV infection lasted, on average, only 8 months (Ho et al., 1998). The duration of infection was longer with older women, and the risk of cervical cancer increases with longer infections.

Vaccine?

As noted, there is a strong association between HPV and cervical cancer. In one study, for example, HPV was found in 93 percent of the 1000 tissue samples from cervical cancer tumors taken from women in 22 different countries (Bosch, Manos, Muñoz, et al., 1995). This finding has led researchers to believe that if they developed a vaccine against HPV, the vaccine might also destroy cervical tumors. Several such vaccines are currently being tested (Hines et al., 1998).

Genital Herpes

Genital herpes is a disease of the genital organs caused by the herpes simplex virus Type II (**HSV**-2) in 90 percent of the cases, and otherwise by the Type I virus (Moreland et al., 1996). HSV-1 usually causes symptoms on nongenital parts of the body; it is responsible for cold sores and fever blisters, for instance. Genital herpes is transmitted by sexual intercourse, although there are people who have it whose only sexual partner does not. HSV-1 may be transmitted to the genitals during oral-genital sex.

Symptoms

The symptoms of genital herpes are small, painful bumps or blisters on the genitals. In women, they are usually found on the vaginal lips; in men, they usually occur on the penis. They may be found around the anus if the person has had anal intercourse. The blisters burst and can be quite painful. Fever, painful urination, and headaches may occur. The blisters heal on their own in about three weeks in the first episode of infection. The virus continues to live in the body, however. It may remain dormant for the rest of the person's life. But the symptoms may recur unpredictably, so that the person repeatedly undergoes 7- to 14-day periods of sores. On the average, herpes patients have four recurrences per year (Benedetti et al., 1994). The disease may also be transmitted from a pregnant woman to the fetus. Some infants recover, but others develop a brain infection that rapidly leads to death. (For excellent reviews, see Corey, 1990; Moreland et al., 1996).

An estimated 45 million Americans—22 percent of persons 12 years of age or older—are infected (Fleming et al., 1997). Infection rates are somewhat

---

Genital herpes (HER-pees): A sexually transmitted disease, the symptoms of which are small, painful bumps or blisters on the genitals.
HSV: The herpes simplex virus.

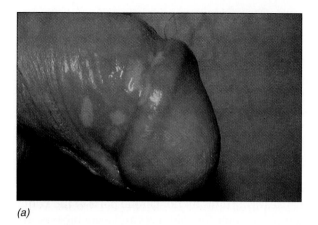

(a)

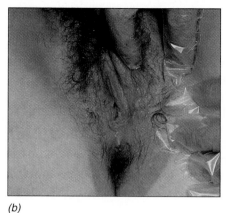

(b)

Figure 20.2   *(a)* Herpes blisters on the penis. *(b)* Herpes infection of the vulva.

higher among women (26 percent) than men (18 percent) and higher among African Americans (46 percent) than whites (18 percent). Of those people who test positive for HSV-2 in their blood, only 10 percent report a history of herpes outbreaks; that is, the great majority are asymptomatic. These persons transmit the disease to others unknowingly.

People with herpes are most infectious when they are having an active outbreak. However, people are infectious even when there is no outbreak (Mertz, 1993). Therefore, there may be no "safe" period.

Treatment
Unfortunately, there is not yet any known drug that kills the virus. That is, there is no cure. Researchers are pursuing two solutions: drugs that would cure symptoms in someone who is already infected, and vaccinations that would prevent herpes. The drug acyclovir prevents or reduces the recurring symptoms, although it does not actually "cure" the disease (Clark et al., 1995; MMWR, 1998). Acyclovir in capsule form controls symptoms and even suppresses recurrences. Other drugs—and even the use of lasers—are also being investigated, as are vaccines.

Long-Term Consequences
Either men or women with recurrent herpes may develop complications such as meningitis or nar-

rowing of the urethra due to scarring, leading to difficulties with urination. However, such complications do not affect the majority of those with herpes. There are two more serious long-term consequences. One is that having a herpes infection increases one's risk of becoming infected with HIV (Mertz, 1993), probably because the open blisters during an outbreak make it possible for HIV to enter the body. Therefore, people who have herpes should be especially careful to use safer sex practices.

The other serious risk involves the transfer of the virus from mother to infant in childbirth, which in some cases leads to serious illness or death in the baby. The risk of transmission to the infant is highest in women who have recently been infected and are having their first outbreak. The risk is less with women who have had the disease longer, and is low if the woman is not having an outbreak (Moreland et al., 1996). C-sections are therefore usually performed on women with an outbreak, but vaginal delivery is possible if there is not an outbreak.

Psychological Aspects: Coping with Herpes
The psychological consequences of herpes need to be taken as seriously as the medical consequences. The range of psychological responses is enormous. At one end of the spectrum are persons with asymptomatic herpes, who are not aware that they have the disease and are happily sexually

active—and at the same time unknowingly spreading the disease to others. At the other end of the spectrum are persons who experience frequent, severe, painful recurrences, who feel stigmatized because of their disease, and who believe that they should abstain from sex in order to avoid infecting others. These difficulties are aggravated by the fact that outbreaks are often unpredictable, and current scientific evidence indicates that people are at least somewhat infectious even when they do not have an active outbreak. A sample of herpes patients reported, not surprisingly, that the disease had the strongest effect on their lives in the area of sexual relations; they reported that it interfered with their sexual freedom, frequency, and spontaneity (Luby & Klinge, 1985). The respondents expressed worries about their ability to establish future intimate relationships, fearing a rejection when the potential partner discovered that they were infected.

On the other hand, many people with herpes are able to cope. For example, in the sample noted above, one-third of the people reported that they had made a successful adjustment to herpes (Luby & Klinge, 1985). In a sample of clinic patients and community volunteers with herpes, 9 percent of the women and 19 percent of the men reported feelings of isolation and loneliness due to the disease (Jadack et al., 1990), but that means that the great majority of HSV-infected people do not! Ironically, the epidemic of herpes in the United States may be a boon to infected persons, because it will become increasingly possible to find a partner who is also infected, so that there are no concerns about transmission. Life goes on.

Psychologists are exploring therapies for herpes patients. One highly effective treatment program consists of a combination of information on herpes, relaxation training, instruction in stress management, and instruction in an imagery technique in which the patient imagines that the genitals are free of lesions and that he or she is highly resistant to the virus (Longo et al., 1988). Patients given six sessions in this program (compared with an untreated control group and a group that participated in a social support group only) showed significantly less depression and anxiety and even had a significant reduction in the average number of herpes outbreaks per year. (See Ebel, 1994, for an excellent book on living with herpes.)

HIV Infection and AIDS

In 1981, a physician in Los Angeles reported a mysterious and frightening new disease identified in several gay men. Within two years, the number of cases had escalated sharply and the gay community had become both frightened and outraged, and within a few more years Washington had funded a major public health effort aimed at understanding and eradicating the disease.

The disease was named **AIDS,** an abbreviation for **acquired immune deficiency syndrome.** As the name implies, the disease destroys the body's natural system of immunity to diseases. Once AIDS has damaged an individual's immune system, opportunistic diseases may take over, and the person usually dies within a few months to a few years. One of these diseases found in AIDS patients is a rare form of cancer that produces purplish lesions on the skin (**Kaposi's sarcoma,** or KS).

A major breakthrough came in 1984 when Robert Gallo, of the National Institutes of Health (NIH), announced that he had identified the virus causing AIDS. A French team, headed by Luc Montagnier of the Pasteur Institute, simultaneously announced the same discovery. The virus is called **HIV,** for human immune deficiency virus. Another strain of the virus, HIV-2, has been identified; it is found almost exclusively in Africa. HIV-1 accounts for almost all infections in North America.

An Epidemic?

As of June, 1998, more than 650,000 persons in the United States had been diagnosed as having AIDS, and more than half of them had died from it. However, public health officials believe that these

---

AIDS (acquired immune deficiency syndrome): A sexually transmitted disease that destroys the body's natural immunity to infection so that the person is susceptible to and may die from a disease such as pneumonia or cancer.

Kaposi's sarcoma (KAP-oh-seez sar-COH-muh): A rare form of skin cancer to which AIDS victims are particularly susceptible.

HIV: Human immune deficiency virus, the virus that causes AIDS.

statistics represent only the tip of the iceberg, for they do not count persons who are just infected with HIV and do not yet have symptoms of full-blown AIDS; nor do they count persons who have mild symptoms of the disease but whose symptoms are not severe enough to be classified as AIDS. It is estimated that 1 million Americans are infected, and experts estimate that 5 to 10 million persons worldwide are infected with HIV, although the majority of them show no symptoms yet and are unaware that they are infected (Liskin & Blackburn, 1986). Thus the term "global epidemic" has been used with reason. The term "pandemic" (a widespread epidemic) has also been used to describe the situation (Stoneburner et al., 1994).

Transmission

When people speak of HIV being transmitted by exchange of body fluids, the body fluids they are referring to are semen and blood and possible secretions of the cervix and vagina. HIV is spread in four ways: (1) by sexual intercourse (either penis-in-vagina intercourse or anal intercourse[4]); (2) by contaminated blood (a risk for people who receive a blood transfusion if the blood has not been screened); (3) by contaminated hypodermic needles (a risk for those who inject drugs or health care workers who receive accidental sticks); and (4) from an infected woman to her baby during pregnancy or childbirth.

Supporting these assertions, statistics for adult and adolescent cases of AIDS in the United States indicate that those infected are from the following exposure categories: (1) men who have sex with

[4]There is also a chance that mouth-genital sex can spread AIDS, particularly if there is ejaculation by an infected person into the mouth.

Figure 20.3 Experts agree that, in the absence of a cure or vaccine for AIDS, the best weapon that we have is education. The New York City Department of Health has produced many posters to promote AIDS education.

men (51 percent); (2) people who inject drugs (21 percent); (3) people who have multiple sources of exposure (13 percent); (4) hemophiliacs who receive transfusions of infected blood products (1 percent); (5) heterosexuals who have sexual contact with an infected person (7 percent); and (6) recipients of contaminated blood transfusions (2 percent) (Spach & Keen, 1996). Although approximately 90 percent of U.S. cases are men, in Africa 35 to 57 percent of cases are women. Worldwide, 70 percent of the cases result from heterosexual transmission (Ehrhardt, 1996).

How great is your risk of contracting AIDS? In essence, it depends on what your sexual practices are (leaving aside issues of drug abuse, which are beyond the scope of this book). Suppose you are a gay man. The San Francisco Men's Health Study provides relevant data (Winkelstein et al., 1987b). The study sampled 1034 single men from neighborhoods in San Francisco where the AIDS epidemic is most serious. Among the homosexual or bisexual participants, 48.5 percent had positive blood tests, meaning they were infected with the virus. Among the heterosexual participants, none had positive blood tests. On the other hand, even among the gay men there was considerable variability. Among those gay men who had had no sexual partners in the previous 2 years, only 18 percent had positive blood tests; in contrast, for those gay men who had had more than 50 partners in the previous 2 years, 71 percent had positive blood tests. One simple lesson is, *the greater your number of sexual partners, the greater your risk of getting infected with HIV.*

The San Francisco Men's Health Study also makes it clear that *the sexual behavior most likely to spread AIDS is anal intercourse, and being the receiving partner puts one most at risk* (Winkelstein et al., 1987b). For example, among those men who had been receiving partners, approximately 50

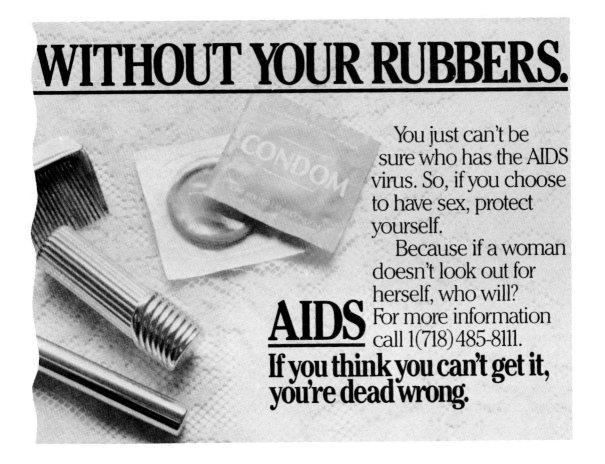

WITHOUT YOUR RUBBERS.

You just can't be sure who has the AIDS virus. So, if you choose to have sex, protect yourself.

Because if a woman doesn't look out for herself, who will?

AIDS For more information call 1(718) 485-8111.

If you think you can't get it, you're dead wrong.

Table 20.1 Risk of Being Infected with an STD as a Result of One Act of Sexual Intercourse with an Infected Person, Using No Condom

	Percent Risk of Transmission from	
	Male to Female	Female to Male
Gonorrhea	50 to 90%	20%
Genital herpes	0.2%	0.05%
HIV	0.1 to 20%	0.01 to 10%

Source: Stone, K. M. (1994). HIV, other STDs, and barriers. In C. Mauck et al. (Eds.). *Barrier contraceptives: Current status and future prospects.* New York: Wiley. Pp. 203–212.

percent tested positive for HIV, compared with 27 percent for those who had been inserters only and 21 percent for those who had not engaged in anal intercourse.

Let's suppose, instead, that you are heterosexual. The range of risk is wide. If you have heterosexual intercourse once with a person chosen at random and you use a condom, your chance of becoming infected is only 1 in 50 million. At the other end of the spectrum, if you have heterosexual intercourse 500 times with an infected person and you don't use condoms, the chances are 2 in 3 that you will become infected (Hearst & Hulley, 1988). *Sexual behavior is riskier if it is with a person who is infected with HIV (seropositive), if the person is in a high-risk group (gay, IV drug user, or hemophiliac), or if condoms are not used.* If you are heterosexual, the risk of transmission depends on whether you are male or female. As Table 20.1 shows, a woman is more likely to get HIV from heterosexual intercourse with an infected person than a man is. This finding was confirmed by a 10-year-long study of the heterosexual transmission of AIDS among 360 couples in which one partner was clearly infected first (Padian et al., 1997). If the male partner was infected, 19 percent of the women became infected, whereas if the woman was infected, only 2.4 percent of the men became infected.

The Virus

HIV is one of a group of retroviruses. Retroviruses reproduce only in living cells of the host species, in this case humans. They invade a host cell, and each time the host cell divides, copies of the virus are produced along with more host cells, each containing the genetic code of the virus. Current research is aimed at finding drugs that will prevent the virus from infecting new cells.

HIV invades particularly a group of white blood cells (lymphocytes) called T-helper cells, or T4 cells. These cells are critical to the body's immune response in fighting off infections. When HIV reproduces, it destroys the infected T cell. Eventually the HIV-positive person's number of T cells is so reduced that infections cannot be fought off.

Scientists have pressed hard to understand the functioning of HIV. Major discoveries were announced in 1996 (Alkhatib et al., 1996; Balter, 1996; Feng et al., 1996). Scientists had identified two *coreceptors* for HIV, fusin and CC-CKR-5, which allow HIV to enter T cells. CC-CKR-5 seems to be the important coreceptor in the early stages of the disease and fusin in later stages. This discovery may lead to advances in treatment if drugs can be used that block these coreceptors.

The Disease

In 1986 the Centers for Disease Control established the following categorization for four broad classes of HIV infection:

1. Initial infection with the virus and development of antibodies to it. Usually people who have been infected show no immediate symptoms, but they do develop antibodies in the blood within 2 to 8 weeks after infection.

2. Asymptomatic carrier state (the person is infected with the virus but shows no symptoms). The mean asymptomatic period in adults is 7 to 9 years (Hirsch, 1990). These asymptomatic carriers can infect other persons, which is a dangerous situation.

3. Lymphadenopathy. This is a more severe condition in which infected persons develop symptoms, which are not immediately life-threatening: swollen lymph nodes, night sweats, fever, diarrhea, weight loss, and fatigue.

4. AIDS. According to the 1986 standards of the U.S. Centers for Disease Control, the diagnosis of AIDS was applied when the person is af-

fected by life-threatening opportunistic infections (infections that occur only in people with severely reduced immunity), such as *Pneumocystis carinii* pneumonia and Kaposi's sarcoma. The diagnosis is also used when other opportunistic infections or cancers of the lymph tissue are present and the person shows a positive test for HIV antibodies. Neurological problems can occur in AIDS patients because the virus can infect the cells of the brain; symptoms may include seizures and mental problems. Other infections in this stage include herpes, candida in the mouth and throat, and human papillomavirus (HPV).

The Centers for Disease Control later refined its definition of AIDS because the original definition required the patient to have an opportunistic infection such as Kaposi's sarcoma (KS) or *Pneumocystis carinii* pneumonia (PCP). Some activists argued that the original definition was particularly inadequate for infected women, who are unlikely to get Kaposi's sarcoma but are likely to develop cervical cancer, precancerous cervical disease, or pelvic inflammatory disease (PID) (Stephens, 1991).

The CDC's new definition includes the criteria listed above, but also takes into account the person's T4 cell count (Steinberg, 1993). The normal count is approximately 1000 cells per cubic millimeter of blood. The early stage of the disease would be defined as infection and lasting as long as the person keeps feeling well and the T4 cell count stays around 1000. In the middle stage, the T4 cell count drops by half, to around 500, but the person still may have no outward symptoms. The immune system is silently failing, however. Treatment with AZT, DDI, and other drugs may begin at this time; there is some evidence that these drug treatments (discussed below) are more effective if begun early. In the late stage, the T4 cell count drops to 200 or less and, although the person may initially have no symptoms, he or she becomes increasingly vulnerable to bacterial, viral, and fungal infections. Early in this stage people may show weight loss, develop diarrhea, and experience fatigue and fevers. Later, serious opportunistic infections may develop, including KS, PCP, and toxoplasmosis, a parasitic infection that attacks the brain. According to the new definition, a T4 cell count below 200 is by itself an indication of AIDS. Other indicators of AIDS have been added for people who are HIV positive, including cervical cancer.

Diagnosis

The blood test that detects the presence of antibodies to HIV uses the ELISA (for enzyme-linked immunosorbent assay) technique. It is easy and cheap to perform. It can be used in two important ways: (1) to screen donated blood; all donated blood in the United States is now screened with ELISA, so that infections because of transfusions should rarely occur, although some did occur before ELISA was developed and a tiny risk remains even with ELISA;[5] and (2) to help people determine whether they are infected (HIV positive) but are asymptomatic carriers. The latter use is important because if people suspect that they are infected and find through the blood test that they are, they should either abstain from sexual activity or, at the very least, use a condom consistently, in order not to spread the disease to others. Only by responsible behavior of this kind can the epidemic be brought under control.

ELISA is a very sensitive test; that is, it is highly accurate in detecting HIV antibodies (it has a very low rate of false negatives, in statistical language). However, it does produce a substantial number of false positives—the test saying HIV antibodies are present when they really are not. Thus positive results on ELISA should always be confirmed by a second, more specific test.

The other major test, using the Western blot or immunoblot method, provides such confirmation. It is more expensive and difficult to perform, so it is not practical for mass screening of blood, as ELISA is. However, it is highly accurate (false positives are rare), and thus it is very useful in confirming or disconfirming a positive test from ELISA.

---

[5]The risk that ELISA will miss occasional cases of infected blood results from the fact that it detects antibodies to HIV, not HIV itself. It takes 6 to 8 weeks for antibodies to form. Thus if a person donates blood within a few weeks of becoming infected and before antibodies form, ELISA will not detect the infection. Usually ELISA is positive within 3 months of infection (Spach & Keen, 1996). It is estimated that there is about 1 case of HIV transmission out of every 550,000 blood donations (Lackritz et al., 1995).

It should be emphasized that both tests detect only the presence of HIV antibodies. They do not predict whether the person will develop symptoms or will progress to the AIDS classification.

Treatment

There is not yet any cure for AIDS. However, some progress is being made in developing treatments to control the disease. One antiviral drug, **AZT** (azidothymidine, also called zidovudine or ZDV), has been used widely. It has the effect of stopping the virus from multiplying. However, even if the virus is stopped, there is still a need to repair the person's badly damaged immune system. Unfortunately, AZT has many side effects and cannot be used by some patients, or can be used for only limited periods of time. Therefore there has been a concerted effort to find new drugs that will slow or stop the progression of the disease.

DDI (dideoxyinosine or didanosine) is one such drug. It was released for clinical trials in late 1989 after less testing than usual, owing to the urgency of finding additional drugs to treat HIV-infected persons. It was made available particularly to those patients who could no longer take AZT. One of its side effects is inflammation of the pancreas. Like AZT, DDI slows the progression of the disease by preventing replication of the virus. DDC (dideoxycytidine) is another drug developed in tandem with DDI; it, too, stops the AIDS virus from replicating. It is helpful for patients who develop problems taking AZT and DDI (Flaskerud & Ungvarski, 1992). D4T is yet another, similar drug.

A major breakthrough came in 1996 with the availability of a new category of drugs, *protease inhibitors* (Ezzell, 1996; Kempf et al., 1995; Vacca et al., 1994). Protease inhibitors attack the viral enzyme protease, which is necessary for HIV to make copies of itself and multiply. Patients take a "drug cocktail" of one of the protease inhibitors (indinavir or ritonavir) combined with AZT and one other anti-HIV drug. Within a year, thrilling reports emerged that the HIV count had become undetectable in the blood of persons taking the drug cocktail (Cohen, 1997). Some believed that the cure had been found. The number of deaths from AIDS declined for the first year since the disease had been identified.

HIV research, unfortunately, is much like a roller coaster, with elated highs followed by plunges to the depths. HIV mutated to drug-resistant forms. And, although HIV had become undetectable in the blood of persons treated with the drug cocktail, HIV was hiding out—in T cells and the lymph nodes and in organs such as the brain, eye, and testes (Cohen, 1998; Finzi et al., 1997; Wong et al., 1997). In short, it is not eradicated by the drug cocktail treatment. The next challenge to scientist is to develop drugs that attack HIV in its hideouts. One team has already reported a combination drug treatment that essentially clears HIV from lymph tissue (Cavert et al., 1997).

On another front, progress is also being made with drugs that prevent the opportunistic infections that strike people with AIDS. The drug pentamidine, for example, in aerosol form, is a standard treatment to prevent *Pneumocystis carinii* pneumonia.

Women, Children, Ethnic Minorities, and AIDS

Approximately 90 percent of the cases of AIDS in the United States have been men, so most attention has been focused on them. However, AIDS infects men and women worldwide nearly equally, and the number of infected women in the United States is rising rapidly (Klirsfeld, 1998). Thus, the urgency of addressing the needs of women with AIDS is increasing.

According to 1996 statistics from the Centers for Disease Control, new cases of women with AIDS fall into the following risk categories: injected drugs (38 percent); had heterosexual contact with an infected man (38 percent); received a contaminated blood transfusion (2 percent); and other, not identified (22 percent). AIDS is one of the five leading causes of death for women in the United States between the ages of 15 and 44; it is the leading cause of death for women in this age group in New York City (Chin, 1990). It is estimated that 9.2 million women worldwide are infected with HIV (Stine, 1996). As noted earlier, some of the diseases that women contract as a result of HIV infection, such as cervical cancer, are different from those of men. The pattern of sexual transmission is differ-

---

AZT: A drug used to treat HIV-infected persons; also called ZDV.

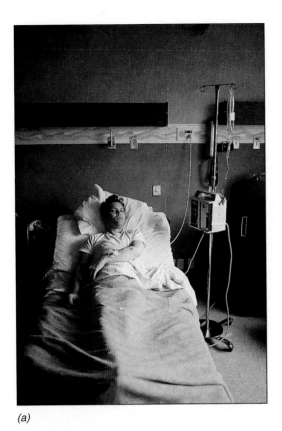

(a)

(b)

Figure 20.4 AIDS is a multicultural disease. *(a)* An American man with AIDS. *(b)* An African woman with AIDS.

ent for women than it is for men. Most women become infected through heterosexual activity, whereas the great majority of men become infected through homosexual activity.

Women need far more recognition in AIDS research (Amaro, 1995; Morokoff et al., 1996; O'Leary & Jemmott, 1995). For example, intervention programs (such as the Harlem AIDS Project) need to be developed tailored to their needs (Deren et al., 1993). Such programs should include sexual assertiveness training, in which women are empowered to insist that their sex partners use condoms. Women also need to be included in clinical trials of drug treatments.

Some of the saddest cases are children with AIDS; these cases are known as pediatric AIDS. Children become HIV-infected either at birth from an infected mother (89 percent of cases) or because of hemophilia (4 percent) or transfusions with contaminated blood (6 percent). Babies born to infected mothers are often, although not always, infected. One bright spot is the finding that treating infected women with AZT during pregnancy

substantially reduces the rate of infection in their babies (Blank et al., 1994).

People of color in the United States—and worldwide, as well—have borne a disproportionate burden of the cases of AIDS. African Americans constitute just 12 percent of the U.S. population, but they account for 30 percent of men with AIDS and 55 percent of women with AIDS (Centers for Disease Control, 1996). Hispanics are also overrepresented; they are 9 percent of the U.S. population, but 17 percent of the cases of AIDS. The incidence of AIDS is low among Asian Americans (less than 1 percent of cases) and Native Americans (less than 1 percent of cases).

In understanding the impact of AIDS on ethnic minorities in the United States, it is important to recognize that some minority groups, particularly African Americans, hold a very different view of AIDS from the white majority. Many African Americans see AIDS as a planned strategy to kill African Americans. In one survey of African American church members in five cities, 34 percent agreed with the statement "I believe that AIDS is a man

Focus 20.1
AIDS in Thailand

Globally, HIV is pandemic; that is, it is an epidemic that has spread around the world. The World Heath Organization estimates that by the year 2000, 30 million to 40 million persons will be infected worldwide, of whom 90 percent will be in developing nations and almost half will be women. This pandemic will produce serious social and economic strains on many countries, not to mention suffering of individuals.

Asia has generally lagged behind other regions in the spread of HIV. Thailand is a notable exception to the pattern: an epidemic is raging there. According to research, depending on the region of the country, 40 to 50 percent of injecting drug users are infected, as are 20 to 45 percent of brothel prostitutes. Approximately 1 million of the 60 million people of Thailand are believed to be infected.

Sociocultural context is the key to understanding why AIDS is devastating Thailand. Two closely linked factors are Thailand's commercial sex industry and its booming tourism. A large commercial sex industry flourishes in Thailand. It is estimated that approximately half a million young women work as prostitutes, and most of them are in the 16 to 24 age group. A wide variety of forms of prostitution exist, catering to every budget, ranging from call girl agencies, executive clubs, and go-go bars to massage parlors, brothels, streetwalkers, and even mobile operations that go out to rural areas.

Why is there such a large commercial sex industry in Thailand? The major factor is economics. Social and income gaps in Thailand are enormous, ranging from rural areas characterized by great poverty to the conspicuous wealth of Bangkok. A young rural girl may go to the city and enter prostitution in order to pay off family debts, for example. Prostitution is very lucrative; it has been estimated that commercial sex workers can earn 25 times as much as young women who work as maids and in other jobs available to them. Another factor is a cultural belief that women are of only two types:

made virus," and an additional 44 percent were unsure about this issue (Thomas & Quinn, 1991). Furthermore, 35 percent agreed with the statement, "I believe AIDS is genocide against the African American race." These beliefs must be understood in cultural context. Most African Americans are well aware of the Tuskegee Syphilis Study, in which black men in Alabama who had syphilis were purposely left untreated for decades in order to see what the long-term consequences of the disease were. That study was appallingly cruel and unethical; no research ethics committee would approve it today. In the wake of that study, African Americans find it hard to trust the white-dominated medical community, or white Americans in general, in matters having to do with sexually transmitted diseases.

There is an urgent need to develop education and prevention programs for the black and Latino communities like those that have been launched in the gay community (e.g., Marin et al., 1993). These programs must be culturally sensitive and should focus on the elimination of needle sharing and unsafe sexual practices.

Psychological Considerations in AIDS
Psychological issues for those infected with HIV and for AIDS patients are profound. Focus 20.2 provides a personal view of the psychological struggles of such a person. There are some analogies to people who receive a diagnosis of an incurable cancer, for AIDS is—at least at present—in-

virtuous women who are virgins until marriage, and prostitutes. Thailand is also a country that historically practiced polygamy and concubinage. And a man who does not use prostitutes is considered something less than a real man. These cultural factors combine to create great demand for prostitutes. Finally, beginning around the time of the Vietnam War, Thailand became a sex playground for foreign tourists and remains so to the present day.

Tourism provides more clients for prostitutes and therefore provides the incentive for more women to enter prostitution. But tourism also plays a major role in the international spread of HIV, bringing it into Thailand and in turn spreading it to the home countries of the tourists.

Public health officials in Thailand are alarmed by the situation and have taken action in cooperation with the government. In 1990 AIDS education was introduced into the schools. In 1991 a "100% condom program" was instituted with the help of the owners of sex parlors and sex workers, encouraging all clients to use condoms. The government supplied 60 million condoms a year to the effort. The results look promising. Research shows that the incidence of extramarital sex and sex with commercial sex workers decreased from 22 per-

cent in 1990 to 10 percent in 1997 (Phoolchareon, 1998). By 1997, the rate of HIV infection in new military draftees had declined substantially from its peak in 1993.

Despite the success of these efforts, the epidemic is probably beyond control by these methods alone. Thai officials, working with researchers in the United States, are therefore considering more daring strategies.

Two promising vaccines against HIV have been developed in the United States, but U.S. officials have decided not to move them into trials in this country. Thai officials, facing a much more desperate situation, are negotiating to conduct the trials there. Ironically, the political complications of the democratic process in the United States, combined with a great concern for safety, slow down developments in the United States—whereas in Thailand, if officials decide to do it, it happens. The strategy would be to vaccinate uninfected volunteers whose sex partners are HIV-positive. Thus, it is possible that Thailand's tragedy will lead to the identification of an effective vaccine that can be used worldwide in stemming the pandemic.

Sources: Balter (1998); Cohen (1995); Ford & Koetsawang (1991); Phoolcharoen (1998); Stoneburner et al. (1994)

---

curable. Many patients experience the typical reactions for such situations, including a denial of the reality, followed by anger, depression, or both. However, the analogy to cancer patients is not perfect, for AIDS is a socially stigmatized disease in a way that cancer is not. Thus the revelation that one has AIDS must often be accompanied by the announcement that one is gay or drug-addicted. Also, as the patient becomes sicker, he or she is unlikely to be able to hold a job, and financial worries become an additional strain.

There is a great need to be sensitive to the psychological needs of AIDS patients. In most cities, support groups for AIDS patients and their families have formed. Social and psychological support from others is essential as people weather this crisis (Pakenham et al., 1994).

In one study HIV-positive men who had developed symptoms were randomly assigned to a 10-week program of cognitive-behavioral stress management therapy or a no-therapy control group (the second group received the therapy as soon as the research was finished). Those in the therapy group showed significant reductions in depressed mood and anxiety compared with the controls (Lutgendorf et al., 1997). The effectiveness of this therapy is important for two reasons. First, it improves the quality of life of affected people. Second, research shows that HIV-positive men who do not have symptoms but are depressed die earlier than similar men who are not depressed (Burack et al., 1993). Psychotherapy is therefore likely to have a positive impact on both mental health and physical health.

Focus 20.2

An HIV-Positive Person Tells His Story

Scott Christensen is 28 and a graduate of the University of Wisconsin. He moved to Madison in 1990 from a small town in Minnesota. In Madison he found his first opportunity to be openly gay and began relationships with two lovers.

In the spring of 1990 he tested positive for HIV. The very next day he began to attend a support group organized by the Madison AIDS Support Network. He had had some early warning about the possible infection because he had been very ill for five weeks, including swollen lymph nodes and night sweats (an unusual occurrence at such an early stage of infection). During those five weeks he came to grips with the idea that he had the disease and decided that he was going to go on living rather than just giving up and dying.

One of his lovers from that time still tests negative for HIV. Scott believes that the other person is positive and knew at the time and did not tell him. Scott quite openly acknowledges that he hates that man.

A physician monitors Scott's health status and prescribes drugs for him. Often Scott hears of a new drug just being tested even before the physician does because of the communication network among HIV-infected people. By January 1991 his T-cell count had dropped to 211, so he began treatment with AZT and continued it for nine months.

He suffered no side effects from it. Then he was given Bactrim for pneumonia prevention. The two drugs interacted and lowered his red blood cell count so far that he had to be hospitalized for transfusions. He stopped all medications at that time. In January 1992, he started treatment with another antipneumonia drug, dapsone, which continues to work well for him.

One month after his diagnosis, he began a relationship that lasted two and a half years and ended in 1992. The relationship was an enormous psychological boost to him, as he came to grips with being infected. His former partner is HIV-negative. Scott let him know that he was infected even before their first date. They practiced safer sex, and there seem to have been no problems. He then started a new relationship, again with an HIV-negative man. This man, too, knew about Scott's infection before they began the relationship. Scott says that he doesn't want a relationship with an HIV-positive person because he is very nurturant and would exhaust himself caring for an infected partner. He has never had a potential partner reject him because he is HIV-positive. Scott believes that everyone, gay and straight, male and female, should practice safer sex because the disease is so widespread.

Recent Progress in AIDS Research

As this discussion makes clear, much more research on AIDS is needed. We need better treatments to control this disease, we need a cure for it, and we need a vaccine against it. Those are pretty tall orders, and it is unlikely that any of them will appear in the next 2 or 3 years.

Vaccine.   Researchers have been working hard to develop a vaccine against HIV, but the job has turned out to be much more difficult than expected. The problem is that HIV actually has many forms and, to make matters worse, it mutates rapidly and

recombines, creating even more forms (Robertson et al., 1995). In effect, the virus doesn't hold still long enough for a vaccine to take effective aim at it.

One strategy in developing a vaccine is to first develop a vaccine that works with monkeys, which can be infected by an analog to HIV called SIV (simian immunodeficiency virus). It has even been suggested that SIV is the same virus as HIV-2 (Gao et al., 1992). Progress has been made toward developing a vaccine that protects rhesus monkeys from infection with SIV (Agnew & Stein, 1990; Nowak, 1991). It awaits clinical trials with humans.

Scott has worked for the Madison AIDS Support Network. He spends much of his time speaking to groups about being HIV-positive. He feels that it is important for him to talk openly with as many people as possible about HIV. He has also been working with a research program that has provided 70 HIV-positive people with access to networked computers in their homes so that they can easily communicate with each other. This allows them to share information rapidly and efficiently. A physician has been placed on the network to answer questions. One strength of this experimental program is that it allows these individuals to be in contact with others while remaining anonymous if they so wish.

When asked how he deals emotionally with his situation and with the future, Scott replied that he recognizes his own mortality, but that he still has many plans for the future. He does best when he concentrates on the present moment—How can I best utilize my time now? Sometimes he feels like he is waiting for the other shoe to drop. He thinks that he has come to grips with his anger and expresses it openly. He feels he needs to come to grips with his sadness, which he doesn't handle as well. In 1992 alone, he attended the funerals of six of his friends. He has benefited a great deal from being in an art therapy program for HIV-positive people (his college major was art). He plans to go to graduate school to become an art therapist so he can give back to the HIV community some of what he has received.

We reinterviewed Scott three years later, in 1995. Not much had changed, but that's good news. His T-cell count is around 190. He is still in the same relationship, which has now lasted four years and is good and stable. His partner remains uninfected with HIV. They practice safer sex and feel that it works. At the same time, his partner believes that he knows the risks he is agreeing to—that is, he has given informed consent—and if he becomes infected, he won't blame anyone.

Because of his earlier bad reaction to AZT, Scott takes no antiretroviral drugs now. (His doctor does not agree with this decision.) He does take two drugs that prevent the opportunistic infections to which AIDS patients are vulnerable: Bactrim (for pneumonia) and rifabutin (for MAI, a disease of the nervous system). He had an outbreak of shingles, which is caused by the herpes virus, and he takes acyclovir to prevent a recurrence.

Scott works for a bookstore chain, continues to do his art, and has been active on the board of the Rodney Scheel House, an apartment complex being built in Madison for people with AIDS. It will offer independent living, with rent adjusted to income, and lots of social support.

Scott feels mentally healthy and said that AIDS has helped him clarify his priorities, so that in some ways he is actually happier. He is careful not to let his stress level get too high. He feels that much of his emotional health is due to all the support he receives from his partner; a stable relationship really helps.

Source: Based on interviews conducted by the authors.

Two other strategies involve developing a vaccine that stimulates the body to form resistance (i.e., antibodies) to HIV, or a vaccine that acts at the cellular level by stimulating the production of specialized T cells that are toxic to HIV (Heilman & Baltimore, 1998). Another possibility is a vaccine that combines both.

Research on Nonprogressors. Some groups of people are being studied for clues to breakthroughs in the war against AIDS. One such group is *nonprogressors* (Barnes, 1995; Buchbinder et al., 1994; Cao et al., 1995; Pantaleo et al., 1995). Approximately 5 percent of HIV-infected people go for 10 years or more without symptoms and with no deterioration of their immune system. Their T-cell count remains higher than 500. Nonprogressors turn out to have less HIV in their bodies, even though they have been infected for over a decade. Why? One possibility is that these people have unusually strong immune systems which have essentially managed to contain the virus. Another is that they were infected with a weak strain of the virus. Some individuals were infected with a genetically defective strain of HIV that does not replicate; this

Figure 20.5 One promising strategy in the search for a cure for AIDS is to study those rare individuals who have been HIV-infected for 10 years or more but have not died or even developed AIDS. Artist Robert Anderson has had HIV for 15 years and is still healthy.

situation is intriguing because it might be useful in producing a vaccine.

Chemokines. Another major scientific advance is the discovery of HIV-suppressor factors or *chemokines* (Balter, 1995; Cocchi et al., 1995; Cohen, 1997). Certain lymphocytes (CD8+ T cells) battle against HIV in the body. They do so by secreting three chemokines (which are molecules), named RANTES, MIP-1α, and MIP-1β. HIV-infected persons who are nonprogressors have high levels of CD8 cells and high levels of chemokines compared with rapid progressors (Haynes et al., 1996). The chemokines can bind to the coreceptor CC CKR 5, blocking HIV from entering cells. Scientists hope

that these discoveries may lead to improved treatments for HIV-infected persons and possibly to vaccines that boost the level of chemokines and therefore boost the body's resistance to HIV infection.

People Who Resist Infection Another intriguing group are some prostitutes in Nairobi, Kenya, who have been repeatedly exposed to HIV because of their work, but who have never become infected (Taylor, 1994). Another research group is studying heterosexual couples in the United States in whom the man is HIV-infected but the woman, despite repeated unprotected intercourse, has not become infected. It seems that some people are resistant to HIV infection and that their immune systems are unusual, seeking and destroying the virus. Unraveling this mystery may also yield clues to a vaccine or a cure.

Mutations. Some researchers believe that HIV's property of rapid mutation could actually be used against it (Chow et al., 1993). They believe that HIV could be forced to mutate to a harmless form. This might be done by exposing the virus to the various drugs used to treat HIV; in response to each, the virus will mutate until eventually it will mutate to a form that cannot replicate itself. This approach, too, holds exciting possibilities.

Gonorrhea

Historical records indicate that **gonorrhea** ("the clap," "the drip") is the oldest of the sexual diseases. Its symptoms are described in the Old Testament, in Leviticus 15 (about 3500 years ago). The Greek physician Hippocrates, 2400 years ago, believed that gonorrhea resulted from "excessive indulgence in the pleasures of Venus," the goddess of love (hence the term "venereal" disease). Albert Neisser identified the bacterium that causes it, the gonococcus *Neisseria gonorrhoeae*, in 1879.

Gonorrhea (gon-uh-REE-uh): A sexually transmitted disease that usually causes symptoms of a puslike discharge and painful, burning urination in the male but is frequently asymptomatic in the female.

Focus 20.3

Safer Sex in the AIDS Era

In this age of AIDS, everyone needs to think about positive health practices that will prevent, or at least reduce the chances of, infection. Technically, these practices are called "safer sex," there being no true safe sex except no sex. But at least health experts agree that the following practices will make sex safer:

1. If you choose to be sexually active (and abstinence is one alternative to consider), have sex only in a stable, faithful, monogamous relationship with an uninfected partner whom you know is uninfected because you both have been tested.

2. If you are sexually active with more than one partner, always use latex condoms. They have a good track record in preventing many sexually transmitted diseases. Laboratory tests indicate that latex condoms are effective protection against HIV. Condoms have a failure rate in preventing disease just as they do in preventing pregnancy, but they are still much better than nothing.

3. If there is any risk that you are infected or that your partner is, abstain from sex, always use condoms, or consider alternative forms of sexual expression such as hand-genital stimulation.

4. Do not have sexual intercourse with someone who has had many previous partners.

5. Do not engage in anal intercourse if there is even the slightest risk that your partner is infected.

Figure 20.6 Quality control of condoms.

6. Remember that both vaginal intercourse and anal intercourse transmit HIV. Mouth-genital sex may, also, particularly if semen enters the mouth.

7. If you think that you may be infected, have a blood test to determine whether you are. If you find that you are infected, at the very least use a condom every time you engage in anal or vaginal intercourse or, preferably, abstain from these behaviors.

8. If you are a woman and you think that you might be infected, think carefully before getting pregnant, because of the risk of transmitting the disease to the baby during pregnancy and childbirth.

Gonorrhea has always been a particular problem in wartime, when it spreads rapidly among the soldiers and the prostitutes they patronize. In the 20th century, a gonorrhea epidemic occurred during World War I, and gonorrhea was also a serious problem during World War II. Then, with the discovery of penicillin and its use in curing gonorrhea, the disease became much less prevalent in the 1950s; indeed, public health officials thought that it would be virtually eliminated.

Figure 20.7 Symptoms of gonorrhea in men include a puslike discharge. About 80 percent of women, however, are asymptomatic.

Then there was a resurgence of gonorrhea. About 700,000 cases per year were reported in the 1990s (Hatcher et al., 1994). It is clear that one of the reasons for the resurgence was the shift in contraceptive practices to the use of the pill, which (unlike the condom) provides no protection from gonorrhea and actually increases a woman's susceptibility.

Symptoms

Most cases of gonorrhea result from penis-in-vagina intercourse. In the male, the gonococcus invades the urethra, producing gonococcal urethritis (inflammation of the urethra). White blood cells rush to the area and attempt to destroy the bacteria, but the bacteria soon win the battle. In most cases, symptoms appear five to eight days after infection, although they may appear as early as the first day or as late as two weeks after infection (Sherrard & Barlow, 1996). Initially a thin, clear mucous discharge seeps out of the meatus (the opening at the tip of the penis). Within a day or so it becomes thick and creamy and may be white, yellowish, or yellow-green (see Figure 20.7). This is often referred to as a "purulent" (puslike) discharge. The area around the meatus may become swollen. About half of infected men experience a painful burning sensation when urinating (Sherrard & Barlow, 1996). The urine may contain pus or

blood, and in some cases, the lymph glands of the groin become enlarged and tender.

Because the early symptoms of gonorrhea in the male are obvious and often painful, most men seek treatment immediately and are cured. If the disease is not treated, however, the urethritis spreads up the urethra, causing inflammations in the prostate (prostatitis), seminal vesicles (seminal vesiculitis), urinary bladder (cystitis), and epididymis (epididymitis). Pain on urination becomes worse and is felt in the whole penis. Then these early symptoms may disappear as the disease spreads to the other organs. If the epididymitis is left untreated, it may spread to the testicles, and the resulting scar tissue may cause sterility.

Asymptomatic gonorrhea (gonorrhea with no symptoms) does occur in males, but its incidence is low (Hook & Handsfield, 1990). In contrast, about 60 to 80 percent of women infected with gonorrhea are asymptomatic during the early stages of the disease. Many women are unaware of their infection unless they are told by a male partner. Therefore, it is extremely important for any male who is infected to inform all his contacts.

The gonorrheal infection in the woman invades the cervix. Pus is discharged, but the amount may be so slight that it is not noticed. When present, it is yellow-green and irritating to the vulva, but it is generally not heavy (it is not to be confused with normal cervical mucus, which is clear or white and nonirritating, or with discharges resulting from the various kinds of vaginitis—discussed in this chapter—which are irritating but white). Although the cervix is the primary site of infection, the inflammation may also spread to the urethra, causing burning pain on urination (not to be confused with cystitis).

If the infection is not treated, the Bartholin glands may become infected and, in rare cases, swell and produce pus. The infection may also be spread to the anus and rectum, either by a heavy cervical discharge or by the menstrual discharge.

Because so many women are asymptomatic in the early stages of gonorrhea, many receive no treatment, and thus there is a high risk of serious complications. In about 20 percent of women who go untreated, the gonococcus moves up into the uterus, often during the menstrual period. From there it infects the fallopian tubes, causing salpingitis (Whittington et al., 1996). The tissues become swollen and inflamed, and thus the condition is also called pelvic inflammatory disease (PID)—although PID can be caused by diseases other than gonorrhea. The major symptom is pelvic pain and, in some cases, irregular or painful menstruation. If the salpingitis is not treated, scar tissue may form, blocking the tubes and leaving the woman sterile. Indeed, untreated gonorrhea is one of the commonest causes of sterility in women. If the tubes are partially blocked, so that sperm can get up them but eggs cannot move down, ectopic pregnancy can result, because the fertilized egg is trapped in the tube.

There are three other major sites for nongenital gonorrhea infection: the mouth and throat, the anus and rectum, and the eyes. If fellatio is performed on an infected person, the gonococcus may invade the throat. (Cunnilingus is less likely to spread gonorrhea, and mouth-to-mouth kissing rarely does.) Such an infection is often asymptomatic; the typical symptom, if there is one, is a sore throat. Rectal gonorrhea is contracted through anal intercourse and thus affects both women in heterosexual relations and, more commonly, men in homosexual relations. Symptoms include some discharge from the rectum and itching, but many cases are asymptomatic. Gonorrhea may also invade the eyes. This occurs only rarely in adults, when they touch the genitals and then transfer the bacteria-containing pus to their eyes by touching them. This eye infection is much more common in newborn infants. The infection is transferred from the mother's cervix to the infant's eyes during birth. For this reason, most states require that silver nitrate, or erythromycin or some other antibiotic, be put in every newborn's eyes to prevent any such infection. If left untreated, the eyes become swollen and painful within a few days, and there is a discharge of pus. Blindness was a common result in the preantibiotic era.

Diagnosis

In males, the physician obtains a sample of the discharge by inserting a swab about ½ inch up the urethra (this causes some pain, but it is not severe). In the most accurate laboratory test, the bacteria are then grown ("cultured") by wiping the swab onto a medium on a culture plate.

If gonorrhea in the throat is suspected, a swab should be taken and cultured using the same technique. People who suspect that they may have rectal gonorrhea should request that a swab be taken from the rectum, since many physicians will not automatically think to do this.

In females, a sample of the cervical discharge is taken from the cervix and cultured. A pelvic examination should also be performed. Pain during this exam may indicate salpingitis. Women who suspect throat or rectal infection should request that samples be taken from those sites as well.

Treatment

The traditional treatment for gonorrhea was a large dose of penicillin, or tetracycline for those who were allergic to penicillin. However, strains of the gonococcus that are resistant to penicillin and tetracycline have become so common that a newer antibiotic, ceftriaxone, must now be used (Whittington et al., 1996). It is highly effective, even against resistant strains.

Syphilis

There has been considerable debate over the exact origins of **syphilis.** The disease, called "the Great Pox," was present in Europe during the 1400s and became a pandemic by 1500.

The bacterium that causes syphilis is called *Treponema pallidum*. It is spiral-shaped and is thus often called a *spirochete*. In 1906, Wassermann, Neisser, and Bruck described a test for diagnosing syphilis; this was known as the *Wassermann test* or *Wassermann reaction*. This test has been replaced by more modern blood tests, but the "Wassermann" label hangs on.

The incidence of syphilis is much less than that of gonorrhea and chlamydia. There were 68,000 reported cases in 1995. It is also true that syphilis rates rose in the 1990s and are at their highest

Syphilis (SIFF-ih-lis): A sexually transmitted disease that causes a chancre to appear in the primary stage.

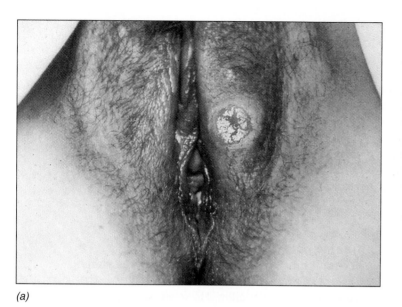

(a)

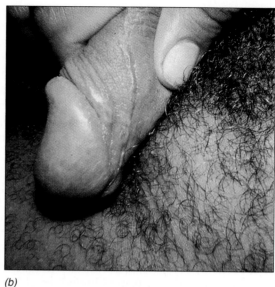

(b)

Figure 20.8   The chancre characteristic of primary-stage syphilis *(a)* on the labia majora and *(b)* on the penis.

levels in 40 years (Sanchez, 1994). One factor in this resurgence is the exchange of sex for drugs, especially crack cocaine.

Although syphilis is not nearly so common as gonorrhea, its effects are much more serious. In most cases, gonorrhea causes only discomfort and, sometimes, sterility; syphilis, if left untreated, can damage the nervous system or even cause death. There are many cases today of coinfection, in which the person is infected with both syphilis and HIV (Sanchez, 1994). Syphilis infection makes one more vulnerable to HIV, and vice versa.

Symptoms

The major early symptom of syphilis is the **chancre.** It is a round, ulcerlike lesion with a hard, raised edge, resembling a crater. One of the distinctive things about the chancre is that although it looks terrible, it is painless. It appears about three weeks (as early as ten days or as late as three months) after intercourse with an infected person. The chancre appears at the point where the bacteria entered the body. Typically, the bacteria enter through the mucous membranes of the genitals as a result of intercourse with an infected person. Thus in men the chancre often appears on the glans or corona of the penis, or it may appear anywhere on the pe-

nis or scrotum. In women, the chancre often appears on the cervix, and thus the woman does not notice it and is unaware that she is infected (nature's sexism again; this may be a good reason for a woman to do the pelvic self-exam with a speculum as described in Chapter 4). The chancre may also appear on the vaginal walls or, externally, on the vulva (see Figure 20.8).

If oral sex or anal intercourse with an infected person occurs, the bacteria can also invade the mucous membranes of the mouth or rectum. Thus the chancre may appear on the lips, tongue, or tonsils or around the anus. Finally, the bacteria may enter through a cut in the skin anywhere on the body. Thus it is possible (though rare) to get syphilis by touching the chancre of an infected person. The chancre would then appear on the hand at the point where the bacteria entered through the break in the skin.

The progress of the disease once the person has been infected is generally divided into four stages: primary-stage syphilis, secondary-stage syphilis, latent syphilis, and late syphilis. The phase de-

> **Chancre (SHANK-er):** A painless, ulcerlike lesion with a hard, raised edge that is a symptom of syphilis.

scribed above, in which the chancre forms, is **primary-stage syphilis.** If left untreated, the chancre goes away by itself within one to five weeks after it appears. This marks the end of the primary stage. However, the disease has not gone away just because the chancre has healed. The disease has only gone underground.

Beginning one to six months after the original appearance of the chancre, a generalized body rash develops, marking the beginning of **secondary-stage syphilis.** The rash is very variable in its appearance, the most distinctive feature being that it does not itch or hurt. Hair loss may also occur during the secondary stage. Usually the symptoms are troublesome enough to cause the person to seek medical help. With appropriate treatment at this stage, the disease can still be cured, and there will be no permanent effects.

Even without treatment, the secondary-stage symptoms go away in two to six weeks, leading people to believe mistakenly that the disease has gone away. Instead, it has entered a more dangerous stage.

After the symptoms of the secondary stage have disappeared, the disease is in the latent stage; **latent syphilis** may last for years. Although there are no symptoms in this stage, *T. pallidum* is busily burrowing into the tissues of the body, especially the blood vessels, central nervous system (brain and spinal cord), and bones. After the first year or so of the latent stage, the disease is no longer infectious, except that a pregnant woman can still pass it on to the fetus.

About half of the people who enter the latent stage remain in it permanently, living out the rest of their lives without further complications. The other half, however, move into the dangerous **late syphilis.** In *cardiovascular late syphilis* the heart and major blood vessels are attacked; this occurs 10 to 40 years after the initial infection. Cardiovascular syphilis can lead to death. In *neurosyphilis* the brain and spinal cord are attacked, leading to insanity and paralysis, which appear 10 to 20 years after infection. Neurosyphilis may be fatal.

If a pregnant woman has syphilis, the fetus may be infected when the bacteria cross the placental barrier, and the child gets **congenital** (meaning present from birth) **syphilis.** The infection may cause early death of the fetus (spontaneous abortion) or severe illness at or shortly after birth. It

may also lead to late complications which show up only at 10 or 20 years of age. Women are most infectious to their baby when they have primary- or secondary-stage syphilis, but they may transmit the infection to the fetus in utero as long as 8 years after the mother's initial infection. If the disease is diagnosed and treated before the fourth month of pregnancy, the fetus will not develop the disease. For this reason, a syphilis test is done as a routine part of the blood analysis in a pregnancy test.

Diagnosis

Syphilis is somewhat difficult to diagnose from symptoms because, as noted above, its symptoms are like those of many other diseases.

The physical exam should include inspection not only of the genitals but also of the entire body surface. Women should have a pelvic exam so that the vagina and the cervix can be checked for chancres. If the patient has had anal intercourse, a rectal exam should also be performed.

If a chancre is present, some of its fluid is taken and placed on a slide for inspection under a dark-field microscope. If the person has syphilis, *T. pallidum* should be present.

The most common tests for syphilis are blood tests, all of which are based on antibody reactions. The VDRL (named for the Venereal Disease Research Laboratory of the U.S. Public Health Service, where the test was developed) is one of these blood tests. It is fairly accurate, cheap, and easy to perform.

Primary-stage syphilis: The first few weeks of a syphilis infection during which the chancre is present.
Secondary-stage syphilis: The second stage of syphilis, occurring several months after infection, during which the chancre has disappeared and a generalized body rash appears.
Latent (LAY-tent) syphilis: The third stage of syphilis, which may last for years, during which symptoms disappear although the person is still infected.
Late syphilis: The fourth and final stage of syphilis, during which the disease does damage to major organs of the body such as the lungs, heart, or brain.
Congenital (kun-JEN-ih-tul) syphilis: A syphilis infection in a newborn baby resulting from transmission from an infected mother.

Treatment

The treatment of choice for syphilis is penicillin. *Treponema pallidum* is actually rather fragile, so large doses are not necessary in treatment; however, the bacteria may survive for several days, and therefore a long-acting penicillin is used. The recommended dose is two shots of benzathine penicillin, of 1.2 million units each, one in each of the buttocks. Latent, later, and congenital syphilis require larger doses.

For those allergic to penicillin, tetracycline or doxycycline is the recommended treatment, but it should not be given to pregnant women (Larsen et al., 1996).

Follow-up exams should be performed to make sure that the patient is completely cured.

Elimination of Syphilis?

In 1998 the U.S. Public Health Service announced that it was targeting syphilis for complete elimination in the United States (St. Louis & Wasserheit, 1998). On the surface, the goal may seem odd, given that syphilis is relatively rare compared with other STDs. But syphilis infection makes a person much more vulnerable to HIV infection because of the open sores produced by syphilis, and a person who is HIV-infected and then gets syphilis may experience immediately life-threatening conditions. For these reasons, syphilis is more serious than the small rate of infection indicates. Moreover, syphilis is completely curable with a single dose of penicillin, it is easily detectable with inexpensive laboratory tests, and it has not developed resistant strains. Experience over the several decades since World War II indicates that, even when syphilis is effectively treated and brought to a low incidence, it still resurges periodically in epidemics. Therefore, total elimination is the best goal and it seems feasible.

Viral Hepatitis

Viral hepatitis is a disease of the liver. One symptom is an enlarged liver that is somewhat tender. The disease can vary greatly in severity from asymptomatic cases to cases in which there is a fever, fatigue, jaundice (yellowish skin), and vomiting, much as one might experience with a serious case of the flu. There are five types of viral hepatitis: hepatitis A, B, C, D, and E. The one that is of most interest in a discussion of sexually transmitted diseases is hepatitis B. C and D (or delta) can also be transmitted sexually, but they are rare compared with B (Shapiro & Alter, 1996).

The virus for hepatitis B (HBV) can be transmitted through blood, saliva, semen, vaginal secretions, and other body fluids. The behaviors that spread it include needle sharing by people who inject drugs, vaginal and anal intercourse, and oral-anal sex. The disease is found among male homosexuals and among heterosexuals. It has many similarities to AIDS, although hepatitis B is more contagious.

Hepatitis B is more common than most people think because it receives relatively little publicity compared with AIDS and herpes. There are approximately 200,000 new cases in the United States every year. People who have had the disease continue to have a positive blood test for it for the rest of their lives. In samples of the general population, 50 percent of homosexual men test positive, compared with 5 to 10 percent of heterosexual men (Lemon & Newbold, 1990; Hatcher et al., 1994). Among heterosexual university students, those who had had more than three sexual partners in the past four months were likelier to test positive (14 percent) than were students with fewer partners (1.5 percent) (Alter et al., 1986). This raises the important point—which is relevant not only to hepatitis B, but also to all sexually transmitted diseases—that having many sexual partners increases your risk of contracting a disease, not only because you have a higher probability of being exposed but also because you are being exposed to a riskier group of partners, namely, people who engage in casual sex with many partners.

Treatment for hepatitis B is complex; there is no cure except rest, and treatments are aimed mainly at relieving the individual symptoms. The good news is that there is a vaccine against hepatitis B. The current recommendation is that all teenagers and infants should be vaccinated. We urge you to be vaccinated if you are a gay man or a heterosexual man or woman who has had a number of partners.

Pubic Lice

Pubic lice ("crabs" or pediculosis pubis) are tiny lice that attach themselves to the base of pubic hairs and there feed on blood from their human host. They are about the size of a pinhead and, under magnification, resemble a crab (see Figure 20.9). They live for about 30 days, but they die within 24 hours if they are taken off a human host. They lay eggs frequently, the eggs hatching in 7 to 9 days. Crabs are transmitted by sexual contact, but they may also be picked up from sheets, towels, sleeping bags, or toilet seats. (Yes, Virginia, there are some things you can catch from toilet seats.)

The major symptom of pubic lice is an itching in the region of the pubic hair, although some people do not experience the itching. Diagnosis is made by finding the lice or the eggs attached to the hairs.

Pubic lice are treated with the drug lindane, which is available by prescription only, as a cream, lotion, or shampoo, under several brand names (e.g., Kwell in the United States, Kwellada in Canada). Rid is also effective and is available without prescription. Both kill the lice. After treatment, the person should put on clean clothing. Since the lice die within 24 hours, it is not necessary to disinfect clothing that has not been used for over 24 hours. However, the eggs can live up to 6 days, and in difficult cases it may be necessary to boil or dry-clean one's clothing or use a spray such as R and C (Billstein, 1990).

Figure 20.9   A pubic louse, enlarged. The actual size is about the same as the head of a pin.

Preventing STDs

While most of the literature one reads concentrates on the rapid diagnosis and treatment of STDs, prevention would be much better than cure, and there are some ways in which one can avoid getting STDs, or at least reduce one's chances of getting them. The most obvious, of course, is limiting yourself to a monogamous relationship with an uninfected person or abstaining from sexual activity. If this strategy is unacceptable to you, there are other techniques you might be willing to follow.

The condom, in addition to being a decent contraceptive, gives some protection against HIV, gonorrhea, herpes, syphilis, and other STDs. With the rise of the STD epidemic, the condom is again becoming popular. The key is to eroticize condom use.

Some simple health precautions are also helpful. Successful prostitutes, who need to be careful about STDs, take such precautions. Washing the genitals before intercourse helps remove bacteria. This may not sound like a romantic prelude to lovemaking, but prostitutes make a sensuous game out of soaping the man's genitals. You can do this as part of taking a shower or bath with your partner. The other important technique is inspecting your partner's genitals. If you see a chancre, a wart, a herpes blister, or a discharge, put on your clothes and leave or at the very least, immediately start a conversation about STD infection status (do not fall for the "it's only a pimple" routine). This technique may sound a little crude or embarrassing, but if you are intimate enough with someone to make love with that person, you ought to be intimate enough to look at her or his genitals. Once again, if you are cool about it, you can make this an erotic part of foreplay.

On the other hand, just because a partner has no obvious symptoms like herpes blisters or warts, don't assume that the person is uninfected. We

Pubic lice: Tiny lice that attach themselves to the base of pubic hairs and cause itching; also called crabs or pediculosis pubis.

Focus 20.4
Cool Lines About Safer Sex

The journal *Medical Aspects of Human Sexuality* (March 1987) thoughtfully published some responses you can give to a partner who is resistant to safer sex practices. Here are some of these ideas.

If the partner says	*You can say*
I'm on the pill; you don't need a condom.	I'd like to use it anyway. We'll both be protected from infections we may not realize we have.
I'll lose my erection by the time I stop and put it on.	I'll help you put it on—that'll help you keep it up.
Condoms are unnatural, fake, a total turnoff.	Please let's try to work this out—an infection isn't great either. So let's give the condom a try. Or maybe we can look for alternatives.
What kind of alternatives?	Maybe we'll just pet, or postpone sex for a while.
None of my other boyfriends uses a condom. A *real* man isn't afraid.	Please don't compare me to them. A real man cares about the woman he dates, himself, and their relationship.
Just this once.	Once is all it takes.

have seen in this chapter how many of these diseases—for example, chlamydia, herpes, and warts—can be asymptomatic. The only way to really know is to have a complete battery of tests for STDs, a choice more and more people are making.

Urinating both before and after intercourse helps to keep bacteria out of the urethra.

Finally, each person needs to recognize that it is his or her ethical responsibility to get early diagnosis and treatment. Probably the most important responsibility is that of informing prospective partners if you have an STD and of informing past partners as soon as you discover that you have one. For example, because so many women are asymptomatic for chlamydia, it is particularly important for men to take the responsibility for informing their female partners if they find that they have the disease. It is important to take care of your own health, but it is equally important to take care of your partner's health.

Other Genital Infections

Vaginitis (vaginal inflammation or irritation) is very common among women and is endemic in college populations. Three kinds of vaginitis, as well as cystitis (inflammation of the urinary bladder) and prostatitis, will be considered here. None of these infections are STDs because they are not transmitted by sexual contact; they are, however, common infections of the sex organs.

Monilia

Monilia (also called *candida, yeast infection, fungus,* and *moniliasis*) is a form of vaginitis caused by

Vaginitis (vaj-in-ITE-is): An irritation or inflammation of the vagina, usually causing a discharge.
Monilia (Moh-NILL-ee-uh): A form of vaginitis causing a thick, white discharge; also called candida or yeast infection.

Focus 20.5
Preventing Vaginitis

The following steps help to prevent vaginitis:

1. Wash the vulva carefully and dry it thoroughly.
2. Do not use feminine hygiene deodorant sprays. They are not necessary for cleanliness, and they can irritate the vagina.
3. Wear cotton underpants. Nylon and other synthetics retain moisture, and vaginitis-producing organisms thrive on moisture. Consider cutting out the center panel in the crotch of panty hose. Even those with cotton crotches do not allow much air to circulate.
4. Avoid wearing pants that are too tight in the crotch; they increase moisture and may irritate the vulva. Loose-fitting clothes are best, since they permit air to circulate.
5. Keep down the amount of sugar and carbohydrates in your diet.
6. Wipe the anus from front to back so that bacteria from the anus do not get into the vagina. For the same reason, never go immediately from anal intercourse to vaginal intercourse. After anal intercourse, the penis must be washed before it is put into the vagina.
7. Douching with a mildly acidic solution (1 or 2 tablespoons of vinegar in a quart of water) can help to prevent vaginitis.

---

the yeast fungus *Candida albicans. Candida* is normally present in the vagina, but if the delicate environmental balance there is disturbed (for example, if the pH is changed), the growth of *Candida* can get out of hand. Conditions that encourage the growth of *C. albicans* include long-term use of birth control pills, menstruation, diabetes or a pre-diabetic condition, pregnancy, and long-term use of antibiotics such as tetracycline. It is not a sexually transmitted disease, although intercourse may aggravate it.

The major symptom is a thick, white, curdlike vaginal discharge, found on the vaginal lips and the walls of the vagina. The discharge can cause extreme itching, to the point where the woman is not interested in having intercourse.

Treatment is by the drug nystatin (Mycostatin or other drugs) in vaginal suppositories. The problem sometimes resists treatment, though. Women, especially those who have frequent bouts of monilia, may want to take some steps to prevent it (see Focus 20.5). Drugs to treat yeast infections are now available in drugstores without prescription.

If a woman has monilia while she is pregnant, she can transmit it to her baby during birth. The baby gets the yeast in its digestive system, a condition known as *thrush*. Thrush can also result from oral-genital sex.

Trichomoniasis

Trichomoniasis ("trich") is caused by a single-celled organism, *Trichomonas vaginalis*. Trichomoniasis can be passed back and forth from a man to a woman, and so it is technically an STD (for this reason, it is important for the man as well as the woman to be treated), but it can also be transmitted by such objects as toilet seats and washcloths. Some women are asymptomatic.

Trichomoniasis (trick-oh-moh-NY-us-is): A form of vaginitis causing a frothy white or yellow discharge with an unpleasant odor.

The symptom is an abundant, frothy, white or yellow vaginal discharge which irritates the vulva and also has an unpleasant smell. There are usually no symptoms in the male.

It is extremely important that accurate diagnosis of the different forms of vaginitis be made, since the drugs used to treat them are different and since the long-term effects of untreated trichomoniasis can be serious. A drop of the vaginal discharge should be put on a slide and examined under a microscope. If the infection is trichomoniasis, then *Trichomonas vaginalis* should clearly be present. Cultures can also be grown.

The treatment of choice is metronidazole (Flagyl) taken orally.

If left untreated for long periods of time, trichomoniasis may cause damage to the cells of the cervix, making them more susceptible to cancer.

Nonspecific Vaginitis

Nonspecific vaginitis occurs when there is a vaginal infection, with accompanying discharge, but it is not a case of monilia or trichomoniasis. This diagnosis can be made only when microscopic examination of the discharge, as well as other laboratory tests, has eliminated all the other common causes of vaginitis (trichomoniasis, monilia, gonorrhea) as possibilities. A distinctive symptom is often the foul odor of the discharge. Treatment is by Flagyl taken orally.

Cystitis

Cystitis is an infection of the urinary bladder that occurs almost exclusively in women. In most cases it is caused by the bacterium *Escherichia coli*. The bacteria are normally present in the body (in the intestine), and in some cases, for unknown reasons, they get into the urethra and the bladder. Sometimes frequent, vigorous sexual intercourse will irritate the urethral opening, permitting the bacteria to get in.

The symptoms are a desire to urinate every few minutes, with a burning pain on urination. The urine may be hazy or even tinged with red; this is caused by pus and blood from the infected bladder. There may also be backache. Diagnosis can usually be made simply on the basis of these symptoms. A urine sample should be taken and analyzed, though, to confirm that *E. coli* is the culprit.

Treatment is usually with a sulfa drug (such as Gantrisin or ampicillin) taken orally. The drug may include a dye that helps relieve the burning sensation on urination; the dye turns the urine bright orange-red. If the cause is a bacterium other than *E. coli*, ampicillin is typically prescribed.

If cystitis is left untreated (which it seldom is, since the symptoms are so unpleasant), the bacteria may get into the kidneys, causing a dangerous kidney infection.

To prevent cystitis or prevent recurring bouts of it, drink lots of water and urinate frequently, especially just before and after intercourse. This will help flush any bacteria out of the bladder and urethra.

Prostatitis

Prostatitis is an inflammation of the prostate gland. As with cystitis in women, the infection is usually caused by the bacterium *E. coli*. The symptoms are fever, chills, pain around the anus and rectum, and a need for frequent urination. It may produce sexual dysfunction, typically painful ejaculation. In some cases, prostatitis may be chronic (long-lasting) and may have no symptoms, or only lower-back pain. Antibiotics are used in treatment.

Nonspecific vaginitis: A form of vaginitis caused by an anaerobic bacterium.

Cystitis (sis-TY-tis): An infection of the urinary bladder in women, causing painful, burning urination.

Prostatitis (pros-tuh-TY-tis): An infection or inflammation of the prostate gland.

SUMMARY

Sexually transmitted diseases (STDs) are at epidemic levels in the United States and worldwide. The three most common STDs among college students are chlamydia, HPV, and herpes.

Chlamydia is often asymptomatic, especially in women. In men, it produces a thin discharge from the penis and mild pain on urination. It is quite curable with antibiotics. If left untreated in women, the possible complications include damage to the cervix, salpingitis, pelvic inflammatory disease, and possibly infertility.

HPV (human papilloma virus) causes genital warts. Sometimes the warts are obvious, but in other cases they are small and may not be visible. HPV infection increases women's risk of cervical cancer.

Genital herpes, caused by the HSV virus, produces bouts of painful blisters on the genitals. These episodes may recur for the rest of the person's life, although some infected persons experience no or only a few outbreaks. Currently there is no cure, although the drug acyclovir minimizes the symptoms. Herpes infection increases one's risk of HIV infection.

AIDS, caused by the virus HIV, is an incurable disease that destroys the body's natural immune system and leaves the person vulnerable to certain kinds of infections and cancer that lead to death. Most HIV-positive people come from three risk groups: gay men, injection drug users, and heterosexual partners of infected persons. Drugs such as AZT and ritonavir (a protease inhibitor) are used to slow the progression of the disease. More attention is needed for the special concerns of HIV-infected women and ethnic minorities. Several strategies for producing a vaccine are being pursued.

The primary symptoms of gonorrhea in the male, appearing five to eight days after infection, are a white or yellow discharge from the penis and a burning pain on urination. The majority of women with gonorrhea are asymptomatic. Gonorrhea is caused by a bacterium, the gonococcus, and is cured with antibiotics. If left untreated, it may lead to infertility.

Syphilis is caused by the bacterium *Treponema pallidum*. The first symptom is a chancre. Penicillin is effective as a cure. If left untreated, the disease progresses through several stages that may lead to death.

Hepatitis B, caused by the virus HBV, is transmitted sexually as well as by needle sharing. The only treatment is rest until the symptoms go away. A vaccine is now available and is being administered widely.

Pubic lice are tiny lice that attach to the pubic hair. They are spread through sexual and other types of physical contact. Shampoos are available for treatment.

Techniques for preventing STDs include thorough washing of both partners' genitals before intercourse; urination both before and after intercourse; inspecting the partner's genitals for symptoms like a wart or urethral discharge; and the use of a condom.

Vaginal infections that are not STDs include monilia, trichomoniasis, and nonspecific vaginitis. Cystitis is an infection of the urinary bladder in women, leading to frequent, burning urination. Prostatitis is inflammation of the prostate gland.

REVIEW QUESTIONS

1. A thin discharge from the penis and mildly painful urination are symptoms of _____.
2. If left untreated, chlamydia in women can cause infertility. True or false?
3. Small, cauliflowerlike growths on the genitals are symptoms of _____.
4. Small, painful bumps or blisters on the genitals are symptoms of _____.
5. Currently there is no known cure for genital herpes. True or false?
6. HIV is transmitted by bodily fluids; the two principal fluids are _____ and _____.
7. Gonorrhea is frequently asymptomatic in women, but it is never asymptomatic in men. True or false?
8. Although syphilis is much less common than gonorrhea, its effects are much more serious. True or false?
9. A vaccine against hepatitis B is available. True or false?
10. Pubic lice are tiny lice that attach themselves to the base of pubic hairs and may cause cancer if left untreated. True or false?

(The answers to all review questions are at the end of the book, after the Glossary.)

QUESTIONS FOR THOUGHT, DISCUSSION, AND DEBATE

1. Contact the student health service on your campus and see whether they are willing to share with you the number of cases of chlamydia, gonorrhea, genital warts, and herpes they diagnose per year; then share the information with your class.
2. Design a program to reduce the number of cases of sexually transmitted disease on your campus.
3. Michael has had a monogamous relationship with Sonya for 6 months. At her annual pelvic exam, Sonya discovers that she has chlamydia and blows up at Michael for giving it to her.

Michael has never been tested for anything. What should Michael and Sonya discuss with each other, what other information do they need, and what should they do?

4. If you were designing an intervention for Latino teenagers to reduce the spread of HIV, how would you make that intervention culturally sensitive? Would the approach be different for women and for men? If you feel that you don't know enough about Latino culture, interview some Latino students or community members to find out more.

SUGGESTIONS FOR FURTHER READING

Ebel, Charles. (1994). *Managing herpes: How to live and love with a chronic STD*. Research Triangle Park, NC: American Social Health Association. People say this is the best herpes book ever.

Jones, James H. (1981). *Bad blood: The Tuskegee syphilis experiment*. New York: Free Press. The shocking story of the study in which black men with syphilis were left untreated so that the course of the disease could be observed.

McIlvenna, Ted (Ed.). (1992). *The complete guide to safer sex*. Ft. Lee, NJ: Barricade Books. An excellent guide to creating a safer sex life.

WEB RESOURCES

Sexually transmitted diseases:

http://www.cdc.gov/nchstp/dstd/dstdp.html
   Centers for Disease Control and Prevention, Center for STD Prevention Homepage; information and data about STDs, STD prevention.

http://www.sexhealth.org/infocenter/STDsFile/stds.htm
   The Sexual Health InfoCenter: STDs Page.

HIV/AIDS:

http://www.cdcnpin.org
   Centers for Disease Control National AIDS Clearinghouse.

http://www.ama-assn.org/special/hiv
   American Medical Association.

http://www.plannedparenthood.org
   Planned Parenthood: Sexual Health pages.

http//www.medscape.com/
   News about a variety of health topics, including AIDS.

http://www.sph.emory.edu/bshe/AIDS/college.html
   Bibliography of articles about HIV/AIDS and college students.

http://www.caps.ucsf.edu/index.html
   Center for AIDS Prevention Studies; information and resources; material regarding AIDS and Hispanics.

Safer Sex:

http://www.sexuality.org/safesex.html
   The Society for Human Sexuality.

http://www.sexhealth.org/infocenter/GuideSS/GuideSS.htm
   The Sexual Health InfoCenter: Guide to Safer Sex Page.

http://www.safersex.org
   The Safer Sex Institute.

chapter
21

ETHICS, RELIGION, AND SEXUALITY

CHAPTER HIGHLIGHTS

A viable sexual theology for our time will affirm that sexuality is always much more than genital expression. Sexuality expresses the mystery of our creation as those who need to reach out for the physical and spiritual embrace of others. It expresses God's intention that we find our authentic humanness not in isolation but in relationship.*

A high school student is in love with her boyfriend and wonders whether they ought to begin sleeping together. A corporation executive hears rumors that one of his employees is gay, and he tries to decide what to do about it. A minister is asked to counsel a husband and wife, one of whom is involved in an affair. A presidential candidate is confronted by a right-to-life group demanding support for a constitutional amendment to ban abortion. All these people are facing the need to make decisions that involve sexuality, and they find that issues of values make the decisions difficult. The two principal conceptual frameworks for dealing with questions of values are religion and ethics, which are therefore the topics of this chapter.

There are two concerns that give force to our consideration of the religious and ethical aspects of human sexuality. First, there is the scientific concern to explain sexual phenomena. Since religion and ethics are important influences on people's behaviors, especially in matters of sex, one cannot fully appreciate why people do what they do without looking at these influences. Second, there is also a personal side to this coin. We are all ethical decision makers; we all have a personal system of values. Each of us must make decisions with respect to our own sexuality. Therefore, we would do well to consider how such decisions are made.

Ethics refers to a system of moral principles, a way of deciding what is right and wrong. We use ethics when there is a conflict between things we prize or desire highly. Sexual gratification may be an important value for one person and something to be avoided for another. However, regardless of the importance we attach to sex, we need a way of integrating our sexuality into our patterns of decision making. To do this we use such categories as "right or wrong," "good or bad," "appropriate or inappropriate," and "moral or immoral." These are the kinds of distinctions made in the field of ethics; since we use them every day, we are all practical ethicists.

Religion enters the picture as a source of values, attitudes, and ethics. For believers, religion sets forth an ethical code and provides sanctions (rewards and punishments) that motivate them to obey the rules. When a particular religion is practiced by many people in a society, it helps create culture, which then influences even those who do not accept the religion. Therefore, it is important to study the relationship of religion to sexuality for two reasons. First, it is a powerful influence on the sexual attitudes of many individuals. And second, as a creator of culture, it often forms a whole society's orientation toward human sexuality.

Let us begin by defining some terms that will be useful in discussing ethics, religion, and sexuality. **Hedonism** and **asceticism** have to do with one's approach to the physical and material aspects of life in general and to sexuality in particular. The word "hedonism" comes from the Greek word meaning "pleasure" and may be used to refer to the belief that the ultimate goal of human life is the pursuit of pleasure, the avoidance of pain, and the fulfillment of physical needs and desires: "Eat,

*James B. Nelson. (1992). *Body theology.* Louisville, KY: Westminster/John Knox Press.

Ethics: A system of moral principles; a way of determining right and wrong.
Hedonism: A moral system based on maximizing pleasure and avoiding pain.
Asceticism: An approach to life emphasizing discipline and impulse control.

drink, and be merry, for tomorrow we die." Asceticism, in contrast, holds that there is more to life than its material aspects, which must be transcended to achieve true humanity. Ascetics are likely to view sexuality as neutral at best and evil at worst; they prize self-discipline, the avoidance of physical gratification, and the cultivation of spiritual values. Orders of monks and nuns, found in Eastern religions as well as in Christianity, are good examples of institutionalized asceticism, with their affirmation of celibacy, virginity, poverty, and so on.

Two other terms, **legalism** and **situationism,** refer to methods of ethical decision making. As an approach to ethics, legalism is concerned with following a moral law, or set of principles, which comes from a source outside the individual, such as nature or religion. Legalistic ethics are focused on the rightness or wrongness of specific acts and set forth a series of rules—"Do this" and "Don't do that"—that persons are to follow. The term "situationism" has been used since it was coined by Joseph Fletcher in his 1966 book *Situation Ethics*. Also called contextual ethics, this approach suggests that although there may be broad general guidelines for ethical behavior, each ethical decision should be made according to the individuals and situations involved. Situationism is based in human experience and, in matters of sexual morality, tends to focus on relationships rather than rules. Whereas legalism deals in universal laws, situationism decides matters on a case-by-case basis, informed by certain guiding principles, such as love. Traditional religious ethical systems (which we might call the Old Morality) have tended to be quite legalistic, and many continue to be today (e.g., Orthodox Judaism, Roman Catholicism, and evangelical Protestantism). However, with the advent of the modern empirical scientific world view, the situationist approach (the New Morality) has attracted many adherents (Nelson, 1978).

Of course, few ethical systems are purely hedonistic or ascetic or entirely legalistic or situationist; most lie between these extremes. However, the terms are useful in pointing to tendencies and will be used for that purpose in this chapter. The concluding section will try to put them into perspective and offer a critique of their strengths and weaknesses as approaches to sexual ethics.

Sexuality in Great Ethical Traditions

With these ways of looking at ethics, especially sexual ethics, as background, let us examine certain great ethical traditions to see how they deal with norms for sexual behavior. Although some attention will be given to non-Western sexual ethics, the focus of this section will be on ethical traditions of Western culture, primarily because this is a text for American undergraduates, who are part of that culture. We are understandably interested in *our* story, in the world of ideals and practice in which *we* live. That culture can, at the risk of oversimplification, be seen as originating in the confrontation of Greek culture, preserved and developed by the Romans, and Jewish tradition, extended by Christianity. From that point on until rather recently, Western culture was Christian, at least officially. Even self-conscious revolts against Christian culture in the West are part of that tradition because of their roots.

Classical Greek Philosophy

During the Golden Age of Greek culture, covering roughly the fifth and fourth centuries B.C.E.,[1] philosophers such as Socrates, Plato, and Aristotle pondered most of the great ethical questions. They regarded the beautiful and good as the chief goal of life, and they admired the figure of the warrior-intellectual, who embodied the virtues of wisdom, courage, temperance, justice, and piety.

While nothing in Greek culture rejected sex as evil—the gods and goddesses of Greek mythology are often pictured enjoying it—the great

[1]Before the Common Era; some people prefer this rather than the Christian-centered Before Christ (B.C.).

Legalism: Ethics based on the assumption that there are rules for human conduct and that morality consists of knowing the rules and obeying them.
Situationism: Ethics based on the assumption that there are no absolute rules, or at least very few, and that each situation must be judged individually.

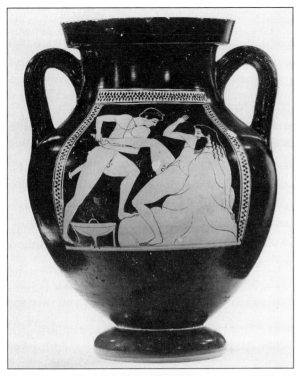

Figure 21.1   The ancient Greeks not only approved of, but idealized, pederasty.

philosophers did develop a kind of asceticism that assumed an important place in Western thought. They thought that virtue resulted from wisdom, and they believed that people would do right if they could, failing to live morally only through ignorance. To achieve wisdom and cultivate virtue, violent passions must be avoided, and these might well include sex. Plato believed that love led toward immortality and was therefore a good thing. However, since this kind of love was rather intellectual and more like friendship than vigorous sexuality, the term "platonic love" has come to mean sexless affection. There was also, among the warrior class, a certain approval of **pederasty** (a sexual relationship between an older man and a younger one), although the practice was far from universal and was frequently ridiculed (Brinton, 1959).

Later, Greek philosophy became even more ascetic than in the Golden Age. Epicurus (341–270 B.C.E) taught that the goal of life was *ataraxia*, a tranquil state between pleasure and pain in which

the mind is unaffected by emotion. He, like other Stoics of the same period, valued detachment from worldly anxieties and pleasures and, indeed, a total indifference to either life or death. Sex was seen not necessarily as evil but as less important than wisdom and virtue, something to be transcended to achieve the beautiful and good.

Judaism

The basic source of the Judeo-Christian tradition, which is the religious foundation of Western culture, is the Hebrew scriptures or the Old Testament of the Bible, which is the basis for Judaism and a major source for Christianity as well. Written between approximately 800 and 200 B.C.E, the Old Testament has a great deal to say about the place of sexuality in human life and society, always seen in religious terms.

The Old Testament view of sexuality is fundamentally positive. In the Genesis myth of creation we read, "So God created man in his own image, in the image of God he created him; male and female he created them" (Genesis 1:27). Human sexual differentiation is not an afterthought or an aberration; it is part and parcel of creation, which God calls "good." Judaism sees sexuality as a gift to be used responsibly and in obedience to God's will, never as something evil in itself. The command to marry and to procreate within marriage is clear (Farley, 1994). Looking at the Old Testament as a whole, we can find three themes in this view of sexuality.

First, sex is seen not as just another biological function but as a deep and intimate part of a *relationship* between two people. The very ancient story of Adam and Eve states that "a man leaves his father and cleaves to his wife and the two become one flesh" (Genesis 2:24). Frequently, biblical Hebrew uses the verb "to know" to mean sexual intercourse (as in "Adam knew Eve and she conceived a child"). It also uses the word "knowledge," with this suggestion of deep intimacy, to describe the

Pederasty: Sex between an older man and a younger man, or a boy; sometimes called "boy love."

relationship between God and his people.[2] The use of sexual imagery in describing both marital and divine-human relationships testifies to the positive view of the Old Testament toward sex.

Second, in the Old Testament, sexuality could never be separated from its *social consequences.* Historically, Israel began as a small group of nomadic tribes fighting to stay alive in the near-desert of the Arabian peninsula. Sheer survival demanded that there be enough children, especially males, so that there would be enough herdsmen and warriors.[3] Thus, nonprocreative sex could not be allowed. Further, since the tribes were small and close-knit, sex had to be regulated to prevent jealousy over sexual partners, which could have divided and destroyed the group. It is not surprising, then, that so much of the Old Testament is concerned with laws regarding people living together in society and that these laws often include the regulation of sexual practices.

Finally, the Old Testament sees sexual behavior as an aspect of *national and religious loyalty.* When the Israelites settled in what is now the state of Israel, about 1200 to 1000 B.C.E., they came into contact with the original inhabitants, whom they called Canaanites. Like many agricultural peoples of the time, the Canaanites sought to encourage the growth of their crops through their religion. In this **fertility cult** Baal, the Sky Father, was encouraged to mate with Asherah (Astarte or Ishtar), the Earth Mother, so that crops would grow. This mating was encouraged by ritual sex, and temple prostitutes (male and female) were very much a part of Canaanite religion. Hebrew religious leaders saw in the fertility cult a threat to their religion, and many sexual practices are forbidden in the Old Testament because they were found among the Canaanites and might lead to infidelity to Israel's God.

The many sexual regulations of the Old Testament need to be seen both in the context of the times and against this historical background. From Israel's struggle for survival during the nomadic period came institutions such as polygamy (many wives) and concubinage (slaves kept for childbearing purposes) designed to produce children in the case of a childless marriage (Farley, 1994), and also laws against illegitimacy and nonprocreative sex. From the confrontation with the fertility cult, Israel derived prohibitions against nakedness, cultic prostitution, and other such typically Canaanite practices. Both themes are present in this passage from Leviticus 20:10–19:

> If a man commits adultery with his neighbor's wife, both adulterer and adulteress shall be put to death. The man who has intercourse with his father's wife has brought shame on his father. They shall both be put to death; their blood shall be on their own heads. . . . A man who has intercourse with any beast shall be put to death, and you shall kill the beast. . . . If a man takes his sister, his father's daughter or his mother's daughter, and they see one another naked, it is a scandalous disgrace. They shall be cut off in the presence of their people. . . . If a man lies with a woman during her monthly period and brings shame upon her, he has exposed her discharge and she has uncovered the source of her discharge; they shall both be cut off from their people.

Adultery and incest are threats to the harmony of the group. Not only is bestiality "unnatural," but it is also nonprocreative and may have been a feature of Canaanite religion. The menstrual taboo is typical of many societies (see Chapter 6).

It should be noted that all societies have had laws regulating sex (Chapter 1) and that these laws, however exotic they may seem to us, made sense in their historical context and were, for the most part, remarkably humane for the time. The Old Testament is also marked by a great regard for married love, affection, and sexuality; this is in marked contrast to, for example, the Greek view of marriage as an institution for breeding and housekeeping. Old Testament Judaism is highly legalistic but not particularly ascetic in its high regard for responsible sexuality as a good and integral part of human life.

---

[2]See, for example, Hosea, the Song of Solomon, and, in the New Testament, Revelation.

[3]Note that the heart of God's promise to the patriarch Abraham was descendants as numberless as the grains of sand or the stars in the sky (Genesis 13:14–17, among many other places).

Fertility cult: A form of nature-religion in which the fertility of the soil is encouraged through various forms of ritual magic, often including ritual sexual intercourse.

Christianity

As our discussion turns to Christianity, which grew in three centuries from an obscure Jewish sect to the dominant religion in the West, the complex conditions of the Mediterranean world between 100 B.C.E and 100 C.E.[4] must be noted. The world in which Christianity developed was one of tremendous ferment in the spheres of philosophy, religion, and morals. Although Stoicism remained popular among intellectuals, ordinary folks preferred various blends of mythology, superstition, and religion. Few people were much concerned with the pursuit of wisdom and virtue. There were many strange cults, often characterized by some sort of **dualism.** This was the notion that body and spirit were unalterably separate and opposed to each other and that the goal of life was to become purely spiritual by transcending the physical and material side of life. Public morals were notably decadent, and even ethical pagans were shocked by a society in which people, at least those who could afford it, prized pleasure above all things.

Revulsion at the excesses of Roman life affected Judaism, which became markedly more dualistic and antisex by the time of Jesus' birth and the growth of the Christian Church. That Church's ethical tradition is rooted in Old Testament Judaism and was given its direction by the teachings of Jesus, the writings of St. Paul, and the theology of the Fathers of early Christianity. From these beginnings, Christian ethics has evolved and developed over 2000 years in many and various ways. This makes oversimplification a real danger, and yet it is possible to speak in general terms of a Christian tradition of sexual ethics and morality.

Christianity is distinctive among the major world religions in insisting on monogamy (Parrinder, 1996). Most other religions permit polygyny, or a man having several wives. The Christian standard of monogamy, which may seem strict by today's standards, may be viewed in another light as a major step toward equality between women and men. Men were no longer permitted to have many wives as "possessions." Similarly, Jesus opposed divorce, which again may seem strict. How-

Figure 21.2   The Virgin Mary. During the Middle Ages, a great devotion to the mother of Jesus developed, emphasizing her perpetual virginity, purity, and freedom from all sin. That devotion lives on in most Latin American countries today.

ever, it reversed the traditional Hebrew rule—and the practice in many other cultures—that a man could divorce his wife simply at will, yet a wife had no similar power.

The New Testament

At the heart of the Christian scriptures are the Gospels, which describe the life and teachings of Jesus. Since Jesus said almost nothing on the subject of sex, it is difficult to derive a sexual ethic from the Gospels alone. Jesus' ethical teaching is based

---

[4]The Common Era, an alternative to Anno Domini (A.D.).

> **Dualism:** A religious or philosophical belief that body and spirit are separate and opposed to each other and that the goal of life is to free the spirit from the bondage of the body; thus a depreciation of the material world and the physical aspect of humanity.

on the tradition of the Old Testament prophets, and his view of sexuality follows in that tradition. He urged his followers to strive for ethical perfection, and he spoke strongly against pride, hypocrisy, injustice, and the misuse of wealth. Toward penitent sinners, including those whose sins were sexual, the Gospels show Jesus as compassionate and forgiving (see, for example, his dealings with "fallen women" in John 4:1–30, John 8:53–9:11, and Luke 7:36–50). He did not put any particular emphasis on sexual conduct, apparently regarding it as a part of a whole moral life based on the love of God and neighbor.

The task of first applying the principles of Jesus in concrete situations fell to St. Paul, documented in his letters. Paul's view of sexuality and women was ambivalent, deriving both from the immorality of much Greco-Roman culture and from the expectation he shared with most early Christians that Jesus would return soon, bringing the world to an end (Parrinder, 1996). Paul advocated celibacy, not necessarily because he was opposed to sex but because marriage might prove a distraction from prayer, worship, and the proclamation of the Gospel. As a Jew, Paul opposed all sexual expression outside marriage and judged sexual immorality harshly. However, he did not single out sexual sin. He condemned the "sins of the flesh," but by this he meant all aspects, such as "immorality, impurity, sorcery, enmity, strife, jealousy, anger, selfishness, party spirit, envy, drunkenness, carousing and the like" (Galatians 5:19–21). Later Christian theologians tended to understand the "sins of the flesh" primarily in sexual terms and thus gave Christianity a bias against sexuality beyond what Paul probably intended.

Other New Testament writings show a concern that Christians behave in a manner which is above reproach. Again, sexual morality is part of this concern for ethical perfection but is not necessarily the most important part. Given the times, it is not surprising that the New Testament is ambivalent about sexuality, probably more so than the Old Testament.

The Early Christian Church

The "Fathers of the Church," such as St. Augustine, wrote roughly between 150 and 600 C.E, and completed the basic theological shape of the Christian faith. During this time, Christian ethics became increasingly ascetic for several reasons: its natural tendencies, the assimilation of often dualistic Greek philosophy (especially Stoicism), the decadence of Roman society, and the conversion of the Roman Emperor Constantine in 325. As the Church became the official religion of the Roman Empire, much of its original fervor was lost and the Church began to grow corrupt and worldly.

Serious Christians revolted against this situation by moving to the desert to become monks and hermits, to fast, to pray, and to practice all sorts of self-denial, including **celibacy.** From this point on, monks and monasticism became a permanent reform movement within the Church, a vanguard of ascetics calling Christians to greater rigor. Their success can be seen in the twelfth-century requirement that all clergy in the West be celibate, a departure from early Church practice.[5] The Fathers of the Church, almost all of whom were celibates, allowed that marriage was good and honorable but thought virginity to be a much superior state.

The Middle Ages

During the Middle Ages, these basic principles continued to be elaborated and extended. The most important figure of the period, and even today the basic source of Catholic moral theology, was St. Thomas Aquinas (1225–1274). His great achievement was the *Summa Theologica,* a synthesis of Christian theology with Aristotelian philosophy which answered virtually any question a Christian might have on any topic. Thomas combined reason with divine revelation to discern the moral law of the universe, which was to be obeyed as God's intention for creation. This "natural law" approach to ethics was normative in Western Christianity for many centuries and remains so for Roman Catholicism.

---

[5]The First Epistle of Timothy, Chapter 3, shows the clear expectation that clergy will be married and father children.

Celibate (SEL-ih-bit): A person who remains unmarried, usually for religious reasons. (Celibacy: the practice of remaining celibate.) Sometimes used to refer to abstaining from sexual intercourse, the correct term for which is "chastity."

Figure 21.3 The most notable of the Western Fathers was St. Augustine (354–430 C.E), who had had a promiscuous youth and overreacted after his conversion to Christianity. For Augustine, sexuality was a consequence of the Fall, and every sexual act was tainted by concupiscence (from the Latin word *concupiscentia,* meaning "lust" or "evil desire of the flesh"). Even sex in marriage was sinful, and in *The City of God,* he wrote that "children could not have been begotten in any other way than they know them to be begotten now, i.e., by lust, at which even honorable marriage blushes" (1950 ed., Article 21). The stature of Augustine meant that his negative view of sexuality was perpetuated in subsequent Christian theology.

Thomas believed that sex was obviously intended for procreation and that, therefore, all nonprocreative sex violates the natural law and is sinful, being opposed to both human nature and the will of God. In the *Summa,* Thomas devoted a chapter to various sorts of lust and condemned as grave sin such things as fornication (premarital intercourse), nocturnal emissions, seduction, rape, adultery, incest, and "unnatural vice," which includes masturbation, bestiality, and homosexuality.

The theology of Aquinas and other moralists was communicated to the ordinary Christian through the Church's canon law, which determined when intercourse was or was not sinful. All sex outside marriage was, by definition, a sin. Even within marriage the Church forbade intercourse during certain times in a woman's physiological cycle (during menstruation, pregnancy, and up to 40 days postpartum) as well as on certain holy days, fast days (such as Fridays), and even during whole liturgical seasons (such as Advent and Lent). These rules were "enforced" through the *Penitentials,* guidebooks for priests hearing confessions, which instructed them on how to judge certain sins and what penances to assign for them (Brundage, 1984). All of this communicated to the ordinary person that the Church regarded sex as basically evil, for procreation only, and probably not something one should enjoy!

The Protestants

The Protestant Reformation in the sixteenth century destroyed the Christian unity of Europe and shook the theological foundations of the Catholic Church. However, in matters of sexual ethics there were few changes. The Protestant churches abandoned clerical celibacy, regarding it as unnatural and the source of many abuses, and placed a higher value on marriage and family life. Reformers nonetheless feared illegitimacy and approved of sexuality only in the confines of matrimony. Even then, they were often ambivalent. For example, Martin Luther, who was the founder of the Reformation and was happily married to a former nun, called marriage "a hospital for the sick" and saw its purpose as being to "aid human infirmity and prevent unchastity" (quoted in Thielicke, 1964, p. 136)—scarcely an enthusiastic approach.

A significant contribution of Reformation Protestantism to Christianity was a renewed emphasis on the individual conscience in matters such as the interpretation of the Bible and ethical decision making. Such an emphasis on freedom and individual responsibility has led to the serious questioning of legalistic ethics and, in part, to today's ethical debates.

The Reformation also gave rise to Puritanism. The Puritans followed Augustine in emphasizing the doctrine of "original sin" and the "total depravity" of fallen humanity. This led them to use civil

Focus 21.1

Dissent over Sexual Ethics in the Roman Catholic Church

In the 1980s and 1990s the Roman Catholic Church in the United States has experienced a number of serious controversies, many of them in the area of sexual ethics. The Church's traditional teaching on sexuality has been vigorously reasserted by the current pope, John Paul II, in the face of calls for a less legalistic and strict approach. John Paul has repeatedly condemned all sexual activity outside marriage and all nonprocreative sex, such as masturbation. In 1983, for example, the Vatican issued *Educational Guidance in Human Love,* a pamphlet for parents and teachers. In this document procreation is seen as the essential purpose of marital sex; masturbation, extramarital sex, and homosexuality are all described as "grave moral disorders." However, such teaching has not always been welcomed either by Catholic ethicists or by ordinary Catholic laypeople. Thus the debate within American Catholicism mirrors the controversy that has been going on in the society at large. At issue are such topics as contraception, abortion, homosexuality, reproductive technologies, and sexual abuse.

Contraception

Although the Vatican and the American hierarchy have not moved from the condemnation of "artificial" birth control as set forth in Pope Paul VI's encyclical *Humanae Vitae* of 1968, there is evidence that many American Catholics ignore the condemnation and use contraceptives, often with the tacit approval of their priests. For example, a 1993 *Newsweek* poll asked American Catholics, "Do you or do other Catholics you know personally use artificial birth control?" and 63 percent responded yes. Moreover, since *Humanae Vitae* was not an infallible teaching, some Catholic ethicists still treat contraception as an open question.

Abortion

There is sharp division among American Catholics on the issue of abortion. A large segment of Catholics actively or passively support the antiabortion, or right-to-life, movement, yet there are many dissenters. The *Newsweek* poll asked, "Is the Catholic Church's position on abortion too conservative, too liberal, or about right?" Forty-one percent thought it was too conservative and 43 percent thought it was about right, showing a nearly even division of opinion.

Homosexuality

In 1976 Jesuit priest and psychotherapist John J. McNeill published *The Church and the Homosexual,* in which he questioned the Church's traditional teaching on homosexuality and its scriptural and theological bases. At the time, Father McNeill was forbidden by his Jesuit superiors from speaking or writing further on the subject. He obeyed the order for 10 years, until his superiors ordered him to cease all ministry to the gay community, at which point he left the Jesuits and the priesthood.

Reproductive Technologies

In 1987, the Congregation for the Doctrine of the Faith issued a document called "Instruction for Human Life in Its Origins and on the Dignity of Procreation," which roundly condemned in vitro fertilization, surrogate motherhood, artificial insemination, and other new reproductive technologies. The document was severely criticized by many American Catholic ethicists, who found it too rigid and ill-informed.

Sexual Abuse by Clergy

Perhaps the most difficult sexual issue that faces the Catholic Church today is that of sexual abuse of children by Catholic priests. One of the most publicized cases was that of Father David Holley, who, after a stay in a treatment center (ordered by his bishop because of allegations), in the 1970s established a pattern of sexual abuse of boys as young as

11. In 1993 Holley, then 65, pleaded guilty and was sentenced to 175 years in prison. By 1995, an estimated 600 priests faced allegations of sexual misconduct. These scandals have caused the Church considerable embarrassment, particularly because critics have charged that bishops for many years did not deal adequately with accused priests, covering up the allegations so that the Church's reputation would not be tarnished. About two-thirds of American Catholics believe that the Church has treated abusive priests too leniently (*Newsweek*, 1993). Perhaps more seriously, this scandal has caused an enormous shift in Catholics' views of priests; they were once regarded as holy because of their renunciation of all sexual activity but are now revealed to be capable of sexual abuse. Two-thirds of American Catholics believe that abusive priests are a serious problem (*Newsweek*, 1993). The scandal has caused a profound shift in authority within the Catholic Church, which once dictated sexual morals for laypeople, but now finds itself defending some of its priests against charges of unethical sexual behavior.

Sources: Sipe (1995), Grammick (1986), Reuther (1985), McNeill (1987), Curran (1988), *Newsweek* (Aug. 16, 1993), Jenkins (1996).

Figure 21.4   Some nuns and other Catholic women challenge teachings of their Church.

law to regulate human behavior in an attempt to suppress frivolity and immorality. As will be seen in the next chapter, this urge to make people good by law has many sexual applications, although the Puritans were probably no more sexually repressive than other Christians of the time. What we often think of as "Puritan" sexual rigidity is probably more properly referred to as "Victorian." During the 60-year reign of Queen Victoria (1819–1901), English society held sexual expression in exaggerated disgust and probably exaggerated its importance. While strict public standards of decency and purity were enforced, many Victorians indulged in private vices of pornography, prostitution, and the rest. It is against this typically Victorian combination of repressiveness and hypocrisy that many people of the twentieth century have revolted, wrongly thinking the Victorian period to be representative of the whole Christian ethical tradition.

Current Trends

Across Western history there has been a fairly stable consensus on the fundamentals of sexual ethics, which some call the Old Morality. Sex has been understood as a good part of divine creation but also as a source of temptation that needs to be controlled. Although at various times chastity has been exalted, marriage and the family have always been held in esteem, and sex outside marriage condemned, in theory if not always in practice. The only approved purpose for sex has been procreation, and nonprocreative sex has been regarded as unnatural and sinful. However, this consensus has largely broken down in this century, and sexual ethics is now a topic of heated debate. Several factors, both within the religious community and outside it, have contributed to this ferment.

The rise of historical-critical methods in biblical scholarship has led to a questioning of the absoluteness of scriptural norms. Many scholars see them as conditioned by the time and culture in which they were written and not necessarily binding today. Furthermore, traditional understandings of biblical statements about sexuality have been questioned as more is learned about the original historical context. The Reformation emphasis on the Bible and individual conscience called the natural-law approach into question. The religious community has also been influenced by the behavioral sciences, which suggest that sexuality is much more complex than had been thought, and ques-

tion older assumptions about what is "natural" or "normal." Technology has made it possible, for the first time in human history, to prevent conception reliably and to terminate pregnancy safely, which, among other things, blunts the force of arguments against premarital intercourse on the basis of the disapproval of illegitimacy. Indeed, technology itself has raised a host of ethical issues as humans gain more and more control over what had once been a matter of "doing what comes naturally."

All this has led religious groups into serious debate and even conflict. Within Judaism, the Orthodox who still live by the rabbinic interpretation of the Old Testament law may be in serious conflict with the Conservative and Reformed groups. Protestants are deeply divided among conservatives who hold to the Old Morality, liberals inclined to the New Morality, and others who come out somewhere in between. Perhaps no single religious community has experienced as much tension over sexual morality as the Roman Catholic Church in the United States (see Focus 21.1). Controversy in matters of sexuality tends to be heated, as we shall see when we examine specific topics.

Humanism

It would be misleading to suggest that within Western culture, at least since the Renaissance, all ethical thinking has been religious in origin. Many ethicists have quite consciously tried to find a framework for moral behavior that does not rely on divine revelation or any direction from a source outside human intellect. Nonreligious ethics covers the whole spectrum; however, we can look at a fairly broad mainstream called **humanism.**

Humanistic ethics accepts no supernatural source for direction and insists that values can be found only in human experience in this world, as observed by the philosopher or social scientist. Most humanists would hold that the basic goals of human life are happiness, self-awareness, the avoidance of pain and suffering, and the fulfillment of human needs. Of course, the individual

Humanism: A philosophical system which denies a divine origin for morality and holds that ethical judgments must be made on the basis of human experience and human reason.

pursuit of these ends must be tempered by the fact that no one lives in the world alone and that some limitation of individual happiness may be required for the common good. Another important humanistic principle is that the individual must make his or her own decisions and accept responsibility for them and their consequences, without appeal to some higher authority, such as God.

In the area of human sexuality, humanism demands a realistic approach to behavior: one that does not create arbitrary or unreasonable standards and expectations. Humanism is very distrustful of the legalistic approach. It seeks real intimacy between persons and condemns impersonal and exploitative relationships, though probably not with the vigor that marks religious ethics. It tends to be tolerant, compassionate, and skeptical of claims of absolute right or wrong.

Sexuality in Other Major Religions

The discussion so far has been mostly concerned with Western culture and the Judeo-Christian tradition. It will broaden our outlook if we consider human sexuality in religious traditions outside dominant American culture. Obviously, this could be the topic for a very large book itself; we will be able to provide only a rather brief look at the three non-Western religions with the largest number of adherents: Islam, Hinduism, and Buddhism.

Islam

Geographically, and in terms of its roots, Islam is the closest faith to the Judeo-Christian heritage. It was founded by the Prophet Muhammad, who lived from 570 to 632 in what is now Saudi Arabia. Its followers are called *Muslims,* and its sacred scripture is the Koran. Classical Islam values sexuality very positively, and Muhammad saw intercourse, especially in marriage, as the highest good of human life. Islam sanctions both polygamy and concubinage, and the Prophet had several wives. In legend, great sexual prowess is attributed to him, although it may be exaggerated (Parrinder, 1996). He opposed celibacy, and Islam has very

Figure 21.5 In a resurgence of Islamic fundamentalism, some Muslims have attacked the liberalization of sexual practices and gender roles. The regime of the Ayatollah Khomeini led this return to fundamentalism. In Iran, there has been great repression of women, symbolized by the requirement that they resume the wearing of veils in public.

little ascetic tradition. A male-dominated faith, Islam has a strong double standard but recognizes a number of rights and prerogatives for women as well. Although sex outside either marriage or concubinage is viewed as a sin, Islam can be ethically quite moderate and rather tolerant of sexual sin, including homosexuality. All these aspects combine to create a "sex-positive" religion (Bullough, 1976, p. 205).

Islam is not a monolithic faith, and there is great variety in the ways in which the laws of the Koran are carried out in societies throughout the Moslem world. Some Arab states (e.g., Saudi Arabia and Iran) are theocracies in which the religious law is enacted in civil law. In these nations sexual offenses are likely to be more stringently punished, and women have very little freedom. Other Islamic countries (such as Egypt and Syria) are secular states in which Western values have been adopted to some extent. In these, women have more rights, and sexual mores are more pluralistic. Moreover, there is considerable variation in the interpretation of the Koran, especially between Sunnis and Shi'ites, the two principal Islamic "denominations."

Hinduism

"Hinduism" is a rather inclusive term that refers to a highly varied complex of mythology and religious practice founded on the Indian subcontinent. Here can be found virtually every approach to sexuality that the human species has yet invented. However, certain themes are worth mentioning. In Hinduism, four possible approaches to life are acceptable: Kama, the pursuit of pleasure; Artha, the pursuit of power and material wealth; Dharma, the pursuit of the moral life; and Moksha, the pursuit of liberation through the negation of the self in a state of being that is known as nirvana. Kama is notable because it has produced an extensive literature on the achievement of sexual pleasure, notably the *Kama Sutra* of Vatsyayana, a masterpiece

Figure 21.6 Buddhism encourages men to live celibate lives as monks.

of erotic hedonism. This book testifies to the highly positive view of sexuality to be found in Hinduism.

In contrast, the ways of Dharma and Moksha can be as rigorously ascetic as anything in Christianity. By avoiding all passions, including sex, the follower of these ways of renunciation seeks to pass out of the cycle of continual rebirth to absorption into the godhead. Part of this is *brahmacarya,* or celibacy, which is to be cultivated at the beginning of life (for the purposes of education and discipline) and at the end of life (for the purpose of finding peace). It is interesting to note that in between it is permissible to marry and raise a family, and thus this form of Hinduism makes active sexuality and asceticism possible in the same lifetime (Noss, 1963).

Buddhism

Buddhism developed out of Hinduism; it originated in the life and thought of Guatama (560–480 B.C.E), the Buddha, and has been elaborated in many forms since then. There is little discussion of sex in the teachings of the Buddha; his way is generally ascetic and concentrates on the achievement of enlightenment and on escape from the suffering of the world. Two main traditions, Therevada and Mahayana, are found in contemporary Buddhism, and they differ greatly. The ethics of Therevada include the strict nonindulgence of the desires that bring joy; understanding, morals, and discipline are emphasized. The ethics of Mahayana are more active and directed toward love of others. Both encourage men to live celibate lives as monks. Originally, Buddha sought a "middle way" between extreme asceticism and extreme hedonism, but today the situation is rather like that of medieval Christianity: The masses live ordinary and usually married lives, while the monks cultivate ascetic wisdom.

Tantric Buddhism, found particularly in Tibet, is a form of Buddhism which is of particular interest. There is a devotion to natural energy *(shakti),* and followers are taught that passion can be exhausted by passion. For example, sexual desire can be overcome while engaging in intercourse, according to occult knowledge. Sexuality can therefore be used as a means of transcending the limitations of human life. This sexual mysticism is by no means common, but it is one of the various forms which Eastern religion may take (Parrinder, 1980).

Contemporary Issues in Sexual Ethics

It cannot be said too frequently that human sexuality is heavily value-laden and, therefore, is likely to be the subject of strongly and emotionally held convictions. It is also likely to be the focal point of conflicts in society, if there is no broad consensus on the norms of sexual behavior. This is clearly the case in contemporary American society. The "sexual revolution" is perceived by many people as a threat to all they hold dear; not surprisingly, they respond with fear and anger. The backlash against the more liberal view of sexuality and the greater freedom of sexual behavior which has come about in the last 40 years, accelerated by the spread of AIDS and herpes, has resulted in explosive public debate, organized attempts at legislating the Old Morality back into force, and a reassertion of a highly legalistic view of the Judeo-Christian ethic. The debate promises to continue for some time and to generate much heat (Nelson, 1978).

This debate over the limits, if any, of individual sexual freedom can be seen as a clash between the New Morality and the Old Morality, but let us propose a more helpful model. The Old Morality is, to a great extent, supported by people who believe that there are clearly and objectively defined standards of right and wrong and that a society has a right to insist that all its members conform to them, at least outwardly. We might call this view **moralism;** it has many proponents in the religious community who see the objective standard of morality as deriving from divine law. Opposed to this view are the proponents of **pluralism;** they see the question of public morality as being much more complex. Pluralists deny that there are

---

Moralism: A religious or philosophical attitude that emphasizes moral behavior, usually according to strict standards, as the highest goal of human life. Moralists tend to favor strict regulation of human conduct to help make people good.

Pluralism: A philosophical or political attitude that affirms the value of many competing opinions and believes that the truth is discovered in the clash of diverse perspectives. Pluralists, therefore, believe in the maximum human freedom possible.

objective standards of morality, and they are likely to contend that truth is to be discovered in the clash of differing opinions and convictions. According to this view, society is wise to allow many points of view to be advocated and expressed. The conscience and rights of the individual are to be stressed over society's needs for order and uniformity. Pluralists are much less likely than moralists to appeal to either law or religion for the enforcement of their views, and they are more likely to allow freedom to individuals even if society might be endangered by their actions. The debate between moralist and pluralist has been going on for a very long time, both in the religious community itself and throughout American history. It will not be settled any time soon.

An illustration of the moralistic view can be found in the "profamily" position which is rooted in religious conservatism and is increasingly attempting to influence the legislative process. Profamily activists are in favor of an absolute constitutional ban on abortion, against any kind of legal tolerance of the cohabitation of unmarried persons, and in favor of legal discrimination against homosexuals in such areas as housing, child custody, and employment.

This position is essentially that of the New Religious Right, a coalition of conservative religious and political groups like the Christian Coalition. Members of this movement, largely but not exclusively fundamentalist Protestants, argue that the New Morality has sapped the moral vigor of American society, leaving the country open to inner decay and divine judgment. Their efforts to enforce their religious convictions by legislation have created one of the most intense church-state controversies of the twentieth century (see Chapter 22). Their position is clearly odious to pluralists and to those who have benefited by the liberalization of laws and attitudes concerning sexuality. These persons and groups will fight to keep what they consider to be gains, while profamily and New Right activists will seek to turn the clock back to what they perceive to have been a healthier and more moral time.

This conflict can be found within most religious communities today. Even liberal "mainline" Protestant groups, which have tended to accommodate at least some of the New Morality, have been under attack from portions of their own membership on such issues as abortion, premarital sex, and homosexuality. Reports in the press of national gatherings of American religious groups reveal a remarkable number of debates related to human sexuality, debates that parallel those in society at large. Here we will illustrate this ferment by discussing the ethical issues posed by sex outside marriage, contraception, abortion, homosexuality, AIDS, and reproductive technologies.

Sex Outside Marriage

The biblical tradition underlying Western ethics has almost always limited sexual intercourse to marriage. This is rooted in a religious understanding of marriage as God's will for most men and women, the way in which sin is avoided and children are cared for. Thus, the tradition has condemned both sex before marriage (which the Bible calls **fornication**) and sex by persons married to others **(adultery).** Today, this position continues to be held by theological conservatives among Jews, Protestants, and Roman Catholics. A Roman Catholic statement on the subject is typical of this position:

> Today there are many who vindicate the right to sexual union before marriage, at least in those cases where a firm intention to marry and an affection which is already in some way conjugal in the psychology of the subjects require this completion which they judge to be connatural. . . . This option is contrary to Christian doctrine, which states that every genital act must be within the framework of marriage. (Sacred Congregation for the Doctrine of the Faith, 1976, p. 11)

However, trends in society have caused many ethicists to reopen the question and to take less dogmatic positions. Among these are the development of safe and reliable contraception, later age at first marriage, the fact that many people experi-

Fornication: The biblical term for sex by unmarried persons and, more generally, all immoral sexual behavior.
Adultery: Voluntary sexual intercourse by a husband or wife with someone other than one's spouse; thus betrayal of one's marriage vows.

ence the singleness of divorce and widowhood, and empirical evidence to suggest widespread sexual activity among adolescents. These ethicists are concerned that people be given more helpful guidance than "thou shalt not." For them, the quality of the relationship is more important ethically than its legal status.

Criteria for judging the morality of nonmarital sexual acts could include the following (Countryman, 1994). First, is there a genuine respect for the personhood of all involved? Virtually all ethicists would agree that sexual exploitation of one person by another (whether married or not)—the use of other human beings merely for one's pleasure—is wrong. Furthermore, most would require genuine affection and serious commitment from both parties. This commitment would be manifested in responsible behavior such as using contraceptives if the couple were not willing or able to have children and taking precautions against STDs. Finally, many ethicists would insist that moral sexual behavior must include genuine openness and honesty between the partners. Public and private institutions, in this view, should be involved in helping people to make good ethical choices about sexual behavior in a culture that tends to glorify and exploit sex (Lebacqz, 1987; Moore, 1987).[6]

Extramarital sex (adultery) has always been regarded as a grave matter in the Judeo-Christian tradition. In the Hebrew Scriptures, the penalty for it was to be stoned to death; in the New Testament, it is the only grounds for divorce allowed by Jesus (Matthew 6:21–22). Adultery has been understood as a serious breach of trust by a spouse, as well as an act of unfaithfulness to God (a violation of religiously significant promises). Few contemporary ethicists would modify this position, but many would argue for a less judgmental, more humane approach to those involved. In this view, people in extramarital relationships should be helped to find the root causes of their infidelity and to move toward a reconciliation with their spouses based on forgiveness and love. This approach suggests that counseling is more helpful than condemnation. Above all, some argue, religious organizations

need to assist people in establishing and maintaining good marriages based on mutual respect, communication, and commitment.

Contraception

Roman Catholics and Orthodox Jews oppose any "artificial" means of contraception; other Jews and most Protestants favor responsible family planning by married couples. Moreover, most ethicists would suggest that unmarried persons who are sexually active ought to be using some means to prevent pregnancy.

Those who oppose birth control for religious reasons see it as being contrary to the will of God, against the natural law, or both. Orthodox Judaism cites the biblical injunction to "be fruitful and multiply" (Genesis 1:26), as God's command to his people, not to be disobeyed in any way. Furthermore, some members of other Jewish communities warn that limiting family size threatens the future existence of the Jewish people, and they call for a return to the traditionally large Jewish family.

The Roman Catholic position is best articulated in Pope Paul VI's 1968 encyclical, *Humanae Vitae:*

> Marriage and conjugal love are by their nature ordained toward the begetting and educating of children. . . . In the task of transmitting life, therefore, they are not free to proceed completely at will, as if they could determine in a wholly autonomous way the honest path to follow, but they must conform their activity to the creative intention of God, expressed in the very nature of marriage and by its acts, and manifested by the constant teaching of the Church. (Pope Paul VI, 1968, p. 20)

The encyclical continued the Church's approval of "natural family planning," that is, abstinence during fertile periods, popularly known as the "rhythm method" or "Vatican roulette." *Humanae Vitae* was not enthusiastically accepted by all Catholics and there is evidence to suggest that many Catholic couples, often with the encouragement or tacit approval of their priest, ignore these teachings and use contraceptives anyway. Nonetheless, the teachings of *Humanae Vitae* have been repeatedly and resoundingly reiterated by the current pope, John Paul II.

Those in the religious community who favor the use of contraceptives do so for a variety of reasons. Many express a concern that all children who are

---

[6]A fine discussion of these issues from different perspectives (liberal Protestant and Roman Catholic, respectively) can be found in Nelson (1978) and Genovesi (1987).

born should be "wanted," and they see family planning as a means to this end. Others, emphasizing the dangers that overpopulation poses to the quality and future of human life, the need for a more equitable distribution of natural resources, and the needs of the emerging nations, call for family planning as a matter of justice. Another point of view regards the use of contraceptives as part of the responsible use of freedom. In this view, any couple who are unwilling or unready to assume the responsibility of children have a duty to use contraceptives. For these groups, the decision to use contraceptives is a highly individual one, and the government must allow each individual the free exercise of his or her conscience (Curran, 1988).

Abortion

One of the most convulsive debates of our time is being waged over the issue of abortion. Prolife and prochoice activists are well organized and deeply convinced of the rightness of their positions. The conflict is a clash of religious belief, political conviction, and world view in the realm of public policy, one that allows for no easy solutions.[7]

Two distinctions should be made at the outset. First, there is no consensus on the relation between abortion and contraception. For the Roman Catholic Church, and others within the prolife movement, the two are the same in intention; indeed, many prolife activists wish to ban all contraception except natural family planning. In the other camp, there are some prochoice advocates who also regard abortion as a variety of contraception—less desirable, perhaps, but better than unwanted pregnancy. However, most centrist ethicists do distinguish between abortion and contraception, typically favoring the latter while raising ethical questions about the former. Second, a distinction is frequently made between therapeutic abortion and elective abortion. Therapeutic abortion is a termination of pregnancy when the life or mental health of the woman is threatened or when there is trauma, such as in cases of rape or incest. Many ethical theorists are

[7]For a thorough and careful study of what prolife and prochoice activists believe they have at stake, see Ginsburg (1989).

willing to endorse therapeutic abortion as the lesser of two evils but do not sanction elective abortion, i.e., abortion whenever requested by a woman for any reason.

The leadership of the antiabortion movement clearly comes from the Roman Catholic Church, for which putting an end to abortion is a major policy goal. For many Catholics, opposition to abortion is seen as part of an overall commitment to respect for life, which Joseph Cardinal Bernardin called the "seamless garment" that includes opposition to capital punishment, euthanasia (mercy killing), and social injustice, and a very positive stance toward peace (Cahill, 1985). Pope John Paul II reaffirmed this position in his 1995 encyclical, *Evangelium Vitae* (Gospel of Life), saying "I declare that direct abortion, that is, abortion willed as an ends and a means, always constitutes a grave moral disorder, since it is the deliberate killing of an innocent human being." He went on to condemn capital punishment and euthanasia. The underlying principle of this position is that all life is a gift from God that human beings are not permitted to take. It is the position of the prolife movement that human life begins at the moment of conception and that the fetus is, from that beginning, entitled to full rights and protections. The Roman Catholic position is shared by Orthodox Jews, Eastern Orthodox Christians, and many conservative, or fundamentalist, Protestants (see Focus 21.2). An end to legalized abortion is at the top of the political agenda of many theologically conservative groups and has been a major issue in presidential and congressional elections and Supreme Court appointments for the past 25 years.

Nonetheless, there has been some significant dissent from this position even within the Catholic Church. Prochoice Catholics point out that for most of its history, the Church accepted Aristotle's teaching, reaffirmed by St. Thomas Aquinas, that "ensoulment," that is, the entry into the fetus of its distinctively human soul, takes place 40 days after conception for a male and 80 or 90 days after conception for a female. Theoretically, this permits abortions at least until the fortieth day. In 1869, Pope Pius IX eliminated the concept of ensoulment, holding that life begins at conception and that all abortion is therefore murder (Luker, 1984). Though regularly denounced by the Vatican and

Figure 21.7   Prolife and prochoice advocates are both adamant about their positions regarding abortion.

the American hierarchy, some Catholic ethicists insist that the Church's position is not unchangeable and argue that the concerns and needs of women should be more carefully considered in the matter (Maguire, 1983; Kolbenschlag, 1985; Reuther, 1985).

The prochoice position takes at least two forms: absolute and modified. The absolute position argues that pregnancy is solely the concern of a woman and that she should have the absolute right to control her own body and determine whether to carry a fetus to term or not. Ethically, this position is based on the conviction that the individual must be free and autonomous in all personal decisions. It is also inspired by feminism, which regards such autonomy as necessary if women are to be truly equal to men. Feminists also observe that, historically, the rules about abortion were made by men, who do not become pregnant, and thus are deeply suspect. Indeed, for many feminists complete access to abortion is an absolute principle for women's liberation. Concerns for autonomy and

individualism have formed a significant part of Western ethics and American social theory for over two centuries.

For those who hold the modified prochoice position—and this includes most liberal Protestants and Jews—the issue is more complex and means balancing several goods against one another. They affirm that human life is good and ought to be preserved but also argue that the quality of life is important. They argue that an unborn child may have a right to life, but ask if it does not also have the right to be wanted and cared for. In high-risk situations, might not the danger to the well-being of a woman already alive take precedence over the well-being of an unborn fetus? Few in this camp regard abortion as a good thing but suggest that there may be many situations in which it is the least bad choice. Moreover, these ethicists tend to observe that since there is no real consensus in society over the morality of abortion, the government ought to keep out and let the individual woman make up her own mind.

Focus 21.2

Religious Position Statements on Abortion: Prolife Versus Prochoice

The following statements come from a variety of major religious organizations. They reflect the nature of the arguments and rhetoric of the abortion debate. Consider these statements in relation to the Gallup poll (Table 21.1) which shows a wide diversity of opinion among the American public.

Prolife Statements

[The] Church has always rejected abortion as a grave moral evil. It has always seen that the child's helplessness, both before and after birth, far from diminishing his or her right to life, increases our moral obligation to respect and to protect that right. . . . The Church also realizes that a society which tolerates the direct destruction of innocent life, as in the current practice of abortion, is in danger of losing its respect for life in all other contexts. (National Conference of Catholic Bishops, 1985)

All human beings ought to value every person for his or her uniqueness as a creature of God, called to be a brother or sister of Christ by reason of the incarnation and universal redemption. For us, the sacredness of human life is based on these premises. And it is on the same premises that there is based our celebration of human life—all human life. This explains our efforts to defend human life

Table 21.1 Gallup Poll Findings on Americans' Attitudes Toward Abortion

	Percentage		
	1975	1992	1998*
Abortion should be:			
Legal under any circumstances	21%	31%	23%
Legal under only certain circumstances*	54%	53%	—
Legal under certain (most) circumstances	—	—	16%
Legal under certain (a few) circumstances	—	—	42%
Illegal under all circumstances	22%	14%	17%
No opinion	3	2	2

*In 1996 the Gallup Organization changed the wording of this question, splitting the older option "Legal under only certain circumstances" into two alternatives: "Legal under certain circumstances (most)" and "Legal under certain circumstances (only a few)."

against every influence or action that threatens or weakens it, as well as our endeavors to make every life more human in all its aspects. (Pope John Paul II, 1979)

The "prolife" position is, as is typical with moralism, much more absolute and apparently simple, while the pluralist "prochoice" position is subtle and differentiated. Both positions agree on the value and dignity of human life, but are sharply divided on when life begins, how various conflicting interests are to be balanced, and how human life is best preserved and enhanced. Several factors ensure that the debate will continue for some time. Advances in neonatal medicine are pushing back the threshold of "viability" (the survival of premature infants), and this may affect the ethical acceptability of second-trimester abortions for some people (Callahan, 1986). The politicalization of the issue will keep it in the public consciousness, and legal challenges will undoubtedly continue (see Chapter 22). Certainly, the intensity is not likely to diminish, as it is a clash about life, law,

Orthodox Christians have always viewed the willful abortion of unborn children as a heinous act of evil. The Church's canonical tradition identifies any action intended to destroy a fetus as the crime of murder. (Orthodox Church in America, 1992)

Abortion is not a moral option, except as a tragically unavoidable by-product of medical procedures necessary to prevent the death of another human being, viz., the mother. (Lutheran Church—Missouri Synod, 1979)

[W]e do affirm our opposition to legalized abortion and our support of appropriate federal and state legislation and/or constitutional amendment which will prohibit abortion except to prevent the imminent death of the mother. (Southern Baptist Convention, 1989)

Prochoice Statements

Therefore, be it resolved, that the Sixteenth General Synod. . . upholds the right of men and women to have access to adequately funded family planning services, and to safe, legal abortions as one option among others. . . and urges pastors, members, local churches, . . . to oppose actively legislation and amendments which seek to revoke or limit access to safe and legal abortions. (United Church of Christ, General Synod, 1987)

Our belief in the sanctity of unborn human life makes us reluctant to approve abortion. But we are equally bound to respect the sacredness of life and well-being of the mother for whom devastating damage may result from an unacceptable pregnancy. In continuity with past Christian teaching, we recognize tragic conflicts of life with life that may justify abortion, and in such cases support the legal option of abortion under proper medical procedures. (United Methodist Church, General Conference, 1988)

The American Jewish Congress has long recognized that reproductive freedom is a fundamental right, grounded in the most basic notions of personal privacy, individual integrity and religious liberty. Jewish religious traditions hold that a woman must be left free to her own conscience and God to decide for herself what is morally correct. The fundamental right to privacy applies to contraception to avoid unintended pregnancy as well as to freedom of choice on abortion to prevent unwanted birth. (American Jewish Congress, Biennial Convention, 1989)

While we acknowledge that in this country it is the legal right of every woman to have a medically safe abortion, as Christians we believe strongly that if this right is exercised, it should be used only in extreme situations. We emphatically oppose abortion as a means of birth control, family planning, sex selection or any reason of mere convenience. (Episcopal Church, General Convention, 1988)

Sources: National Right to Life; Religious Coalition for Reproductive Choice.

freedom, and values, and few people are neutral on these issues.

Homosexuality

Mirroring society as a whole, the religious community has been engaged in a vigorous debate on the subject of homosexuality. Until recently, it was assumed that all homosexual acts and persons were condemned by the Judeo-Christian tradition. However, many contemporary ethicists, and some religious bodies, are reexamining their attitudes toward homosexuality (Siker, 1994). This change has occurred in part because some recent scholarship suggests that the traditional interpretation of the Bible passages on this topic is not accurate, and in part because the impact of social science has led many ethicists to question whether homosexuality is truly abnormal and unnatural and therefore against the will of God. There are three positions, broadly speaking, on the issue: rejection, modified rejection or qualified acceptance, and full acceptance (Nelson, 1978).

Rejectionism

Regarding the *rejectionist position,* it has generally been presumed that the weight of the Judeo-Christian tradition absolutely opposes any sexual acts between persons of the same gender and regards those committing such acts as dreadful sinners, utterly condemned by God. Although there are few references in the Bible, all are negative, the most famous being the passage about the destruction of Sodom (see Figure 21.8) (see also Leviticus 20:13).

In the Hebrew Scriptures, homosexual practices are included on lists of offenses against God. Jesus made no comment on the subject, but St. Paul was unambiguously against homosexual acts, seeing them as perverse behavior by fundamentally heterosexual persons. Thus he included them in lists of sexual sins, along with adultery and fornication, that are in opposition to the will of God and symptomatic of human depravity. However, Paul does not seem to have found homosexuality any more dreadful than other sexual sins.

Homosexuality was not uncommon in the Mediterranean world of the early Church, and the Church condemned it as part of the immoral world in which it found itself. The Church saw it as a crime against nature that might bring down the wrath of God upon the whole community (Kosnick, 1977). In the Middle Ages, Thomas Aquinas stated that "unnatural vice . . . flouts nature by transgressing its basic principle of sexuality and is in this matter the gravest of sin" (1968 ed., II-II, q. 154, a. 12). Thielicke noted a similar attitude among Reformation theologians and quoted a seventeenth-century Lutheran, Benedict Carpzov, as listing the following results of homosexuality: "earthquakes, famine, pestilence, Saracens, floods, and very fat voracious field mice" (1964, p. 276).

The rejectionist position continues to be held by many members of the religious community who condemn homosexual acts and reject homosexual persons unless they repent and become heterosexual. An example of this stance is a 1987 resolution of the Southern Baptist Convention that states: "Homosexuality is a perversion of divine standards

and a violation of nature and of natural affections. . . . [While] God loves the homosexual and offers salvation, homosexuality is not a normal lifestyle and is an abomination in the eyes of God" (Associated Press, June 1987).

Modified Rejection or Qualified Acceptance

Many religious groups would modify this position somewhat, through a distinction between homosexual orientation and behavior. In essence this stance of *modified rejection* or *qualified acceptance* regards homosexual orientation, assuming it cannot be changed, as morally neutral, but rejects homosexual acts. Thus, an ethical homosexual person may be fully obedient to the will of God, as long as she or he remains abstinent. This is the official position of the Roman Catholic Church, reiterated in a Vatican directive entitled "The Pastoral Care of Homosexual Persons," which states in part:

> Although the particular inclination of the homosexual person is not a sin, it is a more or less strong tendency ordered toward an intrinsic moral evil and thus the inclination itself must be seen as an objective disorder. Therefore special concern and pastoral attention should be directed toward those who have this condition, lest they be led to believe that the living out of this orientation in homosexual activity is a morally acceptable option. It is not. (Congregation for the Doctrine of the Faith, 1986, p. 379)

As a result of this instruction, many chapters of Dignity, an organization of lesbian and gay Catholics, were denied the use of church facilities by American bishops. Various Protestant groups have taken the same stance—that is, being gay per se may not be sinful, but homosexual acts are.

Farther down the road to qualified acceptance is the 1985 statement of the Episcopal Church's General Convention, in which that body committed itself

> to find an effective way to foster a better understanding of homosexual persons, to dispel myths and prejudices about homosexuality, to provide pastoral support, and to give life to the claim of homosexual persons upon the love, acceptance, and pastoral care and concern of the Church. (*Journal of the General Convention,* 1985, p. 505)

The Episcopal Church stands, with many other religious groups, for full civil rights and liberties for

Figure 21.8 In this Dürer painting, Lot and his family flee as the city of Sodom burns. In Genesis 19:4–11, God sends two angels to the city of Sodom to investigate its alleged immorality. The angels are granted hospitality by Lot, but his house is surrounded by a crowd of men demanding that he send the angels out, "that we may know them." Lot offers his virgin daughters instead, but the men of Sodom insist, the angels strike them blind, and God destroys the city. This story has been understood to condemn all homosexual acts. However, some modern scholars question this interpretation, noting that at most it condemns homosexual rape. Moreover, scholars point out that in other portions of the Bible and in Jewish history, the sin of Sodom is never seen as homosexuality, but rather as general immorality and lack of hospitality (a serious offense in the ancient Near East) (Furnish, 1994).

homosexual persons, although it is officially opposed to the ordination of practicing homosexuals (see below). Thus, rejection can be modified and acceptance qualified in a variety of ways.

Full Acceptance

At the other end of the spectrum, there are those in the religious community who favor *full acceptance* of lesbian and gay persons, usually basing this on a revisionist view of the Bible and Church tradition. Some scholars question whether the apparent condemnation in the Scriptures is really relevant to

homosexuality as it is generally understood today (Furnish, 1994). New Testament scholar Robin Scroggs, for example, says that the only form of same-gender behavior known to the world of the New Testament involved older men and youths (often prostitutes or slaves), and he concludes that "what the New Testament was against was the image of homosexuality as pederasty and primarily here its more sordid and dehumanizing dimensions" (1983, p. 126). Yale historian John Boswell's detailed research into early and medieval Christianity has led him to conclude that up until about the thirteenth century the Christian Church was relatively neutral toward homosexuality and, when it did see homosexual behavior as sinful, did not regard it as any worse than heterosexual transgressions. Boswell found a gay subculture that flourished throughout this period and argues that it was known to the Church, that clergy and church officials were often part of it, that it was not infrequently tolerated by religious and civil authorities alike, and that same-gender marriages or "unions" occurred (Boswell, 1980; 1994). This sort of reinterpretation has led some theologians, such as the Roman Catholic John McNeil and the Anglican Norman Pittenger, to question whether the tradition has been understood properly and to conclude that sexual relationships which are characterized by mutual respect, concern, and commitment—by love in its fullest sense—are to be valued and affirmed, whatever the gender of the partners (Pittenger, 1970; McNeill, 1987). The revisionist view has not captured the religious community, but it has been articulated well enough to keep the debate going.

Institutional expression of the position of full acceptance has been varied. In 1963 a group of English Friends challenged traditional thinking about sexuality, including homosexuality, in *Toward a Quaker View of Sex*. Since that time, Quakers and Unitarians have been notable for their acceptance not only of the homosexual person but also of her or his sexual behavior, as long as it is conscientious. Within virtually all the mainline churches, gay caucuses and organizations have been formed in an effort to move fellow believers toward greater understanding and tolerance. A considerable number of lesbians and gays have simply left the established religious organizations and founded their own churches, synagogues,

Figure 21.9   The Reverend Troy Perry founded the Universal Fellowship of Metropolitan Community Churches in 1968 as part of his coming out—the story of which is told in his book, *The Lord Is My Shepherd and He Knows I'm Gay.* Providing a home for over 30,000 members, in 1983 the UFMCC applied for membership in the National Council of Churches, the major American organization of Christian bodies. After much debate, its application was placed on permanent hold (Glaser, 1994).

temples, and other groups, of which the largest is the Metropolitan Community Church. On the other hand, many homosexual persons reject all forms of religion as oppressive and invalid, making religion as controversial within the gay community as homosexuality is within religious bodies.

Two issues in particular seem to provoke the most debate: ordination and marriage of homosexual people. Beginning in the 1970s, most major American Protestant denominations debated the appropriateness of ordaining lesbians and gays to the ministry. The debates were emotional and explosive, and nearly all resulted in legislation forbidding homosexual ordination. The 1984 debate at the General Conference of the United Methodist Church was typical. That body required all clergy and candidates for ordination to observe "fidelity in marriage and celibacy in singleness" and added a specific prohibition against the ordination of "self-avowed practicing homosexuals" to its *Book*

of Discipline (*The Christian Century*, 1984, *101*, 565). At present only the Unitarian-Universalist Association, the United Church of Christ (Congregationalists), and the American Union of Hebrew Congregations seem willing to ordain gay and lesbian people openly, and the lines are pretty clearly drawn in other religious groups.

Many who favor full acceptance of homosexual persons have argued for formal recognition of committed relationships along the lines of matrimony. The Metropolitan Community Church frequently performs such "holy unions," but so far no mainline church has officially approved of such a rite, although they do occur without permission of ecclesiastical authorities.

AIDS

AIDS has raised a host of complex and difficult ethical issues for individuals, religious communities, and society as a whole. These issues are being debated in an atmosphere of fear, anger, and ignorance, which is focused on the fatality of the disease and the fact that the vast majority of sufferers in the United States are either homosexual men or injection drug users, two populations about which the society has profound ambivalence. Religious groups, like the rest of society, have struggled to develop effective ways of ministering to persons with HIV infection or AIDS (Jantzen, 1994; Cherry & Mitulski, 1988). Responses have ranged from declaring AIDS to be God's punishment on sinners to actively organizing to minister to persons with AIDS and seeking to educate members of churches and synagogues about the disease and how they may respond compassionately (Godges, 1986; Countryman, 1987; Jantzen, 1994).

Broadly speaking, the major ethical conflicts center on the dignity and autonomy of the person, on the one hand, and the welfare of society on the other. This issue has both personal and public aspects. For the person who has AIDS or who is HIV-positive, a primary issue is confidentiality. Given that disclosure can lead to the loss of job, housing, friends, and family, such persons may well wonder if anyone has a right to know about their condition. However, many people argue that the public, or at least certain groups within society, have a right to know who is infected in order to be protected from them.

It has been proposed that children with HIV infection should not be allowed to attend public schools and that infected persons should be registered for the protection of emergency medical personnel, health care workers, coroners, and morticians. Several large populations, including military personnel, would-be immigrants, and some prisoners have been required to undergo mandatory testing for HIV. Many public health officials and AIDS researchers oppose such measures because they fear that persons at risk would be driven underground. They argue that the public health is best protected by voluntary testing and fairly stringent protection of confidentiality (Levine & Bermel, 1985; Levine & Bermel, 1986).

Ethically, a solid middle position would encourage persons in high-risk categories to take responsibility for themselves by undergoing voluntary testing and practicing safer sex. This position would argue that infected persons have a right to confidentiality but should voluntarily disclose their status to anyone put at risk by it—notably health care personnel and sexual partners. For health care workers, there is the personal ethical problem of whether to treat persons who are HIV-positive. It is probable that there is an ethical obligation to treat, but that there is also an obligation to take appropriate precautions (Zuger, 1987).

Many of the ethical issues of public policy revolve around the very high cost of AIDS—both treating its victims and seeking a medical solution. Who should pay this cost? Insurance companies have sought ways in which to deny health coverage to persons in high-risk groups. People with AIDS often cannot work and so lose job-related insurance benefits. Who should pay for their care—hospitals, cities, states, the federal government? Specialized services and facilities—such as home care and hospices—are needed. Who is responsible for providing them? Given limited funds available for research, what should the focus of that research be? Should resources be concentrated on prevention through a vaccine or on treatment of those who already have the disease? Who should develop treatments, public agencies or private drug companies? And again, who will pay?

Public education for prevention, generally regarded as the only really effective response at present, raises some ethical problems. What kind of education is appropriate? At what age should it begin? How graphic should it be? Does government advocacy of "safer sex" mean endorsement of sexual practices abhorrent to many Americans? Because of its stand on birth control, the Roman Catholic Church absolutely opposes the use of condoms. Should Catholics have to support government programs to make them generally available? Some public health officials, alarmed by the rapid spread of AIDS among intravenous drug users, suggest the reversal of a long-time policy of fighting drug abuse by restricting the availability of syringes. They argue that if addicts could readily obtain clean needles, they would not have to share and risk infection. However, this strategy outrages the sensibilities of many who believe that drug addiction is a grave moral and social evil.

Most of the choices that must be made in dealing with AIDS are unappealing, expensive, or both. American society is being challenged to maintain its own values and to deal compassionately and effectively with what many have described as the greatest health crisis since the bubonic plague.

Technology and Sexual Ethics

A major challenge to ethicists in the late twentieth century is the rapid development of technologies that raise new moral issues before the old ones have been resolved, in matters of human sexuality as in anything else. We have already discussed several issues in which technology has played a major role. Although sex outside of marriage is hardly a problem unique to our time, the availability of reliable birth control techniques has probably increased the incidence of premarital and extramarital sex. The fact that millions of people can enjoy vigorous sex lives without conceiving children unless they choose to has markedly changed the basic moral climate.

The issue of abortion is also intensified by technological advances of the past few decades and will only get more complicated in the future. The developing medical science of neonatology means that fetuses are viable outside the uterus earlier and earlier. Some late-pregnancy abortions produce a fetus that could be kept alive, and hospital staffs are faced with agonizing questions about what should be done in such cases. Traditional ethics suggest that such unwanted children be kept alive, yet this can be incredibly costly and the children are often severely handicapped. Hospitals may limit, or even forbid, second-trimester abortions to deal with the issue (Callahan, 1986).

Another complex of ethical issues arises out of the host of new reproductive technologies which enable people to conceive children outside of the "normal" process of sexual intercourse in what has been called "collaborative parenting." These include artificial insemination, either by husband or donor (AIH or AID), in vitro fertilization (IVF), embryo transfer, and surrogate motherhood.

For many ethicists, these technologies can be tentatively approved because they enable otherwise infertile people to have children with at least one partner's genes (Strong, 1997). Certainly, having children of one's own has had a very high value for most people throughout history, and barrenness has been seen as a curse in most cultures.

These technologies bring with them a number of ethical problems, too, chief among which is that they involve "playing God." That is, they give human beings control over things that are, it is argued, best left up to nature and raise serious problems of who will decide how they are to be used. It is also argued that separating conception from marital intercourse may confuse the parenthood of children and have a negative effect on the family. Another concern is the possibility of the exploitation of others, particularly in the case of surrogacy. Some ethicists, especially those taking a feminist perspective, fear that rich couples will "rent wombs" or "buy babies" from low-income women. Many question the morality of conceiving and/or carrying a child one never intends to raise. Others wonder if AID and IVF might not be used to select the sex of a child or predetermine other characteristics, ushering in a "Brave New World" that is less than human (Krimmel, 1983; McDowell, 1983; Schneider, 1985; Boyd et al., 1986).

Two religious communities have condemned most or all of these technologies, though on somewhat different grounds. Orthodox Judaism might permit the use of techniques which would allow an otherwise infertile couple to have a child if both egg and sperm come from the couple, e.g., in AIH or IVF with implantation in the wife's uterus. Any technique involving a third party is condemned as being de facto adultery and confusing the parent-

age of the child (Rosner, 1983; Green, 1984). The Roman Catholic position was stated clearly in a 1987 statement issued by the Vatican, "Instruction on Respect for Human Life in Its Origin and on the Dignity of Procreation." The statement admitted as an open moral question fertility techniques which remained within the woman's body using her husband's sperm not collected by masturbation. Otherwise all techniques such as AID, IVF, and surrogacy were unequivocally condemned as an assault on the dignity of the embryo (in Catholic theology a human person) and on the sanctity of marriage as the only licit means of procreation (Congregation for the Doctrine of the Faith, 1987).

A centrist position on new reproductive technologies might approve their use in many cases, such as AIH in which a married couple use their own egg and sperm and the woman's uterus and simply accomplish fertilization and implantation by artificial means, perhaps because the wife's fallopian tubes are blocked. At the same time, this centrist position might forbid other practices, such as the use of a stranger for her egg and gestation with only the husband's sperm contributed by the "parental" couple. One question then would be whether technological conception should be regulated according to the mother's age. Rosanna Dalla Corte, a 62-year-old Italian woman, created headlines in 1994 when she gave birth, having used ovum donation because she was postmenopausal (Strong, 1997). She would be 80 at the time of her son's high school graduation.

The announcement, in 1997, that Scottish scientist had successfully cloned a sheep, Dolly, confronted the public with the possibility of human cloning and its ethical issues when most had thought that cloning was at most a science-fiction fantasy. The technique itself is called **somatic cell nuclear transfer** and involves substituting the genetic material from an adult's cell for the nucleus in an egg (Shapiro, 1997). President Clinton swiftly asked the National Bioethics Advisory Commission to report on the ethical and legal issues surrounding human cloning. The group considered a

Somatic cell nuclear transfer: A cloning technique that involves substituting genetic material from an adult's cell for the nucleus of an egg.

Figure 21.10   The successful cloning of Dolly, the sheep, was a technological breakthrough that also raises a series of ethical questions.

number of ethical perspectives (Shapiro, 1997). A child born through cloning might have a diminished sense of individuality and personal autonomy, being genetically identical, for example, to her mother. The practice might open the door to a eugenics movement in which many copies of genetically "desirable" individuals were created while others were not permitted to reproduce. Yet these concerns must be balanced against important principles such as the right to privacy and to personal freedom, as well as the need to pursue scientific research. The commission concluded that, at this time, it would be morally unacceptable for anyone to create a human child using somatic cell nuclear transfer cloning, in part because the evidence indicates that the technique, at this time, is not safe and could introduce serious, unknown risks to a fetus. At the same time, the commission concluded that various religious bodies held divergent views on cloning and that widespread public debate needed to occur in order to refine and reach consensus on these ethical issues.

There is no way to stop the development of technology, or its speed. Nonetheless, it is important to recognize that decisions about human life and reproduction ought not to be made on purely scientific grounds. By definition, they have the deepest moral implications, and these implications must be adequately addressed if the essential values of human life are to be preserved.

Toward an Ethics of *Human* Sexuality

The combined forces of the sexual revolution and the New Morality have attacked the traditional Judeo-Christian sexual ethic as narrow and repressive. This may be true, but it has not yet been proved to everyone's satisfaction that the alternatives proposed are a real improvement upon the Old Morality. Whether the debate will be resolved, and how, remains to be seen. Some of the arguments and possibilities are considered below.

The Old Morality tends to be ascetic and legalistic and, at its worst, reduces ethical behavior to following a series of rules. Its asceticism may downgrade the goodness of human sexuality and negate the very real joys of physical pleasure. A healthy personality needs to integrate the physical side of life and affirm it, and this kind of self-acceptance may be made more difficult by the Old Morality. Furthermore, if morality is simply a matter of applying universal rules, there is no real choice, and human freedom is seriously undermined. In short, opponents of the Old Morality might argue, this approach diminishes the full nature of humanity and impoverishes human life.

On the other hand, the traditional approach deserves a few kind words as well. For one thing, with the traditional morality people almost always know where they stand. Right and wrong and good and bad are clearly, if somewhat inflexibly, spelled out. Moreover, asceticism does bear witness to the fact that the human is more than merely the body.

The New Morality, with its situational approach and tendencies toward hedonism, has its own share of pluses and minuses. It affirms quite positively the physical and sexual side of human nature as an integral part of the individual. This is helpful, but if it is pushed too far, it can leave people under no control and thus less than fully human. Situation ethics quite properly calls for an evaluation of every ethical decision on the basis of the concrete aspects of the personalities involved and the context of the decision. Its broad principles of love, respect, and interpersonal responsibility are sound, but it can be argued that situationism does not take sufficiently into account the problem of human selfishness. Dishonesty about our real motives may blind us to the actual effects of our actions, however sincere we profess to be. Furthermore, situationism is a much less certain guide than the older approach, since so many situations are ambiguous.

There is a middle way between these two extremes, one that may prove to be the synthesis that sexual ethics seems to be searching for. This approach would use the traditional principles (laws) as guidelines for actions while insisting that they must occasionally be reworked in view of certain specific situations. This approach differs from the Old Morality by stating that ethical principles must

be flexible, and it differs from the New Morality by holding that departures from tradition must be based on very strong evidence that the old rules do not apply. For those adopting this position, healthy decision making operates in the tensions between the rigid "thou shalt not" of the legalist and the "do your own thing" of the situationist.

In the specific case of sexual ethics, such a middle-of-the-road approach would affirm the goodness of human sexuality but insist that sexual behavior needs to be responsible, and based on reason, experience, and conscience. It would accept sexuality as a vital part of human personality but not the sum total of who we are. Such an approach to sexual ethics would indeed be consistent with what was shown earlier in this chapter to be the heart of the Judeo-Christian tradition, shorn of some of its rigidity and distrust of sexuality. If, as is sometimes claimed, the sexual revolution is over, and there is a movement toward relationship and commitment, this might prove to be just the sexual morality that people in our time are looking for.

SUMMARY

It is important to study religion and ethics in conjunction with human sexuality because they frequently provide the framework within which people judge the rightness or wrongness of sexual activity. They give rise to attitudes that influence the way members of a society regard sexuality, and they are therefore powerful influences on behavior. Religion and ethics may be hedonistic (pleasure oriented) or ascetic (emphasizing self-discipline). They may be legalistic (operating by rules) or situational (making decisions in concrete situations, with few rules).

In the great ethical traditions, ancient Judaism had a positive, though legalistic, view of sexuality. Christian sources are ambivalent about sexuality, with Jesus saying little on the subject and with St. Paul, influenced by the immorality of Roman culture and his expectation of the end of the world, being somewhat negative. Later, Christianity became much more ascetic; this is reflected in the writings of Augustine and Thomas Aquinas, who also placed Catholic moral theology in the natural-law mold. The Protestant Reformation abolished

clerical celibacy and opened the door to greater individual freedom in ethics. Today, technological development and new forms of biblical scholarship have led to a wide variety of positions on issues of sexual ethics.

Humanistic ethics rejects external authority, replacing it with a person-centered approach to ethics. A variety of approaches to sexuality can be found in Islam, Hinduism, and Buddhism.

Six ethical issues involving human sexuality have provoked lively debate recently. Although the Western ethical tradition opposes sex outside marriage, some liberals are open to sex among the unmarried under certain conditions. Contraception is opposed by Roman Catholicism and Orthodox Judaism on scriptural and natural-law grounds, but it is positively valued by other groups. Abortion provokes a very emotional argument, with positions ranging from condemnation on the grounds that it is murder to a view that asserts the moral right of women to control their own bodies. Although the traditional view condemns homosexuality absolutely, there is some movement toward either a qualified approval of at least civil rights for gay people or a more complete acceptance of their lifestyle. The spread of AIDS poses serious ethical problems, which involve a balancing of individual needs with the welfare of society. Developments in the technology of human reproduction are creating complex ethical issues with few clear norms.

A possible resolution of the conflict between the Old Morality and the New Morality involves an ethics of *human* sexuality, neither hedonistic nor rigidly ascetic, which takes seriously the historical tradition of ethical thinking while insisting that decisions be made on the basis of the specific situation.

REVIEW QUESTIONS

1. The ethics of the Old Morality tend to be _____, and those of the New Morality are often called _____.

2. Jewish teaching, as found in the Old Testament, is uniformly negative about sex. True or false?

3. Much Christian teaching, from the Apostle Paul through the Middle Ages, favored _____ over marriage for ordinary Christians, and required it for the clergy.

4. Engagement in sexual intercourse and family life at one time in life and extreme asceticism at others is characteristic of the _____ religion.

5. A political and religious position found in contemporary society that opposes abortion, women's liberation, and gay rights is often called _____.

6. Roman Catholicism opposes contraception on the basis of _____ law.

7. Abortion is uniformly condemned by all religious groups in the United States. True or false?

8. Areas of debate on homosexuality in many religious groups center on the issues of _____ and _____.

9. Ethical debate over AIDS includes discussion about whether an individual who is HIV-positive should be able to keep that information _____.

10. Which of the following reproductive technologies has been approved for use by Roman Catholics: AID, IVF, surrogate motherhood?

(The answers to all review questions are at the end of the book, after the Glossary.)

QUESTIONS FOR THOUGHT, DISCUSSION, AND DEBATE

1. If you are a member of a religious group, investigate your group's beliefs on the issues discussed in this chapter (sex outside marriage, contraception, abortion, homosexuality, AIDS, and reproductive technology). Do you agree with those positions? If you are not a member of a religious group, see if you can formulate a statement of sexual ethics that is consistent with your philosophy of life.

2. Seek out a person, group, or written material which takes the opposite of your position on abortion and carefully consider those arguments. What effect does this have on your views?

3. Ruth Greenberg is the judge in a most unusual custody case. Chad and Michele are a married couple who had an infertility problem. Medical testing revealed that Chad had an extremely low sperm count that was unlikely to produce successful fertilization. To make matters worse, Michele had a small, undeveloped uterus and probably would always miscarry early in a pregnancy. Yet they desperately wanted children. They contracted with a surrogate, Mary, to provide an egg and gestation, and used sperm from a sperm bank for the fertilization. The baby, a little girl, has been born, but Mary has decided that she wants to keep her. The court battle is between Chad and Michele on the one hand and Mary on the other, as to who should get the baby. What should Judge Greenberg do?

SUGGESTIONS FOR FURTHER READING

Biale, David (1997). *Eros and the Jews.* Berkeley, CA: University of California Press. A history of sex and the Jewish people.

Boswell, John. (1980). *Christianity, social tolerance, and homosexuality.* Chicago: University of Chicago Press. A very sophisticated reassessment of Christian attitudes toward gays and their place in Western society through the Middle Ages.

Countryman, L. William. (1988). *Dirt, greed, and sex: Sexual ethics in the New Testament and their implications for today.* This seminary professor carefully examines biblical statements about sexuality to gain a better understanding of their meaning and their relevance today.

Genovesi, Vincent J. (1987). *In pursuit of love: Catholic morality and human sexuality.* Wilmington, DE: Michael Glazier. Well-written, scholarly work treating the whole field from the mainline Roman Catholic point of view.

Gold, Michael (1992). *Does God belong in the bedroom?* Philadelphia, PA: J.P.S. A Jewish approach to sexuality and sexual ethics.

Nelson, James B. (1978). *Embodiment: An approach to sexuality and Christian theology.* Minneapolis: Augsburg. Well-written, scholarly work treating the whole field from the mainline Protestant point of view.

Nelson, James B., and Longfellow, Sandra P. (Eds.). (1994). *Sexuality and the sacred.* Louisville, KY: Westminster Press. An interesting collection of essays on a variety of topics in the realm of sexual ethics.

Parrinder, Geoffrey. (1996). *Sexual morality in the world's religions.* New York: Oxford University Press. A superb and concise treatment of the variety of religious approaches to human sexuality. The best short book in the field.

WEB RESOURCES

http://www.christusrex.org/www1/CDHN/sexeduc.html
Roman Catholic statement.

http://www.thelutheran.org/9703/page32.html
Lutheran statement.

http://www.siecus.org/pubs/biblio/bibs0010.html
Bibliography on religion and sexuality prepared by SIECUS.

http://library.vanderbilt.edu/carpenter/traditioncon.htm
Statements on gender and sexuality from religious traditions; compiled by the Carpenter Program, Vanderbilt University Divinity School.

chapter
22

SEX AND THE LAW

CHAPTER HIGHLIGHTS

Sex, although considered by many in our culture the quintessential private activity, is blanketed by a staggering number and variety of laws. Such a crazy quilt of state laws means that a perfectly legitimate sexual practice in one state may be a felony in another. By crossing a state boundary one may be stepping into a different moral universe.*

Every day, millions of Americans engage in sexual behaviors that are illegal. This may surprise you, since arrests for criminal sexual conduct are rare in the United States (Posner, 1992). But in many states or cities, sexual activity with someone under 18 (statutory rape), oral and anal sexual intercourse ("sodomy"), sex with someone who is legally married to someone else (adultery), and sexual contact involving two persons of the same gender are crimes. In fact, there are numerous laws telling people, in effect, what they can do and how, where, and with whom they can do it. This chapter will consider why such laws exist, what sorts of behaviors are affected, how these laws are enforced, how they are changing, and what the future prospects for sex-law reform might be.

Why Are There Sex Laws?

To begin, we might well ask why there are laws regulating sexual conduct in the first place. This actually is a very modern question, for throughout most of Western history such laws were taken for granted. Sexual legislation is quite ancient, dating back certainly to the time of the Old Testament (see Chapter 21). Since then, in countries where the Judeo-Christian tradition is influential, attempts to regulate morals have been the rule. Today we are likely to regard sex as a private matter, of concern only to those involved. However, historically it has been seen as a matter that very much affects society and therefore as a fit subject for law. Most societies regulate sexual behavior, both by custom and by law.

Even today, certain kinds of sex laws are probably legitimate and necessary. Packer argues that the following might be rationally included in law: protection "against force and equivalent means of coercion to secure sexual gratification," "protection of the immature against sexual exploitation," and (a somewhat problematic category) "the prevention of conduct that gives offense or is likely to give offense to innocent bystanders" (1968, p. 306). It seems obvious that people ought to be free from sexual assault and coercion and that children should not be sexually exploited; individual rights and the interests of society are here in agreement.

However, sex laws have been designed for other purposes which may be more open to debate. Historically, one rationale was to preserve the family as the principal unit of the social order by protecting its integrity from, for example, adultery or desertion of a spouse. Sex laws also seek to ensure that children have a supportive family by prohibiting conduct such as **fornication,** which is likely to result in out-of-wedlock births. Changing social conditions may call for revision of these statutes, but the principles behind them are understandable.

There is yet another realm of motivation behind sex laws which is highly problematic, and that is the protection of society's morals. The concern for public morality results in laws against nonprocreative sex, for reasons outlined in Chapter 21. Thus there have been laws against homosexual acts, bestiality, and contraception. Religious beliefs as to what is "unnatural," "immoral," or "sinful" have found expression in law, as it was often held that the state had a duty to uphold religion as a pillar of civilized society, using the law to make people good. The example of England is instructive, since American law derives so extensively from English law. Church and state in England have historically

---

*Richard Posner and Katharine Silbaugh. (1996). *A guide to America's Sex Laws.* Chicago, IL: University of Chicago Press.

Fornication: Sex between two unmarried persons.

been seen as identical, and the state had an obligation to protect the interests of the Church. A secular government not tied to the Church, such as the United States, was unthinkable, and an individual's morals were a matter of public concern.

In the United States, in contrast, the Constitution separates church and state in order to prevent one religious group from imposing its beliefs on others. To the extent that today's laws are derived from the Judeo-Christian (or any other) religious tradition, they violate this principle. Moreover, we have witnessed increasing heterogeneity of moral beliefs in recent decades (Posner & Silbaugh, 1996). Even if we accept the use of law to regulate morality, there is no longer (if there ever was) a consensus about which morals should be codified into law.

It can be argued that another principal source of sex laws is *sexism,* which is deeply rooted in Western culture. One scholar has suggested that the history of the regulation of sexual activity could as well be called the history of the double standard. He goes on to note that:

The law of marriage and the law controlling sexual expression are really the same question looked at from different angles. Women have always been looked upon as the property of men—whether fathers or husbands. Marriage has frequently in history been a commercial transaction or a way in which the fabric of society could be maintained. The male insistence on chastity was simply an attempt to regulate social relations, to cement dynasties, to ensure the orderly succession of property (particularly real property) and to perpetuate male domination. (Parker, 1983, p. 190)

It is probably not coincidental that the movement for sex-law reform has gone hand in hand with the movement for the liberation of women.

The American tradition of moralism in politics, the prudery of the Victorian period (during which much of the American legal system came into being), and the zealousness of such individuals as Anthony Comstock (see Focus 22.1) combined to provide the United States with an enormous amount of sexual legislation. This legislation reflects a great deal of conflict in our attitudes toward sex, which is perhaps unsurprising in such a pluralistic society. According to one authority, "The United States criminalizes more sexual conduct than other developed countries do and punishes the sexual conduct that it criminalizes in common with those countries more severely" (Posner, 1992, p. 78). Table 22.1 lists the maximum punishment (years in prison) allowed by the laws of six countries for three sex crimes and three property crimes. For the three sex crimes, the allowable punishment in the United States is among the most severe, whereas the punishment for the three property crimes in the United States is about average for the six countries. Persons convicted of a crime in the United States actually spend much less time in jail, but these maximum sentences tell us which crimes legislators think are most serious.

The U.S. legal tradition assumes both the right of the state to enforce morals and a consensus in society as to which morals are to be enforced.

Table 22.1 Severity of Punishment for Three Sex Crimes and Three Property Crimes, Selected Countries

| Crime | Maximum Prison Sentence (years) in | | | | | |
	England	France	Italy	Japan	Sweden	United States
Rape	30	20	10	15	6	18
Statutory rape	16	10	10	15	4	16
Incest	7	20	6.5	NA	2	10
Arson	30	30	7	15	8	14
Robbery	30	30	10	15	6	14
Larceny	10	10	3	10	2	8

Source: R. A. Posner. (1992). *Sex and Reason.* Cambridge, MA: Harvard University Press, Tables 1 and 3.

Focus 22.1

Anthony Comstock: Crusader Against Vice

In any discussion of laws regulating sexual behavior, the name of Anthony Comstock looms large. His zeal for moral reform is reflected in the use of the term "Comstock laws" for the kinds of restrictive statutes considered here.

Comstock was born in Connecticut in 1844 and was reared as a strict Puritan Congregationalist; he had a well-developed sense of his own sinfulness—and of others' as well. He served in the Union Army during the Civil War and worked as a dry-goods salesperson. While still a young man, he became very active in the Young Men's Christian Association, seeking the arrest and conviction of dealers in pornography. He helped found the Committee for the Suppression of Vice within the YMCA. This later became an independent society, as his efforts gained him national attention.

Comstock's most noteworthy success was probably a comprehensive antiobscenity bill which passed the U.S. Congress in 1873. The law prohibited the mailing of obscene matter within the United States, as well as advertisements for obscenity, which included matter "for the prevention of conception." Comstock initiated the passage of a similar law in New York, making it illegal to give contraceptive information verbally, and many other states followed suit. At the same time, Comstock received an appointment as a special agent of the U.S. Post Office, and this gave him the authority personally to enforce the Comstock Law. He did so with a vengeance, claiming at the end of his career to have been personally responsible for the jailing of over 3600 offenders against public decency.

Comstock's energies were directed not only against pornography but also against abortionists, fraudulent advertisers and sellers of quack medicines, lotteries, saloonkeepers, artists who painted nude subjects, and advocates of free love. Among the objects of his wrath were many of the most famous advocates of unpopular opinions of his day. He carried on a crusade against women's movement pioneers Victoria Woodhull and her sister Tennessee Claflin; he helped jail William Sanger, husband of Margaret Sanger, the birth control crusader; he attacked Robert Ingersoll, the atheist; and he tried to prevent the New York production of George Bernard Shaw's play about a prostitute, *Mrs. Warren's Profession*. For the last of these efforts, Shaw rewarded him by coining the word "Comstockery."

Anthony Comstock was a controversial figure during his own lifetime and has often been blamed for all legislation reflecting his views. However, it is important to understand that he had a great deal of support from the public, without which he could not have jailed his 3600 miscreants. He will probably go down in history as a symbol of the effort to make people moral by legislation. He died in 1915 shortly after representing the United States at an International Purity Congress.

Anthony Comstock may be long dead, but Comstockery has always been a feature of American society and seems to have made a comeback in the 1980s and 1990s. He might very well be a patron saint for the Christian far right.

Source: Heywood Broun and Margaret Leech. (1927). *Anthony Comstock: Roundsman of the Lord*. New York: Boni.

However, some contemporary citizens have come to question the legitimacy of government interference in what they regard as their private affairs. This tension has led to a widespread demand for a radical overhaul of laws that regulate sexual conduct. It all makes for a fascinating, if frustrating, field of study, for law has a way of reflecting the ambiguities and conflicts of society.

What Kinds of Sex Laws Are There?

Cataloging the laws pertaining to sexual conduct would be a difficult project. It is possible that no one really knows how many such laws there are, given the large number of jurisdictions in the American legal system. When one considers federal law, the Uniform Code of Military Justice, state laws, municipal codes, county ordinances, and so on, the magnitude of the problem becomes clear. In addition to *criminal law,* portions of civil law that may penalize certain sexual behaviors—such as licensing for professions, personnel rules for government employees, and immigration regulations—must also be considered. Furthermore, these laws are changing all the time, and any list would become obsolete before it went to press. Therefore, what is offered here is not so much a statistical summary of specific sex laws as a look at the *kinds* of laws that are, or have been, on the statute books.[1] The subheadings, all of which contain the word "crime," have been chosen with care, as a reminder that we are discussing legal offenses which can carry with them the penalty of going to jail, loss of reputation, monetary fines, or all of these. However quaint and amusing some of these laws may seem, they are a serious matter.

Crimes of Exploitation and Force

Recalling our earlier discussion about the kinds of sex laws that seem to make sense in a pluralistic society, let us begin with those seeking to prevent the use of force or exploitation in sexual relations—chiefly laws against rape and sexual relations with children. In the past two decades, there has been a movement toward seeing such crimes not so much as sex crimes but as crimes of violence and victimization, with laws being revised to accommodate this different understanding and to protect the victims (see Chapter 17).

For generations, the classic legal definition of (forcible) **rape** was:

> The act of sexual intercourse with a female person not the wife of or judicially separated from bed and board from, the offender, committed without her lawful consent. Emission is not necessary; and any sexual penetration, however slight, is sufficient to accomplish the crime. (Slovenko, 1965, p. 48)

In trials under this definition, the principal issue was the consent of the victim, and many states allowed the victim's prior sexual activities to be considered as evidence of her consent, in effect putting the victim on trial.

Today, rape is typically defined, following current laws in many states, as "nonconsensual oral, anal, or vaginal penetration, obtained by force, by threat of bodily harm, or when the victim is incapable of giving consent" (Koss, 1993, p. 1062). A victim might be "incapable of giving consent" because of being unconscious or drunk.

At least 22 states have revised their laws so that a husband can be tried for raping his wife (Posner & Silbaugh, 1996). The 1979 Rideout case in Oregon was the first prosecution under such a law, although the husband was acquitted. In 1993, John Bobbitt was charged with raping his wife, Lorena, and she was charged with malicious wounding for cutting off his penis. Both were acquitted. Some states have also revised the language of their statutes to eliminate the word rape, choosing instead "criminal sexual assault," which permits prosecution of men for raping men and even of women for raping men (Searles & Berger, 1987; Jaffe & Becker, 1984).

Laws that seek to prevent the sexual exploitation of children and young people are complicated

---

[1]A general discussion of the kinds of laws may be found in MacNamara & Sagarin (1977). State-by-state listings appear in Bernard et al. (1985), Hunter et al. (1992), and Posner & Silbaugh (1996).

Rape: Nonconsensual oral, anal, or vaginal penetration, obtained by force, by threat of bodily harm, or when the victim is incapable of giving consent.

by the issues of consent, coercion, and immaturity, all of which are rather difficult to define. Most states have laws against *statutory rape,* or carnal knowledge of a juvenile. These laws presume that all intercourse by an adult male (normally one over 17 or 18) with any female under a certain age is, by definition, illicit because she cannot give genuine consent. The age of consent varies from state to state, ranging from 14 to 18; in most states, it is 15 or 16 (Posner & Silbaugh, 1996). Many states have laws that also include a reference to the difference in ages between the male and the female, on the assumption that there is a difference in criminality between a 16-year-old girl having intercourse with her 18-year-old boyfriend and with a man in his thirties or forties (MacNamara & Sagarin, 1977; Mueller, 1980).

There are a great variety of laws against the *sexual abuse of children,* called variously child molestation, carnal abuse of a child, or impairing the morals of a child. These general terms usually cover all sexual contact between adult and child, heterosexual and homosexual, and can include the use of sexual language, exhibitionism, showing pornography to a child, having a child witness intercourse, or taking a child to a brothel or gay bar (MacNamara & Sagarin, 1977). Such statutes attempt to protect children, a reasonable goal, but some are so vague as either to be ineffective or to criminalize innocuous behavior. In order to prevent misuse of these laws, it is important to develop more precise ones.

Finally, every state includes laws against **incest** in its penal code. These laws prohibit sexual relations between children and "biological parents, ancestors, or other siblings;" some also prohibit activity involving step and adoptive parents (Posner & Silbaugh, 1996). Most prosecutions are cases involving children and adult relatives. The nearly universal taboo against incest seems to have as its purpose the guarantee to children that the home will be a place where they can be free from sexual pressure. In many states, the closer the relationship, the more severe the penalties against incest (Mueller, 1980). Incest laws also seek to prevent the alleged genetic problems of inbreeding. Greater public awareness of the extent of incest (see Chapter 17) is likely to lead to a reexamination of laws on the subject.

Criminal Consensual Acts

Although it is not hard to see the logic of laws against force and exploitation of the young, many people are amazed to discover the number of sexual acts that are legally forbidden to consenting adults. These laws have been justified on the grounds of the prevention of illegitimacy, the preservation of the family, the promotion of public health, and the enforcement of morality (Bernard et al., 1985). With respect to heterosexuals, there are a number of laws against fornication, **cohabitation,** and adultery. As of 1994, fornication was illegal in 19 states and the District of Columbia; cohabitation was outlawed in 14 states (Posner & Silbaugh, 1996).

Adultery, intercourse involving persons at least one of whom is married to someone else, is a crime in 24 states and the District of Columbia. It is grounds for divorce in almost every state (see Figure 22.1a). Two issues arise in defining adultery. The first is whether both parties or only the married person are guilty of a crime; in 6 states only the married participant can be charged. The other is what constitutes adultery, a single incident (42 states) or habitual or open conduct (10 states).

Besides specifying with whom one may have sex, laws have attempted to regulate what acts are permissible, even in the case of a legally married couple. Twenty-nine states have laws prohibiting **sodomy** (see figure 22.1b). In 13 states, the laws define sodomy very broadly, as "crimes against nature," or "deviate sexual intercourse." In several court cases, judges have ruled that such laws are unconstitutional because this definition is too vague (Leonard, 1993), but in other cases judges have ruled that past judicial decisions, "case law," provide a clear definition (Posner & Silbaugh, 1996). In 15 states, the laws are much more specific, for example, "contact between the penis and

---

Incest: Sexual relations between persons closely related to each other.
Cohabitation: Unmarried persons living together (with sexual relations assumed).
Adultery: Intercourse involving persons at least one of whom is married to someone else.
Sodomy: Originally "crimes against nature"; in contemporary laws, oral and anal intercourse.

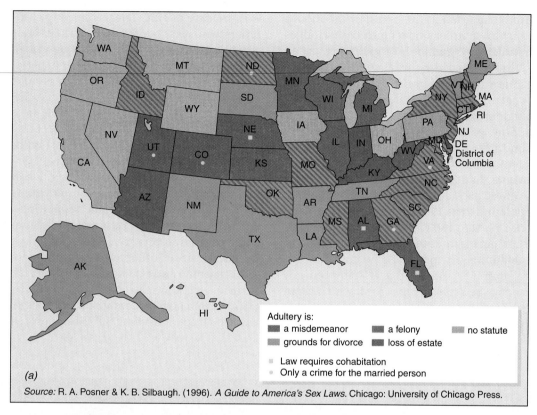

Adultery is:

◼ a misdemeanor   ◼ a felony   ▨ no statute

▨ grounds for divorce   ◼ loss of estate

▪ Law requires cohabitation
• Only a crime for the married person

(a)

Source: R. A. Posner & K. B. Silbaugh. (1996). *A Guide to America's Sex Laws.* Chicago: University of Chicago Press.

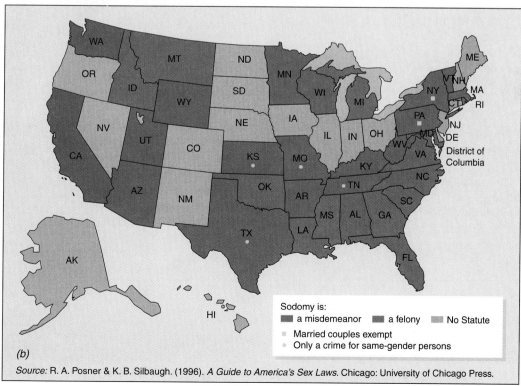

Sodomy is:

◼ a misdemeanor   ◼ a felony   ▨ No Statute

▪ Married couples exempt
• Only a crime for same-gender persons

(b)

Source: R. A. Posner & K. B. Silbaugh. (1996). *A Guide to America's Sex Laws.* Chicago: University of Chicago Press.

Figure 22.1 *(a)* State laws governing adultery. *(b)* State laws governing sodomy.

the anus, the mouth and the penis, or the mouth and the vulva." In at least two court cases, judges ruled that such laws cannot be applied to married persons, arguing that the sexual conduct of married persons is protected by the right to privacy.

One cogent example of specific language used to define sodomy is the Georgia statute, which defines sodomy as "any act involving the mouth or anus of one person, and the sex organs of another." It is a felony and is punishable by not less than 1 year, or more than 20 years, in jail (Hunter et al., 1992, p. 155). This statute was upheld by the U.S. Supreme Court in the 1986 case of *Bowers v. Hardwick* (478 U.S. 186, 1986), in which the Court affirmed the right of states to enact such statutes.

Although in theory most sodomy laws are supposed to prohibit "unnatural acts" between any two persons, when they are enforced, which is not very often, the prosecution is almost invariably against gay people. A widely publicized exception was the case of NBC Television sportscaster Marv Albert, who was charged with forcible sodomy and sexual assault in 1997. A woman described as his mistress charged that he forced her to perform oral sex and bit her. Albert pled guilty to assault (a misdemeanor in Virginia) and in exchange, the prosecutor dropped the sodomy charge (a felony). Sodomy laws are only the tip of the iceberg in a legal structure of discrimination against homosexual persons (Rivera, 1981–1982; Rivera, 1979; Hunter et al., 1992). In most places gay and lesbian persons may be denied private employment. However, the federal civil service regulations forbid such discrimination in public employment, and some state governors have issued executive orders forbidding sexual orientation discrimination. Until 1995, gay men and lesbian women could be denied security clearances, which effectively barred them from certain governmental and other jobs. By Executive Order, President Bill Clinton ordered that

Figure 22.2   President Bill Clinton, together with General Colin Powell, announces the new "Don't ask, don't tell" policy for gays and lesbians in the military.

people no longer be denied clearances solely on the basis of sexual orientation. Gays and lesbians have an ambiguous status with regard to military service. For many years, they could be refused entry to or dismissed from military service. In 1990, the U.S. Supreme Court upheld the military's right to ban gays and lesbians. Recent lower-court cases distinguish between statements of orientation and behavior, and protect the latter (*Meinhold v. U.S.* 34 F 3d 1469, 1994). In 1994, the Clinton administration adopted a "Don't ask, don't tell" policy. Military authorities cannot ask a person about his or her sexual orientation, and nonheterosexuals are not to make their orientation public. Many professional and occupational licensing requirements have "good character" or morality clauses which seem to invite discrimination. For example, a number of public school teachers have been dismissed when their homosexuality became known.

Beyond discrimination in employment, distinctions are drawn in other areas as well. No right to gay or lesbian marriage is supported by statute, and gay and lesbian relationships have very little legal support. This means that the partner of a gay or lesbian is not entitled to social security benefits, although a poll indicates that a majority of Americans support such entitlement (see Table 22.2). Homosexual couples cannot automatically inherit each other's property. The homosexuality of a parent is a serious disadvantage in child custody proceedings. Surveying this landscape, legal scholar Rhonda Rivera concluded that there is

> systematic and pervasive discrimination against homosexual individuals in our courts. . . . Homosexuals are penalized in all aspects of their lives because of their sexual preference. They lose their jobs, their children, and numerous other precious rights as a result of many current judicial policies. (Rivera, 1979, p. 1947)

Gay and lesbian activists have pursued a variety of strategies aimed at overcoming discrimination. They had hoped that the Supreme Court would eventually expand the "right of privacy" (discussed below) to include consensual sodomy and even homosexuality as a lifestyle. The decision in *Bowers v. Hardwick* seems to have closed that door for the present (*Leading Cases,* 1986). Another judicial approach would be to seek to define sexual orientation as a "suspect classification" protected from discrimination under the "equal protection of the laws" clause of the Fourteenth Amendment. However, the courts have shown very little inclination to move in that direction. Seventeen states and 157 municipalities have laws or policies prohibiting discrimination on the basis of sexual orientation, notably in housing and employment (Epstein, 1995). These laws have withstood court tests but are often unpopular and subject to repeal by referendum. Particularly with public sentiment inflamed

Table 22.2 Americans' Attitudes toward Gay Rights

	Percentage who said		
	Should	Should Not	Don't Know
Do you think there should or should not be:			
Equal rights in terms of job opportunities?	84%	13%	3%
Equal rights in terms of housing?	80	16	4
Health insurance and other employee benefits for gay spouses?	59	36	5
Social security benefits for gay spouses?	57	37	6
Special legislation to guarantee equal rights for gays?	47	48	5
Adoption rights for gay spouses?	40	49	11
Equal rights in terms of legally sanctioned gay marriages?	35	56	9

Source: A telephone poll of 753 adults conducted by Princeton Survey Research Associates in June 1997.

by the AIDS crisis, significant increases in legal protection for gays seem unlikely at this time (Conkle, 1987; Note, 1985).

Crimes Against Good Taste

Another broad category of sex offenses can be viewed as crimes against community standards of good taste and delicacy. In this area we find laws against **exhibitionism, voyeurism,** solicitation, disorderly conduct, being a public nuisance, and "general lewdness." These statutes are by and large quite vague and punish acts that are offensive, or *likely* to be offensive, to someone. Forty states have laws prohibiting intentional public nudity and exhibitionism. Twenty-three states have laws declaring sexual contact or activity in a public place a crime (Poser & Silbaugh, 1996). In all but two states these activities are misdemeanors. As will be seen below, unequal enforcement of these laws, their vagueness, and the difference between what is offensive and what is actually criminal make these statutes suspect.

Crimes Against Reproduction

As was noted in Focus 22.1, the Comstock laws included a ban on the giving of information concerning the prevention of conception. Comstock apparently regarded contraception and abortion as identical. These issues will be discussed more fully in the section on the right to privacy, below; here it will be mentioned only that until 1973 abortion was prohibited or severely limited in many jurisdictions, and contraception was prohibited in some. These laws are clear examples of the enshrinement in the statute books of the values of another day. They arise from an understanding of reproduction as the only legitimate purpose of sex and a belief in the necessity of vigorous propagation of the species. Such laws were overturned by Supreme Court action, but continuing agitation, at least in the case of abortion, ensures that public debate will endure for some time.

Criminal Commercial Sex

The law has also deemed it illegal to make money from sex, at least in certain circumstances. It is not illegal to sell products with subtle promises of sex-

Figure 22.3   The oldest profession. A prostitute solicits a potential customer.

ual fulfillment, but it is illegal actually to provide such fulfillment, either in direct form (that is, prostitution) or on paper, as in pornography. Both will be treated in greater detail below; first, however, the kinds of laws on these subjects should be noted.

Prostitution is the exchange of sex for money or other payment such as drugs. Except in Nevada, where counties may allow it, prostitution is illegal in every jurisdiction in the United States, though not in many other countries. In most states, prosti-

Exhibitionism: Showing one's genitals in a public place, to passersby; indecent exposure.
Voyeurism: Secretly watching people who are nude.

tution is a misdemeanor, punishable by a fine or jail sentence. The law also forbids activities related to it, such as solicitation, pandering (pimping, procuring), renting premises for prostitution, and enticing minors into prostitution (Perry, 1980). These activities are felonies in most states. Laws against vagrancy and loitering are also used against prostitutes. Many state laws provide the same penalty for patronizing a prostitute as for prostitution. However, clients are rarely charged. By one estimate, only 10 percent of the 100,000 prostitution-related arrests per year are of clients (Prostitutes Education Network, 1998). In general, all of these laws have proved very difficult to enforce, and so the "oldest profession" goes on unabated.

Obscenity will be discussed in more detail in a later section. Suffice it to say here that in most jurisdictions it is a crime to sell material or to present a play, film, or other live performance that is "obscene." That much is fairly simple. The real problem comes in deciding *what* exactly is obscene and how that will be determined without doing violence to the First Amendment's guarantee of freedom of the press. So far, no satisfactory answer has been found. Obscenity laws seem to have a twofold basis. First, they attempt to prevent the corruption of morals by materials that incite sexual thoughts and desires. Second, they attempt to ensure that no one will profit by the production and distribution of such materials. Whether either can be done, or is worth doing, is a question that will be taken up later in this chapter.

Sex-Law Enforcement

From the foregoing, it is clear that the law has intruded into areas that the reader may well have thought were his or her own business. We can now ask: How are sex laws enforced? The answer is simple: with *great* inconsistency. One authority estimated that "the enforcement rate of private consensual sex offenders must show incredibly heavy odds against arrest—perhaps one in ten million" (Packer, 1968, p. 304). The contrast between the number and severity of the laws themselves and the infrequency and capriciousness of their en-

forcement reflects society's ambivalence toward the whole subject.

This contrast leads to serious abuses and to demands for radical reform of sex laws. A summary of the arguments for reform will be presented in the next section. First, however, it should be noted that as long as the laws are on the books, the *threat* of prosecution, or even of arrest, can exact a great penalty from the "offender." Loss of job, reputation, friendship, family, and so on, can and does in some cases result from the sporadic enforcement of sex laws. For persons engaging in the prohibited acts, especially gays and lesbians, the threat of blackmail is ever present. Of course, for those actually convicted on "morals charges," the situation is even worse. That individuals should be subjected to such punishments for private acts is questionable.

Second, the uneven enforcement of sex laws may have a very bad effect on law enforcement generally. It invites arbitrary and unfair behavior and abuse of authority by police and prosecutors. With regard to prostitution, arrest practices seem highly discriminatory. First, 90 percent of all arrests are of prostitutes. Of these, 50 percent to 80 percent are of minority women, even though most prostitutes (considering all the types discussed in Chapter 18) are white. Finally, a large minority of those sentenced to jail are women of color; white women are more likely to be fined (Prostitutes Education Network, 1998). Another serious abuse is entrapment, in which an undercover police agent, posing, for example, as a gay man or as a prostitute's potential client, actually solicits the commission of a crime. Since a sexual act between consenting parties means that there is no one to report the act to the authorities, undercover agents must create the crime in order to achieve an arrest for it. Such entrapment hardly leads to respect for the law. Moreover, the knowledge that sex laws are violated with impunity creates a general disrespect for the law, particularly among those who know that they are, strictly speaking, "criminals" under it. If nothing else, the failure of Prohibition ought to demonstrate that outlawing activities of which a substantial

Obscenity: That which is offensive to decency or modesty, or calculated to arouse sexual excitement or lust.

proportion of the population approves is bad public policy. It may well be said that more violations of the public good result from the enforcement of sex laws than from the acts they seek to prevent.[2] Keeping this in mind, let us turn to the prospects for the future.

Trends in Sex-Law Reform

It is difficult to specify the number and details of sex laws because change is very rapid in this area. Our distaste for these attempts to regulate human behavior leads us to call this "reform," although many in our society would contend that the change is for the worse. Since sex is a topic heavily laden with values, such reform is not likely to be accomplished without a good deal of conflict. This makes it difficult to predict either the precise directions of change or its speed. However, some important legal principles are used to bring about changes in sex laws; these trends in reform are discussed below.

Efforts at Sex-Law Reform

Following a thorough review of legal practices in the United States, the American Law Institute's Model Penal Code recommended **decriminalization** of many kinds of sexual behavior previously outlawed. Under the section dealing with sexual offenses, it includes as reasonable for the law to regulate only rape, deviate sexual intercourse by force or imposition, corruption and seduction of minors, sexual assault, and indecent exposure (American Law Institute, 1962, Article 213). With the notable exception of prostitution, which it still makes illegal, the American Law Institute follows the principle that private sexual behavior between consenting adults is not really the law's business. The recommendation of the Model Penal Code has been followed by nearly half the states.

While there are exceptions, a state is more likely to reform its sex laws as part of a complete overhaul of its criminal code than it is to make specific repeal of such laws. The reason for this is political and is grounded in the distinction between legal-

ization and decriminalization. If legislators "legalize" unconventional sexual practices, people are likely to become upset and accuse the state of "condoning" them. It is therefore important to note that what is advocated is decriminalization; that is, ceasing to define certain acts as criminal or removing the penalties attached to them. Decriminalization is morally neutral; it neither approves nor disapproves, but simply revises the definitions.

Right to Privacy

A legal principle that has been very important in sex-law reform is the constitutional right to privacy. This has come into play chiefly in attacks on sex laws through the courts. Interestingly enough, although the right to privacy is invoked in connection with an amazing variety of matters—criminal records, credit bureaus and banks, school records, medical information, government files, wiretapping, and the 1974 amendment to the Freedom of Information Act (known as the Privacy Act), to name a few—the definitive articulation of the constitutional principle came in a sex-related case (Brent, 1976).

In 1965 the Supreme Court decided the case of *Griswold v. Connecticut* and invalidated a state law under which a physician was prosecuted for providing a married couple with information and medical advice concerning contraception. Justice William Douglas stated flatly that "we deal with a right of privacy older than the Bill of Rights, older than our political parties, older than our school system" (*Griswold v. Connecticut*, 1965, p. 486). The problem that Douglas, and the six justices who voted with him, faced was finding the specific provisions of the Constitution which guaranteed this right. Douglas found it not in any actual article of the Bill of Rights but in "penumbras, formed by emanations from those guarantees that help give them life and substance" (*Griswold v. Connecticut*, 1965, p. 484). Critics have found this a splendid example of constitutional double-talk, and the debate over the application of the right to privacy continues. Nonetheless, in invalidating the Con-

---

[2]For a good discussion of these and other arguments, see Packer (1968, pp. 301–306).

Decriminalization: Removing an act from those prohibited by law, ceasing to define it as a crime.

necticut law, the Court defined a constitutional right to privacy which was, in this instance, abridged when a married couple was denied access to information on contraception.

While the decision in the Griswold case declared the marriage bed an area of privacy, in the 1972 case of *Eisenstadt v. Baird*, the Court invalidated a Massachusetts law forbidding the dissemination of contraception information to the unmarried. In doing so, the Court stated that "if the right of privacy means anything, it is the right of the individual, married or single, to be free from unwarranted governmental intrusion into matters as fundamentally affecting a person as the decision whether to bear or beget a child" (*Eisenstadt v. Baird*, 1972, p. 453). Other decisions have established one's home as a protected sphere of privacy which the law cannot invade (Brent, 1976).

The right to privacy was also invoked by the Court in 1973 in one of its most controversial cases, *Roe v. Wade*, which invalidated laws prohibiting abortion. Suing under the assumed name of Jane Roe, a Texas resident argued that her state's law against abortion denied her a constitutional right. The Court agreed that "the right of personal privacy includes the abortion decision" (*Roe v. Wade*, 1973, p. 113). However, it held that such a right is not absolute and that the state has certain legitimate interests that it may preserve through law, such as the protection of a viable fetus. Nonetheless, the Court declared that a fetus is not a person and therefore is not entitled to constitutional protection. The effect of the *Roe* case, and of related litigation, was to invalidate most state laws against abortion. The Court limited second- and third-trimester abortions to reasons of maternal health but made a woman's right to a first-trimester abortion nearly absolute. However, the 1989 decision in *Webster v. Reproductive Health Services* and the 1992 decision in *Planned Parenthood v. Casey* changed the shape of abortion laws, as discussed later in this chapter.

The *Roe* decision was based on the right to privacy, but the Supreme Court has not chosen to extend the right of sexual privacy much further. The reasoning in *Bowers v. Hardwick* is interesting in this regard. Michael Hardwick was arrested for sodomy in his own bedroom. Though he was not prosecuted, he sued the state of Georgia, arguing that the existence of the sodomy law violated his right to privacy. The Court, in a 5-to-4 decision, ruled that the Constitution does not confer "a fundamental right upon homosexuals to engage in sodomy." Since a fundamental right was not being violated, the statute would be lawful if it rested on a rational basis, and the Court said that the "majority sentiments" of citizens of the state provide such a basis (Leonard, 1993). The *Hardwick* case suggests that the right to privacy is limited to the traditional categories of marriage, family, and procreation (*Leading Cases*, 1986; Green, 1989).

Equal Protection

Another important legal principle is the right to equal protection of the law. This right is guaranteed by the U.S. Constitution, in the fifth and fourteenth amendments. If law or government action results in disadvantage to some group, that group can seek relief, typically through court action.

Challenges to laws or policies that discriminate against gays, lesbians, prostitutes, and other groups distinguished by sexual conduct have been based on this principle. A series of cases have been brought by gays and lesbians against actions by or policies of the U.S. military, such as policies requiring the discharge of lesbians or gays from the armed forces. Some of these cases have been decided in favor of the plaintiffs, while others have not.

Discrimination against gays and lesbians in employment and in other arenas has also been challenged in court. The most celebrated case is *Evans v. Romer*. In 1992, Amendment 2 to the Constitution of the State of Colorado was put on the ballot; it prohibited the enactment of antidiscrimination laws or policies favoring homosexuals. In November 1992, the amendment was adopted with 54 percent of the votes cast. Nine individuals, three cities, and a school district filed a lawsuit, asking that Amendment 2 be overturned because it violated the equal protection clause of the U.S. Constitution. After a series of decisions by District Judge Jeffrey Bayless and the Colorado Supreme Court, the United States Supreme Court agreed to review Amendment 2. In 1996, the Supreme Court declared Amendment 2 unconstitutional (517 U.S. 620). That is, states may not prohibit antidiscrimination laws.

The equal protection clause is also the basis for challenges to bans on same-gender marriages. The so-called "Hawaiian case," *Baehr v. Lewin*, involves a suit by two women who claim that the state's refusal to marry them constitutes discrimination on the basis of gender. Both the Hawaii Supreme Court and a Circuit Judge have ruled in favor of the two women. But the case triggered a strong backlash. In 1996, President Clinton signed a law which states that the United States government will not recognize same-sex marriages. At least 16 states have passed laws banning such unions.

Victimless Crimes

In the past three decades, a great deal of legislative change has taken place involving the principle of "victimless crimes"—a concept that has broad applicability beyond sexual behavior. It is asserted that when an act does no legal harm to anyone or does not provide a demonstrable victim, it cannot reasonably be defined as a crime. The thrust of the argument is well articulated by former University of Chicago Law School Dean Norval Morris:

> Most of our legislation concerning drunkenness, narcotics, gambling and sexual behavior is wholly misguided. It is based on an exaggerated conception of the capacity of the criminal law to influence men and, ironically, on a simultaneous belief in the limited capacity of men to govern themselves. We incur enormous collateral costs for that exaggeration and we overload our criminal justice system to a degree that renders it grossly defective where we really need protection—from violence and depredations on our property. But in attempting to remedy this situation, we should not substitute a mindless "legalization" of what we now proscribe as crime. Instead, regulatory programs, backed up by criminal sanctions, must take the

Figure 22.4   A cartoonist's view of victimless crime. Reprinted with special permission of King Features Syndicate.

"Be right with you, ma'am, soon as I've brought these lawbreakers to justice."

place of our present unenforceable, crime-breeding and corrupting prohibitions. (1973, p. 11)

The victimless-crime argument should appeal not only to the public's sense of privacy but also to its pocketbooks. Crimes in which there is no readily identifiable victim account for over half the cases handled by U.S. courts (Boruchowitz, 1973). If the court dockets could be cleared and law enforcement officers reassigned, protection against violent crimes would be rendered more efficient and less expensive.

The application of the principle to some of the issues discussed above should be obvious. A sexual act performed by consenting adults produces no legal harm, and neither of the participants are victims. The only conceivable end served by criminalizing such an act is the protection of "public morals," which, in a society with many values, seems an end not worth the cost, if it is even achievable. If sodomy laws, for example, are removed by a state legislature, the victimless-crime argument is likely to be used, by itself or in combination with an appeal to the right to privacy.

The most common reference to the decriminalization of victimless acts is with respect to prostitution. Police efforts at curbing the "oldest profession" seem to be ineffective, open to corruption and questionable practices, and tremendously expensive. Since prosecution is normally of the prostitute and not her customer, there seems to be a clear pattern of discrimination against women which violates the constitutional principle of equal protection. Finally, as all manner of adult consensual behavior is decriminalized, the legitimacy of distinguishing between commercial and noncommercial consensual sex has been questioned (Parnas, 1981). It has been suggested that much of the demonstrable harm associated with prostitution, such as the committing of robbery and other crimes by prostitutes and pimps and the connections with organized crime, has resulted *because* the practice is illegal (Caughey, 1974). Thus, it has been argued that if prostitution were no longer defined as a crime, all would benefit—the prostitute, her patron, the police, and society at large (Parnas, 1981; Rosenbleet & Pariente, 1973).

San Francisco has a long tradition of prostitution, dating back to the 1860s. It also has thousands of people employed in commercial sexual activity.

In March 1994, the Board of Supervisors created a Task Force on Prostitution. After 18 months of hearings and study, the Task Force concluded:

> not only are the current responses ineffective, they are also harmful. They marginalize and victimize prostitutes, making it more difficult for those who want to get out of the industry and more difficult for those who remain in prostitution to claim their civil and human rights. (San Francisco Task Force on Prostitution, 1996)

The Task Force recommended:

> that the City departments stop enforcing and prosecuting prostitution crimes. It further recommends that the departments instead focus on the quality of life infractions about which the neighborhoods complain and redirect funds from prosecution, public defense, court time, legal system overhead, and incarceration towards services and alternatives for needy constituencies.

The argument against the criminalization of prostitution assumes that if prostitution were legal it could be regulated and the problems of crime, public offense, and the spread of sexually transmitted diseases associated with it might be avoided. In this case, then, there would be no victims and no societal need to ban the practice. However, this argument can be countered by suggestions that prostitution may indeed have victims. First, as AIDS spreads to the heterosexual population, prostitutes may be carriers. Second, a feminist analysis of prostitution suggests that the prostitute herself may be a victim of the profession. Some assert that prostitution is inherently degrading and that many prostitutes would welcome alternative means of earning a living.

Other activities by consenting adults may also harm other persons. Adultery, for example, may damage one or two marriages, harming the spouses and children of the participants. In general, the "victimless crime" argument is becoming less persuasive.

The Problem of Obscenity and Pornography

Among the most controversial topics in the area of sexual regulation is obscenity and pornography. A substantial portion of the American populace apparently finds pornography offensive and

Figure 22.5   The production and sale of pornography have become a major commercial enterprise. Most large cities have areas in which such businesses are concentrated, as shown in this photograph taken in San Francisco.

wishes it suppressed. Many others do not share this view and find any form of censorship outrageous and unconstitutional. Antivice crusaders consider "smut" dangerous to the average citizen. Legislators are swamped with demands that something be done, while the courts have labored unsuccessfully for years to balance the First Amendment right of freedom of speech with the desire of some to outlaw, or at least regulate, pornography.

The discussion will begin with a problem of definition. Here it is helpful to distinguish between "pornography" as a popular term and "obscenity" as a legal concept. "Pornography" comes from the Greek work *porneia,* which means, quite simply, "prostitution," and *graphos,* which means "writing." In general usage today, it refers to literature, art, films, speech, and so on, that are intended to be sexually arousing, or presumed to be arousing, in nature. Pornography

may be "soft-core," that is, suggestive, or "hard-core," which usually means that there is explicit depiction of some sort of sexual activity. Pornography, as such, has never been illegal, but obscenity is. The word "obscene" refers to that which is foul, disgusting, or lewd, and it is used as a legal term for that which is offensive to the authorities or to society (Wilson, 1973).

Obscenity has been a legal issue ever since the Supreme Court decided the Roth case in 1957. The Court explicitly stated that obscenity was not protected by the First Amendment—which guarantees freedom of speech and the press and which, by long-recognized extensions, includes films, pictures, literature, and other forms of artistic expression. However, it also ruled that not all sexual expression is obscene, defining obscenity as material "which deals with sex in a manner appealing to prurient interest" (*U.S. v. Roth,* 1957, p. 487). The Roth decision evoked much controversy, both from those who thought the Court had opened the floodgates of pornography and from civil libertarians who found the definition too restrictive. The Court continued to try to refine the test for obscenity. In the 1966 *Memoirs* case, obscenity was additionally defined as that which is "utterly without redeeming social value." In the *Ginzburg* decision the same year, the Court upheld the obscenity conviction of a publisher for "pandering" in his advertising; that is, flagrantly exploiting the sexually arousing nature of his publication. However, none of these tests was persuasive to more than five members of the Court, much less to the public at large.[3]

The current standard definition of obscenity by the Supreme Court came in the 1973 case of *Miller v. California*. Rejecting the "utterly without redeeming social value" test, Chief Justice Burger and the four justices concurring with him proposed the following definition:

> (a) whether "the average person, applying contemporary community standards," would find that the work, taken as a whole, appeals to the prurient interest, (b) whether the work depicts or describes, in a patently offensive way, sexual conduct specifically defined by the applicable state law, and (c) whether the work, taken as a whole, lacks serious literary, artistic, political or scientific value. (*Miller v. California*, 1973, p. 24)

The goals of this decision seem to be to define as obscenity "hard-core pornography" in the popular sense, to require from state statutes precise descriptions of that which is to be outlawed, and to give governments more power to regulate it (Gruntz, 1974). The notable problem with the *Miller* test, at least for civil libertarians, is the "contemporary community standards" provision. This allows the local community to determine what is obscene, rather than using national norms, making it impossible to predict what a given jury in a particular town might find obscene.

One important factor that has affected the law on pornography is the extent to which legislators, law enforcement officers, and courts believe that it causes harm to the general population. For evidence, they have often turned to social science and found a mixture of data and conclusions. The Commission on Pornography appointed by Attorney General Edwin Meese, in its 1986 *Final Report*, using personal testimony and scientific studies, concluded that pornography is harmful and linked it to the abuse of children and women. The commission's controversial recommendations affirmed the existing antiobscenity statutes and urged their vigorous enforcement. It also seemed to approve other strategies for combating pornography (U.S. Department of Justice, 1986).

The least controversial of these strategies is concerned with the problem of child pornography, which is widespread and damaging to the young people involved in it (Burgess, 1984). In the 1982 case of *New York v. Ferber*, the U.S. Supreme Court ruled unanimously that child pornography, whether or not it is obscene under the prevailing legal standards, is not protected by the Constitution. This decision gave states broader latitude in legislation, based on the government's obligation to protect children from abuse (Shewaga, 1983). The Court required states to be precise about whether they were prohibiting the production, processing, or distribution of child pornography, and to develop clear definitions of what was to be outlawed. This has proved difficult, but efforts continue. Another approach to the problem is to make tougher and more precise laws against child sexual abuse, without which "kiddie porn" could not be made. This approach would criminalize the very production of such material and sidestep the more complicated constitutional issue of distribution and sale (Shouvlin, 1981).

Far more problematic is a strategy advocated by some feminists which seeks to define pornography as inherently discriminatory against women, and thus a matter of civil rights. This approach takes the results of studies, such as those cited in Chapter 18, and feminist analysis as a basis for arguing that pornography, especially the more violent kind, is indeed harmful to women. Writer Andrea Dworkin and legal scholar Catherine MacKinnon drafted a model city ordinance which defines pornography as "the graphic sexually explicit subordination of women through pictures and/or words," and goes on to describe this subordination

---

[3]The frustration of defining obscenity was perhaps best expressed by Justice Potter Stewart in *Jacobelis v. Ohio* (1964): "I shall not further attempt to define [hard-core pornography], and perhaps I could not ever succeed in intelligibly doing so. But I know it when I see it" (378 U.S. 197).

(Dworkin, 1985, p. 25). A version of this ordinance was adopted in 1982 in Minneapolis amid much controversy but was vetoed by the mayor. Subsequently, it was adopted by Indianapolis and tested in the courts in the case of *American Booksellers Association v. Hudnut* (475 U.S. 1001, 1986). Feminists filed friend-of-the-court briefs on both sides of the issue. Both the Federal District Court and the U.S. Court of Appeals rejected the ordinance on First Amendment grounds. While recognizing the possible harm to women in pornography, the courts would not accept limitation on free, constitutionally permitted expression as an allowable remedy. In 1986, the Supreme Court affirmed the lower-court opinions without comment, and further efforts to fight pornography in this manner have not been successful (Brest & Vandenberg, 1987; Benson, 1986).

A more fruitful approach for those who oppose pornography has been the attempt to regulate, or eliminate, its sale through zoning. The city of Renton, Washington, in an effort to keep adult movie theaters out of the community, passed an ordinance forbidding adult film theaters within 1000 feet of any residential zone, single- or multiple-family dwelling, church, park, or school. Citing precedents and a municipality's right to prevent crime and protect property values, the Supreme Court upheld the Renton ordinance in 1986. Such zoning laws are content-neutral and thus can avoid First Amendment issues. National City, California, enacted a very restrictive zoning ordinance, which has been upheld by the California Supreme Court (Ragona, 1993).

The Supreme Court has not abandoned the *Miller* "community standards" test, but it seems likely that various attempts will continue to be made to reduce or eliminate the availability of what some see as harmful pornography. The complexity of the legal issues and the continuing debate over what is appropriate for Americans to read and view will undoubtedly keep the matter of pornography, obscenity, and erotica controversial for some time to come. While it is difficult to define obscenity, the courts have consistently ruled that states and local communities have the right to regulate public conduct and public artistic expression (Leonard, 1993).

Contemporary controversy about pornography is centered on the Internet. We described in Chapter 18 the various sex-oriented goods and services available on-line to anyone with a computer and a modem. In the past, a potential user of X-rated books, photographs, videos, or CD-ROMs had to purchase them through a retail store or a mail-order firm. These businesses could, at least in theory, control who purchased these items. Specifically, they could require proof of age, thereby preventing minors from gaining access to such material. There is no person who serves as a gatekeeper and controls access to these items on the Internet.

Existing laws governing the production, distribution, sale, and possession of kiddie porn apply to the Internet, and some law enforcement agencies seek out and arrest offenders. More problematic are materials featuring persons over 18. While adult websites require users to click on a statement certifying that they are 18 or older in order to gain access, there is no practical way to enforce this restriction. Once the user gains access, many adult websites provide free viewing of pornographic photos and stories; users may download the photographs and print the stories. In response to this situation, the U.S. Congress passed the Communications Decency Act in 1996 which made it illegal to distribute via the Internet "indecent material" which a child could access. Several groups immediately challenged the constitutionality of the Act, claiming that it violated individuals' First Amendment rights. The U.S. Supreme Court declared the law unconstitutional in 1997.

At least three states have passed laws banning all state employees (including those who teach and do research in human sexuality) from using government computers to access sexually explicit materials. A number of corporations and many school districts have adopted similar policies.

The Controversy over Reproductive Freedom

An even more convulsive controversy is to be found in the matter of abortion. Although the Supreme Court's decision in the *Roe* case was quite clear, it has been under continuous attack since it was handed down in 1973. Opposition comes from a broad coalition of antiabortion groups that prefer to call themselves "prolife" and include the Roman Catholic Church, Evangelical Protestants, various "New Right" organizations, and the Republican Party. The controversy has been carried on in recent elections, in the courts, in state legislatures, and in Congress. The prolife movement is well organized and well financed and has proved to be an effective lobbying force, instrumental in the defeat of a number of legislators who have not supported the antiabortion cause. Those seeking to preserve the right of women to legal abortions, who call themselves "prochoice," have also organized and been effective as well. In the 1980s and 1990s the prolife movement has used five basic strategies to eliminate or reduce abortions: funding restrictions, parental consent and spousal notification requirements, procedural requirements, the Human Life Amendment, and disruptive action against abortion providers.

The most notable example of funding restrictions is the Hyde amendment, annually proposed by Congressman Henry Hyde (no relation to the coauthor of this text). This is a rider to the appropriations bills forbidding the expenditure of any federal money for abortions under most circumstances. In various years, this restriction has been applied to the Departments of Health and Human Services, Defense, Justice, and the Treasury (Wilcox, Robbennolt, & O'Keeffe, 1998). This approach was ruled constitutional in the 1980 case of *Harris v. McRae* (440 U.S. 297, 1980), and subsequently poor women have been denied this means of obtaining abortions (Milbauer, 1983). Various states have introduced similar restrictions on state money that has been used to cover the gap in federal funding for abortions. Another strategy involves regulations in Title X of the Public Health Act denying funds to organizations, such as family planning agencies, which make referrals for abortions and granting funds to organizations that oppose abortions (Paul & Klassel, 1987).

The second strategy to make abortion more difficult to obtain has been to restrict abortion by requiring *parental consent* for a minor to have an abortion or *notification* of the husband of a married woman seeking an abortion. In the 1992 case of *Planned Parenthood v. Casey* (112 S.Ct. 2791, 120 L.Ed. 674), the Supreme Court ruled that states could require parental consent for unmarried girls under 18 seeking abortions. However, it struck down the requirement that married women must notify their husbands.

A third strategy to restrict abortion by making it more difficult or unpleasant is to require that women seeking an abortion be given certain information, for example, that they be informed about fetal development or the medical or psychological consequences of abortion. Some laws along these lines have specified information that is reasonably accurate scientifically, whereas others specify information that is propaganda with little scientific basis. In 1983 the Supreme Court struck down an Akron, Ohio, ordinance which required information of the propaganda variety (Fox, 1983). However, in the 1992 *Planned Parenthood v. Casey* decision, the Supreme Court upheld Pennsylvania's requirement that women seeking abortion be informed about fetal development during the three trimesters of pregnancy and the possible viability of fetuses during the third trimester.

The waning of the Supreme Court majority recognizing a woman's right to an abortion became clear in the important 1989 decision *Webster v. Reproductive Health Services*. The case concerned a Missouri law that (a) prohibited state employees from assisting in abortions and prohibited abortions from being performed in state-owned hospitals, and (b) banned abortions of "viable" fetuses. The Court upheld the Missouri law. Thus it essentially said that states may pass certain kinds of laws regulating abortions. For example, states may require doctors to perform a "viability test" on any fetus if a woman is believed to be 20 weeks or more pregnant and may make it illegal to perform an abortion if the test shows that the fetus could live (such tests are not 100 percent accurate). Ironically, abortions of genetically defective fetuses might be prevented by this ruling because amniocentesis does not provide results until the eighteenth or nineteenth week of pregnancy. The broader effect of the Court's decision, however, is to throw the hot potato back to state legislatures.

By 1992 the membership of the Supreme Court had shifted to a majority of conservatives as a result of appointments made during the Reagan and Bush administrations. As mentioned previously, the *Planned Parenthood v. Casey* case concerned a Pennsylvania law that placed many procedural obstacles in the way of a woman seeking abortion. She had to be given information about fetal development and then wait 24 hours before the abortion could be performed. Unmarried girls under 18 had to get the consent of at least one parent, or a state judge had to rule that she was mature enough to

Figure 22.6 The murder in 1998 of Dr. Slepian, a physician who performed abortions in a Buffalo, NY, clinic, reflects the increased violence associated with one wing of the antiabortion movement.

make the decision herself (called a judicial bypass). Married women had to notify their husbands before obtaining an abortion. The Court's decision was complex. It did not overturn *Roe v. Wade*, as many had expected. Instead, it reaffirmed a woman's constitutional right to an abortion before the fetus is viable. On the other hand, it upheld all the restrictions in the Pennsylvania law except the one requiring a married woman to notify her husband. The Court was again ruling that states could pass laws placing restrictions on abortion, although abortion itself could not be outlawed, and the laws could not place "undue burden" on a woman seeking an abortion.

In summary, then, *Roe v. Wade* (1973) decriminalized abortion and said that states could not restrict access to first-trimester abortion. Two decisions have chipped away at *Roe v. Wade* without overturning it completely. In *Webster v. Reproductive Health Services* (1989) the Court said that states could restrict abortion in some ways, namely, by forbidding it in state-owned hospitals and by banning abortion of viable fetuses. In *Planned Parenthood v. Casey* (1992) the Court again upheld a state law placing restrictions on abortion, involving parental consent, information, and a waiting period. However, requiring a woman to notify her husband was going too far, according to the decision, and laws could not place an undue burden on women seeking an abortion.

A fourth strategy used by opponents of abortion has been to champion the Human Life Amendment to the Constitution. This amendment, which would prohibit all abortions, was defeated in the Senate in 1983, falling 18 votes short of the necessary two-thirds majority. This amendment or similar legislation has been introduced in every session of Congress since 1983, but no serious effort has been made to bring it to a vote (Wilcox et al., 1998). Legislation designed to restrict access to abortion in various ways also has been introduced in each session. One example are bills prohibiting the use of specific procedures, such as so-called partial birth abortion.

A fifth strategy adopted by members of Operation Rescue and other prolife activists involves disruption, such as picketing and engaging in acts of civil disobedience outside abortion clinics, Planned Parenthood facilities, or the homes of physicians who perform abortions. Some individuals, including members of Defensive Action and the Army of God, have engaged in arson and bombings of clinics where abortions are performed. There have been at least 153 actual or attempted cases of arson and bombings, causing damages of $13 million ("Abortion: Who's Behind the Violence?" 1994). In 1994, Paul Hill, the director of Defensive Action, shot and killed Dr. John Britton, 69, and James Barrett, 74, outside a clinic in Pensacola, Florida. The Army of God has published a handbook on how to conduct violent protests. In response to these incidents, the U.S. Congress passed the Freedom of Access to Clinic Entrances Act in 1994. This law makes it a federal crime to use violence or the threat of violence to interfere with access to a reproductive services provider. The Supreme Court ruled that the act is constitutional in October 1995.

Ethnicity and Sex Laws

Although the Constitution promises equal protection to people of all races, in practice people of color and low-income people are often at a disadvantage, and this is no less true in the area of sexuality than elsewhere. Here we will consider abortion as an example (Nsiah-Jefferson, 1989; Roberts, 1993).

Little information exists on the abortion or the reproduction-related needs of women of color. Until 1990, abortion statistics were published for

Figure 22.7   It is important that women of color have the same access to abortion as middle-class white women.

only two categories of American women: white and black. Data are now available on Hispanic women but not on Native American or Asian American women. Given the different cultural heritages of these ethnic groups, abortion undoubtedly has different meanings for women in these groups; yet we lack data on the specifics.

Although for decades white women have had some control over their reproduction, for many women of color this is a new step. They may be wary because of the history of negative experiences of women of color in this area, such as the experimental work on the introduction of the birth control pill, which was done with poor women in Puerto Rico. Therefore, women of color need far more access to information and education, and it is critical that this information be sensitive to their cultural heritage.

Women of color are more likely to have abortions than are white women. Of the reported abortions performed in 1995, 44.1 percent were for white women (who make up 80 percent of the population), 35 percent were for black women (12 percent of the population), and 15.4 percent were for Latinas (9 percent of the population) (Centers for Disease Control and Prevention, 1997). A significantly higher percentage of women of color obtain abortions after the first trimester than do white women (Nsiah-Jefferson, 1989). Statistics indicate that 10.2 percent of all abortions obtained by white women are done after the first trimester (that is, 89.8 percent were done in the first trimester). By comparison, 14.3 percent of all abortions obtained by nonwhite women are done after the first trimester (Koonin, Smith, & Ramick, 1993). It seems clear that the Hyde Amendment, which prohibits the use of federal Medicaid funds to pay for abortions for low-income women, is a key factor in this difference. Women of color, who are disproportionately represented in the low-income group, must often spend a good deal of time raising the funds for an abortion because Medicaid is denied them, and this process of raising funds delays the abortion until the second trimester.

In some cases, women of color may have significantly less access to abortion (Nsiah-Jefferson, 1989). For example, Native American women living on reservations are denied federal funding for abortions and, to make matters worse, no Indian Health Service clinics or hospitals may perform abortions even when paid for with private funds. Native American women can therefore be literally hundreds of miles away from access to an abortion. The number of abortion providers in the United States has been slowly declining since 1982 (Henshaw & Van Vort, 1994). Eighty-four percent of U.S. counties have no known provider, and 33 percent of metropolitan areas have no provider or one who reports fewer than 50 procedures per year. This means that, for millions of women, access to an abortion requires travel to another city, county, or state. Poor women, who are disproportionately women of color, are less likely to be able to afford such travel.

Abortion is only one example of ways in which people of color are disadvantaged under the present system of sex laws. This example illustrates clearly that efforts at sex-law reform need to include a consideration of the impact on people of color.

Sex and the Law in the Future

Nothing seems riskier than to predict with any degree of confidence how society's views of sex, and the laws that express those views, will develop and change. Thus, any look ahead is at best a guess about what might happen, based on what has happened. Unforeseen events have a way of upsetting our calculations and introducing new variables into the mix. For example, the first edition of this text (1978) predicted that sex-law reform would go on as it had in the 1970s, with the extension of the right to privacy and wide use of the "victimless-crime argument" for decriminalization of various sexual practices. The election of 1980 in which Ronald Reagan became president, the rise of the New Right, and the increasing appeal of its "social agenda" rendered that prophecy wrong. Likewise, the third edition (1986) was written without any attention to the complex legal issues posed by AIDS and the new reproductive technologies. As we focus our attention on these issues, you might well speculate on what will need to be included in the next edition.

Sex-Law Reform and Backlash

At this writing, it appears that the movement toward more permissive sex laws, which probably had its roots in the civil rights movement, the sexual revolution, and feminism, has achieved virtually all the gains it is likely to for the time being. The Supreme Court has limited the right to privacy in sexual matters. The decriminalization of sex offenses had essentially ceased by 1980. New strategies for combating pornography have been tried, and some have succeeded. The New Right and its allies in the conservative evangelical sector of the religious community have, it would seem, taken the initiative away from those who favor less restrictive sex laws. The latter now find themselves seeking to defend and preserve gains already made, rather than extending them.

While the "profamily" coalition has by no means attained all its goals, it has certainly stopped sex-law reform and may be on the way to reversing it. The mood of the nation in the 1990s was far less hospitable to such reform and to the extension of legal protection to sexual minorities. Above all, fear of AIDS (see Focus 22.2) may be the motivating force behind increasingly restrictive laws regulating sexual behavior. This would be very much in the tradition of the American legal system, which has always tended to reflect both the dominant values and the conflicts of the society.

The Legal Challenge of New Reproductive Technologies

Very complex legal questions are being raised by the proliferation of techniques enabling previously infertile people, and others, to have children. These include artificial insemination, in vitro fertilization (IVF), surrogate motherhood, and various kinds of embryo fertilization and transfer. There are few state laws on the subject, with the exception of statutes regulating artificial insemination in 27 states and laws in all jurisdictions against trafficking in children ("baby buying"), which may impinge on surrogacy. There are likewise few federal standards other than regulations on IVF experimentation (Taub, 1987; Shapiro, 1986). However, the issues must be addressed, and very soon.

A fundamental difficulty is that these technologies bring a very public quality to what has always been one of the most private of all human activities, the conception of children. Some of them involve a third party, as a donor of sperm, egg, embryo, or uterus, in what had been a matter solely between a man and a woman. Even the nomenclature is complicated. In this section we adopt the convention of designating as "parent(s)" the person(s) who rear the child and accept legal responsibility for her or him. Those who provide some necessary aspect of the process we call "donors," or in the special case of women in whose wombs embryos are implanted, "surrogates." In the absence of clear legislation or case law, we are mostly able to point to the questions raised.

Perhaps the foremost question is whether or not there is a fundamental right to reproduce (Strong, 1997). If there is, it is hard to argue against the use of any appropriate technique to achieve that end, including third-party participation, or what some call "collaborative conception." On the other hand, if there is no such fundamental right, it may be reasonable to limit or even prohibit the use of such technology. Unfortunately, this question cannot be answered definitively at present. There is a well-established right *not* to conceive under the right to privacy, but the converse has not been established. Since our ancestors did not face this problem, their legacy is not much help.

Closely related is the question of the legal status of an embryo, since in several of these techniques fertilization and conception take place outside the uterus. Those who assert that life begins at conception would accord full personhood and legal rights to the embryo from the first cell division. Many states have enacted laws aimed at protecting embryos. For example, a Louisiana law specifies that an embryo, even if outside anyone's body, is a person and shall not intentionally be destroyed (Andrews, 1989). At the other extreme are those who regard an embryo merely as tissue, to be disposed of at will. This view puts no limits on reproductive technologies. A moderate view sees the embryo as less than a person but more than mere tissue and argues that it should be treated with the respect due potential human life. This view would seem to lead to some regulation of reproductive technology short of prohibition (Robertson, 1986). The question of what can be done with embryos not implanted also arises and is related. An embryo's status—even in the maternal womb, much less outside it—is a question the Supreme Court has explicitly avoided considering in its abortion decisions.

Another complex of questions is to be found in the matter of kinship, parental rights, and responsibilities. When a child is born as a result of these techniques, who exactly are the parents? One commentator notes that there can be five: an egg donor, a sperm donor, the donor of a uterus for all or part of the gestation, and the couple who rear the child (Shapiro, 1986, p. 54). This is a particularly thorny question in the case of a surrogate mother who has donated her egg and her uterus for nine months and then is required to give up the child. Do donors have any legal claim to contact with "their" children after birth?

Focus 22.2
AIDS and the Law

The spread of AIDS (see Chapter 20) has created some challenging legal issues. None of the choices facing legislators and judges is easy. Any reflection on AIDS and the law must take into account two factors, both of which render the issue more complex and emotional. First, AIDS is identified, both statistically and in the popular mind, with outsiders, persons who are already disliked, feared, and discriminated against: gay men and intravenous drug users. Second, at this writing no vaccine or cure for AIDS exists. This combination of factors has led some commentators to write of the "two epidemics": the social epidemic of fear of AIDS and the medical epidemic of the disease.

The existence of the two epidemics makes the legal balancing act even more difficult. Government has two very important responsibilities with respect to HIV infection and those who suffer from it. On the one hand, the state is obligated to protect individual rights and defend its citizens from discrimination and injustice. On the other hand, it is equally obligated to protect the health and welfare of the population. AIDS is a tragic case in which these two obligations are in severe conflict. This may be why the Federal government has failed to develop comprehensive plans for research and treatment of HIV and AIDS. It is difficult to predict how government will resolve this tension, although history suggests that it is more likely to emphasize the general welfare than individual rights, especially in view of the epidemic of fear.

There is significant case law, much of it from the early twentieth century, when cities were swept with epidemics of tuberculosis and other infectious diseases, affirming the right and the responsibility of the state to protect the populace. As one scholar notes, "Courts have traditionally deferred to public health authorities in their struggle to control epidemics, even when these efforts infringed on the constitutional rights of individual citizens" (Nanula, 1987, p. 330). Measures approved include the reporting of cases to local public health officials, mass vaccination programs, quarantine, and other restrictions on known or suspected disease carriers. At the same time, public health officials must demonstrate that any measure is directly aimed at the disease in question and is not arbitrary, capricious, or oppressive (Lazzo & McElgunn, 1986).

There has been much debate about the mandatory testing of certain populations. Testing identifies those who are infected, and it is the gateway to treatment. So far, testing is legally required of all military recruits, Job Corps entrants, would-be immigrants, and, in some states, prison inmates. None of these requirements has been invalidated by the courts. There are proposals to test other high-risk groups, such as people with STDs or all pregnant women. The issue is especially important in the case of pregnant women, because the baby of an HIV-positive woman has a 33 percent probability of being born infected. There is now a treatment regimen which, if started early enough in pregnancy, reduces the infant's risk by two-thirds. However, some fear that widespread mandatory testing—of, for example, all fertile women or all who seek treatment for other STDs—will only serve to keep away those who most need testing and treatment, or drive them underground (Nanula, 1987).

Public health authorities in some cities have closed gay bathhouses and bars where sexual activity was reportedly occurring. Some of these closings have been very unpopular in the gay community. Many other traditional public health measures, such as mass testing in public places and the quarantine of infected persons, are rendered suspect by the number of persons thought to

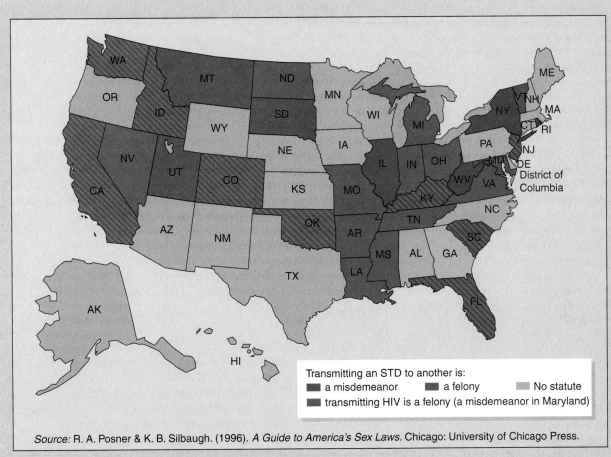

Transmitting an STD to another is:
▮ a misdemeanor ▮ a felony ▮ No statute
▮ transmitting HIV is a felony (a misdemeanor in Maryland)

Source: R. A. Posner & K. B. Silbaugh. (1996). *A Guide to America's Sex Laws*. Chicago: University of Chicago Press.

Figure 22.8 State laws governing disease transmission.

be infected (1 million to 2 million in the United States) and by the relatively restricted ways in which the virus is transmitted.

Protection of individual rights poses a thorny problem. To have AIDS, or even to be thought to have the disease, has caused many individuals to be fired from their jobs, divorced parents to lose custody or visitation rights, people to lose health insurance coverage or even to be deprived of medical care, and children to be barred from attending school; these people have also endured all sorts of informal harassment and discrimination. However, federal regulations, laws in some states, and ordinances in some counties and municipalities

prohibit discrimination against those who have or are thought to have AIDS.

A major issue concerning AIDS is the confidentiality of test results and medical records. Most states have passed laws protecting the confidentiality of antibody test records. Persons with positive tests have a reasonable fear of losing their jobs, and some insurance companies have made efforts to find ways of denying coverage to individuals who are thought to carry the virus, and even to all members of high-risk groups (Schatz, 1987). Some legal scholars argue that the constitutional right to privacy should afford protection to individuals infected with HIV. Still, positive test results need to be disclosed to medical care providers so that they can provide the necessary treatment. In one-half of the states, physicians have the discretion to notify sexual partners of those who test positive for HIV (Burris, 1993). Such contact tracing and partner notification have been important public health tools in dealing with other STDs.

Individuals who experience discrimination because they are HIV-positive or have AIDS can seek protection under several federal laws. The Vocational Rehabilitation Act of 1973 (29 U.S.C.S., 701, *et seq*) prohibits discrimination against disabled persons in employment, transportation, and access to public and health services. The Fair Housing Act of 1988 (42 U.S.C.S., 3601, *et seq*) prohibits discrimination in housing practices against disabled individuals and their families. The Americans with Disabilities Act of 1990 (42 U.S.C.S., 12101, *et seq*) prohibits discrimination in employment and places of public accommodation on the basis of disability. In June 1998, the U.S. Supreme Court ruled that HIV infection is a disability covered by these statutes, even if the person is asymptomatic.

Several hundred people have been prosecuted for putting others at risk of HIV infection. The laws of the states are portrayed in Figure 22.8. About one-half of these cases involve prisoners spitting at or biting law enforcement officers. Such cases usually result in conviction and a severe penalty. The other cases involve people who fail to tell sex partners that they are HIV-positive or have AIDS. The charge may be battery, assault, or attempted murder. In order to obtain a conviction, prosecutors must show that harm occurred, that the accused person caused the harm, and that the accused person intended to cause it. In practice, it has been difficult to prove all three, so such cases are less likely to result in conviction (Dalton, 1993). In New York, Nushawn Reynolds, a 20-year-old drifter, had unprotected sex with dozens of women after he tested positive for HIV. He was charged with sexual assault. In Tennessee, 30-year-old Pamela Wiser was charged with two counts of criminal exposure to HIV, a felony under state law. She initially claimed to have had unprotected sex with 50 men, but later said the number was only five. If convicted, she could be sentenced to 12 years in prison.

In the future, legislatures and courts will struggle to resolve a host of questions, including these: Who has a right to know that a person has tested positive for HIV? Should there be registries of infected persons, and who should have access to such registries? Should people be held legally liable if they knowingly (or even unknowingly) infect another person with the disease? Who is going to pay the enormous costs of this disease? How are the rights of the public and of individuals to be balanced and protected?

Sources: Burris (1993); Dalton (1993); Dalton & Burris (1987); Dennis (1993); Nanula (1987); Schatz (1987).

Surrogacy and surrogacy contracts raise particularly complex issues, especially if the woman has contributed not only her uterus but an egg as well. A man who donates his sperm to a sperm bank renounces any right to further contact with children who may be conceived; however, no equivalent principles have been established for women who donate eggs or a uterus. Some argue that state laws against trafficking in children ("buying babies") make surrogacy contracts, especially surrogacy for pay, illegal. Others have raised the concern that commercializing reproduction is inherently corrupting and should be prohibited because it turns children into commodities. There is a concern that surrogacy leads to the exploitation of low-income women, who will "rent" their wombs because they need the money, despite the possible psychological stresses and health risks of surrogate motherhood (Andrews, 1989; Taub, 1987; O'Brien, 1986).

Taking a broad perspective, there are a number of alternatives for legal approaches to the issue of surrogacy (Andrews, 1989). The most restrictive one is to outlaw all surrogacy contracts for pay, as Michigan did in 1988. A second alternative would be to have courts scrutinize all surrogacy contracts to ensure, for example, that the woman has not been coerced. Some proposed laws require that a mental health professional interview all participants—surrogate, egg donor, sperm donor, potential rearing parents—to be sure that all are truly

Figure 22.9 The law has not yet dealt adequately with the consequences of new reproductive technologies, as this cartoon indicates. "Solomon Updated" by Signe Hutkinson. Reprinted by permission of Cartoonists & Writers Syndicate.

giving informed consent. Another alternative, to address the concern that the surrogate may be exploited, is to require that surrogates must have their own legal counsel when they enter into a contract. Some suggest that all surrogacy arrangements and contracts should be handled by nonprofit agencies, as adoptions are, to discourage profit making. Another possibility is to declare surrogacy to be a special case of adoption. That is, the surrogate is treated the same as any other birth mother who arranges to give her child up for adoption. This ensures that she has a period of 6 weeks or so after the birth to decide whether she wants to keep the baby, and it clearly establishes whose baby it is.

A final possibility, at the other end of the spectrum, is that government stay out of this matter entirely, on the grounds that this is a private matter between the people involved. We don't see this as a viable option, because contested surrogacy cases are already reaching the courts. The government is already involved. It would be far better to have some well-thought-out legislation on this matter to provide guidance and reduce the number of painful contested cases.

Finally, there are a host of procedural issues that will inevitably arise if any of these reproductive techniques are legally permitted and regulated.

What will the standards of confidentiality be, especially with regard to the identity of nonparent donors? If something goes wrong during one of these procedures—and they are risky—who is liable? Who will take responsibility for a defective child born through one of these techniques? Who will bear the expense? Must insurance companies pay for artificial insemination, IVF, embryo transfer, or even surrogacy? Should such procedures be covered under Medicaid for the poor? Many more such questions will develop in the future.

Underlying all these issues are the root questions: What is the government's interest, if any, in human reproduction, and how shall it be expressed? What procedures shall be used to put into law society's concerns, if this is not deemed to be strictly a private matter? And, above all, who is to decide these issues? Legislatures and courts have shown a marked disinclination to enter this field, but evasion cannot remain a viable strategy much longer.[4]

In 1996, a group of medical and social scientists and legal scholars presented a report to the European Commission which discussed many of these

[4]For fuller development of these issues, see Cohen & Taub (1989), Taub (1987), Robertson (1986), Shapiro (1986), O'Brien (1986), Mallory & Rich (1986).

issues (Evans & Evans, 1996). The group agreed that there should be a uniform, *minimum* threshold of laws and regulations governing artificial procreation. The participants reached consensus on a number of issues, including the following:

1. Artificial procreation services should be noncommercial.

2. A licensing system should be created to ensure the quality of clinical services offered to the public.

3. Clinics should obtain advance directives from donors regarding the use or disposal of sperm, ova, and embryos.

4. Time limits should be agreed upon by all parties in advance regarding cryopreservation of gametes and embryos.

5. Access to services should be available to all married or cohabiting couples.

6. Cloning and related technologies should be prohibited.

On a number of other issues, there was not consensus, including the following:

1. Access to services for single women and lesbians.

2. The cryopreservation of human embryos.

3. Allowing children produced by assisted reproduction access to information about their conception and parentage.

4. The permissibility of experimentation with human embryos.

Hopefully, legislatures and policy makers will use this report as a basis for constructing the minimum threshold described in the report.

SUMMARY

Laws to protect adults from coercion, children from sexual exploitation, and the public from offensive behavior are justifiable. However, many laws against sexual conduct originated in a desire to promote public morality and perpetuate sexism, and it is hard to justify them.

The laws governing sexual conduct include laws against crimes of exploitation and force (such as rape, carnal knowledge of a juvenile, and child molestation), against various consensual acts (such as fornication, adultery, and sodomy), against gays and lesbians, against offending public taste (exhibitionism, voyeurism, solicitation, disorderly conduct, lewdness, and the like), against behaviors involved in reproduction (contraception and abortion), and against criminal commercial sex (notably prostitution and obscenity). These laws are often capriciously enforced, and this unequal enforcement has high social costs that may require reform.

Certain trends can be discerned in the reform of such sex laws. The American Model Penal Code included proposals to decriminalize consensual sexual behavior. The legal principle that has ac-

counted for much court action to reform laws against contraception and abortion is the "right to privacy," but it has not been broadened to include consensual acts of adults. The Constitutional principle of equal protection has been used to combat discrimination against groups identified by their sexual conduct, including gays, lesbians, and prostitutes. A factor that has influenced legislators is the movement for the decriminalization of "victimless crimes"; recently, however, critics have challenged the argument that no one is harmed by prostitution or adultery. The issue of pornography and obscenity, which includes such problems as definition, conflicting societal values, and actual demonstration of effects, is a confusing one. The latest controversy is over the availability of X-rated goods and services on the Internet. Abortion remains a volatile and highly controversial matter.

Sex-law reform moved more slowly in the 1990s than it did in the previous decade, and there are signs of a conservative backlash. In the future, the law will need to balance individual rights and the public interest when it comes to issues such as AIDS and new reproductive technologies.

REVIEW QUESTIONS

1. The regulation of sexual conduct by law is a fairly new development in American history. True or false?

2. Until recently a husband could not be prosecuted for raping his wife. True or false?

3. By and large, the law allows consenting adults to do in private whatever they want to do. True or false?

4. Laws which, in theory, may apply to many forms of sexual conduct, such as oral and anal sex, but which are usually enforced only against homosexuals, are _____ laws.

5. The 1965 Supreme Court decision that struck down state laws that had made it illegal for a physician to discuss or prescribe contraceptives for married persons was _____.

6. In legal terms, what is often called pornography is not protected by the Constitution if it can be shown to be _____.

7. The 1992 Supreme Court case that upheld a woman's right to abortion while also upholding a Pennsylvania law's restrictions, including parental notification for minors, was _____.

8. The reform of sex laws and the extension of the right of privacy are likely to continue in the future with little controversy. True or false?

9. The law recognizes the validity of surrogacy contracts. True or false?

10. A woman has an absolute right to an abortion, at any age and with public funds if she cannot pay. True or false?

(The answers to all review questions are at the end of the book, after the Glossary.)

QUESTIONS FOR THOUGHT, DISCUSSION, AND DEBATE

1. Find out what sorts of laws relate to sexual activity in the state in which you live or go to school. Are there any moves to change those laws?

2. What aspects of human sexuality do you think it is reasonable for the law to regulate?

3. How do you think young children can best be protected from sexual abuse and exploitation?

4. If you were a state legislator, what kinds of laws would you favor regarding AIDS and surrogate parenthood?

SUGGESTIONS FOR FURTHER READING

Burris, Scott, Dalton, Harlon L., and Miller, Judith (Eds.). (1993). *AIDS law today.* New Haven, CT: Yale University Press. Produced by the Yale AIDS Law Project, this excellent book discusses all the various legal issues regarding AIDS.

Hunter, Nan, Michaelson, Sherryl, and Stoddard, Thomas. (1992). *The rights of lesbians and gay men: The basic ACLU guide to a gay person's rights.* Carbondale, IL: Southern Illinois University. Prepared by the American Civil Liberties Union, this guide covers most laws relevant to gays and lesbians.

Messer, Ellen, and May, Kathryn E. (1988). *Backrooms: Voices from the illegal abortion era.* New York: St. Martin's Press. This book is based on interviews with 24 women, all of whom had unwanted pregnancies before *Roe v. Wade.* The women tell the stories of their illegal abortions and the consequences. It is important reading, especially for those who have grown up in the era of legal abortions.

Tribe, Laurence H. (1990). *Abortion: The clash of absolutes.* New York: Norton. Constitutional scholar Tribe analyzes the arguments on the two sides of the abortion debate and explains why no progress toward a compromise on this issue has been made.

WEB RESOURCES

http://www.aclu.org/

The American Civil Liberties Union Homepage; includes current information about a variety of legal issues, including reproductive issues, sexual orientation, censorship of the Internet, and HIV/AIDS.

http://www.peacefire.org/

A website devoted to preventing attempts to censor the Internet. Includes information about relevant events, and information about the problems associated with blocking software (for example, Cyber Patrol, CYBERsitter, Surf Watch, and Net Nanny).

chapter

23

SEXUALITY EDUCATION

CHAPTER HIGHLIGHTS

In the Home, in the School, or Somewhere
   Else?

Purposes of Sexuality Education

What to Teach at Different Ages

The Curriculum

The Teacher

Effective Sexuality Education

Effective Multicultural Sexuality Education

That gets me very nervous; that people like this [conservatives] want to take sex education out of the schools. Reagan wanted to take it out. He said he believed sex education *caused* promiscuity: if you had the knowledge, you'd use it. Hey—I took algebra; I *never* do math.*

Children are curious about sex (so are most adolescents and many adults). That is perfectly good and normal, and such curiosity motivates children to learn. The only problem is that adults often do not know what to do about it. This chapter concerns **sexuality education,** the lifelong process of acquiring information about sexual behavior and forming attitudes, beliefs, and values about identity, relationships, and intimacy (SIECUS, 1991). Sex education can occur in many settings: home, school, church or synagogue, youth programs, relationships.

In the Home, in the School, or Somewhere Else?

When parents of children and teenagers get together and urge their school system to adopt a sex education curriculum, invariably some citizens of the community raise a protest. They might say that sex education promotes promiscuity, teenage pregnancy, or AIDS, and they are sure that it should take place only in the home (or possibly the church), but certainly not in the schools.

What these citizens overlook is the realistic alternative to sexuality education in the schools. Among children ages 10 to 12, about half do get information about sex from their parents (see Table 23.1). Among youth ages 13 to 15, the proportion declines to 40 percent. Notice that the most frequently mentioned source in both groups is TV and magazines; among teens, friends are ranked almost as high. In many cases the information provided about sexuality on TV is sensationalized and unrealistic (see Focus 11.1). Relying on friends for information is a classic case of the blind leading the blind. Interestingly, young people would like to

*Elayne Boosler, *Best of Comic Relief 3.*

Table 23.1 Sources of Information About Sex

	Percentage saying "a lot"	
	Children 10–12	Teens 13–15
How much do children your age find out about sex from:		
Parents	54%	40%
Friends	36	60
Entertainment media; TV, magazines	57	61
Schools and teachers	29	45
Churches, religious organizations	29	13

Source: Kaiser Family Foundation and Children Now survey of national random samples of children (n=164) and teens (n=201), October–November 1996. Conducted by Princeton Survey Research Associates.

hear about sex from their parents (Sanders & Mullis, 1988).

The fact is that many children are given no sexuality education in the home. Rather, they learn about sex from TV and peers, and the result is a massive amount of misinformation. Thus, people who say that sex education should be carried out in the home, not in the school, are not making a sensible argument.

There are two reasons why parents do not provide much explicit education to their children. First, many people (and parents *are* people!) are embarrassed about discussing sexuality. We see few models of how to have an explicit, matter-of-

Sexuality education: The lifelong process of acquiring information about sexual behavior and forming attitudes, beliefs, and values about identity, relationships, and intimacy.

Figure 23.1   Although some parents claim that sex education belongs in the home, it is rarely conducted there effectively. Margulies/The Record/Rothco

fact discussion; we are much more likely to see people discussing sex indirectly, with euphemisms and innuendo, or telling dirty jokes. (As a partial corrective, we have tried to write this book in an explicit, straightforward way.) Second, there are many things about sexuality that many adults do not know. They did not have good sexuality education themselves, and they may be painfully aware of their ignorance. Even those who received a good sex education, if they took a human sexuality course before 1980, didn't learn anything about the then-unheard-of disease called AIDS.

Surveys have shown repeatedly that the majority of parents are in favor of sex education in the schools. A survey of 1,011 adults in 1998 found that 80 percent favor sex education in the schools; of those, 82 percent favor teaching children about sex by age 12 (Yankelovich Partners for TIME/CNN, 1998). The results of *Time* magazine's 1986 survey of adult Americans (July 21, 1986) showed that the great majority favored teaching children about AIDS and other sexually transmitted diseases, premarital sex, birth control, abortion, and homosexuality. The great majority believe that school

health clinics should make birth control information available, although only a minority believe that the clinics should provide contraceptives. The point is that there is strong support for detailed sex education in the schools, beginning at least when children are 12 years old (in sixth or seventh grades).

You may be surprised to learn that most adults favor sex education. The media regularly publicize controversies, cases in which parents are protesting sex education in the schools. There are three things to keep in mind about such episodes. First, they are rare. The vast majority of schools with sex education programs have not experienced such conflict. Second, the protesters are usually in a minority. In a case in Wisconsin, protesters packed a school board meeting, and the board voted to delay implementation of a program. A subsequent survey of all the parents in the district found that 71 percent approved and only 18 percent disapproved of the program. Third, the controversy is often not over whether there should be a program, but over the use of a particular curriculum, book, or video.

Purposes of Sexuality Education

The Sexuality Information and Education Council of the United States (**SIECUS**) has been one of the most active groups promoting high-quality sex education. According to SIECUS, the goals of sex education should be:

1. *Information*   To provide accurate information about human sexuality, including: growth and development, human reproduction, anatomy, physiology, masturbation, family life, pregnancy, childbirth, parenthood, sexual response, sexual orientation, contraception, abortion, sexual abuse, HIV/AIDS, and other sexually transmitted diseases.

2. *Attitudes, values, and insights*   To provide an opportunity for young people to question, explore, and assess their sexual attitudes in order to develop their own values, increase self-esteem, develop insights concerning relationships with members of both genders, and understand their obligations and responsibilities to others.

3. *Relationships and interpersonal skills*   To help young people develop interpersonal skills, including communication, decision-making, assertiveness, and peer refusal skills, as well as the ability to create satisfying relationships. Sexuality education programs should prepare students to understand their sexuality effectively and creatively in adult roles. This would include helping young people develop the capacity for caring, supportive, non-coercive, and mutually pleasurable intimate and sexual relationships.

4. *Responsibility*   To help young people exercise responsibility regarding sexual relationships, including addressing abstinence, how to resist pressures to become prematurely involved in sexual intercourse, and encouraging the use of contraception and other sexual health measures. Sexuality education should be a central component of programs designed to reduce the prevalence of sexually related medical problems, including teenage pregnancies, sexually transmitted diseases including HIV infection, and sexual abuse (SIECUS, 1991).

What to Teach at Different Ages

Sexuality education is not something that can be carried out all at once in one week during fifth grade. Like teaching math, it is a process that must begin when children are small. They should learn simple concepts first, progressing to more difficult ones as they grow older. What is taught at any particular age depends on the child's sexual behavior (see Chapter 11), sexual knowledge, and sexual interests at that age. This section will concentrate on theories and research that provide information on these last two points.

Children's Sexual Knowledge

A few researchers have investigated what children know about sex and reproduction at various ages. Children begin to develop an understanding of gender at a very early age. They learn first that gender is constant (a boy will always be a boy) and only later learn that gender is determined by whether one has a penis or a vagina (Gordon & Schroeder, 1995). There is a similar developmental process in children's understanding of pregnancy and birth. Very young children may believe that a baby has always existed, that it existed somewhere else before it got inside the mother. The following dialogue demonstrates this:

> (How did the baby happen to be in your Mommy's tummy?) It just grows inside. (How did it get there?) It's there all the time. Mommy doesn't have to do anything. She waits until she feels it. (You said that the baby wasn't in there when you were there.) Yeah, then he was in the other place . . . in America. (In America?) Yeah, in somebody else's tummy. (Bernstein & Cowan, 1975, p. 86)

As children get older, they develop a more accurate understanding.

Children's knowledge about sex varies by social class. In one study of children 2 to 7 years of age, middle- and upper-class children knew more about body parts and functions and pregnancy than children from lower-class families (Gordon

SIECUS: The Sexuality Information and Education Council of the United States, an organization devoted to fostering sex education.

Figure 23.2 Children often have inaccurate ideas about sex. Dennis the Menace cartoon by T.M. 1983. Dennis the Menace ®: used by permission of Hank Ketcham and © North American Syndicate.

DENNIS the MENACE

"THAT'S FUNNY... MY DAD CAN TELL IF IT'S A BOY OR A GIRL JUST BY LOOKIN' AT THE BOTTOM OF ITS FEET."

et al., 1990). It is possible that these differences reflect parental attitudes rather than social class; lower-class parents tended to have more restrictive attitudes about sexuality, and they reported having given their children less sex education.

By age 7 or 8, children have a more sophisticated understanding of reproduction. They may know that three things are involved in making a baby: a social relationship between two people, such as love or marriage; sexual intercourse; and the union of sperm and egg. At age 12, some children can give a good physiological explanation of reproduction that includes the idea that the embryo begins its biological existence at the moment of conception and is the product of genetic material from both parents. As one preteen explained:

> The sperm encounters one ovum, and one sperm breaks into the ovum which produces, the sperm makes like a cell, and the cell separates and divides. And so it's dividing, and the ovum goes through a tube and embeds itself in the wall of the, I think it's the fetus of the woman. (Bernstein & Cowan, 1975, p. 89)

As we discussed in Chapter 11, research suggests that many children engage in sexual play and exploration. How does sexual behavior relate to sexual knowledge? One review of the literature concludes:

> This work indicates that sexual experience and behavior typically precede knowledge and understanding, at least among younger children. The fact that children engage in sexual behavior before they have a clear understanding of what it is all about places them at high risk for a variety of adverse experiences which can impact negatively on their development. (Gordon & Schroeder, 1995, p. 11)

These findings have important implications for sex education. Educators need to be aware of the level of the child's understanding and should not inundate him or her with information inappropriate for his or her age. Instead, the educator should attempt to clarify misunderstandings in the child's beliefs. For example, if a child believes a baby has always existed, the educator might say, "To make a baby, you need two grown-ups, a man and a woman."

Focus 23.1
Are American Children Sexual Illiterates?

Ronald and Juliette Goldman (1982b) did a massive cross-cultural study of children's understanding of sexual matters. From their results, they concluded that American children are sexual illiterates.

The Goldmans did face-to-face interviews with children aged 5, 7, 9, 11, 13, and 15 in four cultures: Australia, England, North America, and Sweden. A total of 838 children were interviewed. The Swedish sample is particularly interesting because there is compulsory sex education for all children in Swedish schools, beginning at age 8. It is also worth noting that the North American sample was originally planned to be a United States sample, but school officials in the United States were so uncooperative that the Goldmans had to go across the border from upstate New York to Canada (where they obtained more cooperation) in order to complete the sample. Therefore, the North American sample is a mixture of children from Canada and the United States.

The Goldmans were careful to avoid controversial topics such as homosexuality in the interview, and they questioned children only about their understanding of sexual concepts, not about their own sexual behavior. They called the study "Children's Concepts of Development." These precautions were taken in order to produce a high rate of cooperation from parents. Parents in general were cooperative; only 20 percent of parents overall refused to allow their children to participate.

A comparison of the results from the North American children with those from children in the other three cultures led the Goldmans to conclude that American children are strikingly lacking in sexual information. Some of the results are shown in Table 23.2. Notice, for example, that only 23 percent of North American 9-year-olds, but 60 percent of Australian 9-year-olds, know the genital differences between newborn baby boys and girls. The Swedish children are consistently more knowledgeable than the American children, indicating the positive effects of sex education.

Some of the children's responses can only be classified as amusing. In response to the question

Children's Sexual Interests

Children's knowledge of and interest in sex are reflected in the questions they ask. In a study of children's questions about sex, it was found that many questions are asked at around age 5, a time when children are generally asking questions. Boys also tend to ask a lot of questions at around age 9, and girls at ages 9 and 13 (Byler, 1969). The areas of sexual curiosity were (beginning with the most common) the origin of babies, the coming of another baby, intrauterine growth, the process of birth, the organs and functions of the body, physical gender differences, the relation of the father to reproduction, and marriage.

Probably the most comprehensive study of children's sex knowledge is described in Focus 23.1.

It is important that the sex education curriculum for a particular age group address the questions of that age group, rather than questions they haven't thought of yet or questions they thought about and answered long ago.

High school students agree that sex education should begin in early elementary school, and that it should progress from the simple to the complex (Eisenberg et al., 1997). They believe that the "ideal" class should cover a wide range of topics, including reproduction, pregnancy, abortion, birth control options, disease prevention, sexual violence, relationships and gender roles, and values. They would like all of these topics presented by eighth grade.

Table 23.2 Responses of 9-Year-Olds in Four Cultures in the Goldman Study

Concept	Percentage of Correct Answers Among			
	Australians	British	North Americans	Swedish
Knowing physical sex differences of newborn babies	60	35	23	40
Knowing correct terms for the genitals	50	33	20	*
Knowing length of gestation is 8 to 10 months	35	32	30	67
Knowing that one purpose of coitus is enjoyment	6	10	4	60
Knowing the meaning of the term "uterus"	0	0	0	23

*Owing to the difficulties of translating from the Swedish language, this percentage is not available.

Source: Ronald Goldman and Juliette Goldman. (1982). *Children's sexual thinking.* London: Routledge & Kegan Paul, pp. 197, 213, 240, 263, 354.

"How can anyone know a newborn baby is a boy or a girl?" an 11-year-old English boy said, "If it's got a penis or not. If it has it's a boy. Girls have a virginia." And in all cultures there seems to be a lot of confusion about contraception. Here are some responses:

The pill goes down the stomach and dissolves the baby and it goes out in the bowels. You should take three pills a day. (American boy, 7 years old)

If you don't want to start one, you don't get married. There's no other way. (English girl, 7 years old)

The tubes are tied, the vocal cords. (Australian girl, 15 years old)

If the Goldmans' conclusion is right, that American children are sexual illiterates, the remedy seems to be a massive program of sex education in the United States.

Source: Ronald Goldman and Juliette Goldman. (1982). *Children's sexual thinking.* London: Routledge & Kegan Paul.

We can also tell something about children's sexual knowledge and interest by the dirty jokes they tell. Anthropologist Rosemary Zumwalt collected dirty jokes from girls between the ages of 7 and 10 as part of her study of children's folklore (1976). The following is typical of the jokes they told her:

There's this little boy, and he wanted to take a bath with his dad. And his dad said, "If you promise not to look under the curtain." And then he took a shower, and he looked under the curtain. And he said, "Dad, what's that long hairy thing?" And the father says, "That's my banana."

Then he asks his Mom, "Can I take a shower with you, Mom?" She says, "If you promise not to look under the curtain." And they get into the shower, and he looks under the curtain. And he

says, "Mom, what's that thing?" And she says, "That's my fruit bowl." And he says, "Mom, can I sleep with you and Dad?" And she says, "Yes, if you promise not to look under the covers." And he looks under the covers and says, "Mom, Dad's banana is in your fruit bowl!" (Zumwalt, 1976, p. 261)

Children's dirty jokes reflect several themes in their attitudes toward sexuality and in their interactions with their parents on the issue. First, children seem to view their parents as always trying to keep sex a secret from them. The parents consistently tell children not to look under the curtain, for example. Second, the jokes reflect children's fascination with sex, particularly with the penis, the vagina, the breasts, and intercourse. The jokes generally revolve around these topics and children's attempts to find out about them. Third, the jokes seem to satirize adults' use of euphemisms for sexual terms. The joke above hinges on a parent's using the term "banana" instead of "penis." Most frequently, the fanciful names used for the sexual organs involve food (banana, hot dog), power (light bulbs, light sockets), or animals (gorilla). Commenting on the bathtub-shower form of dirty joke, an authority said, "In all forms of the . . . joke, the wonderful humor to the child is the mocking of the parents' evasions, which are somehow so foolishly phrased. . . ."(Legman, 1968, p. 53)

Teenagers have outgrown this sort of joke, but they tell a parallel one:

> This little boy walks into the bathroom, and he catches his mother naked. She was a little embarrassed. He said, "Mommy, what's that?" And she says, "Oh, that's where God hit me with an axe." And the little kid says, "Got you right in the cunt, eh?" (Zumwalt, 1976, p. 267)

Once again, this joke has the theme of a parent's embarrassment and use of evasions and euphemisms when dealing with sex. But now the child (teenager) reflects a sophistication about sex, perhaps even a greater sophistication than the parent has.

Sex educators should remember that children are aware of adults' attempts to "cover up" and of their embarrassment and their use of euphemisms, as these jokes indicate.

The Curriculum

The term "sexuality education" has been used to refer to a wide variety of programs. At one end of the continuum are programs that involve showing youngsters one or two videos and distributing some brochures. At the other end are well-developed curricula which include lectures, books, videos, and classroom discussion presented over four to six weeks. We will focus on the more comprehensive programs.

Pregnancy Prevention Programs

Kirby identifies five "generations" of curricula (Kirby, 1992). The first programs, developed 25 years ago, were concerned with the transmission of knowledge. The goal of these programs was to reduce the number of teen pregnancies. Accordingly, the emphasis was on teaching students about sexual intercourse, pregnancy and birth control, and the consequences of having a baby. These programs were generally not carefully evaluated, but there is some evidence that they did increase knowledge.

Second-generation programs retained the informational content of the first ones, but the emphasis was placed on values clarification and decision-making skills. Proponents of these programs believed that young people engage in sexual risk taking because they are unsure of their values and have difficulty making decisions. These programs also taught skills designed to improve communication with partners. Subsequently, evaluations

> clearly demonstrated that these early sex education programs did not markedly hasten or delay onset of intercourse, reduce sexual risk-taking behavior, or reduce teen-age pregnancy. However, they may have increased slightly the use of birth control or had other positive effects that could not be measured by the evaluation methods employed. (Kirby, 1992, p. 281)

Abstinence-based Programs

The third generation consisted of programs developed out of opposition to the existing curricula. Some people were opposed to any sex education in the schools; others felt that the existing programs were too liberal or permissive. The concerns led to passage by the U.S. Congress of the Adolescent Family Life Act in 1981; the AFLA limited the use of federal funds to programs that "promoted sexual abstinence as the sole means of preventing pregnancy and exposure to sexually transmitted diseases" (Wilcox & Wyatt, 1997, p. 4). Millions of dollars have been spent by state and federal government to support the development and widespread use of these programs. The two most widely known of these curricula are called *Sex Respect* (see Focus 23.2) and *Teen Aid*. These curricula are typically presented in 4 to 10 sessions, lasting one hour each, over a period of one to three months.

In 1997, financial support for these curricula increased dramatically. The national Personal Responsibility and Work Opportunity Act of 1996 (welfare reform) contains a provision that provides $50 million per year to states to support abstinence-only educational programs. For example, Wisconsin will receive about $750,000 per year; the money will be given to 10 agencies or schools that propose to introduce these programs.

So, how effective are these curricula? Researchers who assessed the content of *Sex Respect* concluded that it omits a number of important topics, including sexual anatomy (!), sexual physiology, sexual response, contraception, and abortion (Goodson & Edmundson, 1994). We noted earlier that high school students say that all of these topics should be included in an "ideal" class. As a result of their widespread use, there have been many evaluations of these programs' effects on student attitudes and behavior. A review of 52 evaluations that meet minimal methodological standards concluded:

1. In some schools, these programs have positive effects on knowledge and attitudes.
2. Of the 16 evaluations assessing whether students delayed sexual activity, only three report a positive effect.
3. "None of the best studies [by methodological criteria] found positive changes . . . in age of onset of sexual activity, rates of sexual activity, pregnancies, or STDs." (Wilcox & Wyatt, 1997, p. 13)

Therefore, it is clear to us that the millions of dollars designated for these programs are being woefully misused, a conclusion shared by an Editor of the Journal of the American Medical Association (DiClemente, 1998).

HIV and AIDS Risk Education

In the past decade the focus of sex education has shifted from pregnancy prevention to AIDS (and other STD) prevention. A strong case has been made for comprehensive education about HIV and AIDS in the schools (National Commission on AIDS, 1994). Forty states either require or encourage these programs, and 94 percent of parents approve of them (Kirby, 1992). Two-thirds of the school districts in the United States require HIV education (Robenstine, 1994).

Programs of this type developed independently of the first three generations of sex education programs. They are often sharply focused on disease prevention. They have a variety of goals, including removing myths about AIDS and other STDs, encouraging delay of sexual intercourse, and encouraging condom use or abstinence from unprotected intercourse. Early curricula relied on lecture and classroom discussion. On occasion, someone with AIDS was brought in to talk with the class. These programs are usually short, lasting as few as five class periods.

A review of the effectiveness of these programs found 40 published studies (Kim et al., 1997). Most of the studies report that the educational program improved knowledge significantly. Of those which measured the impact on attitudes (12), more than half (7) found that the program increased positive attitudes toward prevention and negative attitudes toward risk. Finally, 6 of the 10 studies reported positive changes in intention to use condoms. Interventions which addressed more topics, were based on theory, and which incorporated cultural issues were more effective.

Recent national surveys indicate that teenagers are very well informed about HIV transmission. Most know that condoms prevent transmission of AIDS and other STDs. School-based HIV education programs may have contributed to these outcomes.

On the other hand, it must be noted that HIV/AIDS risk reduction education is narrowly focused on behavior. It does not take into account the developmental context, the broad range of social and psychological influences on one's sexual behavior (Ehrhardt, 1996). Thus, these programs are not a substitute for a comprehensive sexuality education program as described earlier in this chapter.

Theoretically-based Programs

The newest programs, the fifth generation, are unique in that they are explicitly based on social science theories of health promotion, including the Health Belief Model, social inoculation theory, and social learning theory. The curriculum described in Focus 23.3 is of this type. The best known of these curricula are called *Postponing Sexual Involvement* and *Reducing the Risk*. These programs include discussion of the social pressures to engage in sex, and ways to resist these influences

Focus 23.2

Conservative Sex Education: The *Sex Respect* Curriculum

The sex education curriculum called *Sex Respect* takes a considerably different approach from the one in this book. *Sex Respect* (which is the most well known of several similar curricula) is a political conservative's approach to sex education. Federally funded, it is targeted at middle school audiences and, as of 1991, 1600 school districts nationwide were using it.

The major goal of this curriculum is to teach that abstinence is the only approach that is moral and safe. The curriculum uses cartoons and other attention-grabbing techniques. There are catchy slogans for children to chant in class, such as

Don't be a louse, wait for your spouse!

Do the right thing, wait for the ring!

Pet your dog, not your date!

There is a "chastity pledge" that all students take, and a chart of physical intimacy in which a prolonged kiss is characterized as the "beginning of danger." The curriculum teaches that condoms can be the road to ruin because many fail and pregnancy results.

Sex Respect throws in a lot of gender-role stereotypes as well, characterizing boys as "sexual aggressors" and girls as "virginity protectors." It presents the two-parent, heterosexual couple as "the sole model of a healthy, 'real' family."

Wisconsin's chapter of the American Civil Liberties Union (ACLU), on behalf of parents who objected to the curriculum, demanded that it be removed from all public schools using it. The ACLU argued that the curriculum amounts to discrimination based on gender, marital status, sexual orientation, and religion—all of which are illegal under Wisconsin laws.

In fairness to *Sex Respect*, it may have some good points in that it teaches students skills in resisting peer pressure. On the other hand, it includes a lot of "facts" that are really misinformation (for example, it says that condoms frequently fail, but they actually have a very low failure rate), and it seems out of touch with today's teenagers.

The widespread adoption of this curriculum points out how important it is for parents to examine the sex education materials being presented to their children.

Sources: Newsweek, June 17, 1991; *Wall Street Journal,* February 20, 1992.

(based on inoculation theory). Social learning theory emphasizes the importance of practicing new skills, so these curricula include rehearsal and role-playing activities.

A large-scale evaluation of *Postponing Sexual Involvement* was carried out in California in 1992–1994. More than 10,600 seventh and eighth graders participated; each was randomly assigned to either the curriculum or a control group (Kirby et al., 1997). The *PSI* program consisted of five sessions, 45 to 60 minutes each; it included discussion, group activities, videos or slides, and role-playing. Survey data were collected before the program, 3 months after, and 17 months after the program. At 3 months, youth who participated in *PSI* showed two positive changes in attitude, endorsed more reasons for not having sex, were more likely to recognize sexual content in the media, and were more confidant they could say no to sex. At 17 months, there were no differences between the

Figure 23.3 Sex education programs such as *Sex Respect* promote abstinence, with slogans such as "Control your urgin'. Be a virgin." Evaluations of these programs indicate that they are not effective in helping teens postpone intercourse.

groups. The program had no effect on several measures of sexual behavior. The review concludes that the program may have been too limited in both scope and duration.

Condom Distribution

In the early 1990s, the most visible conflict was over the distribution of condoms to students. There are condom availability programs in 431 pubic schools in 21 states (Brown et al., 1997). In some schools, condoms are available through the sex education program. In other schools, clinics providing health care services to adolescents dispense birth control, including condoms. In still other schools, condoms are sold in vending machines. Again, data indicate widespread support for the distribution of condoms in the schools. In New York City, where condoms are available in all public high schools, a survey found that 69 percent of the parents of high school students

Focus 23.3
A Sample Sexuality Education Curriculum

This set of curriculum guidelines for sexuality education, developed by SIECUS, is based on teaching about six key concepts, and teaching about each concept at each age level, with age-appropriate material. The age levels are as follows:

- Level 1: ages 5 to 8; early elementary school
- Level 2: ages 9 to 12; upper elementary school
- Level 3: ages 12 to 15; middle school or junior high
- Level 4: ages 15 to 18; high school

The guidelines recommend that the following material be taught:

Key Concept 1: Human Development

Topic 1: Reproductive Anatomy and Physiology

Level 1: Each body part has a correct name and a specific function. Boys and men have a penis, scrotum, and testicles. Girls and women have a vulva, clitoris, vagina, uterus, and ovaries. Both girls and boys have body parts that feel good when touched.

Level 2: The maturation of external and internal reproductive organs occurs during puberty. At puberty, boys begin to ejaculate and girls begin to menstruate.

Level 3: The sexual response system differs from the reproductive system. Some of the reproductive organs provide pleasure as well as reproductive capability.

Level 4: Chromosomes determine whether a developing fetus will be male or female. For both sexes, hormones influence growth and development as well as sexual and reproductive function. A woman's ability to reproduce ceases after menopause; a man can usually reproduce throughout his life. Both men and women can experience sexual pleasure throughout their life. Most people enjoy giving and receiving sexual pleasure.

Topic 2: Reproduction

Level 1: Reproduction requires both a man and a woman. Men and women have reproductive organs that enable them to have a child. Not all men and women decide to have children. When a woman is pregnant, the fetus grows inside her body in her uterus. Babies usually come out of a woman's body through an opening called a vagina. Women have breasts that can provide milk for a baby. Sexual intercourse occurs when a man and a woman place the penis inside the vagina.

---

support the program, although half also felt that they should be able to prevent their child from receiving condoms (Guttmacher et al., 1995). A survey of all of the students in one Denver high school found that 85 percent supported distribution of condoms in their school (Fanburg et al., 1995).

The most visible opposition to condom distribution programs is by the Roman Catholic Church. Church officials, including bishops in New York City and Chicago, oppose such programs because of their religion's ban on all artificial forms of contraception. Others oppose such programs on the grounds that they will encourage sexual intercourse outside of marriage. As one critic put it:

Instead of amoral, secular humanistic sex education and condom distribution in the schools, families, churches, schools, social organizations, and the business community must re-emphasize the teaching, learning, and practice of virtues like courtesy, kindness, honesty, decency, moral courage, integrity, justice, fair play, self-respect, respect for others, and the Golden Rule. (Gow, 1994, p. 184)

Level 2: Sexual intercourse provides pleasure. Whenever genital intercourse occurs, it is possible for the woman to become pregnant. There are ways to have genital intercourse without causing pregnancy.

Level 3: People should use contraception during sexual intercourse unless they want to have a child. Conception is most likely to occur midway between a woman's menstrual periods. Ovulation can occur any time during the month; therefore a woman may become pregnant at any time. When a girl begins to menstruate, she can become pregnant. When a boy produces sperm and can ejaculate, he can cause a pregnancy. An important first sign of pregnancy is a missed menstrual period.

Level 4: Menopause is when a woman's reproductive capacity ceases. Some people are unable to reproduce due to physiological reasons. Medical procedures can help some people with fertility problems. People who cannot reproduce can choose to adopt children. New reproductive technologies, such as artificial insemination, in vitro fertilization, and surrogate motherhood allow people with fertility problems to have children.

Topic 3: Puberty

Level 1: Bodies change as children grow older. People are able to have babies only after they have reached puberty.

Level 2: Puberty begins and ends at different ages for different people. Most changes in puberty are similar for boys and girls. Girls often begin pubertal changes before boys. Early adolescents often feel uncomfortable, clumsy, and/or self-conscious because of the rapid changes in their bodies. The sexual and reproductive systems mature during puberty. During puberty, girls begin to ovulate and menstruate, and boys begin to produce sperm and ejaculate. During puberty, many people begin to develop sexual and romantic feelings.

Level 3: Some people will not reach full puberty until their middle or late teens.

Topic 4: Body Image

Level 1: Individual bodies are different sizes, shapes, and colors. Male and female bodies are equally special. All bodies are special, including those that are disabled. Good health habits, such as diet and exercise, can improve the way a person looks and feels. Each person can be proud of the special qualities of his or her own body.

Level 2: A person's appearance is determined by heredity, environment, and health habits. The media portray beautiful people, but most people do not fit these images. The value of a person is not determined by his or her appearance.

Level 3: The size and shape of the penis or breasts does not affect reproductive ability or the ability to be a good sexual partner. The size and shape of a person's body may affect how others feel about and behave toward that person. People with physical disabilities have the same feelings, needs, and desires as people without disabilities.

Level 4: Physical appearance is only one factor that attracts one person to another. A person who accepts and feels good about his or her body will seem more likeable and attractive to others. Physical attractiveness should not be a major factor in choosing friends or dating partners.

In January 1996, the U.S. Supreme Court rejected a challenge to a condom distribution program in the Falmouth, Massachusetts, public schools.

Condoms are available in all 15 high schools in Seattle (Brown et al., 1997). Forty-eight percent of the students who reported having intercourse during the two years prior to a survey said they had obtained condoms from school. In focus groups, students said that the availability of condoms had not led to an increase in rates of sexual activity. Students preferred that condoms be available in private locations (nurse's office) rather than public ones (vending machines). Students also wanted comprehensive sexuality education programs in conjunction with condoms.

In some cities, school-based clinics appear to have led to a reduction in births to students; they have not led to increased rates of sexual activity. Other types of condom distribution programs are too new to have been carefully evaluated (Kirby, 1992).

Focus 23.3, Cont'd

Topic 5: Sexual Identity and Orientation

Level 1: Everyone is born a boy or a girl. Boys and girls grow up to be men and women. Most men and women are heterosexual, which means they will be attracted to and fall in love with someone of the other gender. Some men and women are homosexual, which means they will be attracted to and fall in love with someone of the same gender. Homosexuals are also known as gay men and lesbian women.

Level 2: Sexual orientation refers to whether a person is heterosexual, homosexual, or bisexual. A bisexual person is attracted to men and women. It is not known why a person has a particular sexual orientation. Homosexual, heterosexual, and bisexual people are alike except for their sexual attraction. Homosexual and bisexual people are often mistreated, called hurtful names, or denied their rights because of their sexual orientation. Some people are afraid to admit they are homosexual because they fear they will be mistreated. Homosexual love relationships can be as fulfilling as heterosexual relationships. Gay men and lesbians can form families by adopting children or having their own children.

Level 3: Many young people have brief sexual experiences (including fantasies and dreams) with the same gender, but they mainly feel attracted to the other gender. When a homosexual person accepts his or her sexual orientation, gains strength and pride as a gay or lesbian person, and tells others, it is known as "coming out." People do not choose their sexual orientation. Sexual orientation cannot be changed by therapy or medicine.

Level 4: One's understanding and identification of one's sexual orientation may change during life. Teenagers who have questions about their sexual orientation should consult a trusted and knowledgeable adult.

The curriculum guidelines continue with equally detailed content for the remaining four key concepts:

Key Concept 2: Relationships

Key Concept 3: Personal Skills

Key Concept 4: Sexual Behavior

Key Concept 5: Sexual Health

Space does not permit us to list all the details under each of these concepts, so interested readers may want to consult the source listed below.

Source: Condensed from SIECUS. (1991). *Guidelines for comprehensive sexuality education.* New York: Sexuality Information and Education Council of the United States.

The Teacher

Suppose you have decided to start a program of sexuality education. You have found a curriculum that is consistent with your objectives, whether those are to promote premarital abstinence or condom use with every act of sexual intercourse. Wherever the program is to be carried out—in the home, the school, the place of worship, or someplace else—the next resource you need is the teacher. There are two essential qualifications: the person must be educated about sexuality, and he or she must be comfortable interacting with learners about sexual topics. High school students in Minneapolis-St. Paul participating in focus groups agreed that these two qualifications are essential (Eisenberg et al., 1997). They also cited the ability of the teacher to relate the material to their lives as important.

Sex education teachers need to be educated about sex. Reading a comprehensive text such as this one or taking a college or university course in sexuality are good ways to acquire the information that is needed. The teacher does not have to have a graduate degree in sexology; the important qualifications are a good basic knowledge, a willingness to admit it when he or she does not know the

Figure 23.4   The distribution of condoms in the public schools has been very controversial. These young people display condoms obtained at their high school in New York City.

answer, and the patience to look things up. In school settings, the preparation of the teachers is an important influence on the success of the program. A survey of a probability sample of school districts found that although two-thirds of the districts provided preparation, the most frequent method was simply to give teachers written material to read. Forty percent of the districts provided no in-service training; in the other 60 percent, the average length of training was just 3 hours (Robenstine, 1994).

Equally important is the teacher's comfort with sexual topics. Even when parents or other adults willingly give factual information about sex to a child or adolescent, they may convey negative attitudes because they become anxious or blush or because they use euphemisms rather than explicit sexual language. According to one 16-year-old girl, "The personal development classes are a joke. Even the teacher looks uncomfortable. There is no

way anybody is going to ask a serious question" (Stodghill, 1998). One evaluation of the conservative *Teen Aid* curriculum studied seventh- and eighth-grade classes in 24 schools; the researchers assessed student outcomes and the teachers' philosophy and level of implementation of the program. There was greater change in students' values, attitudes, and behavioral intention regarding abstinence when the teachers' philosophy was congruent with the abstinence emphasis in the curriculum (de Gaston et al., 1994).

Some people are relaxed and comfortable in discussing sex. Others must work to learn this attitude. There are a number of ways to do this. For example, the teacher can role-play, with another adult, having sexual discussions with children. Some communities periodically offer programs designed to desensitize the sexuality teacher or to enhance awareness of his or her own sexual values and attitudes.

A good teacher is also a good listener who can assess what the learner knows from the questions asked and who can understand what a child really wants to know when she or he asks a question. As one joke had it, little Billy ran into the kitchen one day after kindergarten and asked his mother where he had come from; she gritted her teeth, realized the time had come, and proceeded with a 15-minute discussion of intercourse, conception, and birth, blushing the whole time. Billy listened, but at the end he appeared somewhat confused and walked away shaking his head, saying, "That's funny. Jimmy says he came from Illinois."

Effective Sexuality Education

In light of the continuing high levels of teenage pregnancy (1 million pregnancies per year), the sharp increases in rates of STDs among persons 15 to 24 years of age, and the increasing rate of HIV infection in adolescents, it is imperative that we identify sex education programs that appear to be effective in reducing sexual risk-taking behavior. At the request of the U.S. Centers for Disease Control and Prevention, Kirby and a number of other researchers undertook a thorough review of the research on the effectiveness of school-based programs (Kirby et al., 1994). They identified six characteristics that are associated with delaying the initiation of intercourse, reducing the frequency of intercourse, reducing the number of sexual partners, and increasing the use of condoms and other contraceptives.

Effective programs focus on reducing risk-taking behavior.   Such programs have a small number of specific goals. They do not emphasize general issues such as gender equality and dating.

Effective programs are based on theories of social learning.   Programs that utilize theory in designing the curriculum are more effective than nontheoretical programs. The theories suggest that, to be effective, the program must increase knowledge, elicit or increase motivation to protect oneself, demonstrate that specific behaviors will protect the person, and teach the person how to effectively use those behaviors.

Effective programs teach through experiential activities that personalize the messages.   Such programs avoid lectures and videos; instead they utilize small-group discussions, simulation and games, role playing, rehearsal, and similar educational techniques. Some of these programs rely on peer educators.

Effective programs address media and other social influences that encourage sexual risk-taking behaviors.   Some programs look at how the media use sex to sell products. All the effective programs analyze the "lines" that young people use to try to get someone else to engage in sex, and teach ways of responding to these approaches.

Effective programs reinforce clear and appropriate values.   These programs are not value-free. They emphasize the values of postponing sex and avoiding unprotected sex and high-risk partners. The values and norms must be tailored to the target population. Different programs are needed for middle school students, for white middle-class high school students, and for ethnic minority high school students.

Effective programs enhance communication skills. Such programs provide models of good communication and opportunities for practice and skill rehearsal.

The United Nations Programme on HIV/AIDS commissioned a review of the effectiveness of sexuality education programs, with data from countries as diverse as Mexico, France, and Thailand, in addition to the United States (UNAIDS, 1997). The review focused on studies that measured the impact of educational programs on behavior. Three studies found an increase in sexual behavior following a program. Twenty-two of fifty-three studies reported that the program delayed the initiation of sexual activity, led to a reduction in number of partners, or reduced rates of unplanned pregnancy and STDs. The review concluded that the most effective programs:

1. focus on risk-reduction;
2. are grounded in social learning theories;
3. focus on activities that address media and social influences; and
4. teach and allow for practice in communication and negotiation skills.

Sex education programs that reduce sexual risk-taking behavior by adolescents do exist, and the inclusion of such programs in the schools is supported by a large majority of the parents in every survey. We need to convince school administrators to implement such programs, to provide adequate training and support to the teachers, and to stand firm in the face of opposition from vocal opponents of these programs.

Effective Multicultural Sexuality Education

Much of the discussion in this chapter assumes that the participants in a sexuality education program are homogeneous, that they are all from the same culture. In some situations, that assumption is valid; but in other settings, the learners may be from diverse cultural backgrounds.

Cultures vary in a number of ways that are directly related to the success or failure of a sexuality education program (Irvine, 1995). There are cultural differences in sexual practices; some of these were discussed in Chapters 1 and 11. There are differences in the acceptability of explicit sexual language or of particular types of language, such as street slang. Cultures vary in the meaning they attach to sexuality. White, Euro-American cultures have emphasized sex for the purpose of reproduction, and thus tend to regard vaginal intercourse as the norm (see Chapter 21). Other cultures place greater emphasis on the pleasure that can be derived from sexual stimulation. Finally, cultures vary in the definition of and the roles expected within the family.

Of necessity, sexuality education programs utilize language. Street slang might enhance rapport with black urban youth but deeply offend Latinas. Programs are based on assumptions about the prevalence of specific sexual practices, such as

Figure 23.5   Sex education around the world. At this family planning clinic in India, an educator explains the female reproductive system to these mothers.

Figure 23.6 Research indicates that sex education programs are most effective when they allow students to gain experience through role playing and rehearsing discussions they might have with a potential partner. Here students role-play ways of resisting pressure to engage in unwanted sexual activity.

vaginal and anal intercourse. They implicitly or explicitly identify some practices as desirable, for example, condom use. They reflect assumptions about the purposes of sexual intimacy; for example, abstinence-based programs assume that sexual intercourse is most or only meaningful within marriage.

If sexuality education is to be successful, it must reflect, or at least accept, the culture(s) of the participants. The educator must assess the audience, the intended messages, and the context, and then target the program accordingly (Irvine, 1995). Educators must recognize their own sexual culture, learn about the sexual culture(s) of the participants, and be aware of the power differences between groups in our society. In the classroom, they should use this knowledge to enhance the effectiveness of the presentation. The use of communication styles and media common to the culture(s)

of the participants—for example, certain rap songs that appeal to young urban African Americans—can be a valuable tool. Finally, it is important that the program not advocate beliefs and practices that are incompatible with participants' culture(s). Such programs are doomed to failure.

One attempt to develop a curriculum for African-American adolescents is the "Let the Circle Be Unbroken: Rites of Passage" program (Okwumabua et al., 1998). This program is based on the premise that a successful transition into adolescence requires preparation and celebration. It is presented to 10- to 14-year-olds, lasts four to six months, and involves youth, parents, and friends. Staff are specially trained during an "orientation phase." During the "passage phase," there are weekly programs lasting 60 to 90 minutes; these programs focus on preparation for adult roles, including the making of sexual decisions and behavior. The last four weeks make up the "culminating phase," in which everyone plans for the final celebration, the *rite de passage*. Programs like this one respond to the call for sexuality education that incorporates the family and community context in which our sexuality is grounded (Maddock, 1997; Young, 1996).

SUMMARY

Most children receive their sex education from their peers or other sources, not from their parents. As a result, those who argue for sex education in the home rather than in the school are being unrealistic. Most Americans do favor sex education in the school. Cases of opposition to sex education are rare and involve a small number of people.

The purposes of sexuality education include providing children with adequate knowledge of the physical and emotional aspects of sex, with an opportunity to develop their own values and interpersonal skills, and with the maturity to take responsibility for their sexuality.

What is taught at each age should depend on what children are thinking about at that age. Children pass through various stages in their understanding of sexuality. For example, at first they believe that a baby has always existed. Later, they realize that the parents caused the baby's creation, but they don't know exactly how. Older children acquire a more scientific understanding of reproduction. Children's sexual play seems to precede rather than follow the development of sexual knowledge. Children's dirty jokes reflect their parents' attempts to hide sex from them, their parents' use of euphemisms rather than actual terms, and their great fascination with sexual organs and intercourse.

Sexuality education curricula have evolved a great deal over the past 25 years. The first- and second-generation programs focused on knowledge, values clarification, and decision-making skills. They did not have substantial effects. The third generation consists of the conservative programs such as *Sex Respect,* which emphasize abstinence and are limited in content. Although millions of dollars are spent annually to promote these programs, the evidence indicates they are not effective. HIV education programs are required in many school districts; surveys suggest that they may lead to increased knowledge and more frequent condom use. Contemporary programs are based on social science theory and emphasize the importance of allowing children to practice new behaviors such as communication skills. An evaluation of one of these programs found some changes three months after participation. None of the changes lasted seventeen months.

There is a good deal of conflict over the distribution of condoms in high schools. There is widespread support among both high school students and their parents for such programs, but the Catholic Church, among others, is a vocal opponent.

A good sexuality education instructor must have accurate knowledge about sexuality, must be comfortable discussing it, and must be good at listening to the questions learners ask. Students say that the instructor's ability to relate the material to their lives is also important.

Research suggests that sexuality education programs that are effective at delaying the onset of intercourse, reducing the frequency of intercourse and the number of partners, and increasing condom use share several characteristics. They focus on specific risk-taking behaviors, are based on theory, utilize experiential activities, address social influences on sexual behavior, reinforce values, and provide opportunities to practice new skills.

To be effective, multicultural sexuality education must reflect or be consistent with the culture(s) of participants. It should present messages that are compatible with participants' beliefs and practices. Such programs should utilize language and styles of communication that are appropriate.

REVIEW QUESTIONS

1. For most children, the major source of sex information is the mother. True or false?

2. Approximately 35 percent of American parents favor sex education in the schools. True or false?

3. By age 12, some children can give a good physiological explanation of reproduction. True or false?

4. On the basis of their research, the Goldmans argue that North American children are sexual illiterates who possess less sexual knowledge than their counterparts in other cultures such as Australia and Sweden. True or false?

5. Analyses of children's dirty jokes indicate that children realize their parents would prefer to keep sex a secret from them. True or false?

6. The national Personal Responsibility and Work Opportunity Act of 1996 provides $50 million dollars per year to states to support _____ educational programs.

7. According to an editor of the *Journal of the American Medical Association,* money spent on abstinence-only sexuality education programs is money well spent. True or false?

8. At least two-thirds of parents and high-school students support programs which make condoms available in the schools. True or false?

9. *Sex Respect* is a model sex education curriculum sponsored by SIECUS and favored by most sex education experts. True or false?

10. Research on the effects of sexuality education indicates that those teenagers who had sexuality education in high school are more likely to engage in premarital intercourse and have a higher incidence of premarital pregnancies. True or false?

(The answers to all review questions are at the end of the book, after the Glossary.)

QUESTIONS FOR THOUGHT, DISCUSSION, AND DEBATE

1. Design a sample sexuality education curriculum for the schools, indicating what topics you think would be important to teach at various age levels and what your reasoning is behind your choices.

2. Debate the following topic. Resolved: A sexuality education unit, at least a week long, should be included in all grades in all schools.

3. Data from surveys indicate that mass media are a major source of information about sexuality for children and adolescents. We saw, in Chapter 1 and Focus 11.1, that media portrayals in the United States are generally unrealistic. How would you deal with access to media portrayals of sexuality for your child at ages 5, 10, and 15? What would you do, and what would you tell your child at each age?

SUGGESTIONS FOR FURTHER READING

Donovan, P. (1989). *Risk and responsibility: Teaching sex/education in America's schools today.* New York: Alan Guttmacher Institute. To contact the Institute, look in the Directory of Resources following this chapter.

Gordon, Sol (1986, Oct.). What kids need to know. *Psychology Today.* Reprinted in O. Pocs (Ed.), *Human Sexuality 88/89.* Guilford, CT: Dushkin. Sol Gordon, a leading sex educator, provides a humorous view of what sexuality education ought to be accomplishing.

Mayle, P., Robins, A., and Walter P. (1973). *Where did I come from?* Secaucus, NJ: Lyle Stuart. A delightful sexuality education book for young children. A companion volume—*What's happening to me?*—is for children approaching or experiencing puberty.

SIECUS. (1991). *Guidelines for comprehensive sexuality education: Kindergarten–12th grade.* New York: Sexuality Information and Education Council of the United States. The guidelines outline a comprehensive program, divided into 36 topics, and contain developmentally appropriate messages according to school level. To contact SIECUS, look in the Web Resources list or the Directory of Resources following this chapter.

WEB RESOURCES

http://www.plannedparenthood.org
   Planned Parenthood: Parents pages.

http://www.siecus.org
   Sexuality Information and Education Council of the United States: Parents area.

http://educ.indiana.edu/cas/tt/v1i3/table.html
   An issue of *Teacher Talk* devoted to sex education from a teacher's viewpoint.

http://www.cis.yale.edu/ynhti/curriculum/guides/1991/5/91.05.01.x.html
   Information about sex education in multicultural settings.

For answers to questions about sexuality:

http://www.sfsi.org
   San Francisco Sex Information; telephone information and referral service.

http://www.goaskalice.columbia.edu
   A Health Question & Answer Service by Alice!, Columbia University's Health Education Program.

For general medical information:

http://www.healthfinder.gov/default.htm
   Health information site maintained by the U.S. Department of Health and Human Services.

http://www.medicinenet.com/
   A wealth of medical information, including virtually every known disease.

The McGraw-Hill Sexuality Resources Page:

http://www.mhhe.com/socscience/sex/

Epilogue

In the AIDS era, the challenge for us is to create joyous, intimate, fulfilling sex. We are convinced that there are two courses of action that are absolute folly. One is to be so overwhelmed by the threat of AIDS that sex becomes locked in the icy grip of anxiety. The other absolute folly is to hide our heads in the sand, ignore the threat of AIDS (and herpes and genital warts), and continue with sex as usual. The AIDS epidemic will then only intensify. All of us need to steer a middle course in which we take AIDS and other STDs seriously, while keeping sex joyful and satisfying.

We have no easy answers to these challenges. Sex education is one part of the solution, as is drug education. Continued funding for both medical and psychological research is essential. We might even need to consider wild and crazy lifestyles like monogamy. Meanwhile, we need to remember that sex can and should be joyous.

Directory of Resources in Human Sexuality

I. Health Issues: Pregnancy, Contraception, Abortion, Diseases

American Cancer Society
1599 Clifton Road, N.E.
Atlanta, GA 30329-4251
1-800-227-2345
http://www.cancer.org
Offers up-to-date and accurate information on cancer treatment and support. Funds research on cancer.

American Foundation for the Prevention of Venereal Disease, Inc.
799 Broadway, Suite 638
New York, NY 10003
(212) 759-2069
Provides a complete guide, *STD Prevention,* emphasizing sex education and personal hygiene.

AVSC International
79 Madison Avenue
New York, NY 10016
(212) 561-8000
email: info@avsc.org
http://www.avsc.org
Supports service delivery and offers information regarding family planning and a wide range of reproductive health services.

CDC National AIDS Clearinghouse
P.O. Box 6003
Rockville, MD 20849-6003
1-800-458-5231
Operated by the U.S. Public Health Service, this clearinghouse provides information and referrals on AIDS/HIV.

La Leche League International
1400 N. Meacham Road
P.O. Box 4079
Schaumburg, IL 60168
(847) 519-7730
email: lllhq@llli.org
http://www.lalecheleague.org/
An organization devoted to helping mothers worldwide to breastfeed through mother-to-mother support, encouragement, education, and information.

National Abortion Rights Action League (NARAL)
1156 15th Street N.W., Suite 700
Washington, DC 20005
(202) 973-3000
http://www.naral.org
A political action organization working at both state and national levels, dedicated to preserving a woman's right to safe and legal abortion and also to teaching its members effective use of the political process to ensure abortion rights.

Division of STD/HIV/TB Prevention
National Center for Prevention Services
Centers for Disease Control
Mail Stop E-06
Atlanta, GA 30333
(404) 639-8063
email: nchstp@cdc.gov
http://www.cdc.gov/nchstp/od/nchstp.html
Offers the most up-to-date information on prevention-related issues and sexually transmitted diseases and HIV; administers federal programs for the prevention of STD and HIV infection.

National Lesbian and Gay Health Association
1407 S Street N.W.
Washington, DC 20009
(202) 939-7880
An organization devoted to meeting the health needs of lesbians and gay men.

National Right to Life Committee, Inc.
Suite 500
419 7th Street N.W.
Washington, DC 20004
(202) 626-8800
http://www.nrlc.org
An organization based on the belief that human life begins at conception and that abortions should therefore be opposed.

National Women's Health Resource Center
2425 L Street N.W., 3rd floor
Washington, DC 20037
(202) 293-6045
A national clearinghouse on women's health information. Publishes a newsletter, *National Women's Health Report.*

Planned Parenthood Federation of America (PPFA)
810 Seventh Avenue
New York, NY 10019
(212) 541-7800
http://www.plannedparenthood.org
PPFA is the nation's oldest and largest voluntary family planning agency. Through local clinics (call 1-800-230-PLAN for the clinic nearest you), it offers birth control information and services, pregnancy testing, voluntary sterilization, prenatal care, abortion, pelvic and breast exams, and other reproductive health services, including sexuality education.

Population Information Program
Johns Hopkins Center for Communication Programs
111 Market Place, Suite 310
Baltimore, MD 21202
Fax (410) 659-6266
Publishes *Population Reports,* frequent, up-to-date reports on contraception and family planning with emphasis on developing countries.

II.   Sex Education, Sex Research, and Sex Therapy

The Alan Guttmacher Institute
120 Wall Street
New York, NY 10005
(212) 248-1111
email: info@agi-usa.org
http://www.agi-usa.org
A not-for-profit organization for reproductive health research, policy analysis, and public education. It produces many excellent, informative publications.

American Association of Sex Educators, Counselors, and Therapists (AASECT)
P.O. Box 238
Mount Vernon, IA 52314-0238
(319) 895-8407

email: aasect@worldnet.att.net
http://www.aasect.org
This organization certifies sex educators, sex counselors, and sex therapists and provides other services associated with sexuality education and sex therapy.

Sexuality Information and Education Council of the United States (SIECUS)
130 West 42nd Street, Suite 350
New York, NY 10036
(212) 819-9770
email: siecus@siecus.org
http://www.siecus.org
Provides a library and information service on sexuality education, including curricula. Publishes bibliographies and maintains a database of titles of books and journals on human sexuality, currently consisting of over 8000 entries.

The Society for the Scientific Study of Sexuality
P.O. Box 208
Mount Vernon, IA 52314-0208
(319) 895-8407
email: thesociety@worldnet.att.net
http://www.ssc.wisc.edu/ssss
An organization devoted to promoting quality sex research; publishes the *Journal of Sex Research.*

III.   Lifestyle Issues

Harry Benjamin International Gender Dysphoria Association
P.O. Box 1718
Sonoma, CA 95476
Society for professionals interested in the study and care of transsexualism and gender dysphoria.

Intersex Society of North America
P.O. Box 31791
San Francisco, CA 94131
email: info@isna.org
http://www.isna.org
ISNA is a peer support and advocacy group for intersexuals (persons born with mixed sexual anatomy).

J2CP Information Services
P.O. Box 184
San Juan Capistrano, CA 92693-0184
Information on transsexualism and professional referrals.

Lambda Legal Defense and Education Fund
120 Wall Street, Suite 1500
New York, NY 10005-3904
(212) 809-8585
and
6030 Wilshire Blvd., Suite 200
Los Angeles, CA 90036-3617
(213) 937-2728

email: http://www.lldef.org
http://www.lambdalegal.org
Advances the legal rights of lesbians, gay men, and people with AIDS through test case litigation and public education. Publishes many resource manuals, newsletters, bibliographies, and articles on current topics for lesbians, gay men, and people with HIV/AIDS.

National Gay and Lesbian Task Force (NGLTF)
1734 14th Street N.W.
Washington, DC 20009-3409
(202) 332-6483
The oldest national gay and lesbian civil rights advocacy organization. Lobbying, grassroots organizing, publications (call or write for listing), and referrals.

Society for the Second Self
Box 194
Tulare, CA 93275
(209) 688-9246
email: trichi1@aol.com
http://members.aol.com/chitriess/trisss/
chimain.htm
An organization for heterosexual men who cross-dress and their wives.

IV. Media

Multi-Focus, Inc.
1525 Franklin Street
San Francisco, CA 94109-4592
1-800-821-0514
Has one of the largest selections of sexuality education and sex therapy films and videos available for rental or purchase.

Focus International
1160 East Jericho Turnpike, Suite 15
Huntington, NY 11743
1-800-843-0305
email: Sex_Help@focusint.com
http://www.hip.com/focus
Another organization with a large selection of sexuality education and sex therapy films and videos.

V. Sexual Victimization

Violence and Traumatic Stress Research Branch
National Institute of Mental Health
5600 Fishers Lane, Room 10C-024
Rockville, MD 20857
(301) 443-3728
This branch is the focal point at the National Institute of Mental Health for research on violent behavior, including sexual abuse, sexual assault, and trauma (including PTSD).

VI. Feminism and Gender Issues

National Organization for Women (NOW)
1000 Sixteenth St. N.W., Suite 700
Washington, DC 20036
(202) 331-0066
email: now@now.org
http://www.now.org/
NOW seeks to take action to bring women into full participation in the mainstream of U.S. society, exercising all the privileges and responsibilities thereof in truly equal partnership with men.

VII. Journals

Annual Review of Sex Research
The Society for the Scientific Study of Sexuality
P.O. Box 208
Mount Vernon, IA 52314-0208

Archives of Sexual Behavior
Plenum Publishing Corporation
233 Spring Street
New York, NY 10013-1578

Gender & Society
Sage Publications
2455 Teller Road
Newbury Park, CA 91320

Journal of Child Sexual Abuse
Haworth Press
10 Alice Street
Binghamton, NY 13904-1580

Journal of Gay and Lesbian Psychotherapy
Haworth Press
10 Alice Street
Binghamton, NY 13904-1580

Journal of Homosexuality
Haworth Press
10 Alice Street
Binghamton, NY 13904-1580

Journal of Men's Studies
Men's Studies Press
P.O. Box 32
Harriman, TN 37748-0032

Journal of Sex and Marital Therapy
Brunner/Mazel, Inc.
19 Union Square West
New York, NY 10003
(212) 924-3344

Journal of Sex Education and Therapy
American Association of Sex Educators,
Counselors, and Therapists
P.O. Box 208
Mount Vernon, IA 52314-0208

Journal of Sex Research
The Society for the Scientific Study of Sexuality
P.O. Box 208
Mount Vernon, IA 52314-0208

Journal of the History of Sexuality
University of Chicago Press
P.O. Box 37005
Chicago, IL 60637

Psychology of Women Quarterly
Cambridge University Press
40 West 20th Street
New York, NY 10011

Sex Roles: A Journal of Research
Plenum Publishing Corporation
233 Spring Street
New York, NY 10013

Sexual Abuse: A Journal of Research and Treatment
Plenum Publishing Corporation
233 Spring Street
New York, NY 10013-1578

Sexual Addiction and Compulsivity
Brunner/Mazel, Inc.
19 Union Square West
New York, NY 10003
(212) 924-3344

Sexuality and Disability
Human Sciences Press
233 Spring Street
New York, NY 10013-1578
(212) 620-8000

Bibliography

Abbey, Antonia. (1991). Misperception as an antecedent of acquaintance rape: A consequence of ambiguity in communication between men and women. In A. Parrott & L. Bechhofer (Eds.), *Acquaintance rape: The hidden crime.* New York: Wiley.

Abel, Ernest L. (1984). *Fetal alcohol syndrome and fetal alcohol effects.* New York: Plenum.

Abel, Ernest L. (1980). Fetal alcohol syndrome. *Psychological Bulletin, 87,* 29–50.

Abel, Gene G., Osborn, Candace, Anthony, David, & Gardos, Peter. (1992). Current treatments of paraphiliacs. *Annual Review of Sex Research, 3,* 255–290.

Abel, Gene, & Rouleau, Joanne-L. (1990). The nature and extent of sexual assault. In W. L. Marshall, D. R. Laws, & H. E. Bartarce (Eds.), *Handbook of sexual assault* (pp. 9–21). New York: Plenum.

Abortion: Who's behind the violence? (1994, Nov. 14). *U.S. News and World Report,* 50–55 ff.

Abramowitz, Stephen I. (1986). Psychosocial outcomes of sex reassignment surgery. *Journal of Consulting and Clinical Psychology, 54,* 183–189.

Acker, Michele, & Davis, Mark. (1992). Intimacy, passion and commitment in adult romantic relationships: A test of the triangular theory of love. *Journal of Social and Personal Relationships, 9,* 21–50.

Ackerman, Mark D., & Carey, Michael P. (1995). Psychology's role in the assessment of erectile dysfunction: Historical precedents, current knowledge, and methods. *Journal of Consulting and Clinical Psychology, 63,* 862–876.

ACSF Investigators. (1992). AIDS and sexual behaviour in France. *Nature, 360,* 407–409.

Addiego, Frank, Belzer, E. G., Comolli, J., Moger, W., Perry, J. D., & Whipple, B. (1981). Female ejaculation: A case study. *Journal of Sex Research, 17,* 13–21.

Adler, Nancy E., David, Henry P., Major, Brenda P., Roth, Susan H., Russo, Nancy Felipe, & Wyatt, Gail E. (1992). Psychological factors in abortion: A review. *American Psychologist, 47,* 1194–1204.

Adler, N. E., et al. (1990). Psychological responses after abortion. *Science, 248,* 41–44.

Afriat, Cydney. (1995). Antepartum care. In Donald R. Coustan, Ray V. Hunning, Jr., & Don Singer (Eds.), *Human reproduction: Growth and development* (pp. 213–234). Boston: Little, Brown & Co.

Agnew, Bruce, & Stein, Rob. (1990, Jan.). Getting closer to an AIDS vaccine. *The Journal of NIH Research, 2,* 33.

Alan Guttmacher Institute. (1994). *Sex and America's teenagers.* New York: Author.

Albin, Rochelle S. (1977). Psychological studies of rape. *Signs, 3,* 423–435.

Alexander, Craig J., Sipski, Marca L., & Findley, Thomas W. (1993). Sexual activities, desire, and satisfaction in males pre- and postspinal cord injury. *Archives of Sexual Behavior, 22,* 217–228.

Alexander, Pamela C., & Lupfer, Shirley L. (1987). Family characteristics and long-term consequences associated with sexual abuse. *Archives of Sexual Behavior, 16,* 235–246.

Alkhatib, G., et al. (1996). CC CKR5: A RANTES, MIP1α, MIP-Iβ receptor as a fusin cofactor for macrophage-tropic HIV-1. *Science, 272,* 1955.

Allen, Donald M. (1980). Young male prostitutes: A psychological study. *Archives of Sexual Behavior, 9,* 399–426.

Allen, Katherine R., & Demo, David H. (1995). The families of lesbians and gay men: A new frontier in family research. *Journal of Marriage and the Family, 57,* 111–127.

Allgeier, Elizabeth Rice, & Wiederman, Michael W. (1994). How useful is evolutionary psychology for understanding contemporary human sexual behavior? *Annual Review of Sex Research, 5,* 218–256.

Alter, Miriam J., et al. (1986). Hepatitis B virus transmission between heterosexuals. *Journal of the American Medical Association, 256,* 1307–1310.

Alzate, Heli. (1990). Vaginal erotogeneity, "female ejaculation," and the "Grafenberg spot." *Archives of Sexual Behavior, 19,* 607–609.

Alzate, Heli, & Londono, M. L. (1984). Vaginal erotic sensitivity. *Journal of Sex & Marital Therapy, 10,* 499–506.

Amaro, Hortensia. (1995). Love, sex, and power: Considering women's realities in HIV prevention. *American Psychologist, 50,* 437–447.

Ambrosone, Christine B., et al. (1996). Cigarette smoking, N-Acetyltransferase 2 genetic polymorphisms, and breast cancer risk. *Journal of the American Medical Association, 276,* 1494–1501.

American Cancer Society. (1998). *Cervical cancer, testicular cancer, breast cancer, and prostate cancer.* Atlanta, GA: American Cancer Society.

American Cancer Society. (1995). *Cancer facts and figures—1995.* Atlanta, GA: American Cancer Society.

American Cancer Society. (1988). *Facts on testicular cancer.* New York: American Cancer Society.

American Law Institute. (1962). *Model penal code: Proposed official draft.* Philadelphia, PA: ALI.

American Psychiatric Association. (1994). *Diagnostic and statistical manual of mental disorders,* 4th ed. Washington, D.C.: American Psychiatric Association.

American Psychological Association, Ad Hoc Committee on Ethical Standards in Psychological Research. (1973). *Ethical principles in the conduct of research with human participants.* Washington, D.C.: APA.

Ames, Thomas-Robert. (1991). Guidelines for providing sexuality-related services to severely and profoundly retarded individuals: The challenge for the 1990s. *Sexuality and Disability, 9,* 113–122.

Andersen, Barbara L. (1983). Primary orgasmic dysfunction: Diagnostic considerations and review of treatment. *Psychological Bulletin, 93,* 105–136.

Anderson, E. (1989). Sex codes and family life among poor inner-city youths. *Annals of the American Academy of Political and Social Science, 501,* 59–78.

Andrews, Frank, & Halman, L. Jill. (1992). Infertility and subjective well-being: The mediating roles of self-esteem, interpersonal control, and interpersonal conflict. *Journal of Marriage and the Family, 54,* 408–417.

Andrews, Lori B. (1989). Alternative modes of reproduction. In S. Cohen & N. Taub (Eds.), *Reproductive laws for the 1990s* (pp. 361–404). Clifton, NJ: Humana Press.

Anson, Robert S. (1977, October 25). *San Francisco Chronicle.*

Antle, Katharyn. (1978). Active involvement of expectant fathers in pregnancy: Some further considerations. *Journal of Obstetric, Gynecologic and Neonatal Nursing, 7(2),* 7–12.

Aquinas, St. Thomas. (1968). *Summa theologica* (Vol. 43). (Thomas Gilly, Trans.). New York: McGraw-Hill.

Arafat, Ibtihaj, S., & Cotton, Wayne L. (1974). Masturbation practices of males and females. *Journal of Sex Research, 10,* 293–307.

Aries, Elizabeth. (1996). *Men and women in interaction: Reconsidering the differences.* New York: Oxford University Press.

Ashton, Eleanor. (1983). Measures of play behavior: The influence of sex-role stereotyped children's books. *Sex Roles, 9,* 43–47.

Associated Press. (1998). Sixteen deaths among Viagra users prompt renewed warning. http://www.cnn.com/HEALTH/9806/09/viagra.

Associated Press. (1984, Mar. 25). Baby girl is born from transferred embryo. *New York Times.*

Avis, Nancy E., & McKinlay, Sonja M. (1995, Mar.–Apr.). The Massachusetts Women's Health Study: An epidemiological investigation of the menopause. *JAMWA, 50,* 45–63.

Bach, G., & Wyden, P. (1969). *The intimate enemy: How to fight fair in love and marriage.* New York: Morrow.

Bachmann, G. A., & Leiblum, S. R. (1991). Sexuality in sexagenarian women. *Maturitas, 13,* 43–50.

Badgett, M. V. Lee. (1995). The wage effects of sexual orientation discrimination. *Industrial and Labor Relations Review, 48,* 726–739.

Bailey, J. Michael, Bobrow, David, Wolfe, Marilyn, & Mikach, Sarah. (1995). Sexual orientation of adult sons of gay fathers. *Developmental Psychology, 31,* 124–129.

Bailey, J. Michael, & Pillard, Richard C. (1995). Genetics of human sexual orientation. *Annual Review of Sex Research, 6,* 126–150.

Bailey, J. Michael, & Pillard, Richard C. (1991). A genetic study of male sexual orientation. *Archives of General Psychiatry, 48,* 1089–1096.

Bailey, J. M., Pillard, R. C., Neale, M. C., & Agyei, Y. (1993). Heritable factors influence sexual orientation in women. *Archives of General Psychiatry, 50,* 217–223.

Bailey, J. M., Willerman, L., & Parks, C. (1991). A test of the maternal stress theory of human male homosexuality. *Archives of Sexual Behavior, 20,* 277–294.

Bailey, J. Michael, & Zucker, Kenneth J. (1995). Childhood sex-typed behavior and sexual orientation. *Developmental Psychology, 31,* 43–55.

Baladerian, Nora J. (1991). Sexual abuse of people with developmental disabilities. *Sexuality and Disability, 9,* 323–335.

Baldwin, John D., & Baldwin, Janice I. (1997). Gender differences in sexual interest. *Archives of Sexual Behavior, 26,* 181–210.

Baldwin, John D., & Baldwin, Janice I. (1989). The socialization of homosexuality and heterosexuality in a non-Western society. *Archives of Sexual Behavior, 18,* 13–30.

Baldwin, John D., Whiteley, Scott, & Baldwin, Janice I. (1992). The effect of ethnic group on sexual activities related to contraception and STDs. *Journal of Sex Research, 29,* 189–205.

Balter, Michael. (1998). Impending AIDS vaccine trial opens old wounds. *Science, 279,* 650.

Balter, Michael. (1996). A second coreceptor for HIV in early stages of infection. *Science, 272,* 1740.

Balter, Michael. (1995). Elusive HIV-suppressor factors found. *Science, 270,* 1560–1561.

Bancroft, John. (1987). A physiological approach. In J. H. Geer & W. T. O'Donohue (Eds.), *Theories of human sexuality* (pp. 411–421). New York: Plenum.

Bancroft, John. (1978). The prevention of sexual offenses. In C. B. Qualls et al. (Eds.), *The prevention of sexual disorders* (pp. 95–116). New York: Plenum.

Bancroft, John, & Gutierrez, P. (1996). Erectile dysfunction in men with and without diabetes mellitus. *Diabetic Medicine, 13,* 84–89.

Bandura, Albert J. (1977). *Social learning theory.* Englewood Cliffs, NJ: Prentice-Hall.

Bandura, Albert. (1982). Self-efficacy mechanism in human agency. *American Psychologist, 37,* 122–147.

Bandura, Albert, & Walters, Richard H. (1963). *Social learning and personality development.* New York: Holt.

Banks, Amy, & Gartrell, Nanette K. (1995). Hormones and sexual orientation: A questionable link. *Journal of Homosexuality, 28,* 247–268.

Barash, David P. (1982). *Sociobiology and behavior* (2d ed.). New York: Elsevier.

Barash, David P. (1977a). *Sociobiology and behavior.* New York: Elsevier.

Barash, David P. (1977b). Sociobiology of rape in mallards: Response of the mated male. *Science, 197,* 788–789.

Barbach, Lonnie G. (1993). *The pause: Positive approaches to menopause.* New York: Dutton.

Barbach, Lonnie G. (1975). *For yourself: The fulfillment of female sexuality.* Garden City, NY: Anchor/Doubleday.

Barbaree, H. E., & Marshall, W. L. (1991). The role of male sexual arousal in rape: Six models. *Journal of Consulting and Clinical Psychology, 59,* 621–630.

Barlow, David H. (1986). Causes of sexual dysfunction: The role of cognitive interference. *Journal of Consulting and Clinical Psychology, 54,* 140–148.

Barlow, David. H., Leitenberg, H., Agras, W. S. (1969). Experimental control of sexual deviation through manipulation of noxious scenes in covert sensitization. *Journal of Abnormal Psychology, 74,* 596–601.

Barnes, Deborah. (1995, Feb.). HIV-1-infected long-term nonprogressors. *Journal of NIH Research, 7,* 19–21.

Baron, Larry, & Straus, Murray A. (1989). *Four theories of rape in American society.* New Haven, CT: Yale University Press.

Barr, Helen M., Streissguth, Ann P., Darby, Betty L., & Sampson, Paul D. (1990). Prenatal exposure to alcohol, caffeine, tobacco and aspirin: Effects on fine and gross motor performance in 4-year-old children. *Developmental Psychology, 26,* 339–348.

Barrows, Sydney B. (1986). *Mayflower madam.* New York: Arbor House.

Bart, Pauline B., & O'Brien, P. H. (1985). *Stopping rape: Successful survival strategies.* New York: Pergamon.

Bartell, Gilbert D. (1970). Group sex among the mid-Americans. *Journal of Sex Research, 6,* 113–130.

Baumeister, Roy F. (1988a). Masochism as escape from the self. *Journal of Sex Research, 25,* 28–59.

Baumeister, Roy F. (1988b). Gender differences in masochistic scripts. *Journal of Sex Research, 25,* 478–499.

Bauserman, Robert, & Rind, Bruce. (1997). Psychological correlates of male child and adolescent sexual experiences with adults: A review of the nonclinical literature. *Archives of Sexual Behavior, 26,* 105–142.

Beach, Frank A. (1947). Evolutionary changes in the physiological control of mating behavior in mammals. *Psychological Review, 54,* 297–315.

Beach, Frank A. (Ed.). (1976). *Human sexuality in four perspectives.* Baltimore: Johns Hopkins University Press.

Beach, Frank, & Merari, A. (1970). Coital behavior in dogs. V. Effects of estrogen and progesterone on mating and other forms of social behavior in the bitch. *Journal of Comparative and Physiological Psychology Monograph, 70*(1), Part 2, 1–22.

Beall, Anne, & Sternberg, Robert. (1995). The social construction of love. *Journal of Social and Personal Relationships, 12,* 417–438.

Beck, J. Gayle. (1995). Hypoactive sexual desire disorder: An overview. *Journal of Consulting and Clinical Psychology, 63,* 919–927.

Becker, Judith V. (1989). Impact of sexual abuse on sexual functioning. In S. R. Leiblum & R. C. Rosen (Eds.), *Principles and practice of sex therapy: Update for the 1990s* (pp. 298–318). New York: Guilford.

Beckman, Linda J., & Harvey, S. Marie. (Eds.). (1998). *The new civil war: The psychology, culture, and politics of abortion.* Washington, D.C.: American Psychological Association.

Beier, E. G., & Sternberg, D. P. (1977). Marital communication. *Journal of Communication, 27,* 92–103.

Bell, Alan P. (1974a, Summer). *Childhood and adolescent sexuality.* Address delivered at the Institute for Sex Research, Indiana University.

Bell, Alan P. (1974b). Homosexualities: Their range and character. In *Nebraska symposium on motivation 1973.* Lincoln, NE: University of Nebraska Press.

Bell, Alan P., & Weinberg, Martin S. (1978). *Homosexualities.* New York: Simon & Schuster.

Bell, Alan P., Weinberg, Martin S., & Hammersmith, Sue K. (1981). *Sexual preference.* Bloomington, IN: Indiana University Press.

Belzer, E. G. (1981). Orgasmic expulsions of women: A review and heuristic inquiry. *Journal of Sex Research, 17,* 1–12.

Bem, Daryl J. (1998). Is EBE theory supported by the evidence? Is it androcentric? A reply to Peplau et al. (1998). *Psychological Review, 105,* 395–398.

Bem, Daryl J. (1996). Exotic becomes erotic: A developmental theory of sexual orientation. *Psychological Review, 103,* 320–335.

Bem, Sandra L. (1981). Gender schema theory: A cognitive account of sex typing. *Psychological Review, 88,* 354–364.

Benedetti, Jacqueline, Corey, Lawrence, & Ashley, Rhoda. (1994). Recurrence rates in genital herpes after symptomatic first-episode infection. *Annals of Internal Medicine, 121,* 847–854.

Benson, Rebecca. (1986). Pornography and the first amendment. *Harvard Women's Law Journal, 9,* 153–172.

Bérard, E. J. J. (1989). The sexuality of spinal cord injured women: Physiology and pathophysiology: A Review. *Paraplegia, 27,* 99–112.

Beretta, G., Chelo, E., & Zanollo, A. (1989). Reproductive aspects in spinal cord injured males. *Paraplegia, 27,* 113–118.

Berg, J. H., & Derlega, V. J. (1987). Themes in the study of self-disclosure. In V. J. Derlega & J. H. Berg (Eds.), *Self-disclosure: Theory, research and therapy* (pp. 1–8). New York: Plenum.

Bergen, D. J., & Williams, J. E. (1991). Sex stereotypes in the United States revisited: 1972–1988. *Sex Roles, 24,* 413–423.

Berliner, David L., Jennings-White, Clive, & Lavker, Robert M. (1991). The human skin: Fragrances and pheromones. *Journal of Steroid Biochemistry and Molecular Biology, 39,* 671–679.

Bermant, Gordon, & Davidson, Julian M. (1974). *Biological bases of sexual behavior.* New York: Harper & Row.

Bernard, M., Levine, E., Presser, S., & Stecich, M. (1985). *The rights of single people.* New York: Bantam Books.

Berne, Eric. (1970). *Sex in human loving.* New York: Simon & Schuster.

Bernstein, Anne C., & Cowan, Philip, A. (1975). Children's concepts of how people get babies. *Child Development, 46,* 77–92.

Berrill, K. T. (1992). Antigay violence and victimization in the United States. In G. M. Herek & K. T. Berrill (Eds.), *Hate crimes: Confronting violence against lesbians and gay men* (pp. 259–269). Newbury Park, CA: Sage.

Berscheid, Ellen, & Hatfield, Elaine. (1978). *Interpersonal attraction* (2d ed.). Reading, MA: Addison-Wesley.

Berscheid, Ellen, & Walster, Elaine. (1974a). A little bit about love. In T. L. Huston (Ed.), *Foundations of interpersonal attraction.* New York: Academic.

Berscheid, Ellen, & Walster, Elaine. (1974b). Physical attractiveness. *Advances in Experimental Social Psychology, 7,* 157–215.

Berscheid, Ellen, et al. (1971). Physical attractiveness and dating choice: A test of the matching hypothesis. *Journal of Experimental Social Psychology, 7,* 173–189.

Bess, Barbara E., & Janus, Samuel S. (1976). Prostitution. In B. J. Sadock et al. (Eds.), *The sexual experience.* Baltimore, MD: Williams & Wilkins.

Bibring, Grete, et al. (1961). A study of the psychological processes in pregnancy and of the earliest mother-child relationship. *The psychoanalytic study of the child* (Vol. XVI, pp. 9–72). New York: International Universities Press.

Bieber, Irving, Dain, H. J., & Dince, P. R. (1962). *Homosexuality: A psychoanalytic study.* New York: Basic Books.

Biller, Henry, & Meredith, D. (1975). *Father power.* New York: Anchor Books.

Billings, Andrew. (1979). Conflict resolution in distressed and nondistressed married couples. *Journal of Consulting and Clinical Psychology, 47,* 368–376.

Billstein, Stephan A. (1990). Human lice. In K. Holmes et al. (Eds.), *Sexually transmitted diseases* (2d ed.). New York: McGraw-Hill.

Billy, J. O. G., Tanfer, K., Grady, W. R., & Klepinger, D. H. (1993). The sexual behavior of men in the United States. *Family Planning Perspectives, 25*(2), 52–60.

Binson, Diane, Dolcini, M. Margaret, Pollack, Lance M., & Catania, Joseph. (1993). IV. Multiple sexual partners among young adults in high-risk cities. *Family Planning Perspectives, 25,* 268–272.

Birchler, Gary R., Weiss, R. L., & Vincent, J. P. (1975). Multimethod analysis of social reinforcement exchange between maritally distressed and nondistressed spouse and stranger dyads. *Journal of Personality and Social Psychology, 31,* 349–360.

Bixler, Ray H. (1986). Of apes and men (including females). *Journal of Sex Research, 22,* 255–267.

Blair, C. David, & Lanyon, Richard I. (1981). Exhibitionism: Etiology and treatment. *Psychological Bulletin, 89,* 439–463.

Blanchard, Ray. (1997). Birth order and sibling sex ratio in homosexual versus heterosexual males and females. *Annual Review of Sex Research, 8,* 27–67.

Blanchard, Ray, & Bogaert, Anthony F. (1996a). Biodemographic comparisons of homosexual and heterosexual men in the Kinsey interview data. *Archives of Sexual Behavior, 25,* 551–579.

Blanchard, Ray, & Bogaert, Anthony F. (1996b). Homosexuality in men and number of older brothers. *American Journal of Psychiatry, 153,* 27–31.

Blanchard, Ray, Clemmensen, L. H., & Steiner, B. W. (1983). Gender reorientation and psychosocial adjustment in male-to-female transsexuals. *Archives of Sexual Behavior, 12,* 503–510.

Blanchard, Ray, Dickey, Robert, & Jones, Corey L. (1995). Comparison of height and weight in homosexual versus nonhomosexual gender dysphorics. *Archives of Sexual Behavior, 24,* 543–554.

Blanchard, Ray, & Klassen, Philip. (1997). H-Y antigen and homosexuality in men. *Journal of Theoretical Biology, 185*, 373–378.

Blanchard, Ray, Zucker, Kenneth J., Bradley, Susan J., & Hume, Caitlin S. (1995). Birth order and sibling ratio in homosexual male adolescents and probably prehomosexual boys. *Developmental Psychology, 31*, 22–30.

Blank, Anne, Mofenson, Lynne M., Willoughby, Anne, & Yaffe, Sumner J. (1994). Maternal and pediatric AIDS in the United States: The current situation and future research directions. *Acta Paediatrica*, Supp. 400, 106–110.

Blechman, Elaine A., Clay, Connie J., Kipke, Michele D., & Bickel, Warren K. (1988). The premenstrual experience. In E. Blechman & K. Brownell (Eds.), *Handbook of behavioral medicine for women* (pp. 80–91). New York: Pergamon.

Blee, Kathleen M., & Tickamyer, Ann R. (1995). Racial differences in men's attitudes about women's gender roles. *Journal of Marriage and the Family, 57*, 21–30.

Blumstein, Philip W., & Schwartz, Pepper. (1983). *American couples.* New York: Morrow.

Bodlund, Owe, & Kullgren, Gunnar. (1996). Transsexualism—General outcome and prognostic factors: A five-year follow-up study of 19 transsexuals in the process of changing sex. *Archives of Sexual Behavior, 25*, 303–316.

Bogren, Lennart Y. (1991). Changes in sexuality in women and men during pregnancy. *Archives of Sexual Behavior, 20*, 35–45.

Boles, J., & Ellifson, K. (1994). Risk factors associated with HIV seropositivity in clients of male and female prostitutes. Presented at the Annual Meeting, Society for the Scientific Study of Sexuality, Miami, FL.

Bonaparte, Marie. (1965). *Female sexuality.* New York: Grove. (First published by International Universities Press, 1953).

Booth, Cathryn L., & Meltzoff, Andrew N. (1984). Expected and actual experience in labour and delivery and their relationship to maternal attachment. *Journal of Reproductive and Infant Psychology, 2*, 79–91.

Bootzin, R. R., & Natsoulas, T. (1965). Evidence for perceptual defense uncontaminated by response bias. *Journal of Personality and Social Psychology, 1*, 461–468.

Bornstein, Robert F. (1989). Exposure and affect: Overview and meta-analysis of research, 1968–1987. *Psychological Bulletin, 106*, 265–289.

Boruchowitz, Robert C. (1973). Victimless crimes: A proposal to free the courts. *Judicature, 57*, 69–78.

Bosch, F. X., Manos, M. M., Muñoz, N. et al. (1995). Prevalence of human papillomavirus in cervical cancer: A worldwide perspective. *Journal of the National Cancer Institute, 87*, 796–802.

Boston Women's Health Book Collective. (1996). *The new our bodies, ourselves: A book by and for women.* New York: Simon & Schuster.

Boston Women's Health Book Collective. (1992). *The new our bodies, ourselves.* New York: Simon & Schuster.

Boswell, John. (1994). *Same-sex unions in premodern Europe.* New York: Villard Books.

Boswell, John. (1980). *Christianity, social tolerance, and homosexuality.* Chicago: University of Chicago Press.

Boyd, K., Callaghan, B., & Shotter, E. (1986). *Life before birth.* London: SPCK.

Bradford, John M. W., & Greenberg, D. M. (1996). Pharmacological treatment of deviant sexual behaviour. *Annual Review of Sex Research, 7*, 283–306.

Bradford, John M. W., & Pawlak, Anne. (1993a). Double-blind placebo crossover study of cyproterone acetate in the treatment of the paraphilias. *Archives of Sexual Behavior, 22*, 383–402.

Bradford, John M. W., & Pawlak, Anne. (1993b). Effects of cyproterone acetate on sexual arousal patterns of pedophiles. *Archives of Sexual Behavior, 22*, 629–642.

Brady, Katherine. (1978). *Father's days.* New York: Dell.

Braun, Stephen. (1996). New experiments underscore warnings on maternal drinking. *Science, 273*, 738–739.

Brecher, Edward M. (1984). *Love, sex, and aging.* Mount Vernon, NY: Consumers Union.

Breitman, P., Knutson, K., & Reed, P. (1987). *How to persuade your lover to use a condom . . . and why you should.* Rocklin, CA: Prima Publishing.

Brent, Jonathan. (1976). A general introduction to privacy. *Massachusetts Law Quarterly, 61*, 10–18.

Brenton, Myron. (1972). *Sex talk.* New York: Stein and Day.

Breslow, N., Evans, L., & Langley, J. (1985). On the prevalence and roles of females in the sadomasochistic subculture: Report of an empirical study. *Archives of Sexual Behavior, 14*, 303–318.

Brest, Paul, & Vandenberg, Ann. (1987). Politics, feminism, and the Constitution. *Stanford Law Review, 39*, 607–671.

Breton, Sylvie, Smith, Peter J. S., Lut, Ben, & Brown, Dennis. (1996). Acidification of the male reproductive tract by a proton pumping (H^+)-ATPase. *Nature Medicine, 2*, 470–472.

Bretschneider, Judy G., & McCoy, Norma L. (1988). Sexual interest and behavior in healthy 80- and 102-year-olds. *Archives of Sexual Behavior, 17*, 109–130.

Brewster, Karin. (1994). Race differences in sexual activity among adolescent women: The role of neighborhood characteristics. *American Sociological Review, 59*, 408–424.

Brim, Orville G. (1976). Theories of the male mid-life crisis. *Counseling Psychologist, 6*, 2–9.

Brinton, Crane. (1959). *A history of Western morals.* New York: Harcourt Brace Jovanovich.

Brinton, Louise A., & Schairer, Catherine. (1997). Postmenopausal hormone-replacement therapy—time for a reappraisal? *New England Journal of Medicine, 336*, 1821–1822.

Broderick, Carlfred B. (1966a). Sexual behavior among preadolescents. *Journal of Social Issues, 22*(2), 6–21.

Broderick, Carlfred B. (1966b). Socio-sexual development in a suburban community. *Journal of Sex Research, 2*, 1–24.

Brosius, Hans-Berad, Weaver, James B. III, & Staab, Joachim. (1993). Exploring the social and sexual "reality" of contemporary pornography. *Journal of Sex Research, 30*, 161–170.

Brotman, Harris. (1984, Jan. 8). Human embryo transplants. *New York Times Magazine*, 42ff.

Brown, Jane, & Steele, Jeanne R. (1996). Sexuality and the mass media: An overview. *SIECUS Report, 24*(4), 3–9.

Brown, Nancy L., Pennylegion, Michelle, & Hillard, Pamela. (1997). A process evaluation of condom availability in the Seattle, Washington public schools. *Journal of School Health, 67*, 336–340.

Brownmiller, Susan. (1975). *Against our will: Men, women, and rape.* New York: Simon & Schuster.

Bryant, J., & Rockwell, S.C. (1994). Effects of massive exposure to sexually oriented prime-time television programming on adolescents' moral judgment. In D. Zillman, J. Bryant, and A.C. Houston (Eds), Media, children, and the family: Social, scientific, psychodynamic, and clinical perspectives (pp. 183–195). Hillsdale, NJ: Lawrence Erlbaum.

Bryant, Wayne M. (1997). *Bisexual characters in film.* New York: Haworth.

Buchanan, K. M. (1986). *Apache women warriors.* El Paso, TX: Texas Western Press.

Buchbinder, Susan P., et al. (1994). Long-term HIV-1 infection without immunologic progression. *AIDS, 8*, 1123–1128.

Bullough, Bonnie, & Bullough, Vern. (1997). Are transvestites necessarily heterosexual? *Archives of Sexual Behavior, 26*, 1–12.

Bullough, Vern L. (1994). *Science in the bedroom: A history of sex research.* New York: Basic Books.

Bullough, Vern L. (1976). *Sexual variance in society and history.* New York: Wiley.

Bumpass, Larry L., Sweet, James A., & Cherlin, Andrew. (1991). The role of cohabitation in declining rates of marriage. *Journal of Marriage and the Family, 53,* 913–927.

Burack, J. H., et al. (1993). Depressive symptoms and CD4 lympho-cyte decline among HIV-infected men. *Journal of the American Medical Association, 270,* 2568–2573.

Bureau of the Census. (1997). *Statistical abstract of the United States 1997.* Washington, D.C.: Bureau of the Census.

Bureau of the Census. (1990). Marital status and living arrangements: March 1990. *Current Population Reports,* Series P-20, No. 450.

Burger, H. G. (1993). Evidence for a negative feedback role of inhibin in follicle stimulating hormone regulation in women. *Human Reproduction, 8,* Suppl. 2, 129–132.

Burgess, Ann W. (1984). *Child pornography and sex rings.* Lexington, MA: Lexington Books (D.C. Heath).

Burgess, Ann W., & Holmstrom, Lynda L. (1974a). Rape trauma syndrome. *American Journal of Psychiatry, 131,* 981–986.

Burgess, Ann W., & Holmstrom, Lynda L. (1974b). *Rape: Victims of crisis.* Bowie, MD: Robert J. Brady.

Burleson, Brant, & Denton, Wayne. (1997). The relationship between communication skill and marital satisfaction: Some moderating effects. *Journal of Marriage and the Family, 59,* 884–902.

Burris, Scott. (1993). Testing, disclosure, and the right to privacy. In S. Burris, H. L. Dalton, & J. L. Miller (Eds.), *AIDS law today* (pp. 115–149). New Haven, CT: Yale University Press.

Burt, Martha R., & Estep, Rhoda E. (1981). Apprehension and fear: Learning a sense of sexual vulnerability. *Sex Roles, 7,* 511–522.

Burton, Frances D. (1970). Sexual climax in Macaca Mulatta. *Proceedings of the Third International Congress on Primatology, 3,* 180–191.

Buss, Arnold. (1966). *Psychopathology.* New York: Wiley.

Buss, David M. (1994). *The evolution of desire: Strategies of human mating.* New York: Basic Books.

Buss, David M. (1989). Sex differences in human mate preferences: Evolutionary hypotheses tested in 37 cultures. *Behavioral and Brain Sciences, 12,* 1–49.

Buss, David M. (1988). The evolution of human intra-sexual competition: Tactics of mate attraction. *Journal of Personality and Social Psychology, 54,* 616–628.

Buss, David, Larsen, Randy, & Westen, Drew. (1996). Sex differences in jealousy: Not gone, not forgotten, and not explained by alternative hypotheses. *Psychological Science, 7,* 373–375.

Buss, David M., & Schmitt, David P. (1993). Sexual strategies theory: An evolutionary perspective on human mating. *Psychological Review, 100,* 204–232.

Buss, David, & Shackelford, Todd. (1997a). From vigilance to violence: Mate retention tactics in married couples. *Journal of Personality and Social Psychology, 72,* 346–361.

Buss, David, & Shackelford, Todd. (1997b). Susceptibility to infidelity in the first year of marriage. *Journal of Research in Personality, 31,* 193–221.

Buunk, Bram, Angleitner, Alois, Oubaid, Viktor, & Buss, David. (1996). Sex differences in jealousy in evolutionary and cultural perspective: Tests from the Netherlands, Germany, and the United States. *Psychological Science, 7,* 359–363.

Buvat, Jacques, et al. (1990). Recent developments in the clinical assessment and diagnosis of erectile dysfunction. *Annual Review of Sex Research, 1,* 265–308.

Byard, Roger, Hucker, Stephen, & Hazelwood, Robert. (1993). Fatal and near-fatal autoerotic asphyxial episodes in women. *The American Journal of Forensic Medicine and Pathology, 14,* 70–73.

Byers, E. Sandra. (1996). How well does the traditional sexual script explain sexual coercion? Review of a program of research. *Journal of Psychology and Human Sexualtiy, 8,* 7–25.

Byler, Ruth V. (1969). *Teach us what we want to know.* New York: Mental Health Materials Center (for the Connecticut State Board of Education).

Byne, William. (1995). Science and belief: Psychobiological research on sexual orientation. *Journal of Homosexuality, 28,* 303–344.

Byne, William, & Parsons, Bruce. (1993). Human sexual orientation: The biologic theories reappraised. *Archives of General Psychiatry, 50,* 228–239.

Byrne, Donn. (1997). An overview (and underview) and research and theory within the attraction paradigm. *Journal of Social and Personal Relationships, 14,* 417–431.

Byrne, Donn. (1983). Sex without contraception. In D. Byrne & W. A. Fisher (Eds.), *Adolescents, sex, and contraception.* Hillsdale, NJ: Lawrence Erlbaum.

Byrne, Donn. (1977). Social psychology and the study of sexual behavior. *Personality and Social Psychology Bulletin, 3,* 3–30.

Byrne, Donn. (1971). *The attraction paradigm.* New York: Academic.

Byrne, Donn, Ervin, C. E., & Lamberth, J. (1970). Continuity between the experimental study of attraction and real-life computer dating. *Journal of Personality and Social Psychology, 16,* 157–165.

Cabaj, Robert P., & Stein, Terry S. (Eds.). (1996). *Textbook of homosexuality and mental health.* Washington, D.C.: American Psychiatric Press.

Cado, Suzanne, & Leitenberg, Harold. (1990). Guilt reactions to sexual fantasies during intercourse. *Archives of Sexual Behavior, 19,* 49–64.

Cahill, Lisa Sowle. (1985). The "seamless garment": Life in its beginnings. *Theological Studies, 46,* 64–80.

Calderwood, Deryck. (1987, May). The male rape victim. *Medical Aspects of Human Sexuality,* 53–55.

Calhoun, Karen S., Atkeson, B. M., & Resick, P. A. (1982). A longitudinal examination of fear reactions in victims of rape. *Journal of Counseling Psychology, 29,* 655–661.

Call, Vaughn, Sprecher, Susan, & Schwartz, Pepper. (1995). The incidence and frequency of marital sex in a national sample. *Journal of Marriage and the Family, 57,* 639–652.

Callahan, Daniel. (1986). How technology is reframing the abortion debate. *Hastings Center Report,* February 1986, 33–42.

Callahan, Sidney, & Callahan, Daniel. (1984). *Abortion: Understanding differences.* New York: Plenum.

Canary, Daniel J., & Dindia, Kathryn. (Eds.). (1998). *Sex differences and similarities in communication.* Mahwah, NJ: Erlbaum.

Canary, Daniel J., & Hause, Kimberley S. (1993). Is there any reason to research sex differences in communication? *Communication Quarterly, 41,* 129–144.

Cao, Yunzhen, et al. (1995). Virologic and immunologic characterization of long-term survivors of human immunodeficiency virus Type 1 infection. *New England Journal of Medicine, 332,* 201–208.

Carani, Cesare, et al. (1990). Effects of androgen treatment in impotent men with normal and low levels of free testosterone. *Archives of Sexual Behavior, 19,* 223–234.

Carey, Michael P., & Johnson, Blair T. (1996). Effectiveness of yohimbine in the treatment of erectile disorder: Four meta-analytic integrations. *Archives of Sexual Behavior, 25,* 341–360.

Carmichael, Marie S., Warburton, Valerie L., Dixen, Jean, & Davidson, Julian M. (1994). Relationships among cardiovascular, muscular, and oxytocin responses during human sexual activity. *Archives of Sexual Behavior, 23,* 59–80.

Carroll, Janell, Volk, K., & Hyde, J. S. (1985). Differences between males and females in motives for engaging in sexual intercourse. *Archives of Sexual Behavior, 14,* 131–139.

Carter, C. Sue. (1992). Hormonal influences on human sexual behavior. In J. B. Becker et al. (Eds.), *Behavioral endocrinology* (pp. 131–142). Cambridge, MA: MIT Press.

Cass, Vivienne C. (1979). Homosexual identity formation: A theoretical model. *Journal of Homosexuality, 4,* 219–235.

Castaneda, Donna. (1993). The meaning of romantic love among Mexican-Americans. *Journal of Social Behavior and Personality, 8,* 257–272.

Catalona, William J., et al. (1991). Measurement of prostate-specific antigen in serum as a screening test for prostate cancer. *New England Journal of Medicine, 324,* 1156–1161.

Catania, Joseph A., et al. (1992). Prevalence of AIDS-related risk factors and condom use in the United States. *Science, 258,* 1101–1106.

Catania, Joseph A., Binson, Diane, Van Der Straten, Ariane, & Stone, Valerie. (1995). Methodological research on sexual behavior in the AIDS era. *Annual Review of Sex Research, 6,* 77–125.

Catania, Joseph A., Gibson, D., Marin, B., Coates, T., & Greenblatt, R. (1990a). Response bias in assessing sexual behaviors relevant to HIV transmission. *Evaluation and Program Planning, 13,* 19–29.

Catania, Joseph A., Gibson, David R., Chitwood, Dale D., & Coates, Thomas J. (1990b). Methodological problems in AIDS behavioral research: Influences on measurement error and participation bias in studies of sexual behavior. *Psychological Bulletin, 108,* 339–362.

Catania, Joseph, Coates, Thomas, Peterson, John, Dolcini, Margaret, Kegeles, Susan, Siegel, David, Golden, Eve, & Fullilove, Mindy Thompson. (1993). Changes in condom use among black, Hispanic, and white heterosexuals in San Francisco: The AMEN cohort survey. *The Journal of Sex Research, 30,* 121–128.

Caughey, Madeline S. (1974). The principle of harm and its application to laws criminalizing prostitution. *Denver Law Journal, 51,* 235–262.

Cavert, Winston, et al. (1997). Kinetics of response in lymphoid tissues to antiretroviral therapy of HIV-1 infection. *Science, 276,* 960–964.

Centers for Disease Control and Prevention. (1997). Abortion surveillance: Preliminary analysis—United States, 1995. *Morbidity and Mortality Weekly Report, 46,* 1133–1137.

Centers for Disease Control and Prevention. (1997). 1995 Assisted reproductive technology success rates: National summary. Located at www.cdc.gov/nccdphp/dvh/arts.

Chalker, Rebecca. (1995, Nov./Dec.) *Sexual pleasure unscripted.* Ms, 49–52.

Chan, Connie S. (1995). Issues of sexual identity in an ethnic minority: The case of Chinese American lesbians, gay men, and bisexual people. In A. R. D'Augelli & C. J. Patterson (Eds.), *Lesbian, gay, and bisexual identities over the lifespan* (pp. 87–101). New York: Oxford University Press.

Chapman, Audrey. (1997). The black search for love and devotion: Facing the future against all odds. In H. P. McAdoo (Ed.), *Black families,* 3d ed. Thousand Oaks, CA: Sage.

Chapman, Heather, Hobfoll, Stevan, & Ritter, Christian. (1997). Partners' stress underestimations lead to women's distress: A study of pregnant inner-city women. *Journal of Personality and Social Psychology, 73,* 418–425.

Cherukuri, R., Minkoff, H., Feldman, J., Parekh, A., & Glass, L. (1988). A cohort study of alkaloidal cocaine ("crack") in pregnancy. *Obstetrics and Gynecology, 72,* 147–151.

Cherry, Kittredge, & Mitulski, James. (1988). We are the church alive, the Church with AIDS. *Christian Century, 105,* 85–88.

Chesler, Ellen. (1992). *Woman of valor: Margaret Sanger and the birth control movement.* New York: Simon & Schuster.

Chin, J. (1990). Current and future dimensions of the HIV/AIDS pandemic in women and children. *Lancet, 336,* 221–224.

Chow, J. M., et al. (1990). The association between chlamydia trachomatis and ectopic pregnancy: A matched-pair, case-control study. *Journal of the American Medical Association, 263,* 3164.

Chow, Yung-Kang, et al. (1993). Use of evolutionary limitations of HIV-1 multidrug resistance to optimize therapy. *Nature, 361,* 650–654.

Christensen, Cornelia V. (1971). *Kinsey: A biography.* Bloomington, IN: Indiana University Press.

Christopher, F. Scott, Owens, Laura, & Stecker, Heidi. (1993). Exploring the dark side of courtship: A test of a model of male premarital sexual aggressiveness. *Journal of Marriage and the Family, 55,* 469–479.

Cimbalo, R. S., Faling, B., & Mousaw, P. (1976). The course of love: A cross-sectional design. *Psychological Reports, 38,* 1292–1294.

Clanton, Gordon, & Smith, Lynn G. (Eds.). (1986). *Jealousy.* Lanham, MD: University Press of America.

Clark, Jacquelyn L., Taum, Nancy O., & Noble, Sara L. (1995). Management of genital herpes. *American Family Physician, 51,* 175–182.

Clarren, S. K., & Smith, D. W. (1978). The fetal alcohol syndrome. *New England Journal of Medicine, 298,* 1063–1067.

Clement, Ulrich. (1990). Surveys of heterosexual behavior. *Annual Review of Sex Research, 1,* 45–74.

Clemente, Carmine D. (1987). *Anatomy: A regional atlas of the human body,* 3d ed. Baltimore, MD: Urban & Schwarzenberg.

Cocchi, Fiorenzo, et al. (1995). Identification of RANTES, MIP-1α, and MIP-1β as the major HIV-suppressive factors produced by CD8$^+$ T cells. *Science, 270,* 1811–1815.

Cochran, Susan D., Mays, V. M., & Leung, L. (1991). Sexual practices of heterosexual Asian American young adults: Implications for risk of HIV infection. *Archives of Sexual Behavior, 20,* 381–392.

Cochran, Susan D., & Peplau, L. Anne. (1985). Value orientations in heterosexual relationships. *Psychology of Women Quarterly, 9,* 477–488.

Cochran, W. G., Mosteller, F., & Tukey, J. W. (1953). Statistical problems of the Kinsey report. *Journal of the American Statistical Association, 48,* 673–716.

Cogan, Jeanine C. (1996). The prevention of anti-lesbian/gay hate crimes through social change and empowerment. In E. Rothblum & L. A. Bond (Eds.), *Preventing heterosexism and homophobia* (pp. 219–238). Thousand Oaks, CA: Sage.

Cohen, Dale J., & Bruce, Kathleen E. (1997). Sex and mortality: Real risk and perceived vulnerability. *Journal of Sex Research, 34,* 279–291.

Cohen, I., et al. (1990). Improved pregnancy outcome following successful treatment of chlamydial infection. *Journal of the American Medical Association, 263,* 3160.

Cohen, Jon. (1998). Exploring how to get at—and eradicate—hidden HIV. *Science, 279,* 1854–1855.

Cohen, Jon. (1997). Exploiting the HIV-chemokine nexus. *Science, 275,* 1261–1264.

Cohen, Jon. (1996). Results on new AIDS drugs bring cautious optimism. *Science, 271,* 755–756.

Cohen, Jon. (1995). Thailand weighs AIDS vaccine tests. *Science, 270,* 904–907.

Cohen, Sherrill, & Taub, Nadine. (1989). *Reproductive laws for the 1990's.* Clifton, NJ: Humana Press.

Cohn, Lawrence. (1983, Nov. 16). Pix less able but porn is stable. *Variety, 313*(3), 1–2.

Cole, Steve W., Kemeny, Margaret E., Taylor, Shelley E., & Visscher, Barbara R. (1996). Elevated physical health risk among gay men who conceal their homosexual identity. *Health Psychology, 15,* 243–251.

Cole, Theodore M., & Cole, S. (1978). The handicapped and sexual health. In A. Comfort (Ed.), *Sexual consequences of disability.* Philadelphia, PA: G. F. Stickley.

Coleman, Eli. (1991). Compulsive sexual behavior: New concepts and treatments. *Journal of Psychology and Human Sexuality, 4*(2), 37–51.

Coleman, Eli. (1982). Developmental stages of the coming-out process. In W. Paul et al. (Eds.), *Homosexuality: Social, psychological, and biological issues.* Beverly Hills, CA: Sage.

Coleman, Eli, & Reece, Rex. (1988). Treating low sexual desire among gay men. In S. R. Leiblum & R. C. Rosen (Eds.), *Sexual desire disorders.* New York: Guilford.

Coleman, S., Piotrow, P. T., & Rinehart, W. (1979, Mar.). Tobacco: Hazards to health and human reproduction. *Population Reports,* Series L(1).

Coles, Robert, & Stokes, Geoffrey. (1985). *Sex and the American teenager.* New York: Harper & Row.

Coley, Rebekah, & Chase-Lansdale, P. Lindsay. (1998). Adolescent pregnancy and parenthood: Recent evidence and future directions. *American Psychologist, 53,* 152–166.

Collins, James. (1997, July 7). Throwing away the key. *Time,* 29.

Collins, Nancy L., & Miller, Lynn C. (1994). Self-disclosure and liking: A meta-analytic review. *Psychological Bulletin, 116,* 457–475.

Comas-Diaz, Lillian. (1987). Feminist therapy with mainland Puerto Rican women. *Psychology of Women Quarterly, 11,* 461–474.

Comfort, Alex. (1991). *The new joy of sex: A gourmet guide to lovemaking for the nineties.* New York: Crown.

Commission on Obscenity and Pornography (1970). *The report of the commission on obscenity and pornography.* New York: Bantam Books.

Congregation for the Doctrine of the Faith. (1986). The pastoral care of homosexual persons. *Origins, 26,* 378–382.

Congregation for the Doctrine of the Faith. (1987). Instruction on respect for human life in its origin and on the dignity of procreation. *Origins, 16,* 198–211.

Conkle, D. O. (1987). The second death of substantive due process. *Indiana Law Journal, 62,* 215–242.

Consumer Reports Staff. (1995, May). How reliable are condoms? *Consumer Reports,* 320–324.

Coombs, N.R. (1974). Male prostitution: A psychosocial view of behavior. *American Journal of Orthopsychiatry, 44,* 782.

Corey, Lawrence. (1990). Genital herpes. In K. Holmes et al. (Eds.), *Sexually transmitted diseases* (pp. 391–414). New York: McGraw-Hill.

Cosentino, Clare E., Meyer-Bahlburg, Heino, Alpert, Judith L., Weinberg, Sharon L., & Gaines, Richard. (1995). Sexual behavior problems and psychopathology symptoms in sexually abused girls. *Journal of the American Academy of Child and Adolescent Psychiatry, 34,* 1033–1042.

Council on Scientific Affairs. (1995). Female genital mutilation. *Journal of the American Medical Association, 274,* 1714–1716.

Countryman, L. William. (1994). New Testament sexual ethics and today's world. In J. B. Nelson & S. P. Longfellow (Eds.), *Sexuality and the sacred* (pp. 28–53). Louisville, KY: Westminster/John Knox Press.

Countryman, L. William. (1987). The AIDS crisis: Theological and ethical reflections. *Anglican Theological Review, 69,* 125–134.

Coustan, Donald. (1995a). Obstetric analgesia and anesthesia. In Donald R. Coustan, Ray V. Hunning, Jr., & Don Singer (Eds.), *Human reproduction: Growth and development* (pp. 327–340). Boston: Little, Brown & Co.

Coustan, Donald. (1995b). Obstetric complications. In Donald R. Coustan, Ray V. Hunning, Jr., & Don Singer (Eds.), *Human reproduction: Growth and development* (pp. 431–455). Boston: Little, Brown & Co.

Coustan, Donald R. (1995c). Prescribing during pregnancy. In Donald R. Coustan, Ray V. Hunning, Jr., & Don Singer (Eds.), *Human reproduction: Growth and development.* Boston: Little, Brown & Co.

Coustan, Donald, & Angelini, Diane. (1995). The puerperium. In Donald R. Coustan, Ray V. Hunning, Jr., & Don Singer (Eds.), *Human reproduction: Growth and development* (pp. 341–358). Boston: Little, Brown & Co.

Couzinet, Beatrice, et al. (1986). Termination of early pregnancy by the progesterone antagonist RU486 (mifepristone). *New England Journal of Medicine, 315,* 1565–1569.

Cowley, J. J., & Brooksbank, B. W. L. (1991). Human exposure to putative pheromones and changes in aspects of social behavior. *Journal of Steroid Biochemistry and Molecular Biology, 39,* 647–659.

Cox, Daniel J. (1988). Incidence and nature of male genital exposure behavior as reported by college women. *Journal of Sex Research, 24,* 227–234.

Cranston-Cuebas, Margaret A., & Barlow, David H. (1990). Cognitive and affective contributions to sexual functioning. *Annual Review of Sex Research, 1,* 119–162.

Crawford, June, Kippax, Susan, & Waldby, Catherine. (1994). Women's sex talk and men's sex talk: Different worlds. *Feminism and Psychology,* 571–587.

Creighton, James. (1992). *Don't go away mad.* New York: Doubleday.

Crenshaw, Theresa L., & Goldberg, James P. (1996). *Sexual pharmacology: Drugs that affect sexual function.* New York: Norton.

Crowe, L. C., & George, W. H. (1989). Alcohol and human sexuality. *Psychological Bulletin, 105,* 374–386.

Cunningham, F. Gary, MacDonald, Paul C., & Gant, Norman F. (1989). *Williams obstetrics* (18th ed.). Norwalk, CT: Appleton and Lange.

Cunningham, F. Gary, MacDonald, Paul C., Gant, Norman F., Leveno, Kenneth J., & Gilstrap, Larry C. III. (1993). *Williams Obstetrics* (19th ed.). Norwalk, CT: Appleton and Lange.

Cunningham, Michael, Roberts, Alan, Wu, Chang-Huan, Burbee, Anita, & Druen, Perry. (1995). "Their ideas of beauty are, on the whole, the same as ours": Consistency and variability in the cross-cultural perception of female physical attractiveness. *Journal of Personality and Social Psychology, 68,* 261–279.

Curran, Charles E. (1988). Roman Catholic sexual ethics: A dissenting view. *Christian Century, 105,* 1139–1142.

Cutler, Winnifred B., Friedmann, Erika, & McCoy, Norma L. (1998). Phenomenal influences on sociosexual behavior in men. *Archives of Sexual Behavior, 27,* 1–14.

Cutler, Winnifred B., Preti, George, et al. (1986). Human axillary secretions influence women's menstrual cycles: The role of donor extract from men. *Hormones and Behavior, 20,* 463–473.

Dalton, Harlon. (1993). Communal law. In S. Burris, H. L. Dalton, & J. L. Miller (Eds.), *AIDS law today* (pp. 242–262). New Haven, CT: Yale University Press.

Dalton, Harlon L., & Burris, Scott. (Eds.). (1987). *AIDS and the law: A guide for the public.* New Haven, CT: Yale University Press.

Dalton, Katharina. (1979). *Once a month.* Ramona, CA: Hunter House.

Darling, C. A., Davidson, J. K., & Conway-Welch, C. (1990). Female ejaculation: Perceived origins, the Gräfenberg spot/area, and sexual responsiveness. *Archives of Sexual Behavior, 19,* 29–48.

Darling, C. A., Davidson, J. K., Sr., & Parish, W. E., Jr. (1989). Single parents: Interaction of parenting and sexual issues. *Journal of Sex and Marital Therapy, 15,* 227–245.

Darling, Carol A., Davidson, J. Kenneth, & Jennings, Donna A. (1991). The female sexual response revisited: Understanding the multiorgasmic experience in women. *Archives of Sexual Behavior, 20,* 527–540.

Darney, Philip D., et al. (1990). Acceptance and perceptions of Norplant among users in San Francisco, USA. *Studies in Family Planning, 21,* 152–160.

D'Augelli, Anthony R. (1992). Lesbian and gay male undergraduates' experiences of harassment and fear on campus. *Journal of Interpersonal Violence, 7,* 383–395.

D'Augelli, Anthony R., & Garnets, Linda D. (1995). Lesbian, gay, and bisexual communities. In A. D'Augelli & C. Patterson (Eds.), *Lesbian, gay, and bisexual identities over the lifespan.* (pp. 293–320). New York: Oxford University Press.

D'Augelli, Anthony R., & Patterson, Charlotte J. (Eds.). (1995). *Lesbian, gay, and bisexual identities over the lifespan.* New York: Oxford University Press.

David, H. P., Dytrych, Z., Matejcek, Z., & Schuller, V. (Eds.). (1988). *Born unwanted: Developmental effects of denied abortion.* New York: Springer.

David, Henry P. (1992). Born unwanted: Long-term developmental effects of denied abortion. *Journal of Social Issues, 48*(3), 163–181.

David, Henry P., & Matejcek, Z. (1981). Children born to women denied abortion: An update. *Family Planning Perspectives, 13,* 32–34.

Davidson, J. Kenneth, & Darling, C. A. (1988). The stereotype of single women revisited: Sexual practices and sexual satisfaction among professional women. *Health Care for Women International, 9,* 317–336.

Davis, Clive M., & Bauserman, R. (1993). Exposure to sexually explicit materials: An attitude change perspective. *Annual Review of Sex Research, 4,* 121–210.

Davis, J. A., & Smith, T. (1991). *General social surveys, 1972–1991.* Storrs, CT: University of Connecticut, Roper Center for Public Opinion Research.

Davison, Gerald C. (1982). Politics, ethics, and therapy for homosexuality. In W. Paul et al. (Eds.), *Homosexuality: Social, psychological, and biological issues.* Beverly Hills, CA: Sage.

Day, Nancy L., & Richardson, Gale. (1994). Comparative teratogenicity of alcohol and other drugs. *Alcohol Health and Research World,* 42–48.

Day, Randal. (1992). The transition to first intercourse among racially and culturally diverse youth. *Journal of Marriage and the Family, 54,* 749–762.

Dean, Karol E., & Malamuth, Neil M. (1997). Characteristics of men who aggress sexually and of men who imagine aggressing: Risk and moderating variables. *Journal of Personality and Social Psychology, 72,* 449–455.

DeBuono, B. A., Zinner, S. H., Daamen, M., & McCormack, W. M. (1990). Sexual behavior of college women in 1975, 1986, and 1989. *New England Journal of Medicine, 322,* 821–825.

DeLamater, John. (1987). A sociological perspective. In J. H. Geer & W. T. O'Donohue (Eds.), *Theories of human sexuality* (pp. 237–256). New York: Plenum.

DeLamater, John. (1982). Response effects of question content. In W. Dijkstra & J. Van derZouwen (Eds.), *Response behavior in the survey—interview.* London, England: Academic.

DeLamater, John. (1981). The social control of sexuality. *Annual Review of Sociology, 7,* 263–290.

DeLamater, John, & MacCorquodale, Patricia. (1979). *Premarital sexuality: Attitudes, relationships, behavior.* Madison: University of Wisconsin Press.

DeLamater, John, Wagstaff, David, & Havens, Kayt K. (1994). The impact of a health-promotion intervention on condom use by Black male adolescents. Madison, WI: Center for Demography and Ecology, Working Paper 94-26.

Delmas, Pierre D., et al. (1997). Effects of raloxifene on bone mineral density, serum cholesterol concentrations, and uterine endometrium in postmenopausal women. *New England Journal of Medicine, 337,* 1641–1647.

de Gaston, Jacqueline F., Jensen, Larry, Weed, Stan, & Tanas, Raja. (1994). Teacher philosophy and program implementation and the impact of sex education outcomes. *The Journal of Research and Development in Education, 27,* 265–270.

D'Emilio, John, & Freedman, Estelle B. (1988). *Intimate matters: A history of sexuality in America.* New York: Harper & Row.

Dennis, Donna. (1993). HIV screening and discrimination: The federal example. In S. Burris, H. L. Dalton, & J. L. Miller (Eds.), *AIDS law today* (pp. 187–215). New Haven, CT: Yale University Press.

Deren, Sherry, Tortu, Stephanie, & Davis, W. Rees. (1993). An AIDS risk reduction project with inner-city women. In C. Squire (Ed.), *Women and AIDS: Psychological perspectives* (pp. 73–89). London: Sage.

Derlega, Valerian J. (Ed.). (1984). *Communication, intimacy, and close relationships.* New York: Academic.

Derlega, Valerian J., Metts, Sandra, & Margulis, Stephen T. (1993). *Self-disclosure.* Newbury Park, CA: Sage.

Devor, Holly. (1997). *FTM: Female-to-male transsexuals in society.* Bloomington, IN: Indiana University Press.

Dewsbury, Donald A. (1981). Effects of novelty on copulatory behavior: The Coolidge effect and related phenomena. *Psychological Bulletin, 89,* 464–483.

de Zalduondo, Barbara O. (1991). Prostitution viewed cross-culturally: Toward recontextualizing sex work in AIDS intervention research. *Journal of Sex Research, 28,* 223–248.

Diamond, Milton. (1997). Sexual identity and sexual orientation in children with traumatized or ambiguous genitalia. *Journal of Sex Research, 34,* 199–211.

Diamond, Milton. (1996). Prenatal predisposition and the clinical management of some pediatric conditions. *Journal of Sex and Marital Therapy, 22,* 139–147.

Diamond, Milton. (1993). Homosexuality and bisexuality in different populations. *Archives of Sexual Behavior, 22,* 291–310.

Diamond, Milton, & Sigmundson, H. Keith. (1997). Sex reassignment at birth: Long-term review and clinical implications. *Archives of Pediatric and Adolescent Medicine, 151,* 298–304.

Dickinson, Robert L. (1949). *Atlas of human sex anatomy.* Baltimore: Williams & Wilkins.

DiClemente, R. (1998). Preventing sexually transmitted infections among adolescents: A clash of ideology and science. *Journal of the American Medical Association, 279,* (19), 1574–1575.

Dindia, Kathryn, & Allen, Michael. (1992). Sex differences in self-disclosure: A meta-analysis. *Psychological Bulletin, 112,* 106–124.

Dion, Karen K. (1973). Young children's stereotyping of facial attractiveness. *Developmental Psychology, 9,* 183–188.

Dion, Karen K., & Dion, Kenneth L. (1993b). Individualistic and collectivistic perspectives on gender and the cultural content of love and intimacy. *Journal of Social Issues, 49,* 53–69.

Dion, Kenneth (1977). The incentive value of physical attractiveness for young children. *Personality and Social Psychology Bulletin, 3,* 67–70.

Dion, Kenneth L., & Dion, Karen K. (1993a). Gender and ethnocultural comparisons in styles of love. *Psychology of Women Quarterly, 17,* 463–474.

Ditman, Keith. (1964). Inhibition of ejaculation by chlorprothixenes. *American Journal of Psychiatry, 120,* 1004–1005.

Dixon, Joan K. (1984). The commencement of bisexual activity in swinging married women over age thirty. *Journal of Sex Research, 20,* 71–90.

Dixson, Alan F. (1990). The neuroendocrine regulation of sexual behavior in female primates. *Annual Review of Sex Research, 1,* 197–226.

Djerassi, Carl. (1989). The bitter pill. *Science, 245,* 356–361.

Dobert, Margarete. (1975). Tradition, modernity, and woman power in Africa. In M. S. Mednick, S. Tangri, & L. Hoffman (Eds.), *Women and achievement: Social and motivational analysis.* Washington, D.C.: Hemisphere.

Docter, Richard F., & Prince, Virginia. (1997). Transvestism: A survey of 1032 cross-dressers. *Archives of Sexual Behavior, 26,* 589–606.

Dodson, Betty. (1987). *Sex for one: The joy of self-loving.* New York: Harmony Books (Crown).

Doering, Charles, et al. (1978). Plasma testosterone levels and psychologic measures in men over a 2-month period. In R. Friedman et al. (Eds.), *Sex differences in behavior.* Huntington, NY: Krieger.

Doll, Lynda S., Peterson, Lyle R., White, Carol R., et al. (1992). Homosexually and non-homosexually identified men who have sex with men: A behavioral comparison. *Journal of Sex Research, 29,* 1–14.

Donahey, Karen M., & Carroll, Richard A. (1993). Gender differences in factors associated with hypoactive sexual desire. *Journal of Sex & Marital Therapy, 19,* 25–40.

Donnelly, D. A. (1993). Sexually inactive marriages. *Journal of Sex Research, 30,* 171–179.

Donnerstein, Ed. (1980). Aggressive-erotica and violence against women. *Journal of Personality and Social Psychology, 39,* 269–277.

Donnerstein, E., Linz, D., & Penrod, S. (1987). *The question of pornography: Research findings and policy implications.* New York: Free Press.

Donovan, Basil, Bassett, I., & Bodsworth, N. J. (1994). Male circumcision and common sexually transmissible diseases in a developed nation setting. *Genitourinary Medicine, 70,* 317–320.

Donovan, P. (1989). *Risk and responsibility: Teaching sex/education in America's schools today.* New York: Alan Guttmacher Institute.

Doran, Terence A. (1990). Chorionic villus sampling as the primary diagnostic tool in prenatal diagnosis. *Journal of Reproductive Medicine, 35,* 935–940.

d'Oro, Luca C., Parazzini, Fabio, Naldi, Luigi, & LaVecchia, Carlo. (1994). Barrier methods of contraception, spermicides, and sexually transmitted diseases: A review. *Genitourinary Medicine, 70,* 410–417.

Doshi, Mary L. (1986). Accuracy of consumer performed in-home tests for early pregnancy detection. *American Journal of Public Health, 76,* 512–514.

Douglas, Mary. (1970). *Purity and danger: An analysis of concepts of pollution and taboo.* Baltimore: Penguin.

Doyle, James A. (1989). *The male experience,* 2d ed. Dubuque, IA: Wm. C. Brown.

Driscoll, James P. (1971, Mar.–Apr.). Transsexuals. *Transaction,* 28–31.

Driscoll, R., Davis, K. E., & Lipetz, M. E. (1972). Parental interference and romantic love: The Romeo and Juliet effect. *Journal of Personality and Social Psychology, 24,* 1–10.

Dryer, P. Christopher, & Horowitz, Leonard. (1997). When do opposites attract? Interpersonal complementarity vs. similarity. *Journal of Personality and Social Psychology, 72,* 592–603.

Duncan, Richard. (1993). Who wants to stop the church?: Homosexual rights legislation, public policy, and religious freedom. *Notre Dame Law Review, 69,* 393–445.

Duncombe, J. A., & Marsden, D. (1994). Whose orgasm is this anyway? "Sex work" and "emotion work" in long-term couple relationships. Presented at American Sociological Association meetings.

Dunn, H. G., et al. (1977). Maternal cigarette smoking during pregnancy and the child's subsequent development: II. Neurologic and intellectual maturation to the age of 6½ years. *Canadian Journal of Public Health, 68,* 43–50.

Dunn, Marian E., & Trost, Jan E. (1989). Male multiple orgasms: A descriptive study. *Archives of Sexual Behavior, 18,* 377–388.

Dutton, Donald G., & Aron, Arthur P. (1974). Some evidence for heightened sexual attraction under conditions of high anxiety. *Journal of Personality and Social Psychology, 30,* 470–517.

Dworkin, Andrea. (1985). Against the flood. *Harvard Women's Law Journal, 8,* 1–29.

Eagly, Alice H., Ashmore, R. D., Makhijani, M. G., & Longo, L. (1991). What is beautiful is good, but . . . : A meta-analytic review of research on the physical attractiveness stereotype. *Psychological Bulletin, 110,* 109–128.

Earls, Christopher M., & David, Helene. (1989). A psychosocial study of male prostitution. *Archives of Sexual Behavior, 18,* 401–420.

East, Patricia. (1998). Racial and ethnic differences in girl's sexual, marital, and birth expectations. *Journal of Marriage and the Family, 60,* 150–162.

Ebel, Charles. (1994). *Managing herpes: How to live and love with a chronic STD.* Research Triangle Park, NC: American Social Health Association.

Ehrhardt, Anke. (1996). Sexual behavior among heterosexuals. In J. Mann & D. Tarantola (Eds.), *AIDS in the world II* (pp. 259–263). New York: Oxford University Press.

Ehrhardt, Anke A., Yingling, Sandra, & Warne, Patricia A. (1991). Sexual behavior in the era of AIDS: What has changed in the United States? *Annual Review of Sex Research, 2,* 25–48.

Eisenberg, M. E., Wagenaar, A., Neumark-Sztainer, D. (1997). Viewpoints of Minnesota students on school-based sexuality education. *Journal of School Health, 67,* 322–326.

Elder, Glen. (1969). Appearance and education in marriage mobility. *American Sociological Review, 34,* 519–533.

Elias, James, & Gebhard, Paul. (1969). Sexuality and sexual learning in childhood. *Phi Delta Kappan, 50,* 401–405.

Elias, Marilyn. (1997, August 14). Modern matchmaking. *USA Today,* D1–D2.

Elliott, Ann N., & O'Donohue, William T. (1997). The effects of anxiety and distraction on sexual arousal in a nonclinical sample of heterosexual women. *Archives of Sexual Behavior, 26,* 607–624.

Ellis, Bruce J., & Symons, Donald. (1990). Sex differences in sexual fantasy: An evolutionary psychological approach. *Journal of Sex Research, 27,* 527–555.

Ellis, Henry Havelock. (1939). *My life.* Boston: Houghton Mifflin.

Ellis, Lee. (1996). The role of perinatal factors in determining sexual orientation. In R. C. Savin-Williams & K. M. Cohen (Eds.), *The lives of lesbians, gays, and bisexuals* (pp. 35–70). Fort Worth, TX: Harcourt Brace.

Ellis, Lee, & Ames, M. Ashley. (1987). Neurohormonal functioning and sexual orientation: A theory of homosexuality-heterosexuality. *Psychological Bulletin, 101,* 233–258.

Ellis, L., Ames, M. A., Peckham, W., & Burke, D. (1988). Sexual orientation of human offspring may be altered by severe maternal stress during pregnancy. *Journal of Sex Research, 25,* 152–157.

Elmer-Dewitt, P. (1993, May 31). Orgies on-line. *Time,* 61.

English, Abigail. (1992). Expanding access to HIV services for adolescents: Legal and ethical issues. In R. DiClemente (Ed.), *Adolescents and AIDS: A generation in jeopardy* (pp. 262–283). Newbury Park, CA: Sage.

Enquist, Roy J. (1983). The churches' response to abortion. *Word and World, 5,* 414–425.

Entwisle, Doris R., & Doering, Susan G. (1981). *The first birth: A family turning point.* Baltimore, MD: Johns Hopkins University Press.

Epstein, Aaron. (1995, Oct. 8). *Court joins gay rights debate.* Wisconsin State Journal, 1B ff.

Erikson, Erik H. (1968). *Identity: Youth and crisis.* New York: Norton.

Erikson, Erik H. (1950). *Childhood and society.* New York: Norton.

Erlichman, Howard, & Eichenstein, Rosalind. (1992). Private wishes: Gender similarities and differences. *Sex Roles, 26,* 399–422.

Ernst, Frederick A., et al. (1991). Condemnation of homosexuality in the black community: A gender-specific phenomenon? *Archives of Sexual Behavior, 20,* 579–585.

Ernulf, Kurt E., & Innala, Sune M. (1995). Sexual bondage: A review and uobtrusive investigation. *Archives of Sexual Behavior, 24,* 631–654.

Espin, Oliva. (1987). Issues of identity in the psychology of Latina lesbians. In Boston Lesbian Psychologies Collective, *Lesbian psychologies.* Urbana, IL: University of Illinois Press.

Esterberg, Kristin G. (1996). Gay cultures, gay communities: The social organization of lesbians, gay men, and bisexuals. In R. C. Savin-Williams & K. M. Cohen (Eds.), *The lives of lesbians, gays, and bisexuals* (pp. 377–392). Fort Worth, TX: Harcourt Brace.

Evans, Donald, & Evans, Marilyn. (1996). Fertility, infertility and the human embryo: Ethics, law, and the practice of human mitoficial prevention. *Human Reproduction Update, 2,* 208–224.

Evans, Harriet. (1995). Defining differences: The "scientific" construction of sexuality and gender in the People's Republic of China. *Signs, 20,* 357–394.

Everett, Guy M. (1975). Amyl nitrate ("poppers") as an aphrodisiac. In M. Sandler and G. L. Gessa (Eds.), *Sexual behavior: Pharmacology and biochemistry.* New York: Raven.

Everitt, Barry J., & Bancroft, John. (1991). Of rats and men: The comparative approach to male sexuality. *Annual Review of Sex Research, 2,* 77–118.

Ezzell, Carol. (1996, Jan.). Emergence of the protease inhibitors: A better class of AIDS drugs? *Journal of NIH Research, 41,* 41–43.

Ezzell, Carol. (1996, May). Gene-therapy trial using BRCA1 to begin with ovarian cancer. *Journal of NIH Research, 8,* 24–25.

Ezzell, Carol. (1995, Apr.). Test may detect heritable breast-cancer mutations. *Journal of NIH Research, 7,* 42–44.

Ezzell, Carol. (1994, Oct.). Breast cancer genes. *Journal of NIH Research, 6,* 33–35.

Fagot-Diaz, Jose G. (1988). Employment discrimination against AIDS victims: Rights and remedies available under the Federal Rehabilitation Act of 1973. *Labor Law Journal, 39,* 148–166.

Fanburg, Jonathan Thomas, Kaplan, David, & Naylor, Kelly. (1995). Student opinion of condom distribution at a Denver, Colorado high school. *Journal of School Health, 65,* 181–185.

Farkas, G. M., Sine, L. G., & Evans, I. M. (1978). Personality, sexuality, and demographic differences between volunteers and nonvolunteers for a laboratory study of male sexual behavior. *Archives of Sexual Behavior, 7,* 513–520.

Farley, Margaret A. (1994). Sexual ethics. In J. B. Nelson & S. P. Longfellow (Eds.), *Sexuality and the sacred* (pp. 54–67). Louisville, KY: Westminster/John Knox Press.

Fasteau, Marc F. (1974). *The male machine.* New York: McGraw-Hill.

Fay, R. E., Turner, C. F., Klassen, A. D., & Gagnon, J. H. (1989). Prevalence and patterns of same-gender sexual contact among men. *Science, 243,* 338–348.

Feder, H. H. (1984). Hormones and sexual behavior. *Annual Review of Psychology, 35,* 165–200.

Federal Bureau of Investigation. (1993). *Uniform crime reports.* Washington, D.C.: U.S. Government Printing Office.

Feingold, Alan. (1988). Matching for attractiveness in romantic partners and same-sex friends: A meta-analysis and theoretical critique. *Psychological Bulletin, 104,* 226–235.

Feingold, Allen. (1990). Gender differences in effects of physical attractiveness on romantic attraction. *Journal of Personality and Social Psychology, 59,* 981–993.

Feinleib, Joel A., & Michael, Robert. (1999). Reported changes in sexual behavior in response to AIDS in the U.S. In E. O. Laumann, & R. Michael (Eds.), *The social organization of sexuality: Further studies,* in press.

Feldman-Summers, Shirley, & Pope, Kenneth S. (1994). The experience of "forgetting" childhood abuse: A national survey of psychologists. *Journal of Consulting and Clinical Psychology, 62,* 636–639.

Felton, Gary, & Segelman, Florrie. (1978). Lamaze childbirth training and changes in belief about person control. *Birth and the Family Journal, 5,* 141–150.

Feng, Y., Broder, C. C., Kennedy, P. E., & Berger, E. A. (1996). HIV-1 entry cofactor: Functional cDNA cloning of a seven-trans membrane G protein-coupled reception. *Science, 272,* 872.

Finkelhor, David. (1994). Current information on the scope and nature of child sexual abuse. *Future of Children, 4*(2), 31–53.

Finkelhor, David (1984). *Child sexual abuse: New theory and research.* New York: The Free Press.

Finkelhor, David, & Russell, D. (1984). Women as perpetrators: Review of the evidence. In D. Finkelhor (Ed.), *Child sexual abuse: New theory and research.* New York: The Free Press.

Finkelhor, David. (1980). Sex among siblings: A survey on prevalence, variety and effects. *Archives of Sexual Behavior, 9,* 171–194.

Finzi, Diana, et al. (1997). Identification of a reservoir for HIV-1 in patients on highly active antiretroviral therapy. *Science, 278,* 1295–1300.

Fisher, Helen. (1992). *Anatomy of love: The mysteries of mating, marriage and why we stray.* New York: Ballantine Books.

Fisher, W. A., & Grenier, G. (1994). Violent pornography, antiwoman thoughts, and antiwoman acts: In search of reliable effects. *Journal of Sex Research, 31,* 23–38.

Fisher, William A., Byrne, D., White, L. A., & Kelley, K. (1988). Erotophobia-erotophilia as a dimension of personality. *Journal of Sex Research, 25,* 123–151.

Fisher, William A., Byrne, D., & White, L. A. (1983). Emotional barriers to contraception. In D. Byrne and W. A. Fisher (Eds.), *Adolescents, sex, and contraception.* Hillsdale, NJ: Lawrence Erlbaum.

Fiske, Susan T., & Glick, Peter. (1995). Ambivalence and stereotypes cause sexual harassment: A theory with implications for organizational change. *Journal of Social Issues, 51*(1), 97–115.

Fitzgerald, Louise F. (1993). Sexual harassment. *American Psychologist, 48,* 1070–1076.

Flaks, David K., et al. (1995). Lesbians choosing motherhood: A comparative study of lesbian and heterosexual parents and their children. *Developmental Psychology, 31,* 105–114.

Flaskerud, Jacquelyn H., & Ungvarski, Peter J. (1992). *HIV/AIDS: A guide to nursing care.* Philadelphia: Saunders.

Fleming, Alison, Ruble, Diane, Krieger, Howard, & Wong, P. Y. (1997). Hormonal and experiential correlates of internal responsiveness during pregnancy and the puerperium in human mothers. *Hormones and Behavior, 31,* 145–158.

Fleming, C., & Ingrassia, M. (1993, Aug. 16). "The Heidi Chronicles." *Newsweek,* 50ff.

Fleming, Douglas T., et al. (1997). Herpes simplex virus type 2 in the United States, 1976 to 1994. *The New England Journal of Medicine, 337,* 1105–1111.

Fleming, M., Steinman, C., & Boeknok, G. (1980). Methodological problems in assessing sex reassignment surgery: A reply to Meyer and Reter. *Archives of Sexual Behavior, 9,* 451–456.

Foa, E. B., Steketee, G., & Olasov, B. (1989). Behavioral/cognitive conceptualization of post-traumatic stress disorder. *Behavior Therapy, 20,* 155–176.

Ford, Clellan S., & Beach, Frank A. (1951). *Patterns of sexual behavior.* New York: Harper & Row.

Ford, Kathleen, & Norris, Anne. (1997). Sexual networks of African-American and Hispanic youth. *Sexually Transmitted Diseases, 24,* 327–333.

Ford, Kathleen, & Norris, Anne. (1991). Methodological considerations for survey research on sexual behavior: Urban African American and Hispanic youth. *Journal of Sex Research, 28,* 539–555.

Ford, Nicholas, & Koetsawang, Suporu. (1991). The socio-cultural context of the transmission of HIV in Thailand. *Social Science and Medicine, 33,* 405–414.

Forrest, Bruce. (1991). Women, HIV, and mucosal immunity. *Lancet, 337,* 835.

Forrest, Jacqueline D., & Singh, Susheela. (1990). The sexual and reproductive behavior of American women, 1982–1988. *Family Planning Perspectives, 22*(5), 206–214.

Fox, Laura. (1983). The 1983 abortion decisions. *University of Richmond Law Review, 18,* 137–159.

Fox, Ronald C. (1995). Bisexual identities. In A. R. D'Augelli & C. J. Patterson (Eds.), *Lesbian, gay, and bisexual identities over the lifespan.* New York: Oxford University Press.

Frank, E., Anderson, C., & Rubinstein, D. (1978). Frequency of sexual dysfunction in "normal" couples. *New England Journal of Medicine, 299*(3), 111–115.

Frank, L. K. (1961). *The Conduct of Sex.* New York: Morrow.

Frayser, Suzanne G. (1994). Defining normal childhood sexuality: An anthropological approach. *Annual Review of Sex Research, 5,* 173–217.

Frayser, Suzanne G. (1985). *Varieties of sexual experience: An anthropological perspective on human sexuality.* New Haven, CT: Human Relations Area Files Press.

Freud, S. (1948). The psychogenesis of a case of homosexuality in a woman (1920). In *The collected papers* (Vol. II, pp. 202–231). London: Hogarth.

Freud, Sigmund. (1924). *A general introduction to psychoanalysis.* New York: Permabooks, 1953. (Boni & Liveright edition, 1924).

Freund, M., Lee, N., & Leonard, T. (1991). Sexual behavior of clients with street prostitutes in Camden, NJ. *Journal of Sex Research, 28,* 579–591.

Freyd, Jennifer J. (1996). *Betrayal trauma theory.* Cambridge, MA: Harvard University Press.

Friday, Nancy. (1975). *Forbidden flowers: More women's sexual fantasies.* New York: Simon & Schuster.

Friday, Nancy. (1973). *My secret garden: Women's sexual fantasies.* New York: Simon & Schuster.

Fried, Peter A. (1986). Marijuana use in pregnancy. In I. J. Chasnott (Ed.), *Drug use in pregnancy: Mother and child.* Boston: MTP Press.

Friedrich, W. N., Beilke, R. L., & Urquiza, A. J. (1988). Behavior problems in young sexually abused boys. *Journal of Interpersonal Violence, 3,* 1–12.

Frieze, Irene H. (1983). Causes and consequences of marital rape. *Signs, 8,* 532–553.

Frisch, R. E., & McArthur, J. W. (1974). Menstrual cycles: Fatness as a determinant of minimum weight for height necessary for their maintenance or onset. *Science, 185,* 949–951.

Frishman, G. (1995). Abortions, miscarriages, and ectopic pregnancies. In D. R. Coustrin, R. V. Hunning, Jr., & D. B. Singer (Eds.), *Human reproduction: Growth and development.* Boston: Little, Brown, and Company.

Fromm, Erich. (1956). *The art of loving.* New York: Harper & Row.

Fuleihan, Ghada. (1997). Tissue-specific estrogens—The promise for the future. *New England Journal of Medicine, 337,* 1686–1687.

Fullilove, Mindy T., Fullilove, R. E., Haynes, K., & Gross, S. (1990). Black women and AIDS prevention: A view towards understanding the gender rules. *Journal of Sex Research, 27,* 47–64.

Fullilove, Robert E. (1988). Minorities and AIDS: A review of recent publications. *Multicultural Inquiry and Research on AIDS Newsletter, 2,* 3–5.

Furby, L., Weinrott, M. R., & Blackshaw, L. (1989). Sex offender recidivism: A review. *Psychological Bulletin, 105,* 3–30.

Furnish, Victor P. (1994). The Bible and homosexuality: Reading the texts in context. In J. S. Siker (Ed.), *Homosexuality in the church* (pp. 18–35). Louisville, KY: Westminster/John Knox Press.

Furstenberg, Frank F., Brooks-Gunn, J., & Morgan, S. P. (1987). *Adolescent mothers in later life.* New York: Cambridge University Press.

Fyfe, B. (1983). "Homophobia" or homosexual bias reconsidered. *Archives of Sexual Behavior, 12,* 549–554.

Gabelnick, Henry L. (1998). Future methods. In R. Hatcher et al. (Eds.), *Contraceptive technology,* 17th ed., (pp. 615–622). New York: Ardent Media.

Gage, A. J. (1995). *An assessment of the quality of data on age at first union, first birth, and first sexual intercourse.* Calverton, MD: Macro International.

Gagnon, John H. (1990). The explicit and implicit use of the scripting perspective in sex research. *Annual Review of Sex Research, 1,* 1–44.

Gagnon, John H. (1977). *Human sexualities.* Glenview, IL: Scott, Foresman.

Gagnon, John H., & Simon, William. (1973). *Sexual conduct: The social origins of human sexuality.* Chicago: Aldine.

Gangestad, Steven W., & Buss, David M. (1993). Pathogen prevalence and human mate preferences. *Ethology and Sociobiology, 14,* 89–96.

Gangestad, Steven W., & Thornhill, R. (1997). Human sexual selection and developmental stability. In J. A. Simpson & D. T. Kenrick (Eds.), *Evolutionary social psychology* (pp. 169–195). Mahwah, NJ: Lawrence Erlbaum Associates.

Gao, F., et al. (1992). Human infection by genetically diverse SIV-SM-related HIV-2 in West Africa. 495–499. *Nature, 358,* 495.

Garcia, Luis T. (1982). Sex-role orientation and stereotypes about male-female sexuality. *Sex Roles, 8,* 863–876.

Garcia-Velasco, Jose, & Mondragon, Manuel. (1991). The incidence of the vomeronasal organ in 1000 human subjects and its possible clinical significance. *Journal of Steroid Biochemistry and Molecular Biology, 39,* 561–563.

Gay, Peter. (1984). *The bourgeois experience: Victoria to Freud.* New York: Oxford University Press.

Gebhard, Paul H. (1976). The Institute. In M. S. Weinberg (Ed.), *Sex research: Studies from the Kinsey Institute.* New York: Oxford University Press.

Gebhard, P. H., Gagnon, J. H., Pomeroy, W. B., & Christenson, C. V. (1965). *Sex offenders: An analysis of types.* New York: Harper & Row.

Geer, James H., & Bellard, Heidi S. (1996). Sexual content induced delays in unprimed lexical decisions: Gender and context effects. *Archives of Sexual Behavior, 25,* 379–396.

Geer, James H., & Manguno-Mire, Gina M. (1996). Gender differences in cognitive processes in sexuality. *Annual Review of Sex Research, 7,* 90–124.

Genovesi, Vincent J. (1987). *In pursuit of love: Catholic morality and human sexuality.* Wilmington, DE: Michael Glazier.

George, Linda K., & Weiler, Stephen J. (1981). Sexuality in middle and late life. *Archives of General Psychiatry, 38,* 919–923.

Georges, Eugenia. (1996). Abortion policy and practice in Greece. *Social Science & Medicine, 42,* 509–519.

Giami, Alain, & Schlitz, Marie-Ange. (1996). Representations of sexuality and relations between partners: Sex research in France in the era of AIDS. *Annual Review of Sex Research, 7,* 125–157.

Gijs, Luk, & Gooren, Louis. (1996). Hormonal and psychopharmacological interventions in the treatment of paraphilias. *Journal of Sex Research, 33,* 273–290.

Gil, Vincent E. (1991). An ethnography of HIV/AIDS and sexuality in The People's Republic of China. *Journal of Sex Research, 28,* 521–537.

Gilligan, Carol. (1982). *In a different voice: Psychological theory and women's development.* Cambridge, MA: Harvard University Press.

Gilmartin, Brian G. (1975). The swinging couple down the block. *Psychology Today, 8*(9), 54.

Ginsburg, Faye. (1989). *Contested lives: The abortion debate in an American community.* Berkeley, CA: University of California Press.

Gladue, Brian A., Green, Richard, & Helman, R. E. (1984). Neuroendocrine response to estrogen and sexual orientation. *Science, 225,* 1496–1498.

Glascock, Jack, & LaRose, Robert. (1993). Dial-a-porn recordings: The role of the female participant in male sexual fantasies. *Journal of Broadcasting & Electronic Media,* 313–324.

Glaser, Chris. (1994). The love that dare not pray its name: The gay and lesbian movement in America's churches. In J. S. Siker (Ed.), *Homosexuality in the church* (pp. 150–157). Louisville, KY: Westminster/John Knox Press.

Glass, S. J., & Johnson, R. W. (1944). Limitations and complications of organotherapy in male homosexuality. *Journal of Clinical Endocrinology, 4,* 540–544.

Godges, John. (1986). Religious groups meet the San Francisco AIDS challenge. *Christian Century, 103,* 771–775.

Gold, Alice R., & Adams, David B. (1981). Motivational factors affecting fluctuations of female sexual activity at menstruation. *Psychology of Women Quarterly, 5,* 670–680.

Gold, S. N., Hughes, D., & Hohnecker, L. (1994). Degrees of repression of sexual abuse memories. *American Psychologist, 49,* 441–442.

Goldberg, Daniel C., et al. (1983). The Grafenberg spot and female ejaculation. A review of initial hypotheses. *Journal of Sex and Marital Therapy, 9,* 27–37.

Goldberg, M. (1987). Understanding hypersexuality in men and women. In G. R. Weeks & L. Hof (Eds.), *Integrating sex and marital therapy.* New York: Brunner–Mazel.

Goldberg, Susan. (1983). Parent-infant bonding: Another look. *Child Development, 54,* 1355–1382.

Goldfoot, D. A., Westerberg-van Loon, W., Groeneveld, W., & Koos Slob, A. (1980). Behavioral and physiological evidence of sexual climax in the female stump-tailed macaque (Macaca arctoides). *Science, 208,* 1477–1478.

Goldman, Ronald J., & Goldman, Juliette D. G. (1982). *Children's sexual thinking.* London: Routledge & Kegan Paul.

Goldstein, Bernard. (1976). *Human sexuality*. New York: McGraw-Hill.

Goldstein, Irwin, et al. (1998). Oral sildenafil in the treatment of erectile dysfunction. *New England Journal of Medicine, 338,* 1397–1404.

Golombok, Susan, & Tasker, Fiona. (1996). Do parents influence the sexual orientation of their children? Findings from a longitudinal study of lesbian families. *Developmental Psychology, 32,* 3–11.

Golub, Sharon. (1992). *Periods: From menarche to menopause.* Newbury Park, CA: Sage.

Gonsiorek, John C. (1996). Mental health and sexual orientation. In R. C. Savin-Williams & K. M. Cohen (Eds.), *The lives of lesbians, gays, and bisexuals* (pp. 462–478). Fort Worth, TX: Harcourt Brace.

Goodchilds, Jacqueline, & Zellman, Gail. (1984). Sexual signaling and sexual aggression in adolescent relationships. In N. Malamuth & E. Donnerstein (Eds.), *Pornography and sexual aggression.* New York: Academic.

Goodson, Patricia, & Edmundson, Elizabeth. (1994). The problematic promotion of abstinence: An overview of Sex Respect. *Journal of School Health, 64,* 205–210.

Gooren, L., Fliers, E., & Courtney, K. (1990). Biological determinants of sexual behavior. *Annual Review of Sex Research, 1,* 175–196.

Gordon, B. N., Schroeder, C. S., & Abrams, J. M. (1990). Age and social-class differences in children's knowledge of sexuality. *Journal of Clinical Child Psychology, 19,* 33–43.

Gordon, Betty, & Schroeder, Carolyn. (1995). *Sexuality: A developmental approach to problems.* New York: Plenum Press.

Gordon, Sol (1986, Oct.). What kids need to know. *Psychology Today.* Reprinted in O. Pocs (Ed.), *Human Sexuality 88/89.* Guilford, CT: Dushkin. Sol Gordon, a leading sex educator, provides a humorous view of what sexuality education ought to be accomplishing.

Gosselin, Chris, & Wilson, Glenn. (1980). *Sexual variations: Fetishism, sadomasochism, transvestism.* New York: Simon & Schuster.

Gottman, J., Markman, H., & Notarius, C. (1977). The topography of marital conflict: A sequential analysis of verbal and nonverbal behavior. *Journal of Marriage and the Family, 39,* 461–478.

Gottman, J., Notarius, C., Gonso, J., & Markman, H. (1976). *A couple's guide to communication.* Champaign, IL: Research Press.

Gottman, John, Conn, James, Carrere, Sybil, & Swanson, Catherine. (1998). Predicting marital happiness and stability from newlywed interactions. *Journal of Marriage and the Family, 60,* 5–22.

Gottman, John M. (1994). *Why marriages succeed or fail.* New York: Simon & Schuster.

Gottman, John M., & Porterfield, A. L. (1981). Communicative competence in the nonverbal behavior of married couples. *Journal of Marriage and the Family, 43,* 817–824.

Gottman, John S. (1990). Children of gay and lesbian parents. In F. W. Bozett & M. B. Sussman (Eds.), *Homosexuality and family relations* (pp. 177–196). New York: Harrington Park.

Gould, Stephen J. (1987). *An urchin in the storm.* New York: Norton.

Gow, Haven Bradford. (1994). Condom distribution in high school. *The Clearing House, 67,* 183–184.

Gower, D. B., & Ruparelia, B. A. (1993). Olfaction in humans with special reference to odorous 16-androstenes: Their occurrence, perception and possible social, psychological and sexual impact. *Journal of Endocrinology, 137,* 167–187.

Graber, Benjamin. (1993). Medical aspects of sexual arousal disorders. In W. O'Donohue & J. H. Geer (Eds.), *Handbook of sexual dysfunctions* (pp. 103–156). Boston: Allyn & Bacon.

Grady, W. R., Klepinger, D. H., Billy, J. O. G., & Tanfer, K. (1993). Condom characteristics: The perceptions of men in the United States. *Family Planning Perspectives, 25*(2), 67–73.

Graham, Cynthia A. (1991). Menstrual synchrony: An update and review. *Human Nature, 2,* 293–311.

Graham, Cynthia A., et al. (1995). The effects of steroidal contraceptives on the well-being and sexuality of women. *Contraception, 52,* 363–369.

Grammick, Jeannine. (1986). The Vatican's battered wives. *Christian Century, 103,* 17–20.

Grau, Ina, & Kimpf, Martin. (1993). Love, sexuality, and satisfaction: Interventions of men and women. *Zeitschrift fur Sozial Psychologie, 24,* 83–93.

Greeley, Andrew. (1994). Review of the Janus Report on Sexual Behavior. *Contemporary Sociology, 23,* 221–223.

Green, Cynthia P. (1992). The environment and population growth: Decade for action. *Population Reports,* Series M, No. 10, 1–31.

Green, Richard. (1975). Adults who want to change sex; adolescents who cross-dress; and children called "sissy" and "tomboy." In R. Green (Ed.), *Human sexuality: A health practitioner's text.* Baltimore, MD: Williams & Wilkins.

Green, Richard, & Fleming, Davis T. (1990). Transsexual surgery follow-up: Status in the 1990s. *Annual Review of Sex Research, 1,* 163–174.

Green, Richard. (1989). Griswold's legacy: Fornication and adultery as crimes. *Ohio Northern University Law Review, 16,* 545–549.

Green, Richard. (1987). *The Sissy Boy Syndrome and the development of homosexuality.* New Haven: Yale University Press.

Green, Richard. (1978). Intervention and prevention: The child with cross-sex identity. In C. B. Qualls et al. (Eds.), *The prevention of sexual disorders* (pp. 75–94). New York: Plenum.

Green, Ronald M. (1984). Genetic medicine in Jewish legal perspective. *The Annual of the Society of Christian Ethics,* 249–272.

Greenberg, Bradley, Brown, Jane D., & Buerkel-Rothfuss, Nancy L. (Eds.). (1993). *Media, sex, and the adolescent.* Cresskill, NJ: Hampton Press.

Greenberg, Bradley S., & Busselle, Rick. (1996). What's old, what's new: Sexuality on the soaps. *SIECUS Report, 24*(5), 14–16.

Greene, Sheila, Nugent, J. Kevin, Wieczorek-Deering, Dorit E., O'Mahoney, Paul, et al. (1991). The patterning of depressive symptoms in a sample of first-time mothers. *Irish Journal of Psychology, 12,* 263–275.

Greenwald, Evan, & Leitenberg, Harold. (1989). Long-term effects of sexual experiences with siblings and nonsiblings during childhood. *Archives of Sexual Behavior, 18,* 389–400.

Greer, Arlette E., & Buss, David M. (1994). Tactics for promoting sexual encounters. *Journal of Sex Research, 31,* 185–201.

Gregersen, Edgar. (1983). *Sexual practices: The story of human sexuality.* New York: Franklin Watts.

Gregoire, Alain. (1992). New treatments for erectile impotence. *British Journal of Psychiatry, 160,* 315–326.

Gregor, Thomas. (1985). *Anxious pleasures: The sexual lives of an Amazonian people.* Chicago: University of Chicago Press.

Grenier, Guy, & Byers, E. Sandra. (1997). The relationships among ejaculatory control, ejaculatory latency, and attempts to prolong heterosexual intercourse. *Archives of Sexual Behavior, 26,* 27–48.

Grenier, Guy, & Byers, E. Sandra. (1995). Rapid ejaculation: A review of conceptual, etiological, and treatment issues. *Archives of Sexual Behavior, 24,* 447–474.

Griffin, Susan. (1981). *Pornography and silence.* New York: Harper & Row.

Griffit, W. (1970). Environmental effects on interpersonal affective behavior: Ambient effective temperature and attraction. *Journal of Personality and Social Psychology, 15,* 240–244.

Grodstein, Francine, et al. (1997). Postmenopausal hormone therapy and mortality. *New England Journal of Medicine, 336,* 1769–1775.

Gross, Alan E., & Bellew-Smith, Martha. (1983). A social psychological approach to reducing pregnancy risk in adolescence. In D. Byrne & W. A. Fisher (Eds.), *Adolescents, sex, and contraception.* Hillsdale, NJ: Lawrence Erlbaum.

Gruntz, Louis, G., Jr. (1974). Obscenity 1973: Remodeling the house that Ruth built. *Loyola Law Review, 20,* 159–174.

Gunderson, B. H., et al. (1981). Sexual behavior of preschool children. In L. L. Constantine & F. M. Martinson (Eds.), *Children and sex* (pp. 45–62). Boston: Little, Brown.

Gursoy, Akile. (1996). Abortion in Turkey: A matter of state, family, or individual decision. *Social Science & Medicine, 42*, 531–542.

Gutek, Barbara A. (1985). *Sex and the workplace.* San Francisco: Jossey-Bass.

Guttmacher, Sally, Lieberman, L., Ward, David, Radosh, Alice, Rafferty, Yvonne, & Freudenberg, Nicholas. (1995). Parents' attitudes and beliefs about HIV/AIDS prevention with condom availability in New York City public high schools. *Journal of School Health, 65*, 101–106.

Guyer, Bernard, Martin, Joyce, MacDorman, Marion, Anderson, Robert, & Strobino, Donna. (1997). Annual summary of vital statistics—1996. *Pediatrics, 100*, 905–918.

Hahn, S. R., & Paige, K. E. (1980). American birth practices: A critical review. In J. E. Parsons (Ed.), *The psychobiology of sex differences and sex roles.* New York: McGraw-Hill, Hemisphere.

Halbreich, Uriel. (1996). Reflections on the cause of premenstrual syndrome. *Psychiatric Annals, 26*, 581–585.

Haldeman, Douglas C. (1994). The practice and ethics of sexual orientation conversion therapy. *Journal of Consulting and Clinical Psychology, 62*, 221–227.

Hall, Gordon C. Nagayama. (1997). Prevention of sexual aggression: Sociocultural risk and protective factors. *American Psychologist, 52*, 5–14.

Hall, Gordon C. Nagayama. (1996). *Theory-based assessment, treatment, and prevention of sexual aggression.* New York: Oxford University Press.

Hall, Gordon C. Nagayama. (1995). Sexual offender recidivism revisited: A meta-analysis of recent treatment studies. *Journal of Consulting and Clinical Psychology, 63*, 802–809.

Hall, Gordon C. Nagayama & Barongan, Christy (1997). Prevention of sexual aggression: *American Psychologist, 52*, 5–14.

Hamer, Dean, Hu, Stella, Magnuson, Victoria L., Hu, Nan, Pattatucci, Angela M. L. (1993). A linkage between DNA markers on the X chromosome and male sexual orientation. *Science, 261*, 321–327.

Hanash, Kamal A. (1997). Comparative results of goal-oriented therapy for erectile dysfunction. *Journal of Urology, 157*, 2135–2138.

Handler, A., Kistin, N., Davis, J., & Ferre, C. (1991). Cocaine use during pregnancy: Perinatal outcomes. *American Journal of Epidemiology, 133*, 818–825.

Hanneke, Christine R., Shields, Nancy M., & McCall, George J. (1986). Assessing the prevalence of marital rape. *Journal of Interpersonal Violence, 1*, 350–362.

Hanson, R. Karl, & Bussiere, Monique T. (1998). Predicting relapse: A meta-analysis of sexual offender recidivism studies. *Journal of Consulting and Clinical Psychology, 66*, 348–362.

Harbison, R. D., & Mantilla-Plata, B. (1972). Prenatal toxicity, maternal distribution and placental transfer of tetrahydrocannabinol. *Journal of Pharmacology and Experimental Therapeutics, 180*, 446–453.

Harlow, Harry F. (1959, June). Love in infant monkeys. *Scientific American, 200*, 68–70.

Harlow, Harry F., Harlow, M. K., & Hause, F. W. (1963). The maternal affectional system of rhesus monkeys. In H. L. Rheingold (Ed.), *Maternal behavior in mammals.* New York: Wiley.

Harris, G. W., & Levine, S. (1965). Sexual differentiation of the brain and its experimental control. *Journal of Physiology, 181*, 379–400.

Harrison, Albert. (1977). Mere exposure. In L. Berkowitz (Ed.), *Advances in experimental social psychology* (Vol. 10). New York: Academic.

Hart, Linda L. (1990). Accuracy of home pregnancy tests. *Annals of Pharmacotherapy, 24*, 712–713.

Hartman, William, & Fithian, Marilyn. (1984). *Any man can: The multiple orgasmic technique for every loving man.* New York: St. Martin's Press.

Hartman, William E., & Fithian, Marilyn A. (1974). *Treatment of sexual dysfunction.* New York: Jason Aronson.

Hass, Aaron. (1979). *Teenage sexuality.* New York: Macmillan.

Hatcher, Robert A., et al. (1998). *Contraceptive technology* (17th ed.). New York: Ardent Media.

Hatcher, Robert A., et al. (1994). *Contraceptive technology* (16th ed.). New York: Irvington.

Hatcher, Robert A., et al. (1976). *Contraceptive technology, 1976–1977* (8th ed.). New York: Irvington.

Hatfield, Elaine. (1978). Equity and extramarital sexuality. *Archives of Sexual Behavior, 7*, 127–141.

Hatfield, Elaine, & Rapson, Richard. (1993a). Historical and cross-cultural perspectives on passionate love and sexual desire. *Annual Review of Sex Research, 4*, 67–97.

Hatfield, Elaine, & Rapson, Richard. (1993b). *Love, sex, and intimacy.* New York: HarperCollins.

Hatfield, Elaine, & Sprecher, Susan. (1986a). Measuring passionate love in intimate relations. *Journal of Adolescence, 9*, 383–410.

Hatfield, Elaine, & Sprecher, Susan. (1986b). *Mirror, mirror . . . : The importance of looks in everyday life.* Albany: State University of New York Press.

Hatfield, Elaine, & Walster, G. William. (1978). *A new look at love.* Reading, MA: Addison-Wesley.

Hatfield, Elaine, Walster, G. W., & Berscheid, E. (1978). *Equity theory and research.* Boston: Allyn and Bacon.

Hausknecht, Richard U. (1995). Methotrexate and misoprostol to terminate early pregnancy. *New England Journal of Medicine, 333*, 537–540.

Hawton, Keith. (1992). Sex therapy research: Has it withered on the vine? *Annual Review of Sex Research, 3*, 49–72.

Haynes, Barton F., Pantaleo, Giuseppe, & Fauci, Anthony S. (1996). Toward an understanding of the correlates of protective immunity to HIV infection. *Science, 271*, 324–328.

Hays, Constance L. (1990, Sept. 30). Reports of assaults on homosexuals increase. *New York Times.*

Hazan, C., & Shaver, P. (1987). Love conceptualized as an attachment process. *Journal of Personality and Social Psychology, 52*, 511–524.

Hearst, Norman, & Hulley, Stephen B. (1988). Preventing the heterosexual spread of AIDS: Are we giving our patients the best advice? *Journal of the American Medical Association, 259*, 2428–2432.

Heath, Robert C. (1972). Pleasure and brain activity in man. *Journal of Nervous and Mental Disease, 154*, 3–18.

Hedricks, Cynthia A. (1994). Sexual behavior across the menstrual cycle: A biopsychosocial approach. *Annual Review of Sex Research, 5*, 122–172.

Heilman, Carole A., & Baltimore, David. (1998). HIV-vaccines—where are they going? *Nature Medicine Vaccine Supplement, 4*, 532–534.

Heiman, Julia R. (1975). The physiology of erotica: Women's sexual arousal. *Psychology Today, 8*(11), 90–94.

Heiman, Julia, LoPiccolo, Leslie, & LoPiccolo, Joseph. (1976). *Becoming orgasmic: A sexual growth program for women.* Englewood Cliffs, NJ: Prentice-Hall.

Heiman, Julia R., & Meston, Cindy M. (1997). Empirically validated treatment for sexual dysfunction. *Annual Review of Sex Research, 8*, 148–194.

Heiman, Julia R., et al. (1991). Psychophysiological and endocrine responses to sexual arousal in women. *Archives of Sexual Behavior, 20*, 171–186.

Heise, Lori. (1993). Violence against women: The hidden health burden. *World Health Statistics Quarterly, 46*, 78–85.

Hellerstein, Herman K., & Friedman, Ernst H. (1969, Mar.). Sexual activity and the post-coronary patient. *Medical Aspects of Human Sexuality, 3*, 70–74.

Henahan, John. (1984). Honing the treatment of early breast cancer. *Journal of the American Medical Association, 251*, 309–310.

Hendrick, Susan. (1981). Self-disclosure and marital satisfaction. *Journal of Personality and Social Psychology, 40*, 1150–1159.

Hendrick, Susan, & Hendrick, Clyde. (1992). *Liking, loving, and relating* (2d ed.). Pacific Grove, CA: Brooks/Cole.

Hendrick, S. S., Hendrick, Clyde, & Adler, N. L. (1988). Romantic relationships: Love, satisfaction, and staying together. *Journal of Personality and Social Psychology, 54*, 980–988.

Henshaw, Stanley K. (1997). Teenage abortion and pregnancy rates by state, 1992. *Family Planning Perspectives, 29,* 115–122.

Henshaw, Stanley K. (1990). Induced abortion: A world review, 1990. *Family Planning Perspectives, 22,* 76–89.

Henshaw, Stanley, & Van Vort, Jennifer. (1994). Abortion services in the United States, 1991 and 1992. *Family Planning Perspectives, 26,* 100–106, 112.

Herbert, J. (1966). The effect of estrogen applied directly to the genitalia upon the sexual attractiveness of the female rhesus monkey. *Exerpta Medica International Congress Series, 3,* 212.

Herbst, A. (1972). Clear cell adenocarcinoma of the genital tract in young females. *New England Journal of Medicine, 287*(25), 1259–1264.

Herdt, Gilbert. (1990). Mistaken gender: 5-alpha reductase hermaphroditism and biological reductionism in sexual identity reconsidered. *American Anthropologist, 92,* 433–446.

Herdt, Gilbert H. (1984). *Ritualized homosexuality in Melanesia.* Berkeley, CA: University of California Press.

Herdt, Gilbert. (1994). *Guardians of the flutes: Idioms of masculinity.* Chicago: University of Chicago Press.

Herek, Gregory M. (1993). Sexual orientation and military service: A social science perspective. *American Psychologist, 48,* 538–549.

Herek, Gregory M. (1990). Gay people and government security clearances: A social science perspective. *American Psychologist, 45,* 1035–1042.

Herek, Gregory M., & Capitanio, John P. (1995). Black heterosexuals' attitudes towards lesbians and gay men in the United States. *The Journal of Sex Research, 32,* 95–105.

Herek, Gregory M., & Glunt, Eric K. (1993). Interpersonal contact and heterosexuals' attitudes toward gay men: Results from a national survey. *The Journal of Sex Research, 30,* 239–244.

Herek, Gregory M., Kimmel, Douglas C., Amaro, Hortensia, & Melton, Gary (1991). Avoiding heterosexist bias in psychology research. *American Psychologist, 46,* 957–963.

Herman, Judith L. (1981). *Father-daughter incest.* Cambridge, MA: Harvard University Press.

Hersey, R. B. (1931). Emotional cycles in man. *Journal of Mental Science, 77,* 151–169.

Hershberger, Scott L., & D'Augelli, Anthony R. (1995). The impact of victimization on the mental health and suicidality of lesbian, gay, and bisexual youths. *Developmental Psychology, 31,* 65–74.

Heyl, Barbara Sherman. (1979). *The madam as entrepreneur.* New Brunswick, NJ: Transaction Books.

Hicks, Esther K. (1996). *Infibulation: Female mutilation in Islamic Northeastern Africa.* New Brunswick, NJ: Transaction.

Higham, Eileen. (1980). Sexuality in the infant and neonate: Birth to two years. In B. B. Wolman & J. Money (Eds.), *Handbook of human sexuality.* Englewood Cliffs, NJ: Prentice-Hall.

Hill, C. T., Rubin, Z., & Peplau, L. A. (1976). Breakups before marriage: The end of 103 affairs. *Journal of Social Issues, 32*(1).

Hines, Jeffrey F., Ghim, Shin-je, & Jenson, A. Bennett. (1998). Prospects for human papillomavirus vaccine development emerging HPV vaccines: *Current Opinion in Infectious Diseases, 11,* 57–61.

Hines, Melissa, & Collaer, Marcia (1993). Gonadal hormones and sex differences. *Annual Review of Sex Research, 4,* 1–48.

Hirsch, Martin S. (1990). Clinical manifestations of HIV infection in adults in industrialized countries. In K. Holmes et al. (Eds.), *Sexually transmitted diseases* (pp. 331–342). New York: McGraw-Hill.

Hite, Shere. (1981). *The Hite report on male sexuality.* New York: Alfred Knopf.

Hite, Shere. (1976). *The Hite report.* New York: Macmillan.

Ho, Gloria Y. F., et al. (1998). Natural history of cervicovaginal papillomavirus infection in young women. *New England Journal of Medicine, 338,* 423–428.

Hobart, C. Q. (1958). The incidence of romanticism during courtship. *Social Forces, 36,* 364.

Hobfoll, Stevan, Ritter, Christian, Lavin, Justin, Hulsizer, Michael, et al. (1995). Depression prevalence and incidence among inner-city pregnant and postpartum women. *Journal of Consulting and Clinical Psychology, 63,* 445–453.

Hoff, Gerard A., & Schneiderman, Lawrence J. (1985, Dec.). Having babies at home: Is it safe? Is it ethical? *Hastings Center Report,* 19–27.

Hofferth, Sandra L. (1990). Trends in adolescent sexual activity, contraception, and pregnancy in the United States. In J. Bancroft & J. Reinisch (Eds.), *Adolescence and puberty* (pp. 217–233). New York: Oxford University Press.

Hollander, D. (1996). Programs to bring down cesarean section rate prove to be successful. *Family Planning Perspectives, 28,* 182–185.

Holmes, King K., et al. (Eds.). (1990). *Sexually transmitted diseases* (2d ed.). New York: McGraw-Hill.

Holmstrom, Lynda Lytle, & Burgess, Ann Wolbert. (1980). Sexual behavior of assailants during reported rapes. *Archives of Sexual Behavior, 9,* 427–440.

Holroyd, Jean C., & Brodsky, Annette M. (1977). Psychologists' attitudes and practices regarding erotic and nonerotic physical contact with patients. *American Psychologist, 34,* 843–849.

Hong, Lawrence K. (1984). Survival of the fastest: On the origin of premature ejaculation. *Journal of Sex Research, 20,* 109–122.

Hook, Edward W., & Handsfield, H. Hunter. (1990). Gonococcal infections in adults. In K. Holmes et al. (Eds.), *Sexually transmitted diseases* (pp. 149–160). New York: McGraw-Hill.

Hoon, P. W., Bruce, K., & Kinchloe, B. (1982). Does the menstrual cycle play a role in sexual arousal? *Psychophysiology 19,* 21–26.

Hopkins, J., Marcues, M., & Campbell, S. B. (1984). Postpartum depression: A critical review. *Psychological Bulletin, 95,* 498–515.

Hopwood, Nancy J., et al. (1990). The onset of human puberty: Biological and environmental factors. In J. Bancroft & J. M. Reinisch (Eds.), *Adolescence and puberty.* New York: Oxford University Press.

Horney, Karen. (1973). The flight from womanhood (1926). In K. Horney, *Feminine psychology.* New York: Norton.

Horowitz, Carol R., & Jackson, T. Carey. (1997). Female "circumcision": African women confront American medicine. *Journal of General Internal Medicine, 12,* 491–499.

Horrocks, R. (1997). *An introduction to the study of sexuality.* New York: St. Martin's Press.

House, Carrie. (1997). Navajo warrior women: An ancient tradition in a modern world. In S. Jacobs, et al. (Eds.), *Two-spirit people* (pp. 223–227). Urbana, IL: University of Illinois Press.

Houston, L. N. (1981). Romanticism and eroticism among black and white college students. *Adolescence, 16,* 263–272.

Hoyenga, Katherine B., & Hoyenga, Kermit T. (1993). *Gender-related differences: Origins and outcomes.* Boston: Allyn and Bacon.

Hsu, Bing, Kling, Arthur, Kessler, Christopher, Knapke, Kory, Difenbach, Pamela, & Elias, James. (1994). Gender differences in sexual fantasy and behavior in a college population: A ten-year replication. *Journal of Sex and Marital Therapy, 20,* 103–118.

Hucker, Stephen, & Blanchard, Ray. (1992). Death scene characteristics in 118 fatal cases of autoerotic asphyxia compared with suicidal asphyxia. *Behavioural Sciences and the Law, 10,* 509–523.

Hudson, Walter W., & Ricketts, Wendell A. (1980). A strategy for the measurement of homophobia. *Journal of Homosexuality, 5,* 357–372.

Hughes, Jean O., & Sandler, Bernice R. (1987). *"Friends" raping friends: Could it happen to you?* Washington, D.C.: Association of American Colleges.

Hulter, Birgitta, & Lundberg, B. O. (1994). Sexual function in women with hypothalamo-pituitary disorders. *Archives of Sexual Behavior, 23,* 171–184.

Humphreys, Laud. (1970). *Tearoom trade: Impersonal sex in public places.* Chicago: Aldine.

Hunt, Morton. (1974). *Sexual behavior in the 1970s*. Chicago: Playboy Press.

Hunt, Morton, & Hunt, Bernice. (1977). *The divorce experience*. New York: McGraw-Hill.

Hunter, Nan, Michaelson, Sherryl, & Stoddard, Thomas. (1992). *The rights of lesbians and gay men: The basic ACLU guide to a gay person's rights*. Carbondale, IL: Southern Illinois University Press.

Huston, T. L., & Levinger, G. (1978). Interpersonal attraction and relationships. In M. R. Rosenzweig & L. W. Porter (Eds.), *Annual Review of Psychology* (Vol. 29). Palo Alto, CA: Annual Reviews.

Hutchins, Loraine, & Kaahumanu, Lani. (Eds.). (1991). *Bi any other name: Bisexual people speak out*. Boston: Alyson.

Hutchinson, Karen A. (1995). Androgens and sexuality. *American Journal of Medicine, 98* (Suppl. 1A), 1A111S–1A115S.

Hyde, Janet S. (1996). *Half the human experience: The psychology of women* (5th ed.). Boston: Houghton-Mifflin.

Hyde, Janet S. (1984). How large are gender differences in aggression? A developmental meta-analysis. *Developmental Psychology, 20*, 722–736.

Hyde, Janet S., DeLamater, John, & Hewitt, Erri. (1998). Sexuality and the dual-earner couple: Multiple rules and sexual functioning. *Journal of Family Psychology, 12*, 354–368.

Hyde, Janet S., DeLamater, John, Plant, E. Ashby, and Byrd, Janis M. (1996). Sexuality during pregnancy and the year postpartum. *Journal of Sex Research, 33*, 143–151.

Icard, Larry D. (1996). Assessing the psychosocial well-being of African American gays. In J. F. Longres (Ed.), *Men of color: A context for service to homosexually active men*. New York: Haworth.

Idanpaan-Heikkila, J., et al. (1969). Placental transfer of tritiated-1 tetrahydrocannabinol. *New England Journal of Medicine, 281*, 330.

Imperato-McGinley, J., et al. (1974). Steroid 5 reductase deficiency in man: An inherited form of male pseudohermaphroditism. *Science, 186*, 1213–1215.

Innala, Sune M., & Ernulf, Kurt E. (1989). Asphyxiophilia in Scandinavia. *Archives of Sexual Behavior, 18*, 181–190.

Irvine, Janice M. (1995). *Sexuality education across cultures: Working with differences*. San Francisco: Jossey-Bass.

Jacob, Kathryn A. (1981). The Mosher report. *American Heritage*, 57–64.

Jacobs, Sue-Ellen, Thomas, Wesley, & Lang, Sabine. (Eds.). (1997). *Two-spirit people*. Urbana, IL: University of Illinois Press.

Jacobson, Sandra W., Jacobson, Joseph L., & Sokol, Robert J. (1994). Effects of fetal alcohol exposure on infant reaction time. *Alcoholism: Clinical and Experimental Research, 18*, 1125–1132.

Jacobson, Sandra W., Jacobson, Joseph L., Sokol, Robert J., Martier, Susan S., & Ager, Joel W. (1993). Prenatal alcohol exposure and infant information processing ability. *Child Development, 64*, 1706–1721.

Jadack, Rosemary A., Keller, Mary L., & Hyde, Janet S. (1990). Genital herpes: Gender comparisons and the disease experience. *Psychology of Women Quarterly, 14*, 419–434.

Jaffe, A., & Becker, R. E. (1984). Four new basic sex offenses: A fundamental shift in emphasis. *Illinois Bar Journal, 72*, 400–403.

Janssen, Erick, & Everaerd, Walter. (1993). Determinants of male sexual arousal. *Annual Review of Sex Research, 4*, 211–246.

Jantzen, Grace. (1994). AIDS, shame, and suffering. In J. B. Nelson & S. P. Longfellow (Eds.), *Sexuality and the sacred* (pp. 305–313). Louisville, KY: Westminster/John Knox Press.

Janus, Samuel S., & Janus, Cynthia L. (1993). *The Janus report on sexual behavior*. New York: Wiley.

Jay, Karla, & Young, Allen (1979). *The gay report*. New York: Summit Books.

Jemail, Jay Ann, & Geer, James. (1977). Sexual scripts. In R. Gemme & C. C. Wheeler (Eds.), *Progress in sexology*. New York: Plenum.

Jenkins, J. S., & Nussey, S. S. (1991). The role of oxytocin: Present concepts. *Clinical Endocrinology, 34*, 515–525.

Jenkins, Philip. (1996). *Pedophiles and priests: Anatomy of a contemporary crisis*. New York: Oxford University Press.

Jenks, Richard J. (1985). Swinging: A replication and test of a theory. *Journal of Sex Research, 21*, 199–210.

Jenks, Richard J. (1985). Swinging: A test of two theories and a proposed new model. *Archives of Sexual Behavior, 14*, 517–528.

Jenny, Carole, Roesler, Thomas A., & Poyer, Kimberly A. (1994). Are children at risk for sexual abuse by homosexuals? *Pediatrics, 94*, 41–44.

Jensen, Gordon D. (1976a). Adolescent sexuality. In B. J. Sadock et al. (Eds.), *The sexual experience*. Baltimore: Williams & Wilkins.

Jensen, Gordon D. (1976b). Cross-cultural studies and animal studies of sex. In B. J. Sadock et al. (Eds.), *The sexual experience*. Baltimore, MD: Williams & Wilkins.

John, E. M., Savitz, D. A., & Sandler, D. P. (1991). Prenatal exposure to parents' smoking and childhood cancer. *American Journal of Epidemiology, 133*, 123–132.

Johnson, Anne M., Wadsworth, J., Wellings, K., Bradshaw, S., & Field, J. (1992). Sexual lifestyles and HIV risk. *Nature, 360*, 410–412.

Johnson, Anne M., Wadsworth, Jane, Wellings, Kaye, & Field, Julia. (1994). *Sexual attitudes and lifestyles*. London: Blackwell.

Johnson, Brooke R., Horga, Mihai, & Andronache, Laurentia. (1996). Women's perspectives on abortion in Romania. *Social Science & Medicine, 42*, 521–530.

Johnson, Catherine B., Stockdale, M. S., & Saal, F. E. (1991). Persistence of men's misperceptions of friendly cues across a variety of interpersonal encounters. *Psychology of Women Quarterly, 15*, 463–475.

Jones, James H. (1997). *Alfred C. Kinsey: A public/private life*. New York: Norton.

Jones, Jennifer C., & Barlow, David H. (1990). Self-reported frequency of sexual urges, fantasies, and masturbatory fantasies in heterosexual males and females. *Archives of Sexual Behavior, 19*, 269–280.

Jorgensen, S. R. (1980). Contraceptive attitude-behavior consistency in adolescence. *Population and Environment, 3*, 174–194.

Kabalin, John N., & Kuo, Jeffrey C. (1997). Long-term follow-up of and patient satisfaction with the Dynaflex self-contained inflatable penile prosthesis. *Journal of Urology, 158*, 456–469.

Kafka, Martin P. (1997a). A monoamine hypothesis for the pathophysiology of paraphilic disorders. *Archives of Sexual Behavior, 26*, 343–358.

Kafka, Martin P. (1997b). Hypersexual desire in males: An operational definition and clinical implications for males with paraphilias and paraphilia-related disorders. *Archives of Sexual Behavior, 26*, 505–526.

Kalick, S. Michael, Zebowitz, Leslie, Langlois, Judith, & Johnson, Robert. (1998). Does human facial attractiveness honestly advertise health? Longitudinal data on an evolutionary question. *Psychological Science, 9*, 8–13.

Kalil, Kathleen, Gruber, James, Conley, Joyce, & Sytniac, Michael. (1993). Social and family pressures on anxiety and stress during pregnancy. *Pre- and Perinatal Psychology Journal, 8*, 113–118.

Kanin, E. J., Davidson, K. D., & Scheck, S. R. (1970). A research note on male-female differentials in the experience of heterosexual love. *Journal of Sex Research, 6*, 64–72.

Kanin, Eugene J. (1985). Date rapists: Differential sexual socialization and relative deprivation. *Archives of Sexual Behavior, 14*, 219–232.

Kantner, John F., & Zelnik, Melvin. (1973). Contraception and pregnancy: Experience of young unmarried women in the United States. *Family Planning Perspectives, 5*(1), 21–35.

Kantner, John F., & Zelnik, Melvin. (1972). Sexual experience of young unmarried women in the United States. *Family Planning Perspectives, 4*(4), 9–18.

Kaplan, Helen S. (1995). *The sexual desire disorders: Dysfunctional regulation of sexual motivation*. New York: Brunner/Mazel.

Kaplan, Helen S. (1974). *The new sex therapy*. New York: Brunner/Mazel.

Kaplan, Helen S., & Owett, Trude. (1993). The female androgen deficiency syndrome. *Journal of Sex and Marital Therapy, 19*, 3–25.

Kaplan, Helen S., & Sager, C. J. (1971, June). Sexual patterns at different ages. *Medical Aspects of Human Sexuality,* 10–23.

Kaplan, Helen Singer. (1979). *Disorders of sexual desire.* New York: Simon & Schuster.

Karr, Rodney K. (1978). Homosexual labeling and the male role. *Journal of Social Issues, 34*(3), 73–83.

Kashon, Michael L., Ward, O. Byron, Grisham, William, & Ward, Ingeborg L. (1992). Prenatal, β-endorphin can modulate some aspects of sexual differentiation in rats. *Behavioral Neuroscience, 106,* 555–562.

Katz, Kathryn D. (1982). Sexual morality and the Constitution: People v. Onofre. *Albany Law Review, 46,* 311–362.

Kaunitz, Andrew M. (1997). Reappearance of the intrauterine device: A "user-friendly" contraceptive. *International Journal of Fertility, 42,* 120–127.

Kegel, A. H. (1952). Sexual functions of the pubococcygeus muscle. *Western Journal of Surgery, 60,* 521–524.

Kempf, D. J., et al. (1995). ABT-538 is a potent inhibitor of human immunodeficiency virus protease. *Proceedings of the National Academy of Sciences, 92,* 2484.

Kempton, Winifred, & Kahn, Emily. (1991). Sexuality and people with intellectual disabilities: A historical perspective. *Sexuality and Disability, 9,* 93–111.

Kendall-Tackett, K., Williams, L., & Finkelhor, D. (1993). Impact of sexual abuse on children: A review and synthesis of recent empirical studies. *Psychological Bulletin, 113,* 164–180.

Kenrick, D., et al. (1980). Sex differences, androgyny and approach responses to erotica: A new variation on an old volunteer problem. *Journal of Personality and Social Psychology, 38,* 517–524.

Kessler, Suzanne J. (1998). *Lessons from the intersexed.* Piscataway, NJ: Rutgers University Press.

Kessler, Suzanne. (1997). Creating good-looking genitals in the service of gender. In M. Duberman (Ed.), *A queer world* (pp. 153–176). New York: New York University Press.

Kessler, Suzanne. (1990). The medical construction of gender: Case management of intersexed infants. *Signs: Journal of Women in Culture and Society, 16,* 3–26.

Kestelman, Philip, & Trussell, James. (1991, Oct.). Efficacy of the simultaneous use of condoms and spermicides. *Family Planning Perspectives, 23,* 226–232.

Kiecolt, K. J., Fossett, M. A., & Smith, W. (1995). Mate availability and marriage among African-Americans: Aggregate- and individual-level analyses. In M. B. Tucker & C. Mitchell-Kerum (Eds.), *The decline in marriage among African-Americans: Causes, consequences, and policy implications* (pp. 103–116). New York: Russell Sage Foundation.

Kilmartin, Christopher T. (1994). *The masculine self.* New York: Macmillan.

Kim, Elaine H. (1990, Winter). "Such opposite creatures": Men and women in Asian American literature. *Michigan Quarterly Review, 29,* 68–93.

Kim, N., Stanton, B., Li, X., Dickersin, K., & Galbraith, J. (1997). Effectiveness of the 40 adolescent AIDS-risk reduction interventions: A quantitative review. *Journal of Adolescent Health, 20,* 204–215.

Kimlicka, Thomas, Cross, H., & Tarnai, J. (1983). A comparison of androgynous, feminine, masculine, and undifferentiated women on self-esteem, body satisfaction, and sexual satisfaction. *Psychology of Women Quarterly, 1,* 291–294.

King, Michael, & Woollett, Earnest. (1997). Sexually assaulted males: 115 men consulting a counseling service. *Archives of Social Behavior, 26,* 579–588.

Kingman, Sharon. (1995, July). Human papilloma virus vaccine tested in cervical cancer. *Journal of NIH Research, 7,* 46–47.

Kinsey, Alfred C., Pomeroy, Wardell B., & Martin, Clyde E. (1948). *Sexual behavior in the human male.* Philadelphia, PA: Saunders.

Kinsey, Alfred C., Pomeroy, Wardell B., Martin, Clyde E., & Gebhard, Paul H. (1953). *Sexual behavior in the human female.* Philadelphia, PA: Saunders.

Kiragu, Karungari. (1995, Oct.). Female genital mutilation: A reproductive health concern. *Population Reports* (Supplement), Series J, No. 41, Vol. 23.

Kirby, Douglas. (1992). School-based programs to reduce sexual risk-taking behavior. *Journal of School Health, 62,* 281–287.

Kirby, Douglas. (1977). The methods and methodological problems of sex research. In J. S. DeLora & C. A. B. Warren (Eds.), *Understanding sexual interaction.* Boston: Houghton Mifflin.

Kirby, D., Koupi, M., Barth, R. P., & Cagampang, H. H. (1997). The impact of the Postponing Sexual Involvement curriculum among youths in California. *Family Planning Perspectives, 29,* 100–108.

Kirby, Douglas, Short, Lynn, Collins, Janet, Rugg, Deborah, Kolbe, Lloyd, Howard, Marion, Miller, Brent, Sonenstein, Freya, & Zubin, Laurie. (1994). School-based programs to reduce sexual risk behaviors: A review of effectiveness. *Public Health Reports, 109,* 339–360.

Kirkpatrick, Lee, & Davis, Keith. (1994). Attachment style, gender, and relationship stability: A longitudinal analysis. *Journal of Personality and Social Psychology, 66,* 502–512.

Kirkpatrick, Martha. (1996). Lesbians as parents. In R. P. Cabaj & T. S. Stein (Eds.), *Textbook of homosexuality and mental health.* Washington, D.C.: American Psychiatric Press.

Kirkpatrick, M., Smith, C., & Roy, R. (1981). Lesbian mothers and their children: A comparative survey. *American Journal of Orthopsychiatry, 51,* 545–551.

Kiselica, Mark, & Scheckel, Steve. (1995). The couvade syndrome (sympathetic pregnancy) and teenage fathers: A brief primer for counselors. *School Counselor, 43,* 42–51.

Kitzinger, Celia, & Wilkinson, Sue. (1995). Transitions from heterosexuality to lesbianism: The discursive production of lesbian identities. *Developmental Psychology, 31,* 95–104.

Kiviat, Nancy B., et al. (1989). Prevalence of genital papillomavirus infection among women attending a college student health clinic or a sexually transmitted disease clinic. *The Journal of Infectious Diseases, 159,* 293–302.

Klaus, Marshall, & Kennell, John. (1976). Human maternal and paternal behavior. In M. Klaus & J. Kennell (Eds.), *Maternal infant bonding.* St. Louis, MO: Mosby.

Klebanov, Pamela K., & Jemmott, John B. (1992). Effects of expectations and bodily sensations on self-reports of premenstrual symptoms. *Psychology of Women Quarterly, 16,* 289–310.

Klepinger, D. H., Billy, J. O. G., Tanfer, K., & Grady, W. R. (1993). Perceptions of AIDS risk and severity and their association with risk-related behavior among U.S. men. *Family Planning Perspectives, 25*(2), 74–82.

Klirsfeld, Dava. (1998). HIV disease and women. *Medical Clinics of North America, 82,* 335–357.

Kolbenschlag, Madonna. (1985). Abortion and moral consensus: Beyond Solomon's choice. *Christian Century, 102,* 179–183.

Kolker, Aliza. (1989). Advances in prenatal diagnosis. *International Journal of Technology Assessment in Health Care, 5,* 601–617.

Kolodny, R. C., Masters, W. H., Kolodny, R. M., & Toro, G. (1974). Depression of plasma testosterone levels after chronic intensive marihuana use. *New England Journal of Medicine, 290,* 872–874.

Kolodny, Robert C., et al. (1979). Chronic marihuana use by women. Cited in Rosen, R. C. (1991). Alcohol and drug effects on sexual response. *Annual Review of Sex Research, 2,* 119–180.

Kon, Igor S. (1995). *The sexual revolution in Russia: From the age of Czars to today.* New York: Free Press.

Kon, Igor S. (1987). A sociocultural approach. In J. H. Geer & W. T. O'Donohue (Eds.), *Theories of human sexuality* (pp. 257–286). New York: Plenum.

Koonin, Lisa M., Atrash, H. K., Lawson, H. W., & Smith, J. C. (1991a, July). Maternal mortality surveillance, United States, 1979–1986. *Morbidity and Mortality Weekly Report, 40,* No. SS-1, 1–13.

Koonin, Lisa M., Kochanek, K. D., Smith, J. C., & Ramick, M. (1991b, July). Abortion surveillance, United States, 1988. *Morbidity and Mortality Weekly Report, 40*, No. SS-1, 15–42.

Koonin, Lisa M., Smith, Jack C., & Ramick, Merrell. (1995, May 5). Abortion surveillance—United States, 1991. *Morbidity and Mortality Weekly Report, 44*, No. SS-2, 23–53.

Koonin, Lisa, Smith, Jack, & Ramick, Merrell. (1993). Abortion surveillance—United States, 1990. In CDC surveillance summary, December 17, 1993, *Morbidity and Mortality Weekly Report, 42* (No. SS-6), 29–58.

Korff, Janice, & Geer, James H. (1983). The relationship between sexual arousal experience and genital response. *Psychophysiology, 20*, 121–127.

Kosnick, Anthony, et al. (1977). *Human sexuality: New directions in American Catholic thought.* New York: Paulist Press.

Koss, M. P., Dinero, T. E., Siebel, C. A., & Cox, S. L. (1988). Stranger and acquaintance rape: Are there differences in the victim's experience? *Psychology of Women Quarterly, 12*, 1–24.

Koss, M. P., Gidycz, C. A., & Wisniewski, N. (1987). The scope of rape: Incidence and prevalence in a national sample of higher education students. *Journal of Consulting and Clinical Psychology, 55*, 162–170.

Koss, Mary P. (1993). Rape: Scope, impact, interventions, and public policy responses. *American Psychologist, 48*, 1062–1069.

Koss, Mary P., & Cook, Sarah L. (1994). Facing the facts: Date and acquaintance rape are widespread forms of violence. In M. Koss et al. (Eds.), *No safe haven*. Washington, D.C.: American Psychological Association.

Koss, Mary P., Goodman, Lisa A., Browne, Angela, Fitzgerald, Louise F., Russo, Nancy F., & Keita, Gwendolyn P. (1994). *No safe haven: Male violence against women at home, at work, and in the community.* Washington, D.C.: American Psychological Association.

Koss, Mary P., & Heslet, Lynette. (1992). Somatic consequences of violence against women. *Archives of Family Medicine, 1*, 53–59.

Koss, Mary P., Koss, Paul G., & Woodruff, W. Joy. (1991). Deleterious effects of criminal victimization on women's health and medical utilization. *Archives of Internal Medicine, 151*, 342–347.

Kothari, P. (1984). For discussion: Ejaculatory disorders—a new dimension. *British Journal of Sexual Medicine, 11*, 205–209.

Kreiss, J., et al. (1992). Efficacy of nonoxynol-9 contraceptive sponge use in preventing heterosexual acquisition of HIV in Nairobi prostitutes. *Journal of the American Medical Association, 268*, 477–482.

Krimmel, Herbert T. (1983, Oct.). The case against surrogate parenting. *Hastings Center Report*, 35–39.

Kroll, Ken, et al. (1995). *Enabling romance: A guide to love, sex, and relationships for the disabled (and the people who care for them).* Bethesda, MD: Woodbine House.

Kumar, R., Brant, H. A., & Robson, K. M. (1981). Childbearing and maternal sexuality: A prospective survey of 119 primiparae. *Journal of Psychosomatic Research, 25*, 373–383.

Kunkel, D., Cope, K. M., & Colvin, C. (1996). *Sexual messages on family hour television: Content and context.* Menlo Park, CA: Kaiser Family Foundation.

Kurdek, Lawrence A. (1995a). Developmental changes in relationship quality in gay and lesbian cohabiting couples. *Developmental Psychology, 31*, 86–94.

Kurdek, Lawrence A. (1995b). Lesbian and gay couples. In A. R. D'Augelli & C. J. Patterson (Eds.), *Lesbian, gay, and bisexual identities over the lifespan* (pp. 243–261). New York: Oxford.

Kurdek, Lawrence A. (1994). Areas of conflict for gay, lesbian and heterosexual couples: What couples argue about influences relationship satisfaction. *Journal of Marriage and the Family, 56*, 923–934.

Laan, Ellen, & Everaerd, Walter. (1995). Determinants of female sexual arousal: Psychophysiological theory and data. *Annual Review of Sex Research, 6*, 32–76.

Laan, Ellen, Everaerd, Walter, van Bellen, Gerdy, & Hanewald, Gerrit. (1994). Women's sexual and emotional responses to male- and female-produced erotica. *Archives of Sexual Behavior, 23*, 153–170.

Lackritz, Eve M., et al. (1995). Estimated risk of transmission of the human immunodeficiency virus by screened blood in the United States. *New England Journal of Medicine, 333*, 1721–1725.

Ladas, A. K., Whipple, B., & Perry, J. D. (1982). *The G spot and other recent discoveries about human sexuality.* New York: Holt, Rinehart, Winston.

Ladner, Joyce A. (1971). *Tomorrow's tomorrow: The black woman.* Garden City, NY: Doubleday.

LaFromboise, Theresa D., Heyle, Anneliese M., & Ozer, Emily J. (1990). Changing and diverse roles of women in American Indian cultures. *Sex Roles*, 455–476.

Laird, Joan, & Green, Robert-Jay. (Eds.). (1996). *Lesbians and gays in couples and families: A handbook for therapists.* San Francisco: Jossey-Bass.

Lalumieve, M. L., & Quinsey, V. L. (1998). Pavlovian conditioning of sexual interests in human males. *Archives of Sexual Behavior, 27*, 241–252.

Lamb, Michael. (1982, April). Second thoughts on first touch. *Psychology Today*, 9–10.

Lamb, Michael E., & Hwang, C. (1982). Maternal attachment and mother-neonate bonding: A critical review. In M. E. Lamb & A. L. Brown (Eds.), *Advances in developmental psychology* (Vol. 2). Hillsdale, NJ: Lawrence Erlbaum.

Lamberts, Steven W. J., et al. (1997). The endocrinology of aging. *Science, 278*, 419–424.

Lande, Robert E. (1995). New era for injectables. *Population Reports*, Series K, No. 5.

Langer, Ellen J., & Dweck, Carol S. (1973). *Personal politics: The psychology of making it.* Englewood Cliffs, NJ: Prentice-Hall.

Langfeldt, Thore. (1981). Childhood masturbation. In L. L. Constantine & F. M. Martinson (Eds.), *Children and sex* (pp. 63–74). Boston: Little Brown.

Lantz, H. R., Keyes, J., & Schultz, H. (1975). The American family in the preindustrial period: From baselines in history to change. *American Sociological Review, 40*, 21–36.

Larsen, Sandra A., et al. (1996). Syphilis. In S. Morse et al. (Eds.), *Atlas of sexually transmitted diseases and AIDS* (pp. 21–46). London: Mosby-Wolfe.

Latty-Mann, Holly, & Davis, Keith. (1996). Attachment theory and partner choice: Preference and actuality. *Journal of Social and Personal Relationships, 13*, 5–23.

Laumann, Edward O., Gagnon, John H., Michael, Robert T., & Michaels, Stuart. (1994). *The social organization of sexuality: Sexual practices in the United States.* Chicago: University of Chicago Press.

Laurenceau, J-P., Feldman, Barrett, & Pietromonaco, P. R. (1998). Intimacy as an interpersonal process: The importance of self-disclosure, partner disclosure, and perceived partner responsiveness in interpersonal exchanges. *Journal of Personality and Social Psychology, 74*, 1238–1251.

Lawrence, Kelli-An, & Byers, E. Sandra. (1995). Sexual satisfaction in long-term heterosexual relationships: The interpersonal exchange model of sexual satisfaction. *Personal Relationships, 2*, 267–285.

Leading cases: Right to privacy. (1986). *Harvard Law Review, 100*, 200–220.

Leavitt, Fred. (1974). *Drugs and behavior.* Philadelphia: Saunders.

Leavitt, Gregory C. (1989). Disappearance of the incest taboo: A cross-cultural test of general evolutionary hypotheses. *American Anthropologist, 91*, 116–131.

Lebacqz, Karen. (1987). Appropriate vulnerability: A sexual ethic for singles. *Christian Century, 104*, 435–438.

Lebeque, Breck. (1991). Paraphilias in U.S. pornography titles: "Pornography made me do it" (Ted Bundy). *Bulletin of the American Academy of Psychiatry and Law, 19,* 43–48.

Lederer, Laura (Ed.). (1980). *Take back the night: Women on pornography.* New York: Morrow.

Lee, J. A. (1979). The social organization of sexual risk. *Alternative Lifestyles, 2,* 69–100.

Legman, Gershon. (1968). *Rationale of the dirty joke.* New York: Grove.

Leiblum, Sandra R. (1993). The impact of infertility on sexual and marital satisfaction. *Annual Review of Sex Research, 4,* 99–120.

Leiblum, Sandra R., & Rosen, Raymond C. (1989). *Principles and practice of sex therapy* (2d ed.). New York: Guilford.

Leifer, Myra. (1980). *Psychological effects of motherhood: A study of first pregnancy.* New York: Praeger.

Leigh, Barbara C. (1989). Reasons for having and avoiding sex: Gender, sexual orientation, and relationship to sexual behavior. *Journal of Sex Research, 26,* 199–209.

Leitenberg, Harold, Detzer, Mark J., & Srebnik, Debra. (1993). Gender differences in masturbation and the relation of masturbation experience in preadolescence and/or early adolescence to sexual behavior and sexual adjustment in young adulthood. *Archives of Sexual Behavior, 22,* 87–98.

Leitenberg, H., Greenwald, E., & Tarran, M. J. (1989). The relation between sexual activity among children during preadolescence and/or early adolescence and sexual behavior and sexual adjustment in young adulthood. *Archives of Sexual Behavior, 18,* 299–314.

Leitenberg, Harold, & Henning, Kris. (1995). Sexual fantasy. *Psychological Bulletin, 117,* 469–496.

LeMagnen, J. (1952). Les pheromones olfactosexuals chez le rat blanc. *Archives des Sciences Physiologiques, 6,* 295–332.

Lemon, Stanley H., & Newbold, John E. (1990). Viral hepatitis. In K. Holmes et al. (Eds.), *Sexually transmitted diseases* (2d ed.). New York: McGraw-Hill.

Lemp, George, Hirozawa, Anne, et al. (1994). Seroprevalence of HIV and risk behaviors among young homosexual and bisexual men. *Journal of the American Medical Association, 272,* 449–454.

Leonard, Arthur S. (1993). *Sexuality and the law: An encyclopedia of major legal cases.* New York: Garland Publishing.

Lerman, Hannah. (1986). From Freud to feminist personality theory. *Psychology of Women Quarterly, 10,* 1–18.

Lerner, Harriet G. (1989). *The dance of intimacy.* New York: Harper & Row.

LeVay, Simon. (1996). *Queer science: The use and abuse of research into homosexuality.* Cambridge, MA: MIT Press.

LeVay, Simon. (1991). A difference in hypothalamic structure between heterosexual and homosexual men. *Science, 253,* 1034–1037.

Lever, Janet, Kanouse, David E., Rogers, William H., et al. (1992). Behavior patterns and sexual identity of bisexual males. *Journal of Sex Research, 29,* 141–167.

Levine, Carol, & Bermel, Joyce. (Eds.). (1986, Dec.). *AIDS: Public health and civil liberties.* Hastings Center Report Special Supplement, 1–36.

Levine, Carol, & Bermel, Joyce. (Eds.). (1985, Aug.). *AIDS: The emerging ethical dilemmas.* Hastings Center Report Special Supplement, 1–31.

Levine, E. M., Gruenewald, D., & Shaiova, C. H. (1976). Behavioral differences and emotional conflict among male-to-female transsexuals. *Archives of Sexual Behavior, 5,* 81–86.

Levine, R., Sato, S., Hashimoto, T., & Verma, J. (1995). Love and marriage in eleven cultures. *Journal of Cross-Cultural Psychology, 26,* 554–571.

Levinson, D. R., Johnson, M. L., & Devaney, D. M. (1988). *Sexual harassment in the federal government: An update.* Washington, D.C.: U.S. Merit Systems Protection Board.

Levinson, Daniel J. (1978). *The seasons of a man's life.* New York: Ballantine.

Levitt, Eugene, Moser, Charles, & Jamison, Karen. (1994). The prevalence and some attributes of females in the sadomasochistic subculture: A second report. *Archives of Sexual Behavior, 23,* 465–473.

Levitt, Eugene E. (1983). Estimating the duration of sexual behavior. A laboratory analog study. *Archives of Sexual Behavior, 12,* 329–336.

Levitt, Eugene E., & Mulcahy, John J. (1995). The effect of intracavernosal injection of papaverine hydrochloride on orgasm latency. *Journal of Sex & Marital Therapy, 21,* 39–41.

Liebmann-Smith, Joan. (1987). *In pursuit of pregnancy: How couples discover, cope with, and resolve their fertility problems.* New York: Newmarket Press.

Liebowitz, Michael. (1983). *The chemistry of love.* Boston: Little, Brown, and Company.

Lief, Harold I., & Hubschman, Lynn. (1993). Orgasm in the postoperative transsexual. *Archives of Sexual Behavior, 22,* 145–156.

Light, J. Keith. (1997). Editorial: Impotence. *Journal of Urology, 157,* 2139.

Lightfoot-Klein, Hanny. (1993). Disability in female immigrants with ritually inflicted genital mutilation. *Women & Therapy, 14,* 187–194.

Lightfoot-Klein, Hanny. (1989). *Prisoners of ritual: An odyssey into female genital circumcision in Africa.* New York: Haworth.

Lindsey, Robert. (1988, Feb. 1). Circumcision under criticism as unnecessary to newborn. *New York Times,* A1.

Linz, D., Donnerstein, E., & Penrod, S. (1987). The findings and recommendations of the Attorney General's Commission on pornography: Do the psychological "facts" fit the political fury? *American Psychologist, 42,* 946–953.

Linz, Daniel. (1989). Exposure to sexually explicit materials and attitudes toward rape: A comparison of study results. *Journal of Sex Research, 26,* 50–84.

Liskin, Laurie. (1985, Nov.–Dec.). Youth in the 1980s: Social and health concerns. *Population Reports,* XIII, No. 5, M350–M388.

Liskin, Laurie, Beroit, E., & Blackburn, R. (1992). Vasectomy: New opportunities. *Population Reports,* Series D, No. 5. Baltimore: Johns Hopkins University Population Information Program.

Liskin, Laurie, & Blackburn, Richard. (1986, July–Aug.). AIDS: A public health crisis. *Population Reports,* Series L, Number 6, L193–L228.

Liskin, Laurie, Wharton, C., & Blackburn, R. (1990). Condoms—now more than ever. *Population Reports,* Series H, No. 8, 1–35.

Liu, Peter, & Chan, Connie S. (1996). Lesbian, gay, and bisexual Asian Americans and their families. In J. Laird & R. Green (Eds.), *Lesbians and gays in couples and families.* San Francisco: Jossey-Bass.

Lodl, K., McGettigan, & Bucy, J. (1984). Women's responses to abortion: Implications for post-abortion support groups. *Journal of Social Work and Human Sexuality,* 119–132.

Loftus, Elizabeth F. (1993). The reality of repressed memories. *American Psychologist, 48,* 518–537.

Loftus, Elizabeth F., Polonsky, Sara, & Fullilove, Mindy T. (1994). Memories of childhood sexual abuse: Remembering and repressing. *Psychology of Women Quarterly, 18,* 67–84.

Longo, D. J., Clum, G. A., & Yaeger, N. J. (1988). Psychosocial treatment for recurrent genital herpes. *Journal of Consulting and Clinical Psychology, 56,* 61–66.

LoPiccolo, Joseph, & Stock, Wendy E. (1986). Treatment of sexual dysfunction. *Journal of Consulting and Clinical Psychology, 54,* 158–167.

LoPiccolo, Leslie. (1980). Low sexual desire. In S. R. Leiblum & L. A. Pervin (Eds.), *Principles and practice of sex therapy.* New York: Guilford Press.

Lorenz, Konrad. (1966). *On aggression.* New York: Harcourt Brace Jovanovich.

Louis, R. (1997). *Sexpectations: Women talk candidly about sex and dating.* Madison, WI: MPC Press.

Louv, W. C., et al. (1989). Oral contraceptive use and risk of chlamydial and gonococcal infections. *American Journal of Obstetrics and Gynecology, 160*, 396.

Lowenthal, M. F., et al. (1975). *Four stages of life: A comparative study of women and men facing transitions.* San Francisco: Jossey-Bass.

Lowery, Shearon, & Wetli, Charles. (1982). Sexual asphyxia: a neglected area of study. *Deviant Behavior, 3*, 19–39.

Lowry, D. T., & Snidler, J. A. (1993). Prime-time TV portrayals of sex, "safe sex" and AIDS: A longitudinal analysis. *Journalism Quarterly, 70*, 628–637.

Lozoff, B., Jordan, B., & Malone, S. (1995). Childbirth in cross-cultural perspective. In B. M. du Toit (Ed.), *Human sexuality: Cross cultural readings* (pp. 80–97). New York: McGraw-Hill.

Luby, Elliot C., & Klinge, Valerie. (1985). Genital herpes: A pervasive psychosocial disorder. *Archives of Dermatology, 121*, 494–497.

Luke, Barbara. (1994). Nutritional influences on fetal growth. *Clinical Obstetrics and Gynecology, 37*, 538–549.

Luker, Kristin. (1984). *Abortion and the politics of motherhood.* Berkeley: University of California Press.

Luker, Kristin. (1975). *Taking chances: Abortion and the decision not to contracept.* Berkeley: University of California Press.

Lumley, Judith, & Astbury, J. (1989). Advice for pregnancy. In I. Chalmers, M. Enkin, & M. J. N. C. Keirse (Eds.), *Effective care in pregnancy and childbirth, Vol. I: Pregnancy* (pp. 237–254). New York: Oxford University Press.

Lundberg, Per Olov. (1992). Sexual dysfunction in patients with neurological disorders. *Annual Review of Sex Research, 3*, 121–150.

Lutgendorf, Susan K., et al. (1997). Cognitive-behavioral stress management decreases dysphoric mood and herpes simplex virus—Type 2 antibody titers in symptomatic HIV-seropositive gay men. *Journal of Consulting and Clinical Psychology, 65*, 31–43.

Lytton, Hugh, & Romney, David M. (1991). Parents' differential socialization of boys and girls: A meta-analysis. *Psychological Bulletin, 109*, 267–296.

Maccoby, Eleanor E., & Jacklin, Carol N. (1974). *The psychology of sex differences.* Stanford, CA: Stanford University Press.

MacDonald, A. P. (1982). Research on sexual orientation: A bridge that touches both shores but doesn't meet in the middle. *Journal of Sex Education and Therapy, 8*(1), 9–13.

MacDonald, P. T., et al. (1988). Heavy cocaine use and sexual behavior. *Journal of Drug Issues, 18*, 437–455.

MacDougald, D. (1961). Aphrodisiacs and anaphrodisiacs. In A. Ellis & A. Abarbanel (Eds.), *The encyclopedia of sexual behavior* (Vol. I). New York: Hawthorn.

MacFarlane, J. A., et al. (1978). The relationship between mother and neonate. In S. Kitzinger & J. A. Davis (Eds.), *The place of birth.* Oxford: Oxford University Press.

MacKinnon, Catharine A. (1979). *Sexual harassment of working women.* New Haven: Yale University Press.

MacLean, Paul. (1962). New findings relevant to the evolution of psychosexual functions of the brain. *Journal of Nervous and Mental Disease, 135*, 289–301.

MacNamara, Donald E. J., & Sagarin, Edward. (1977). *Sex, crime, and the law.* New York: The Free Press.

Maddock, J. W. (1997). Sexuality education: A history lesson. In J. W. Maddock (Ed.), *Sexuality education in post-secondary and professional training settings* (pp. 1–22). Binghamton, NY: The Haworth Press.

Magana, J. R., & Carrier, J. M. (1991). Mexican and Mexican American male sexual behavior and spread of AIDS in California. *Journal of Sex Research, 28*, 425–441.

Maguire, Daniel C. (1983). Abortion: A question of Catholic honesty. *Christian Century, 100*, 803–807.

Mahay, Jenna, Michaels, Stuart, & Laumann, Edward O. (1999). Race, gender, and class in sexual scripts. In E. O. Laumann, & R. T. Michael (Eds.), *The social organization of sexuality in the United States: Further studies.* Chicago: University of Chicago Press.

Mahoney, E. R. (1983). *Human sexuality.* New York: McGraw-Hill.

Malamuth, Neil. (in press). Pornography. In L. Kurtz (Ed.), *Encyclopedia of violence, peace and conflict.* San Diego, CA: Academic Press.

Malamuth, Neil M. (1998). The confluence model as an organizing framework for research on sexually aggressive men: Risk moderators, imagined aggression and pornography consumption. In R. Geen & E. Donnerstein (Eds.), *Aggression: Theoretical and empirical reviews.* New York: Academic Press.

Malamuth, Neil M., & Brown, Lisa M. (1994). Sexually aggressive men's perceptions of women's communications. *Journal of Personality and Social Psychology, 67*, 699–712.

Malamuth, Neil M., Sockloskie, R. J., Koss, M. P., & Tanaka, J. S. (1991). Characteristics of aggressors against women: Testing a model using a national sample of college students. *Journal of Consulting and Clinical Psychology, 59*, 670–781.

Maletzky, B. M. (1980). Assisted covert sensitization. In D. J. Cox & R. J. Daitzman (Eds.), *Exhibitionism: Description, assessment, and treatment.* New York: Garland.

Maletzky, B. M. (1977). "Booster" sessions in aversion therapy: The permanency of treatment. *Behavior Therapy, 8*, 460–463.

Maletzky, B. M. (1974). "Assisted" covert sensitization in the treatment of exhibitionism. *Journal of Consulting and Clinical Psychology, 42*, 34–40.

Malinowski, Bronislaw. (1929). *The sexual life of savages.* New York: Harcourt Brace Jovanovich.

Mallory, Tammie E., & Rich, Katherine E. (1986). Human reproductive technologies: An appeal for brave new legislation in a brave new world. *Washburn Law Journal, 25*, 458–504.

Matz, Wendy, & Boss, Suzie. (1997). *In the garden of desire. The intimate world of women's sexual fantasies.* New York: Broadway Books.

Marin, Barbara V., Tschann, Jeanne M., Gomez, Cynthia A., & Kegels, Susan M. (1993). Acculturation and gender differences in sexual attitudes and behaviors: Hispanic vs. non-Hispanic white unmarried adults. *American Journal of Public Health, 83*, 1759–1765.

Markman, Howard J. (1981). Prediction of marital distress: A 5-year follow-up. *Journal of Consulting and Clinical Psychology, 49*, 760–762.

Markman, Howard J. (1979). Application of a behavioral model of marriage in predicting relationship satisfaction of couples planning marriage. *Journal of Consulting and Clinical Psychology, 47*, 743–749.

Markman, Howard J., & Floyd, Frank. (1980). Possibilities for the prevention of marital discord: A behavioral perspective. *American Journal of Family Therapy, 8*, 29–48.

Markman, Howard, & Kadushin, Frederick. (1986). Preventive effects of human training for first-time parents: A short-term longitudinal study. *Journal of Consulting and Clinical Psychology, 54*, 872–874.

Marquette, Catherine M., et al. (1995). Vasectomy in the United States, 1991. *American Journal of Public Health, 85*, 644–649.

Marquis, J. N. (1970). Orgasmic reconditioning: Changing sexual object choice through controlling masturbation fantasies. *Journal of Behavior Therapy and Experimental Psychiatry, 1*, 263–272.

Marshall, Donald C. (1971). Sexual behavior on Mangaia. In D. S. Marshall & R. C. Suggs (Eds.), *Human sexual behavior.* New York: Basic Books.

Marshall, Eliot. (1995). NIH's "Gay Gene" study questioned. *Science, 268*, 1841.

Marshall, W. L. (1993). A revised approach to the treatment of men who sexually assault adult females. In G. Nagayama Hall et al. (eds.), *Sexual aggression* (pp. 143–165). Washington, D.C.: Taylor & Francis.

Marshall, W. L., & Pithers, W. D. (1994). A reconsideration of treatment outcome with sex offenders. *Criminal Justice and Behavior, 21,* 10–27.

Marsiglio, William. (1993). Attitudes toward homosexual activity and gays as friends: A national survey of heterosexual 15- to 19-year-old males. *Journal of Sex Research, 30,* 12–17.

Marsiglio, William, & Diekow, Douglas. (1998). Men and abortion: The gender politics of pregnancy resolution. In L. J. Beckman & S. M. Harvey (Eds.), *The new civil war* (pp. 269–284). Washington, D.C.: American Psychological Association.

Martin, Carol L., & Halverson, C. F. (1983). The effects of sex-typing schemas on young children's memory. *Child Development, 54,* 563–574.

Martinson, Floyd M. (1994). *The sexual life of children.* Westport, CT: Bergin & Garvey.

Marx, Jean. (1995a). New clue to prostate cancer spread. *Science, 268,* 799–800.

Marx, Jean. (1995b). Sharing the genes that divide the sexes for mammals. *Science, 269,* 1824–1827.

Marx, Jean L. (1988). The AIDS virus can take on many guises. *Science, 241,* 1039–1040.

Mason, Carolyn, & Elwood, Robert. (1995). Is there a physiological basis for the couvade and onset of paternal care? *International Journal of Nursing Studies, 32,* 137–148.

Masters, W. H., Johnson, V. E., & Kolodny, R. C. (1982). *Human sexuality.* Boston: Little, Brown.

Masters, W., Johnson, V., & Kolodny, R. (1988). *Crisis: Heterosexual behavior in the age of AIDS.* New York: Grove.

Masters, William H., & Johnson, Virginia. (1979). *Homosexuality in perspective.* Boston: Little, Brown.

Masters, William H., & Johnson, Virginia. (1970). *Human sexual inadequacy.* Boston: Little, Brown.

Masters, William H., & Johnson, Virginia. (1966). *Human sexual response.* Boston: Little, Brown.

Masterton, Graham. (1993). *Drive him wild: A hands-on guide to pleasuring your man in bed.* New York: Signet Books.

Matsumoto, David. (1994). *Cultural influences on research methods and statistics.* Pacific Grove, CA: Brooks/Cole.

Matteson, David R. (1985). Bisexual men in marriage: Is a positive homosexual identity and stable marriage possible? In F. Klein & T. J. Wolf (Eds.), *Bisexualities: Theory and research.* New York: Haworth.

Maurer, Harry. (1994). *Sex: Real people talk about what they really do.* New York: Penguin Books.

May, Rollo. (1974). *Love and will.* New York: Dell Books.

Maybach, Kristine L., & Gold, Steven R. (1994). Hyperfemininity and attraction to macho and non-macho men. *The Journal of Sex Research, 31,* 91–98.

Mayle, P., Robins, A., and Walter P. (1973). *Where did I come from?* Secaucus, NJ: Lyle Stuart. A delightful sexuality education book for young children. A companion volume—*What's happening to me?*—is for children approaching or experiencing puberty.

Mays, Vickie M., Cochran, Susan D., & Rhue, Sylvia. (1993). The impact of perceived discrimination on the intimate relationships of Black lesbians. *Journal of Homosexuality, 25,* 1–14.

Mazur, Allan. (1986). U.S. trends in feminine beauty and overadaptation. *Journal of Sex Research, 22,* 281–303.

McBride, Arthur F., & Hebb, D. O. (1948). Behavior of the captive bottlenose dolphin, Tursiops truncatus. *Journal of Comparative and Physiological Psychology, 41,* 111–123.

McCance, Dennis J. (1994). Human papillomaviruses. *Infectious Disease Clinics of North America, 8,* 751–767.

McCarthy, Barry W. (1990). Treating sexual dysfunction associated with prior sexual trauma. *Journal of Sex & Marital Therapy, 16,* 142–146.

McCarthy, Barry W. (1989). Cognitive-behavioral strategies and techniques in the treatment of early ejaculation. In S. R. Leiblum & R. C. Rosen (Eds.), *Principles and practice of sex therapy* (2d ed.). New York: Guilford.

McClintock, Martha K. (1971). Menstrual synchrony and suppression. *Nature, 229,* 244–245.

McClintock, Martha, & Herdt, Gilbert. (1996). Rethinking puberty: The development of sexual attraction. *Current Directions in Psychological Science, 5,* 178–183.

McClure, Robert, & Brewer, R. Thomas. (1980). Attitudes of new parents towards child and spouse with Lamaze or non-Lamaze methods of childbirth. *Journal of Human Behavior, 17,* 45–48.

McConaghy, Nathaniel. (1987). A learning approach. In J. H. Geer & W. T. O'Donohue (Eds.), *Theories of human sexuality* (pp. 287–334). New York: Plenum.

McCoy, Norma L. (1997). Sexual issues for postmenopausal women. *Topics in Geriatric Rehabilitation, 12,* 28–39.

McCoy, Norma L. (1996). Menopause and sexuality. In M. K. Beard (Ed.), *Optimizing hormone replacement therapy: Estrogen-androgen therapy in postmenopausal women* (pp. 32–36). Minneapolis: McGraw-Hill Healthcare.

McCoy, Norma L., & Matyas, Joseph R. (1996). Oral contraceptives and sexuality in university women. *Archives of Sexual Behavior, 25,* 73–90.

McDowell, Janet Dickey. (1983). Ethical implications of in vitro fertilization. *The Christian Century, 100,* 936–938.

McEwen, B. S. (1997). Meeting report—Is there a neurobiology of love? *Molecular Psychiatry, 2,* 15–16.

McFarlane, Jessica, Martin, Carol L., & Williams, Tannis M. (1988). Mood fluctuations: Women versus men and menstrual versus other cycles. *Psychology of Women Quarterly, 12,* 201–224.

McFarlane, Jessica M., & Williams, Tannis M. (1994). Placing premenstrual syndrome in perspective. *Psychology of Women Quarterly, 18,* 339–374.

McGuire, R. J., Carlisle, J. M., & Young, B. G. (1965). Sexual deviations as conditioned behavior: A hypothesis. *Behavioral Research and Therapy, 2,* 185–190.

McGuire, Terry R. (1995). Is homosexuality genetic? A critical review and some suggestions. *Journal of Homosexuality, 28,* 115–145.

McKeganey, N. (1994). Why do men buy sex and what are their assessments of the HIV-related risks when they do? *AIDS Care, 6,* 289–301.

McKelvie, Melissa, & Gold, Steven R. (1994). Hyperfemininity: Further definition of the construct. *Journal of Sex Research, 31,* 219–228.

McKinlay, Sonja M., Brambilla, D. J., & Posner, J. G. (1992). The normal menopause transition. *American Journal of Human Biology, 4,* 37–46.

McKirnan, David J., Stokes, Joseph P., Doll, Lynda, & Burzette, Rebecca G. (1995). Bisexually active men: Social characteristics and sexual behavior. *Journal of Sex Research, 32,* 65–76.

McMillen, Curtis, Zuravin, Susan, & Rideout, Gregory. (1995). Perceived benefit from child sexual abuse. *Journal of Consulting and Clinical Psychology, 63,* 1037–1043.

McNeill, John J. (1987). Homosexuality: Challenging the Church to grow. *Christian Century, 104,* 242–246.

McWhirter, David P., & Mattison, Andrew M. (1984). *The male couple: How relationships develop.* Englewood Cliffs, NJ: Prentice-Hall.

McWhirter, David P., & Mattison, Andrew M. (1980). Treatment of sexual dysfunction in homosexual male couples. In S. R. Leiblum & L. A. Pervin (Eds.), *Principles and practice of sex therapy.* New York: Guilford.

McWilliams, Elaine. (1994). The association of perceived support with birthweights and obstetric complications: Piloting prospective identification and the effects of counseling. *Journal of Reproductive and Infant Psychology, 12,* 115–122.

Mead, Margaret. (1935). *Sex and temperament in three primitive societies.* New York: Morrow.

Mead, Margaret, & Newton, Niles. (1967). Fatherhood. In S. A. Richardson & A. F. Guttmacher (Eds.), *Childbearing: Its social and psychological aspects.* Baltimore: Williams & Wilkins.

Meana, Marta, & Binik, Yitzchak, M. (1994). Painful coitus: A review of female dyspareunia. *Journal of Nervous and Mental Disease, 182,* 264–272.

Meischke, Hendrika. (1995). Implicit sexual portrayals in the movies: Interpretations of young women. *The Journal of Sex Research, 32,* 29–36.

Melman, A. (1992). Neural and vascular control of erection. In R. Rosen & S. Leiblum (Eds.), *Erectile disorders: Assessment and treatment.* New York: Guilford.

Melman, A., & Tiefer, L. (1992). Surgery for erectile disorders: Operative procedures and psychological issues. In R. C. Rosen & S. R. Leiblum (Eds.), *Erectile disorders* (pp. 255– 282). New York: Guilford.

Merrick, E. N. (1995). Adolescent childbearing as career "choice": Perspective from an ecological context. *Journal of Counseling and Development, 73,* 288–295.

Mertz, Gregory J. (1993). Epidemiology of genital herpes infection. In M. Cohen, E. W. Hook, & P. J. Hitchcock (Eds.), *Infectious disease clinics of North America* (Vol. 7, pp. 825–840). Philadelphia: Harcourt, Brace, Jovanovich.

Messe, Madelyn R., & Geer, James H. (1985). Voluntary vaginal musculature contractions as an enhancer of sexual arousal. *Archives of Sexual Behavior, 14,* 13–28.

Messenger, John C. (1993). Sex and repression in an Irish folk community. In D. N. Suggs & A. W. Miracle (Eds.), *Culture and human sexuality.* Pacific Grove, CA: Brooks/Cole.

Meston, Cindy M., Trapnell, Paul D., & Gorzalka, Boris B. (1996). Ethnic and gender differences in sexuality: Variations in sexual behavior between Asian and non-Asian university students. *Archives of Sexual Behavior, 25,* 33–72.

Metz, Michael E., et al. (1997). Premature ejaculation: A psychophysiological review. *Journal of Sex & Marital Therapy, 23,* 3–23.

Meyer, Ilan H. (1995). Minority stress and mental health in gay men. *Journal of Health and Social Behavior, 36,* 38–56.

Meyer, J. K. (1979). Sex reassignment. *Archives of General Psychiatry, 36,* 1010–1015.

Meyer-Bahlburg, Heino F. L. (1998). Gender assignment in intersexuality. *Journal of Psychology and Human Sexuality.*

Meyer-Bahlburg, Heino F. L. (1997). The role of prenatal estrogens in sexual orientation. In L. Ellis & L. Ebertz (Eds.), *Sexual orientation: Toward biological understanding.* Westport, CT: Praeger.

Meyer-Bahlburg, Heino, et al. (1995). Prenatal estrogens and the development of homosexual orientation. *Developmental Psychology, 31,* 12–21.

Meyerowitz, Beth E. (1980). Psychosocial correlates of breast cancer and its treatments. *Psychological Bulletin, 87,* 108–131.

Michael, Richard P., & Keverne, E. B. (1968). Pheromones in the communication of sexual status in primates. *Nature, 218,* 746–749.

Michael, Robert T., Gagnon, John H., Laumann, Edward O., & Kolata, Gina. (1994). *Sex in America: A definitive survey.* Boston: Little, Brown.

Miki, Yoshio, et al. (1994). A strong candidate for the breast and ovarian cancer susceptibility gene BRCA1. *Science, 226,* 66–71.

Milan, Richard J., & Kilmann, Peter R. (1987). Interpersonal factors in premarital contraception. *Journal of Sex Research, 23,* 289–321.

Milbauer, Barbara. (1983). *The law giveth: Legal aspects of the abortion controversy.* New York: Atheneum.

Miller, Eleanor M. (1986). *Street woman.* Philadelphia: Temple University Press.

Miller, L. C., & Fishkin, S. A. (1997). On the dynamics of human bonding and reproductive success: Seeking windows on the adapted-for-human-environmental interface. In J. A. Simpson & D. T. Kenrick (Eds.), *Evolutionary social psychology* (pp. 197–235). Mahwah, NJ: Lawrence Erlbaum Associates.

Miller, Neil. (1992). *Out in the world: Gay and lesbian life from Buenos Aires to Bangkok.* New York: Random House.

Miller, Rickey S., & Lefcourt, Herbert M. (1982). The assessment of social intimacy. *Journal of Personality Assessment, 46,* 514–518.

Miller, S., Corrales, R., & Wachman, D. B. (1975). Recent progress in understanding and facilitating marital communication. *The Family Coordinator, 24,* 143–152.

Millett, Kate. (1969). *Sexual politics.* New York: Doubleday.

Mitchell, James, & Popkin, Michael. (1983). The pathophysiology of sexual dysfunction associated with antipsychotic drug therapy in males: A review. *Archives of Sexual Behavior, 12,* 173–183.

MMWR. (1998). Genital herpes simplex virus (HSV) infection. *Archives of Dermatology, 134,* 650–652.

Moffatt, Michael. (1989). *Coming of age in New Jersey.* New Brunswick, NJ: Rutgers University Press.

Molitch, Mark E. (1995). Neuroendocrinology. In P. Felig et al. (Eds.), *Endocrinology and metabolism.* New York: McGraw-Hill.

Money, John. (1987). Sin, sickness, or status: Homosexual gender identity and psychoneuroendocrinology. *American Psychologist, 42,* 384–399.

Money, John, & Ehrhardt, Anke. (1972). *Man and woman, boy and girl.* Baltimore: Johns Hopkins. Reissued in a facsimile edition by Jason Aronson, Northvale, NJ, 1996.

Monzon, O. T., & Capellan, J. M. B. (1987, July). Female-to-female transmission of HIV. *Lancet.*

Moodbidri, S. B., et al. (1980). Measurement of inhibin. *Archives of Andrology, 5,* 295–303.

Moore, Allen J. (1987). Teenage sexuality and public morality. *Christian Century, 104,* 747–750.

Moore, T. (1994). Porn shop enjoys brisk business year-round. Madison, WI: *The Capital Times,* 5A–6A.

Morales, A., et al. (1998). Clinical safety of oral sildenafil (Viagra) in the treatment of erectile dysfunction. *International Journal of Impotence Research, 10,* 69–74.

Moreland, Adele A., Majmudar, Bhagirath, & Vernon, Suzanne D. (1996). Genital human papilloma-virus infections. In S. A. Morese et al. (Eds.), *Atlas of sexually transmitted diseases and AIDS* (pp. 225–240). London: Mosby-Wolfe.

Moreland, Adele A., Shafran, Steve D., Bryan, John, & Pellett, Phillip. (1996). Genital herpes. In S. A Morese et al. (Eds.), *Atlas of sexually transmitted diseases and AIDS* (pp. 207–224). London: Mosby-Wolfe.

Morell, V. (1998). A new look at monogamy. *Science, 281,* 1982–1983.

Morgan, Robin. (1980). Theory and practice: Pornography and rape. In L. Lederer (Ed.), *Take back the night: Women on pornography.* New York: Morrow.

Morgan, Robin. (1978, Nov.). How to run the pornographers out of town (and preserve the first amendment). *Ms. 55,* 78–80.

Morgan, Ted. (1991, May). The gross product. *Across the Board,* 25–26.

Morin, Jack. (1981). *Anal pleasure and health.* Burlingame, CA: Down There Press.

Morin, Stephen F., & Rothblum, Esther D. (1991). Removing the stigma: Fifteen years of progress. *American Psychologist, 46,* 947–949.

Moritz, Rob. (1998, August). Cosmopolitan lust survey. *Cosmopolitan,* 124–128.

Morley, J. E., et al. (1997). Testosterone and frailty. *Clinics in Geriatric Medicine, 13,* 685–695.

Morokoff, Patricia J. (1993). Female sexual arousal disorder. In W. O'Donohue & J. H. Geer (Eds.), *Handbook of sexual dysfunctions* (pp. 157–199). Boston: Allyn and Bacon.

Morokoff, Patricia J. (1988). Sexuality in perimenopausal and postmenopausal women. *Psychology of Women Quarterly, 12,* 489–511.

Morokoff, Patricia J. (1986). Volunteer bias in the psychophysiological study of female sexuality. *Journal of Sex Research, 22,* 35–51.

Morokoff, Patricia J., Harlow, Lisa L., & Quina, Kathryn. (1996). Women and AIDS. In A. L. Stanton & S. J. Gallant (Eds.), *Women's health.* Washington, D.C.: American Psychological Association.

Morrell, Martha J., Dixen, Jean M., Carter, Sue, & Davidson, Julian. (1984). The influence of age and cycling status on sexual arousability in women. *American Journal of Obstetrics and Gynecology, 148,* 66–71.

Morris, Norval J. (1973, Apr. 18). The law is a busy-body. *The New York Times Magazine,* 58–64.

Morrison, Diane M. (1985). Adolescent contraceptive behavior: A review. *Psychological Bulletin, 98,* 538–568.

Morrison, Eleanor S., et al. (1980). *Growing up sexual.* New York: Van Nostrand.

Morse, Stephen A., Moreland, Adele A., & Holmes, King K. (Eds.). (1996). *Atlas of sexually transmitted diseases and AIDS.* London: Mosby-Wolfe.

Mortola, Joseph F. (1998). Premenstrual syndrome—pathophysiologic considerations. *New England Journal of Medicine, 338,* 256–257.

Moser, Charles, & Levitt, Eugene E. (1987). An exploratory-descriptive study of a sadomasochistically oriented sample. *Journal of Sex Research, 23,* 322–337.

Moses, Stephen, et al. (1990). Geographical patterns of male circumcision practices in Africa: Association with HIV seroprevalence. *International Journal of Epidemiology, 19,* 693–697.

Mosher, Donald, & MacIan, Paula. (1994). College men and women respond to X-rated videos intended for male or female audiences: Gender and sexual scripts. *Journal of Sex Research, 31,* 99–113.

Mosher, William D. (1990, Oct.). Contraceptive practice in the United States, 1982–1988. *Family Planning Perspectives, 22,* 198–205.

Moss, B. F., & Schwebel, A. I. (1993). Marriage and romantic relationships: Defining intimacy in romantic relationships. *Family Relations, 42,* 31–37.

Muehlenhard, Charlene, Friedman, D. E., & Thomas, C. M. (1985). Is date rape justifiable? The effects of dating activity, who paid, and men's attitudes toward women. *Psychology of Women Quarterly, 9,* 297–310.

Muehlenhard, Charlene L. (1988). Misinterpreted dating behaviors and the risk of date rape. *Journal of Social and Clinical Psychology, 6,* 20–37.

Muehlenhard, Charlene L., & Cook, Stephen W. (1988). Men's self-reports of unwanted sexual activity. *Journal of Sex Research, 24,* 58–72.

Muehlenhard, Charlene L., Harney, Patricia A., & Jones, Jayne M. (1992). From "victim-precipitated rape" to "date rape": How far have we come? *Annual Review of Sex Research, 3,* 219–254.

Muehlenhard, Charlene L., & McCoy, Marcia L. (1991). Double standard/double bind: The sexual double standard and women's communication about sex. *Psychology of Women Quarterly, 15,* 447–462.

Mueller, G. O. W. (1980). *Sexual conduct and the law* (2d ed.). Dobbs Ferry, NY: Oceana Publications.

Muller, James, Mittleman, Murray, Maclure, Malcolm, Sherwood, Jane, & Tofler, Geoffrey. (1996). Triggering myocardial infarction by sexual activity. *Journal of the American Medical Association, 275,* 1405–1409.

Murnen, Sarah K., Perot, Annette, & Byrne, D. (1989). Coping with unwanted sexual activity: Normative responses, situational determinants, and individual differences. *Journal of Sex Research,* 85–106.

Murnen, Sarah K., & Stockton, Mary. (1997). Gender and self-reported sexual arousal in response to sexual stimuli: A meta-analytic review. *Sex Roles, 37,* 135–154.

Murphy, Timothy F. (1992). Redirecting sexual orientation: Techniques and justifications. *Journal of Sex Research, 29,* 502–523.

Myers, Barbara J. (1984). Mother-infant bonding: The status of this critical-period hypothesis. *Developmental Review, 4,* 240–274.

Nanda, Serena. (1997). The Hijras of India. In M. Duberman (ed.), *A queer world* (pp. 82–86). New York: New York University Press.

Nanula, Peter J. (1987). Protecting confidentiality in the effort to control AIDS. *Harvard Journal of Legislation, 24*(1), 315–349.

Narod, Steven A., et al. (1988). Human mutagens: Evidence from paternal exposure? *Environmental and Molecular Mutagenesis, 11,* 401–415.

National Cancer Institute. (1994). *What you need to know about cancer of the cervix.* Bethesda, MD: National Cancer Institute.

National Commission on AIDS. (1994). Preventing HIV/AIDS in adolescents. *Journal of School Health, 64,* 39–51.

Neiger, S. (1968). Sex potions. *Sexology,* 730–733.

Nelson, Adie, & Robinson, Barrie. (1994). *Gigolos and madames bountiful: Illusions of gender, power and intimacy.* Toronto: University of Toronto Press.

Nelson, James B. (1978). *Embodiment: An approach to sexuality and Christian theology.* Minneapolis, MN: Augsburg.

Nelson, Joan A. (1986). Incest: Self-report findings from a nonclinical sample. *Journal of Sex Research, 22,* 463–477.

Newton, Niles A. (1972). Childbearing in broad perspective. In Boston Children's Medical Center. *Pregnancy, birth and the newborn baby.* New York: Delacorte Press.

Nichols, Margaret. (1989). Sex therapy with lesbians, gay men, and bisexuals. In S. R. Leiblum & R. C. Rosen (Eds.), *Principles and practice of sex therapy* (2d ed.). New York: Guilford.

Niemann, Yolanda F., Jennings, Leilani, Rozelle, Richard M., Baxter, James C., & Sullivan, Elroy. (1994). Use of free responses and cluster analysis to determine stereotypes of eight groups. *Personality and Social Psychology Bulletin, 20,* 379–390.

Noller, P. (1984). *Nonverbal communication and marital interaction.* New York: Pergamon.

Norris, Anne, Ford, Kathleen, Shyu, Yu, & Schork, M. Anthony. (1996). Heterosexual experiences and partnerships of urban, low income African-American and Hispanic youth. *Journal of Acquired Immune Deficiency Syndromes and Human Retrovirology, 11,* 288–300.

Norton, Arthur J. (1987, July–Aug.). Families and children in the year 2000. *Children Today,* 6–9.

Noss, John B. (1963). *Man's religions* (3d ed.). New York: Macmillan.

Note: The constitutional status of sexual orientation. (1985). *Harvard Law Review, 98,* 1285–1309.

Notzon, Francis C. (1990). International differences in the use of obstetric interventions. *Journal of the American Medical Association, 263,* 3286–3291.

Novak, E. R., Jones, G. S., & Jones, H. W. (1975). *Novak's textbook of gynecology* (9th ed.). Baltimore, MD: Williams & Wilkins.

Novak, Emil, & Novak, Edmund R. (1952). *Textbook of gynecology.* Baltimore: Williams & Wilkins.

Nowak, Rachel. (1991, Sept.). AIDS vaccines: Key questions still unanswered. *The Journal of NIH Research, 3,* 37–39.

Nsiah-Jefferson, Laurie. (1989). Reproductive laws, women of color, and low-income women. In S. Cohen & N. Taub (Eds.), *Reproductive laws for the 1990s* (pp. 23–68). Clifton, NJ: Humana Press.

Nulman, Irena, Rovet, Joanne, et al. (1997). Neurodevelopment of children exposed in utero to antidepressant drugs. *New England Journal of Medicine, 336,* 258–262.

O'Brien, Shari. (1986). Commercial conceptions: A breeding ground for surrogacy. *North Carolina Law Review, 65,* 127–153.

Obzrut, L. (1976). Expectant fathers' perceptions of fathering. *American Journal of Nursing, 76,* 1440–1442.

O'Connell, Helen E., et al. (1998). Anatomical relationship between urethra and clitoris. *Journal of Urology, 159,* 1982–1897.

O'Connor, Art. (1987). Female sex offenders. *British Journal of Psychiatry, 150,* 615–620.

O'Connor, Mary J., Sigman, Marian, & Kasari, Connie. (1993). Interactional model for the association among maternal alcohol use, mother-infant interaction, and infant cognitive development. *Infant Behavior and Development, 16,* 177–192.

O'Donohue, William, Dopke, Cynthia A., & Swingen, Diane N. (1997). Psychotherapy for female sexual dysfunction: A review. *Clinical Psychology Review, 17,* 537–566.

Oesterling, Joseph E. (1995). Benign prostatic hyperplasia. *New England Journal of Medicine, 332,* 99–109.

O'Hara, Michael W., & Swain, Annette M. (1996). Rates and risk of postpartum depression: A meta-analysis. *International Review of Psychiatry, 8,* 37–54.

Okami, Paul. (1995). Childhood exposure to parental nudity, parent-child co-sleeping, and "primal scenes": A review of clinical opinion and empirical evidence. *Journal of Sex Research, 32,* 51–64.

Okami, Paul, Olmstead, Richard, & Abramson, Paul. (1997). Sexual experiences in early childhood: 18-year longitudinal data from the UCLA Family Lifestyles Project. *Journal of Sex Research, 34,* 339–347.

Okwumabua, T. M., Okwumabua, J. O., & Elliott, V. (1998). "Let the circle be unbroken" helps African-Americans prevent teen-pregnancy. *SIECUS Report, 26,* 12–17.

Olds, James. (1956). Pleasure centers in the brain. *Scientific American, 193,* 105–116.

Olds, James, & Milner, Peter. (1954). Positive reinforcement produced by electrical stimulation of the septal area and other regions of the rat brain. *Journal of Comparative and Physiological Psychology, 47,* 419–427.

Olds, Sally, & Eiger, M. S. (1973). *The complete book of breastfeeding.* New York: Bantam.

O'Leary, Ann, & Jemmott, Loretta S. (Eds.). (1995). *Women at risk: Issues in the primary prevention of AIDS.* New York: Plenum.

Oliver, Mary Beth, & Hyde, Janet S. (1993). Gender differences in sexuality: A meta-analysis. *Psychological Bulletin, 114,* 29–51.

Olson, Beth, & Douglas, William. (1997). The family on television: Evaluation of gender roles in situation comedy. *Sex Roles, 36,* 409–427.

Olson, Heather Carmichael, Sampson, Paul D., Barr, Helen, Streissguth, Ann P., & Bookstein, Fred L. (1992). Prenatal exposure to alcohol and school problems in late childhood: A longitudinal prospective study. *Development and Psychopathology, 4,* 341–359.

Oriel, David. (1990). Genital human papillomavirus infection. In K. Holmes et al. (Eds.), *Sexually transmitted diseases* (pp. 433–442). New York: McGraw-Hill.

O'Shea, P. A. (1995). Congenital defects and their causes. In D. R. Constan, R. V. Haning, Jr., & D. B. Singer (Eds.), *Human reproduction: Growth and development.* Boston: Little, Brown.

Østensten, Monika. (1994). Optimisation of antirheumatic drug treatment in pregnancy. *Clinical Pharmacokinetics, 27,* 486–503.

Otis, Melanie D., & Skinner, William F. (1996). The prevalence of victimization and its effect on mental well-being among lesbian and gay people. *Journal of Homosexuality, 30,* 93–117.

Otto, H. A. (1963). Criteria for assessing family strengths. *Family Process, 2,* 329–337.

Packer, H. L. (1968). *The limits of the criminal sanction.* Stanford, CA: Stanford University Press.

Padian, Nancy S., et al. (1997). Heterosexual transmission of human immunodeficiency virus (HIV) in Northern California: Results from a ten-year study. *American Journal of Epidemiology, 146,* 350–357.

Page, David C., et al. (1987). The sex-determining region of the human Y chromosome encodes a finger protein. *Cell, 51,* 1091–1104.

Paige, Karen E. (1973). Women learn to sing the menstrual blues. *Psychology Today, 7*(4), 41.

Paige, Karen E. (1971). Effects of oral contraceptives on affective fluctuations associated with the menstrual cycle. *Psychosomatic Medicine, 33,* 515–537.

Pakenham, Kenneth I., Dadds, Mark R., & Terry, Deborah J. (1994). Relationships between adjustment to HIV and both social support and coping. *Journal of Consulting and Clinical Psychology, 62,* 1194–1203.

Palace, Eileen M. (1995a). A cognitive-physiological process model of sexual arousal and response. *Clinical Psychology: Science and Practice, 2,* 370–384.

Palace, Eileen M. (1995b). Modification of dysfunctional patterns of sexual response through autonomic arousal and false physiological feedback. *Journal of Consulting and Clinical Psychology, 63,* 604–615.

Palmer, Craig T. (1991). Human rape: Adaptation or by-product? *Journal of Sex Research, 28,* 365–386.

Pantaleo, Giuseppe, et al. (1995). Studies in subjects with long-term nonprogressive human immunodeficiency virus infection. *New England Journal of Medicine, 332,* 209–216.

Paredes, Raul G., & Baum, Michael J. (1997). Role of the medial preoptic area/anterior hypothalamus in the control of masculine sexual behavior. *Annual Review of Sex Research, 8,* 68–101.

Parker, Graham. (1983). The legal regulation of sexual activity and the protection of females. *Osgoode Hall Law Journal, 21,* 187–244.

Parlee, Mary B. (1978, April). The rhythms in men's lives. *Psychology Today,* 82–91.

Parlee, Mary Brown. (1973). The premenstrual syndrome. *Psychological Bulletin, 80,* 454–465.

Parnas, Raymond I. (1981). Legislative reform of prostitution laws: Keeping commercial sex out of sight and out of mind. *Santa Clara Law Review, 21,* 669–696.

Parrinder, Geoffrey. (1996). *Sexual morality in the world's religions.* Oxford: Oneworld.

Parrinder, Geoffrey. (1980). *Sex in the world's religions.* New York: Oxford University Press.

Patterson, Charlotte. (1992). Children of lesbian and gay parents. *Child Development, 63,* 1025–1042.

Patterson, Charlotte J. (1996). Lesbian mothers and their children: Findings from the Bay Area Families Study. In J. Laird & R. Green (Eds.), *Lesbians and gays in couples and families* (pp. 420–438). San Francisco: Jossey-Bass.

Patterson, Charlotte J. (1995). Families of the lesbian baby boom: Parents' division of labor and children's adjustment. *Developmental Psychology, 31,* 115–123.

Patterson, Charlotte J., & D'Augelli, Anthony. (1995). *Lesbian, gay, and bisexual identities over the lifespan.* New York: Oxford University Press.

Patton, Wendy, & Mannison, Mary. (1995). Sexual coercion in dating situations among university students: Preliminary Australian data. *Australian Journal of Psychology, 47,* 66–72.

Paul, Eva W., & Klassel, Dara. (1987). Minors' rights to confidential contraceptive services. *Women's Rights Law Reporter, 10,* 45–64.

Peirce, Kate. (1993). Socialization of teenage girls through teen magazine fiction: The making of a new woman or an old lady? *Sex Roles, 29,* 59–68.

Pennisi, Elizabeth. (1996). Homing in on a prostate cancer gene. *Science, 274,* 1301.

Peplau, L. Anne, Cochran, Susan D., & Mays, Vickie M. (1997). A national survey of the intimate relationships of African American lesbians and gay men. In B. Greene (Ed.), *Ethnic and cultural diversity among lesbians and gay men* (pp. 11–38). Thousand Oaks, CA: Sage.

Peplau, L. Anne, Garnets, L. D., Spalding, L. R., Conley, T. D., & Veniegas, R. C. (1998). A critique of Bem's "Exotic becomes erotic" theory of sexual orientation. *Psychological Review, 105,* 387–394.

Peplau, L. Anne, Veniegas, Rosemary C., & Campbell, Susan M. (1996). Gay and lesbian relationships. In R. C. Savin-Williams & K. M. Cohen (Eds.), *The lives of lesbians, gays, and bisexuals* (pp. 250–273). Fort Worth, TX: Harcourt Brace.

Perkins, D. F., Luster, T., Villarruel, F. A., & Small, S. (1998). An ecological risk-factor examination of adolescents' sexual activity in three ethnic groups. *Journal of Marriage and the Family, 60,* 660–673.

Perkins, Roberta, & Bennett, Garry. (1985). *Being a prostitute: Prostitute women and prostitute men.* London: Allen & Unwin.

Perlman, Daniel, & Fehr, B. (1987). The development of intimate relationships. In D. Perlman & S. Duck (Eds.), *Intimate relationships: Development, dynamics, and deterioration.* Newbury Park, CA: Sage.

Perper, T. (1985). *Sex signals: The biology of love.* Philadelphia: ISI Press.

Perry, C. D. (1980). Right of privacy challenges to prostitution statutes. *Washington University Law Quarterly, 58,* 439–480.

Perry, John D., & Whipple, Beverly. (1981). Pelvic muscle strength of female ejaculators: Evidence in support of a new theory of orgasm. *Journal of Sex Research, 17,* 22–39.

Persson, Goran. (1980). Sexuality in a 70-year-old urban population. *Journal of Psychosomatic Research, 24,* 335–342.

Peterson, J., & Marin, G. (1988). Issues in the prevention of AIDS among Black and Hispanic men. *American Psychologist, 43,* 871–877.

Peterson, J. L., et al. (1992). High-risk sexual behavior and condom use among gay and bisexual African American men. *American Journal of Public Health, 82,* 1490–1494.

Peterson, Maxine E., & Dickey, Robert. (1995). Surgical sex reassignment: A comparative survey of international centers. *Archives of Sexual Behavior, 24,* 135–156.

Pfeiffer, E., Verwoerdt, A., & Wang, H. S. (1968). Sexual behavior in aged men and women. *Archives of General Psychiatry, 19,* 753–758.

Pfeiffer, Eric (1975). Sex and aging. In L. Gross (Ed.), *Sexual issues in marriage.* New York: Spectrum.

Phibbs, C. S., Bateman, D. A., & Schwartz, R. M. (1991). The neonatal costs of maternal cocaine use. *Journal of the American Medical Association, 266,* 1521–1526.

Phoenix, C. H., Goy, R. W., Gerall, A. A., & Young, W. C. (1959). Organizing action of prenatally administered testosterone propionate on the tissues mediating mating behavior in the female guinea pig. *Endocrinology, 65,* 369–382.

Phoolcharoen, Wiput. (1998). HIV/AIDS prevention in Thailand: Success and challenges. *Science, 280,* 1873–1874.

Piccinino, Linda J., & Mosher, William D. (1998). Trends in contraceptive use in the United States: 1982–1995. *Family Planning Perspectives, 30,* 4–10.

Pillard, Richard C. & Weinrich, James D. (1987). Periodic table model of transpositions. *Journal of Sex Research, 23,* 425–454.

Pithers, W. D. (1993). Treatment of rapists. In G. Nagayama Hall et al. (eds.), *Sexual aggression* (pp. 167–196). Washington, DC: Taylor & Francis.

Pittenger, W. Norman. (1970). *Making sexuality human.* Philadelphia: Pilgrim Press.

Pittman, Frank III. (1993, May–June). Beyond betrayal: Life after infidelity. *Psychology Today,* 32–38, ff.

Plant, T. M., Winters, S. J., Attardi, B. J., & Majumdar, S. S. (1993). The follicle stimulating hormone—inhibin feedback loop in male primates. *Human Reproduction, 8,* Suppl. 2, 41–44.

Pleck, Joseph H. (1981). *The myth of masculinity.* Cambridge, MA: MIT Press.

Pomeroy, Wardell B. (1975). The diagnosis and treatment of transvestites and transsexuals. *Journal of Sex and Marital Therapy, 1,* 215–224.

Pomeroy, Wardell B. (1972). *Dr. Kinsey and the Institute for Sex Research.* New York: Harper & Row.

Pope Paul VI. (1968, July 30). *Humanae vitae.* (English text in *The New York Times,* 20.)

Population Information Program. (1983). Vasectomy—Safe and simple. *Population Reports,* Series D, No. 4, D61–D100.

Posner, Richard. (1992). *Sex and reason.* Cambridge, MA: Harvard University Press.

Posner, Richard, & Silbaugh, Katherine. (1996). *A guide to America's sex laws.* Chicago: University of Chicago Press.

Powdermaker, Hortense. (1933). *Life in Lesu.* New York: Norton.

Preti, George, Cutler, Winnifred B., et al. (1986). Human axillary secretions influence women's menstrual cycles: The role of donor extract of females. *Hormones and Behavior, 20,* 474–482.

Price, E. O., et al. (1988). The relationship of male-male mounting to the sexual preferences of young rams. *Applied Animal Behaviour Science, 21,* 347–355.

Price, James, Allensworth, Diane, & Hillman, Kathleen. (1985). Comparison of sexual fantasies of homosexuals and heterosexuals. *Psychological Reports, 57,* 871–877.

Price, James H. (1981). Toxic shock syndrome—An update. *Journal of School Health, 51,* 143–145.

Pritchard, J. A., & MacDonald, P. C. (1980). *Williams Obstetrics* (16th ed.). New York: Appleton-Century-Crofts.

Pritchard, Jack A., MacDonald, P. C., & Grant, N. F. (1985). *Williams Obstetrics* (17th ed.). Norwalk, CT: Appleton-Century-Crofts.

Prostitutes Education Network. (1998). Prostitution in the United States—the statistics. @ www.bayswan.org/stats.html.

Purnine, Daniel, & Carey, Michael. (1997). Interpersonal communication and sexual adjustment: The roles of understanding and agreement. *Journal of Consulting & Clinical Psychology, 65,* 1017–1025.

Quadagno, D., et al. (1991). The menstrual cycle: Does it affect athletic performance? *The Physician & Sports Medicine, 19,* 121–124.

Quadagno, David, Nation, Ann Jo, et al. (1995). Cardiovascular disease and sexual functioning. *Applied Nursing Research, 8,* 143–146.

Quadagno, David, Sly, David, Harrison, Dianne, Eberstein, Isaac, & Soler, Hosunna. (1998). Ethnic differences in sexual decisions and sexual behavior. *Archives of Sexual Behavior, 27,* 57–75.

Qualls, C. B., Wincze, J. P., & Barlow, D. H. (1978). *The prevention of sexual disorders.* New York: Plenum.

Quittner, Joshua. (1997, April 14). Divorce, internet style. *Time,* 72.

Rachman, S. (1966). Sexual fetishism: An experimental analogue. *Psychological Record, 16,* 293–296.

Radlove, Shirley. (1986). Sexual response and gender roles. In E. R. Allgeier & N. B. McCormick (Eds.), *Changing boundaries: Gender roles and sexual behavior.* Palo Alto, CA: Mayfield.

Ragona, Steven. (1993). City of National City v. Wiener: The further erosion of First Amendment protection for adult businesses. *Loyola of Los Angeles Entertainment Law Journal, 14,* 331–355.

Rakic, Zoran, et al. (1997). Testosterone treatment in men with erectile disorder and low levels of total testosterone in serum. *Archives of Sexual Behavior, 26,* 495–504.

Ramey, Estelle. (1972, Spring). Men's cycles. *Ms.,* 8–14.

Rasmussen, Stephanie J. (1998). Chlamydia immunology. *Current Opinion in Infectious Diseases, 11,* 37–41.

Rawicki, H. B., & Hill, S. (1991). Semen retrieval in spinal cord injured men. *Paraplegia, 29,* 443–446.

Ray, Amy L., & Gold, Steven R. (1996). Gender roles, aggression, and alcohol use in dating relationships. *Journal of Sex Research, 33,* 47–55.

Reamy, Kenneth J., & White, Susan E. (1987). Sexuality in the puerperium: A review. *Archives of Sexual Behavior, 16,* 165–186.

Reinharz, Shulamit. (1992). *Feminist methods in social research.* New York: Oxford University Press.

Reisenzein, Rainer. (1983). The Schachter theory of emotion: Two decades later. *Psychological Bulletin, 94,* 239–264.

Reiss, Ira L. (1986). *Journey into sexuality: An exploratory voyage.* Englewood Cliffs, NJ: Prentice-Hall.

Reiss, Ira L. (1967). *The social context of premarital sex permissiveness.* New York: Holt.

Reiss, Ira L. (1960). *Premarital sexual standards in America.* New York: Free Press.

Renaud, Cheryl, & Byers, E. Sandra. (1997). Sexual and relationship satisfaction in mainland China. *Journal of Sex Research, 34,* 399–410.

Renne, Elisha P. (1996). The pregnancy that doesn't stay: The practice and perception of abortion by Ekiti Yoruba women. *Social Science & Medicine, 42,* 483–494.

Repke, John T. (1994). Calcium and vitamin D. *Clinical Obstetrics and Gynecology, 37,* 550–557.

Resick, Patricia A. (1983). The rape reaction: Research findings and implications for intervention. *The Behavior Therapist, 6,* 129–132.

Reuther, Rosemary Radford. (1985). Catholics and abortion: Authority vs. dissent. *Christian Century, 102,* 859–862.

Rice, Berkeley. (1974). Rx: Sex for senior citizens. *Psychology Today, 8*(1), 18–20.

Rich, Melissa K., & Cash, Thomas F. (1993). The American image of beauty: Media representations of hair color for four decades. *Sex Roles, 29,* 113–124.

Richardson, J. Derek. (1991). I. Medical causes of male sexual dysfunction. *The Medical Journal of Australia, 155,* 29–33.

Riddle, Dorothy I. (1978). Relating to children: Gays as role models. *Journal of Social Issues, 34*(3), 38–58.

Rigdon, Susan M. (1996). Abortion law and practice in China: An overview with comparisons to the United States. *Social Science & Medicine, 42,* 543–560.

Rimm, Marty. (1995). Marketing pornography on the information superhighway: A survey of 917,410 images. *The Georgetown Law Journal, 83,* 1849–1925.

Riportella-Muller, Roberta. (1989). Sexuality in the elderly: A review. In K. McKinney & S. Sprecher (Eds.), *Human sexuality: The societal and interpersonal context* (pp. 210–236). New York: Ablex.

Rivera, Rhonda. (1982). Homosexuality and the law. In W. Paul et al. (Eds.), *Homosexuality: Social, psychological, and biological issues.* Beverly Hills, CA: Sage.

Rivera, Rhonda. (1981–1982). Recent developments in sexual preference law. *Drake Law Review, 30,* 311–346.

Rivera, Rhonda. (1979). Our straight-laced judges. The legal position of homosexual persons in the U. S. *Hastings Law Journal, 30,* 799–955.

Robenstine, Clark. (1994). HIV/AIDS education for adolescents: Social policy and practice. *The Clearing House, 67,* 229–232.

Roberson, Bruce, & Wright, Rex. (1994). Difficulty as a determinant of interpersonal appeal: A social-motivational application of energization theory. *Basic and Applied Social Psychology, 15,* 373–388.

Roberto, Laura G. (1983). Issues in diagnosis and treatment of transsexualism. *Archives of Sexual Behavior, 12,* 445–473.

Roberts, Dorothy E. (1993). Crime, race, and reproduction. *Tulane Law Review, 67,* 1945–1977.

Robertson, David L., Hahn, Beatrice, & Sharp, Paul M. (1995). Recombination in AIDS viruses. *Molecular Evolution, 40,* 249–259.

Robertson, John A. (1986). Embryos, families and procreative liberty: The legal structure of the new reproduction. *Southern California Law Review, 59,* 942–1041.

Robertson, John A. (1983). Surrogate mothers: Not so novel after all. *Hastings Center Report,* October 1983, 28–34.

Robinson, D., & Rock, J. (1967). Intrascrotal hyperthermia induced by scrotal insulation: Effect on spermatogenesis. *Obstetrics and Gynecology, 29,* 217.

Robinson, I., Ziss, K., Ganza, B., & Katz, S. (1991). Twenty years of the sexual revolution, 1965–1985: An update. *Journal of Marriage and the Family, 53,* 216–220.

Rodriguez, Ned, Ryan, Susan W., Van de Kamp, Hendrika, & Foy, David W. (1997). Posttraumatic stress disorder in adult female survivors of childhood sexual abuse: A comparison study. *Journal of Consulting and Clinical Psychology, 65,* 53–59.

Rogers, Susan M., & Turner, Charles F. (1991). Male-male sexual contact in the U.S.A.: Findings from five sample surveys, 1970–1990. *Journal of Sex Research, 28,* 491–519.

Romer, Daniel et al. (1997). "Talking computers": A reliable and private method to conduct interviews on sensitive topics with children. *Journal of Sex Research, 34,* 3–9.

Rosaldo, Michelle A. (1974). Woman, culture, and society: A theoretical overview. In M. S. Rosaldo & L. Lamphere (Eds.), *Woman, culture, and society.* Stanford, CA: Stanford University Press.

Rosario, Margaret, Meyer-Bahlburg, Heino, Hunter, Joyce, Exner, Theresa, Swadz, Marya, & Keller, Auden. (1996). The psychosexual development of urban lesbian, gay and bisexual youths. *Journal of Sex Research, 33,* 113–126.

Roscoe, Bruce, Kennedy, Donna, & Pope, Tony. (1987). Adolescents' views of intimacy: Distinguishing intimate from nonintimate relationships. *Adolescence, 22,* 511–516.

Roscoe, Will. (1997). Gender diversity in Native North America. In M. Duberman (Ed.), *A queer world* (pp. 65–81). New York: New York University Press.

Rose, S., & Frieze, I. H. (1993). Young singles' contemporary dating scripts. *Sex Roles, 28,* 499–509.

Rosen, David H. (1974). *Lesbianism: A study of female homosexuality.* Springfield, IL: Charles C. Thomas.

Rosen, Raymond. (1998). Sildenafil: Medical advance or media event? *The Lancet, 351,* 1599–1600.

Rosen, Raymond C. (1991). Alcohol and drug effects on sexual response: Human experimental and clinical studies. *Annual Review of Sex Research, 2,* 119–180.

Rosen, Raymond C., & Ashton, Adam K. (1993). Prosexual drugs: Empirical status of the "new aphrodisiacs." *Archives of Sexual Behavior, 22,* 521–543.

Rosen, Raymond C., & Beck, J. Gayle. (1988). *Patterns of sexual arousal.* New York: Guilford.

Rosen, Raymond C., & Leiblum, Sandra R. (1995a). Hypoactive sexual desire. *Psychiatric Clinics of North America, 18,* 107–121.

Rosen, Raymond C., & Leiblum, Sandra R. (1995b). Treatment of sexual disorders in the 1990s: An integrated approach. *Journal of Consulting and Clinical Psychology, 63,* 877–890.

Rosen, Raymond C., & Leiblum, Sandra R. (1992). *Erectile disorders.* New York: Guilford.

Rosen, Raymond C., Leiblum, Sandra R., & Spector, Ilana P. (1994). Psychologically-based treatment for male erectile disorder: A cognitive-interpersonal model. *Journal of Sex & Marital Therapy, 20,* 67–85.

Rosenberg, Lynn. (1993). Hormone replacement therapy: The need for reconsideration. *American Journal of Public Health, 83,* 1670–1673.

Rosenbleet, C., & Pariente, B. J. (1973). The prostitution of the criminal law. *American Criminal Law Review, 11,* 373–427.

Rosler, Ariel, & Witztum, Eliezer. (1998). Treatment of men with paraphilia with a long-lasting analogue of gonadotropin-releasing hormone. *New England Journal of Medicine, 338,* 416–422.

Rosner, Fred. (1983). In vitro fertilization and surrogate motherhood: The Jewish view. *Journal of Religion and Health, 22,* 139–160.

Ross, David, & Stevenson, John. (1993, Nov.–Dec.). HRT and cardiovascular disease. *British Journal of Sexual Medicine,* 10–13.

Ross, Michael N., Paulsen, J. A., & Stalstrom, O. W. (1988). Homosexuality and mental health: A cross-cultural review. *Journal of Homosexuality, 15,* 131–152.

Rotello, Gabriel. (1997). *Sexual ecology: AIDS and the destiny of gay men.* New York: Dutton.

Rothbaum, B. O., Foa, E. B., Riggs, D. S., Murdock, T., & Walsh, W. (1992). A prospective examination of post-traumatic stress disorder in rape victims. *Journal of Traumatic Stress, 5,* 455–475.

Rothblum, Esther D. (1994). "I only read about myself on bathroom walls": The need for research on the mental health of lesbians and gay men. *Journal of Consulting and Clinical Psychology, 62,* 213–220.

Rothblum, Esther D. (1990). Depression among lesbians: An invisible and unresearched phenomenon. *Journal of Gay and Lesbian Psychotherapy, 1,* 67–87.

Rothblum, Esther D., & Bond, Lynne A. (Eds.). (1996). *Preventing heterosexism and homophobia.* Thousand Oaks, CA: Sage.

Rousseau, S., et al. (1983). The expectancy of pregnancy for "normal" infertile couples. *Fertility & Sterility, 40,* 768–772.

Rowland, David A., & Slob, A. Koos. (1997). Premature ejaculation: Psychophysiological considerations in theory, research, and treatment. *Annual Review of Sex Research, 8,* 224–253.

Ruan, F., Bullough, V. L., & Tsai, Y. (1989). Male transsexualism in mainland China. *Archives of Sexual Behavior, 18,* 517–522.

Ruan, Fang-fu. (1991). *Sex in China.* New York: Plenum.

Rubenstein, Carin. (1983, July). The modern art of courtly love. *Psychology Today,* 40–49.

Rubin, Isadore. (1966). Sex after forty—and after seventy. In Ruth Brecher & Edward Brecher (Eds.), *An analysis of human sexual response.* New York: Signet Books, New American Library.

Rubin, Isadore. (1965). *Sexual life after sixty.* New York: Basic Books.

Rubin, Lillian B. (1979). *Women of a certain age: The midlife search for self.* New York: Harper & Row.

Rubin, Robert T., Reinisch, J. M., & Haskett, R. F. (1981). Postnatal gonadal steroid effects on human behavior. *Science, 211,* 1318–1324.

Rubin, Zick. (1973). *Liking and loving: An invitation to social psychology.* New York: Holt.

Rubin, Zick, et al. (1980). Self-disclosure in dating couples: Sex roles and the ethic of openness. *Journal of Marriage and the Family, 42,* 305–317.

Ruble, Diane N. (1977). Premenstrual symptoms: A reinterpretation. *Science, 197,* 291–292.

Ruble, Diane N., & Stangor, Charles. (1986). Stalking the elusive schema: Insights from developmental and social-psychological analyses of gender schemas. *Social Cognition, 4,* 227–261.

Russell, Diana E. H. (1990). *Rape in marriage.* (rev. ed.). Bloomington, IN: Indiana University Press.

Russell, Diana E. H. (1983). *Rape in marriage.* New York: Macmillan.

Russell, Diana E. H. (1980). Pornography and violence: What does the new research say? In L. Lederer (Ed.), *Take back the night: Women on pornography.* New York: Morrow.

Ruth, Sheila. (Ed.). (1990). *Issues in feminism: An introduction to women's studies* (2d ed.). Mountain View, CA: Mayfield.

Ryan, Alan S., et al. (1991). Recent declines in breast-feeding in the United States, 1984 through 1989. *Pediatrics, 88,* 719–727.

Rylko-Bauer, Barbara. (1996). Abortion from a cross-cultural perspective. *Social Science & Medicine, 42,* 479–482.

Sabatelli, R. M., Buck, R., & Dreyer, A. (1982). Nonverbal communication accuracy in married couples: Relationships with marital complaints. *Journal of Personality and Social Psychology, 43,* 1088–1097.

Sacred Congregation for the Doctrine of the Faith (1976, Jan. 16). Declaration on certain questions concerning sexual ethics. (English text in the *New York Times,* 2).

Sadock, Benjamin J., & Sadock, Virginia A. (1976). Techniques of coitus. In B. J. Sadock et al. (Eds.), *The sexual experience.* Baltimore, MD: Williams & Wilkins.

Saegert, S., Swap, W., & Zajonc, R. B. (1973). Exposure, context, and interpersonal attraction. *Journal of Personality and Social Psychology, 25,* 234–242.

Sagarin, Edward. (1973). Power to the peephole. *Sexual Behavior, 3,* 2–7.

St. Louis, Michael F., & Wasserheit, Judith N. (1998). Elimination of syphilis in the United States. *Science, 281,* 353–354.

Salamon, Edna. (1989). The homosexual escort agency: Deviance disavowal. *British Journal of Sociology, 40,* 1–21.

San Francisco Task Force on Prostitution. (1996). Final report. http://www.bayswan.org/2.

Sanchez, Miguel R. (1994). Infectious syphilis. *Seminars in Dermatology, 13,* 234–242.

Sanday, Peggy R. (1990). *Fraternity gang rape.* New York: New York University Press.

Sanday, Peggy R. (1981). The socio-cultural context of rape: A cross-cultural study. *Journal of Social Issues, 37*(4), 5–27.

Sanders, G., & Mullis, R. (1988). Family influences on sexual attitudes and knowledge as reported by college students. *Adolescence, 23,* 837–845.

Santen, Richard J. (1995). The testis. In P. Felig, J. D. Baxter, & L. A. Frohman (Eds.), *Endocrinology and metabolism* (3rd ed.). New York: McGraw-Hill.

Sarrel, Lorna, & Sarrel, Philip. (1984). *Sexual turning points: The seven stages of adult sexuality.* New York: Macmillan.

Sarrel, Philip, & Masters, William. (1982). Sexual molestation of men by women. *Archives of Sexual Behavior, 11,* 117–132.

Savage, Olayinka M. N., & Tchombe, Therese M. (1994). Anthropological perspectives on sexual behaviour in Africa. *Annual Review of Sex Research, 5,* 50–72.

Savin-Williams, Ritch. (1995). An exploratory study of pubertal maturation timing and self-esteem among gay and bisexual male youths. *Developmental Psychology, 31,* 56–64.

Savin-Williams, Ritch C., & Cohen, K. M. (Eds.). (1996). *The lives of lesbians, gays, and bisexuals: Children to adults.* New York: Harcourt Brace.

Schachter, Julius, & Barnes, Robert C. (1996). Chlamydia. In S. Morese et al. (Eds.), *Atlas of sexually transmitted diseases and AIDS* (pp. 65–86). London: Mosby-Wolfe.

Schachter, Stanley. (1964). The interaction of cognitive and physiological determinants of emotional state. In L. Berkowitz (Ed.), *Advances in experimental social psychology* (Vol. I). New York: Academic.

Schacter, Stanley, & Singer, J. F. (1962). Cognitive, social, and physiological determinants of emotional state. *Psychological Review, 69,* 379–399.

Schaefer, Mark T., & Olson, David H. (1981). Assessing intimacy: The PAIR Inventory. *Journal of Marital and Family Therapy,* 47–60.

Schaffer, H. R., & Emerson, Peggy E. (1964). Patterns of response to physical contact in early human development. *Journal of Child Psychology and Psychiatry, 5,* 1–13.

Scharfe, Elaine, & Bartholomew, Kim. (1995). Accommodation and attachment representations in young couples. *Journal of Social and Personal Relationships, 12,* 389–401.

Schatz, B. (1987). The AIDS insurance crisis: Underwriting or overreaching? *Harvard Law Review, 100*(7), 1782–1805.

Schenker, J. G., & Evron, S. (1983). New concepts in the surgical management of tubal pregnancy and the consequent postoperative results. *Fertility and Sterility, 40,* 709–723.

Schewe, Paul A., & O'Donohue, William. (1996). Rape prevention with high-risk males: Short-term outcome of two interventions. *Archives of Sexual Behavior, 25,* 455–472.

Schiavi, Raul C. (1990). Sexuality in aging men. *Annual Review of Sex Research, 1,* 227–250.

Schiavi, Raul C., Mandeli, John, & Schreiner-Engel. (1994). Sexual satisfaction in healthy aging men. *Journal of Sex and Marital Therapy, 20,* 3–13.

Schiavi, Raul C., White, Daniel, Mandeli, John, & Levine, Alice C. (1997). Effect of testosterone administration on sexual behavior and mood in men with erectile dysfunction. *Archives of Sexual Behavior, 26,* 231–242.

Schieffelin, E. L. (1976). *The sorrow of the lonely and the burning of the dancers.* New York: St. Martin's Press.

Schlenker, Jennifer A., Caron, Sandra L., & Halteman, William A. (1998). A feminist analysis of *Seventeen* magazine: Content analysis from 1945 to 1995. *Sex Roles, 38,* 135–150.

Schmidt, Madeline H. (1970). Superiority of breast-feeding: Fact or fancy? *American Journal of Nursing, 70,* 1488–1493.

Schmitt, David, & Buss, David. (1996). Strategic self-promotion and competitor derogation: Sex and content effects on the perceived effectiveness of mate attraction tactics. *Journal of Personality and Social Psychology, 70,* 1185–1204.

Schneider, Edward D. (Ed.). (1985). *Questions about the beginning of life.* Minneapolis, MN: Augsburg.

Schofield, Alfred T., & Vaughan-Jackson, Percy. (1913). *What a boy should know*. New York: Cassell.

Schoof-Tams, K., Schlaegel, J., & Walczak, L. (1976). Differentiation of sexual morality between 11 and 16 years. *Archives of Sexual Behavior, 5*, 353–370.

Schott, Richard. (1995). The childhood and family dynamics of transvestites. *Archives of Sexual Behavior, 24*, 309–327.

Schow, Douglas A., Redmon, Bruce, & Pryor, Jon L. (1997). Male menopause: How to define it, how to treat it. *Postgraduate Medicine, 101*, 62–79.

Schroeder, Patricia. (1994). Female genital mutilation—a form of child abuse. *New England Journal of Medicine, 331*, 739–740.

Schultz, W. C. M., et al. (1989). Vaginal sensitivity to electric stimuli: Theoretical and practical implications. *Archives of Sexual Behavior, 18*, 87–96.

Schultz, Willibrord C. M. Weijmar, van de Wiel, Harry B. M., Hah, Daniela E. E., & van Driel, Mels F. (1992). Sexuality and cancer in women. *Annual Review of Sex Research, 3*, 151–200.

Schwartz, Lisa Barrie. (1997, Dec. 20/27). Understanding human parturition. *The Lancet, 350*, 1792–1793.

Sciarra, John J. (1991). Infertility: A global perspective on the role of infection. *Annals of the New York Academy of Sciences, 626*, 478–483.

Scott, John Paul. (1964). The effects of early experience on social behavior and organization. In W. Etkin (Ed.), *Social behavior and organization among vertebrates*. Chicago: University of Chicago Press.

Scott, Joseph E., & Cuvelier, Steven J. (1993). Violence and sexual violence in pornography: Is it really increasing? *Archives of Sexual Behavior, 22*, 357–372.

Scroggs, Robin. (1983). *The New Testament and homosexuality*. Philadelphia, PA: Fortress.

Seal, David W. (1997). Interpartner concordance of self-reported sexual behavior among college dating couples. *Journal of Sex Research, 34*, 39–55.

Searles, Patricia, & Berger, Ronald J. (1987). The current status of rape reform legislation. *Women's Rights Law Reporter, 10*, 25–44.

Segraves, R. Taylor. (1989). Effects of psychotropic drugs on human erection and ejaculation. *Archives of General Psychiatry, 46*, 275–284.

Segraves, R. Taylor. (1988). Drugs and desire. In S. R. Leiblum & R. C. Rosen (Eds.), *Sexual desire disorders* (pp. 313–347). New York: Guilford.

Sell, Randall L. (1997). Defining and measuring sexual orientation: A review. *Archives of Sexual Behavior, 26*, 643–658.

Sell, Randall L., Wells, James A., & Wypij, David. (1995). The prevalence of homosexual behavior and attraction in the United States, the United Kingdom and France. *Archives of Sexual Behavior, 24*, 235–248.

Serrin, W. (1981, Feb. 9). Sex is a growing multimillion dollar business. *New York Times*, B1–B6.

Service, Robert F. (1994). Contraceptive methods go back to the basics. *Science, 266*, 1480–1481.

The Sex Industry. (1998, February 14). *The Economist*, 21–23.

Shackelford, Todd, & Buss, David. (1997). Cues to infidelity. *Personality and Social Psychology Bulletin, 23*, 1034–1045.

Shandera, K. C., & Thompson, I. M. (1994). Urologic prostheses. *Emergency Medicine Clinics of North America, 12*, 729–748.

Shapiro, Craig N., & Alter, Miriam J. (1996). Syphilis. In S. A. Morse et al. (Eds.), *Atlas of sexually transmitted diseases* (pp. 241–268). London: Mosby-Wolfe.

Shapiro, E. Donald. (1986). New innovations in conception and their effects upon our law and morality. *New York Law Review, 21*, 37–59.

Shapiro, Harold T. (1997). Ethical and policy issues in human cloning. *Science, 277*, 195–196.

Sharpstein, Don J., & Kirkpatrick, Lee. (1997). Romantic jealousy and adult romantic attachment. *Journal of Personality and Social Psychology, 72*, 627–640.

Shattuck-Eidens, Donna, et al. (1995). A collaborative survey of 80 mutations in the BRCA1 breast and ovarian cancer susceptibility gene. *Journal of the American Medical Association, 273*, 535–541.

Shaw, Jeanne. (1989). The unnecessary penile implant. *Archives of Sexual Behavior, 18*, 455–460.

Shelp, Earl E., & Sunderland, Ronald H. (1987). The challenge of AIDS to the church. *St. Luke's Journal of Theology, 30*, 273–280.

Sherfey, Mary Jane. (1966). The evolution and nature of female sexuality in relation to psychoanalytic theory. *Journal of the American Psychoanalytic Association, 14*, 28–128.

Sherman, Karen J., et al. (1990). Sexually transmitted diseases and tubal pregnancy. *Sexually Transmitted Diseases, 17*, 115–121.

Sherrard, J., & Barlow, D. (1996). Gonorrhoea in men: Clinical and diagnostic aspects. *Genitourinary Medicine, 72*, 422–426.

Sherwin, Barbara B. (1991). The psychoendocrinology of aging and female sexuality. *Annual Review of Sex Research, 2*, 181–198.

Shewaga, Duane. (1983). Note on New York v. Ferber. *Santa Clara Law Review, 23*, 675–684.

Shields, W. M., & Shields, L. M. (1983). Forcible rape: An evolutionary perspective. *Ethology and Sociobiology, 4*, 115–136.

Shoemaker, Donald J. (1977). The teeniest trollops: "Baby pros," "chickens," and child prostitutes. In C. D. Bryant (Ed.), *Sexual deviancy in social context* (pp. 241–254). New York: Franklin Watts.

Shostak, Arthur B. (1984). *Men and abortion: Lessons, losses, and love*. New York: Praeger.

Shouvlin, David P. (1981). Preventing the sexual exploitation of children: A model act. *Wake Forest Law Review, 17*, 535–560.

Shusterman, L. R. (1979). Predicting the psychological consequences of abortion. *Social Science and Medicine, 13*, 683–689.

SIECUS. (1991). *Guidelines for comprehensive sexuality education*. New York: Sexuality Information and Education Council of the United States.

Siegel, Karolynn, & Glassman, Marc. (1989). Individual and aggregate level change in sexual behavior among gay men at risk for AIDS. *Archives of Sexual Behavior, 18*, 335–348.

Siegel, Karolynn, Krauss, Beatrice J., & Karus, Daniel. (1994). Reporting recent sexual practices: Gay men's disclosure of HIV risk by questionnaire and interview. *Archives of Sexual Behavior, 23*, 217–230.

Signorielli, Nancy. (1990). Children, television, and gender roles. *Journal of Adolescent Health Care, 11*, 50–58.

Signorile, Michelangelo. (1997). *Life on the outside: The Signorile report on gay men*. New York: HarperCollins.

Siker, Jeffrey S. (Ed.). (1994). *Homosexuality in the church: Both sides of the debate*. Louisville, KY: Westminster/John Knox Press.

Silvestre, Louise, et al. (1990). Voluntary interruption of pregnancy with mifepristone (RU-486) and a prostaglandin analogue: A large-scale French experience. *New England Journal of Medicine, 322*, 645.

Simon, William, & Gagnon, John H. (1986). Sexual scripts: Permanence and change. *Archives of Sexual Behavior, 15*, 97–120.

Simpson, J. A. (1990). Influence of attachment styles on romantic relationships. *Journal of Personality and Social Psychology, 59*, 971–980.

Simpson, J. A., Campbill, B., & Berscheid, E. (1986). The association between romantic love and marriage. *Personality and Social Psychology Bulletin, 12*, 363–372.

Singer, D. B. (1995). Human embryogenesis. In D. R. Coustan, R. V. Haning, Jr., & D. B. Singer (Eds.), *Human reproduction: Growth and development*. Boston: Little, Brown, and Company.

Siosteen, A., et al. (1990). Sexual ability, activity, attitudes and satisfaction as part of adjustment in spinal cord-injured subjects. *Paraplegia, 28*, 285–295.

Sipe, A. W. Richard. (1995). *Sex priests, and power: Anatomy of a crisis.* New York: Brunner/Mazel.

Sipski, Marca L., & Alexander, Craig J. (1995). Spinal cord injury and female sexuality. *Annual Review of Sex Research, 6,* 224–244.

Sipski, Marca L., & Alexander, Craig J. (Eds.). (1997). *Sexual function in people with disability and chronic illness.* Gaithersburg, MD: Aspen.

Slater, Suzanne. (1994). *The lesbian family life cycle.* New York: Free Press.

Slob, A. K., et al. (1991). Menstrual cycle phase and sexual arousability in women. *Archives of Sexual Behavior, 20,* 567–578.

Slovenko, Ralph. (1965). *Sexual behavior and the law.* Springfield, IL: Charles C. Thomas.

Small, Meredith F. (1993). *Female choices: Sexual behavior of female primates.* Ithaca, NY: Cornell University Press.

Smith, George, Frankel, Stephen, & Yarnell, John. (1997). Sex and death: Are they related? Findings from the Caerphilly cohort study. *British Medical Journal, 315,* 1641–1645.

Smith, Jeffrey R., et al. (1996). Major susceptibility locus for prostate cancer on chromosome 1 suggested by a genome-wide search. *Science, 274,* 1371–1373.

Smith, R. Spencer. (1976). Voyeurism: A review of literature. *Archives of Sexual Behavior, 5,* 585–608.

Smith, Tom W. (1994). *The demography of sexual behavior.* Menlo Park, CA: Kaiser Family Foundation.

Smith, Tom W. (1991). Adult sexual behavior in 1989: Number of partners, frequency of intercourse and risk of AIDS. *Family Planning Perspectives, 23*(3), 102–107.

Sobo, E. J. (1996). Abortion traditions in rural Jamaica. *Social Science & Medicine, 42,* 495–508.

Soley, Lawrence C., & Kurzbard, Gary. (1986). Sex in advertising: A comparison of 1964 and 1984 magazine advertisements. *Journal of Advertising, 15*(3), 46–54.

Sommers-Flanagan, Rita, Sommers-Flanagan, John, & Davis, Britta. (1993). What's happening on music television? A gender role content analysis. *Sex Roles, 28,* 745–753.

Sonenstein, Freya, Pleck, Joseph H., & Ku, Leighton C. (1989). Sexual activity, condom use and AIDS awareness among adolescent males. *Family Planning Perspectives, 21*(4), 152–158.

Song, Y. I. (1991). Single Asian women as a result of divorce: Depressive affect and changes in social support. *Journal of Divorce and Remarriage, 14,* 219–230.

Sorensen, Robert C. (1973). *Adolescent sexuality in contemporary America.* New York: World.

Sorenson, Susan B., & Siegel, Judith M. (1992). Gender, ethnicity, and sexual assault: Findings from a Los Angeles study. *Journal of Social Issues, 48*(1), 93–104.

Spaccarelli, Steve. (1994). Stress, appraisal, and coping in child sexual abuse: A theoretical and empirical review. *Psychological Bulletin, 116,* 340–362.

Spach, David, & Keen, Peggy. (1996). HIV and AIDS. In S. Morse et al. (Eds.), *Atlas of sexually transmitted diseases and AIDS* (pp. 165–206). London: Mosby-Wolfe.

Spahn, William C. (1988). The moral dimension of AIDS. *Theological Studies, 49,* 88–110.

Spalding, Leah R., & Peplau, L. Anne. (1997). The unfaithful lover: Heterosexuals' perceptions of bisexuals and their relationships. *Psychology of Women Quarterly, 21,* 611–625.

Specter, Michael. (1998, January 11). Contraband women: A special report. *New York Times, 1,* 6.

Spector, Ilana P., & Carey, Michael P. (1990). Incidence and prevalence of the sexual dysfunctions: A critical review of the empirical literature. *Archives of Sexual Behavior, 19,* 389–408.

Spitz, Irving M., et al. (1998). Early pregnancy termination with mifepristone and misoprostol in the United States. *New England Journal of Medicine, 338,* 1241–1247.

Spitz, Rene A. (1949). Autoeroticism: Some empirical findings and hypotheses on three of its manifestations in the first year of life. *The Psychoanalytic Study of the Child* (Vols. III–IV, pp. 85–120). New York: International Universities Press.

Sprecher, Susan. (1987). The effects of self-disclosure given and received on affection for an intimate partner and stability of the relationship. *Journal of Social and Personal Relationships, 4,* 115–127.

Sprecher, Susan, Barbee, Anita, & Schwartz, Pepper. (1995). "Was it good for you, too?": Gender differences in first sexual intercourse experiences. *The Journal of Sex Research, 32,* 3–15.

Sprecher, Susan, & Hatfield, Elaine. (1996). Premarital sexual standards among U.S. college students: Comparison with Russian and Japanese students. *Archives of Sexual Behavior, 25,* 261–288.

Sprecher, Susan, & McKinney, Kathleen. (1993). *Sexuality.* Newbury Park, CA: Sage.

Sprecher, Susan, Sullivan, Quintin, & Hatfield, Elaine. (1994). Mate selection preferences: Gender differences examined in a national sample. *Journal of Personality and Social Psychology, 66,* 1074–1080.

Spring-Mills, E., & Hafez, E. S. (1980). Male accessory sexual organs. In E. S. Hafey (Ed.), *Human reproduction* (pp. 60–90). New York: Harper & Row.

Squire, Corrine. (1993). *Women and AIDS: Psychological perspectives.* London: Sage.

Srivastava, A., Borries, C., & Sommer, V. (1991). Homosexual mounting in free-ranging female Hanuman langurs. *Archives of Sexual Behavior, 20,* 487–512.

Stack, Steven, & Gundlach, James H. (1992). Divorce and sex. *Archives of Sexual Behavior, 21,* 359–368.

Stamm, Walter E., & Holmes, King K. (1990). Chlamydia trachomatis infections of the adult. In K. K. Holmes (Ed.), *Sexually transmitted diseases* (pp. 181–194). New York: McGraw-Hill.

Staples, Robert. (1978). Masculinity and race: The dual dilemma of black men. *Journal of Social Issues, 34*(1), 169–183.

Starks, Kay J., & Morrison, Eleanor S. (1996). *Growing up sexual* (2d ed.). New York: HarperCollins.

St. Augustine. (1950). *The city of God.* (Marcus Dods, Trans.). New York: Modern Library.

Steege, J. F., Stout, A. L., & Carson, Culley C. (1986). Patient satisfaction in Scott and Small-Carrion penile implant recipients. *Archives of Sexual Behavior, 15,* 393–400.

Steinberg, Jennifer. (1993, Feb.). CDC broadens AIDS definition. *Journal of NIH Research, 5,* 32.

Steinberg, Karen K., et al. (1991). A meta-analysis of the effect of estrogen replacement therapy on the risk of breast cancer. *Journal of the American Medical Association, 265,* 1985–1990.

Steinman, Debra L., et al. (1981). A comparison of male and female patterns of sexual arousal. *Archives of Sexual Behavior, 10,* 529–548.

Stephens, Tim. (1992, Feb.). RU 486: New studies, same old politics. *The Journal of NIH Research, 4,* 44–46.

Stephens, Tim. (1991, Feb.). AIDS in women reveals health-care deficiencies. *The Journal of NIH Research, 3,* 27–30.

Stern, Kathleen, & McClintock, Martha K. (1998). Regulation of ovulation by human pheromones. *Nature, 392,* 177–179.

Sternberg, Robert. (1998). *Love is a story: A new theory of relationships.* New York: Oxford University Press.

Sternberg, Robert. (1997). Construct validation of a triangular love scale. *European Journal of Social Psychology, 27,* 313–335.

Sternberg, Robert. (1996). Love stories. *Personal Relationships, 3,* 59–79.

Sternberg, Robert J. (1987). Liking versus loving: A comparative evaluation of theories. *Psychological Bulletin, 102,* 331–345.

Sternberg, Robert J. (1986). A triangular theory of love. *Psychological Review, 93,* 119–135.

Stevens-Simon, Catherine, & White, Marguerite M. (1991). Adolescent pregnancy. *Pediatric Annals, 20*(6), 322–331.

Stevenson, Michael R. (1995). Searching for a gay identity in Indonesia. *Journal of Men's Studies, 4*, 93–108.

Stiles, William B., Walz, Nicolay, Schroeder, Michelle, Williams, Laura, & Ickes, William. (1996). Attractiveness and disclosure in initial encounters of mixed-sex dyads. *Journal of Social and Personal Relationships, 13*, 303–312.

Stine, Gerald J. (1996). *Acquired Immune Deficiency Syndrome: Biological, medical, social, and legal issues.* Englewood Cliffs, NJ: Prentice Hall.

Stodghill II, Ron. (1998). Where'd you learn that? *Time* 151 (no. 23), June 15, 52–59.

Stokes, Joseph P., McKirnan, David, & Bumette, Rebecca. (1993). Sexual behavior, condom use, disclosure of sexuality and stability of sexual orientation in bisexual men. *The Journal of Sex Research, 30*, 203–213.

Stone, Katherine M. (1994). HIV, other STDs, and barriers. In C. K. Mauck et al. (Eds.), *Barrier contraceptives: Current status and future prospects* (pp. 203–212). New York: Wiley.

Stoneburner, Raud L., Sato, Paul, Burton, Anthony, & Mertens, Thierry. (1994). The global HIV pandemic. *Acta Paediatrica,* Suppl. 400, 1–4.

Storms, Michael D. (1981). A theory of erotic orientation development. *Psychological Review, 88*, 340–353.

Storms, Michael D. (1980). Theories of sexual orientation. *Journal of Personality and Social Psychology, 38*, 783–792.

Strassberg, Donald S., & Lowe, Kristi. (1995). Volunteer bias in sex research. *Archives of Sexual Behavior, 24*, 369–382.

Streissguth, A. P., Sampson, P., & Barr, H. M. (1989). Neurobehavioral dose-response effects of prenatal alcohol exposure in humans from infancy to adulthood. *Annals of the New York Academy of Sciences, 562*, 145–158.

Strickland, Bonnie R. (1995). Research on sexual orientation and human development. *Developmental Psychology, 31*, 137–140.

Striegel-Moore, Ruth, Goldman, Susan, Garvin, Vicki, & Rodin, Judith. (1996). A prospective study of somatic and emotional symptoms of pregnancy. *Psychology of Women Quarterly, 20*, 393–408.

Strong, Carson. (1997). *Ethics in reproductive and perinatal medicine.* New Haven: Yale University Press.

Struckman-Johnson, Cindy. (1988). Forced sex on dates: It happens to men, too. *Journal of Sex Research, 24*, 234–241.

Struckman-Johnson, Cindy, et al. (1996). Sexual coercion reported by men and women in prison. *Journal of Sex Research, 33*, 67–76.

Struewing, Jeffery P., et al. (1997). The risk of cancer associated with specific mutations of BRCA1 and BRCA2 among Ashkenazi Jews. *New England Journal of Medicine, 336*, 1401–1408.

Sudarkasa, Niara. (1997). African American families and family values. In H. P. McAdoo (Ed.), *Black families,* 3d ed. (pp. 9–40). Thousand Oaks, CA: Sage.

Suitor, J. Jill, & Reavis, Rebel. (1995). Football, fast cars, and cheerleading: Adolescent gender norms. *Adolescence, 30*, 265–273.

Sunstein, Cass R. (1986). Pornography and the first amendment. *Duke Law Journal, 1986*, 589–629.

Swaab, D. F., Gooren, L. J. G., & Hofman, M. A. (1995). Brain research, gender, and sexual orientation. *Journal of Homosexuality, 28*, 283–301.

Symons, Donald. (1987). An evolutionary approach: Can Darwin's view of life shed light on human sexuality? In J. H. Geer & W. T. O'Donohue (Eds.), *Theories of human sexuality* (pp. 91–126). New York: Plenum.

Symons, Donald. (1979). *The evolution of human sexuality.* New York: Oxford University Press.

Syrjanen, Kari, et al. (1990). Prevalence, incidence, and estimated life-time risk of cervical human papillomavirus infections in a nonselected Finnish female population. *Sexually Transmitted Diseases, 17*, 15–19.

Szasz, Thomas S. (1980). *Sex by prescription.* Garden City, NY: Anchor Press/Doubleday.

Szasz, Thomas S. (1965). Legal and moral aspects of homosexuality. In J. Marmor (Ed.), *Sexual inversion: The multiple roots of homosexuality.* New York: Basic Books.

Taberner, Peter V. (1985). *Aphrodisiacs: The science and the myth.* Philadelphia: University of Pennsylvania Press.

Taffel, Selma M., et al. (1991, June). 1989 U.S. cesarean section rate steadies—VBAC rate rises to nearly one in five. *Birth, 18*, 73–77.

Tafoya, Terry, & Wirth, Douglas A. (1996). Native American two-spirit men. In J. F. Longres (Ed.), *Men of color* (pp. 51–67). New York: Haworth.

Talamini, John T. (1982). *Boys will be girls: The hidden world of the heterosexual male transvestite.* Washington, D.C.: University Press of America.

Tamir, Lois M. (1982). *Men in their forties: The transition to middle age.* New York: Springer.

Tanfer, Koray, Grady, W. R., Klepinger, D. H., & Billy, J. O. G. (1993). Condom use among U.S. men, 1991. *Family Planning Perspectives, 25*(2), 61–66.

Tanfer, Koray, & Schoorl, Jeannette J. (1992). Premarital sexual careers and partner change. *Archives of Sexual Behavior, 21*, 45–68.

Tangri, Sandra, Burt, M. R., & Johnson, L. B. (1982). Sexual harassment at work: Three explanatory models. *Journal of Social Issues, 38*(4), 33–54.

Tannen, Deborah. (1991). *You just don't understand: Women and men in conversation.* New York: William Morrow.

Tannen, Deborah. (1986). *That's not what I meant! How conversational style makes or breaks relationships.* New York: Ballantine Books.

Tanner, James M. (1967). Puberty. In A. McLaren (Ed.), *Advances in reproductive physiology* (Vol. II). New York: Academic.

Tasker, Fiorna L., & Golombok, Susan. (1997). *Growing up in a lesbian family: Effects on child development.* New York: Guilford.

Taub, Nadine. (1987). Amicus brief: In the matter of Baby M. *Women's Rights Law Reporter, 10*, 7–24.

Tavris, Carol. (1977). Masculinity. *Psychology Today, 10*(8), 34.

Taylor, Robert. (1994, Apr.). Quiet clues to HIV-1 immunity: Do some people resist infection? *Journal of NIH Research, 6*, 29–31.

Templeman, Terrel L., & Stinnett, Ray D. (1991). Patterns of sexual arousal and history in a "normal" sample of young men. *Archives of Sexual Behavior, 20*, 137–150.

Terman, Lewis, et al. (1938). *Psychological factors in marital happiness.* New York: McGraw-Hill.

Terman, Lewis M. (1948). Kinsey's *Sexual Behavior in the Human Male:* Some comments and criticisms. *Psychological Bulletin, 45*, 443–459.

Tewksbury, R. (1990). Patrons of porn: Research notes on the clientele of adult bookstores. *Deviant Behavior, 11*, 259–271.

Thielicke, Helmut. (1964). *The ethics of sex.* New York: Harper & Row.

Thomas, David J. (1982). San Francisco's 1979 White Night riot. In W. Paul et al. (Eds.), *Homosexuality: Social, psychological, and biological issues.* Beverly Hills, CA: Sage.

Thomas, J. J. (1992). *Informal economic activity.* Ann Arbor: University of Michigan Press.

Thomas, S., & Quinn, S. (1991). The Tuskegee Syphilis Study 1932–1972: Implications for HIV education and AIDS risk education programs in the African American community. *American Journal of Public Health, 81*, 1498–1505.

Thompson, Anthony P. (1983). Extramarital sex: A review of the research literature. *Journal of Sex Research, 19*, 1–22.

Thorne, Barrie. (1993). *Gender play: Girls and boys in school.* New Brunswick, NJ: Rutgers University Press.

Thornton, Michael C., & Wason, Suzanne. (1995). Intermarriage. In D. Levinson (Ed.), *Encyclopedia of marriage and the family* (Vol. 2, pp. 396–402). New York: Macmillan.

Thorpe, L. P., Katz, B., & Lewis, R. T. (1961). *The psychology of abnormal behavior.* New York: Ronald Press.

Tiefer, Leonore. (1995). *Sex is not a natural act, and other essays.* Boulder, CO: Westview.

Tiefer, Leonore. (1994). Three crises facing sexology. *Archives of Sexual Behavior, 23,* 361–374.

Tiefer, Leonore. (1991a). Historical, scientific, clinical, and feminist criticisms of "The Human Sexual Response Cycle" model. *Annual Review of Sex Research, 2,* 1–24.

Tiefer, Leonore. (1991b). New perspectives in sexology: From rigor (mortis) to richness. *Journal of Sex Research, 28,* 593–602.

Toubia, Nahid. (1995). *Female genital mutilation: A call for global action.* New York: Women Ink.

Toubia, Nahid. (1994). Female circumcision as a public health issue. *New England Journal of Medicine, 331,* 712–716.

Touchette, Nancy. (1991, July). HIV-1 link prompts circumspection on circumcision. *Journal of NIH Research, 3,* 44–46.

Townsend, John M. (1995). Sex without emotional involvement: An evolutionary interpretation of sex differences. *Archives of Sexual Behavior, 24,* 173–206.

Traven, Sheldon, Cuyllen, Ken, & Protter, Barry. (1990). Female sexual offenders: Severe victims and victimizers. *Journal of Forensic Sciences, 35,* 140–150.

Treiman, Katherine, et al. (1995). IUDs—An update. *Population Reports,* Series B, No. 6. Baltimore, MD: Johns Hopkins School of Public Health.

Triandis, H. C., McCusker, C., & Hui, C. H. (1990). Multimethod probes of individualism and collectivism. *Journal of Personality and Social Psychology, 59,* 1006–1020.

Trussell, James, & Vaughan, Barbara. (1991). *Selected results concerning sexual behavior and contraceptive use from the 1988 National Survey of Family Growth and the 1988 National Survey of Adolescent Males.* Working Paper 91-12. Princeton, NJ: Office of Population Research.

Tsai, Mavis, & Uemura, Anne. (1988). Asian Americans: The struggles, the conflicts, and the successes. In P. Bronstein & K. Quina (Eds.), *Teaching a psychology of people.* Washington, D.C.: American Psychological Association.

Tserotas, K., & Merino, G. (1998). Andropause and the aging male. *Archives of Andrology, 40,* 87–93.

Tullman, Gerald M., et al. (1981). The pre- and post-therapy measurement of communication skills of couples undergoing sex therapy at the Masters & Johnson Institute. *Archives of Sexual Behavior, 10,* 95–109.

Turkle, S. (1995). *Life on the screen: Identity in the age of the Internet.* New York: Simon and Schuster.

Turner, C. F., et al. (1998). Adolescent sexual behavior, drug use, and violence: Increased reporting with computer survey technology. *Science, 280,* 867–873.

Turner, R. (1990). Vaginal ring is comparable in safety and efficacy to other low-dose, progestogen-only methods. *Family Planning Perspectives, 22*(5), 236–237.

Turner, William J. (1995). Homosexuality, type 1: An Xq28 phenomenon. *Archives of Sexual Behavior, 24,* 109– 134.

Tutin, C. E. G., & McGinnis, P. R. (1981). Chimpanzee reproduction in the wild. In C. E. Graham (Ed.), *Reproductive biology of the great apes* (pp. 239–264). New York: Academic Press.

Tutuer, W. (1984). Dangerousness of peeping toms. *Medical Aspects of Human Sexuality, 18,* 97.

Udry, J. Richard. (1993). The politics of sex research. *Journal of Sex Research, 30,* 103–110.

Udry, J. Richard. (1988). Biological predispositions and social control in adolescent sexual behavior. *American Sociological Review, 53,* 709–722.

Udry, J. Richard, et al. (1985). Serum androgenic hormones motivate sexual behavior in adolescent boys. *Fertility and Sterility, 43,* 90–94.

Udry, J. Richard, & Eckland, Bruce K. (1984). Benefits of being attractive: Differential payoffs for men and women. *Psychological Reports, 54,* 47–56.

Udry, J. Richard, & Morris, N. M. (1968). Distribution of coitus in the menstrual cycle. *Nature, 220,* 593–596.

Ullman, Sarah E., & Knight, Raymond A. (1993). The efficacy of women's resistance strategies in rape situations. *Psychology of Women Quarterly, 17,* 23–38.

Ulmann, A., Teutsch, G., & Philibert, D. (1990, June). RU-486. *Scientific American, 262,* 42–48.

Ulrich's. (1995). *International periodical directory* (34th ed.). New Providence, NJ: R. R. Bowker.

UNAIDS—United Nations Program on HIV/AIDS. (1997). *Impact of HIV and sexual health education on the sexual behavior of young people.* Geneva, SU.

UPI (1981, Nov. 5). Toxicologist warns against butyl nitrite. *Delaware Gazette,* p. 3.

U.S. Bureau of the Census. (1997). *Statistical abstract of the United States: 1997* (117th ed.). Washington, D.C.: U.S. Bureau of the Census.

U.S. Department of Justice. (1986). *The Attorney General's Commission on Pornography: Final Report.* Washington, D.C.: U.S. Department of Justice.

U.S. Department of Labor. (1994). *Employment and earnings, April, 1994.* Washington, D.C.: U.S. Department of Labor.

Vacca, J. P., et al. (1994). L-735, 524: An orally bioavailable human immunodeficiency virus Type I protease inhibitor. *Proceedings of the National Academy of Sciences USA, 91,* 4096.

Vance, Ellen B., & Wagner, Nathaniel N. (1976). Written descriptions of orgasm: A study of sex differences. *Archives of Sexual Behavior, 5,* 87–98.

Van der ven, Paul, Bornholt, Laurel, & Bailey, Michael. (1996). Measuring cognitive, affective, and behavioral components of homophobic reaction. *Archives of Sexual Behavior, 25,* 155–180.

Van Dis, H., & Larsson, K. (1971). Induction of sexual arousal in the castrated male rat by intracranial stimulation. *Physiology & Behavior, 6,* 85–86.

Van Goozen, Stephanie H. M., et al. (1997). Psychoendocrinological assessment of the menstrual cycle: The relationship between hormones, sexuality, and mood. *Archives of Sexual Behavior, 26,* 359–382.

Van Kesteren, Paul J., Gooren, Louis J., & Megans, Jos A. (1996). An epidemiological and demographic study of transsexuals in the Netherlands. *Archives of Sexual Behavior, 25,* 589–600.

Van Lankveld, Jacques. (1998). Bibliotherapy in the treatment of sexual dysfunctions: A meta-analysis. *Journal of Consulting and Clinical Psychology, 66,* 702–708.

Van Look, Paul, & von Hertzen, Helena. (1995). Clinical uses of antiprogestogens. *Human Reproduction Update, 1,* 19–34.

Vanwesenbeeck, Ine. (1994). *Prostitutes' well-being and risk.* Amsterdam: VU University Press.

Vasey, P. L. (1995). Homosexual behavior in primates: A review of evidence and theory. *International Journal of Primatology, 61,* 173–204.

Vasquez, Melba, & Barron, Augustine. (1988). The psychology of the Chicano experience: A sample course structure. In P. Bronstein & K. Quina (Eds.), *Teaching a psychology of people.* Washington, D.C.: American Psychological Association.

Ventura, S. J., Martin, J. A., Cursten, S. C., & Mathews, T. J. (1997). Report of final natality statistics, 1995. *Monthly Vital Statistics Report, 45*(11, Suppl. 2). Hyattsville, MD: National Center for Health Statistics.

Veronesi, Umberto, et al. (1981). Comparing radical mastectomy with quadrantectomy, axillary dissection, and radiotherapy in patients with small cancers of the breast. *New England Journal of Medicine, 305,* 6–11.

"Video turns big profit for porn products." (1982, March 10). *Variety, 306*, 35.

Vincent, J. P., Friedman, L. C., Nugent, J., & Messerly, L. (1979). Demand characteristics in observations of marital interaction. *Journal of Consulting and Clinical Psychology, 47*, 557–566.

Vogel, Gretchen. (1997). Cocaine wreaks subtle damage on developing brains. *Science, 278*, 38–39.

Von Hertzen, Helena, & Van Look, Paul. (1996). Research on new methods of emergency contraception. *Family Planning Perspectives, 28*, 52–57.

Von Krafft-Ebing, Richard. (1886). *Psychopathia sexualis.* (Reprinted by Putnam, New York, 1965).

Wabrek, Alan J., & Burchell, R. Clay. (1980). Male sexual dysfunction associated with coronary heart disease. *Archives of Sexual Behavior, 9*, 69–75.

Wagner, Goran, & Kaplan, Helen S. (1993). *The new injection treatment for impotence: Medical and psychological aspects.* New York: Brunner/Mazel.

Waite, Linda, & Joyner, Kara. (1999). Men's and women's general happiness and sexual satisfaction in marriage, cohabitation, and single living. In E. O. Laumann & R. Michael (Eds.), *The social organization of sexuality: Further studies,* in press.

Walen, Susan R., & Roth, David. (1987). A cognitive approach in J. H. Geer & W. T. O'Donohue (Eds.), *Theories of human sexuality.* New York: Plenum.

Walfish, Steven, & Myerson, Marilyn. (1980). Sex role identity and attitudes toward sexuality. *Archives of Sexual Behavior, 9*, 199–204.

Wallen, Kim, & Parsons, William A. (1997). Sexual behavior in same-sexed nonhuman primates: Is it relevant to understanding human homosexuality? *Annual Review of Sex Research, 8*, 195–223.

Wallerstein, Edward. (1980). *Circumcision: An American health fallacy.* New York: Springer.

Wallin, Paul. (1949). An appraisal of some methodological aspects of the Kinsey report. *American Sociological Review, 14*, 197–210.

Walling, M., Andersen, B. L., & Johnson, S. R. (1990). Hormonal replacement therapy for postmenopausal women: A review of sexual outcomes and related gynecologic effects. *Archives of Sexual Behavior, 19*, 119–138.

Walser, Robyn D., & Kern, Jeffrey M. (1996). Relationships among childhood sexual abuse, sex guilt, and sexual behavior in adult clinical samples. *Journal of Sex Research, 33*, 321–326.

Walster, Elaine. (1978). Equity and extramarital sexuality. *Archives of Sexual Behavior, 7*, 127–141.

Walster, Elaine, et al. (1973). "Playing hard-to-get": Understanding an elusive phenomenon. *Journal of Personality and Social Psychology, 26*, 113–121.

Wampler, Karen S. (1982). The effectiveness of the Minnesota Couple Communication Program: A review of research. *Journal of Marital and Family Therapy,* 345–355.

Ward, Ingeborg, et al. (1988). Transmission of human immunodeficiency virus (HIV) by blood transfusions screened as negative for HIV antibody. *New England Journal of Medicine, 318*, 473–478.

Ward, Ingeborg L., & Reed, J. (1985). Prenatal stress and prepubertal social rearing conditions interact to determine sexual behavior in male rats. *Behavioral Neuroscience, 99*, 301–309.

Ward, Ingeborg L., & Stehm, Kathleen E. (1991). Prenatal stresses feminize juvenile play patterns in male rats. *Physiology & Behavior, 50*, 601–605.

Warr, M. (1985). Fear of rape among urban women. *Social Problems, 32*, 239–250.

Wass, Debbie M., et al. (1991). Completed follow-up of 1,000 consecutive transcervical chorionic villus samplings performed by a single operator. *Australia New Zealand Journal of Obstetrics and Gynecology, 31*, 240–245.

Weber, Robert P. (1990). *Basic content analysis.* (2d ed.) Newbury Park, CA: Sage.

Weideger, Paula. (1976). *Menstruation and menopause.* New York: Knopf.

Weinberg, Martin, Williams, Colin, & Calhan, Cassandra. (1995). "If the shoe fits . . .": Exploring male homosexual foot fetishism. *The Journal of Sex Research, 32*, 17–27.

Weinberg, Martin S., & Williams, Colin. (1974). *Male homosexuals: Their problems and adaptations.* New York: Oxford University Press.

Weinberg, Martin S., & Williams, Colin J. (1988). Black sexuality: A test of two theories. *Journal of Sex Research, 25*, 197–218.

Weinberg, Martin S., Williams, Colin J., & Pryor, Douglas W. (1994). *Dual attraction: Understanding bisexuality.* New York: Oxford University Press.

Weinberg, Samuel K. (1955). *Incest behavior.* New York: Citadel Press.

Weinberg, Thomas S. (1994). Research in sadomasochism: A review of sociological and social psychological literature. *Annual Review of Sex Research, 5*, 257–279.

Weinberg, Thomas S. (1987). Sadomasochism in the United States: A review of recent sociological literature. *Journal of Sex Research, 23*, 50–69.

Weinhardt, Lance S., & Carey, Michael P. (1996). Prevalence of erectile disorder among men with diabetes mellitus. *Journal of Sex Research, 33*, 205–214.

Weinstock, Hillard, Dean, Deborah, & Bolan, Gail. (1994). Chlamydia trachomatis infections. *Infectious Disease Clinics of North America, 8*, 797–819.

Weisberg, D. Kelly. (1985). *Children of the night: A study of adolescent prostitution.* Lexington, MA: Lexington Books.

Weissman, Myrna, & Klerman, G. (1977). Sex differences and the epidemiology of depression. *Archives of General Psychiatry, 34*, 98–111.

Weller, Leonard, Weller, Aron, & Avinir, Ohala. (1995). Menstrual synchrony: Only in roommates who are close friends? *Physiology & Behavior, 58*, 883–889.

Wertz, R. W., & Wertz, D. C. (1977). *Lying-in: A history of childbirth in America.* New York: Free Press.

Westoff, Charles F., Blanc, A. K., & Nyblade, L. (1994). *Marriage and entry into parenthood.* Calverton, MD: Macro International.

Wharton, Chris, & Blackburn, Richard. (1988). Lower-dose pills. *Population Reports,* Series A, No. 7, 1–31.

Whatley, Mark A. (1993). For better or worse: The case of marital rape. *Violence and Victims, 8*, 29–39.

Wheeler, Garry D., et al. (1984). Reduced serum testosterone and prolactin levels in male distance runners. *Journal of the American Medical Association, 252*, 514–516.

Whipple, Beverly, Gerdes, Carolyn A., & Komisaruk, Barry R. (1996). Sexual response to self-stimulation in women with complete spinal cord injury. *Journal of Sex Research, 33*, 231–240.

Whipple, Beverly, Ogden, Gina, & Komisanak, Barry. (1992). Physiological correlates of imagery-induced orgasm in women. *Archives of Sexual Behavior, 21*, 121–133.

Whitam, Frederick L. (1983). Culturally invariable properties of male homosexuality: Tentative conclusions from cross-cultural research. *Archives of Sexual Behavior, 12*, 207–226.

Whitam, Frederick L. (1977). The homosexual role: A reconsideration. *Journal of Sex Research, 13*, 1–11.

Whitam, Frederick L., Daskalos, Chrisopher, Sobolewski, Curt G., & Padilla, Peter. (1998). The emergence of lesbian sexuality and identity cross-culturally: Brazil, Peru, the Philippines, and the United States. *Archives of Sexual Behavior, 27*, 31–56.

Whitam, Frederick L., Diamond, Milton, & Martin, James. (1993). Homosexual orientation in twins. *Archives of Sexual Behavior, 22*, 187–206.

Whitam, Frederick L., & Mathy, Robin M. (1991). Childhood cross-gender behavior of homosexual females in Brazil, Peru, the Philippines, and the United States. *Archives of Sexual Behavior, 20*, 151–170.

White, Gregory L., Fishbein, S., & Rutstein, J. (1981). Passionate love and the misattribution of arousal. *Journal of Personality and Social Psychology, 41*, 56–62.

White, Gregory L., & Knight, T. D. (1984). Misattribution of arousal and attraction: Effects of salience of explanations for arousal. *Journal of Experimental Social Psychology, 20*, 55–64.

White, Gregory L., & Mullen, Paul E. (1989). *Jealousy: Theory, research, and clinical strategies.* New York: Guilford.

Whitley, Bernard, Jr. (1993). Reliability and aspects of the construct validity of Sternberg's Triangular Love Scale. *Journal of Social and Personal Relationships, 10*, 475–480.

Whittington, William, Ison, Catherine, & Thompson, Sumner. (1996). Gonorrhea. In S. Morse et al. (Eds.), *Atlas of sexually transmitted diseases and AIDS* (pp. 99–118). London: Mosby-Wolfe.

Wickler, Wolfgang. (1973). *The sexual code.* New York: Anchor Books. (Original in German, 1969).

Wiederman, Michael W. (1993). Demographic and sexual characteristics of nonresponders to sexual experience items in a national survey. *Journal of Sex Research, 30*, 27–35.

Wiederman, Michael W., Weis, David L., & Allgeier, Elizabeth R. (1994). The effect of question preface on response rates to a telephone survey of sexual experience. *Archives of Sexual Behavior, 23*, 203–216.

Wielandt, Hanne, et al. (1989). Age of partners at first intercourse among Danish males and females. *Archives of Sexual Behavior, 18*, 449–454.

Wilcox, A. J., Weinberg, C. R., & Baird, D. D. (1995). Timing of sexual intercourse in relation to ovulation. *New England Journal of Medicine, 333*, 1517–1521.

Wilcox, Brian L., Robbenoit, J. K., & O'Keefe, J. E. (1998). Federal abortion policy and politics: 1973 to 1996. In L. J. Beckman & S. M. Harvey (Eds.), *The new civil war: The psychology, culture and politics of abortion* (pp. 3–24). Washington, D.C.: American Psychological Association.

Wilcox, Brian L., & Wyatt, J. (1997). Adolescent abstinence education programs: A meta-analysis. Presented at the annual meeting, Society for the Scientific Study of Sexuality, Arlington, VA.

Wilkinson, Ross. (1995). Changes in psychological health and the marital relationship through child bearing: Transition or process as stressor. *Australian Journal of Psychology, 47*, 86–92.

Williams, Linda M. (1994). Recall of childhood trauma: A prospective study of women's memories of child abuse. *Journal of Consulting and Clinical Psychology, 62*, 1167–1176.

Williams, Lindy, & Sobieszczyk, Teresa. (1997). Attitudes surrounding the continuation of female circumcision in the Sudan: Passing tradition to the next generation. *Journal of Marriage and the Family, 59*, 966–981.

Williams, Martin H. (1992). Exploitation and inference: Mapping the damage from therapist-patient sexual involvement. *American Psychologist, 47*, 412–421.

Wilson, E. O. (1975). *Sociobiology: The new synthesis.* Cambridge, MA: Harvard University Press.

Wilson, Glenn D. (1987). An ethological approach to sexual deviation. In G. D. Wilson (Ed.), *Variant sexuality: Research and theory.* Baltimore, MD: Johns Hopkins University Press.

Wilson, R. D., et al. (1991). Chorionic villus sampling: Analysis of fetal losses to delivery, placental pathology, and cervical microbiology. *Prenatal Diagnosis, 11*, 539–550.

Wilson, W. Cody. (1973). Pornography: The emergence of a social issue and the beginning of psychological study. *Journal of Social Issues, 29*, 7–17.

Wincze, John P., Albert, Alexa, & Bansal, Sudhir. (1993). Sexual arousal in diabetic females: Physiological and self-report measures. *Archives of Sexual Behavior, 22*, 587–602.

Wincze, John P., & Carey, Michael P. (1991). *Sexual dysfunction: A guide for assessment and treatment.* New York: Guilford.

Winick, Charles, & Kinsie, Paul M. (1971). *The lively commerce.* New York: Quadrangle.

Winkelstein, Warren, et al. (1987a). The San Francisco Men's Health Study: III. Reduction in human immunodeficiency virus transmission among homosexual/bisexual men, 1982–86. *American Journal of Public Health, 76*, 685–688.

Winkelstein, Warren, et al. (1987b). Sexual practices and risk of infection by the human immunodeficiency virus. The San Francisco Men's Health Study. *Journal of the American Medical Association, 257*, 321–325.

Winkelstein, Warren, et al. (1987c). Selected sexual practices of San Francisco heterosexual men and risk of infection by the human immunodeficiency virus. *Journal of the American Medical Association, 257*, 1470.

Winn, Rhonda L., & Newton, Niles. (1982). Sexuality in aging: A study of 106 cultures. *Archives of Sexual Behavior, 11*, 283–298.

Winter, Jeremy S. D., & Couch, Robert M. (1995). Sexual differentiation. In P. Felig, J. D. Baxter, & L. A. Frohman (Eds.), *Endocrinology and metabolism* (3d ed., pp. 1053–1104). New York: McGraw-Hill.

Wise, Phyllis M., Krajnak, Kristine M., & Kashon, Michael L. (1996). Menopause: The aging of multiple pacemakers. *Science, 273*, 67–70.

Wiswell, Thomas E., Enzenauer, R. W., Cornish, J. D., & Hawkins, C. T. (1987). Declining frequency of circumcision: Implications for changes in the absolute incidence and male to female sex ratio of urinary tract infections in early infancy. *Pediatrics, 79*, 338–342.

Wolchik, Sharlene A., Spencer, S. L., & Lisi, I. S. (1983). Volunteer bias in research employing vaginal measures of sexual arousal. *Archives of Sexual Behavior, 12*, 399–408.

Wolf, Timothy J. (1985). Marriages of bisexual men. In F. Klein & T. J. Wolf (Eds.), *Bisexualities: Theory and research.* New York: Haworth.

Wong, Joseph K. et al. (1997). Recovery of replication-competent HIV despite prolonged suppression of plasma viremia. *Science, 278*, 1291–1295.

Wood, Julia T. (1994). *Gendered lives: Communication, gender, and culture.* Belmont, CA: Wadsworth.

Woods, James D. (1993). *The corporate closet.* New York: Free Press.

Woollett, Anne, Dosaujh, Neelam, et al. (1995). The ideas and experiences of pregnancy and childbirth of Asian and non-Asian women in East London. *British Journal of Medical Psychology, 68*, 65–84.

Worthington, Everett, Martin, Glenn, Shumate, Michael, & Carpenter, Johnice. (1983). The effect of brief Lamaze training and social encouragement on pain endurance in a cold pressor task. *Journal of Applied Social Psychology, 13*, 223–233.

Wyatt, Gail E. (1992). The sociocultural context of African American and White American women's rape. *Journal of Social Issues, 48*(1), 77–92.

Wyatt, Gail E. (1991). Child sexual abuse and its effects on sexual functioning. *Annual Review of Sex Research, 2*, 249–266.

Wyatt, Gail E., Peters, S. D., & Guthrie, D. (1988). Kinsey revisited, Part I: Comparisons of the sexual socialization and sexual behavior of white women over 33 years. *Archives of Sexual Behavior, 17*, 201–240.

Wylie, Kevan R. (1997). Treatment outcome of brief couple therapy in psychogenic male erectile disorder. *Archives of Sexual Behavior, 26*, 527–546.

Wysocki, Charles J., & Lepri, John J. (1991). Consequences of removing the vomeronasal organ. *Journal of Steroid Biochemistry and Molecular Biology, 39*, 661–669.

Yalom, Irvin D. (1960). Aggression and forbiddenness in voyeurism. *Archives of General Psychiatry, 3*, 317.

Yarbro, E. Scott, & Howards, Stuart S. (1987). Vasovasostomy. *Urologic Clinics of North America, 14*, 515–526.

Young, Ian. (1996). Education for sexuality—the role of the school. *Journal of Biological Education, 30*, 250–255.

Zabin, L. S., Hirsch, M. B., Smith, E. A., & Hardy, J. B. (1984). Adolescent sexual attitudes and behaviors: Are they consistent? *Family Planning Perspectives, 4*, 181–185.

Zaslow, Martha J., et al. (1985). Depressed mood in new fathers: Association with parent-infant interaction. *Genetic, Social, and General Psychology Monographs, 111*(2), 133–150.

Zaviačič, M. (1994). Sexual asphyxiophilia (Koczwarism) in women and the biological phenomenon of female ejaculation. *Medical Hypotheses, 42*, 318–322.

Zax, M., Sameroff, A., & Farnum, J. (1975). Childbirth education, maternal attitude and delivery. *American Journal of Obstetrics and Gynecology, 123*, 185–190.

Zelnik, Melvin, & Kantner, John F. (1980). Sexual activity, contraceptive use and pregnancy among metropolitan-area teenagers: 1971–1979. *Family Planning Perspectives, 12*(5), 230–237.

Zelnik, Melvin, & Kantner, John F. (1977). Sexual and contraceptive experience of young unmarried women in the United States, 1976 and 1971. *Family Planning Perspectives, 9*(2), 55–71.

Zhang, Heping, & Bracken, M. B. (1995). Tree-based risk factor analysis of preterm delivery and small-for-gestational-age birth. *American Journal of Epidemiology, 141*, 70–78.

Zilbergeld, Bernie. (1992). *The new male sexuality.* New York: Bantam Books.

Zilbergeld, Bernie. (1978). *Male sexuality.* Boston: Little, Brown.

Zilbergeld, Bernie, & Ellison, Carol Rinklieb. (1980). Desire discrepancies and arousal problems in sex therapy. In S. R. Leiblum & L. A. Pervin (Eds.), *Principles and practice of sex therapy* (pp. 65–104). New York: Guilford.

Zilbergeld, Bernie, & Evans, Michael. (1980, Aug.). The inadequacy of Masters and Johnson. *Psychology Today, 14*(3), 28–43.

Zillmann, Dolf, Schweitzer, Karla J., & Mundorf, Norbert. (1994). Menstrual cycle variations of women's interest in erotica. *Archives of Sexual Behavior, 23*, 579–598.

Zimmer, D. (1983). Interaction patterns and communication skills in sexually distressed and normal couples: Two experimental studies. *Journal of Sex and Marital Therapy, 9*, 251–265.

Zimmerman, Rick S., & Langer, Tilly M. (1995). Improving estimates of prevalence rates of sensitive behaviors: The randomized lists technique. *Journal of Sex Research, 32*, 107–118.

Zimmerman-Tansella, Christa, Bertagni, Paolo, Siani, Roberta, & Micciolo, Rocco. (1994). Marital relationships and somatic and psychological symptoms in pregnancy. *Social Science and Medicine, 38*, 559–564.

Zoldbrod, Aline P. (1993). *Men, women, and infertility: Intervention and treatment strategies.* New York: Lexington Books.

Zoucha-Jensen, J. M., & Coyne, A. (1993). The effects of resistance strategies on rape. *American Journal of Public Health, 83*, 1633–1634.

Zuger, Abigail. (1987, June). AIDS on the wards: A residency in medical ethics. *Hastings Center Report,* June 16–20.

Zumpe, Doris, & Michael, R. P. (1968). The clutching reaction and orgasm in the female rhesus monkey (Macaca mulatta). *Journal of Endocrinology, 40*, 117–123.

Zumwalt, Rosemary. (1976). Plain and fancy: A content analysis of children's jokes dealing with adult sexuality. *Western Folklore, 35*, 258–267.

Zussman, Leon, Zussman, S., Sunley, R., & Bjornson, E. (1981). Sexual response after hysterectomy-oophorectomy. *American Journal of Obstetrics and Gynecology, 140*, 725–729.

Glossary

Abortion The ending of a pregnancy and the expulsion of the contents of the uterus; may be spontaneous or induced by human intervention.

Abstinence (sexual) Not engaging in sexual activity.

Acculturation The process of incorporating the beliefs and customs of a new culture.

Acquired Immune Deficiency Syndrome (AIDS) A sexually transmitted disease that destroys the body's natural immunity to infection.

Adrenal gland An endocrine gland located just above the kidney; in the female, it is the major producer of androgens.

Adrenogenital syndrome See *congenital adrenal hyperplasia*.

Adultery Sexual intercourse between a married person and someone other than her or his spouse.

Afterbirth The placenta and amniotic sac, which come out after the baby during childbirth.

AIDS See *Acquired Immune Deficiency Syndrome*.

Ambisexual See *bisexual*.

Amenorrhea The absence of menstruation.

Amniocentesis A test done to determine whether a fetus has birth defects; done by removing amniotic fluid from the pregnant woman's uterus.

Amniotic fluid The watery fluid surrounding a developing fetus in the uterus.

Amyl nitrate A drug, usually inhaled, that some people use to prolong or intensify orgasm.

Anal intercourse Sexual behavior in which one person's penis is inserted into another's anus.

Analogous organs Organs in the male and female that have similar functions.

Anaphrodisiac A substance that decreases sexual desire.

Androgens "Male" sex hormones, produced in the testes; an example is testosterone. In females, the adrenal glands produce androgens.

Androgyny Having both feminine and masculine characteristics.

Anilingus Mouth-anus stimulation.

Anorgasmia The inability of a woman to orgasm; a sexual disorder.

Antigay prejudice Negative attitudes and behaviors toward gays and lesbians.

Anus The opening of the rectum, located between the buttocks.

Aphrodisiac A substance that increases sexual desire.

Areola The dark circular area of skin surrounding the nipple of the breast.

Artificial insemination Artificially putting semen into the vagina or uterus for the purpose of inducing pregnancy.

Asexual Without sexual desires.

Asphyxiophilia A desire for a state of oxygen deprivation to enhance sexual arousal or orgasm.

Asymptomatic Having no symptoms.

Attachment A psychological bond that forms between an infant and the mother, father, or other care giver.

Autoeroticism Sexual self-stimulation; masturbation is one example.

Axillary hair Underarm hair.

Barr body A small, black dot appearing in the cells of genetic females; it represents an inactivated X chromosome.

Bartholin glands Two tiny glands located on either side of the vaginal entrance.

Basal body temperature method A method of rhythm birth control in which the woman determines when she ovulates by keeping track of her temperature.

Behavior therapy A system of therapy based on learning theory and focusing on the problem behavior, not the unconscious.

Bestiality Sexual contact with an animal; also called *zoophilia*.

Bisexual A person whose sexual orientation is toward men and women.

Blastocyst A small mass of cells that results after several days of cell division by the fertilized egg.

Bondage A type of sadomasochism in which sexual pleasure is derived from feeling restricted, usually by being tied up with ropes.

Braxton-Hicks contractions Contractions of the uterus during pregnancy that are not part of actual labor.

Breech presentation Birth of a baby with buttocks or feet first.

Brothel   A house of prostitution.

Buccal smear   A test of genetic gender.

Bulbourethral glands   See *Cowper's glands*.

Butch   A very masculine lesbian; may also refer to a very masculine gay man.

Candida albicans   A yeast or fungus in the vagina; if its growth gets out of control, it causes vaginitis, or irritation of the vagina, with an accompanying discharge.

Carpopedal spasm   A spastic contraction of the hands or feet which may occur during orgasm.

Castration   The removal (usually by means of surgery) of the gonads (the testes in men or the ovaries in women).

Celibate   Unmarried; also used to refer to someone who abstains from sexual activity.

Cervical cap   A birth control device similar to the diaphragm.

Cervix   The lower part of the uterus; the part next to the vagina.

Cesarean section   Surgical delivery of a baby through an incision in the abdominal and uterine walls.

Chancre   A painless open sore with a hard ridge around it; it is an early symptom of syphilis.

Chancroid   A sexually transmitted disease.

Chastity   Sexual abstinence.

Chlamydia   An organism causing a sexually transmitted disease.

Chorionic villus sampling (CVS)   A technique for early detection of birth defects in the fetus.

Cilia   Tiny hairlike structure lining the vas deferens and the fallopian tubes.

Circumcision   Surgical removal of the foreskin of the penis.

Classical conditioning   A learning process whereby a previously neutral stimulus (CS) is repeatedly paired with an unconditioned stimulus (US) that reflexively elicits an unconditioned response (UR). Eventually the CS will evoke the response.

Climacteric   See *menopause*.

Climax   An orgasm.

Clitoridectomy   Removal of the clitoris.

Clitoris   A small, highly sensitive sexual organ in the female, located in front of the vaginal entrance.

Cloning   Producing genetically identical individuals from a single parent.

Cognitive   Relating to mental activity, such as thought, perception, understanding.

Cohabitation   Unmarried persons living together (with sexual relations assumed).

Coitus   Sexual intercourse, insertion of the penis into the vagina.

Coitus interruptus   See *withdrawal*.

Coitus reservatus   Sexual intercourse in which the man intentionally refrains from ejaculating.

Colostrum   A watery substance that is secreted from the breast at the end of pregnancy and during the first few days after delivery.

Coming out   The process of acknowledging to oneself, and then to others, that one is gay or lesbian.

Companionate love   A feeling of deep attachment and commitment to a person with whom one has an intimate relationship.

Conceptus   The product of conception; sometimes used to refer to the embryo or fetus.

Condom   A sheath placed over the penis; used in the prevention of pregnancy and of sexually transmitted diseases.

Congenital adrenal hyperplasia (CAH)   A condition in which a genetic female has an abnormally functioning adrenal gland that produces an excess of androgens so that she is born with genitals that look like a male's. Also called adrenogenital syndrome.

Contraceptive technique   A method of preventing conception.

Coprophilia   A sexual variation in which arousal is associated with defecation or feces.

Copulation   Sexual intercourse.

Corona   The rim of tissue between the glans and the shaft of the penis.

Corpora cavernosa   Two cylindrical masses of erectile tissue running the length of the penis; also present in the clitoris.

Corpus luteum   The mass of cells remaining after a follicle has released an egg; it secretes progesterone.

Corpus spongiosum   A cylinder of erectile tissue running the length of the penis.

Correlation   A number that measures the relationship between two variables.

Couvade   The experiencing of the symptoms of pregnancy and childbirth by a male.

Cowper's glands   A pair of glands that secrete substances into the male's urethra.

Crabs   See *Pediculosis pubis*.

Cramps   Painful menstruation, or dysmenorrhea.

Cremaster muscle   A muscle in the scrotum.

Cryptorchidism   Undescended testes.

Cul-de-sac   The end of the vagina, past the cervix.

Culdoscopy   A female sterilization procedure.

Cunnilingus   Mouth stimulation of the female genitals.

Cystitis   Inflammation of the urinary bladder; the major symptom is a burning sensation while urinating.

Dartos muscle   A muscle in the scrotum.

Decriminalization   Removing criminal penalties for an activity that was previously defined as illegal.

Defloration   The rupture of a virgin's hymen, through intercourse or other means.

Depo-Provera   A drug containing progestin; used as a form of birth control in women, as well as a treatment for male sex offenders.

Deprivation homosexuality   Homosexual activity that occurs in certain situations, such as prisons, when people are deprived of their regular heterosexual activity.

Detumescence   The return of an erect penis to the flaccid (unaroused) state.

Diaphragm   A cap-shaped rubber contraceptive device that fits inside a woman's vagina over the cervix.

Diethylstilbestrol (DES)   A potent estrogen drug that was used for a "morning-after" pill.

Dilate   To enlarge; used to refer to the enlargement of the cervical opening during labor.

Dildo   An artificial penis.

Don Juanism   Hypersexuality in a male.

Double standard   A standard in which premarital intercourse is considered acceptable for males but not for females.

Douche   To flush out the inside of the vagina with a liquid.

Drag queen   A male homosexual who dresses in women's clothing.

Ductus deferens   See *vas deferens*.

Dysmenorrhea   Painful menstruation.

Dyspareunia   Painful intercourse.

Ectopic pregnancy   A pregnancy in which the fertilized egg implants somewhere other than the uterus.

Edema   An excessive fluid retention and swelling.

Editing   In communication, censoring or not saying things that would be deliberately hurtful to your partner or that are irrelevant.

Effacement   Thinning out of the cervix during labor.

Ejaculation   The expulsion of semen from the penis, usually during orgasm.

Electra complex   In Freudian theory, a little girl's sexual desires for her father; the female analogue of the Oedipus complex.

Embryo   In humans, the term used to refer to the unborn young from the first to the eighth week after conception.

Embryo transfer   A technique in which a fertilized, developing egg (embryo) is transferred from the uterus of one woman to the uterus of another woman.

Endocrine gland   A gland that secretes substances (hormones) directly into the bloodstream.

Endometriosis   A condition in which the endometrium grows in some place other than the uterus, such as the fallopian tubes.

Endometrium   The inner lining of the uterus.

Epididymis   Highly coiled tubules located on the edge of the testes; the site of sperm maturation.

Episiotomy   An incision that is sometimes made at the vaginal entrance during childbirth.

Erectile disorder   The inability to get or maintain an erection.

Erection   An enlargement and hardening of the penis which occurs during sexual arousal.

Erogenous zones   Areas of the body that are particularly sensitive to sexual stimulation.

Erotica   Sexually arousing material that is not degrading to women, men, or children.

Erotophilia   Feeling comfortable with sex, the opposite of erotophobia.

Erotophobia   Feeling guilty and fearful about sex.

Estrogens   A small group of "female" sex hormones; also produced in smaller quantities in males.

Estrus   The period of ovulation and sexual activity in nonhuman female mammals.

Ethnocentrism   The tendency to regard one's own ethnic group and culture as superior to others and to believe that its customs and way of life are the standards by which other cultures should be judged.

Eunuch   A castrated male.

Exhibitionist   A person who derives sexual gratification from exposing his or her genitals to others.

Extramarital sex   Sexual activity by a married person with someone other than her or his spouse.

Fallopian tube   The tube extending from the uterus to the ovary; also called the oviduct.

Fellatio   Mouth stimulation of the male genitals.

Female circumcision   Amputation of the clitoris.

Female orgasmic disorder   A sexual disorder in which the woman is unable to have an orgasm.

Female sexual arousal disorder   A lack of responding to sexual stimulation by a woman.

Femme   A feminine lesbian.

Fertilization   The union of sperm and egg, resulting in conception.

Fetal alcohol syndrome (FAS)   Serious growth deficiency and malformations in a child born to a woman who abuses alcohol.

Fetishism   A sexual variation in which a person is sexually aroused by and attaches great erotic significance to an inanimate object.

Fetus   In humans, the term used to refer to the unborn young from the third month after conception until birth.

Fimbriae   Fingerlike projections at the end of the fallopian tube near the ovary.

Fitness   In evolutionary theory, an individual's reproductive success.

Flaccid   Not erect.

Follicle   The capsule of cells surrounding an egg in the ovary.

Follicle-stimulating hormone (FSH)   A hormone secreted by the pituitary; it stimulates follicle development in females and sperm production in males.

Follicular phase   The first phase of the menstrual cycle, beginning just after menstruation.

Foreskin   The sheath of skin covering the tip of the penis in an uncircumcised male.

Fornication   Sexual intercourse between two unmarried people.

Fourchette   The place where the inner lips come together behind the vaginal opening.

Frenulum   A highly sensitive area of skin on the underside of the penis next to the glans.

Frigidity   Lack of sexual response in a woman.

Gamete intra-fallopian transfer (GIFT)   A procedure in which sperm and eggs are collected and then inserted together into the fallopian tube.

Gametes   Sperm or eggs.

Gay   Homosexual; particularly a male homosexual.

Gender   The state of being male or female.

Gender dysphoria   See *transsexual.*

Gender identity   The psychological sense of one's own maleness or femaleness.

Gender role   A cluster of socially defined expectations that people of one gender are expected to fulfill.

Genitals   The sexual or reproductive organs.

Genital warts   A sexually transmitted disease causing warts on the genitals, caused by human papillomavirus (HPV).

Gerontophilia   Sexual attraction to the elderly.

Gestation   The period of pregnancy; the time from conception until birth.

GIFT   See *Gamete intra-fallopian transfer.*

Gigolo   A male who sells his sexual services to women.

Glans   The tip of the penis or clitoris.

Gonadotropin-releasing hormone (Gn-RH)   A hormone secreted by the hypothalamus that regulates the pituitary's secretion of hormones.

Gonadotropins   Pituitary hormones (FSH, LH) that stimulate the activity of the gonads.

Gonads   The ovaries or testes.

Gonorrhea   A common sexually transmitted disease.

Gossypol   A substance used as a male contraceptive in China.

Gräfenberg spot (G spot)   A hypothesized small gland on the front wall of the vagina, emptying into the urethra, which may be responsible for female ejaculation.

Granuloma inguinale   A rare sexually transmitted disease.

Gynecomastia   Temporary enlargement of a male's breasts during puberty.

Hedonism A moral system based on maximizing pleasure and avoiding pain.

Hegar's sign A sign of pregnancy based on a test done by a physician, in which a softening of the uterus is detected.

Hepatitis B A disease that can be transmitted by anal intercourse or oral-anal sex.

Hermaphrodite A person with both male and female sex glands, that is, both ovaries and testicular tissue (see also *pseudohermaphrodite*).

Herpes genitalis A disease characterized by painful bumps on the genitals.

Heterosexual A person who is sexually attracted to members of the other gender.

HIV Human Immune Deficiency Virus, the virus that causes AIDS.

Homologous organs Organs in the male and female that develop from the same embryonic tissue.

Homophily The tendency to have contact with people equal in social status.

Homophobia Strong, irrational fear of homosexuality.

Homosexual A person whose sexual orientation is toward members of her or his own gender.

Homosocial A pattern of social grouping in which boys associate with other boys and girls associate with other girls.

Hormones Chemical substances secreted by the endocrine glands into the bloodstream.

Human chorionic gonadotropin (HCG) A hormone produced by the placenta; HCG is what is detected in most pregnancy tests.

Human papillomavirus (HPV) The organism that causes genital warts.

Hustlers Male prostitutes who sell their services to other males.

Hyaluronidase An enzyme secreted by the sperm that allows one to penetrate the egg.

Hymen A membrane that partially covers the vaginal opening.

Hypersexuality An excessive, insatiable sex drive.

Hypoactive sexual desire A sexual disorder in which there is a lack of interest in sexual activity.

Hypothalamus A part of the brain which is important in regulating certain body functions including sex hormone production.

Hysterectomy Surgical removal of the uterus.

Hysterotomy A method of abortion sometimes used during the second trimester.

Id In Freudian theory, the part of the personality containing the libido or sex drive.

Imperforate hymen A condition where the hymen is unusually thick and covers the vaginal entrance completely.

Implantation The burrowing of the fertilized egg into the lining of the uterus.

Impotence See *erectile disorder*.

Impregnate To make pregnant.

Incest Sexual activity between close relatives, such as a brother and sister.

Incest taboo A regulation prohibiting sexual activity between blood relatives.

Incidence The percentage of people giving a particular response.

Infertility A woman's inability to conceive and give birth to a child, or a man's inability to impregnate a woman.

Infibulation A ritual practice of cutting off the inner lips of the vagina and sewing together the outer lips, making intercourse impossible.

Informed consent An ethical principle in research, in which participants have a right to be informed, before participating, of what they will be asked to do.

Inguinal canal In the male, the passageway from the abdomen to the scrotum through which the testes usually descend shortly before birth.

Inhibin Substance produced by the testes and ovaries, which regulates FSH levels.

Inner lips Thin folds of skin on either side of the vaginal entrance.

Intercourse (sexual) Sexual activity in which the penis is inserted into the vagina; coitus (see also *anal intercourse*).

Interfemoral intercourse Sexual activity in which the penis moves between the partner's thighs.

Interstitial cells Cells in the testes which manufacture male sex hormones; also called *Leydig cells*.

Intimacy The level of commitment and positive affective, cognitive, and physical closeness one experiences with a partner.

Intrauterine device (IUD) A plastic or metal device that is inserted into the uterus for contraceptive purposes.

Introitus Entrance to the vagina.

Intromission Insertion of the penis into the vagina.

In vitro fertilization (IVF) A technique in which sperm and egg are mixed outside the body in a laboratory dish, so that conception can occur.

John Slang term for a prostitute's customer.

Kaposi's sarcoma A rare form of skin cancer to which AIDS patients are susceptible.

Kegel exercises Exercises to strengthen the muscle surrounding the genitals.

Kiddie porn Pictures or films of sexual acts with children.

Labia majora See *outer lips*.

Labia minora See *inner lips*.

Labor The series of stages involved in giving birth.

Lactation Secretion of milk from the female's breasts.

Lamaze method A method of "prepared" childbirth involving relaxation and controlled breathing.

Laparoscopy A method of female sterilization.

Latinos People of Latin American origin.

Lesbian A female homosexual.

Leveling In communication, telling your partner what you are feeling by stating your thoughts clearly, simply, and honestly.

Leydig cells See *interstitial cells*.

Libido The sex drive.

Limbic system A set of structures in the interior of the brain, including the amygdala, hippocampus, and fornix; believed to be important for sexual behavior in both animals and humans.

Lochia A discharge from the uterus and vagina that occurs during the first few weeks after childbirth.

Low sexual desire See *hypoactive sexual desire*.

Lumpectomy A surgical treatment for breast cancer in which only the lump and a small bit of surrounding tissue are removed.

Luteal phase The third phase of the menstrual cycle, following ovulation.

Luteinizing hormone (LH) A hormone secreted by the pituitary. In females, it causes ovulation.

Lymphogranuloma venereum (LGV) A virus-caused disease affecting the lymph glands in the genital region.

Madam A woman who manages a brothel.

Male orgasmic disorder See *retarded ejaculation*.

Mammary gland The milk-producing part of the breast.

Mammography X-rays for diagnosing breast cancer.

Masochism A sexual variation in which the person derives sexual pleasure from experiencing physical or mental pain.

Mastectomy Surgical removal of the breast.

Masturbation Self-stimulation of the genitals with the hand or some object.

Matching phenomenon The tendency for men and women to choose as partners people who match them, that is, who are similar in attitudes, intelligence, and attractiveness.

Mean The average of respondents' scores.

Menage à trois A sexual relationship involving three people.

Menarche The first menstruation.

Menopause The gradual cessation of menstruation in a woman, generally at around age 50.

Menses The menstrual flow.

Menstruation A bloody discharge of the lining of the uterus, generally occurring about once a month in women.

Mere-exposure effect The tendency to like someone more if we have been exposed to them repeatedly.

Midwife A person (often a nurse) trained as a birth attendant.

Mind reading Making assumptions about what your partner thinks or feels.

Mini-pill A birth control pill containing a low dose of progesterone and no estrogen.

Miscarriage A pregnancy that terminates on its own; spontaneous abortion.

Miscegenation Sex between two people of different races.

Mittelschmerz Abdominal cramps at the time of ovulation.

Monilia A yeast infection of the vagina.

Monogamy The pairing of one person with just one other person in a long-term relationship in which neither engages in sexual activity with anyone else.

Mons pubis The fatty pad of tissue under the pubic hair; also called the *mons* or *mons veneris*.

Morning-after pill A pill which can be used in emergency situations for preventing pregnancy after intercourse has occurred.

Mucosa Mucous membrane.

Müllerian ducts In the embryo, a pair of ducts that eventually become part of the female reproductive system.

Multiparous A term used to refer to a woman who has had more than one baby.

Myotonia Muscle contraction.

Necrophilia A sexual variation involving having sexual contact with a dead person.

Nipples The pigmented tip of the breast, through which milk goes when a woman is breast-feeding.

Nocturnal emission Involuntary orgasm and ejaculation while asleep.

Nongonococcal urethritis An inflammation of the male's urethra not caused by gonorrhea.

Norplant An implanted progestin-only contraceptive for women.

Nulliparous A term used to refer to a woman who has never given birth to a baby.

Nymphomania An excessive, insatiable sex drive in a woman.

Obscenity Something that is offensive according to accepted standards of decency, the legal term for pornography.

Oedipus complex In Freudian theory, the sexual attraction of a little boy to his mother.

Onanism Withdrawal of the penis from the vagina before ejaculation; sometimes also used to refer to masturbation.

Oophorectomy Surgical removal of the ovaries.

Operant conditioning The process of changing the frequency of a behavior (the operant) by following it with reinforcement or punishment.

Operational definition Defining a concept or term by how it is measured.

Oral-genital sex Sexual activity in which the mouth is used to stimulate the genitals.

Orchidectomy Surgical removal of the testes.

Orgasm An intense sensation that occurs at the peak of sexual arousal and is followed by release of sexual tension.

Orgasmic platform The thickening of the walls of the outer third of the vagina that occurs during sexual arousal.

Outer lips The fatty pads of tissue lying on either side of the vaginal opening and inner lips.

Ovaries The paired sex glands in the female which produce ova (eggs) and sex hormones.

Oviduct Fallopian tube.

Ovulation Release of an egg by the ovaries.

Ovum Egg.

Oxytocin A hormone secreted by the pituitary which stimulates the contractions of the uterus during childbirth; also involved in breast-feeding.

Pander To produce a prostitute for a client; sometimes used to mean any catering to another's sexual desires.

Pap test The test for cervical cancer.

Paraphilia A sexual variation; erotic attraction to unusual things or behaviors.

Participant-observer study A research method in which the scientist becomes part of the community to be studied and makes observations from inside the community.

Parturition Childbirth.

Passionate love A state of intense longing for union with another person and of intense physiological arousal.

Pederasty Sexual relations between a man and a boy; sometimes also used to mean anal intercourse.

Pediculosis pubis Lice attaching themselves to the roots of the pubic hair; crabs.

Pedophilia A sexual variation in which an adult is sexually attracted to children; child molesting.

Pelvic inflammatory disease Infection of the pelvic organs such as the fallopian tubes.

Penile strain gauge A device used to measure physiological sexual arousal in the male.

Penis A male sexual organ, which functions both in sexual activity and urination.

Perineum The area between the vaginal opening and the anus.

Period The menstrual period.

Perversion A sexual deviation.

Phallus Penis.

Pheromones Chemical substances secreted outside the body that are important in communication between animals and may serve as sex attractants.

Phimosis A condition in which the foreskin of the penis is so tight that it cannot be pulled back.

Photoplethysmograph A device used to measure physiological sexual arousal in the female.

Pimp A prostitute's protector, companion, and master.

Pituitary gland A gland located on the lower surface of the brain; it secretes several hormones important to sexual and reproductive functioning.

Placenta An organ formed on the wall of the uterus through which the fetus receives oxygen and nutrients and gets rid of waste products.

Plateau phase Masters and Johnson's term for the second phase of sexual response, occurring just before orgasm.

Polygamy Marriage in which a man has more than one wife or a woman has more than one husband.

Population A group of people a researcher wants to study and make conclusions about.

Pornography Sexually arousing art, literature, or films.

Postpartum The period of time following childbirth.

Postpartum depression Mild to moderate depression in women following the birth of a baby.

Post-traumatic stress disorder (PTSD) Long-term psychological distress suffered by someone who has experienced a terrifying event.

Preeclampsia A serious disease of pregnancy, marked by high blood pressure, severe edema, and proteinuria.

Premarital intercourse Intercourse before marriage.

Premature ejaculation A sexual disorder in which the male ejaculates too soon.

Premenstrual syndrome (PMS) A combination of severe physical and psychological symptoms (such as depression and irritability) occurring in some women just before menstruation.

Prenatal Before birth.

Prepuce Foreskin.

Preterm birth An infant born weighing only 2500 grams (5½ pounds) or less.

Priapism A rare condition in which erections are long-lasting and painful.

Primipara A woman having her first baby.

Progesterone A female sex hormone produced by the corpus luteum in the ovary.

Prolactin A hormone secreted by the pituitary; it is involved in lactation.

Promiscuous A term used to refer to someone who engages in sexual activity with many different people.

Prophylactic A drug or device used to prevent disease, often specifically sexually transmitted disease; often used to mean "condom."

Prostaglandins Chemicals that stimulate contractions of the muscles of the uterus; the likely cause of painful menstruation.

Prostate The gland in the male, located below the bladder, that secretes some of the fluid in semen.

Prostatitis An infection or inflammation of the prostate gland.

Prostitution Indiscriminate sexual activity in return for money or drugs.

Pseudocyesis False pregnancy, in which the woman displays the signs of pregnancy but is not pregnant.

Pseudohermaphrodite An individual who has a mixture of male and female reproductive structures, so that it is not clear whether the individual is a male or a female.

Psychoanalytic theory A psychological theory originated by Freud; its basic assumption is that part of the human psyche is unconscious.

Puberty The period of time during which the body matures from that of a child to that of an adult capable of reproducing.

Pubic hair Hair on the lower abdomen and genital area, appearing at puberty.

Pubic lice See *Pediculosis pubis*.

Pubococcygeal muscle A muscle around the vaginal entrance.

Pudendum The external genitals of the female.

Purposeful distortion Purposely giving false information on a survey.

Radical mastectomy A surgical treatment for breast cancer in which the entire breast, as well as underlying muscle and lymph nodes, are removed.

Random sampling An excellent method of sampling in research, in which each member of the population has an equal chance of being included in the sample.

Rape Nonconsenting oral, anal, or vaginal penetration obtained by force, by threat of bodily harm, or when the victim is incapable of giving consent.

Refractory period The period following orgasm during which the person cannot be sexually aroused.

Resolution phase Masters and Johnson's term for the last phase of sexual response, in which the body returns to the unaroused state.

Retarded ejaculation A sexual disorder in which the male cannot have an orgasm even though he is highly aroused. Also called male orgasmic disorder.

Retrograde ejaculation A condition in which orgasm in the male is not accompanied by an external ejaculation; instead, the ejaculate goes into the urinary bladder.

Rhythm method A method of birth control that involves abstaining from sexual intercourse during the fertile days of the woman's menstrual cycle.

RU-486 A new pill that produces a very early abortion.

Sadism A sexual variation in which the person derives sexual pleasure from inflicting pain on someone else.

Salpingectomy See *tubal ligation*.

Salpingitis Infection of the fallopian tubes.

Sample A part of a population.

Satyriasis An excessive, insatiable sex drive in a male.

Schema A general knowledge framework a person has about a given topic, e.g., a gender schema.

Scoptophilia A sexual variation in which the person becomes sexually aroused from viewing others' sexual acts.

Scripts What we have learned to be appropriate sequences of behavior.

Scrotum The pouch of skin that contains the testes.

Secondary sex characteristics The physical characteristics, other than the sex organs, that distinguish the male from the female; examples are the woman's breasts and the man's beard.

Self-disclosure Telling personal things about yourself.

Semen The fluid that is ejaculated from the penis during orgasm; it contains sperm.

Seminal vesicles The two organs lying on either side of the prostate, which secrete much of the fluid in semen.

Seminiferous tubules Highly coiled tubules in the testes that manufacture sperm.

Sensate focus exercises Exercises prescribed by sex therapists to increase sexual response.

Sex-change operation The surgery done on transsexuals to change their anatomy to the other gender.

Sex-determining Region, Y chromosome (SRY)   A region on the Y chromosome that causes testes to differentiate prenatally. Also called *testis-determining factor (TDF)*.

Sex flush   A rashlike condition on the skin that occurs during sexual arousal.

Sexual disorder   A problem with sexual responding that causes a person mental distress; examples are erectile disorder in men and anorgasmia in women; also called *sexual dysfunction*.

Sexual dissatisfaction   Masters and Johnson's term for homosexuals who are unhappy with their sexual orientation and seek therapy to become heterosexual.

Sexual harassment   Unwanted imposition of sexual requirements in the context of a relationship of unequal power, such as an employer and an employee.

Sexual identity   A person's sense of his or her own sexual orientation, whether heterosexual, homosexual, or bisexual.

Sexual orientation   A person's erotic and emotional orientation toward members of his or her own gender or members of the other gender.

Sexual selection   An evolutionary theory proposed to explain gender differences in processes of mate selection.

Sexuality education   The lifelong process of acquiring information about sexual behavior and forming attitudes, beliefs, and values about identity, relationships, and intimacy.

Sixty-nining   Simultaneous mouth-genital stimulation.

Skene's glands   Glands opening into the urethra.

Smegma   A cheesy substance formed under the foreskin of the penis.

Sociobiology   A theory that applies evolutionary biology to understanding the social behavior of animals, including humans.

Sodomy   Originally "crimes against nature"; in contemporary laws, oral and anal intercourse.

Spectatoring   Acting as an observer or judge of one's own sexual performance.

Sperm   The mature male reproductive cell, capable of fertilizing an egg.

Spermatogenesis   The production of sperm.

Sperm bank   A place that stores frozen sperm for later use.

Spermicide   A substance that kills sperm.

Spirochete   A spiral-shaped bacterium; one kind causes syphilis.

Squeeze technique   A form of therapy for premature ejaculation.

Statutory rape   Sexual relations with a person who is below the legal "age of consent."

Stereotype   A generalization about a group of people (e.g., men) that distinguishes them from others (e.g., women).

Sterile   Incapable of reproducing.

Sterilization technique   A procedure by which an individual is made incapable of reproducing.

Steroids   A group of chemical substances including the sex hormones estrogen, progesterone, and testosterone.

Straight   Heterosexual.

Structural functionalism   A sociological theory that views society as an interrelated set of structures that function together to maintain that society.

Superego   According to Freud, the part of the personality containing the conscience.

Surrogate   A member of a sex therapy team who serves as a sexual partner for the client while in therapy.

Surrogate mother   A woman who, through artificial insemination or in vitro fertilization, gestates a fetus for someone else.

Swinging   A form of extramarital sex in which married couples exchange partners.

Syphilis   A sexually transmitted disease.

Tearoom trade   Impersonal sex in public places such as restrooms.

Teratogen   A substance that produces defects in the fetus.

Test-retest reliability   A method for determining whether self-reports are reliable or accurate; people are tested and then tested a second time some time later to determine whether their answers are the same both times.

Testes   The sex glands of the male, located in the scrotum; they manufacture sperm and sex hormones.

Testicle   A testis.

Testis-determining factor (TDF)   A gene on the Y chromosome that causes testes to differentiate prenatally.

Testosterone   A hormone secreted by the testes in the male; it maintains secondary sex characteristics. It is also secreted in smaller amounts in the female.

Toxemia   A dangerous disease of pregnancy.

Toxic shock syndrome   A sometimes fatal disease associated with tampon use.

Transgender   A transsexual.

Transsexual   A person who feels that he or she is trapped in the body of the wrong gender; a person who undergoes a sex-change operation.

Transvestism   Dressing in the clothing of the other gender.

Tribadism   A sexual technique in which one woman lies on top of another, moving rhythmically to produce sexual pleasure.

Trichomoniasis   A vaginal infection.

Trimester   Three months.

Troilism   A sexual variation in which three people engage in sexual activity together.

Tubal ligation   A surgical method of female sterilization; also called *salpingectomy*.

Tumescence   Swelling due to congestion with body fluids; erection.

Tyson's glands   Glands under the foreskin of the penis that secrete a cheesy substance called smegma.

Umbilical cord   The tube that connects the fetus to the placenta.

Urethra   The tube through which urine leaves the bladder and passes out of the body; in males, also the tube through which semen is discharged.

Urophilia (or urolagnia)   A sexual variation in which the person derives sexual pleasure from urine or urination.

Uterus   The organ in the female in which the fetus develops.

Vacuum aspiration   A method of abortion that is performed during the first trimester.

Vagina   The barrel-shaped organ in the female into which the penis is inserted during intercourse and through which a baby passes during birth.

Vaginal ring   An experimental device for contraception.

Vaginismus   A strong, spastic contraction of the muscles around the vagina, closing off the vaginal entrance and making intercourse impossible.

Vaginitis   An inflammation or irritation of the vagina, usually due to infection.

Varicocele   Essentially, varicose veins in the testes; may be related to infertility in men.

Vas deferens   The ducts through which sperm pass on their way from the testes to the urethra.

Vasectomy   A surgical procedure for male sterilization involving severing of the vas deferens.

Vasocongestion   An accumulation of blood in the blood vessels of a region of the body, especially the genitals; a swelling or erection results.

Venereal disease   A disease transmitted primarily by sexual contact.

Viral hepatitis   See *hepatitis B*.

Virgin   A person who has never had sexual intercourse.

Volunteer bias   A problem in sex research caused by some people refusing to participate, so that those who are in the sample are volunteers who may in some ways differ from those who refuse to participate.

Voyeurism   A sexual variation in which the person derives sexual pleasure from watching nudes; also called *peeping Tomism*. See also *scoptophilia*.

Vulva   The collective term for the external genitals of the female; includes the mons, clitoris, inner and outer lips, and vaginal and urethral openings.

Warts   See *genital warts*.

Wassermann test   A blood test for syphilis.

Wet dream   See *nocturnal emission*.

Withdrawal   A method of birth control in which the male withdraws his penis from the vagina before he ejaculates.

Wolffian ducts   Embryonic ducts which form part of the male's reproductive system.

Womb   See *uterus*.

Zoophilia   See *bestiality*.

Zygote   The fertilized egg.

Zygote intra-fallopian transfer (ZIFT)   A procedure in which an egg fertilized by sperm in a laboratory dish (zygote) is place into the fallopian tube.

Answers to Review Questions

Chapter 1. Sexuality in Perspective

1. False
2. heterosexual intercourse between unmarried persons
3. False
4. True
5. True
6. False
7. True
8. gender
9. True
10. True

Chapter 2. Theoretical Perspectives on Sexuality

1. Sociobiology
2. libido
3. False
4. True
5. False
6. True
7. True
8. Macro level, subcultural level, interpersonal level, individual level
9. True
10. self-disclosure

Chapter 3. Sex Research

1. Volunteer bias
2. True
3. Direct observation
4. False
5. False
6. False
7. content analysis.
8. Participant-observer study
9. False
10. False

Chapter 4. Sexual Anatomy

1. clitoris
2. Infibulation
3. False
4. endometrium
5. estrogen and progesterone
6. Circumcision
7. interstitial cells
8. True
9. 80
10. True

Chapter 5. Sex Hormones and Sexual Differentiation

1. hypothalamus
2. pituitary gland
3. True
4. Wolffian
5. True
6. ovaries
7. True
8. False
9. False
10. True

Chapter 6. Menstruation and Menopause

1. luteal or secretory phase
2. True
3. True
4. Prostaglandins
5. True
6. False
7. True
8. Estrogen-replacement therapy (or hormone-replacement therapy)
9. True
10. True

Chapter 7. Conception, Pregnancy, and Childbirth

1. False
2. False
3. placenta
4. human chorionic gonadotropin (HCG)
5. True
6. True
7. Lamaze
8. False
9. in vitro fertilization
10. Gamete intra-fallopian transfer (GIFT)

Chapter 8. Contraception and Abortion

1. estrogen and progestin (progesterone)
2. True
3. False
4. progestin; 3 months
5. False
6. True
7. RU-486
8. rhythm
9. condoms; vasectomy
10. False

Chapter 9. The Physiology of Sexual Response

1. excitement, plateau, orgasm, resolution
2. vasocongestion
3. cognitive
4. False
5. sexual desire, vasocongestion, muscular contractions
6. True
7. Grafenberg spot (G-spot)
8. organizing
9. Pheromones
10. False

Chapter 10. Techniques of Arousal and Communication

1. False
2. True
3. False
4. cunnilingus
5. False
6. True
7. True
8. Leveling
9. False
10. validating

Chapter 11. Sexuality and the Life Cycle: Childhood and Adolescence

1. birth or even prenatally
2. False
3. True
4. True
5. True
6. True
7. False
8. testosterone
9. False
10. True

Chapter 12. Sexuality and the Life Cycle: Adulthood

1. True
2. True
3. True
4. False
5. False
6. Swinging
7. equity theory
8. True
9. True
10. regular sexual expression

Chapter 13. Attraction, Intimacy, and Love

1. False
2. True
3. reinforcements
4. False
5. intimate
6. passion, intimacy
7. True
8. False
9. physiological arousal
10. collectivist

Chapter 14. Gender Roles, Female Sexuality, and Male Sexuality

1. gender role
2. True
3. False
4. masturbation; attitudes toward casual sex
5. 63; 42
6. penile strain gauge
7. False
8. True
9. transsexual (or transgender)
10. gender dysphoria

Chapter 15. Sexual Orientation: Gay, Straight, or Bi?

1. lesbian, gay, bisexual
2. False
3. False
4. True
5. False
6. attachment, autonomy, equality
7. 2; 1
8. False
9. True
10. True

Chapter 16. Variations in Sexual Behavior

1. True
2. Fetishism
3. True
4. masochist
5. False
6. hypersexuality
7. sexual addiction
8. False
9. True
10. False

Chapter 17. Sexual Coercion

1. False
2. victim-precipitated
3. post-traumatic stress disorder (PTSD)
4. True
5. False
6. True
7. False
8. False
9. False
10. True

Chapter 18. Sex for Sale

1. True
2. In-call service
3. pimp
4. Gigolos
5. True
6. True
7. True
8. False
9. True
10. False

Chapter 19. Sexual Disorders and Sex Therapy

1. Erectile disorder
2. Primary orgasmic disorder
3. True
4. True
5. False
6. cognitive interference
7. True
8. cognitive-behavioral
9. erectile disorder
10. False

Chapter 20. Sexually Transmitted Diseases

1. chlamydia
2. True
3. genital warts
4. herpes
5. True
6. Blood, semen
7. False
8. True
9. True
10. False

Chapter 21. Ethics, Religion, and Sexuality

1. legalistic; situation ethics
2. False
3. celibacy
4. Hindu
5. profamily
6. natural
7. False
8. ordination; marriage of homosexuals
9. confidential
10. None

Chapter 22. Sex and the Law

1. False
2. True
3. False
4. sodomy
5. *Griswold v. Connecticut*
6. obscene/obscenity
7. *Planned Parenthood v. Casey*
8. False
9. False
10. False

Chapter 23. Sexuality Education

1. False
2. False
3. True
4. True
5. True
6. abstinence-only
7. False
8. True
9. False
10. False

Acknowledgments

PHOTOGRAPHS

Chapter 1

1.1 (a) AP/Wide World Photos; (b) Culver
1.2 Reprinted by permission of The Kinsey Institute for Research in Sex, Gender and Reproduction, Inc., photo by Bill Dellenback
1.4 Photofest
1.5 AP/Wide World Photos
1.6 (a) Stephen McBrady/PhotoEdit; (b) Robert A. Isaacs/Photo Researchers
1.7 (a) Stephen McBrady/PhotoEdit; (b) Lawrence Migdale/Stock, Boston
1.8 (a) © Meredith F. Small; (b) Photograph by Frans de Waal.

Chapter 2

2.1 (a) J.H. Robinson/Photo Researchers; (b) Spencer Grant/Stock, Boston
2.2 Barbara Campbell/Gamma Liaison
2.5 Bob Daemmrich/Stock, Boston
2.6 © Jim Vecchione/Liaison International
2.7 Suzanne Arm/The Image Works.

Chapter 3

3.1 D. Young-Wolff/PhotoEdit
3.3 Reprinted by permission of The Kinsey Institute for Research in Sex, Gender and Reproduction, Inc., photo by Bill Dellenback
3.4 Courtesy Edward Laumann
3.5 Bob Kalman/The Image Works
3.6 © Summer Productions
3.7 Photo by Russell D. Curtis/Photo Researchers, Inc.

Chapter 4

4.4 Lori Grinker/Contact Press Images
4.5 Kathy Bendo

4.12 (a,b) Joel Gordon
4.14a © Taeke Henstra/Petit Format/Photo Researchers, Inc.
4.14b © David Parker/SPL/Photo Researchers, Inc.
4.15 © Biophoto Associates/Science Source/Photo Researchers, Inc.
4.17 © Carey/The Image Works

Chapter 5

5.6 © Michael Geissinger
5.8 (a) Elizabeth Crews/The Image Works; (b) Joe Sohm/The Image Works
5.9 (1) Jason Laure/Woodfin Camp; (2) Blair Seitz/Photo Researchers, Inc.

Chapter 6

6.2 Dr. Landrum B. Shettles
6.6 © Don Smetzer/Tony Stone Images, Inc.
6.7 David Lissy/Index Stock Imagery

Chapter 7

7.3 Dr. Landrum B. Shettles
7.4 (a) Petit Format/Nestle/Science Source/Photo Researchers; (b-d) Lennart Nilsson, from *A Child is Born*, Dell Publishing Company
7.5 Spencer Grant/Monkmeyer
7.6 (a) Streissguth, A.P, Landesman-Dwyer, S., Martin, J.C., & Smith, D.W. (1980). "Teratogenic effects of alcohol in humans and laboratory animals," *Science*, 209, 353-361; (b) John Chiasson/Gamma Liaison
7.7 (a,b) © Scott Camazine/Sue Trainor/Photo Researchers, Inc.
7.10 Lawrence Migdale/Stock, Boston
7.11 (a,b) © Monkmeyer/Byron
7.12 Mary M. Thacher/Photo Researchers
7.15 (a,b) Hank Morgan/Photo Researchers.

Chapter 8

8.2 UPI/Corbis/Bettmann
8.3 Tony Freeman/PhotoEdit
8.4 Joel Gordon
8.5 Gyno Pharma, Inc.
8.8 © Joel Gordon
8.9 Tony Freeman/PhotoEdit
8.10 J & M Studios/Gamma Liaison
8.11 The Children's Defense Fund
8.15 (b) J. Berry/Gamma Liaison
8.16 © The McGraw-Hill Companies, Inc./Bob Coyle, photographer.

Chapter 9

9.4 UPI/Bettmann Newsphotos
9.6 Leonard Freed/Magnum
9.8 (a) Joel Gordon; (b) AP/Wide World Photos
9.12 (Box 9.3) Bob Daemmrich/Stock Boston
9.13 (a) Guy Gilette/Photo Researchers; (b) D. Young-Wolff/PhotoEdit.

Chapter 10

10.1 (a,b) © Joel Gordon
10.2 © Mark Antman/The Image Works
10.4 © Luiz C. Marico/Peter Arnold, Inc.
10.11 (a) © Bachmann/Photo Researchers, Inc.; (b) © Dion Ogust/The Image Works
10.12 © Esbin-Anderson/The Image Works.

Chapter 11

11.1 © Amy C. Etra/PhotoEdit
11.2 Tana Hoban
11.3 © Frank Siteman/Stock Boston
11.4 © Bob Daemmrich/Stock, Boston
11.6 © Monkmeyer/Grantpix
11.8 © Stuart Cohen/Comstock
11.10 © B. Daemmrich/The Image Works.

Chapter 12

12.1 © Monkmeyer/Grantpix
12.3 (a) © A. Albert/The Image Works; (b) © Jon Feingersh/Stock, Boston
12.4 (a) © Esbin-Anderson/The Image Works; (b) © SuperStock
12.5 © Spencer Grant/Index Stock Imagery
12.6 © Bruce Davidson/Magnum
12.9 © Network Prod./The Image Works.

Chapter 13

13.1 © SuperStock
13.2 © R. Lord/The Image Works
13.6 Guy Le Querrec/Magnum
13.7 ©1999 Summer Productions
13.8 Jerry Cooke/Photo Researchers.

Chapter 14

14.1 Rod Rolle/Gamma Liaison
14.2 Photofest
14.3 F.A. Rinehart for B.A.E./Smithsonian Institute
14.4 (a) Ulrike Welsch/Photo Researchers; (b) Victor Friedman/Photo Researchers
14.5 Photo courtesy of J.R. Heiman
14.9 (a,b) AP/Wide World Photos.

Chapter 15

15.1 (a) Joyce R. Wilson/Photo Researchers, Inc.
15.2 AP/Wide World Photos
15.3 Joel Gordon
15.4 David Young-Wolff/PhotoEdit
15.6 (a) Chris Maynard/Gamma Liaison; (b) S. Gazin/The Image Works
15.8 © Bob Daemmrich/Stock Boston
15.9 Photofest.

Chapter 16

16.1 Herb Snitzer/Stock, Boston
16.3 (a) Photofest; (b) Lee Snider/The Image Works
16.4 Jacques Prayer/Gamma Liaison
16.5 Randy Matusow
16.6 © Christie's Images/SuperStock
16.8 © David Harry Stewart/Tony Stone Images.

Chapter 17

17.1 Rhoda Sidney/PhotoEdit
17.3 Yvonne Hemsey/Gamma Liaison
17.4 Courtesy National Victim Center
17.5 (a,b) Brad Markel/Gamma Liaison
17.6 © B. Daemmrich/The Image Works.

Chapter 18

18.1 Kent Knudson/Stock, Boston
18.2 © Joel Gordon
18.3 © Richard B. Levine
18.4 © Joel Gordon
18.5 Photofest
18.6 Joel Gordon
18.7 Barbara Nitke/Femm Distribution, Inc.
18.8 (a) Charles Gatewood/The Image Works; (b) Myrleen Ferguson/PhotoEdit.

Chapter 19

19.2 (a) Paula A. Scully/Gamma Liaison; (b) Rob Nelson/Stock, Boston
19.6 © Larry Mulvehill/The Image Works
19.9 © 1999 Jack Ziegler from cartoonbank.com. All Rights Reserved.

Chapter 20

20.1 (a) Marc S. Berger/The Stock Shop; (b) © Biophoto Associates/Photo Researchers, Inc.
20.2 (a,b) © Biophoto Associates/Photo Researchers, Inc.

20.3   New York City Department of Health
20.4   (a) Thomas Bowman/PhotoEdit; (b) Claude Poulet/Gamma Liaison
20.5   AP/Wide World Photos
20.6   Stuart Franklin/Magnum
20.7, 20.8   (a,b) Courtesy Centers for Disease Control
20.9   © E. Gray/Science Photo Library/Photo Researchers, Inc.

Chapter 21

21.1   The Metropolitan Museum of Art, Rogers Fund, 1941 (41.162.101) Neg. #177754
21.2   Hermitage Museum, St. Petersburg, Russia/Super-Stock
21.3   Ognissanti Church, Florence, Italy/Bridgeman Art Library, London/SuperStock
21.4   Frances M. Roberts
21.5   Marc Riboud/Magnum
21.6   Sujoy Das/Stock, Boston
21.7   Joel Gordon
21.8   National Gallery of Art, Washington, D.C./Super-Stock
21.9   © Rick Gerharter/Impact Visuals
21.10   © Najlah Feanny/Stock Boston.

Chapter 22

22.2   Ira Schwarz/Reuters/Bettmann
22.3   Michael Newman/PhotoEdit
22.5   M. Siluk/The Image Works
22.6   Reuters/Joe Trave/Archive Photos
22.7   © Joel Gordon.

Chapter 23

23.3   James Schnepf/Gamma Liaison
23.4   Renato Rotolo/Gamma Liaison
23.5   Robert Nickelsberg/Gamma Liaison
23.6   Steve McCurry/Magnum Photos.

LINE ART AND TEXT

Chapter 1

Excerpt   Edward O. Laumann et al., (1994). *The social organization of sexuality: Sexual practices in the United States*. Reprinted by permission of The University of Chicago Press and Edward O. Laumann.

Table 1.1   Edward O. Laumann et al., (1994). *The social organization of sexuality: Sexual practices in the United States*. Reprinted by permission of The University of Chicago Press and Edward O. Laumann.

Table 1.2   Edward O. Laumann et al., (1994). *The social organization of sexuality: Sexual practices in the United States*. Reprinted by permission of The University of Chicago Press and Edward O. Laumann.

Chapter 2

Figure 2.4   John DeLamater (1987). A sociological approach to understanding human sexuality for a sociological perspective. In J.H. Geer and W.T. O'Donohue (eds.), *Theories of Human Sexuality*, pp. 237–256, 1987. Reprinted by permission of Plenum Publishing Corporation and John DeLamater.

Lyrics   Cole Porter (1953). "Too Darn Hot" from *Kiss Me, Kate*. Music Lyrics by Cole Porter; book by Sam and Bella Spewack. © 1949 (Renewed) Chappell and Company. (ASCAP). All Rights Reserved. Used by Permission. Warner Bros. Publications U.S. Inc., Miami, FL 33014

Chapter 3

Table 3.1   From "The Janus Report" by Andrew M. Greeley in *Contemporary Sociology*, Vol 23, pp. 221–223. Reprinted by permission of the American Sociological Association and Andrew M. Greeley.

Chapter 4

Focus 4.2   "The Breast Self-Exam." Reprinted by permission of the American Cancer Society.

Focus 4.3   "The Testicular Self-Exam."

Chapter 5

Table 5.2   Bernard Goldstein (1976). *Introduction to Human Sexuality*. New York: McGrawHill, pp. 80–81. Reprinted by permission of Bernard Goldstein.

Poem   "This Way" by Sherri Groveman from *Hermaphrodites with Attitude*, 1995, p. 2. Reprinted by permission of the author.

Excerpt   John Money (1987). Sin, sickness, or status: Homosexual gender identity and psychoneuroendocrinology. *American Psychologist*, 42, pp. 384–399. Copyright © 1996 by the American Psychological Association. Reprinted by permission.

Chapter 6

Excerpt 6-1   From the Revised Standard Version of the Bible, copyright © 1946, 1952, 1971 by the Division of Christian Education of the National Council of the Churches of Christ in the USA. Used by permission.

Chapter 7

Table 7.1   R. Kumar, H.A. Brant and K.M. Robson (1981). Childbearing and maternal sexuality: A prospective study of 119

Table 7.2 — primiparas. *Journal of Psychosomatic Research*, 25, 373–383. Copyright 1981 by Elsevier Science Ltd. Reprinted by permission of Elsevier Science Inc. Republished with permission of The Society for the Scientific Study of Sexuality, PO Box 208, Mount Vernon, IA 52314. *Sexuality During Pregnancy and the Year Postpartum* (Table), J.S. Hyde, J.D. DeLamater, E.A. Plant, & J.M. Byrd, *The Journal of Sex Research*, 33 (1996). Reproduced by permission of the publisher via Copyright Clearance Center, Inc.

Chapter 8

Figure 8.1 — Cynthia P. Green. The environment and population growth: Decade for action. Population Reports, Series M. No. 10. Baltimore, Johns Hopkins School of Public Health, Population Information Program, May 1992. Reprinted by permission.

Figure 8.15a — © 1993 Time Inc. Reprinted by permission.

Focus 8.3 — Robert Hatcher et al. (1976). *Contraceptive Technology*, 1976–1977, 8/e. Reprinted with permission of Irvington Publishers, Inc.

Table 8.3 — Reproduced with the permission of The Alan Guttmacher Institute from Henshaw SK. Induced abortion: a world review, 1990. *Family Planning Perspectives*. 1990, 22(2): 76–89.

Table 8.4 — Robert A. Hatcher et al (1994). *Contraceptive Technology, 16th edition.* Adapted by permission of Irvington Publishers, Inc.

Chapter 9

Excerpt — Excerpt from *Drive Him Wild* by Graham Masterton. Copyright © 1993 by Graham Masterton. Used by permission of Dutton Signet, a division of Penguin Putnam Inc.

Chapter 10

Excerpt — From *Sex: an oral history* by Harry Maurer. Copyright © 1994 by Harry Maurer. Used by permission of Viking Penguin, a division of Penguin Putnam Inc. and International Creative Management, Inc.

Focus 10.3 — "How Solid is Your Relationship?" Reprinted with the permission of Simon & Schuster and John Gottman from *Why Marriages Succeed or Fail* by John Gottman. Copyright © 1994 by John Gottman.

Chapter 11

Selection 11-1 — Floyd M. Martinson (1994). *The sexual life of children*, pp. 37, 59, 62. Copyright © 1994. Reprinted with permission of Greenwood Publishing Group, Inc., Westport, Connecticut.

Excerpt — Gail E. Wyatt et al. (1988). Kinsey revisited. Part I: Compairisons of the suxual sociolization and sexual behavior of white women over 33 years. *Archives of Sexual Behavior*, 17, pp. 20140. Excerpted and reprinted by permission of Plenum Publishing Corporation and the author.

Table 11.3 — Randal D. Day (1992). The transition to first intercourse among racially and culturally diverse youth. *Journal of Marriage and the Family*, 54, pp. 749–762. Copyright 1992 by the National Council on Family Relations, 3989 Central Avenue NE, Suite 550, Minneapolis, MN 55421. Reprinted by permission of the National Council on Family Relations.

Excerpt — Reprinted with permission from Playboy Enterprises, Inc., from *Sexual Behavior in the 1970's* by Morton Hunt. Copyright 1974 by Morton Hunt.

Excerpt — Robert T. Michael et al. (1994). *Sex in America: A definitive survey*. Copyright © 1994. Reprinted by permission of Little, Brown and Company.

Chapter 12

Excerpt — From *The Janus report on sexual behavior* by Samuel S. Janus and Cynthia L. Janus, pg. 383, 1993. Reprinted by permission of John Wiley & Sons, Inc.

Excerpt — From *The Janus report on sexual behavior* by Samuel S. Janus and Cynthia L. Janus, pg. 191, 1993. Reprinted by permission of John Wiley & Sons, Inc.

Excerpt — From *The Janus report on sexual behavior* by Samual S. Janus and Cynthia L. Janus, pg. 8. Reprinted by permission of John Wiley & Sons, Inc.

Table 12.3 — From *Love, Sex and Aging* by Edward M. Brecher, et al. Copyright © 1984 by Consumers Union of the United States. Reprinted by permission of Little, Brown and Company.

Chapter 13

Table 13.1 — Edward O. Laumann et al., (1994). *The social organization of sexuality: Sexual practices in the United States*. Reprinted by permission of The University of Chicago Press and Edward O. Laumann.

Table 13.2 — Data from Robert Levine et al., (1995). "Love and marriage in eleven cultures." *Journal of Cross-Cultural Psychology*, 26, pp. 554–571. Used by permission of Robert Levin. Reprinted in E. Hatfield and R.L. Rapson (1996). Love and sex: Cross-cultural perspectives. Needham Heights Massachusetts: Allyn & Bacon.

Chapter 14

Table 14.1   Yolanda F. Niemann et al. (1994). Use of free responses and cluster analysis to determine the stereotypes of eight groups. *Personality and Social Psychology Bulletin*, Vol. 20, pp. 379–390. Reprinted by Permission of Sage Publications, Inc

Excerpt   Reprinted with permission from Playboy Enterprises, Inc., from *Sexual Behavior in the 1970's* by Morton Hunt. Copyright 1974 by Morton Hunt.

Chapter 15

Table 15.1   J.A. Davis and T. Smith (1973, 1996). The National Opinion Research Center, General Social Survey, 1973, 1996. Reprinted by permission of the National Opinion Research Center.

Excerpt   From "Reports of assaults...," by Constance L. Hays in *The New York Times*, 9/30/90. Copyright © 1990 by The New York Times. Reprinted by permission.

Excerpt   Esther D. Rothblum (1993). I only read about myself on bathroom walls: The need for research on the mental health of lesbians and gay men. *Journal of Consulting and Clinical Psychology*, 62, 1993. Copyright © 1993 by the American Psychological Association. Reprinted by permission.

Table 15.3   Edward O. Laumann et al., (1994). *The social organization of sexuality: Sexual practices in the United States*. Reprinted by permission of The University of Chicago Press and Edward O. Laumann.

Figure 15.7   Daryl J. Bem (1996). Exotic becomes erotic: A developmental theory of sexual orientation. *Psychological Review*, 103, p. 321. Copyright © 1996 by the American Psychological Association. Reprinted by permission.

Focus 15.4   Lisa Yost (1991). Bisexual tendencies. In L. Hutchins and L. Kaahumanu (eds.), *Bi any other name: Bisexual people speak out*. Reprinted with permission, Alyson Publications 1991 by Loraine Hutchins and Lani Kaahumanu.

Chapter 16

Excerpt   Reprinted with permission of The Society for the Scientific Study of Sexuality. PO Box 208, Mount Vernon, IA 52314. *If The Shoe Fits...Exploring Male Homosexual Foot Fetishes* (excerpt), Weinberg, Williams & Calhan, *The Journal of Sex Research*, 32, 1995. Reproduced by permission of the publisher via Copyright Clearance Center, Inc.

Excerpt   Richard Green (1978). Intervention and prevention: The child with cross-sex identity. In C.B. Qualls et al. (eds.), *The prevention of sexual disorders*, p. 88. Reprinted by permission of Plenum Publishing Corporation.

Poem   "Rape poem" from *Circles on the Water* by Marge Piercy. Copyright © 1982 by Marge Piercy. Reprinted by permission of Alfred A. Knopf, Inc. Also from *Living in the Open* by Marge Piercy. Copyright © 1976 by Marge Piercy and Middlemarsh, Inc. Used by permission of the Wallace Literary Agency.

Excerpt   From Jean O. Hughes & Bernice R. Sandler. (1987). "Friends" raping friends: Could it happen to you? Washington, DC: Association of American Colleges. Used with permission.

Chapter 17

Focus 17.2   Condensed from Jean O. Hughes & Bernice R. Sandler. (1987). "Friends" raping friends: Could it happen to you? Washington, DC: Association of American Colleges. Used with permission.

Excerpt   Reprinted with permission of The Society for the Scientific Study of Sexuality, PO Box 208, Mount Vernon, IA 52314. *Forced Sex on Dates: It Happens to Men Too* (excerpt), Cindy Struckman-Johnson, *The Journal of Sex Research*, 24, 1988. Reproduced by permission of the publisher via Copyright Clearance Center, Inc.

Table 17.1   Edward O. Laumann et al., (1994). *The social organization of sexuality: Sexual practices in the United States*. Reprinted by permission of The University of Chicago Press and Edward O. Laumann.

Chapter 18

Focus 18.1   From *Being a Prostitute* by Roberta Perkins and Garry Bennett, 1985. Reprinted by permission of Allen & Unwin Australia Pty. Ltd.

Table 18.1   Martin Rimm (1995). Marketing pornography on the information superhighway. *Georgetown Law Journal*, Vol. 83, p. 1891. Reprinted with permission of the publisher, Georgetown University and Georgetown Law Journal. Copyright © 1995.

Excerpt   From the *San Francisco Chronicle*, October 25, 1977. Copyright © San Francisco Chronicle. Reprinted by permission.

Selection 18-24   T. Moore. "Porn Shop Enjoys Brisk Business Year Round." *The Capital Times*, 1994, pp. 5A–6A. Reprinted by permission of The Capital Times, Madison, WI.

Focus 18.2   "Ernie: A Pedophile and Child Pornographer" from *Child pornography and sex rings* by Ann W. Burgess, pp. 26–27, 1984, Lexington Books. Reprinted by permission of the author.

Chapter 19

Figure 19.3   David H. Barlow (1986). Causes of sexual dysfunction: The role of cognitive interference. *Journal of Consulting and Clinical Psychology*, 54. Copyright © 1986 by the American Pschological Association. Reprinted by permission.

Chapter 20

Table 20.1   From *Barrier contraceptives: Current status and future prospects* by Katherine M. Stone, pg. 203–212, 1994. Reprinted by permission of John Wiley & Sons, Inc.

Focus 20.4   Reprinted by permission of the publisher from Patient Guide: "How To Use A Condom," *Medical Aspects of Human Sexuality*, 21 (7): 74–75. Copyright 1987 by Quadrant Healthcom Inc.

Chapter 21

Focus 21-1   United Church of Christ, 1987 General Synod. Reprinted by permission.

Focus 21-2   Biennial Convention, 1989. Reprinted with permission of the American Jewish Congress, New York.

Focus 21.2   General Convention of the Episcopal Church, 1988. Reprinted with permission of the Episcopal Church Center, New York.

Excerpt   Congregation for the Doctrine of the Faith, 1976, p. 11. Reprinted by permission.

Chapter 22

Table 22.1   From the Gallup Poll on Americans on their attitudes toward abortion (1975, 1977, 1980, 1992). Reprinted by permission of The Gallup Organization.

Excerpt   Congregation for the Doctrine of the Faith. The pastoral care of homosexual persons. *Origins*, 16, 1986, p. 379. Reprinted by permission.

Excerpt   R.A. Poser & K.B. Silbaugh. (1996). *A Guide to America's Sex Laws*. Chicago: University of Chicago Press. Reprinted by permission.

Table 22.1   R.A. Posner (1992). *Sex and reason*, tables 1 & 3. Copyright © 1992 by the President and Fellows of Harvard College. Reprinted by permission of Harvard University Press, Cambridge, Massachusetts.

Excerpt   From "The Legal Regulation of Sexual Activity and the Protection of Females" by Graham Parker from *Osgoode Hall Law Journal*, 21, 187–244, (1983). Reprinted by permission of Université 82 York University and Graham Parker.

Figure 22.1a   R.A. Poser & K.B. Silbaugh. (1996). *A Guide to America's Sex Laws*. Chicago: University of Chicago Press. Reprinted by permission.

Figure 22.1b   R.A. Poser & K.B. Silbaugh. (1996). *A Guide to America's Sex Laws*. Chicago: University of Chicago Press. Reprinted by permission.

Table 22.2   A telephone poll of 753 adults conducted by Princeton Survey Research Associates in June 1997 for *Newsweek*.

Figure 22.8   R.A. Poser & K.B. Silbaugh. (1996). *A Guide to America's Sex Laws*. Chicago: University of Chicago Press. Reprinted by permission.

Chapter 23

Excerpt 23-4   From "Sex Education." Reproduced with permission of the Sexuality Information and Education Council of the United States, Inc. (SIECUS). Copyright © SIECUS, 130 West 42nd Street, Suite 350, New York, NY 10036.

Selection 23-4   Anne C. Bernstein and Philip A. Cowan (1975). Children's concepts of how people get babies. *Child Development*, 46, pp. 77–92. © The Society for Research in Child Development, Inc.

Excerpt 23-6   Rosemary Zumwalt (1976). Plain and fancy: A content analysis of children's jokes dealing with adult sexuality. *Western Folklore*, 35, pp. 261, 267. Reprinted by permission of the California Folklore Society.

Table 23a   From *Children's Sexual Thinking* by Ronald and Juliette Goldman, pgs 197, 213, 240, 263, and 354, 1982. Reprinted by permission of Routledge.

Excerpt   Douglas Kirby. School-based programs to reduce sexual risktaking behavior. *Journal of School Health*, Vol. 62, No. 7, September 1992, p. 281. Reprinted with permission from American School Health Association, Kent, Ohio.

Excerpt   Hoven Bradford Gow (1994). 1994 Condom distribution in high schools. *The Clearing House*, Vol. 67, pp. 183–184. Copyright © 1994. Reprinted with permission of The Helen Dwight Reid Educational Foundation. Published by Heldref Publications, 1319 18th Street N.W. Washington, DC 20036-1802.

Index